THE ROUTLEDGE CO. PHILOSOPHY OF MEDICINE

The Routledge Companion to Philosophy of Medicine is a comprehensive guide to topics in the fields of epistemology and metaphysics of medicine. It examines traditional topics such as the concept of disease, causality in medicine, the epistemology of the randomized controlled trial, the biopsychosocial model, explanation, clinical judgment, and phenomenology of medicine, and emerging topics, such as philosophy of epidemiology, measuring harms, the concept of disability, nursing perspectives, race and gender, the metaphysics of Chinese medicine, and narrative medicine. Each of the 48 chapters is written especially for this volume and with a student audience in mind. For pedagogy and clarity, each chapter contains an extended example illustrating the ideas discussed. This text is intended for use as a reference for students in courses in philosophy of medicine and philosophy of science, and pairs well with *The Routledge Companion to Bioethics* for use in medical humanities and social science courses.

Miriam Solomon is Professor and Chair in the Philosophy Department at Temple University, and Affiliated Professor at the Center for Bioethics, Urban Health, and Policy at the Lewis Katz School of Medicine at Temple University. She works in the areas of philosophy of science, philosophy of medicine, epistemology, and feminist philosophy. She is the author of *Social Empiricism* (2001), *Making Medical Knowledge* (2015), and numerous articles. She is Fellow of the College of Physicians of Philadelphia.

Jeremy R. Simon, MD, PhD, is an emergency physician. He is Associate Professor of Medicine at Columbia University Medical Center, an attending physician in the New York-Presbyterian Emergency Medicine residency, and a member of the Ethics Consultation Service at New York-Presbyterian/CUMC. His primary academic research is in philosophy of medicine, and he also writes on medical ethics.

Harold Kincaid is Professor of Economics at the University of Cape Town, South Africa. He is the author of numerous books, book chapters, and articles in the philosophy of science. Among his many books is the most recent, *Classifying Psychopathology: Mental Illness and Natural Kinds* (2014).

ROUTLEDGE PHILOSOPHY COMPANIONS

Routledge Philosophy Companions offer thorough, high-quality surveys and assessments of the major topics and periods in philosophy. Covering key problems, themes, and thinkers, all entries are specially commissioned for each volume and written by leading scholars in the field. Clear, accessible, and carefully edited and organized, *Routledge Philosophy Companions* are indispensable for anyone coming to a major topic or period in philosophy, as well as for the more advanced reader.

For a full list of published *Routledge Philosophy Companions*, please visit https://www.routledge.com/series/PHILCOMP.

The Routledge Companion to Islamic Philosophy
Edited by Richard C. Taylor and Luis Xavier López-Farjeat

Forthcoming:

The Routledge Companion to Sixteenth Century Philosophy
Edited by Benjamin Hill and Henrik Lagerlund

The Routledge .Companion to Seventeenth Century Philosophy
Edited by Dan Kaufman

The Routledge Companion to Philosophy of Literature
Edited by Noël Carroll and John Gibson

The Routledge Companion to Medieval Philosophy
Edited by Richard Cross and JT Paasch

The Routledge Companion to Philosophy of Race
Edited by Paul C. Taylor, Linda Martín Alcoff, and Luvell Anderson

The Routledge Companion to Environmental Ethics
Edited by Benjamin Hale and Andrew Light

The Routledge Companion to Free Will
Edited by Meghan Griffith, Neil Levy, and Kevin Timpe

The Routledge Companion to Philosophy of Technology
Edited by Joseph Pitt and Ashley Shew Helfin

The Routledge Companion to Philosophy of Medicine
Edited by Miriam Solomon, Jeremy R. Simon, and Harold Kincaid

The Routledge Companion to Feminist Philosophy
Edited by Ann Garry, Serene Khader, and Alison Stone

The Routledge Companion to Philosophy of Social Science
Edited by Lee McIntyre and Alex Rosenberg

THE ROUTLEDGE COMPANION TO PHILOSOPHY OF MEDICINE

Edited by
Miriam Solomon, Jeremy R. Simon,
and Harold Kincaid

LONDON AND NEW YORK

First published 2017
by Routledge

2 Park Square, Milton Park, Abingdon, Oxfordshire OX14 4RN
52 Vanderbilt Avenue, New York, NY 10017

Routledge is an imprint of the Taylor & Francis Group, an informa business

First issued in paperback 2020

Copyright © 2017 Taylor & Francis

The right of the editors to be identified as the authors of the editorial material, and of the authors for their individual chapters, has been asserted in accordance with sections 77 and 78 of the Copyright, Designs and Patents Act 1988.

All rights reserved. No part of this book may be reprinted or reproduced or utilised in any form or by any electronic, mechanical, or other means, now known or hereafter invented, including photocopying and recording, or in any information storage or retrieval system, without permission in writing from the publishers.

Notice:
Product or corporate names may be trademarks or registered trademarks, and are used only for identification and explanation without intent to infringe.

Library of Congress Cataloging-in-Publication Data
Names: Solomon, Miriam, editor. | Simon, Jeremy R., editor. | Kincaid, Harold, 1952– editor.
Title: The Routledge companion to philosophy of medicine / edited by Miriam Solomon, Jeremy R. Simon, and Harold Kincaid.
Description: 1 [edition]. | New York : Routledge-Taylor & Francis, 2017.
Identifiers: LCCN 2016011922 | ISBN 9781138846791 (hardback) | ISBN 9781315720739 (e-book)
Subjects: LCSH: Medicine—Philosophy. | Medical ethics.
Classification: LCC R723 .R68 2017 | DDC 610.1—dc23
LC record available at https://lccn.loc.gov/2016011922

ISBN: 978-1-138-84679-1 (hbk)
ISBN: 978-0-367-36036-8 (pbk)

Typeset in Goudy
by Apex CoVantage, LLC

CONTENTS

List of Contributors   ix

Introduction   1
MIRIAM SOLOMON, JEREMY R. SIMON, AND HAROLD KINCAID

PART I
General Concepts   3

1. The Concept of Disease   5
DOMINIC SISTI AND ARTHUR L. CAPLAN

2. Disease, Illness, and Sickness   16
BJØRN HOFMANN

3. Health and Well-Being   27
DANIEL M. HAUSMAN

4. Disability and Normality   36
ANITA SILVERS

5. Mechanisms in Medicine   48
PHYLLIS ILLARI

6. Causality and Causal Inference in Medicine   58
JULIAN REISS

7. Frequency and Propensity: The Interpretation of Probability in Causal Models for Medicine   71
DONALD GILLIES

8. Reductionism in the Biomedical Sciences   81
HOLLY K. ANDERSEN

9. Realism and Constructivism in Medicine   90
JEREMY R. SIMON

PART II
Specific Concepts — 101

10 Birth — 103
CHRISTINA SCHÜES

11 Death — 115
STEVEN LUPER

12 Pain, Chronic Pain, and Suffering — 124
VALERIE GRAY HARDCASTLE

13 Measuring Placebo Effects — 134
JEREMY HOWICK

14 The Concept of Genetic Disease — 144
JONATHAN MICHAEL KAPLAN

15 Diagnostic Categories — 156
ANNEMARIE JUTEL

16 Classificatory Challenges in Psychopathology — 170
HAROLD KINCAID

17 Classificatory Challenges in Physical Disease — 180
MATHIAS BROCHHAUSEN

PART III
Research Methods

(a) Evidence in Medicine — 193

18 The Randomized Controlled Trial: Internal and External Validity — 195
ADAM LA CAZE

19 The Hierarchy of Evidence, Meta-Analysis, and Systematic Review — 209
ROBYN BLUHM

20 Statistical Evidence and the Reliability of Medical Research — 218
MATTIA ANDREOLETTI AND DAVID TEIRA

21 Bayesian Versus Frequentist Clinical Trials — 228
CECILIA NARDINI

22 Observational Research — 237
OLAF M. DEKKERS AND JAN P. VANDENBROUCKE

23 Philosophy of Epidemiology — 248
ALEX BROADBENT

24 Complementary/Alternative Medicine and the Evidence Requirement KIRSTEN HANSEN AND KLEMENS KAPPEL	257

PART III
Research Methods

(b) Other Research Methods	269
25 Models in Medicine MICHAEL WILDE AND JON WILLIAMSON	271
26 Discovery in Medicine BRENDAN CLARKE	285
27 Explanation in Medicine MAËL LEMOINE	296
28 The Case Study in Medicine RACHEL A. ANKENY	310
29 Values in Medical Research KIRSTIN BORGERSON	319
30 Outcome Measures in Medicine LEAH MCCLIMANS	330
31 Measuring Harms JACOB STEGENGA	342
32 Expert Consensus MIRIAM SOLOMON	353

PART IV
Clinical Methods

	361
33 Clinical Judgment ROSS UPSHUR AND BENJAMIN CHIN-YEE	363
34 Narrative Medicine DANIELLE SPENCER	372
35 Medical Decision Making: Diagnosis, Treatment, and Prognosis ASHLEY GRAHAM KENNEDY	383

PART V
Variability and Diversity

	395
36 Personalized and Precision Medicine ALEX GAMMA	397

37 Gender in Medicine INMACULADA DE MELO-MARTÍN AND KRISTEN INTEMANN	408
38 Race in Medicine SEAN A. VALLES	419
39 Atypical Bodies in Medical Care ELLEN K. FEDER	432

PART VI
Perspectives 443

40 The Biomedical Model and the Biopsychosocial Model in Medicine FRED GIFFORD	445
41 Models of Mental Illness JACQUELINE SULLIVAN	455
42 Phenomenology and Hermeneutics in Medicine HAVI CAREL	465
43 Evolutionary Medicine MICHAEL COURNOYEA	475
44 Philosophy of Nursing: Caring, Holism, and the Nursing Role(s) MARK RISJORD	487
45 Contemporary Chinese Medicine and Its Theoretical Foundations JUDITH FARQUHAR	497
46 Double Truths and the Postcolonial Predicament of Chinese Medicine ERIC I. KARCHMER	508
47 Medicine as a Commodity CARL ELLIOTT	519
Index	529

CONTRIBUTORS

Holly K. Andersen is an Associate Professor at Simon Fraser University in Burnaby, British Columbia. Her work focuses on causation and explanation, ranging from metaphysical foundations to methodology and issues in application in the sciences.

Mattia Andreoletti is a PhD student in Foundations of Life Sciences at European Institute of Oncology (Milan). He has a background in philosophy of science, and now he is working on the epistemology of novel designs of cancer clinical trials.

Rachel A. Ankeny is a Professor in the School of Humanities at the University of Adelaide, Australia. Among other topics, her research explores the history and philosophy of the biological and biomedical sciences, particularly case-based reasoning and experimental organisms.

Robyn Bluhm is an Associate Professor in the Department of Philosophy and Lyman Briggs College at Michigan State University. She has written extensively about philosophical questions arising from medical research.

Kirstin Borgerson is an Associate Professor of Philosophy at Dalhousie University in Canada. Dr. Borgerson researches and teaches in medical epistemology and medical ethics.

Alex Broadbent is Professor of Philosophy and Executive Dean of the Faculty of Humanities at the University of Johannesburg, South Africa. He works in the areas of causation, statistical reasoning, and epidemiology. He is the author of *Philosophy of Epidemiology* (2013) and *Philosophy for Graduate Students* (Routledge, 2016).

Mathias Brochhausen is at the Department of Biomedical Informatics, University of Arkansas for Medical Sciences, Little Rock. His research centers on biomedical classification systems, in particular formal ontologies and their use in managing biomedical data.

Arthur L. Caplan, PhD, is the Drs. William F. and Virginia Connolly Mitty Professor and the founding director of the Division of Medical Ethics in the New York University Langone Medical Center's Department of Population Health.

Havi Carel is Professor of Philosophy at the University of Bristol, UK. She researches the experience of illness. She holds a Senior Investigator Award from the Wellcome Trust. Her publications include *Illness* (2013), *Life and Death in Freud and Heidegger* (2006), and *Phenomenology of Illness* (2016).

Benjamin Chin-Yee is a medical student at the University of Toronto. His research includes topics in the philosophy of medicine and biology, as well as the medical humanities. He is particularly interested in the intersection between philosophy and clinical practice.

Brendan Clarke is lecturer in history and philosophy of medicine at the Department of Science and Technology Studies, University College London.

CONTRIBUTORS

Michael Cournoyea is a doctoral candidate at the University of Toronto's Institute for the History and Philosophy of Science and Technology. His research focuses on explanatory strategies in biomedicine, particularly medically unexplained physical symptoms, evolutionary medicine, and network medicine.

Olaf M. Dekkers was trained in medicine, epidemiology, and philosophy. He is working in internal medicine at the Leiden University Medical Centre and is involved in many observational studies in clinical medicine.

Carl Elliott is Professor in the Center for Bioethics at the University of Minnesota. His books include A *Philosophical Disease* (Routledge, 2008), *Better than Well* (2003), and *White Coat, Black Hat* (2010).

Judith Farquhar is Max Palevsky Professor Emeritus in the Department of Anthropology of the University of Chicago. Her research on contemporary China spans three decades, focusing on theories and practices of modern traditional Chinese medicine, everyday life and embodiment, popular culture and media, post-Mao and post-socialist micropolitics, and, most recently, national movements to systematize the traditional medicine practices of China's ethnic minorities. She is the author of *Knowing Practice: The Clinical Encounter of Chinese Medicine* (1994); *Appetites: Food and Sex in Post-Socialist China* (2004); and *Ten Thousand Things: Nurturing Life in Contemporary Beijing* (2012).

Ellen K. Feder is William Fraser McDowell Professor of Philosophy at American University in Washington, DC. She is author of *Making Sense of Intersex: Changing Ethical Perspectives in Biomedicine* (2014).

Alex Gamma is a neurobiologist at the University of Zurich, Switzerland. He has worked in experimental brain research in humans, psychiatric epidemiology, and applied statistics. From 2010–2013, he pursued his interest in the philosophy of biology at the Federal Institute of Technology (ETH), Zurich.

Fred Gifford is Professor in the Department of Philosophy and Faculty Associate in the Center for Ethics and Humanities in the Life Sciences at Michigan State University. His research and teaching interests include philosophy of science and ethics, especially topics at their intersection, such as in medical research and practice. He edited *Philosophy of Medicine: Handbook of the Philosophy of Science* (2011).

Donald Gillies worked in the University of London from 1971 until his retirement in 2009. He is currently Emeritus Professor of Philosophy of Science and Mathematics in University College London.

Kirsten Hansen is a special advisor at the Danish Health Authority. She holds an MA and PhD in philosophy from the University of Copenhagen, Denmark.

Valerie Gray Hardcastle is Professor of Philosophy, Psychology, and Psychiatry and Behavioral Neuroscience at the University of Cincinnati. She studies the nature and structure of interdisciplinary theories in the cognitive sciences and has focused primarily on developing a philosophical framework for understanding conscious phenomena, like pain, that are responsive to neuroscientific, psychiatric, and psychological data.

Daniel M. Hausman is the Herbert A. Simon and Hilldale Professor at the University of Wisconsin-Madison. His research lies at the boundaries of economics and philosophy, and he is one of the founding editors of the journal *Economics and Philosophy*.

CONTRIBUTORS

Bjørn Hofmann is a professor at the Norwegian University of Science and Technology Gjøvik and an adjunct professor at the Centre for Medical Ethics at the University of Oslo, Norway. He holds a PhD in philosophy of medicine and is trained both in the natural sciences and in the humanities. His main research interests are philosophy of medicine, philosophy of science, technology assessment, and bioethics. Hofmann teaches ethics, philosophy of science, science and technology studies, and philosophy of medicine at the BA, MA, and PhD levels. He has been a researcher at The Norwegian Knowledge Centre for the Health Services and a Harkness fellow (Commonwealth Fund) at Dartmouth College, New Hampshire.

Jeremy Howick received his PhD from the London School of Economics in 2008. He was a lecturer at University College London, and he is currently a senior researcher at the University of Oxford, where he does both philosophical and epidemiological research.

Phyllis Illari is a Lecturer in Philosophy of Science in the Science and Technology Studies department at University College London. Her current interests are mechanisms, causality, and information, and how they impact on evidence assessment in biomedical sciences. With Clarke, Gillies, Russo, and Williamson, she is just beginning an Arts and Humanities Research Council project entitled "Evaluating Evidence in Medicine," led by the University of Kent and University College, London, and collaborating with the National Institute for Health and Care Excellence and the International Agency for Research on Cancer.

Kristen Intemann is an Associate Professor of Philosophy at Montana State University, specializing in philosophy of science and feminist philosophy of science. She has published in a variety of journals, including *Philosophy of Science*, *Hypatia*, *Synthese*, *Philosophy & Biology*, *European Journal of Epidemiology*, and *EMBO Reports*.

Annemarie Jutel is a sociologist and clinician who focuses on the critical study of diagnosis. She is Professor of Health at Victoria University (NZ). Her books include *Putting a Name to It: Diagnosis in Contemporary Society* and *Social Issues in Diagnosis*.

Jonathan Michael Kaplan is a Professor of Philosophy in the School of History, Philosophy, and Religion at Oregon State University.

Klemens Kappel is Associate Professor and Chair of the Section of Philosophy at Copenhagen University, Denmark. His research interests are social epistemology and political philosophy.

Eric I. Karchmer is a medical anthropologist, teaching at Appalachian State University, North Carolina. His research explores the various ways in which Chinese medicine has changed through the encounter with biomedicine. As part of his fieldwork, he earned a Bachelor's of Medicine from the Beijing University of Chinese Medicine and has been practicing Chinese medicine for more than 15 years.

Ashley Graham Kennedy, PhD, is Assistant Professor of Philosophy in the honors college and Assistant Professor of Clinical Biomedical Science (secondary) in the medical college of Florida Atlantic University. Her research focuses on philosophical issues in diagnostic and clinical reasoning. Ashley also teaches philosophy of medicine, biomedical ethics, and logic to undergraduate and medical students.

Harold Kincaid is Professor of Economics at the University of Cape Town, South Africa, and Visiting Professor at the Finnish Center of Excellence for Philosophy of Social Science at the University of Helsinki. He is the author of numerous books, book chapters, and articles in the philosophy of science. Among his many books is the most recent, *Classifying Psychopathology: Mental Illness and Natural Kinds* (2014).

Adam La Caze's research focuses on philosophical approaches to the problems that arise in health care, including the appropriate interpretation and application of clinical research. He is a Lecturer in the School of Pharmacy at The University of Queensland, Australia.

Maël Lemoine teaches at Tours' Medical School (France). He is the author of *La désunité de la médecine*, an essay on medical explanations, and has written several articles in the philosophy of medicine. He is preparing a book about the naturalization of depression.

Steven Luper is Professor of Philosophy at Trinity University, Texas. His books include *Cambridge Companion to Life and Death* (2014), *The Philosophy of Death* (2009), *Invulnerability: On Securing Happiness* (1996), and *The Possibility of Knowledge* (1987).

Leah McClimans is an Associate Professor at the University of South Carolina. Her publications and research interests are in philosophy of medicine, philosophy of science, and medical ethics.

Inmaculada de Melo-Martín is Professor of Medical Ethics at Weill Cornell Medical College, Cornell University. Her research interests include bioethics and philosophy of science, and she has published widely on both areas. She is the author of *Making Babies* (1998) and *Taking Biology Seriously* (2005).

Cecilia Nardini graduated in Theoretical Physics with a thesis on social networks dynamics. She then obtained a PhD in "Foundation and Ethics of the Life Sciences" from the University of Milan, with a thesis on the statistics of clinical trials. She is currently employed in the industry.

Julian Reiss is Professor of Philosophy at Durham University, UK, and Co-Director of the Centre for Humanities Engaging Science and Society (CHESS). His research focuses on methodological, conceptual, and ethical issues in the philosophy of science, especially in economics and medicine. He is the author of *Causation, Evidence, and Inference* (Routledge, 2015).

Mark Risjord is a Professor of Philosophy at Emory University, Georgia, and he is affiliated faculty of the Nell Hodgson Woodruff School of Nursing. His books include *Nursing Knowledge: Science, Practice, and Philosophy* (2010) and *Philosophy of Social Science: A Contemporary Introduction* (Routledge, 2014).

Christina Schües is Professor of Anthropology and Ethics in the Institute for the History of Medicine and Science Studies at the University of Lübeck, and Adjunct Professor in Philosophy at the Institute for Philosophy and Art Sciences at the Leuphana University, Lüneburg (Germany).

Anita Silvers, PhD, is Professor and Chair of Philosophy at San Francisco State University, California. She is a community member of the San Francisco General Hospital Ethics Committee. Silvers has been awarded the American Philosophical Association Quinn Prize for service to philosophy and philosophers and the Phi Beta Kappa Society Lebowitz Prize for philosophical achievement and contribution and was appointed by the President of the United States to serve on the National Endowment for the Humanities' National Council.

Jeremy R. Simon, MD, PhD, is an emergency physician. He is an Associate Professor of Medicine at Columbia University Medical Center (CUMC), an attending physician in the New York-Presbyterian Emergency Medicine residency, and a member of the Ethics Consultation Service at NewYork-Presbyterian/CUMC.

Dominic Sisti, PhD, is Director of the Scattergood Program for Applied Ethics of Behavioral Health Care and Assistant Professor in the Department of Medical Ethics and Health Policy at the Perelman School of Medicine at the University of Pennsylvania.

Miriam Solomon is Professor of Philosophy at Temple University, Pennsylvania. She works in the areas of philosophy of science, philosophy of medicine, epistemology, and feminist philosophy. She is the author of *Social Empiricism* (2001) and *Making Medical Knowledge* (2015).

Danielle Spencer is a faculty member of the Program in Narrative Medicine at Columbia University as well as the Einstein-Cardozo Master of Science Program in Bioethics in New York. She is a co-author of *The Principles and Practice of Narrative Medicine* (2017).

Jacob Stegenga is an Assistant Professor in the Department of Philosophy at the University of Victoria, British Columbia. His area of research is philosophy of science, including methodological problems of medical research, conceptual questions in evolutionary biology, and fundamental topics in reasoning and rationality.

Jacqueline Sullivan is an Assistant Professor of Philosophy, a member of the Rotman Institute of Philosophy, and an associate member of the Brain and Mind Institute (BMI) at the University of Western Ontario. She has published numerous articles on topics in philosophy of neuroscience and philosophy of psychiatry. She is co-editor with Harold Kincaid of *Classifying Psychopathology: Mental Kinds and Natural Kinds* (2014).

David Teira is Associate Professor of Philosophy at the National University of Distance Education (UNED, Madrid). He currently leads a research project on the correction of subjective biases in social and medical experiments.

Ross Upshur is a Professor in the Department of Family and Community Medicine and Dalla Lana School of Public Health, and an affiliate member of the Institute of the History and Philosophy of Science and Technology at the University of Toronto.

Sean A. Valles is Associate Professor at Michigan State University, with a dual appointment in the Lyman Briggs College and Department of Philosophy.

Jan P. Vandenbroucke was trained in medicine and epidemiology. He is mainly involved in the conduct of observational studies and the application of epidemiologic methods to problems of etiology and pathogenesis as they are investigated in academic medical centers.

Michael Wilde is a Research Fellow in Medical Methodology in the Department of Philosophy at the University of Kent, UK. He works mainly in epistemology and its application to medicine.

Jon Williamson is Professor of Reasoning, Inference, and Scientific Method at the University of Kent, UK. He works on causality, probability, logics, and reasoning, and their application in the sciences.

INTRODUCTION

Miriam Solomon, Jeremy R. Simon, and Harold Kincaid

Philosophy of medicine is an emerging field. Although medical ethics established itself as a discipline in the 1970s, the epistemology and metaphysics of medicine—the central topics of this volume—have come to the fore more recently. In the United States, we look back to 2005 and 2008, when Harold Kincaid organized conferences at the University of Alabama at Birmingham that are now regarded as the first and second conferences of the International Philosophy of Medicine Roundtable (https://philosmed.wordpress.com/). Since then, four conferences were held: Rotterdam 2009, San Sebastian 2011, New York City 2013, and Bristol 2015; the seventh Roundtable is being planned for 2017 in Toronto.

Many of the members of the International Philosophy of Medicine Roundtable (now over 250) teach courses on the philosophy of medicine to undergraduates, graduate students, and medical students. Although the field began with such topics as the definition of disease and the epistemology of evidence-based medicine (with classic papers such as Christopher Boorse's "Health as a Theoretical Concept" [1977] and John Worrall's "*What* Evidence in Evidence-Based Medicine" [2002]), it has now expanded to include a wide range of topics reflecting the latest initiatives in medical science, medical humanities, public health, nursing, and decision sciences. There is a need for a teaching resource that will encourage the next generation of students and scholars to use philosophical tools in these areas.

We intentionally took on a wide and comprehensive range of topics, some of which have little prior mention in the philosophical literature. This is because we wanted to emphasize the continuity between philosophical and other literatures and stimulate more interdisciplinary work. As a result, some of the entries represent first attempts at defining the issues and do not report on well-developed debates. We aimed to err on the side of being maximally inclusive of both topics and philosophical traditions. We have also emphasized topics that have practical implications for medical research and clinical practice. We intend this book to be useful to the medical community, not only the philosophy community.

Philosophy of psychiatry has been partially covered with two articles addressing core questions of ontology: "Psychiatric Classification" and "Models of Mental Illness." We included the ways in which philosophy of psychiatry is continuous with the more general philosophy of medicine literature, rather than address questions more particular to the domain of psychiatry, such as the nature of delusion. Philosophy of psychiatry, like medical ethics, has a much longer history than the rest of philosophy of medicine, and several excellent volumes are available for pedagogical use, such as *The Philosophy of Psychiatry: A Companion* (Radden 2004) and *The Oxford Handbook of Philosophy and Psychiatry* (Davies 2013), to which we refer the interested reader.

In our choices of topics and authors, we were guided by an Advisory Committee of prominent researchers in the fields of philosophy of science, medical ethics, and medical humanities, as well as philosophy of medicine, some of whom also contributed chapters: Rachel Ankeny,

Alexander Bird, Alexander Broadbent, Arthur Caplan, Havi Carel, Tod Chambers, Alice Dreger, Fred Gifford, Trisha Greenhalgh, Brian Hurwitz, Rebecca Kukla, Hilde Lindemann, Kathryn Montgomery, Julian Reiss, David Teira, and John Worrall. We thank the Advisory Committee for their advice about topics and authors to include.

We asked authors to write a general overview piece for their topic and—for pedagogical reasons—to feature centrally when possible a specific case exemplifying the conceptual or epistemological issues in the topic. Apart from these instructions, we gave authors leeway in structuring their chapters, in writing style, and in points of view. The result is considerable variety, which we think adds liveliness to the text. We have managed the contents so that there is minimum overlap among chapters but many cross-references and interconnections. Authors were also asked to provide up to five recommendations for further reading; these should guide students and instructors who want to pursue a topic in more depth.

We have attempted to include a variety of philosophical methodologies—from conceptual analysis (e.g., "Death") to naturalistic inquiry (e.g., "Pain and Suffering") to phenomenological exploration (e.g., "Birth"), though the predominant thread is the philosophy of medicine treated as part of the Anglophone philosophy of science tradition. In some cases—for example "Death" and "Birth"—this means that apparently complementary topics are not treated symmetrically. We had room for more than one philosophical approach, but not room to include more than one essay on each topic. We hope that this stimulates readers to develop the philosophical approaches further, and perhaps to combine them when appropriate.

Even with almost 50 chapters, some topics have had to be omitted or treated indirectly. For example, we do not have a chapter on translational medicine, a research initiative that emerged in the early 2000s (for an account of translational medicine, see Solomon (2015), Chapter 7). And we do not have a chapter on the "art of medicine," instead exploring that elusive concept in several chapters, including "Clinical Judgment," "Narrative Medicine," "Medical Decision Making," "The Biomedical Model and the Biopsychosocial Model," and "Phenomenology and Hermeneutics." A chapter on "Social Determinants of Health" did not materialize at the last minute; we refer the reader to other texts for treatment of this popular topic.

Our division of the text into six parts—General Concepts, Specific Concepts, Research Methods, Clinical Methods, Variability and Diversity, and Perspectives—is an attempt to group the chapters in a meaningful way. The chapters are all self-contained and do not presuppose any philosophy or medicine, so they may be used in any order.

We thank the contributors for working with us, sometimes through several drafts and rounds of editing, to produce accessible yet scholarly chapters. It has been a pleasure to bring together our authors' talents to produce this volume.

Finally, we thank Andy Beck of Routledge Press for conceiving the project and encouraging us throughout the process, Elizabeth Vogt of Routledge Press for working with us on the editorial processes, and Jennifer Bonnar of Apex CoVantage for her project management.

References

Boorse, C. 1977, "Health as a Theoretical Concept," *Philosophy of Science*, vol. 44, no. 4, pp. 542–573.
Davies, M. 2013, *The Oxford handbook of philosophy and psychiatry*, Oxford University Press, Oxford.
Radden, J. 2004, *The philosophy of psychiatry: A companion*, Oxford University Press, Oxford.
Solomon, M. 2015, *Making medical knowledge*, Oxford University Press, Oxford.
Worrall, J. 2002, "What Evidence in Evidence-Based Medicine?" *Philosophy of Science*, vol. 69, no. S3, pp. 316–330.

Part I
GENERAL CONCEPTS

1
THE CONCEPT OF DISEASE

*Dominic Sisti and
Arthur L. Caplan*

Introduction

"What is disease?"

This simple question has vexed philosophers of medicine, yet has been overlooked entirely by bioethicists, physicians, and epidemiologists, with the exception of a rare few. On the one hand, dozens of philosophical theories paint starkly different pictures of the essential nature of disease. On the other hand, most bioethicists, policy makers, and clinicians take disease to be a given—the things we call diseases exist as such, they are bad, and they should, if possible, be treated or eliminated. Conversely, health is, quite simply, often thought to be the absence of disease.

It is across this landscape—from fine-grained analyses to straightforward pragmatism—that a multitude of perspectives on the nature and importance of the concept of disease coexist. This chapter offers a survey of this landscape and specific examples that illustrate the confusing nature and ambiguous applications of the concept of disease. We begin with a historical sketch and then examine contemporary philosophical theories of disease. To demonstrate the way these theories shape empirical realities, we offer cases across a range of medical specialties to illustrate theoretical perspectives.

From there we discuss both the clinical and health policy dimensions of the concept of disease. The concept of disease serves many purposes—identifying behavior that requires control, supporting excuses for absence from work or school, creating eligibility for benefits, and serving as the basis for exculpation in the eyes of the law. We pay particular attention to the complicated nature of psychiatric disease or disorder and lay out the significant practical stakes that depend upon the underlying theoretical orientation of disease.

Historical Thumbnails

The quest to "carve nature at its joints" and provide answers to the titular question of this chapter may be found in the metaphysics and epistemology of classifications of antiquity. Various approaches to categorizing the maladies of human existence are found in the Hippocratic Corpus and in the writings of Plato, Aristotle, and Galen (Hippocrates, 1780, Jowett, 1925, Chadwick, 1983). For the better part of two millennia, classical humoral explanations prevailed. Imbalances in the four basic substances of the body—blood, black and yellow bile, and phlegm—caused disease and disability. The medieval theological constructions of disease by the likes of Maimonides and Aquinas both drew upon the classicists and additionally characterized various diseases in terms of moral failing (Caplan et al., 2004).

The Renaissance polymath and occultist, Paracelsus, rejected humoral models of disease and instead developed a nosology based on external substance or "poisonous" causes (Caplan et al., 1981, Porter and Rousseau, 2004). Descartes, whose anatomical studies complemented his conception of the human organism as an ensouled machine, reframed the concept of disease in a more modern way. Disease was considered a mechanical dysfunction or external attack on the body and a deviation in the teleological functioning of the automatic motions of the body.

Throughout history, the question about the nature of psychiatric disease and mental disorder preoccupied physicians and philosophers alike. Phillipe Pinel aimed to parallel his methodical systematizing of medical and mental illness with that of Hippocrates. He drew upon the philosophies of Locke and Condillac in developing a system of psychopathology that was based upon empirical observation and fact-finding (Charland, 2010). To Pinel, the various mental disorders emerged from a single kind of mental alienation. He believed that sudden, often unexpected reversals in life arising "from the pleasure of success to an overwhelming idea of failure, from a dignified state—or the belief that one occupies one—to a state of disgrace and being forgotten" could bring on "mental alienation" or mania—what we might term today depression or anxiety (Gerard, 1997).

Around the same time, an exceptionally detailed and empirically based psychiatric nosology is presented in Kant's *Anthropology from a Pragmatic Point of View* (Kant, 2006). In *Anthropology*, Kant constructs a highly accessible nosology that was not geared for physicians but might have been useful to philosophers and individuals interested in self-treatment (Frierson, 2009). He defines two general forms of mental disorder that affect the cognitive faculty: melancholia and derangement. The first is a milder form of mental illness that is treatable, whereas the latter seems to include severe psychotic syndromes, bipolar illness, and egosyntonic disorders (Sisti, 2012).

Formal systems of post-humoral nosology find their origin in the work of Thomas Sydenham; disease classification and diagnosis slowly evolves toward what is basically the mainstream view today—an understanding that infectious agents or internal mechanisms cause localized acute or chronic dysfunctions. William Harvey's cardiovascular findings and Morgangni's nosology set up the foundations of understanding diseases as tissue and organ specific—a way of thinking that would prove exceptionally valuable with the later advent of germ theory (Baronov, 2008).

The 19th century saw the emergence of scientific medicine and careful observation in both empirical research and in the categorization of ailments. Claude Bernard, Rudolf Vichow, and Walter Cannon developed theories of disease based on physiological findings related to inflammation, cancerous cell growth, and the loss of homeostasis.

The renewal of interest in the concept of disease in the second half of the 20th century may have in part emerged out of a new and applied ethics and philosophy of science that used biomedicine as both a source of puzzling cases and as a professional substrate (Toulmin, 1982). Some argued that a clarification of the basic concepts of medicine—disease, illness, health, disability, etc.—would be necessary for sorting out clinical and health policy problems (Daniels, 1985).

Clinicians who examined the concepts of health and disease began to recognize deep problems in the ways clinicians equivocated in their use of these terms. Members of the first generation of biomedical ethicists sought to build a new philosophy of medicine within which a more coherent theory of disease and health served a critical role. Edmund Pellegrino, for example, wrote early nosological tracts on osseous lesions that foreshadowed over a half-century of scholarship in the philosophy of medicine and bioethics (Pellegrino et al., 1971). Pellegrino's conceptual work on the concept of disease fit within his overall project on defining the special, if not sacred, relationship between the virtuous physician and his patient (Pellegrino, 2008). He states that "clarification of medicine's basic concepts is as much a moral as an intellectual obligation. . . . [C]onfusion about the nature of health and disease is ultimately confusion about the concept of medicine itself" (Pellegrino, 2004).

Today there are dozens of philosophical theories on the concept of disease, from which has emerged a complex classification-of-classifications debate. The most common is to distinguish between naturalism and normativism. For example, Hofmann illustrates that across the complex array of these many accounts, theories on the concept of disease fall within two broad categories—real essence and nominal essence accounts (Hofmann, 2001). Ereshefsky helpfully adds a third category and groups the theories into the broad categories of naturalist, normativist, and hybrid (Ereshefsky, 2009). This is the tripartite division we will follow in our discussion.

Naturalism

Disease naturalism is the general view that the concept of disease reflects an objective reality about cell, organ, or system function or dysfunction. Physician J.G. Scadding introduced a foundational naturalistic theory grounded upon a biostatistical concept of disease. To Scadding,

> A disease is the sum of the abnormal phenomena displayed by a group of living organisms in association with a specified common characteristic or set of characteristics by which they differ from the norm for their species in such a way as to place them at a biological disadvantage.
>
> (Scadding, 1968)

Christopher Boorse, echoing Scadding, developed and defended the most widely discussed contemporary naturalist theory. Boorse claims that the concept of disease is grounded in the

> autonomous framework of medical theory, a body of doctrine that describes the functioning of a healthy body, classifies various deviations from such functioning as diseases. . . . This theoretical corpus looks in every way continuous with theory in biology and other natural sciences, and [is] value-free.
>
> (Boorse, 1975)

Disease is defined by Boorse as

> a type of internal state which is either an impairment of normal functional ability, i.e. a reduction of one or more functional abilities below typical efficiency, or a limitation on functional ability caused by environmental agents.
>
> (Boorse, 1977)

As such, Boorse argues that diseases are recognizable against the objective backdrop of species-typical function—a concept he borrowed and refined from Scadding. Thus, the epistemic core of Boorse's theory of disease is statistical—determining species typicality is an empirical question. Boorse has labeled his particular brand of naturalism the "biostatistical theory." Biological dysfunction is both necessary and sufficient for defining disease.

Boorse draws a distinction between the concepts of disease and illness. He defines the concept of illness as a subclass of disease: those diseases that carry with them "certain normative features reflected in the instructions of medical practice" are considered illnesses (Boorse,

1975). To support the claim that the concept of disease is value-free, Boorse reminds us that we typically do not claim that plants or animals suffer from an illness. Rather, we describe plants and animals as simply afflicted by a disease. Potato blight, for example, renders its host diseased, not ill. One might ask, "What about cases of animal companions who are 'sick' or 'ill'?" Boorse might respond that in those cases where we refer to animals, such as pets, as being sick or ill, it is in the context of a personal relationship, where we, acting as a companion and caregiver, bring to bear certain values we hold about the animal—for example, that it deserves treatment by a veterinarian.

Second, Boorse recognized that the ascription of illness grants the sufferer "special treatment and diminished moral accountability." Thus, illness is a morally laden concept, whereas the concept of disease is, he maintains, completely value-free. According to Boorse (1975), "A disease is an illness only if it is serious enough to be incapacitating, and therefore is: (1) undesirable for its bearers; (2) a title to special treatment; and (3) a valid excuse for normally criticizable behavior" (Boorse, 1975).

Since Boorse's theory was advanced in the 1970s, a plethora of objections have appeared. Some accuse Boorse of covert normativism by questioning the "evaluative" nature of his terminology—that the concept of function itself smuggles values because it is the product of choice on a range of things to examine (Fulford, 2001). Others deny the possibility that objective evaluations of species-typical functioning are possible.

Key objections turn on the possibility of asymptomatic individuals who present with dysfunctions or infectious agents. Are infertile individuals who do not wish to have children diseased or in any way disordered? Should individuals who have rare mutations in a suite of genes, but who are not experiencing any effects of those mutations, be considered to have a disease? Should carriers of HIV, who experience no symptoms of HIV/AIDS, be thought to have a disease (Wakefield, 2014)? Boorse has replied comprehensively and repeatedly to these and other objections (Boorse, 1997, Boorse, 2014).

Like Boorse, Lennart Nordenfelt argues that the word "disease" is simply an empirical statement and not to be taken as an evaluation of a person's state. But, Nordenfelt's holistic theory of health is different in the way it treats disease. Whereas Boorse works his way upward from diseases defined in terms of biostatistical deviations, Nordenfelt argues that we should work in reverse by first recognizing the suffering and the lived experience of illness and then move to examine the underlying cause of such suffering to reveal the disease state (Nordenfelt, 2007). This approach is resonant of Canguilhem's theory of health, disease, and illness (Nordenfelt, 2007).

Clouser, Culver, and Gert provide a distinctly different account of disease, substituting the concept of "malady" for disease. According to the authors,

> a person has a malady if and only if he or she has a condition, other than a rational belief or desire, such that he or she is suffering, or at increased risk of suffering, an evil (death, pain, disability, loss of freedom or opportunity, or loss of pleasure) in the absence of a distinct sustaining cause.
>
> (Clouser et al., 1981)

The idea here is that certain ontic evils—the authors draw explicitly on Aquinas—are objectively and universally bad. This malady account includes values but only insofar as those values are considered objective and universal; rational persons would agree that suffering, pain, injury, and death are bad and ought to be avoided. But how can an assessment of objective values be reasonably made? Who are these so-called rational persons, and how can their values be identified in a pluralistic world? Should certain maladies that predispose one to self-injury, seizure, or hallucination—such as schizophrenia or epilepsy—but that are valued in a particular culture for spiritual reasons count as objectively evil? Clouser, Culver, and Gert's account is very close to toppling over into normativism.

Normativism

The key feature of normativist theories of disease is that they all reject the possibility that objective necessary and sufficient conditions can be found to identify diseases as such. To express this position positively, normativist theories all consider the concepts of disease and health to be influenced by subjective human values to some extent. Subjective human values are beliefs, preferences, and goals that individuals or communities might reasonably hold as important to achieving their version of the good life. For a normativist, when we signify something as a disease, we are marking out something that is subjectively disvalued by society, culture, or individual preferences. Instead of appealing to purportedly objective, biologically discoverable concepts like "dysfunction," disease signifies something that has compromised an individual's values—i.e., some or all of those things that contribute to the well-being of a person.

Reznick, for example, argues that biological malfunction need not be involved in ascriptions of disease, and that any search for a biological cause of pathology presupposes a subjective judgment about what counts as pathological (Reznek, 1987). Such judgments may be ultimately based upon socio-political values. The normativist view on the concept of disease is now widely held. In fact, the World Health Organization's definition of health reflects this view: "Health is a state of complete physical, mental and social well-being and not merely the absence of disease or infirmity" (World Health Organization, 1948).

Normativist theorists point to historical examples to make a compelling case that our concept of disease is inextricably tied to social, political, and religious values. One need only note the accepted antebellum disease of *drapetomania*—the disease that caused slaves to abscond—to see an example of how medicine and politics amplify one another through deeply flawed nosological constructs (Cartwright, 2004). Ostensibly disordered sexual behaviors—such as masturbation—are often held up as examples in support of the normativist claim that the concept of disease is subjective and in flux (Engelhardt, 1974). Homosexuality is a stark example of how moral values penetrate the scientific endeavor to categorize mental disorder and disease. It was not until 1980 that the continued medicalization of homosexuality was officially rejected by American psychiatry (Spitzer, 1981).

One response of the naturalist to these historical examples is to say they were simply wrong. We now know they were faux diseases that were artifacts of a particular ideology or historical moment. But how do we know this? Because, a naturalist might argue, we can now see they lacked the objective criteria necessary for defining disease and instead reflect bad science. Thus, blatantly erroneous historical examples actually support the naturalist cause—without an objective ideal of the concept of disease, how can we know when we were wrong about labeling something a disease?

The normative position is often found in the writings of medical and psychiatric critical theorists. Extending the position of his mentor and colleague, George Canguilhem, Michel Foucault developed a critical philosophical-historical theory on the use of medical diagnosis. He showed how the medical enterprise was a form of social control meant to cordon off individuals whose sickness or, in the case of mental illness, "unreason" rendered them useless to society (Canguilhem and Delaporte, 2000, Foucault, 2013).

Similarly, we see across the corpus of papers and books by Thomas Szasz a vociferous repudiation of his own profession's most fundamental concept: mental illness. In his classic, *The Myth of Mental Illness*, Szasz argues that mental disorders are whole-cloth fictions created to either marginalize eccentric behaviors or exculpate criminal acts (Szasz, 2010). In one sense, Szasz is a strict naturalist because he argues that the only legitimate mental disorders are actually diseases of the brain, which can be observed in the form of lesions or other "external" causes. In rejecting the conventional categories of mental disorder, he maintains they are indefensibly subjective. R.D. Laing also rejected the psychiatric orthodoxy arguing that mental illnesses are really just instances of mismatch and friction between an individual's and society's values about what constitutes the good life.

A strong normativist perspective is also reflected in scholarship on the nature of disability, where the social model of disability seems to have emerged victorious in a theoretical and political contest with the biomedical model (Sisti, 2014). The social model considers disabilities to be reflective of social and cultural barriers, and that there is nothing intrinsically dysfunctional about conditions like blindness, deafness, dwarfism, or paraplegia. Thus, it rejects naturalist definitions of dysfunction (Amundson, 1992). Although the social model has enjoyed success as reflected in landmark policies such as the Americans with Disabilities Act, some disability scholars think it has outlived its usefulness. For example, Shakespeare has argued that the social model is too blunt an instrument to provide a realistic account of disability, and that more nuanced appraisals of disability that recognize the continuum of function along which we all exist are needed (Shakespeare, 2006).

Another important political offshoot of the normative position is the ongoing examination of how ties to biomedical and pharmaceutical corporations influence the expanding broadness of the concept of disease. The basic concern here is that expert panels responsible for creating or revising nosologies often have ties to industry, and these ties influence their decisions to expand the criteria of those who are diseased or to create new diseases, which hold promise for additional profit. This latter phenomenon, now known as "disease mongering," has been a topic of recent health policy, ethics, and sociological scholarship, though the idea that particular stakeholders endorse and motivate the creation of particular disease categories is not new (Conrad, 1975, Moynihan and Henry, 2006, Moynihan et al., 2013). Examples of purportedly manufactured or intentionally over-diagnosed diseases include mental illnesses such as depression, bipolar II, attention deficit disorder, and premenstrual dysphoric disorder (Conrad, 1975, Batstra and Frances, 2012, Chrisler and Gorman, 2015). Physical illnesses such as erectile dysfunction and high blood pressure have been singled out as examples of diagnostic expansion propelled by pharmaceutical companies' drug sales (Carpiano, 2001). Even the most mundane of daily experiences—such as how and when one sleeps and wakes—are within the purview of medicine. Medical treatments and drugs are increasingly marketed to the public to treat "shift-work sleep disorder" and other conditions that have been created by industrial, political, and economic structures and systems.

Hybrid Models

Efforts have been made to take the best parts of both naturalistic theories and normative theories and graft them together to form hybrid models of the concept of disease and disorder. Jerome Wakefield has developed one of the most important hybrid theories—the harmful dysfunction model (Wakefield, 1992). For a condition to be considered a disorder, it must meet two criteria. First, is the objective biological criteria defined in terms of a failure or dysfunction of an organ or body part to meet its evolutionarily determined function (Wakefield, 2007)? Second, does the dysfunction cause a particular social harm? In contrast to Boorse, biological dysfunction is necessary but *not* sufficient to defining disease or disorder (Wakefield, 2014).

The importance of a more holistic view of disease finds traction in George Engel's classic papers on the biopsychosocial model of illness (Engel, 1978). He sees disease as both a biological reality—a dysfunction in the objective sense—that is fully recognized and treated only by attending to the broader context of the patient. Disease is seen as a point on a continuum of an individual's existence. The physician who is competent in applying the biopsychosocial model would view a myocardial infarction not simply as a malfunction of the heart, but as potentially a result of the patient's personality structure, outside stressors, environmental insults, or the result of a significant personal loss. Thus we can think of the concept of disease in Engel's writings as something slightly different than either a natural kind or a social construction, but as a real thing that manifests within the rich context of human existence.

One of us (Caplan) has argued that a middle ground between naturalism and normativism is both possible and already intuitively exists in the way medicine is practiced. In the cases of mosaicism, hirsuteness, or albinism, one might say that there are biological abnormalities, but because there is relatively little dysfunction—correctible eyesight issues, melanoma risk for albinism—it is arguable whether these conditions ought to be considered diseases. Although there may be social harm—i.e., the significant stigma associated with being albino in many cultures leads to the condition being seen as abhorrent—the biological abnormality causes only manageable, mild impairment. Albinism is seen as unusual in many societies but not really a disease any more than a near-sighted, very light-skinned person would be viewed as diseased. To constitute a disease, both species atypicality and more than minimal dysfunction are necessary conditions (Caplan, 1992).

We might also invoke a form of pluralistic realism, such as that described by philosophers of biology in parsing the species concept, which allows for both recognition of biologically real entities and pragmatic usage of differing definitions to meet particular needs (Ereshefsky, 1992). For example, there is ongoing debate among neurologists about the definition of epilepsy, and one flashpoint is how to accurately define seizures (Walker and Kovac, 2015). It turns out that, for clinicians, the version of "seizure" that makes the most practical sense includes criteria related to outward signs and behavioral changes. This definition is of little use to basic researchers who must detect pre-ictal seizures, those not yet causing overt behavior, in electroencephalogram tracings to develop preventative medications. Such is also the case in psychotic disorders, including schizophrenia, where debate continues about the edges of clinical high-risk states, attenuated psychosis, and the fully developed syndrome (Haroun et al., 2006).

Risk states, such as high blood pressure and hypercholesterolemia, have converged with the disease they portend (Aronowitz, 2009). With new, more precise detection technologies, such as functional magnetic resonance imaging (fMRI), prodromal states of serious neurological and mental illnesses, such as Alzheimer's and schizophrenia, are quickly changing from at-risk states into bona fide chronic illnesses. In some cases, individuals may be completely asymptomatic, but because of genetic or physiological biomarkers, risk states may be labeled as disease states (Karlawish, 2014). As such, these states must be carefully monitored, managed, and, if possible, treated lest the full-blown disease emerge. We can therefore see the concept of disease shifts according to our ability to detect and prevent diseases—the concept is a moving target tethered to the technology of the day.

Diseases, Disorders, and the Future of Psychiatric Nosology

As the previous examples illustrate, psychiatry presents us with some of the most challenging cases when trying to define disease. A large part of the problem is that there are, as yet, no specific biological structures or identifiable causes of mental disorders. We do know that particular parts of the brain and neurotransmitters are implicated in some symptoms of diseases, such as schizophrenia, bipolar disorder, autism, and depression. But direct causal connections—such as we have between insulin and diabetes or fever and infection, for example—remain elusive. To some extent this is a function of definition. When a genetic cause, brain pathology, or lesion is identified as causing a mental disease—such as in the case of Rett syndrome—those disorders are redacted from the official psychiatric nosology and assigned physical disease status.

We have endorsed a hybrid model of disease, disability, and mental disorder, which recognizes that biological dysfunctions are knowable realities but that values infuse both the appraisal of those dysfunctions and the way one responds to them. In the case of mental disorder, it is not at all incoherent to argue that the function of the brain-mind is to provide the individual with both the intra- and inter-personal capabilities to flourish. How flourishing cashes out will vary from person to person, but most would agree that it would include a degree of free thought, relational abilities, independence, and the ability to discern prudential preferences (Sisti et al., 2013).

We also recognize that nosologies reflect certain practical concerns and many entities that are not "true" diseases. There may be times when application of the concept of disease is legitimate but practically pointless, such as in the labeling of conditions for which there is little understanding or cure. Other times it may make sense to call something a disease that is really not a "true" one, but it seems that medicine has the capacity to relieve the human suffering this state causes. For example, migraines may be considered a chronic disease rather than an unfortunate condition, which affords sufferers access to treatment coverage and protection under disability laws. In other words, medicalization is not always bad, and sometimes cure and treatment is possible without clarity about whether what is being treated is a disease.

One case where medicalization might be acceptable is in the description of complicated grief as a mental disorder. Although it has been simmering for decades, the debate about the status of medical grief and bereavement reached a crescendo in the run-up to the publication of the *Diagnostic and Statistical Manual of Mental Disorders, 5th Edition (DSM-5)*—the nosological schedule used by thousands of mental health professionals worldwide (American Psychiatric Association, 2013). It was proposed that the so-called bereavement exclusion to major depression—the stipulation that depression ought not be diagnosed within two months of a significant loss—be eliminated. This opened the door to the idea that a person could be usefully treated and medicated immediately after suffering a serious loss under the aegis of depression.

Protest ensued. On the one hand, grief or bereavement seems like a natural response to a significant loss such as the death of a spouse, child, or loved one, or the loss of worldly possessions after a natural disaster, life savings in a stock market crash, or employment (Horwitz and Wakefield, 2007). In such cases, it seemed to be stretching credulity to view grief as a part of depression. After all, are not grief and loss a normal part of life? On the other hand, serious grief would often present in ways similar to depression, it could be incapacitating, and it often responds positively to antidepressants. Why not treat it as disease in its own right?

The sustained examination of the concept of disease will continue into the future as researchers and clinicians continue to unravel causes of psychiatric disorders that are now only defined by outward behaviors. The National Institute of Mental Health (NIMH), for example, has moved to redefine the concept of mental disorder according to more granular models of genetics, neurochemistry, and neurological systems and circuitry. As previously mentioned, a diagnosis of schizophrenia may become possible in the preclinical or prodromal stages, before any obvious symptoms emerge. This brings us back to the question of asymptomatic carriers—should they be considered to be diseased and offered the entitlements of sick individuals, such as increased access to preventative treatment, social allowances, and particular accommodations? These questions have yet to be answered but are ultimately rooted in the simple question, "What is disease?"

References

AMERICAN PSYCHIATRIC ASSOCIATION. 2013. *Diagnostic and statistical manual of mental disorders (DSM-5)*, Washington, DC, American Psychiatric Association.
AMUNDSON, R. 1992. Disability, handicap, and the environment. *Journal of Social Philosophy*, 23, 105–119.
ARONOWITZ, R. A. 2009. The converged experience of risk and disease. *Milbank Quarterly*, 87, 417–442.
BARONOV, D. 2008. Biomedicine: An ontological dissection. *Theoretical Medicine and Bioethics*, 29, 235–254.
BATSTRA, L. & FRANCES, A. 2012. Diagnostic inflation: Causes and a suggested cure. *The Journal of Nervous and Mental Disease*, 200, 474–479.
BOORSE, C. 1975. On the distinction between disease and illness. *Philosophy & Public Affairs*, 5(1), 49–68.
BOORSE, C. 1977. Health as a theoretical concept. *Philosophy of Science*, 44(4), 542–573.
BOORSE, C. 1997. A rebuttal on health. In J. M. Humber & R. F. Almeder, *What is disease?* Towtowa, NJ: Humana Press, pp. 3–134.
BOORSE, C. 2014. A second rebuttal on health. *Journal of Medicine and Philosophy*, 39, 683–724.
CANGUILHEM, G. & DELAPORTE, F. 2000. *A Vital Rationalist: Selected Writings from Georges Canguilhem*, Brooklyn, NY: Zone Books.
CAPLAN, A. L. 1992. "Is aging a disease?" In A. L. CAPLAN, *If I were a rich man could I buy a pancreas?* Bloomington: Indiana University Press, pp. 195–209.
CAPLAN, A. L., ENGELHARDT, H. T. & MCCARTNEY, J. J. 1981. *Concepts of health and disease: Interdisciplinary perspectives*, Reading, MA, Addison-Wesley Publishing.
CAPLAN, A. L., MCCARTNEY, J. & SISTI, D. 2004. *Health, disease, and illness: Concepts in medicine*, Washington, DC: Georgetown University Press.
CARPIANO, R. M. 2001. Passive medicalization: The case of Viagra and erectile dysfunction. *Sociological Spectrum*, 21, 441–450.
CARTWRIGHT, S. A. 2004. Report on the diseases and physical peculiarities of the Negro Race. In: ARTHUR CAPLAN, JAMES MCCARTNEY & DOMINIC SISTI (eds.). *Health, disease, and illness: Concepts in medicine*, Washington, DC: Georgetown University Press, pp. 28–39.
CHADWICK, J. 1983. *Hippocratic writings*, London: Penguin Classics.
CHARLAND, L. C. 2010. Science and morals in the affective psychopathology of Philippe Pinel. *History of Psychiatry*, 21, 38–53.
CHRISLER, J. C. & GORMAN, J. A. 2015. The medicalization of women's moods: Premenstrual syndrome and premenstrual dysphoric disorder. In: *The Wrong Prescription for Women: How Medicine and Media Create a "Need" for Treatments, Drugs, and Surgery*, Santa Barbara, CA: Praeger Press, pp. 77–98.
CLOUSER, K. D., CULVER, C. M. & GERT, B. 1981. Malady: A new treatment of disease. *The Hastings Center Report*, 11, 29–37.
CONRAD, P. 1975. Discovery of hyperkinesis: Notes on the medicalization of deviant behavior. *Social Problems*, 23, 12–21.
DANIELS, N. 1985. *Just health care*, Cambridge, UK: Cambridge University Press.
ENGEL, G. L. 1978. The biopsychosocial model and the education of health professionals. *Annals of the New York Academy of Sciences*, 310, 169–181.
ENGELHARDT, H. T. 1974. The disease of masturbation: Values and the concept of disease. *Bulletin of the History of Medicine*, 48, 234.
ERESHEFSKY, M. 1992. Eliminative pluralism. *Philosophy of Science*, 59, 671–690.
ERESHEFSKY, M. 2009. Defining 'health' and 'disease'. *Studies in History and Philosophy of Science Part C: Studies in History and Philosophy of Biological and Biomedical Sciences*, 40, 221–227.

FOUCAULT, M. 2013. *History of madness*, London/New York: Routledge.
FRIERSON, P. 2009. Kant on mental disorder: Part 1: An overview. *History of Psychiatry*, 20, 267–289.
FULFORD, K. W. M. 2001. What is (mental) disease?: An open letter to Christopher Boorse. *Journal of Medical Ethics*, 27, 80–85.
GERARD, D. L. 1997. Chiarugi and Pinel considered: Soul's brain/person's mind. *Journal of the History of the Behavioral Sciences*, 33, 381–403.
HAROUN, N., DUNN, L., HAROUN, A. & CADENHEAD, K. S. 2006. Risk and protection in prodromal schizophrenia: Ethical implications for clinical practice and future research. *Schizophr Bull*, 32, 166–178.
HIPPOCRATES. 1780. *The history of epidemics*. Translated into English from the Greek, with notes and observations, and a preliminary dissertation on the nature and cause of infection. By Samuel Farr, M.D. F.R.S. In: S. FARR (ed.). *The history of epidemics*, London: Printed for T. Cadell.
HOFMANN, B. 2001. Complexity of the concept of disease as shown through rival theoretical frameworks. *Theoretical Medicine and Bioethics*, 22, 211–236.
HORWITZ, A. V. & WAKEFIELD, J. C. 2007. *The loss of sadness: How psychiatry transformed normal sorrow into depressive disorder*, New York: Oxford University Press.
JOWETT, B. 1925/1892. *The dialogues of Plato*, London: Oxford University Press.
KANT, I. 2006. *Kant: Anthropology from a pragmatic point of view*, Cambridge, UK: Cambridge University Press.
KARLAWISH, J. 2014. How are we going to live with Alzheimer's disease? *Health Affairs*, 33, 541–546.
MOYNIHAN, R. N., COOKE, G. P. E., DOUST, J. A., BERO, L., HILL, S. & GLASZIOU, P. P. 2013. Expanding disease definitions in guidelines and expert panel ties to industry: A cross-sectional study of common conditions in the United States. *PLoS Med*, 10, e1001500.
MOYNIHAN, R. N. & HENRY, D. 2006. The fight against disease mongering: Generating knowledge for action. *PLoS Med*, 3, e191.
NORDENFELT, L. 2007. The concepts of health and illness revisited. *Medicine, Health Care and Philosophy*, 10, 5–10.
PELLEGRINO, E. D. 2004. Forward: Renewing medicine's basic concept. In: A. CAPLAN, J. J. MCCARTNEY & D. A. SISTI (eds.). *Health, disease, and illness: Concepts in medicine*, Washington, DC: Georgetown University Press, pp. xi–xiv.
PELLEGRINO, E. D. 2008. *The philosophy of medicine reborn: A Pellegrino reader*, Notre Dame, IN: University of Notre Dame Press.
PELLEGRINO, E. D., BILTZ, R. M. & PELLEGRINO, M. J. 1971. A microradiographic study of bone disease in uremia. *Transactions of the American Clinical and Climatological Association*, 82, 177–187.
PORTER, R. & ROUSSEAU, G. S. 2004. Prometheus's Vulture: The Renaissance Fashioning of Gout. In: ARTHUR CAPLAN, JAMES MCCARTNEY & DOMINIC SISTI (eds.). *Health, disease, and illness: Concepts in medicine*, Washington, DC: Georgetown University Press, pp. 11–28.
REZNEK, L. 1987. *The nature of disease*, London: Routledge & Kegan Paul.
SCADDING, J. G. 1968. The clinician and the computer. *The Lancet*, 291, 139–140.
SHAKESPEARE, T. 2006. The social model of disability. *The Disability Studies Reader*, 2, 197–204.
SISTI, D. A. 2012. Was Kant a normativist or naturalist for mental illness? *Journal of Ethics in Mental Health*, 7, 1–7.
SISTI, D. A. 2014. Naturalism and the social model of disability: Allied or antithetical? *Journal of Medical Ethics*, 41, 553–556.
SISTI, D. A., YOUNG, M. & CAPLAN, A. 2013. Defining mental illnesses: Can values and objectivity get along? *BMC Psychiatry*, 13, 346.
SPITZER, R. L. 1981. The diagnostic status of homosexuality in DSM-III: A reformulation of the issues. *American Journal of Psychiatry*, 138, 210–215.
SZASZ, T. S. 2010. *The myth of mental illness: Foundations of a theory of personal conduct*, New York: Harper Perennial.
TOULMIN, S. 1982. How medicine saved the life of ethics. *Perspectives in Biology and Medicine*, 25(4), 736–750.

WAKEFIELD, J. C. 1992. The concept of mental disorder: On the boundary between biological facts and social values. *American Psychologist*, 47, 373.
WAKEFIELD, J. C. 2007. The concept of mental disorder: Diagnostic implications of the harmful dysfunction analysis. *World Psychiatry*, 6, 149–156.
WAKEFIELD, J. C. 2014. The biostatistical theory versus the harmful dysfunction analysis, part 1: Is part-dysfunction a sufficient condition for medical disorder? *Journal of Medicine and Philosophy*, 39, 648–682.
WALKER, M. C. & KOVAC, S. 2015. Seize the moment that is thine: How should we define seizures? *Brain*, 138, 1127–1128.
WORLD HEALTH ORGANIZATION. 1948. Preamble to the Constitution of the World Health Organization as adopted by the International Health Conference, New York. Official Records of the World Health Organization.

Further Reading

ARONOWITZ, R. A. 1998. *Making sense of illness: Science, society and disease*, Cambridge, UK: Cambridge University Press.
CAPLAN, A. L., ENGELHARDT, H. T. & MCCARTNEY, J. J. 1981. *Concepts of health and disease: Interdisciplinary perspectives*, Reading, MA: Addison-Wesley Publishing.
CAPLAN, A. L., MCCARTNEY, J. & SISTI, D. 2004. *Health, disease, and illness: Concepts in medicine*, Washington, DC: Georgetown University Press.
ENGELHARDT, H. T. 1996. *The foundations of bioethics*, Oxford, UK: Oxford University Press.
FOUCAULT, M. 1988. *Madness and civilization: A history of insanity in the age of reason*, New York: Vintage Books.
FULFORD, B., THORNTON, T. & GRAHAM, G. 2006. *Oxford textbook of philosophy and psychiatry*, Oxford, UK: Oxford University Press.
GHAEMI, S. 2003. *The concepts of psychiatry*, Baltimore: John Hopkins University.
HORWITZ, A. V. & WAKEFIELD, J. C. 2007. *The loss of sadness: How psychiatry transformed normal sorrow into depressive disorder*, Oxford, UK: Oxford University Press.
HUMBER, J. M. & ALMEDER, R. F. 1997. *What is disease?* Towtowa, NJ: Humana Press.
MARGOLIS, J. 1986. Thoughts on definitions of disease. *Journal of Medicine and Philosophy*, 11, 233–236.
NELSON, J. L. & NELSON, H. L. 1999. *Meaning and medicine: A reader in the philosophy of health care*, New York/London: Routledge/Psychology Press.
PORTER, R. & ROUSSEAU, G. S. 2000. *Gout: The patrician malady*, New York/London: Yale University Press.
RADDEN, J. 2002. *The nature of melancholy: From Aristotle to Kristeva*, Oxford, UK: Oxford University Press.
SHELL, S. M. 1996. *The embodiment of reason: Kant on spirit, generation, and community*, Chicago, IL: University of Chicago Press.
SZASZ, T. 1974. *The myth of mental illness: Foundations of a theory of personal conduct*, revised edition, New York: Perennial.

2
DISEASE, ILLNESS, AND SICKNESS

Bjørn Hofmann

Background

There are many heated debates about the concept of disease: Is aging a disease? What about obesity, electromagnetic hypersensitivity, insomnia, and grief? How can we understand myalgic encephalomyelitis, chronic fatigue syndrome, and Lyme disease? In such cases, we often confer with definitions of disease to decide. However, as shown in Chapter 1, it is much more difficult to define disease (the general concept) than we might think at first sight. (It is also difficult to define particular diseases, and this is the topic of Chapters 16 and 17.) There are many diverging definitions of *disease*, and it is difficult to make a coherent and consistent synthesis. One of the reasons for this may be that our conception of disease is complex, comprising various dimensions of human malady (Hofmann, 2001). For example, it is covered by a personal perspective—i.e., how it feels to be ill (illness); a professional perspective—i.e., how health care professionals define, detect, predict, and handle disease entities (disease); and a social perspective—i.e., how a person's social role is defined or changed by social norms and institutions (sickness). These perspectives focus on different phenomena and entities, they comprise different types of knowledge, and they call for different actions from health care professionals. This can explain some of the controversies over the concept of disease (see Chapter 1) and help to reduce the complexity. At the same time, the perspectives have challenges of their own. The goal of this chapter is to introduce a more precise sub-classification of the general concept of disease into three parts: disease (strictly speaking), illness, and sickness. *Malady* will be used as a generic term covering a wide range of terms, such as disease, illness, sickness, but also injury, lesion, defect, disorder, deformity, disability, impairment, infirmity, etc. (Clouser et al., 1997, Sadegh-Zadeh, 1981, 2000). Accordingly, the objective of this chapter is to explore the three prominent perspectives on human malady: disease, illness, and sickness.

Defining the Triad *Disease, Illness,* and *Sickness*

The distinction between *illness* and *disease* has been noted in the theoretical literature on medicine since the 1950s (Parsons, 1951, 1958, 1964). The sociologist Andrew Twaddle was the first to elaborate on the distinction among disease, illness, and sickness in his doctoral dissertation defended in 1967 (Twaddle, 1968, 1994a, 1994b). The distinction has since become commonplace in medical sociology, medical anthropology, and philosophy of medicine. (For the literature on various perspectives on human malady, see for example, Boyd, 2000, Engelhardt & Wildes, 2004, Fabrega, 1972, 1979, King, 1954, Marinker, 1975, Parsons,

1951, Rothschuh, 1972, Sedgwick, 1973, Susser, 1990, Taylor, 1979, Von Engelhardt, 1995, and Young, 1995.) The triad *disease*, *illness*, and *sickness* has been elaborated and more strictly defined (Susser, 1990, Twaddle, 1994a, 1994b), but also challenged (Nordenfelt, 2007, Twaddle, 1994b).

As can be seen in Chapter 1, there is still debate on how to define *disease*. The same goes for *illness* and *sickness*. However, there is substantial agreement that physiological, biochemical, genetic, and mental entities and events are the basic phenomena of *disease*, and most definitions contend that *disease* can be observed, examined, mediated, and measured, and is objective in this sense. It is also the target of health professionals who want to classify, detect, control, and treat *disease*, ultimately in order to cure.

Illness, on the other hand, has emotions and experience, such as anxiety, fear, pain, and suffering, as its basic phenomena. Although *illness* is in this sense subjective, it can be argued that we can have access to another person's illness through his or her verbal reports of introspection (phenomenology; Carel, 2013, 2014a, Svenaeus, 2014, Toombs, 1990), through a common language (the philosophy of language; Hofmann, 2015), or through brain states (Daniel, 1991). For example, illness is characterized in terms of bodily and/or mental awareness and a feeling of estrangement, unpleasantness, or uncanniness (Carel, 2014b). Health professionals' aim with regard to *illness* is comfort, care, and/or relief of suffering.

Sickness, on the other hand, has expectations, conventions, policies, and social norms and roles as basic phenomena (Susser, 1990). Its criteria are discovered through social interaction, participation, and social studies. Accordingly, the knowledge about *sickness* is inter-subjective (i.e., it is knowledge shared by a social group). *Sickness* determines whether a person is entitled to treatment and economic rights, exemption from social duties, such as work (sick leave), but also whether a person is legally accountable for his or her actions. In such matters, *sickness* is established and governed by formal structures, such as social institutions (including laws). However, *sickness* can also be framed by overt or covert norms, which give diseases different prestige and render them stigmatizing or discriminating. For example, myocardial infarction has a higher prestige than fibromyalgia among medical doctors (Album and Westin, 2008); the fact that homosexuality was classified as a disease for many years was experienced as stigmatizing; and obesity can lead to discrimination in health care and society at large (Hofmann, 2010, Puhl and Heuer, 2009, Wilt et al., 2010).

All aspects of *sickness* do not apply all the time (e.g., hypertension found by a health check qualifies for treatment and additional check-ups, but not for sick leave). As *sickness* is constituted by social norms, *sickness* may vary from place to place. For example, pelvic girdle pain qualifies for sick leave in Norway but not in other countries (Dorheim et al., 2013). Moreover, although *sickness* is a societal construct, it is important to notice that both patients and health professionals are social agents influencing *sickness*. For example, health professionals are involved as gatekeepers and sometimes as "rescuers." At the beginning of the 20th century, health professionals argued that making homosexuality a disease would save it from being a social stigma (Hofmann, 2014).

Hence, the concepts of *disease*, *illness*, and *sickness* highlight different perspectives on important aspects of human life. These concepts reflect medical professional, personal, and social perspectives, respectively. Furthermore, *disease*, *illness*, and *sickness* are primarily negative notions (i.e., they refer to occurrences in human life of negative value). They may, however, also have positive aspects, such as increased attention, justified work absence, or economic support (Hofmann, 2014), but the positive aspects of human malady are beyond the scope of this chapter.

Moreover, each concept calls for action. *Disease* calls for actions by health professionals with the goals of identifying, treating, and handling the entities and events and to care for the person. *Illness* changes the self-concept, relationships, and activities of the individual (e.g.,

Table 2.1 Different perspectives of human malady: Disease, illness, and sickness investigated

	Disease	*Illness*	*Sickness*
Field, Area, Primary agents/ stakeholders	Profession, medical and other health care professions	Personal, (experiential, existential)	Society, social institutions, health policy makers, lawyers
Basic phenomena	Physiological, mental, genetic, environmental entities or events	Subjective experience, first-person negative experience, suffering, pain	Social conventions, norms, roles (including social prejudice)
Access to phenomena through:	Observations, examinations, measurements (by the natural sciences and by the use of technology)	Introspection, intuition (phenomenology), interaction (language), mental states (psychology)	Participation, interaction, social (science) studies
Knowledge status	Objective	Subjective	Inter-subjective
Altruistic approach	Cure	Care	Resource allocation, justice
Entitles to: Results in:	Examination, diagnostics, treatment	Attention, support, moral and social excuse, reduced accountability	Economic support and compensation, sick leave, but may also result in discrimination and stigmatization

making the person call for help). *Sickness* is a determination of the social status of the person being sick, in particular, with regard to entitlement to treatment and economic rights and exemption from social duties, such as work (sick leave).

Table 2.1 gives a summary of the main features of the three perspectives on human malady (i.e., most definitions of *disease*, *illness*, and *sickness* include such features).

Applying the Triad to Cases from Clinical Practice

Disease, *illness*, and *sickness* are neither mutually exclusive nor exhaustive. They are interrelated and partly interdependent, but there are no necessary connections among them. Although they frequently occur conjoint, they all can occur without the others. Figure 2.1 illustrates the relationship among the three concepts of the triad.

This triad can help clarify some of the issues in the philosophy of medicine (e.g., to address some challenges with naturalist and normativist definitions of disease). Descriptive or naturalist theories of disease have been accused of making pregnancy, excellence, and homosexuality into diseases (Boorse, 1975, Cooper, 2005, Zachar and Kendler, 2012). On the other hand, normativist or nominalist theories are charged with making ageing, shyness, and sadness into diseases (Horwitz and Wakefield, 2007). In the following section, some specific cases will be addressed to illustrate the potential fruitfulness of the distinctions among *disease*, *illness*, and *sickness*.

The paradigm case in health care is when a person feels *ill*, the medical profession is able to detect and treat *disease*, and society attributes to him the status of being *sick*. *Illness* alters the person's situation, explains it to himself, and calls for care, *disease* permits medical explanation,

DISEASE, ILLNESS, AND SICKNESS

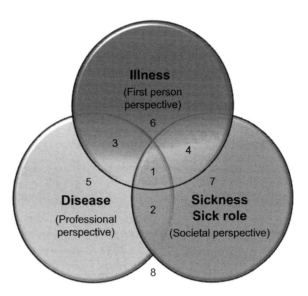

Figure 2.1 Visual outline of the triad disease, illness, and sickness (Hofmann, 2002)

attention, and action and *sickness* frees him from ordinary duties of work and gives him the right to economic support (case 1 in Figure 2.1). Examples of such conditions are numerous. Ischemic heart disease, stroke, chronic obstructive pulmonary disease (COPD), and lower respiratory infections are but three examples that top the World Health Organization's list of leading causes of deaths worldwide (2000 and 2012). There is no disagreement on such cases. Here negative bodily occurrences as conceived of by the individual correspond with negative bodily occurrences recognized by the medical profession and by relevant social institutions. Hence, cases of *disease*, *illness*, and *sickness* are paradigm cases of human malady and of health care.

Less agreement may occur when only two of the perspectives coincide (e.g., instances of conditions that satisfy the criteria for both *disease* and *sickness*, but not for *illness*; case 2 in Figure 2.1). For example, there are conditions in which certain signs or (bio)markers are identified by the medical profession before the patient experiences any illness and which leads to an entitlement to treatment and economic support (*sickness*). High blood pressure (without symptoms), human papilloma virus (HPV), pre-diabetes, biomarkers for Alzheimer's disease, ductal carcinoma in situ (DCIS), and other conditions found by screening, predictive testing, or as incidentalomas belong to this group. The professionals are confident that they are dealing with *disease*, social institutions designate the person in question as *sick*, but the person is (initially) not *ill*. The same situation can be found when patients are unconscious or have impairments recognized by the medical profession and by society, but not by the person in question.

We also have cases with instances of both *disease* and *illness*, but not of *sickness* (case 3 in Figure 2.1). Examples are the common cold and a headache after drinking alcohol. The medical profession is able to recognize these conditions as *disease* by various diagnostics, and the person in question certainly experiences them as negative, but it does not qualify as *sickness* for all, as they are expected to work (if they are not drunk).

Furthermore, there are conditions that are both *illness* and *sickness* but not *disease* (case 4 in Figure 2.1). Some cases of chronic Lyme disease, whiplash, candida, and irritable bowel syndrome (IBS) may be examples of conditions where the person certainly feels ill and society (in many countries) entitles the person to have the status of being *sick*, but where the medical profession is not always able to identify or detect *disease*. It has been extensively debated in the medical literature and in the health insurance setting in many countries whether whiplash, as well as cases where persons are bitten by ticks but do not test positive on validated borrelia tests, have a disease. Correspondingly, pregnancy is commonly not conceived of as *disease* by the medical profession, although it might be experienced by many women as *illness* and accepted by society as a reason for sick leave (*sickness*).

Moreover, asymptomatic instances of hyperglycemia, hypertension (low or moderate), and various genetic mutations are examples of *disease* that are neither *illness* nor *sickness* (case 5 in Figure 2.1). The medical profession conceives of these and diagnoses them as *disease*, but the person does not experience them as such, and they do not normally qualify as *sickness*.

Correspondingly, instances of *illness* that are neither *sickness* nor *disease* (case 6 in Figure 2.1) represent cases that are experienced by the person as negative, but are neither recognized as *sickness* by society nor as *disease* by the medical profession. An intense feeling of fatigue, dissatisfaction, unpleasantness, incompetence, anxiety, or sadness might be examples. Decades ago, fibromyalgia, chronic fatigue syndrome (CFS), and myalgic encephalomyelitis (ME) also belonged to this group. Now they are commonly regarded as diseases, even though we still do not understand the mechanisms underlying them.

Cases of *sickness*, which are neither *disease* nor *illness* (case 7 in Figure 2.1) are also of great interest. Delinquency, dissidence, homosexuality, drapetomania, and masturbation may count as (historical) examples of cases in which social institutions have designated people as *sick*, but the person has not felt *ill*, and the medical profession has not diagnosed any negative bodily correlates, although some tried very hard to do so. The examples of *sickness* but not *illness* or *disease* are mainly historical examples, as we like to believe that today's society is free of such repressive actions. However, in China the members of Falun Gong have been hospitalized on the basis of their religion. Moreover, in many countries, prominent ears in children are treated without the persons feeling *ill* and the professionals necessarily thinking that they have a *disease*. Many instances of attention deficit/hyperactivity disorder (AD/HD) may also be deemed to be so in the future. Many children with this diagnosis do not feel *ill*. However, social norms for education and conduct make them *sick*. Health professionals may not find anything wrong in the children's organs or functioning, beyond their social behavior (defined by society, i.e., *sickness*).

These cases illustrate that the triad can account for controversial cases discussed in the literature. Note that these are not the only possible examples and that the cases may be interpreted otherwise (e.g., due to variations in professional knowledge and social norms with place and time). The point is that the concepts of *disease*, *illness*, and *sickness* represent a framework for analyzing controversial cases and for explaining controversies (i.e., several of the controversies result from conflicting perspectives of important stakeholders).

Epistemic and Normative Consequences

Accordingly, cases incorporating *disease*, *illness*, and *sickness* (case 1 in Figure 2.1) are not epistemically (or normatively) problematic. The person has some negative bodily or mental experience making him/her request help, the medical profession recognizes certain signs and (frequently) knows what can be done, and society and its institutions entitle him/her to treatment, economic support, and a release from certain obligations like work.

The three kinds of cases, in which only one member of the triad is applicable (cases 5, 6, and 7 in Figure 2.1) are quite challenging. First, when the medical profession classifies a condition as a *disease*, but the patient does not feel any *illness* and society does not find any reason to change his/her social status (case 5), both epistemic and normative difficulties result. How can we know that people with asymptomatic diseases will actually develop symptoms and become ill? Should people with low or moderate hypertension be subject to extensive treatment? Is sickle cell trait a matter of the same medical concern in areas with malaria as without? Is a person with lactose intolerance only sick if he/she lives in areas where dairy products are part of the traditional diet? Can it be right to treat polydactylism and obesity if it does not bother the person?

Situations in which a person experiences *illness*, but no *disease* has been found, and where there is no change in one's social status (case 6 in Figure 2.1), also are challenging. Epistemically, it is difficult for the medical profession to uncover the cause of the person's suffering. Normatively, it is hard to see what society ought to do in such situations. A general feeling of dissatisfaction does not normally qualify the person for medical care or economic support. On the other hand, medicine has been criticized for medicalizing a wide range of everyday experiences. Moreover, aspirations to handle all cases of illness are also limited by resource allocation and prioritization.

Cases of *sickness* that are neither *disease* nor *illness* (case 7) are challenging and may even be dangerous. Drapetomania (a "disease" that made slaves run away; Bynum, 2000), masturbation, homosexuality, and political dissidence are crude examples where society (with or without its institutions) has deemed the conditions as *sickness*. There appears to be no professionally accepted diagnostic criteria in these cases (any longer). The epistemic and moral norms that entitled a person to be *sick* in these cases have later been changed.

Hence, cases where only one member of the triad applies certainly call for special attention. However, cases in which two of the three apply (cases 2, 3, and 4 in Figure 2.1) may be epistemically and normatively challenging as well. First, cases of both *disease* and *illness* but not of *sickness* (case 3) are subject to pressure from professionals and patient interest groups (and industry) for support. There may be several reasons why the status of *sickness* is not granted, even though the person has both *disease* and *illness*, such as lack of resources, commonness, or where no treatment is available. Myopia and tooth decay are examples of cases that are not conceived of as *sickness* in many countries with "universal coverage," but are acknowledged to be *disease* by the medical profession and are experienced negatively by persons with these conditions. The epistemic challenge is to find effective and efficient cures, whereas the normative challenges are connected to questions of priority setting and to cases in which people are not able to pay for health care services themselves.

Second, cases of both *illness* and *sickness* but not of *disease* (case 4) put pressure on the medical research community to find mechanisms and causes of these occurrences, which are both personally experienced and economically supported. Low back pain, medically unexplained physical symptoms (MUPS), and sorrow may serve as examples. The etiology of and treatment for these conditions are not commonly agreed upon. They have, however, been accepted in various countries as *sickness*, and people certainly claim to experience them as *illness*. There is pressure on the medical establishment to see these conditions as *disease* as well. There is both an epistemic challenge to establish etiology and a normative challenge to find a treatment, since such conditions ought to be treated.

Third, cases of both *disease* and *sickness* that are not *illness* (case 2) generate some profound challenges. Epistemically, we are challenged by the question of whether a person will actually become *ill* when test results indicate *disease*. Many cases defined as *disease* will not develop to *illness* if left untreated. Ductal carcinoma in situ (DCIS) is but one example. In such cases of overdiagnosis, the person may die with the condition rather than from it (Welch et al., 2011). Normatively, we are faced with a series of questions: How are we to handle the results from screening and predictive testing? Are there limits to the treatment of asymptomatic diseases? Are we going to tell the patient about the findings? The discussion on genetic testing (including incidental finding of uncertain significance), hypercholesterolemia, and hypertension illustrates some of these normative challenges (Fanu, 1999). How far can we go in treatment of cases in which the patient is not *ill*? How is patient autonomy preserved? Who is to balance the risks and benefits of such treatment?

Hence, cases that fall outside of case 1 in Figure 2.1 (i.e., where only one or two of the triad's concepts apply) represent epistemic and normative challenges. Moreover, it may be argued that cases that belong to only one of the spheres of the triad may be more challenging than cases that belong to two. Cases are less controversial if they are recognized by two of the agents as being both *disease* and *sickness* (case 2 in Figure 2.1), both *disease* and *illness* (case 3), or both *illness* and *sickness* (case 4) than if they are only recognized by one of the agents as *disease* (case 5), *illness* (case 6), or *sickness* (case 7). That is, we appear to be more challenged by medical treatment of incompetence, dissatisfaction, homosexuality, dissidence, and low or moderate hyperglycemia than we are by the treatment of asymptomatic breast cancer, common colds, and seasickness. The pressure on medicine to accept an occurrence as *disease* is strong when it is recognized both as *illness* and *sickness*. In the same way, there is pressure on society to provide necessary resources and to admit that an occurrence is *sickness* when it is recognized both as *disease* and *illness*.

In cases of only *illness*, the ill person has to convince both the medical profession and social institutions about his or her situation. Many have found media to be helpful in this regard. Similarly, social institutions have to convince both the medical profession and the person in cases of *sickness* alone. This can be done through funding, regulation, and education. In cases of *disease* alone, both society and the person have to be persuaded (e.g., through scientific and popular publications). Thus, cases in which only one member of the triad is applicable appear to be challenging. When more perspectives coincide, the cases become less controversial.

The Dynamics Among Disease, Illness, and Sickness

The concepts of *disease*, *illness*, and *sickness* are not independent. Making something a *disease* (e.g., by making something subject to medical attention and manipulation) influences the attribution of a social status to the condition (i.e., making it *sickness*). *Infertility*, which was traditionally not considered a *sickness* in many countries, now qualifies for economic support because it became treatable as a *disease*. Conversely, many conditions, including sorrow, pregnancy, and obesity, have become *disease* because they have gained social attention (e.g., by being covered by health insurances).

Similarly, the experience of *illness* is affected by medical knowledge. The personal experience of ailment is influenced by the medical terminology (e.g., a soccer player might state that he has some pain in his *meniscus*, or a patient can feel her "large intestines a bit bound") (Nessa and Malterud, 1998). New imaging technology may also influence both *illness* and *disease* (McCabe and Castel, 2008).

Conversely, the experience of *illness* influences the activities of the medical profession. Research into lower back pain, whiplash, and myalgic encephalomyelitis (ME) was initiated by people's suffering and need for help. The status of pregnancy and childbirth as *illnesses* and *sicknesses* has made the medical establishment hospitalize pregnant women as if they were having *disease*.

Correspondingly, professionals preoccupied with *disease* are influenced by the social status and prestige of *sickness*. As already mentioned, disease entities vary greatly in their prestige

(Album and Westin, 2008). The search for a causal explanation for fibromyalgia is supported by its status as *illness* and *sickness*. On the other hand, cases of the common cold are not always accepted as *sickness* (Copeland, 1977). Furthermore, the social sphere to a large extent governs medical education and research, and the social and psychological influences on the concept of disease are clearly reflected in the influential biopsychosocial model of disease (Engel, 1977).

Moreover, the class membership of the areas may vary over time. To a person, none, one, or more of the triad's concepts may apply at the same time (Twaddle, 1994a). Even more, the membership may be complex and change with time. Both the medical professionals and ill people are members of society, and thus all influence the sphere of *sickness*. In particular, in many countries, the physician is society's gatekeeper and manages both *disease* and *sickness* at the same time. Furthermore, all members of society, whether medical professionals or not, may become *ill*. Social and behavioral criteria may also feed into professional criteria for *disease*.

The concept of disease changes with time, depends on practice, and influences medical taxonomy. Diseases are defined according to abnormalities of morphology, physiological aberrations, biochemical defects, genetic abnormalities, ultrastructural abnormalities, and etiologic agents (Copeland, 1977), technology (Hofmann, 2013), and behavioral criteria. Hence, it has been difficult to provide a consistent medical taxonomy. There is no unified nosology (Hofmann, 2013), and the taxonomy seems to be more influenced by prognostic and therapeutic capacity than by formal definitions (Scadding, 1967). Figure 2.2 illustrates the dynamic relationship among *disease, illness,* and *sickness*, and Figure 2.3 illustrates the influence from various stakeholders.

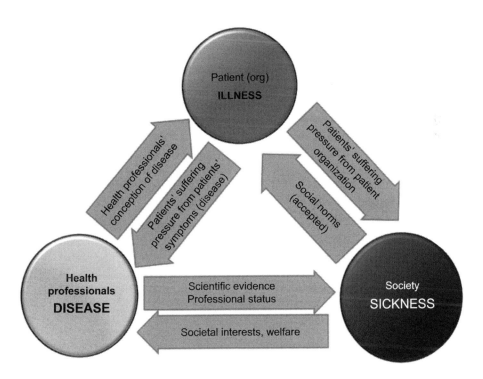

Figure 2.2 Sketch of the dynamic relationship among disease, illness, and sickness (Hofmann, 2001)

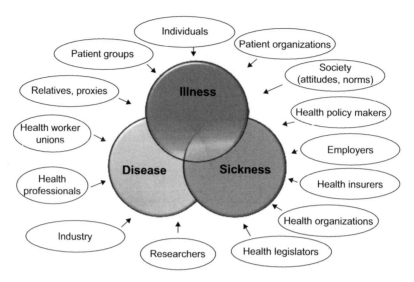

Figure 2.3 Overview of some of the actors influencing the conceptions of disease, illness, and sickness (Hofmann, 2001)

Other Perspectives

Although *disease, illness,* and *sickness* have been sustainable theoretical concepts referring to pertinent perspectives and providing fruitful frameworks for analyzing and addressing hard cases, it is far from obvious that they are the only relevant and fruitful concepts. One obvious candidate to add to the model is disease *risk*. Although it has been argued that risk has converged with disease, it can also be argued that disease risk is so special and influential that it warrants to be differentiated from the existing concept of *disease*.

Moreover, it may also be argued that existential aspects of human malady are special and different from *illness*, and therefore warrant a specific concept in the model of human malady. For example, Seneca wrote in his *Epistulae morales ad Lucilium* about loss of joy (intermission voluptatum) and fear of death (metis mortis).

Furthermore, it can be argued that sickness is too broad of a concept, as it includes both formalized norms, such as structures and regulations for sick leave, as well as informal norms, such as status, prestige, prejudice, stigmatization, and discrimination. It could be argued that these aspects should be differentiated into two perspectives (e.g., "warranted sickness" and "non-warranted sickness"). However, what in a specific time in history is considered to be (warranted) *sickness* by its institutions may in other times not be (warranted) *sickness*.

Although these and many other perspectives readily can be added to better explain certain aspects and challenges with human malady, they also add complexity, potentially making the model less (theoretically and practically) useful.

Conclusion

Disease, illness, and *sickness* are three interrelated concepts that refer to three pertinent perspectives of human malady (i.e., the professional, the personal, and the societal perspectives). They provide a fruitful framework for explaining and addressing several of the epistemic and moral challenges in the philosophy of medicine and in clinical practice.

References

ALBUM, D. & WESTIN, S. 2008. Do diseases have a prestige hierarchy? A survey among physicians and medical students. *Social Science & Medicine*, 66, 182–188.

BOORSE, C. 1975. On the distinction between disease and illness. *Philosophy and Public Affairs*, 5, 49–68.

BOYD, K. M. 2000. Disease, illness, sickness, health, healing and wholeness: Exploring some elusive concepts. *Med Humanit*, 26, 9–17.

BYNUM, B. 2000. Drapetomania. *Lancet*, 356, 1615.

CAREL, H. H. 2013. Illness, phenomenology, and philosophical method. *Theor Med Bioeth*, 34, 345–357.

CAREL, H. H. 2014a. Illness and Its Experience: The Perspective of the Patient. *In:* EDWARDS, S. & SCHRAMME, T. (eds.) *Handbook of concepts in the philosophy of medicine.* Dordecht, Netherlands: Springer.

CAREL, H. H. 2014b. The philosophical role of illness. *Metaphilosophy*, 45, 20–40.

CLOUSER, K., CULVER, C. & GERT, B. 1997. Malady. *In:* HUMBER, J. & ALMEDER, R. (eds.) *What is disease?* Totowa, NJ: Humana Press, pp. 173–218.

COOPER, R. 2005. *Classifying madness*, Dordrecht, The Netherlands: Springer.

COPELAND, D. D. 1977. Concepts of disease and diagnosis. *Perspect Biol Med*, 20, 528–538.

DANIEL, D. 1991. *Consciousness explained.* Boston, MA: Little Brown.

DORHEIM, S. K., BJORVATN, B. & EBERHARD-GRAN, M. 2013. Sick leave during pregnancy: A longitudinal study of rates and risk factors in a Norwegian population. *Bjog*, 120, 521–530.

ENGEL, G. L. 1977. The need for a new medical model: A challenge for biomedicine. *Science*, 196, 129–136.

ENGELHARDT, H. & WILDES, K. 2004. Health and Disease—Philosophical Perspectives. *In:* REICH, W. (ed.) *Encyclopedia of bioethics* (3rd ed.). New York: MacMillan, pp. 1075–1081.

FABREGA, H., JR. 1972. Concepts of disease: Logical features and social implications. *Perspect Biol Med*, 15, 583–616.

FABREGA, H., JR. 1979. The scientific usefulness of the idea of illness. *Perspect Biol Med*, 22, 545–558.

FANU, J. L. 1999. *The rise and fall of modern medicine*, London: Little Brown.

HOFMANN, B. 2001. Complexity of the concept of disease as shown through rival theoretical frameworks. *Theor Med Bioeth*, 22, 211–236.

HOFMANN, B. 2002. On the triad disease, illness, and sickness. *Journal of Medicine and Philosophy*, 27, 651–673.

HOFMANN, B. 2010. Stuck in the middle: The many moral challenges with bariatric surgery. *American Journal of Bioethics*, 10, 3–11.

HOFMANN, B. 2013. The Technological Invention of Disease. *In:* E. CARAYANNIS (ed.) *Encyclopedia of creativity, invention, innovation and entrepreneurship*, pp. 1786–1796. New York: Springer.

HOFMANN, B. 2014. *Hva er sykdom?* Oslo: Gyldendal Akademisk.

HOFMANN, B. 2015. Suffering: Harm to Bodies, Minds, and Persons. *In:* EDWARDS, S. & SCHRAMME, T. (eds.) *Handbook of concepts in the philosophy of medicine.* Berlin: Springer.

HORWITZ, A. & WAKEFIELD, J. 2007. *The loss of sadness: How psychiatry transformed normal sorrow into depressive disorder.* New York: Oxford University Press.

KING, L. 1954. What is a disease? *Philosophy of Science*, 21, 193–203.

MARINKER, M. 1975. Why make people patients? *J Med Ethics*, 1, 81–84.

MCCABE, D. P. & CASTEL, A. D. 2008. Seeing is believing: The effect of brain images on judgments of scientific reasoning. *Cognition*, 107, 343–352.

NESSA, J. & MALTERUD, K. 1998. Feeling your large intestines a bit bound: Clinical interaction—talk and gaze. *Scand J Prim Health Care*, 16, 211–215.

NORDENFELT, L. 2007. The concepts of health and illness revisited. *Med Health Care Philos*, 10, 5–10.

PARSONS, T. 1951. *The social system.* London: Routledge & Kegan Paul.

PARSONS, T. 1958. Definitions of Health and Illness in the Light of American Values and Social Structure. *In:* JACO, E. G. (ed.) *Patients, physicians and illness: Sourcebook in behavioral science and medicine*, pp. 165–187. New York: Free Press.

PARSONS, T. 1964. *Social structure and personality.* New York: Free Press.

PUHL, R. M. & HEUER, C. A. 2009. The stigma of obesity: A review and update. *Obesity*, 17, 941–964.
ROTHSCHUH, K. 1972. Der Krankheitsbegriff. (Was ist Krankheit?). *Hippokrates*, 43, 3–17.
SADEGH-ZADEH, K. 1981. Normative systems and medical metaethics: Part I: Value kinematics, health, and disease. *Metamedicine*, 2, 75–119.
SADEGH-ZADEH, K. 2000. Fuzzy health, illness, and disease. *J Med Philos*, 25, 605–638.
SCADDING, J. G. 1967. Diagnosis: The clinician and the computer. *Lancet*, 2, 877–882.
SEDGWICK, P. 1973. Illness—mental and otherwise. *Stud Hastings Cent*, 1, 19–40.
SUSSER, M. 1990. Disease, illness, sickness: Impairment, disability and handicap. *Psychol Med*, 20, 471–473.
SVENAEUS, F. 2014. The phenomenology of suffering in medicine and bioethics. *Theor Med Bioeth*, 35, 407–420.
TAYLOR, F. 1979. *The concepts of illness, disease and morbus*. Cambridge: Cambridge University Press.
TOOMBS, S. K. 1990. The temporality of illness: Four levels of experience. *Theor Med*, 11, 227–241.
TWADDLE, A. 1968. *Influence and illness: Definitions and definers of illness behavior among older males in Providence, Rhode Island*. PhD dissertation, Brown University, Providence, RI.
TWADDLE, A. 1994a. Disease, Illness and Sickness Revisited. In: TWADDLE, A. & NORDENFELT, L. (eds.) *Disease, illness and sickness: Three central concepts in the theory of health*, pp. 1–18. Linköping: Linköping University Press.
TWADDLE, A. 1994b. Disease, Illness, Sickness and Health: A Response to Nordenfelt. In: TWADDLE, A. & NORDENFELT, L. (eds.) *Disease, illness and sickness: Three central concepts in the theory of health*, pp. 37–53. Linköping: Linköping University Press.
VON ENGELHARDT, D. 1995. Health and Disease—History of the Concepts. In: REICH, W. (ed.) *Encyclopedia of bioethics*, pp. 1057–1085. New York: MacMillan.
WELCH, H. G., SCHWARTZ, L. & WOLOSHIN, S. 2011. *Overdiagnosed: Making people sick in the pursuit of health*. Boston, MA: Beacon Press.
WILT, G. J. V. D., BRAUNACK-MAYER, A., BOMBARD, Y. & HOFMANN, B. 2010. Emerging and Evolving Methods for Ethics in HTA. Health Technology Assessment International, 7th Annual Meeting, Dublin, Ireland, Maximising the Value of HTA, Book of Abstracts.
WORLD HEALTH ORGANIZATION. 2000/2012. The top 10 causes of death. http://www.who.int/mediacentre/factsheets/fs310/en/
YOUNG, A. 1995. Health and Disease—Anthropological Perspectives. In: REICH, W. (ed.) *Encyclopedia of bioethics*, pp. 1097–1101. New York: MacMillan.
ZACHAR, P. & KENDLER, K. S. 2012. The removal of Pluto from the class of planets and homosexuality from the class of psychiatric disorders: A comparison. *Philosophy, Ethics, and Humanities in Medicine*, 7, 1–7.

Further Reading

Brown, W. M. 1985. On defining 'disease'. *Journal of Medicine and Philosophy*, 10, 311–328.
Fulford, K. W. M. 1993. Praxis makes perfect: Illness as a bridge between biological concepts of disease and social conceptions of health. *Theoretical Medicine*, 14, 305–320.
Kovács, J. 1998. The concept of health and disease. *Medicine, Health Care and Philosophy*, 1, 31–39.
Margolis, J. 1976. The concept of disease. *Journal of Medicine and Philosophy*, 1, 238–255.
Reznek, L. 1987. *The nature of disease*. New York: Routledge & Keagen Paul.

3
HEALTH AND WELL-BEING
Daniel M. Hausman

Health is both a cause and constituent of well-being. A stronger view, which I shall contest, is that health is simply one kind of well-being and that the value of health consists of its contribution to well-being. Section 1 briefly defends the view that health is both a cause and constituent of well-being and distinguishes that view from the stronger claim that health is a kind of well-being. Section 2 considers whether contributing to well-being or possessing some other evaluative property is essential to health. Section 3 explores further implications of the view that the value of health consists in its contribution to well-being and lays out arguments in its defense. Section 4 criticizes those arguments, and section 5 points out contrasts between health and well-being. Throughout this chapter, I shall attempt to bring my abstract claims down to earth by linking them to a particular health state: deafness.

1. Health as a Cause and Constituent or as a Species of Well-Being

There is, to say the least, no consensus among philosophers concerning what constitutes well-being—that is, living a good life. So there is no feasible way to show that aspects of health and its consequences meet the established criteria for promoting well-being. Yet there is a broad consensus among both philosophers and non-philosophers concerning many of the typical constituents of well-being. That greater health typically promotes well-being and that *some* ways of being healthier are themselves ways of being better off are part of that consensus. But merely to point to this fact is not much of an argument against someone who disagrees. Fortunately, other platitudes concerning well-being reinforce the connection between health and well-being. For example, even those who reject hedonism usually agree that alleviating pain, nausea, or vertigo makes people better off (other things being equal), and lessening these conditions is, of course, a health improvement. Most people agree that succeeding in worthwhile projects—from child-rearing to treaty-making—enhances well-being, and better mental and physical functioning promote these successes. Without further belaboring of the obvious, we can take it as established that better health typically promotes well-being and that, in some cases at least, health improvements constitute improvements in well-being.

A more controversial view maintains that health is a kind of well-being. One expression of that view can be found in the definition of *health* given by the World Health Organization in 1947 as "a state of complete physical, mental and social well-being and not merely the absence of disease or infirmity" (1948, p. 100). If health is identical to a state of well-being, then it is a necessary, rather than a contingent, truth that health promotes well-being. Most of those who hold such a view have taken the necessity to be conceptual. They have taken health, like well-being, to be an evaluative notion. If they are right and an increase in well-being is entailed by an improvement in health, then if Jack is healthier than Jill and their non-health circumstances are identical, it must necessarily be the case that Jack is better off than Jill. On this view, comparisons of health are not purely naturalistic inquiries into bodily and mental functioning.

Even if it is a conceptual truth that health promotes well-being, it has other properties and consequences beyond its contribution to well-being, and it could turn out to be more convenient to measure health by those other properties or consequences than by measuring well-being. But given the enormous heterogeneity among health states, it would seem to be most direct and natural to measure health or to assign values to health states by measuring their contribution to well-being. Taking the measure of health to be a measure of well-being may seem like an obvious step to make. How else could one make sense of the claim that someone who is deaf is in better health than someone with severe angina? But it is a fateful step as well, because it implies that if those who have a disability such as deafness are not thereby worse off than those who can hear, then either they are not in fact disabled after all or disability is not a form of diminished health. I shall return to this conundrum later.

Most attempts to quantify overall health purport to measure "health-related quality of life" [see Kind (1996), Kaplan et al. (1998, p. 509), and http://www.fhs.mcmaster.ca/hug/]. But it is hard to know what to make of this statement, because most of the health-measurement literature leaves the meaning of "health-related quality of life" completely obscure. Bogner and Hirose are an exception. They describe health-related quality of life as "that fraction of overall well-being that is determined by health" (2014, p. 31). In practice, health economists measure health-related quality of life by eliciting preferences. If one believes that preference satisfaction constitutes well-being or, more reasonably, that the extent to which preferences are satisfied constitutes good evidence concerning well-being, then it appears that health economists are taking the value of health states to consist in the contribution that health makes to well-being.

2. Is Health an Evaluative or a Naturalistic Concept?

There is a heated literature devoted to the question in the section heading, which I shall not attempt to survey. (See also Chapter 1 of this book.) There is one influential naturalistic view, which identifies human health with the efficiency with which the parts and processes of the body are functioning, where (very roughly) efficiency is a matter of the extent to which the parts and processes contribute to the survival and reproduction of the person (Boorse 1977, 1997). On the evaluative side, one finds many different views, which agree that in order for a physical or mental state to constitute a pathology, it must have some evaluative feature such as making people worse off, diminishing their opportunities, calling for medical treatment, excusing behavior that would ordinarily be condemned, or calling for sympathy and care from others (Cooper 2002, Engelhardt 1974, Fulford 1989, Nordenfelt 2000, Reznek 1987, Venkatapuram 2011, Wakefield 1992). Both sides agree that health contributes to well-being both causally and constitutively. The issue is whether this contribution to well-being defines, in part, what constitutes health.

Although the naturalistic view faces detailed criticisms (Hausman 2012a, Kingma 2007, 2010, Schwarz 2007), I think that two of the arguments against evaluative views of health are decisive. And if health is not an evaluative notion, it cannot be identical to a state of well-being. First, evaluative views face counterexamples. Some states of health, such as a malfunction in just one kidney, diminish health without diminishing well-being. The bearing of other health states on well-being is sometimes variable. For example, infertility in premenopausal women constitutes a health problem. This health problem sometimes diminishes well-being, but it sometimes enhances it. To maintain that infertility is a health problem only when it is unwanted squarely contradicts the usage of pathologists, and it would permit us to cure health problems by convincing people to be happy with their physical or mental states or by changing social values. Another response to counterexamples like infertility might be to maintain that it is a conceptual truth that greater health *typically* goes with greater well-being, not that it *always* does. That is an odd conceptual truth, and it's not obvious that it matters whether anything rides on whether one grants it or denies it.

The second, and to my mind more decisive, reason to reject an evaluative notion of health is that the contribution that a given health state makes to well-being and to other values depends on the geographical, technological, and cultural environment and on the objectives and values of individuals, whereas the state of someone's health is largely independent of these other factors. Two people who are both deaf and whose physical and mental health is otherwise the same are in the same state of health. But the contribution that their health makes to their well-being depends on the accommodations that their societies provide, the risks in their respective environments, the attitudes of the communities in which they live, their activities and objectives, and so forth. Nor can one separate the differences in their well-being into a common health-related component and a differing non-health-related component. As John Broome argues (2002), the contribution that health makes to well-being interacts with the contribution of other things. One cannot add them up. Maintaining a sharp distinction between the health of these two deaf individuals and their health-related well-being is crucial if one wishes to describe their situations sensibly.

So one should reject the view that health is defined, at least in part, by its value, whether that value lies in its contribution to well-being or in some other evaluative aspect. To the extent that the view that health is a species of well-being rests on the conceptual claim, one has reason to deny that view as well. Of course, health typically has great value, but that value depends on facts about people and their environments that are separate from health. Health is a matter of how adequately the parts and processes in the body and mind are carrying out their functions.

3. Measuring and Valuing Health by Well-Being

Even though health is not defined by well-being, it may be that *what matters about health* for practical purposes is its contribution to well-being, and there are apparently powerful arguments in defense of valuing health by its impact on well-being. I shall consider three of these arguments. The first points out simply that there are practical methods for measuring well-being by means of measuring preferences or by means of measuring subjective experience. These methods are imperfect, but at least they exist. Their feasibility is not in question. What feasible alternative is there?

The second argument for valuing health by its contribution to well-being may be formulated as follows:

1. Subject to the constraints imposed by other moral considerations, such as fairness or respect for individual rights, health policy should aim to minimize the loss of well-being due to ill health.
2. To minimize the loss of well-being due to ill health (subject to appropriate constraints), those who make health policy need to know the impact of ill health on well-being.
3. Those who make health policy need to know the impact of ill health on well-being.

This argument is rarely stated explicitly, but it is nevertheless important. Its first premise—that social policy, including health policy, should aim to promote well-being—is however contestable, and as I shall suggest following in section 4, there is good reason why health policy should have other aims. Moreover, the argument does not establish that other information is not also of value to those making health policy.

The third argument appears, at first glance, to provide even stronger support for valuing health by well-being. The argument is implicit in the following passage from Bognar and Hirose (2014, p. 30):

> For, ultimately, we do not much care about health itself. What we do care about is its value for us: the way it affects our well-being or quality of life.... Consequently, when we allocate health care resources, we should be interested in their impact on quality of life. In other words, what matters is the impact of health on well-being.

In much the same spirit, Broome (2002, p. 94) maintains that health states should be evaluated so as "to measure how good a person's health is for the person, or how bad her ill-health.... That is to say, it aims to measure the contribution of health to well-being." The argument that is implicit in these quotations can be stated as follows:

1. The value of a health condition for an individual in that condition is the same thing as how good the health condition is for that individual.
2. How good something is for an individual is the extent to which it promotes or diminishes that individual's well-being.
3. Thus, the value of a health condition for an individual in that condition is the extent to which it promotes or diminishes that individual's well-being.

This argument is shot through with ambiguities and possible equivocations. It might seem that nothing could be more obvious than that the value of something for an individual is how good it is for the individual. But the value of something for me may lie in its promoting things that I care about that have nothing to do with my well-being. Those things that make states of affairs good or bad for people need not increase or decrease their well-being. Losing my hearing might be bad for me because of its effect on how well my life goes or because it prevents me from promoting other objectives that I care about. The range of things that affect the choice-worthiness of alternatives is wider than the range of things that affect someone's well-being, and the ranking (and consequently the measure) of alternatives in terms of choiceworthiness is not the same as the ranking or measure of alternatives in terms of well-being (Scanlon 1998, chapter 3).

4. Against Valuing Health by Its Contribution to Well-Being

The most important objections to identifying the values of health states with their contributions to well-being are at the same time critiques of the three arguments in favor of doing so. First, well-being is hard to measure, and the methods of measuring well-being are seriously flawed. So identifying the value of health with its contribution to well-being gives one little empirical grasp of the value of health. Preferences are a good indicator of well-being only under very restrictive conditions (Hausman 2012b), and methods to elicit preferences are faulty. Whether subjective states are good indicators of well-being is controversial, and measurements of subjective states are sensitive to irrelevant features of the measurement context. Second, it is questionable whether the objective of state policy, especially in a liberal state, should be directly to enhance the well-being of the citizenry. The agents of the state do not possess the knowledge or sensitivity to accomplish such a mandate, and it arguably better serves well-being by delegating pursuit of well-being to the individuals themselves and confining state policy to providing them with favorable conditions in which to form and pursue worthwhile objectives. On this view, what matters about health from the perspective of liberal social policy is mainly the extent to which it limits what objectives individuals can pursue rather than how much it diminishes well-being.

The third argument against identifying the values of health states with their contributions to well-being derives from the recognition (which is implicit in the critique of the third argument in favor of measuring health by well-being) that people care about many aspects of health in addition to its contribution to well-being. For example, health problems may limit autonomy. Having suffered an injury that limits my physical abilities to take care of myself, I may be forced to limit my activities to conform to the preferences of my caregivers. Whether these limits make me worse off (i.e., reduce my well-being) is a separate question from whether they limit my ability to govern my own life. Indeed, I may be better off if governed by others. Similarly, a disability such as deafness that need not make me worse off, nevertheless closes off some careers and activities. Health affects freedom, and freedom matters whether or not it has an impact on well-being. That I have more alternatives available to me may make a state of affairs more worthy of choice, but it does not necessarily make my life better. Having more choices, I may make a hash of things.

As this last point implies, identifying the value of health with well-being has seriously mistaken and harmful implications concerning disabilities. As people adapt to a condition such as deafness, their well-being improves more than their physical capacities (Salomon and Murray 2002). Some people are inclined to deny that deafness is a health problem at all, and there is a usage of "healthy" and "sick" according to which we would distinguish among those who are deaf those who are healthy and those who are sick (with the flu, perhaps). But on the view of health as a matter of how well the parts and processes of the body and mind carry out their functions, deafness is a malfunction and a decrement in health. Even those who reject the view of health as functional efficiency ought to concede the point, because otherwise they will find themselves forced to deny that those factors that threaten people's hearing, such as infections, tumors, or injuries, are threats to health.

Much of the resistance to regarding deafness as a disability (i.e., a form of ill-health) is, I conjecture, a consequence of the mistaken identification of how good someone's health is with how well his or her life is going. If the measure of health is well-being, then those who are living the best lives must be fully healthy, regardless of whether they can hear. If the measure of health is well-being, then calling deafness a disability is equivalent to asserting that those who are deaf cannot live truly excellent lives, which is of course false. How good the lives of the deaf turn out to be depends on the accommodations that society is willing to make and (like the well-being of individuals who can hear) on the character and good fortune of individuals.

Consider the following statement from the National Association of the Deaf's position paper on cochlear implants:

> Many within the medical profession continue to view deafness essentially as a disability and an abnormality and believe that deaf and hard of hearing individuals need to be "fixed" by cochlear implants. This pathological view must be challenged and corrected by greater exposure to and interaction with well-adjusted and successful deaf and hard of hearing individuals.
>
> (National Association for the Deaf 2000)

This quotation assumes that calling deafness a disability implies that the deaf are not "well-adjusted and successful." If it follows from granting that deafness is a disability that the lives of the deaf are inferior to the lives of those who can hear, then representatives of the deaf community should *of course* deny that deafness is a disability. Given the mistaken identification of disability with diminished well-being, that denial is entirely justified.

Instead of denying that deafness is a disability, one should deny that disability necessarily diminishes well-being. One should understand the claim that deafness is a significant disability as asserting only that it involves a dysfunction in the auditory system and that this dysfunction has significant consequences. These claims are platitudes and do not challenge the value of the lives of those who are deaf. Not being able to hear increases the risks of injury, and it limits the projects people can pursue and which enjoyments they can have. Some central features of human cultures are inaccessible to the deaf. For these reasons, deafness is a significant disability, but the fact that the range of activities and enjoyments is narrower does not imply that activities and enjoyments within that range are in any way inferior. The fact that deafness is in this sense a disability is consistent with the possibility that the lives of the deaf are every bit as rich and fulfilling as the lives of those with good hearing. Well-being is not a good measure of the value of health.

5. Health Is Not a Kind of Well-Being

As I said at the beginning, this chapter denies both that the value of health is its contribution to well-being and that health is a kind of well-being. In fact, health differs in significant ways from well-being. These differences cast doubt on the prospects of measuring health by its impact on well-being, and they suggest that it should be easier to assess people's health than to measure their well-being or to measure the contribution of health to their well-being.

The first difference between well-being and health and its value is that among people of the same sex and roughly the same age there is comparatively little variation in what counts as good health, while utterly different lives may be good lives and good, in part, because of their differences rather than despite them. To exaggerate the point, one might say that there is one way to be healthy, while there are many ways to have a good life. The good life for some people consists in taking risks, whereas others thrive in quiet comfort. Some people flourish by pursuing their ambitions, but others focus on friends and family. Good lives are almost as diverse as people's objectives, but good health is much the same for everyone.

A second contrast helps to explain the first. One reason why well-being is diverse, while good health is uniform, is that what is good for me depends heavily on my goals and values, while my self-definition is less relevant to how good my health is. This is why the diminished health of someone who is deaf does not imply diminished well-being. By valuing those aspects of life to which deafness is irrelevant, there is no limit to how good the life of someone who is deaf can be. This contrast requires some qualifications. The value of health often depends on the technological, geographical, and cultural environment, but within any given environment, the value of health for the most part varies little from one individual to the next. Well-being, on the other hand, depends heavily on individual values and goals. If, as Joseph Raz (1986), T. M. Scanlon (1998), and others argue, succeeding in the pursuit of one's own worthwhile goals is central to well-being, then one would expect that the details of good lives will be diverse.

A third difference between well-being and health is that interpersonal comparisons of well-being pose serious problems, while interpersonal health comparisons are not substantially more difficult than intrapersonal comparisons. When talking about how healthy someone is, it is largely irrelevant who the somebody is. Comparisons of which of two people is in better health, like comparisons of whether a person at one time is in better health than that same person at another time, are comparisons of the health conditions. The features of the person who experiences those health conditions are typically of little importance.

On the other hand, when one is concerned with well-being or how health states bear on well-being, the standards of comparison vary widely across people. Unlike the activities of a single individual, whose values can often be compared with respect to the individual's unchanged aims, and unlike the mental states of an individual, which are experienced by a single subject, it is unclear how to compare the contributions to well-being of the activities of separate individuals and how to compare the quality of their mental states. For policy purposes, it may be convenient to suppose that people in similar circumstances are equally well-off, but this supposition is at best an extremely rough simplification. In the same circumstances, including the same health state, one person may be thriving, while another is miserable. It is very hard to specify a method for making interpersonal comparisons of well-being that is not subject to serious ethical criticisms (Hausman 1995).

A fourth contrast concerns the objects of appraisal. In assessing the well-being of individuals who are deaf (like the well-being of individuals who can hear), we think primarily of their whole lives, and our appraisal of how well their lives are going during a limited period often depends on what their lives are like before or after that period. On the other hand, when considering people's health, we think mainly about how healthy someone is during some period. Our appraisal of someone's health during a period does not depend on comparing it to health in previous or succeeding periods or on placing it within the narrative of a whole life.

The temporal separability of health does not rule out an evaluation of lifetime health, and the temporal inseparability of well-being does not prevent one from judging people's well-being during one period or another. The difference is that the lifetime health appraisal is less informative than the time-limited appraisals it summarizes, whereas an evaluation of how well whole lives have gone is more informative than judgments concerning their well-being during various periods. How good a life is cannot be determined by adding up or averaging how good it is during separate periods (Griffin 1986, pp. 34–35), whereas how healthy someone's life as a whole has been is precisely such a sum or average. The trajectory and narrative of a life are crucial elements of well-being, but they are irrelevant to the appraisal of health.

These four contrasts between well-being and health do not imply that health cannot be valued by its bearing on well-being, but they undermine the motivation for attempting to value health this way. What counts pretheoretically as good health (or as better health) is reasonably uniform, interpersonally comparable, largely independent of individual aims and values, and concerned mainly with limited periods within people's lives. Why, then, attempt to value health in terms of something that is diverse and hard to compare across individuals, that depends on individual aims and values, and that is mainly concerned with whole lives?

6. Conclusions

Health is not a species of well-being, and it is not defined by well-being. Unless well-being is defined as consisting of everything that matters to individuals, what matters about health and constitutes its value is not only its contribution to well-being. As the example of deafness illustrates, substantial dysfunctions that limit what people can do, what they can enjoy, and what risks in their environments they can respond to need not lessen well-being. Measures of health in terms of well-being are unreliable, and these measures lead to deep confusions concerning the concept of a disability and to needless offense to those who experience disabilities.

In discussing the relations between the value of health and well-being, I inevitably made a number of remarks about the measurement of health, and implicit in this chapter is an argument that health ought not to be measured by its contribution to well-being. Before drawing that conclusion, however, one needs to consider whether there is any better alternative than measuring health by well-being. That is, however, a question for another occasion. Regardless of how that question is answered, health is not a species of well-being. Although both a cause and a constituent of well-being, its value cannot be measured by its contribution to well-being.

References

Bognar, Greg, and Iwao Hirose. 2014. *The Ethics of Health Care Rationing: An Introduction*. London: Routledge.
Boorse, Christopher. 1977. "Health as a Theoretical Concept." *Philosophy of Science* 44: 542–73.
Boorse, Christopher. 1997. "A Rebuttal on Health." In J. M. Humber and R. F. Almeder, eds. *What Is Disease?* Totowa, NJ: Humana Press, pp. 1–134.
Broome, John. 2002. "Measuring the Burden of Disease by Aggregating Well-Being." In Christopher Murray, Joshua Salomon, Colin Mathers, and Alan Lopez, eds. *Summary Measures of Population Health: Concepts, Ethics, Measurement and Applications*. Geneva: World Health Organization, pp. 91–113.
Cooper, R. 2002. "Disease." *Studies in History and Philosophy of Biological and Biomedical Science* 33: 263–82.
Engelhardt, H. T., Jr. 1974. "The Disease of Masturbation: Values and the Concept of Disease." *Bulletin of the History of Medicine* 48: 234–48.
Fulford, K. W. M. 1989. *Moral Theory and Medical Practice*. New York: Cambridge University Press.
Griffin, James. 1986. *Well-Being: Its Meaning, Measurement and Moral Importance*. Oxford: Clarendon Press.
Hausman, Daniel. 1995. "The Impossibility of Interpersonal Utility Comparisons." *Mind* 104: 473–90.
Hausman, Daniel. 2012a. "Health, Naturalism, and Functional Efficiency." *Philosophy of Science* 74: 519–41.
Hausman, Daniel. 2012b. *Preference, Value, Choice and Welfare*. Cambridge: Cambridge University Press.
Kaplan, Robert, Theodor Ganiats, William Sieber, and John Anderson. 1998. "The Quality of Well-Being Scale: Critical Similarities and Differences with the SF-36." *International Journal for Quality in Health Care* 10: 509–20.
Kind, Paul. 1996. "The EuroQoL Instrument: An Index of Health-Related Quality of Life." In B. Spiker, ed. *Quality of Life and Pharmacoeconomics in Clinical Trials*. 2nd ed. Philadelphia: Lippincott-Raven, pp. 191–201.
Kingma, Elselijn. 2007. "What Is It to Be Healthy?" *Analysis* 67: 128–33.
Kingma, Elselijn. 2010. "Paracetamol, Poison, and Polio: Why Boorse's Account of Function Fails to Distinguish Health and Disease." *British Journal for the Philosophy of Science* 61: 241–64.
National Association for the Deaf. 2000. "Position Statement on Cochlear Implants." http://www.nad.org/issues/technology/assistive-listening/cochlear-implants
Nordenfelt, Lennart. 1995. *On the Nature of Health: An Action-Theoretic Approach* (Vol. 26). New York: Springer Science & Business Media.
Nordenfelt, Lennart. 2000. *Action, Ability and Health: Essays in the Philosophy of Action and Welfare*. Dordrecht: Kluwer.
Raz, Joseph. 1986. *The Morality of Freedom*. New York: Oxford University Press.
Reznek, Lawrie. 1987. *The Nature of Disease*. London: Routledge and Kegan Paul.
Salomon, Joshua, and Christopher Murray. 2002. "A Conceptual Framework for Understanding Adaptation, Coping and Adjustment in Health State Valuations." In Christopher Murray, Salomon Joshua, Mathers Colin, and Lopez Alan, eds. *Summary Measures of Population Health: Concepts, Ethics, Measurement and Applications*. Geneva: World Health Organization, pp. 619–26.
Scanlon, Thomas. 1998. *What We Owe to Each Other*. Cambridge: Harvard University Press.
Schwartz, Peter. 2007. "Defining Dysfunction: Natural Selection, Design, and Drawing a Line." *Philosophy of Science* 74: 364–85.
Venkatapuram, Sridhar. 2011. *Health Justice: An Argument from the Capabilities Approach*. London: Polity Press.
Wakefield, J. C. 1992. "The Concept of Mental Disorder: On the Boundary between Biological Facts and Social Values." *American Psychologist* 47: 373–88.

World Health Organization. 1948. "Preamble to the Constitution of the World Health Organization." In *Official Records of the World Health Organization*, no. 2. Geneva: World Health Organization, p. 100.

Further Reading

Bognar, Greg, and Iwao Hirose. 2014. *The Ethics of Health Care Rationing: An Introduction*. London: Routledge.

Boorse, Christopher. 1997. "A Rebuttal on Health." In J. M. Humber and R. F. Almeder, eds. *What Is Disease?* Totowa, NJ: Humana Press, pp. 1–134.

Hausman, Daniel. 2015. *Valuing Health: Well-Being, Freedom, and Suffering*. New York: Oxford University Press.

Murray, Christopher. 1996. "Rethinking DALYs." In Christopher Murray and Alan Lopez, eds. *The Global Burden of Disease: A Comprehensive Assessment of Mortality and Disability from Diseases, Injuries, and Risk Factors in 1990 and Projected to 2020*. Boston: Harvard School of Public Health, pp. 1–98.

Nord, Erik. 1999. *Cost-Value Analysis in Health Care: Making Sense Out of QALYs*. Cambridge: Cambridge University Press.

4
DISABILITY AND NORMALITY
Anita Silvers

About one in five U.S. residents is disabled, a proportion that according to the U.S. Census Bureau has remained steady through the last two census counts (U.S. Census Bureau 2012). Given the abundance of disabled individuals in the population, each of us most likely encounters several disabled people every day. But how should it be decided that an individual is disabled, and what are the consequences of being included in this group?

The idea of disability is more problematic, and identification of people who are disabled more challenging, than the large number found among us may suggest. Medical diagnosis is central to determining who is disabled and who is not. Yet attributions of disability are not like typical diagnoses of medical conditions. Moreover, just how identifying disability goes beyond diagnosing is not clear.

Current practice in ascribing disability has been portrayed as especially frustrating for both physicians and patients (Carey and Hadler 1986; Sokas et al. 1995; O'Fallon and Hillson 2005). One reason for concern about whether disability assessment is inherently dubious comes from physicians' variability in distinguishing people who are disabled from those who are not. To illustrate, 73 physicians participating in a study of the practice of determining disability differed significantly as to the degree of disablement from which the same patient with chronic back pain suffered: 20% concluded that the patient was totally disabled, 40% concluded that the patient was subject to marked activity limitation due to partial disability, and the remaining 40% concluded that the same patient had no disability at all (O'Fallon and Hillson 2005). Such sharp variations of opinion among medical professionals raise questions about whether their divergent assessments of disablement refer, even roughly, to the same thing.

The nature of the linkage of disability to disadvantage is another problematic aspect of disability ascription. Some people cannot even imagine that an individual could be disabled, yet also be no more disadvantaged than similar nondisabled individuals are. Is the tie so tight that deciding about disability is inherently normative, so that we cannot even conceive of individuals being disabled without also thinking this must be a damaging aspect of their lives? To securely capture a normative thrust, some theorists have urged that the concept of disability be explicitly revised. They propose crafting a more determinate definition according to which disability is ineluctably bad for well-being and therefore is necessarily disadvantageous. To illustrate, Savulesco and Kahane (2011) offer a proposal according to which disablement, by definition, makes life worse.

Adopting this stipulated definition, they believe, is commendable, for doing so builds a reason for societal intervention to address disability's difficulties into the very meaning of the term (Savulescu and Kahane 2011: 46). To revise the idea of disability so that disadvantage is inescapable in this way makes it inconceivable that disabled individuals can ever pursue opportunity on a level playing field. Freeing them from disadvantage would require nothing less than eliminating their disablement.

Savulescu and Kahane call their approach "revisionary" because it gives "disability" a new, more determinate meaning. A further strength of their proposal, in their view, is that it is overtly normative. There will be no equivocating: to identify an individual as disabled will express an assessment that the person is markedly deficient or deprived (Savulescu and Kahane 2011: 45). Alluding to disabled people's privations as an element of the expression used to refer to disablement will, they contend, make the case for assisting them a matter of logic rather than of the heart.

Other theorists have taken a diametrically different approach to revising the concept, proposing instead to strip away the familiar stereotyping that stubbornly binds being disabled to disadvantage (e.g., Silvers 2003; Barnes 2016). With their approach, people can be judged to be disabled without having disadvantage necessarily attributed to them. Disablement is understood to bear only a contingent connection to detriment, so as to leave open to discovery in each particular case whether contingencies call for a neutral or instead a value-propelled response to concluding that the person is disabled (Silvers 2003). Of course, to hold this view is not to deny that people with disabilities often experience disadvantage, nor that they may be especially vulnerable to having disadvantage imposed on them, nor that they deserve resources that are effective given their circumstances for acquiring equitable access to personal welfare, but it is to think people who are disabled need not be disadvantaged as well.

Being Disabled and Not Disabled—No Contradiction Here!

Yet another puzzle about the idea of disability is that a person may, without contradiction, identify or be identified as both disabled and not disabled at the same time. A key to elucidating the perplexing issues about normativity and consensus encountered in conceptualizing disability may be found by exploring how this escape from contradiction has come to be. Doing so may suggest a propitious approach to the link between disability and disadvantage.

In 1994 Carolyn Cleveland suffered a stroke that impaired her language capability to speak, read, and write, as well as to concentrate when executing these functions and to bring up in memory information conveyed in words. Three weeks after onset she filed for benefits under the U.S. Social Security Disability Insurance (SSDI) program, with medical documentation that she was totally disabled supporting her claim. But three months later, withdrawing her SSDI claim, she returned to her job.

After another three months had elapsed, however, she was fired for failing to meet performance standards. She then reinstated her application for the disability insurance payments based on her physician's testimony attributing total disability. Nevertheless, in a lawsuit filed simultaneously against the employer under the Americans with Disabilities Act (ADA), Cleveland also claimed that the employer discriminated by firing her even though she would have been able to execute the essential elements of her work, an assessment also supported by her physician's medical testimony. The employer asked the court to issue a summary judgment against Cleveland, preventing her ADA complaint from being tried. The employer argued that Cleveland contradicted herself by (a) presenting medical testimony that she was totally disabled to substantiate the SSDI claim that she could not work, while also (b) presenting medical testimony that she could execute the essential functions of her job in support of her ADA complaint (Kohrman and Berg 2005).

Initially puzzling as Cleveland's seemingly conflicting testimony may be, in their decision in *Cleveland v. Policy Management Systems Corporation* (1999), the Supreme Court justices found unanimously in her favor. It was the Court's opinion that Ms. Cleveland might be both disabled and nondisabled, so her case was remanded back to the lower court for trial on the facts. This precedent now governs what U.S. courts must allow about people with disabilities. But what can it mean for individuals to be disabled and at the same time not be so?

Neither Cleveland nor her physician could claim without contradiction both that she had suffered a stroke but also had not suffered one. Nor could her physician diagnose without contradiction that aphasia both had and had not been a sequela of the stroke. The physician's testimony she submitted invoked the stroke and resulting aphasia in identifying her as totally disabled. But of course neither Cleveland nor her physician had denied these medically described facts for purposes of the ADA civil rights complaint, after affirming them for the SSDI insurance benefits application.

In the ADA complaint, Cleveland asserted, and the physician affirmed, that she suffered a stroke and resulting aphasia but could have executed the essential work of her position but for the employer's discrimination. In the Court's view, plaintiff's and physician's testimony affirming, and their testimony denying, that the same individual is disabled need not be contradictory. The Court's unanimous opinion explained that attributions of disability, such as were at issue in this case, are not about such purely factual matters as "The light is red/green" or "I can/cannot raise my arm above my head."

According to the Court, something more than assertions of pure fact appear to be conveyed by testifying that individuals are disabled. But if attributions of disability are not purely claims of fact, what more is meant or otherwise conveyed when people are identified as disabled? And to what standards ought medical testimony attributing disability be held if contemporaneous declarations can legitimately be made by the same expert that a patient is totally disabled, yet not so?

Disability—Heritage of the Idea

The idea that individuals with biologically based physical, sensory, cognitive, or emotional deprivations can be collectivized for medical or public policy purposes into a class of "the disabled" is a relatively recent invention. Prior to the 19th century, classification occurred in terms of physical, sensory, psychiatric, or cognitive diagnoses, such as being crippled, deaf, blind, mad, or feebleminded. Then "disability," a status that already had a clear meaning under the law, was given a new usage as a term of art crafted to link judgments made in medicine with responses mandated by various kinds of public policy. Understanding how the concept of disability evolved is a step toward understanding the judgments its origin continues to shape today.

Originally, referring to individuals as "disabled" signified only that the person bore a statutory incapacity or lacked a legal qualification to do something. Individuals with disabilities suffered from legally imposed restrictions on their political participation. In the domain of law, little pretense is made that such legal disabilities, however convincingly justified, originate in anything more than social convention.

To illustrate, married women, but not unmarried ones, were disabled from owning property. Theories explaining, as a way of justifying, the deprivation typically turned on considerations of the stability of prevailing social arrangements. The leading explanation was expressed as an hypothesis that successful domestic arrangement required husband and wife to be as one person, permitting only one procurer of income per household (Blackstone 1765: 442–445). Given their domestic role, married women must refrain from disrupting the household by ceding over management rights to their own property. In contrast, unmarried women could retain oversight and decision-making rights to what they owned.

Such paternalistic convention maintained powerful sway long after 1920, when the 19th Amendment gave U.S. women the right to vote. To illustrate, in *Goesaert v. Cleary* (1948), the Supreme Court refused to overturn a Michigan law allowing women to serve drinks only in public houses over which their husbands or fathers exercised property rights. The Court readily acknowledged that, according to science, some women were brawny enough to safeguard themselves from problems bar life could pose for their sex, but for the typical woman and therefore for women generally, oversight by a proprietary male was required to ensure against societal disruption.

Medicine shares with law the understanding that a disability is disadvantageous to whoever is classified as such, but the presumed reason for disadvantage differs in the two domains. In its original usage in the law, disability is understood to be a condition assigned as a socially constructed response when a kind or group of persons appears to be socially dysfunctional by not complying with ordinary societal convention.

The Medical Model: Naturalizing the Disadvantage of Disability

In medicine, however, disability has come to be equated with biological dysfunction, caused by an inability to maintain one or more species-typical biological processes. The medical idea of disability, which began to develop in the middle of the 19th century, differs in an important respect from the older one found in the law. Legal disability is understood to be a social artifact, whereas in medicine disability has been portrayed as a natural fact.

In the latter context, disablement usually refers to being in a biological state that scientific discovery discerns to be dysfunctional. For example, the current edition of *The American Heritage Medical Dictionary* defines *disability* as a "disadvantage or deficiency, especially *a physical or mental impairment* [emphasis added] that prevents or restricts normal achievement." In both domains, attributions of disability are normative, but they are not normative in the same way, for medicine naturalizes the association between disability and disadvantage.

In the 19th century, proposals about how to apply biological statistics to represent the average human individual forged ahead, propelled by methodological commitment to the usefulness of such mathematical understandings of human beings in preventative and therapeutic medicine. Confidence abounded that statistical research could distinguish between health and illness in living organisms. Bolstered by the presumption that humans have proven to be successful as species go, the method called for discovering how most people function, followed by the inference that these modes are not merely typical for our species but are the species norms as well. The improbability that a species whose members are mostly unsound could survive seemed to underwrite equating the most frequently seen kinds of humans with healthy ones, while anomalous individuals were subjected to suspicion about their impact on social stability and continued species success.

As medicine pursued becoming more scientific in this way, the older era's interpretation of certain physical, sensory, psychological, and cognitive anomalies being punishments for people's moral transgressions—an account now generalized as the moral model of disability—was superseded by a model of disability meant to offer a medical rather than a moral explanation. Positing biological failure rather than social transgression as the cause of individuals' physical or mental dysfunction offered a new, scientific route for averting or ameliorating the disadvantage from which persons who are disabled suffer, as well as forestalling the burdens these individuals are envisioned as imposing on their families and the community. To the extent promoting these techniques is beneficial to both the targeted persons and the general community, the advent of the medical model marked a progressive trend.

Medicine offers benign ways of preventing healthy individuals from experiencing disabling injuries or illnesses, and of mitigating disabling outcomes of injuries or illnesses. Yet not all biological conditions designated as disabilities are subject to effective medical measures. The medicalization of disability magnified concern about the continued existence of people with apparently irremediable dysfunctions, individuals whose conditions regrettably appeared to defy scientific efforts to elevate the species' achievement of health. As the science of causes and cures of disabling conditions grew, the deployment of societal resources to address disablement was more pressingly perceived as affecting community interests as well as personal ones. Dealing with disablement came to be viewed as a proper purpose for public policy that aimed at safeguarding individuals diagnosed with dysfunctions who otherwise could be overwhelmed by their disablement, but also safeguarding families and other caregiving groups, and the human species generally, from stressful interaction with members of the disabled minority. The older charitable efforts to offer incurables assistance in forms that improved moral character now were augmented by public welfare services ranging from institutionalization for medical care to disability pensions and segregated education supposed to be specially suited for the societal roles disabled children were expected to play.

To inhibit these measures from having an inflationary effect on the size of the disabled population, the medical model offered intervention in the reproductive process. Eugenics programs, aimed at wiping apparently dysfunctional variation out of the human gene pool by destroying, sterilizing, or otherwise impeding their carriers from reproducing, were pursued vigorously throughout most of the 20th century in the name of the common good to prevent the creation of more supposedly burdensome sufferers. Their targets were primarily intellectually disabled individuals, but also deaf, blind, seizure-prone, psychiatrically ill, and crippled and maimed people. Accurate knowledge about how biological inheritance may result in these very different impairments was not readily available, so some of the people swept up and bundled into the targeted class had experienced disablement due to non-inheritable injury or infectious disease. They too were sterilized, or in some places even euthanized, in the name of saving the human species from inheriting defects and society from noncontributing or burdensome members (Kevles 1985; Proctor 1988; Lombardo n.d.).

Even people whose behavior violated mere social norms were sterilized, based on false beliefs that their transgressive conduct was inheritable. Compulsory sterilization began in the United States in Indiana in 1907, followed by California and Washington two years later. In the infamous *Buck v. Bell*, 274 U.S. 200 (1927), for example, Justice Holmes upheld Virginia's sterilization of a teenager who had been raped because her mother and grandmother also had been unmarried mothers. In *Skinner v. State of Oklahoma, ex. rel. Williamson*, 316 U.S. 535 (1942), the U.S. Supreme Court ruled that sterilization could not be used as a form of punishment, but hastened to observe explicitly that this ruling did not similarly constrain sterilization for eugenic reasons, as the Court had approved that practice in *Buck v. Bell*, 274 U.S. 200 (1927).

Concern remains today that contemporary versions of the old eugenics programs continue to flourish (Holmes 1991; National Institute for Health and Care Excellence 2008; Chew 2013). Prenatal testing and subsequent termination of pregnancy if the fetus is found to be at risk of or have certain disabling conditions are common. Selective abortion prompted by prenatal testing seems to some to be propelled by the familiar stigmatization of people with disabilities.

Disability advocates usually do not challenge women's right to choose, but many object to the frequency with which termination of pregnancy is influenced by assumptions that children who fall outside of the normal ability range cannot live good lives. That such bias, whether overt or implicit, continues to affect how medical technology is used is evidenced by, for example, the uproar that has greeted the rare instances in which prospective parents with disabilities—persons who themselves lead satisfactory lives—propose to employ reproductive medicine's technology to bear children who, like themselves, do not satisfy standards of normality (Scully 2008). Some writers present or even press intuitions assigning preference to the lives of patients whose impairments can be fixed over the lives of those who are stable but not curable. (See Savulescu and Kahane 2011; Kamm 2013; Bognar 2014; Kamm 2015; Bognar 2016 for different views on the epistemic strength of such intuitions about the comparative value of nondisabled and disabled people's lives.) To be made persuasive, this kind of bioethics narrative often is illustrated by examples of rare illnesses or injuries that are characterized by irremediable anguish. Bias must be suspected when such worst-case scenarios are rolled out to represent the lives of the entire membership of the disability class. Nussbaum's trail-blazing book, *Frontiers of Justice* (Nussbaum 2006), examines the question from a wider angle, asking whether theories of justice that consider only people with species-typical capabilities to represent subjects of justice generally can be fair.

The medical explanation of the disadvantage to which disabled people are vulnerable improves over the moral account for several reasons. Medicine offers a larger, more effective repertoire of preventative measures or other progressively supportive responses to medical conditions associated with disability. Although the medical model resembles the moral model in locating the cause of disability-produced disadvantage in flaws within disabled individuals themselves, occasions to blame the victims for being so are reduced, although not completely eliminated. But medicalization's focus on personal biological deficits creates pressures to acquiesce to risky medical treatments with low rates of providing relief (Andrews 2011), as well as to allow segregated public services or overt removal of disabled people from the community.

Sameness and Singularity

Disability as discussed on the terms the medical model supplies invites dividing the human population into two parts, four-fifths of whom function similarly and thus count as normal, while the remaining one-fifth function disparately due to dysfunctions traceable to anomalous biological conditions. (See U.S. Census 2012 for the proportion of disabled people in the population.) But as the philosopher of science and physician Georges Canguilhem observed, the human species persists because our biological system is capable of very diverse modes of functioning executed under a great variety of conditions, including unusual ones that call for physiological or psychological adaptation (Canguilhem 1978; Canguilhem 1991). Canguilhem (1991: 196) argues that anomalous individuals should be diagnosed as diseased or disabled only in relation to specific contexts that affect people's functional success. Health care policy should proceed according to research models that can acknowledge as valuable statistically anomalous modes of functioning that may be useful, even if only under conditions that may be rare.

Functional determinism supposes that there is a natural, species-typical mode of carrying out each human function. It is by invoking this supposed standard that the medical model magnifies diagnoses of specific illnesses and injuries into attributions of functional disablement. Philosopher of science and disability studies scholar Ronald Amundson (2000) calls for abandoning this misleading doctrine of functional determinism to which, he contends, medical thinking continues to cling.

Amundson points out that for a species, variation in how essential tasks are executed is not unusual and can be eminently useful. The data about species persistence made available to us by biological science do not substantiate claims that average or species-typical modes of functioning always are preeminently effective or otherwise superior. A related mistake is that evolution results in the normalization of species by fixing certain traits through the process of selection, which subsequently remain set in their role in speciation. Biologically adaptive plasticity, such as occurs with the human brain, is well-known. Amundson illustrates with the case of a student with only 10% of the usual human brain tissue owing to subclinical hydrocephaly whose IQ and social life were indiscernible from that of other people.

Since Amundson published his influential article, advances in molecular medicine have revealed more and more ways in which people who appear to be average individuals at the level of the phenotype can vary enormously at the level of the genome. Among the three billion base pairs in which a human individual's chromosomes consist, there is an enormous number of distinct variations that can make this individual more risk prone or more resistant to different diseases and disabilities than most others (Rose 2009). Conceivably, the rarest states of body or mind could be the healthiest because the most effectively adaptive or resilient, which may be especially advantageous if the physical or social conditions in which species members must function undergo change.

Amundson therefore rejects conceptualizing human health as if the difference between species-typical and less-common modes of functioning is that the latter also are less good. That a biological state is found in a majority of humans does not make it a healthy state. Yet, as he notes, despite the well-known facts of functional variation and genetic variation within the human species, "the notion of a fixed species design with determinate limits on functional potential still plays a dominant role in health care" (cf. Boorse 1975; Wachbroit 1994; Boorse 1997). In health care, as well as other areas of policy, social norms that impose conformity on modes of functioning often have been camouflaged as biological norms by casting familiar modes as normal and thereupon naturally suited to promote the species' success.

The explanatory framework that assumes functional determinism is true has been used to justify political approaches and social arrangements that exclude kinds of people who, according to that framing theory, are not normal. Based on the mistaken theory that species-typical modes of functioning are the most valuable, anomalous biological functional conditions such as blindness, deafness, or missing limbs are taken to permit or even call for societal arrangements that disadvantage atypical individuals. For Amundson, an illustrative case of illegitimately using such a normative assumption can be found in bioethicist Norman Daniels's influential proposal about the proper policy for distributing health care resources (Daniels 1985; Daniels 1987), which relies on the assumption that functional determinism is true.

According to Daniels, preserving or restoring species-typical functioning is a primary goal of health care. He stipulates that "[T]he kinds of [health care] needs picked out by reference to normal species functioning are objectively important because they meet this high-order interest persons have in maintaining a normal range of opportunities" (Daniels 1987: 301). The normative power accorded to species-typical modes of functioning is supposed to be confirmed by the frequency with which these modes are exhibited in the general population.

That these occur so often in the population is taken to show that nature favors these familiar functional modes and, further, to explain why they are taken to be norms. Consequently, allocation of health care resources to people whom medical intervention cannot normalize should have the lowest priority, because even with treatment they will never be able to take advantage of the normal range of opportunities (Daniels 1985: 48). This policy conclusion is the outcome of invoking the hypothesis of functional determinism in order to explain and then manage disability by naturalizing it.

The Social Model: Addressing the Disadvantage of Disability

In the last part of the 20th century, a social model of disability was advanced to compete with the medical model. Disability advocates contended that shaping social arrangements so as to be receptive to average or normal people, but inflexibly intolerant when it comes to anomalous individuals, restricts disabled people's well-being more than their biological anomalies do. In other words, the expectation that the normality standard should govern the way things are done narrows the opportunity range for atypically functioning individuals.

Applying a social model perspective, Silvers criticizes Daniels's placing greater value on conforming to species-typical modes of functional performance than on effectively executing the function in whatever way the atypical individual can (Silvers 1998: 101; Silvers 2003). People who cannot function effectively by using the same modes of performing that species-typical individuals do nevertheless may achieve workplace and daily living aims if their alternative modes of functioning are allowed. This insight is appreciated by promoters of the social model of disability, which applies materialist analysis to explain the disadvantage of living with disabling biological conditions when the social arrangements that are circumstances of contemporary life are organized to provide only for species-typical people. The social model captures a significant aspect of disabled people's experience, namely, the striking improvement in their functionality that an accommodating environment often makes, and the precipitous deterioration if accommodations are withdrawn. In particular, the social model accounts for the frequent connection of disablement with social exclusion, as their familiar modes of executing actions may seem more comfortable, more efficient, or otherwise preferable to the nondisabled majority (O'Brien 2001).

In Carolyn Cleveland's case, the accommodations she needed were never made. Factually, therefore, she was totally disabled because she was unable to execute her job without accommodation. A counterfactual claim also may be true, however: had the employer not denied her reasonable accommodation request for such accommodations as an augmented communication device, a computer program with word prediction, word search, and speech output software, training, and additional time for some work tasks, she could have functioned effectively in her job. In other words, Cleveland's claim was that she was totally disabled only because her world was one where the employer maintained an unaccommodating work environment that was hostile to her atypical modes of functioning.

It is as if in the SSDI claim Cleveland and her physician were speaking in a world in which functional determinism is true. Were that so, inability to function in the species-typical way would leave people with biological deficits no option other than to abandon employment for subsistence on disability benefits. On the other hand, the ADA claim spoke of a world where the norms endorsed by the outdated theory of functional determinism have not continued to influence medical practice. In that world, employers would appreciate rather than disdain accommodating differences in employees' modes of functioning. Individuals who in the SSDI world count as totally disabled could, in the ADA world, remain on the job.

As neither the medical nor the social model is meant to be a definition of disability, neither should be used to distinguish who is disabled from who is not. Rather, the models provide different explanations for understanding the connection between disability and disadvantage. The social model differs from the medical one as to the cause of the exceptional disadvantage to which disabled people are exposed, but it also diverges from the medical model in regard to a remedy. The social model cites social facts or conditions as the relevant kind of restricting cause. The theme here is not to reject medical ministration categorically but to focus on nonmedical factors that not only are alterable but that, if altered, can ameliorate the social exclusion that is so prominent a part of many disabled people's lives (Silvers 2003: 476).

Different versions propose social structures, prejudicial or fearful attitudes, economic interests, or similar societal phenomena as sources of the barriers disabled people face. All are subject to amelioration through political action aimed at rescuing people with disabilities from social isolation and integrating them into the community. The social model originally was devised to be of practical political service in propelling rights-driven, liberating social progress for disabled people. It was aimed initially at revising UK social support programs that required disabled people to live in nursing homes and to enable them to reside in, and enjoy the opportunities offered by, the wider community.

The course of action the social model recommends is to pursue rights claims to access and inclusion, accommodating practice, and other kinds of societal change that will allow people constrained by illness or injury more and better opportunity. Locating the problem of disability in correctable social biases rather than in disabled persons whose bodies or minds need correction has proven to possess persuasive power (Silvers 2011). Evidence of the social model's capacity for stimulating action is confirmed by the achievements of the global disability movement in codifying disabled people's rights, not only through national legislation such as the ADA in 1990 but also in the United Nations' Convention on the Rights of People with Disabilities (CRPD) in 2006, as well as in legislation in many nations and in programs adopted by regional organizations such as the European Union.

Multiple Meanings

In a 2007 report called *The Future of Disability in America*, the U.S. Institute of Medicine (IOM) designated disability as a pressing problem for public health. Extolling the effectiveness of medical technology to prevent or remedy disability, the report applauds the reduction of activity-limiting biological dysfunction in older adults but warns that increases in physical inactivity, diabetes, and obesity place younger and middle-aged adults at growing risk of disability.

Identifying such biological conditions with disability is typical of the medical model. Nonetheless, the IOM's press release announced the report's findings with words that appear to promote the social model as well:

> Increasingly, scientific evidence reveals that disability results, in large part, from actions society and individuals take [and is] the result of interactions between people and their physical and social environments. Many aspects of the environment contribute to limitations associated with disability—for example, inaccessible transportation systems and workplaces, restrictive health insurance policies, and telecommunications and computer technologies that do not consider people with vision, hearing, or other disabilities.
> (National Academies of Sciences, Engineering, Medicine 2007)

The IOM appears agreeable to a multivariant explanatory model of disadvantage based on disability. Despite the *Cleveland* case, tension between these different normative directions is hard to eliminate completely, even though no contradiction between the assertions they propel exists (Kohrman and Berg 2005). How comfortably this tandem strategy of advocating medical technology while also invoking anti-discrimination law can be pursued remains to be seen. This is, of course, not the only area of current medical practice where picturing the patient as a defective dependent now must be integrated with the patient's status as a respected bearer of rights.

In 2012, a U.S. Census Bureau report commented:

> Because health professionals, advocates, and other individuals use the same term in different contexts, disability does not often refer to a single definition. . . . As a demographic category, disability is an attribute with which individuals may broadly iden-

tify, similar to race or gender. In contrast, certain federal programs narrowly define disability.... The agencies and organizations that provide benefits to, advocate for, or study these populations, each refer to their targeted group as people with disabilities; but because of the differences in definitions, an individual may be considered to have a disability under one set of criteria but not by another.

(Brault 2012)

Theories about disablement can neither engage each other nor be robust if they do not refer to the same thing. After the ADA became law, implementation often was undercut by this problem, despite the Supreme Court's attempt in *Cleveland* to evade it (Kohrman and Berg 2005). To counter medical testimony that a plaintiff's condition satisfies criteria for disablement, defendants would appeal to some other account of that state and insist that the plaintiff did not qualify for protection. Eventually, in 2008, an almost unanimous Congress amended the ADA to reduce the importance of determining whether individuals seeking redress for discrimination based on disability actually are disabled (Stein, Silvers, Areheart, and Francis, 2014). But variability in attributions of disability identity—whether by medical professionals, policy makers, or disabled people themselves—is unlikely to resolve the lack of consensus among policy makers, lay people, and disabled people about the purpose of identifying individuals as disabled.

References

Amundson, R. (2000) "Against Normal Function." *Stud. Hist. Phil. Biol. & Biomed. Sci.* 31(1), 33–53.

Andrews, Erin E. (2011) 'Pregnancy with a Physical Disability: One Psychologist's Journey: American Psychological Association,' *Spotlight on Disability Newsletter* [online] December, Available from: https://web.archive.org/web/20150716062712/ http://www.apa.org/pi/disability/resources/publications/newsletter/2011/12/pregnancy-disability.aspx

Barnes, E. (2016) *The Minority Body: A Theory of Disability*. Oxford: Oxford University Press.

Blackstone, W. (1765) *Commentaries on the Laws of England*. Vol, 1. pp 442–445 on Blackstone Commentaries, Women and the Law. Available from: http://womenshistory.about.com/cs/lives19th/a/blackstone_law.htm

Bognar, Greg. (presented at the SPAWN conference, 2014) "Fairness and the Puzzle of Disability."

———. (2016) "Reproductive Ethics Is Disability Mere Difference?" *J Med Ethics* 42(1), 46–49.

Boorse, C. (1975) "On the Distinction between Disease and Illness." *Philosophy and Public Affairs* 5, 49–68.

———. (1997) "A Rebuttal on Health," in J. M. Humber and R. F. Almeder (eds.), *What Is Disease*. Totowa, NJ: Humana Press, pp. 3–134.

Brault, R. (2012) *Americans With Disabilities: 2010 —Household Economic Studies*. Washington, DC: U.S. Department of Commerce, Economics and Statistics Administration, Census Bureau. Available at http://www.census.gov/prod/2012pubs/p70-131.pdf

Buck v. Bell (1927) 274 U.S. 200.

Canguilhem, G. (1978) *On the Normal and the Pathological*. Dordrecht: Reidel.

———. (1991) *The Normal and the Pathological*, trans. C. R. Fawcett & R. S. Cohen. New York: Zone Books.

Carey, T. and N. Hadler. (1986) "The Role of the Primary Physician in Disability Determination for Social Security Insurance and Workers' Compensation." *Ann Intern Med.* 104(5), 706–710.

Chew, K. (2013) "Would You Abort a Disabled Child?" *The Guardian* 22 April, Available from: http://www.theguardian.com/commentisfree/2013/apr/22/abort-down-syndrome-child-society-shares-blame

Cleveland v. Policy Mgmt. Sys., 120 F.3d 513, 517 (5th Cir. 1997), vacated, 526 U.S. 795 (1999).

Daniels, N. (1985) *Just Health Care*. Cambridge: Cambridge University Press.

———. (1987) "Justice and Health Care," in D. Van deVeer and T. Regan (eds.), *Health Care Ethics*. Philadelphia: Temple University Press, pp. 290–325.

Goesaert v. Cleary (1948) 335 U.S. 464.

Holmes, S. (1991) "Abortion Issue Divides Advocates for Disabled," *The New York Times* 7 July, Available from: http://groups.csail.mit.edu/mac/users/rauch/nvp/consistent/nyt_disabled.html

Kamm, F. M. (2013) "Rationing and the Disabled: Several Proposals," in Nir Eyal, Samia Hurst, Ole F. Norheim, and Daniel Wikler (eds.), *Inequalities in Health: Concepts, Measures, and Ethics*. New York: Oxford University Press, pp. 240–259.

———. (2015) "Cost Effectiveness Analysis and Fairness." *Journal of Practical Ethics* 3(1), 1–14.

Kevles, D. (1985). *In the Name of Eugenics: Genetics and the Uses of Human Heredity*. New York: Knopf.

Kohrman, D. and K. Berg. (2005) "Reconciling Definitions of 'Disability': Six Years Later, Has Cleveland v. Policy Management Systems Lived Up to Its Initial Reviews as a Boost for Workers' Rights." *Marquette Elder's Advisor* 7, 29–67.

Lombardo, P. n.d. Eugenic Sterilization Laws. Available from: http://www.eugenicsarchive.org/html/eugenics/essay8text.html

National Academies of Sciences, Engineering, Medicine. (2007) Press release announcing publication of *The Future of Disability in America*. Available from: http://www8.nationalacademies.org/onpinews/newsitem.aspx?RecordID=04242007

National Institute for Health and Care Excellence. (2008) NICE Clinical Guideline. March. Available from: http://www.nice.org.uk/

Nussbaum, Martha. (2006) *Frontiers of Justice: Disability, Nationality, Species Membership*. Cambridge, MA: Harvard University Press.

O'Brien, Ruth. (2001) *Crippled Justice: The History of Modern Disability Policy in the Workplace*. Chicago: University of Chicago Press.

O'Fallon, E. and S. Hillson. (2005) "Brief Report: Physician Discomfort and Variability with Disability Assessments." *J Gen Intern Med*. 20(9), 852–854.

Proctor, R. (1988) *Racial Hygiene: Medicine Under the Nazis*. Cambridge, MA: Harvard University Press.

Rose, N. (2009) "Normality and Pathology in a Biomedical Age." *Sociological Review* 57(Suppl.), 66–83.

Savulescu, J. and G. Kahane. (2011) "Disability: A Welfarist Approach." *Clinical Ethics* 6(1), 45–51.

Scully, J. L. (2008) *Disability Bioethics: Moral Bodies, Moral Difference: Disability Bioethics: Moral Bodies, Moral Difference (Feminist Constructions)*. Lanham, MD: Rowman & Littlefield.

Silvers, A. (1998) "A Fatal Attraction to Normalizing," in Erik Parens (ed.), *Enhancing Human Traits: Ethical and Social Implications*. Washington, DC: Georgetown University Press, pp. 95–123.

———. (2003) "On the Possibility and Desirability of Constructing a Neutral Conception of Disability." *Theoretical Medicine & Bioethics* 24, 471–487.

———. (2011) "From the Crooked Timber of Humanity, Something Beautiful Should be Made!" *American Philosophical Association Newsletter on Philosophy and Medicine* 10(2), 1–5.

Skinner v. State of Oklahoma, ex. rel. Williamson (1942) 316 U.S. 535.

Sokas, R. K., L. S. Kolb, L. S. Welch, L. Chang, B. C. Horowitz, and J. el-Bayoumi. (1995) "A Single-Session Exercise to Address Medical Residents' Attitudes Toward Work Disability Evaluations." *Acad Med*. 70, 167.

Stein, M., A. Silvers, B. Areheart, and L. Francis. (2014) "Accommodating Every Body." *University of Chicago Law Review* 81(2), 689–757.

United States Census Bureau. (2012) Nearly 1 in 5 People Have a Disability in the U.S. Available from: https://www.census.gov/newsroom/releases/archives/miscellaneous/cb12-134.html

Wachbroit, R. (1994) "Normality as a Biological Concept." *Philosophy of Science* 61, 579–591.

Further Reading

Amundson, Ron. (2000) "Against Normal Function." *Stud. Hist. Phil. Biol. & Biomed. Sci*. 31(1), 33–53.

Barnes, Elizabeth. (2016) *The Minority Body: A Theory of Disability*. Oxford: Oxford University Press.

Nussbaum, Martha. (2006) *Frontiers of Justice: Disability, Nationality, Species Membership.* Cambridge, MA: Harvard University Press.

Silvers, Anita. (2003) "On the Possibility and Desirability of Constructing a Neutral Conception of Disability." *Theoretical Medicine & Bioethics* 24, 471–487.

Silvers, Anita, David Wasserman, and Mary Mahowald. (1998) *Disability, Difference, Discrimination: Perspectives on Justice in Bioethics and Public Policy.* Lanham, MD: Rowman and Littlefield. Reprint with additions forthcoming 2017.

5
MECHANISMS IN MEDICINE
Phyllis Illari

What Is a Mechanism?

The idea of mechanisms, and how they are used in the sciences, is something that has been explored since 1993 in what is often regarded as the core mechanisms literature, focusing on the life sciences and the sciences of mind and brain (Bechtel & Richardson 1993; Glennan 1996; Machamer, Darden & Craver 2000), and since 1983 in the philosophy of social sciences (Elster 1983; Hedström & Swedberg 1998).

Probably the most well-known current characterization of a mechanism is: "Mechanisms are entities and activities organized such that they are productive of regular changes from start or set-up to finish or termination conditions" (Machamer et al. 2000: 3). The paradigm mechanism that Machamer, Darden, and Craver discuss is that of protein synthesis, which explains how cells make proteins. For this case, they say, entities include "the cell membrane, vesicles, microtubules, molecules, and ions," and activities include "biosynthesis, transport, depolarization, insertion, storage, recycling, priming, diffusion, and modulation" (Machamer et al. 2000: 8). There have been various alternative characterizations (Glennan 2002; Bechtel & Abrahamsen 2005), and Illari and Williamson offered the following simplified account: "A mechanism for a phenomenon consists of entities and activities organized in such a way that they are responsible for the phenomenon" (2012: 120). Illari and Williamson argue that this definition captures an emerging consensus within the mechanisms literature, and they show how the account can be interpreted fairly broadly to avoid excluding key mechanisms (compare the closely similar formulations of Craver 2007; Glennan 2008, 2009).

Simply put, then, finding a mechanism explains a phenomenon by finding local parts and discovering what they do and how they are organized to produce the phenomenon. In medicine, we seek to discover mechanisms of disease and of cure. A great deal of this work depends on our existing knowledge of bodily mechanisms, of which protein synthesis is an important example.

What Is Evidence of Mechanism in Medicine?

To explain disease, we seek at least some *evidence* of mechanisms of disease and of cure. A paradigm kind of evidence of mechanism is probably use of breakthrough technologies, such as Franklin's X-ray crystallography photograph of DNA, which help us study entities we could not study directly before and understand their activities and how they are organized. Franklin's work was famously used by Watson and Crick to describe the structure of DNA, which is now central to our understanding of thousands of cellular mechanisms. In general, finding and understanding new entities and their activities, of which bacteria and viruses may well be the most famous examples, has been important in the history of medicine, and no doubt will continue to be so.

Finding entities might be important, but a huge variety of other scientific work might count as finding evidence of mechanism: anything that helps us identify or better understand entities, activities, and their organization, including simulating organization of complex mechanisms, might count as finding evidence of mechanism. (See Bechtel & Richardson 1993; Darden & Craver 2013 for extensive discussion.) It is worth noting that the literature often recognizes two kinds of mechanisms. The first kind is mechanisms that consist of entities and activities *underlying* a phenomenon of interest (called "constitutive" by Craver 2007 and "vertical" by Kincaid 2011, 2012). For example, we observe that cells make proteins and identify the activities of DNA, RNA, and ribosomes as the mechanism underlying this phenomenon. The second kind is mechanisms that consist of intermediate *links* between a posited cause and its effect (called "etiological" by Craver 2007 and "horizontal" by Kincaid 2011, 2012). Taking the example of protein synthesis again, we might suspect that an alteration in DNA affects which proteins are produced, and try to identify the intervening entities and activities by which the cause leads to the effect. This distinction in kinds of mechanism can be important for some purposes (see Kincaid 2011, 2012), but in some cases it might not much matter: if we cannot really tell what kind of mechanism we are examining given what we know. Suppose we are investigating little-known mechanisms of protein synthesis, trying to identify new entities or activities, then we might make the same methodological choices whether we think of ourselves as identifying a constitutive/vertical mechanism underlying this phenomenon, or finding an etiological/horizontal mechanism linking cause and effect. It might only be possible to use this distinction in practice for cases where the mechanisms in question are unusually well known.

In the face of the variety of kinds of evidence of mechanism, I will simplify this short chapter by focusing on a single case: that of breast cancer. This case is unusual, in that evidence of mechanism is used very intensively, but, for this reason, it serves well to illustrate the points, so that they can be applied to thinking about less-clear cases.

Although the breast cancer population is large, it is non-homogenous. Not all breast cancers develop in the same way or respond to treatment in the same way. There is no single mechanism of disease. There are important differences among cancerous cells, differences among tumors as to how they develop and grow, and, as is common, differences in individual patients' positive response to treatment versus toleration of toxicity. These differences are crucial to our understanding of the disease, to understanding that there are multiple related mechanisms of breast cancer, with different prognoses.

We can now get effective evidence of these different mechanisms at the cellular level, even in individual patients. Most notably, we can take biopsies and type breast cancer cells according to the proteins expressed on their surfaces. This has been enormously helpful in understanding possible mechanisms of *cure* (or inhibition). Designing drugs that target particular proteins has led to enormously improved treatment in 80–85% of cases of breast cancer, because it helps to get drugs to harm cancerous cells and not healthy cells. The protein targets become a crucial access point for an effective mechanism of cure.

The other side of this success story is the remaining 15–20% of cases, called triple-negative breast cancer (TNBC). Clinically, "[t]riple-negative breast cancers are defined as tumors that lack expression of estrogen receptor . . ., progesterone receptor . . ., and HER2" (Foulkes et al., 2010: 1938). Although TNBC is more common in younger women, it has a poor prognosis, as we have no drugs that will effectively target it. The search to better understand these mechanisms of disease, and design effective mechanisms of cure, will continue. Improving treatment of breast cancer is dependent on this ongoing, extensive, largely laboratory-based determination of mechanism.

Integrating Evidence of Mechanism with Other Evidence in Medicine

Explaining disease is of course intellectually valuable, but it is not the only aim of medicine. As in breast cancer, in wider medicine we want to act. In this section, I will draw attention to the many things we want to know in medicine in order to act effectively, and begin to illustrate how this leads us to use evidence of mechanism alongside other evidence, most notably randomized controlled trials (RCTs, see Chapter 18).

In medicine, we want to know what causes disease, certainly, but this is not simple, nor is it the only task (see interesting details in Dawid 2000; Kincaid 2011, 2012). Notice that we also want to know how we can treat people, to prevent disease or cause cure, which is not quite the same as knowing what causes disease. Furthermore, we frequently want to know *how* a particular agent causes a disease, or how a drug works, which is often explicitly described as finding the mechanism of disease causation, or of cure.

We also want to know other things, which may be more or less important under varying circumstances. In many cases, quantities matter. For example, we need to know effective dosage of drugs or what levels of radiation are safe. We also want to know various things about the human population we intend to treat. Are some people more at risk? For example, some drugs must not be given to pregnant women. Ultimately, patients and their doctors have to assess what is the best treatment for one particular person.

Mechanisms are directly involved in explaining how disease is caused, but knowledge of mechanisms can also help with these other tasks. To succeed in these tasks, we use evidence of mechanism alongside other sources of evidence. Many philosophers have advocated something like a requirement of "total evidence" (Carnap 1962; Hempel 1965) in science more broadly. This requires us to bring to bear all available evidence on scientific problems and acknowledge that, in science, any study is fallible. All studies have some weaknesses. This means that the best we can do in investigating any particular disease, or possible treatment, is use a variety of kinds of studies, kinds of experimental work, and different sources of evidence. We hope that, at the very least, they have *different* kinds of weaknesses and should not all display the *same bias*. When this is true, and a variety of different studies agree, we can have much more confidence in the outcome (Haack 1993; Wimsatt 2007).

This requirement of total evidence does not seem to accord with privileging a single kind of tool or study. Randomized controlled trials (see Chapter 18), where the study population is randomized to treatment and control groups, are often treated as our gold standard of medical evidence (Cartwright & Munro 2010). RCTs are rightly valued for careful protocols, which, if carefully followed, restrict the influence of bias and unknown confounders on the outcome, both accidental (perhaps due to doctors and patients having faith in the new treatment) and non-accidental (such as fraud). RCTs are an important tool, although, for practical or ethical reasons, they cannot be used in many cases.

Important tool that they are, as we noted above, RCTs, like any evidence, should not be used alone—and indeed they are not. Whether running an RCT (or an observational study), you have to make a variety of decisions that cannot be decided by an observational or experimental clinical study. You have to decide what variables you will include in your study and how you will measure them—many medical studies depend heavily on decisions about diagnosis of disease and judgment of cure or of disease progression. You have to decide how long to run a trial for: 12 weeks, 6 months, or 5 years? When these questions are raised, they are often treated as either simply part of the RCT or as based merely on "preclinical trials" or on "background knowledge." But we can do better than this—we can recognize that justified choices of these kinds can be based on our best evidence for the mechanism of disease causation and possible mechanism of cure. That understanding should inform diagnosis, judgment of cure or disease progression, and decisions about how long patients need to be followed up for.

To illustrate, these kinds of decisions had to be made to run a phase 2 trial of iniparib as an initially promising treatment for TNBC. Diagnosis of disease was performed by typing cells from biopsies, but judgment of successful treatment was difficult for this trial. Sadly, unambiguous complete cure was not expected, which meant the trial designers had to define what they would count as relevant outcomes. The authors explain:

> Primary end points were the rate of clinical benefit (defined as the percentage of patients who had a complete response, a partial response, or stable disease for at least 6 months), as well as safety and tolerability of iniparib. Secondary end points were the overall rate of response and progression-free survival, defined as the time from randomization to confirmation of disease progression or death. Overall survival (defined as the time from randomization until the date of death) was not prespecified as an end point but was analyzed to explore the potential effect of iniparib on survival.
> (O'Shaugnessy et al. 2011: 207)

The study required, of course, a way of assessing progression-free survival: "Tumor response was based on investigator assessment of target and nontarget lesions and was assessed by means of computed tomography or magnetic resonance imaging at baseline and every 6 weeks thereafter, in the absence of clinically evident disease progression" (O'Shaugnessy et al. 2011: 207). The variety of outcomes specified was based on our knowledge of the usual mechanisms of disease progression and what could reasonably be hoped for from the treatment being tested. Furthermore, the trial used what technologies it could to track the underlying mechanisms of disease progression, such as the growth of the tumor, by examining the tumor using computed tomography or magnetic resonance imaging to give an assessment of patient response that was as objective as possible. Evidence of mechanism of disease progression was used here on an individual patient basis.

Evidence of mechanism can also be very helpful for an important group of tasks once trials have been completed: extending the results of studies to non-study populations. This can include wider patient populations, overseas populations, important sub-populations, or even decisions about one single case. An observational study or even RCT does not in itself contain information about *who else* the results of that study might—or might not—apply to. (Steel 2007 discusses the "extrapolator's circle"; Cartwright 2012 presses the worry of "external validity"; see Chapter 18). Evidence of mechanism cannot solve these problems entirely, but it can help us make judgments.

For example, in the case of breast cancer, evidence of mechanism tells us how crucial the cellular differences are to treating sub-populations of cases. This lets us know that typing the cells from a biopsy is likely to be more important to whether a patient will respond to a treatment designed to target that particular protein than whether a patient lives in the United Kingdom or the United States. Of course, this does not exclude other differences being relevant to patient response.

In sum, there are multiple tasks in assessing disease causation in medicine and in choosing sensible treatments. Evidence of mechanism is frequently helpful. In general, it is not possible to make really justified choices for all of these vital decisions without investigating what evidence there may be of mechanism of disease causation or cure.

Assessing Evidence of Mechanism

All of these tasks are involved when putting together kinds of evidence to decide, overall, what to conclude on the basis of multiple studies about any possible treatment, which is the responsibility of many public bodies worldwide. For example, in the United States, the Food and Drug Administration is the agency responsible for assessing drugs for safety and efficacy to decide whether their sale should be permitted in the United States. The National Institute for Care and Health Excellence in the United Kingdom is responsible for assessing treatments for effectiveness versus cost, to say which treatments will be paid for by the UK National Health Service (see National Institute for Care and Health Excellence 2014). The International Agency for Research on Cancer is tasked with categorizing carcinogens (see International Agency for Research on Cancer 2006). These and many other agencies convene panels or committees to assess all the evidence available on particular medical treatments.

Knowledge of mechanisms can help us perform this kind of assessment, but we do not simply either know a complete mechanism or know nothing whatsoever about the mechanism. We need to assess in a more nuanced way than this what kind of evidence of mechanism we have. In this section I will illustrate how we can have different amounts of evidence of mechanism by describing three different levels of evidence of mechanism one might have. I draw heavily on collaborative work with Brendan Clarke, Donald Gillies, Federica Russo, and Jon Williamson (summarized in Clarke et al. 2014).

The main point is that *evidence* of mechanism must amount to more than a hypothesis about a possible mechanism, but need not amount to complete knowledge of a mechanism. Suppose you are on an evidence panel assessing whether a particular drug (*Drug*) causes a cure (*Cure*) of a particular disease. Consider that there are at least three different levels of evidence of mechanism you might have:

1. Evidence that there is a particular mechanism explaining a link between *Drug* and *Cure*
2. Evidence that there is some kind of mechanism or other explaining a link between *Drug* and *Cure*
3. Evidence that there is no mechanism explaining a putative link between *Drug* and *Cure*

We are very seldom in the position where we know the complete mechanism in medicine. However, all three levels above are feasible, and they can all be incredibly useful.

In reverse order, homeopathy is often considered a case of level 3. Ultimately, our major reason for considering homeopathy ineffective is that everything that we understand of the physical chemistry of water suggests that homeopathy's posited mechanism of action cannot work; that water does not have a "memory" in the relevant sense. Notice that this evidence is not narrowly of the posited mechanism, but is much broader, as the physical chemistry of water structures what we know about many other mechanisms in the domain. Ultimately, of course, this is an application of other scientific knowledge, which should not be treated as permanently infallible. We cannot know how our understanding of the physical chemistry of water will develop in the future.

A particularly important case of level 2 is the discovery of the mechanism of the disease tetanus. Tetanus was puzzling, as the disease seemed to involve infection at a wound site, causing effects distant in the body. When we discovered that the mechanism of action was due to a protein that traveled around the body, and isolated the protein, it was an enormous leap forward, even if we did not know the full mechanism. Faced with a different puzzling disease, it is then sensible to ask whether there might be another mediating protein, and search in turn for evidence of its action. This is reasoning from analogy. Isolating that protein, and identifying other entities and activities in the mechanism, shifts the evidence from level 2 to level 1.

Whether you require evidence of level 3, 2, or 1 is highly contextual. It depends heavily on how much you know about the domain in question and on your purposes. In a public health emergency involving a largely unknown domain, a public body might regard level 3 as enough to proceed to trials. When considering a very well-known domain, a drug company might require something closer to level 1, such as identification of a drug target in a well-known mechanism of disease causation, and laboratory evidence that the drug is effective against that target, before funding expensive trials.

Ultimately, both scientists and evidence panels put evidence of mechanism together with evidence from clinical trials to make decisions about treatment. I will illustrate the use of evidence of mechanism throughout the rest of this chapter by looking in more depth at two trials of iniparib as a treatment for TNBC: the earlier trial seemed promising (O'Shaughnessy et al. 2011), whereas a later and larger trial showed no significant benefit (O'Shaughnessy et al. 2014).

The reason to test iniparib as a possible treatment was dependent on lab-based evidence of mechanism: in this case, pretty well-known general mechanisms of cancer causation, combined with reasoning from analogy from the mechanisms successfully designed to target 80% of cases of breast cancer, to search for a drug that may be effective on at least some cases of TNBC. The broad outline of cancer causation is well known. DNA is constantly subject to damage, but cells have multiple DNA repair mechanisms. The ultimate defense against cancer is when unrepairable cells undergo programmed cell death. Only when this, too, fails, do cells begin replicating out of control, which is to say, become cancerous.

The standard treatment used for TNBC is chemotherapy, which aims to kill cancerous cells but has significant toxic effects on healthy cells. Iniparib was designed as a possible new mechanism of cure, to try to increase the effectiveness of chemotherapy. Cancerous cells also have to repair their own DNA to stay alive when they are under attack by treatments such as chemotherapy. Iniparib inhibits PARP (poly(adenosine diphosphate–ribose) polymerase), and inhibiting PARP was found to increase the frequency of double-strand breaks in DNA (Carey & Sharpless 2011: 277). TNBC cells were thought to be particularly susceptible to double-strand breaks killing them (O'Shaughnessy et al. 2011: 205). Finally, lab evidence suggested that iniparib would work well with chemotherapy drugs: "Although the full mechanism of its antitumor activity is still under investigation, iniparib enhances the antiproliferative and cytotoxic effects of . . . [chemotherapy drugs] in in vitro models of triple-negative breast cancer" (O'Shaughnessy et al. 2011: 206).

What was hypothesized here was a way to interrupt a known mechanism of disease causation (level 1), piggybacking on known mechanisms of cure (level 2–1), using a new substance, iniparib, the action of which had been confirmed in lab-based preclinical trials (level 2). Put simply, available evidence and reasoning about mechanism were used to design iniparib to set up the cancerous cells so that the chemotherapy would wipe them out—making the chemotherapy far more toxic to cancerous cells than to healthy cells.

This illustrates, then, how useful evidence of mechanism can be, in this case for the task of continuing the search for treatments of a disease that currently has a very poor prognosis. It also illustrates how evidence of a solitary mechanism is not generally used in an isolated way, but integrated with other available mechanistic knowledge. In the final section, we will turn to examining problems of evidence of mechanism, and looking more carefully at how evidence of mechanism is put together with evidence generated by clinical trials. Iniparib will also be used to illustrate this point, because promising as it seems, as of 2014, iniparib does not work in clinical trials.

Problems with Evidence of Mechanism

So far in this chapter we have seen how mechanisms are used to explain disease and developed thinking about how evidence of mechanism can be useful for many of the different tasks of medicine. In this final section, I will sound a note of caution. Some might be tempted to regard evidence of mechanism as conclusive about whether, say, *Drug* causes *Cure*. However, evidence of mechanism is itself fallible (see Howick 2011). In this section I will examine two common problems with evidence of mechanism and argue that evidence of mechanism is best used in integration with other sorts of evidence, such as that coming from clinical studies such as RCTs.

Recall that I claimed in the previous section that we seldom if ever have complete knowledge of a mechanism. The three levels of evidence I have described and defended as useful are all levels of less than complete knowledge. It is in this, the usual context, where two common problems of evidence of mechanism arise. Again, suppose we are assessing whether *Drug* causes *Cure*:

1. **Complexity**: The mechanism explaining the link between *Drug* and *Cure* is too complex to allow us to estimate an overall, average effect.
2. **Masking**: The mechanism explaining the link between *Drug* and *Cure* is masked by an undiscovered mechanism that also links *Drug* and *Cure*. The undiscovered mechanism may have the *opposite* effect to the known mechanism.

Suppose you have level 1 evidence of a mechanism that explains a link between *Drug* and *Cure*. Perhaps you have isolated some crucial entities on the pathway from *Drug* to *Cure* and established novel activities that explain the action of *Drug*. It can be tempting to suppose that you can now be certain that *Drug* causes *Cure*.

Notice, however, what you most want to know given the task of medicine: you want to know that the drug will work to cure disease in real human populations. Real populations, the human body, and even individual cells, are very complex, interactive entities. This means that identifying one mechanism connecting *Drug* and *Cure* does not tell you what else might connect or otherwise affect *Drug* and *Cure*. You might have a problem of complexity, which means that you have pretty secure lab-based or even *in vivo* evidence of mechanism but are unable to estimate effect size attributable to that mechanism, perhaps even be unable to decide whether its overall effect will be positive or negative. Furthermore, you might have a masking problem: however detailed your knowledge of the known mechanism is, other mechanisms may exist connecting *Drug* and *Cure*, which are completely unknown to you and mask (or alternatively enhance) the action of the known mechanism.

To address these problems, to help decide what will work in real human populations, we need clinical studies. These studies cannot entirely solve the problems at hand, but they can be very helpful. On the one hand, if they are also positive, we have excellent evidence for the effectiveness of *Drug*. On the other hand, it is quite common for a new treatment that appears promising in the lab or in animal studies to turn out to be ineffective in human populations—due perhaps to toxicity, but also sometimes due to not working, for no reason that we can detect.

I will illustrate the importance of using trials alongside evidence of mechanism by examining how work on iniparib proceeded at the trial stage. The later 2014 trial concludes: "The efficacy results from the phase III study did not confirm the promising results of the phase II [2011] study" (O'Shaughnessy et al. 2014: 3846). So the first thing to note is that, however good the lab studies of iniparib looked, the phase 3 trial did not support its effectiveness in a real human population.

I will now suggest that evidence integration goes beyond simple ideas of weighing evidence. It is not the case that positive preclinical evidence is a plus, while negative RCTs or other trial evidence is a minus, and we generate the overall assessment of evidence by simply adding up the pluses and subtracting the minuses. In the case of iniparib, the negative phase 3 trial trumps the positive lab evidence for decisions about treatment, while the lab evidence remains strong enough to support further lab work. Evidence of mechanism and evidence from RCTs inform each other in important ways. This is because, used in combination, evidence of mechanism and evidence of correlations in populations found in trials can help with the weaknesses of each other, as I will now show. Consider the following lists, which illustrate the relation between the advantages of each kind of evidence and the weaknesses of the other (updated from Illari 2011).

Evidence of C-E correlation (such as from clinical studies):
- *Weaknesses*: confounding, non-causal correlations, and bias
- *Advantages*: can reveal masking and can help assess the net effect of a complex mechanism

Evidence of C-E mechanism (such as from lab or animal studies):
- *Weaknesses*: masking and being too complex to assess a net effect
- *Advantages*: can reveal confounding and non-causal correlations

On the one hand, suppose we have identified—even at level 1—a mechanism explaining the link between *Drug* and *Cure*, but we are concerned that its action will be masked, either in a real human body or in a real human population. Then our best evidence that we can put aside this concern is evidence of a correlation between treatment with *Drug* and *Cure* in a real human population, coming from an RCT or other trial. On the other hand, if we are concerned that the correlation found in a trial is due to bias or confounding by a common cause of treatment and recovery, evidence that the mechanism of action of the treatment really works *in vitro* or in animal studies helps ameliorate that concern. In these ways, evidence of mechanism, and evidence of correlation in real human populations, are both essential in assessing total evidence of whether medical treatments really cause cure.

The intertwining of evidence is important to our assessment of iniparib. First, we do not know why it did not work in the 2014 trial. The trial report says: "There is not a strong biologic hypothesis that can explain the discrepancy in the iniparib efficacy results in the first- versus second-/third-line cohorts." The trial designers could find no difference in patient populations, either (O'Shaughnessy et al. 2014: 3846). It is not clear whether the mechanism of action of iniparib did not work or was masked. An editorial comment published at the same time as the 2011 trial illuminates how much we do not know: "We cannot tell whether the benefit from the PARP inhibitor accrued to all triple-negative tumors equally or whether the benefit preferentially accrued to a subgroup of BRCA-deficient tumors, with less effect in those without the deficiency" (Carey & Sharpless 2011: 278).

Specifically, Carey and Sharpless seem to be suggesting that there may be relevant but unknown sub-populations in TNBC, and iniparib might work only in a sub-population we cannot yet clearly define. Recall that, to run any clinical study, you have to decide what counts as disease and what counts as cure or desirable outcome. You have to choose how long to run the trial for and how to *measure* both disease and outcome. I claimed previously that justified decisions on these issues should be based on evidence of the mechanisms of disease and of cure.

To find effective treatments for breast cancer, it is crucial to recognize that there is not one homogenous disease "breast cancer." The classification "triple-negative breast cancer" is already very interesting. It is a diagnostically important classification because it affects prognosis and identifies some treatments as useless. Yet the classification TBNC depends on what we know of the disease mechanisms—and also on known mechanisms of cure.

Carey and Sharpless are drawing our attention to the fact that sub-populations are likely to be central to ongoing development of breast cancer treatments, and hopefully identifying these populations will lead to effective treatments for cases that are now classified together simply as triple-negative. In sum, breast cancer illustrates how evidence of both mechanism and of correlation are used in assessing overall evidence of medical treatments. As we have seen, the complete mechanism is seldom known, but at least three levels of evidence of mechanism may be useful. We have also discussed the problems of masking and of complexity for interpreting evidence of mechanism, and noted that what level of evidence of mechanism is needed depends both on what else is known and the purposes of inquiry.

We have seen throughout this chapter how important evidence of mechanism has been in our much-improved ability to treat 80–85% of cases of breast cancer, even when gold standard RCTs are being used. Well-known mechanisms of cancer causation and progression are used as a background to design new targeted mechanisms of cure. In treatments of other diseases, the use of such evidence may be far less obvious. However, in so far as any trial has to decide what variables to include, how to measure them, how long to run, and what populations the trial results apply to, all trials have to be run based on decisions that the trial itself cannot validate alone. These decisions thus must often be based on evidence of mechanisms.

References

Bechtel, W. and Abrahamsen, A. (2005): "Explanation: A Mechanist Alternative," *Studies in the History and Philosophy of the Biological and Biomedical Sciences*, 36: 421–441.
Bechtel, W. and Richardson, R. (1993): *Discovering Complexity*, Princeton University Press: Princeton.
Carey, L. A. and Sharpless, N. E. (2011): "PARP and Cancer—If It's Broke, Don't Fix It," *New England Journal of Medicine*, 364(3): 277–279.
Carnap, R. (1962): *Logical Foundations of Probability* (2nd ed.), University of Chicago Press: Chicago.
Cartwright, N. (2012): "Will This Policy Work for You?" *Philosophy of Science*, 9(5): 973–989.
Cartwright, N. and Munro, E. (2010): "The Limitations of Randomized Controlled Trials in Predicting Effectiveness," *Journal of Evaluation in Clinical Practice*, 16(2): 260–266.
Clarke, B., Gillies, D., Illari, P., Russo, F., and Williamson, J. (2014): "Mechanisms and the Evidence Hierarchy," *Topoi* (online first), DOI 10.1007/s11245-013-9220-9.
Craver, C. (2007): *Explaining the Brain*, Clarendon Press: Oxford.
Darden, L. and Craver, C. (2013): *In Search of Mechanisms*, University of Chicago Press: Chicago.
Dawid, A. P. (2000): "Causal Inference Without Counterfactuals (With Comments and Rejoinder)," *Journal of the American Statistical Association*, 95: 407–448.
Elster, J. (1983): *Explaining Technological Change*, Cambridge University Press: Cambridge.
Foulkes, W. D., Smith, I. E., and Reis-Filho, J. S. (2010): "Triple-Negative Breast Cancer," *New England Journal of Medicine*, 363: 1938–1948.
Glennan, S. (1996): "Mechanisms and the Nature of Causation," *Erkenntnis*, 44(1): 49–71.
Glennan, S. (2002): "Rethinking Mechanistic Explanation," *Philosophy of Science*, 69: S342–S353.
Glennan, S. (2008): "Mechanisms," S. Psillos and M. Curd (eds.) *Routledge Companion to the Philosophy of Science*, Routledge: Oxford, 420–428.
Glennan, S. (2009): "Mechanisms," H. Beebee, C. Hitchcock and P. Menzies (eds.) *The Oxford Handbook of Causation*, Oxford University Press: Oxford, 315–325.
Haack, S. (1993): *Evidence and Inquiry: Towards Reconstruction in Epistemology*, Blackwell: Oxford.
Hedström, P. and Swedberg, R. (1998): *Social Mechanisms: An Analytical Approach to Social Theory*, Cambridge University Press: Cambridge.
Hempel, C. G. (1965): *Aspects of Scientific Explanation*, Free Press: New York.
Howick, J. (2011): *The Philosophy of Evidence-Based Medicine*. BMJ Books: Chichester, West Sussex, UK.
Illari, P. M. (2011): "Mechanistic Evidence: Disambiguating the Russo-Williamson Thesis," *International Studies in the Philosophy of Science*, 25(2): 139–157.
Illari, P. M., and Williamson, J. (2012): "What Is a Mechanism? Thinking about Mechanisms Across the Sciences," *European Journal for Philosophy of Science*, 2(1), 119–135.

International Agency for Research on Cancer: *Preamble to the IARC Monographs* (amended Jan 2006), available at: http://monographs.iarc.fr/ (accessed November 2015).

Kincaid, H. (2011): "Causal Modelling, Mechanism, and Probability in Epidemiology," P. Illari, F. Russo and J. Williamson (eds.) *Causality in the Sciences*, Oxford University Press: Oxford, 70–90.

Kincaid, H. (2012): "Mechanisms, Causal Modelling, and the Limitations of Traditional Multiple Regression," H. Kincaid (ed.) *The Oxford Handbook of the Philosophy of Social Science*, Oxford University Press: Oxford, 46–64.

Machamer, P. Darden, L., and Craver, C. (2000): "Thinking about Mechanisms," *Philosophy of Science*, 67: 1–25.

National Institute for Care and Health Excellence. (2014): *Developing NICE guidelines: The manual*, available at: http://www.nice.org.uk/media/default/about/what-we-do/our-programmes/developing-nice-guidelines-the-manual.pdf (accessed November 2015).

O'Shaughnessy, J., Osborne, C., Pippen, J. E., Yoffe, M., Patt, D., Rocha, C., Koo, I. C., Sherman, B. M., and Bradley, C. (2011): "Iniparib Plus Chemotherapy in Metastatic Triple-Negative Breast Cancer," *New England Journal of Medicine*, 364: 205–214.

O'Shaughnessy, J., Schwartzberg, L., Danso, M. A., Miller, K. D., Rugo, H. S., Neubauer, M., Robert, N., Hellerstedt, B., Saleh, M., Richards, P., Specht, J. M., Yardley, D. A., Carlson, R. W., Finn, R. S., Charpentier, E., Garcia-Ribas, I., and Winer, E. P. (2014): "Phase III Study of Iniparib Plus Gemcitabine and Carboplatin Versus Gemcitabine and Carboplatin in Patients With Metastatic Triple-Negative Breast Cancer," *Journal of Clinical Oncology*, 32: 3840–3847.

Steel, D. (2007): *Across the Boundaries: Extrapolation in Biology and Social Science*, Oxford University Press: New York.

Wimsatt, W. (2007): *Re-Engineering Philosophy for Limited Beings*, Harvard University Press: Harvard.

Further Reading

Bechtel, William, and Richardson, Robert (1993): *Discovering Complexity*, Princeton University Press: Princeton. (Classic statement of modern mechanism and extended exploration of mechanism discovery.)

Cartwright, Nancy (2012): "Will This Policy Work for You?" *Philosophy of Science*, 9(5): 973–989. (Recent statement of Cartwright's long-standing concerns about applying results of randomized controlled trials to other populations.)

Clarke, Brendan, Gillies, Donald, Illari, Phyllis, Russo, Federica, and Williamson, Jon (2014): "Mechanisms and the Evidence Hierarchy," *Topoi* (online first), DOI 10.1007/s11245-013-9220-9. (Summary of how mechanisms can be used in evidence integration in medicine.)

Howick, Jeremy (2011): *The Philosophy of Evidence-Based Medicine*, Chichester, West Sussex, UK: BMJ Books. (Extensive discussion of evidence-based medicine, including criticism of uncritical use of mechanistic reasoning.)

Steel, D. (2007): *Across the Boundaries: Extrapolation in Biology and Social Science*, Oxford University Press: New York. (Classic recent treatment of mechanisms and the extrapolation problem.)

Related Chapters

Holly Andersen: Reduction in the biomedical sciences (Chapter 8)
Julian Reiss: Causality and causal inference in medicine (Chapter 6)
Adam La Caze: The randomized controlled trial: internal and external validity (Chapter 18)
Robyn Bluhm: The hierarchy of evidence, meta-analysis, and systematic review (Chapter 19)
Mael Lemoine: Explanation in medicine (Chapter 27)

6
CAUSALITY AND CAUSAL INFERENCE IN MEDICINE

Julian Reiss

1. Inferring and Interpreting Causal Claims

Causal claims such as (1) "HPV types 16 and 18 are responsible for 70% of cervical cancers," (2) "Antipsychotic medications are effective in treating acute psychosis and reducing the risk of future psychotic episodes," (3) "Keeping weight under control, exercising, eating healthily, and not smoking prevent diabetes type-2," or (4) "Hal's exposure to vinyl chloride monomer caused his angiosarcoma" are widely used in medicine, epidemiology, public health, and elsewhere. Arguably, this is for good reason: causal claims allow us to *explain* medical outcomes, both at the population level (claim 1) and the individual level (claim 4); they can help to *predict* outcomes (claim 2); they *underwrite treatment decisions* (claim 2) and *public health policies* (claim 3); and they can help to *attribute responsibility* (join claim 4 to the claim that Hal's employer behaved negligently by exposing him)—all important aims of the health sciences and beyond.

Knowledge of causal claims is a good thing, then. However, their usefulness depends on the extent to which we can reliably learn causal relationships from data, the extent to which what we can learn from the data is unambiguous, and the extent to which causal knowledge actually does help to realize these more ultimate purposes. This chapter will examine these three issues in turn, starting with the learning of causal relationships from data, then moving on to the interpretation of causal claims, and finally addressing the usefulness issue. Throughout, I will use vinyl chloride carcinogenicity as a central example. Vinyl chloride is a compound widely used in the production of polyvinyl chloride (PVC). Before it was classified as a human carcinogen by the International Agency for Research on Cancer (IARC), the U.S. Environmental Protection Agency (EPA), and similar bodies in the 1970s, and regulations were tightened in response, workers in PVC manufacturing plants were often exposed to the substance over long periods of time and suffered adverse health consequences, including the very rare condition angiosarcoma of the liver.

2. Inferring Causal Claims

It would be as imprudent as it would be uncouth to talk about causal inference in medicine without beginning the discussion with the so-called Hill criteria for causation, named after Sir Austin Bradford Hill (1897–1991), a British epidemiologist and statistician. He was concerned with how to distinguish genuinely causal from spurious associations. Medical researchers at the time were fully aware that an association between a risk factor such as diet and a medical outcome such as diabetes could be due to numerous reasons other than the risk factor causing the disease. Hill proposed the following criteria for causation (Hill 1965; the descriptions are paraphrased rather than directly quoted):

(1) *Strength*. The stronger the association, the more likely it is causal.
(2) *Consistency*. If the association has repeatedly been observed by different persons, in different places, circumstances, and times, it is more likely to be causal.
(3) *Specificity*. The more specifically the risk factor, outcome, or both can be defined, the more likely an association between them is causal.
(4) *Temporality*. The more closely the temporal dimension of an association is aligned with our expectations, the more likely it is causal.
(5) *Biological gradient*. If there is a dose-response curve, the association is more likely to be causal.
(6) *Plausibility*. If the causal relation appears plausible on the basis of our background knowledge at the time, it is more likely to be causal.
(7) *Coherence*. The causal interpretation of the association should be coherent with general known facts about the natural history of the disease and its biology.
(8) *Experiment*. If the association has been established in an experiment or quasi-experiment, it is more likely to be causal.
(9) *Analogy*. If there is a strong analogy between the risk factor and a known cause of the disease, it is more likely to be causal.

None of these "viewpoints," as Hill sometimes calls them, is either necessary or sufficient for causality, nor is the conjunction of all nine sufficient. A strong association can be due to a confounder and a weak one causal. Plausibility is, as duly noted by Hill, relative to the biological knowledge of the day, which may be imperfect. There may not exist any analogies or experimental evidence. Nevertheless, each "viewpoint" can be regarded as a fallible indicator—rather than a strict criterion—of causality.

To see how the items on Hill's list can perform the role of fallible indicators of causality, it is useful to distinguish two competing approaches to causal reasoning in the biomedical sciences: the experimentalist on the one hand and the inferentialist on the other (cf. Parascandola 2004). The experimentalist approach maintains that randomized experiments are the "gold standard" of causal inference and discounts evidence about causal claims from other sources. Evidence-based medicine and other movements that carry the "evidence-based" label are rooted in the experimentalist approach. Inferentialism, by contrast, holds that causal claims are inferred from diverse bodies of evidence—bioassays, laboratory experiments with animal models, cohort and case-controlled studies, case reports, clinical trials—using pragmatic guidelines such as Hill's. The approach is very widely used in biomedical research, but inferentialists tend to be less vocal than their evidence-based colleagues.

One way to defend experimentalism is to assume a specific interpretation of causality and then proceed to show that under that interpretation of causality, positive results of certain kinds of experiments guarantee the truth of the associated causal claim. A view of causality that has been very popular recently is James Woodward's, according to which one variable X (directly) causes another variable Y if and only if there is a possible intervention (intervention variable I) on X that changes Y or its likelihood of occurring (Woodward 2003: 55). Intervention variable I, in turn, has the following characteristics (Woodward 2003: 98):

I1. I causes X.
I2. I acts as a switch for all the other variables that cause X. That is, certain values of I are such that when I attains those values, X ceases to depend on the values of other variables that cause X and instead depends only on the value taken by I.

I3. Any directed path from I to Y goes through X. That is, I does not directly cause Y and is not a cause of any causes of Y that are distinct from X except, of course, for those causes of Y, if any, that are built into the I-X-Y connection itself; that is, except for (a) any causes of Y that are effects of X (i.e., variables that are causally between X and Y) and (b) any causes of Y that are between I and X and have no effect on Y independently of X.

I4. I is (statistically) independent of any variable Z that causes Y and that is on a directed path that does not go through X.

In a biomedical experiment, the intervention is the assignment of a member of a test population (e.g., a population of animal models) to a treatment or control group. In one experiment, 360 Swiss mice were exposed to different concentrations of vinyl chloride (VC) 4 hours daily on 5 days per week for 30 weeks (Maltoni and Lefemine 1974). The assignment to a treatment group causes the level of exposure (I1). The level of exposure to VC is controlled by the experiment so that it no longer depends on other variables (e.g., proximity to a PVC manufacturing plant; I2). The assignment to a treatment group does not cause cancer through a mechanism that bypasses exposure to VC (I3). This condition would be violated, for example, if different treatment groups received different diets, which in turn affected cancer rates. Finally, the assignment to a treatment group is statistically independent of other causes of cancer (I4). This condition would be violated, for example, if different strains of mice, which have different degrees of cancer susceptibility, were used in different treatment groups. The four conditions are illustrated in Figure 6.1.

In clinical trials on human subjects, randomization is an intervention in Woodward's sense. If X is treatment status (with values x_t = test treatment and x_c = control), and Y a variable measuring the difference in medical outcome between treatment and control group, then randomization causes X, treatment status, to assume its value x_t or x_c: the outcome of the randomization process determines whether a patient will be in the treatment or the control group (I1). Clinical control will ensure that only patients who are in the treatment group will receive the treatment (I2). The outcome of the randomization process will not have a direct effect on the value of Y because Y is defined as the *difference* in medical outcome between the two groups and all participants of the trial—patients, doctors, nurses, analysts—are blinded with respect to treatment status (I3). Finally, successful randomization will guarantee (at least for large samples) statistical independence from other variables responsible for the medical outcome (I4). Thus, under a Woodwardian conception of causality, if in a randomized experiment that fulfills criteria (I1)–(I4) X and Y are associated, then it must be the case that X causes Y. This would appear to support experimentalism.

Critics of experimentalism say two things in response. First, few real clinical trials strictly fulfill criteria (I1)–(I4). Compliance (whether or not a patient takes the assigned treatment) is rarely perfect, and occasionally treatments are shared between the two groups. Withholding treatment status from participants is often not possible or, if it is possible, treatment status

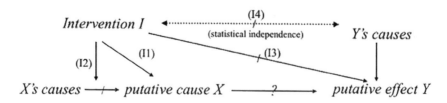

Figure 6.1 Woodward's characterization of an intervention variable

might be revealed if the treatment is actually effective. Randomization guarantees statistical independence of other causes of the medical outcome only in infinitely large samples, and test populations are often small and never infinite. The critics charge that while experimentalists have a good answer why ideal randomized experiments guarantee true causal conclusions, they remain silent about the conditions under which reliable causal judgments can be made under realistic conditions when experiments cannot be implemented ideally or when the causal claim is not open to experimental test. Second, they point out that the bulk of causal knowledge in medicine has been established non-experimentally. As one article published in the *British Medical Journal* put it ironically, we do not need a randomized controlled trial "to determine whether parachutes are effective in preventing major trauma related to gravitational challenge" (Smith and Pell 2003). Critics maintain that experimentalism makes it mysterious that other sources of evidence such as observational studies should reliably support causal claims.

Inferentialists regard causal inference as analogous to medical diagnosis. The presence of most diseases cannot be directly observed but only inferred on the basis of symptoms. Causality can analogously be inferred from "symptoms," albeit, in general, on the population level rather than the individual level. An association between a risk factor and a medical outcome can thus be regarded as a symptom of the existence of a causal relation between factor and outcome. Medical symptoms rarely point unequivocally to a single disease. Likewise, the symptoms of causality have alternative explanations such as confounding (the production of an association by a variable associated with risk factor and medical outcome), selection bias (the determination of treatment status by the patient), experimenter bias (the determination of treatment status by the experimenter), attrition bias (patients' premature exit from the trial at different rates between treatment and control arm), diagnostic error/mismeasurement, and so on. Thus, like a medical diagnostician, the causal researcher comes to a judgment about the hypothesis at stake only after looking at a diverse body of evidence, one function of which is to rule out such alternative explanations.

Just as there is no "golden symptom"—a type of test that reliably indicates the presence of a disease no matter what the disease is—there is no gold standard of evidence according to the inferentialist. There are only bodies of evidence that, if the parts fit together in the right way, can make a convincing case for or against a causal claim.

Hill's viewpoints can play the role of pragmatic criteria that help the inferentialist to come to a judgment concerning the causal claim. For example, a strong association [Hill's no. (1)] certainly does not prove a causal relationship—a confounded relationship may well be strong. (To use a philosopher's favorite example, drops in barometer readings are strongly associated with the occurrence of storms; the relationship is nevertheless confounded by atmospheric pressure.) However, if it is known that the most likely confounder cannot (or most likely does not) produce an association of that size, this alternative can be ruled out. This consideration helped to eliminate R. A. Fisher's "constitutional hypothesis" in the 1950s. According to this hypothesis, a single genetic factor predisposed people to both lung cancer and taking up smoking, thus accounting for the association between smoking and lung cancer without smoking necessarily being a cause of lung cancer. Although genetic factors were known to play a role in cancer susceptibility, they could not explain the 60-fold increased risk observed in the data.

On the other hand, weak associations do not disprove a causal relationship either. Vinyl chloride was classified as carcinogenic by the IARC in 1974 (IARC 1974a), but despite a surge in industrial use of the compound, only very few additional cases of cancer were observed. Carcinogenicity could nevertheless be established because exposed individuals developed angiosarcomas of the liver. These are so rare that chance or other confounders can hardly account for the coincidence of this specific condition [Hill's no. (3)] among exposed individuals.

Experimentation [Hill's no. (8)] plays a role in the inferentialist approach, but like the other items on Hill's list, it is neither necessary nor sufficient for establishing causality. Clinical trials cannot be used to establish carcinogenicity for ethical reasons. Vinyl chloride had been established experimentally to cause cancer in animal models before epidemiological studies confirmed its carcinogenicity in humans (IARC 2014), but even though all human carcinogens have some animal models, there is no guarantee that a substance that causes cancer in animals is also harmful to humans.

Although a well-designed randomized trial eliminates a host of confounders all at once, according to the inferentialist, no feature of experimental design can guarantee that all potential errors, including those having to do with the measurement of outcomes, with data analysis and reporting, with publication and many other aspects of the inference, have been ruled out. Whether or not they have been ruled out remains a judgment that can only be made after the entire body of evidence has been consulted. To give an example, for some time there was a discrepancy between randomized and observational studies in the effect of hormone replacement therapy on breast cancer, with the randomized studies showing a smaller risk than the observational studies. The reason was not, however, that the observational studies were inherently less reliable whereas the randomized studies got it right. The difference lay instead in the timing of the studies: the women in the randomized studies had on average been longer in menopause before starting the treatment. Reanalyzing the data from the randomized trials by adjusting for this temporal gap, the results fell in line and confirmed those of the observational studies (Vandenbroucke 2009). This outcome has nothing to do with the design of either trial or observational study and could only be reached by a systematic review and analysis of all the evidence.

An important issue that has been widely discussed among philosophers of the biomedical sciences concerns the role that evidence about mechanisms plays in causal inference. Risk factors do not cause medical outcomes across spatio-temporal gaps but through continuous biological pathways or "mechanisms" (sometimes also called "modes of action"). There is no doubt that understanding these mechanisms greatly enhances biomedical knowledge and is useful for numerous purposes, including the explanation of medical outcomes, improving intervention strategies, more accurate prognosis, and many more. According to one view, evidence about mechanisms is an important ingredient in successful causal inference (Russo and Williamson 2007). One reason to maintain this view is that causal conclusions can be regarded as always underdetermined by evidence about population-level associations because confounders cannot conclusively be ruled out. As confounders cannot always reliably be measured, and it is always possible that there are unanticipated confounders, causal conclusions should not rest on evidence about correlations alone.

Prima facie, the view discussed in the last paragraph is opposed to experimentalism but can be supported by inferentialism. Experimentalists maintain that confounders are ruled out by the design of a randomized trial. Accordingly, evidence about mechanisms plays only a small role if any in the so-called hierarchies of evidence used in evidence-based medicine. Inferentialists, by contrast, make causal judgments on the basis of evidence from a variety of sources, and evidence about mechanisms naturally fits into a diverse body of evidence. (For a discussion of the diverse roles that evidence about mechanisms can play in causal inference, see Clarke et al. 2014.)

Things are not quite so simple, however. On the one hand, it can be argued that knowledge about mechanisms is necessary in the planning and design of a randomized trial as well as in the analysis and interpretation of data (La Caze 2011). Accordingly, even if the immediate basis for a causal inference is an association generated by the experiment, the inference would not be reliable unless made against a backdrop of knowledge about mechanisms. On the other hand, inferentialists need not require that this kind of knowledge be part of their diverse body of evidence. They can argue that while evidence about mechanisms can play a role in eliminating alternative hypotheses, there is no guarantee that it is a necessary ingredient (Reiss 2012).

Evidence about mechanisms is closely related to Hill's criterion (6), "plausibility." About it he says, "this is a feature I am convinced we cannot demand. What is biologically plausible depends upon the biological knowledge of the day" (Hill 1965). It is a simple fact about the history of medicine that our knowledge about biological pathways has increased dramatically since Hill wrote this. Examining the *IARC Monographs on the Evaluation of Carcinogenic Risk* over time, for instance, we find that in the 1970s there are practically no mechanistic considerations (e.g., IARC 1974b), whereas in the more recent edition the volume of reported data on mechanisms is larger than either that of epidemiological data or of data on animal models (e.g., IARC 2012). Concomitantly, we find an increasing number of substances that have been reclassified mainly on the basis of evidence about mechanisms. At the same time, at least for now it seems false to say that evidence about mechanisms is necessary for reliable causal inference. Many substances have been classified as carcinogenic for decades, and the original judgments were made without the benefit of data on mechanisms. It is not frequently the case that such judgments are overturned once the mechanistic data is in.

3. Interpreting Causal Claims

Once a causal claim has been established, what has been learned? What do we mean when we say that some risk factor causes a medical outcome? Philosophers distinguish five broad families of theories of causation: regularity, probabilistic, counterfactual, interventionist, and mechanistic. These theories provide interpretations of causal claims because they define the term "cause" that occurs in them. All five theories play, or have played, important roles in medicine.

Under the regularity view, a factor *causes* an outcome if and only if it is a necessary condition, a sufficient condition, both a necessary and sufficient condition, or an insufficient but non-redundant part of an unnecessary but sufficient (INUS) condition for the outcome. According to K. Codell Carter, early 19th-century medicine took a leap forward by adopting the regularity view in the guise of an "etiological viewpoint," the belief that

> diseases are best controlled and understood by means of causes, and in particular, by causes that are *natural* (that is, they depend on forces of nature as opposed to the willful transgression of moral or social norms), *universal* (that is, the same cause is common to every instance of a given disease), and *necessary* (that is, a disease does not occur in the absence of its cause).
>
> (Carter 2003: 1; emphasis in original)

The causes of large numbers of bacterial, viral, and deficiency diseases can be understood this way. Koch's postulates, which originate in the late 19th century, still embody the etiological viewpoint:

(i) The microorganism must be found in abundance in all organisms suffering from the disease, but should not be found in healthy organisms.
(ii) The microorganism must be isolated from a diseased organism and grown in pure culture.
(iii) The cultured microorganism should cause disease when introduced into a healthy organism.

(iv) The microorganism must be reisolated from the inoculated, diseased experimental host and identified as being identical to the original specific causative agent.

Postulate (iii) makes the presence of microorganism also sufficient for the disease. This basic understanding of the regularity view as a sufficient understanding of medical causation came under pressure as medical progress and an emerging interest in chronic (as opposed to infectious) diseases around the beginning of the 20th century challenged the understanding of cause as a necessary and/or sufficient condition. Koch knew that postulate (iii) is too strong (when read as a sufficient condition), because he found asymptomatic carriers of cholera (Koch 1893). Moreover, most causes of chronic diseases such as cancers or cardiovascular diseases are neither necessary nor sufficient conditions. The IARC classifies vinyl chloride as a human carcinogen, but exposure to the compound does not always lead to angiosarcomas (or other cancers), nor are all angiosarcomas (much less all cancers) caused by it.

The contemporary version of the regularity view maintains that causes are INUS conditions, which deals with some of the earlier problems. In contemporary epidemiology, for instance, causes are sometimes represented by the use of pie charts in which every wedge represents a condition that is necessary in the circumstances for disease; the entire pie represents the complex of conditions that are jointly sufficient (Rothman 1976). According to the INUS view, causes are neither necessary nor sufficient for their effects. Causes produce their outcomes only in conjunction with additional factors. Smoking, say, on its own is not followed by cancer. At minimum, a smoker has to be genetically susceptible and live long enough for the cancer to develop. Nor are most causal factors found in every instance of the disease as there are numerous alternative causes (each of which requires additional factors to produce the effect).

Both the original etiological viewpoint and the contemporary understanding of causes as INUS conditions are wedded to determinism: they assume that if all the conditions for an outcome are in place, the outcome will happen. Developments in the foundations of physics that occurred in the early 20th century led many researchers to abandon universal determinism and influenced thinking about causality in the biomedical sciences. Causes were no longer understood as sufficient for their effects (singly or jointly with additional factors) but rather as affecting merely the chances of outcomes, and thus the regularity view is no longer adequate.

According to the probabilistic view of causality, a cause is a factor that raises the probability of its effect in a causally homogenous population. The qualification is needed to distinguish between direct and confounded causal relations. The latter are cases where both the apparent cause and effect are in fact independent effects of a common cause. For example, under Fisher's constitutional hypothesis, smoking is not a cause of lung cancer but rather a byproduct of a common genetic factor. To distinguish between the two cases, we divide the population into two groups, one in which the genetic factor is present and one in which it is absent. If smoking raises the probability of lung cancer in both groups (assuming that the genetic factor is the only potential common cause), then it is a causal factor. A population is thus said to be causally homogenous whenever there is no variation among the causes of an outcome of interest in the population. If sex is a causally relevant factor, a causally homogenous population is one in which every member is a woman or one in which every member is a man; if age is a causally relevant factor, a causally homogenous population is one in which every member is in the same age group and so on.

It is important to note that the adequacy of the probabilistic view does not depend on whether factors such as Fisher's genetic condition are known or measurable. The question is whether a factor in fact raises the probability in a causally homogeneous population, not whether there are means to test this. A conceptual rather than practical or epistemic question is, however, whether to demand that a cause raise the probability in *all* causally homogenous populations, in *some* populations, or *on average*. Nancy Cartwright has defended a requirement of "contextual unanimity" according to which only those factors that raise the probability in *all* causally homogenous populations are causes (Cartwright 1979); Brian Skyrms has a slightly weaker requirement according to which the causal factors raise the probability of the effect in *some* populations but do not lower it in any (Skyrms 1980); and John Dupré has argued that factors that raise the probability of their effects *on average* should be called causes (Dupré 1984). Contextual unanimity is a very strong requirement. If, say, there is a gene that makes some people immune to vinyl chloride, then the substance is not to be regarded as carcinogenic even if it raises the probability of cancer for most people. This point, and a look to biomedical practice, led Dupré to abandon it for a focus on average probability increases. What randomized trials establish, according to Dupré, are average causal effects, not contextually unanimous causes.

On the other hand, a disadvantage of calling a factor that raises the probability of their effects only on average a cause is that the status of a factor as cause depends on the actual distribution of factors in a population. Suppose that although VC exposure increases the risk of developing angiosarcoma in most people, a cancer immunization gene *lowered* the probability that those people exposed to VC who have the gene will develop angiosarcomas below that of the general population. That is, for these people, exposure actually protects them against cancer. Dupré would call VC carcinogenic only as long as relatively few people in the population had that gene; if more people had it, the substance would cease to be a cause or even become a preventer of cancer, even for those people without the protective gene. If one believes that whether or not VC is carcinogenic has to do with its intrinsic properties and the intrinsic properties of the person exposed, then this is an unwelcome result. Dupré's interpretation of causal claims is also prone to yield bad advice. A treatment that is effective on average may be harmful to some. If so, to learn that the treatment causes relief is misleading for those subpopulations whose members are harmed by it.

A third view of causality starts from individuals rather than populations. It maintains essentially that the treatment or risk factor is a cause of the medical outcome whenever the outcome would not have occurred if it had not been for the treatment or risk factor. For example, exposure to VC is the cause of a worker's angiosarcoma because he would not have developed the disease had he not been exposed to the substance. In philosophy, this counterfactual theory of causality, which originates in David Lewis' (1973) paper, has received much attention. Lewis and his followers have never found an empirical measure to determine the truth value of such counterfactuals, however. This job has been left to biostatisticians, who have developed the so-called potential-outcomes framework of causality. In that framework, the most fundamental quantity is the individual causal effect (ICE), which is defined as:

(ICE) $Y_t(u) - Y_c(u)$,

where Y measures the medical outcome of u (the patient), and t and c refer to treatment and control status, respectively. Thus, the individual causal effect measures the difference between the value the outcome variable would have assumed had the subject been treated and the value the variable would have assumed had the (same) subject not been treated. In theory, both values are counterfactual in nature. In practice, a patient can either be treated or not, but not both treated and not treated. Therefore, for any given subject u, we can observe only either $Y_t(u)$ or $Y_c(u)$. Much of the literature on this framework develops strategies for identifying the individual causal effect or related quantities from observable data (e.g., Imbens and Rubin 2015).

The fourth, interventionist, account of causation was characterized in the previous section, when we discussed Woodward. Despite its enormous influence in philosophy, it has not been applied much in the biomedical sciences except in the area of psychology (see, for instance, the papers in Part I of Gopnik and Schulz 2007).

The final view of causality I will discuss is the mechanistic account, according to which a risk factor causes a medical outcome if, and only if, it produces the outcome by a mechanism of the appropriate kind. Many definitions of a mechanism have been advanced; according to one (Machamer et al. 2000: 3): "mechanisms are entities and activities organized such that they are productive of regular changes from start or set-up conditions to finish or termination conditions." "Being productive of changes" is itself a causal notion, which is why some commentators have questioned the usefulness of this kind of account for understanding causality (e.g., Parascandola and Weed 2001). But arguably, understanding is provided by accounts of the specific entities of which mechanisms are composed and the activities in which they are engaged (Machamer 2004). At any rate, according to the mechanistic view, causal relations do not necessarily all share characteristics such as regular co-occurrence, probability raising, or counterfactual dependence, which is why proponents of the mechanistic view refuse to define causality in these terms.

Given this multiplicity of accounts of causation, a researcher who seeks guidance in interpreting causal claims is spoilt for choice, so much is clear. What is dubious, to say the least, is whether this is a good state of affairs. After all, confusion and worse can arise when different researchers interpret causal claims differently, especially when these claims are used for further research, treatment decisions, and health policy making. For example, if a team of researchers establishes that VC is a cause of angiosarcomas in the probabilistic sense, but the claim is mistaken by a jury or judge to be a counterfactual causal claim, then a firm that exposed workers to VC, some of whom subsequently developed the disease, may be held responsible because the jury or judge thinks that without having been exposed to the substance, they would not have gotten ill. The counterfactual claim, however, does not follow from the probabilistic causal claim.

There are at least three strategies to deal with this plurality of causal interpretations. The first is to try to reduce one conception to another. If, say, it can be proved that everything that is true about causality under a probabilistic view is equally true under an interventionist view and vice versa, then we can safely ignore one of these views. Unfortunately, all attempts to do so have proved unsuccessful. (See, e.g., Hausman and Woodward 1999; Cartwright 2006.) The second strategy is to put one's foot down and maintain that one's favorite account exhausts the meaning of "cause." The main problem with this strategy is that there are counterexamples to each account; that is, there are bona fide cases of causation that do not count as such under the given account and cases the account regards as causal that are not accepted as causal by common intuitions or scientific practice. Counterexamples to the regularity account were discussed above; similar cases can be found for each of the other accounts (for a detailed review, see Reiss 2015):

- (Probabilistic) Not all causes raise the probability of their effects because a cause can be connected to its effect through more than one mechanism in such a way that positive and negative influences cancel on balance. Birth control pills are a cause of deep vein thrombosis (DVT) but also prevent pregnancy, itself a cause of DVT. On average, the two routes (positive contribution and prevention) might exactly cancel so that women on the pill have the same chance as those who are not to develop DVT.

- (Counterfactual) When two or more causes compete to bring about an effect, there may be causation without counterfactual dependence. A smoker who was also exposed to asbestos may develop lung cancer due to his smoking. However, had he not smoked, he might have developed the disease anyway, because of asbestos exposure.
- (Interventionist) Not all causal relations are invariant to interventions. Interventions may sometimes change the causal structure on the basis of which a higher-level relationship holds. Antibiotics are effective in the treatment of bacterial infections. Used too often, populations can become resistant. In this case, the intervention destroys the causal relation that was aimed to be used to bring about an effect.
- (Mechanistic) Some outcomes are caused by absences. Vitamin-D deficiency can cause multiple sclerosis. However, absences are not mechanistically connected to their effects. (This does not mean that mechanisms are not used in the *explanation* of the onset of a disease caused by an absence, but the absence cannot be the starting point of a causal process that terminates in the effect.)

The third strategy is to maintain that causality is not a feature of the external world but rather one of reasoning agents such as scientific researchers. Information about regularities, probabilistic dependencies, results of experiments, and the like provides reasons to believe causal relationships, but causal claims do not (or need not) represent anything specific in the world. According to one such "subjective" theory of causality, epistemic causality, "[c]ausal relationships are to be identified with the causal beliefs of an omniscient rational agent" (Russo and Williamson 2007: 168). Thus, causality depends on what an agent believes in the ideal case in which he/she is in the possession of all relevant evidence (though it is acknowledged that real agents are rarely in that situation). Another subjective theory, the inferentialist theory (which is closely related to inferentialism about evidence that was discussed above):

> maintains that the meaning of causal claims is given by their inferential connections with other claims. In particular, causal claims are inferentially related to *evidential claims*—the claims from which a causal claim can be inferred—as well as to claims about future events, explanatory claims, claims attributing responsibility, and counterfactual claims (claims predicting "what would happen if")—the claims that can be inferred from a causal claim.
>
> (Reiss 2015a: 20; emphasis in original)

Subjective theories do not suffer from the drawbacks of the other two strategies. They neither try to reduce one feature of causal relationships to another, nor do they suffer from obvious counterexamples (see, for instance, Reiss 2015a, Chapter 5, about how the inferentialist theory deals with cases of redundant causation which pose problems for many standard theories). They also address the original challenge adequately. The epistemic theory holds that causal claims have, in principle, one unique meaning, namely, what an omniscient rational agent believes to be true. The inferentialist theory can be understood as giving up looking for a definition of cause, instead allowing that there may be multiple notions, and asking about what kind of practices provide evidence for causal claims, what inferences from this evidence are justified, and what purposes knowledge of causal claims in medicine serves. It is a pragmatic theory that starts with a medical or policy problem, asks what kinds of causal knowledge are relevant for addressing the problem and what kinds of evidence are needed to substantiate this knowledge without assuming that meaning necessarily carries over from one context to the next.

4. On the Usefulness of Causality in Medicine

In the introductory section, I suggested that biomedical researchers seek knowledge of causal claims because it is useful in attaining the discipline's more ultimate purposes such as explanation, prediction, and making successful treatment and health policy decisions. We may ask, however, whether knowledge of causal claims really does promote these more ultimate purposes. In particular, we may ask whether knowledge of causal claims is necessary or sufficient (or both) for explanation, prediction, and decision making.

It turns out that the answer is ambiguous because it depends on context and on the kind of causal claim that is being used in that context. It is certainly true that causal claims can help explain outcomes, but their ability to do so depends on the precise nature of the explanatory interest and the causal claim at hand. "Vinyl chloride causes cancer of the liver" may be cited to explain a particular clustering of incidences (along with information about exposure), but it hardly explains why this worker rather than that one developed the disease, or why many workers who were exposed do not develop it or some non-exposed people do. Without information about the biological pathways through which they produce neoplasms, causal explanations are very thin at best.

The relation between causality and prediction is even more tenuous. When relationships are stable, we can successfully predict on the basis of correlations—knowledge of the true causal structure is not necessary. On the other hand, when relations are not stable (for instance, because the composition of causal mechanisms changes, some mechanisms operate indeterministically, or interferences occur), knowledge of causal relations does not help much to improve predictability. If, say, exposure to VC produces a stable number of liver cancers in a population and gives all exposed workers a specific set of symptoms, such as peripheral neuropathy and pain in the fingers, we can use information about the symptoms (a correlate) to make a prediction about the chances of developing liver cancer. If, on the other hand, the relationships are not stable, making a prediction on the basis of observing the real cause is prone to be unreliable. It is certainly possible that the relationship between some indicator, which is not directly causally connected to an outcome, and the outcome is more stable than the relationship between a cause and its effect. The point is that for prediction, stability is important, not causality.

Lastly, both of these problems occur in the case of decision making. For good decisions, we first need the right kind of causal knowledge. If some treatment does not operate in a "contextually unanimous" fashion (see previous section), then knowing that it is effective at the population level will not be a good basis for an individual treatment recommendation, even if that individual is a member of that population. Similarly, as Alex Broadbent has pointed out, that C (e.g., some risk factor) causes E (e.g., an adverse medical outcome) does not mean that reducing C *will* reduce E (Broadbent 2013). A ban on sugary drinks will not necessarily lead to lower obesity or diabetes rates because it depends on what people do instead. If, as does not seem implausible, people substitute an equally or more risky behavior, rates may stay put or increase even though the policy was based on a genuine causal relationship. Once again, in the policy context we want stable relationships between a policy variable and an outcome, and what causal relationships, if any, these are based on is immaterial.

Broadbent argues that epidemiologists are primarily interested in explanation and prediction, not in causation (Broadbent 2013). He bemoans that philosophers of science have basically ignored prediction as a topic of methodological analysis. The above-mentioned considerations support this view.

This is not to argue, of course, that knowledge of causal claims is not useful. An important take-home lesson, however, is that biomedical researchers are seldom interested in causal claims in their own right but rather because, and to the extent to which, they help to attain more ultimate purposes. Whether they do so or not is a contextual matter that has to do with the precise nature of the purpose pursued as well as the causal claim in question.

References

Broadbent, Alex (2013). *The Philosophy of Epidemiology*. London: Palgrave Macmillan.
Carter, K. Codell (2003). *The Rise of Causal Concepts of Disease: Case Histories*. Aldershot: Ashgate.
Cartwright, Nancy (1979). "Causal Laws and Effective Strategies." *Noûs* 13:419–437.
——— (2006). "From Metaphysics to Method: Comments on Manipulability and the Causal Markov Condition." *British Journal for Philosophy of Science* 57(1):197–218.
Clarke, Brendan, Donald Gillies, Phyllis Illari, Federica Russo and Jon Williamson (2014). "Mechanisms and the Evidence Hierarchy." *Topoi* 33:339–360.
Dupré, John (1984). "Probabilistic Causality Emancipated." In *Midwest Studies in Philosophy IX*, ed. P. French, T. Uehling Jr and H. Wettstein. 169–175. Minneapolis: University of Minnesota Press.
Gopnik, Alison and Laura Schulz (2007). *Causal Learning: Psychology, Philosophy, and Computation*. New York: Oxford University Press.
Hausman, Daniel and James Woodward (1999). "Independence, Invariance and the Causal Markov Condition." *British Journal for the Philosophy of Science* 50:521–583.
Hill, Austin Bradford (1965). "The Environment and Disease: Association or Causation?" *Proceedings of the Royal Society of Medicine* 58(5):295–300.
IARC (1974a). *IARC Monographs on the Evaluation of Carcinogenic Risk of Chemicals to Man*. Volume 7: Some anti-thyroid and related substances, nitrofurans and industrial chemicals. Lyon: International Agency for Research on Cancer.
——— (1974b). *IARC Monographs on the Evaluation of Carcinogenic Risk of Chemicals to Man*. Volume 14: Asbestos. Lyon: International Agency for Research on Cancer.
——— (2012). *IARC Monographs on the Evaluation of Carcinogenic Risk to Humans*. Volume 106: Trichloroethylene, tetrachloroethylene, and some other chlorinated agents. Lyon: International Agency for Research on Cancer.
——— (2014). *IARC Monographs on the Evaluation of Carcinogenic Risks to Humans*. Volume 106: Trichloroethylene, tetrachloroethylene, and some other chlorinated agents. Lyon: International Agency for Research on Cancer.
Imbens, Guido W. and Donald Rubin (2015). *Causal Inference for Statistics, Social, and Biomedical Sciences: An Introduction*. New York (NY): Cambridge University Press.
Koch, Robert (1893). "Über den augenblicklichen Stand der bakteriologischen Choleradiagnose." *Zeitschrift für Hygiene und Infektionskrankheiten* 14:319–333.
La Caze, Adam (2011). "The Role of Basic Science in Evidence-Based Medicine." *Biology and Philosophy* 26(1):81–98.
Lewis, David (1973). *Counterfactuals*. Cambridge (MA): Harvard University Press.
Machamer, Peter (2004). "Activities and Causation: The Metaphysics and Epistemology of Mechanisms." *International Studies in the Philosophy of Science* 18(1):27–39.
Machamer, Peter, Lindley Darden and Carl Craver (2000). "Thinking About Mechanisms." *Philosophy of Science* 67:1–25.
Maltoni, C. and G. Lefemine (1974). "Carcinogenicity Bioassays on Vinyl Chloride: I. Research Plan and Early Results." *Environmental Research* 7:387–405.
Parascandola, Mark (2004). "Two Approaches to Etiology: The Debate Over Smoking and Lung Cancer in the 1950s." *Endeavour* 28(2):81–86.
Parascandola, M., and Weed, D. L. (2001). "Causation in epidemiology." *Journal of Epidemiology and Community Health*, 55(12), 905–912.
Reiss, Julian (2012). "Third Time's a Charm: Wittgensteinian Pluralisms and Causation." In *Causality in the Sciences*, ed. P. McKay Illari, F. Russo and J. Williamson. 907–927. Oxford: Oxford University Press.
——— (2015). *Causation, Evidence, and Inference*. New York (NY): Routledge.

Rothman, Kenneth (1976). "Causes." *American Journal of Epidemiology* 104(6):587–592.
Russo, Federica and Jon Williamson (2007). "Interpreting Causality in the Health Sciences." *International Studies in the Philosophy of Science* 21(2):157–170.
Skyrms, Brian (1980). *Causal Necessity: Pragmatic Investigation of the Necessity of Laws*. New Haven (CT): Yale University Press.
Smith, Gordon and Jill Pell (2003). "Parachute Use to Prevent Death and Major Trauma Related to Gravitational Challenge: Systematic Review of Randomised Controlled Trials." *British Medical Journal* 327:1459–1461.
Vandenbroucke, Jan P. (2009). "The HRT Controversy: Observational Studies and RCTs Fall in Line." *Lancet* 373:1233–1235.
Woodward, James (2003). *Making Things Happen*. Oxford: Oxford University Press.

Further Reading

Agassi, Joseph (1976). "Causality and Medicine." *Journal of Medicine and Philosophy* 1(4):301–317.
Reiss, Julian (2015). "A Pragmatist Theory of Evidence." *Philosophy of Science* 82(3):341–362.
Worrall, John (2007). "Evidence in Medicine and Evidence-Based Medicine." *Philosophy Compass* 2:981–1022.

7
FREQUENCY AND PROPENSITY: THE INTERPRETATION OF PROBABILITY IN CAUSAL MODELS FOR MEDICINE

Donald Gillies

1. Causality in Medicine

The title of this chapter involves the terms "interpretation of probability" and "causal model," which will not be familiar to many readers. However, the meaning of these expressions will be explained as we go along. To give these explanations, we must begin by making some general observations about the notion of causality. Causality is a key concept in medicine, but, to analyze this concept, it will be useful to begin by distinguishing between theoretical and clinical medicine. What could be called *theoretical medicine* consists of a body of laws and theories, many of them involving causality, which have been discovered and then confirmed by medical research. A typical accepted causal law is the following:

The varicella zoster virus (VZV) causes chickenpox (1)

This kind of causal claim is described as *generic*, because it covers many cases.

In *clinical medicine*, however, a doctor examines a particular patient and has to find out what causes that patient's symptoms. A doctor may, for example, decide that the rash of Miss Anne Smith, aged 4, is chickenpox and so caused by VZV. This is an instance of *single-case*, rather than generic, causality. Some authors speak of type/token causality instead of generic/single-case causality, where type = generic, and token = single-case.

A second important distinction is between *deterministic* and *indeterministic* causality. A causal claim of the form A causes B is deterministic, only if, *ceteris paribus*, whenever A occurs, it is followed by B. Otherwise, the claim is indeterministic.

Deterministic causality is the traditional concept of causality, which is analyzed by 18th- and 19th-century philosophers such as Hume and Kant. Indeed, Kant says in *The Critique of Pure Reason* (1787: B5):

" . . . the very concept of a cause . . . manifestly contains the concept of a necessity of connection with an effect and of the strict universality of the rule . . ."

So, according to Kant, if A causes B, and A occurs, then B is sure to follow. This is what we have called deterministic causality.

Nineteenth- and early-20th-century scientific medicine, as it was developed by Pasteur, Koch, and others, used a deterministic notion of causality. An attempt was made to show that each disease had a single cause, which was both necessary and sufficient for the occurrence of that disease (Codell Carter 2003). So, for example, tuberculosis was caused by a sufficiently large number of tubercle bacilli in the patient.

In the 20th century, however, a new concept of causality emerged in medicine, particularly in connection with medical epidemiology. If a causal claim of the form A causes B is indeterministic, then A can occur without always being followed by B. One of the first examples of indeterministic causality in medicine was the claim that

Smoking causes lung cancer (2)

In 1947, when hypothesis (2) was first seriously investigated in England, it was regarded as unlikely to be true. The standard view was that lung cancer was caused by general urban atmospheric pollution. Nowadays, after much controversy, (2) is almost universally accepted as being true. Curiously, few participants in the heated discussion of the truth of (2) pointed out that it involved a new notion of causality. Yet, this is clearly the case.

To show this, I will quote statistics to be found in Doll and Peto (1976). These are concerned with a sample of 34,440 male doctors in the UK. The 1976 paper deals with the mortality rates of the doctors over the 20 years from 1 November 1951 to 31 October 1971. During that time, 10,072 of those who had originally agreed to participate in the survey had died, and 441 of these had died of lung cancer. As about 83% of the doctors sampled were smokers, this means that only about 5% of these smokers died of lung cancer. So, although smoking causes lung cancer, smoking is not always followed by lung cancer. But although smoking is not invariably followed by lung cancer, smoking definitely increases the probability of getting lung cancer. Doll and Peto calculated the annual death rate for lung cancer per 100,000 men standardized for age. The results in various categories were as follows (1976: 1527):

Non-smokers	10
Smokers	104
1–14 gms tobacco per day	52
14–24 gms tobacco per day	106
25 gms tobacco per day or more	224

(A cigarette is roughly equivalent to 1 gm of tobacco)

If we take these frequencies as good estimates of the underlying probabilities, then we have that P(lung cancer | smoker) [read as: the probability of lung cancer given that the person is a smoker] is about 10 times greater than P(lung cancer | non-smoker), while, if we define a heavy smoker as someone who smokes more than 25 gms of tobacco per day, then P(lung cancer | heavy smoker) is more than 22 times P(lung cancer | non-smoker).

The notion of indeterministic causality has, since 1950, become ubiquitous in medicine. Eating fast food is recognized as causing heart disease, having a gene or genes of a particular kind is recognized as causing Alzheimer's, infections by certain viruses are recognized as causing cancers, and so on. In all of these cases, we are dealing with indeterministic causality.

When indeterministic causes are used in medicine, they often appear in what could be called a *multi-causal fork*. I will describe such forks in the next section.

2. Multi-Causal Forks

As we have seen, from the 1950s on, medicine has had to introduce an indeterministic notion of causality. This goes hand in hand with explaining diseases as caused by the conjunction of several causes acting together. We can illustrate this by developing our earlier example of smoking causing lung cancer. Since not all smokers get lung cancer, we might postulate that those who do have a genetic susceptibility for the disease. We now have two causes which operate: smoking and genetic susceptibility. If we use deterministic causality, we can confine ourselves to one cause, since that is sufficient to produce the disease. This is called *mono-causality*. If we use indeterministic causality and postulate several causes, this is called *multi-causality*. The various different causes, which act together to produce the disease, will be called *causal factors*. The use of multi-causality, or several causal factors, leads to the introduction of *multi-causal forks*, as shown in Figure 7.1.

Here Z, a disease, has a finite number n of causal factors $X_1, X_2, \ldots, X_n$. Perhaps the most important case of a multi-causal fork in contemporary medicine is the case of heart disease (see Levy and Brink 2005). In the last 60 years, quite a number of causal factors for heart disease have been discovered, including smoking, eating fast food, high blood pressure, diabetes, and obesity. In the last few years, investigations have begun into possible genetic causal factors. As heart disease is still the number-one killer in the developed world, this case is obviously an important one. A fully developed causal model for heart disease would have to include all the factors just mentioned and perhaps others as well. However, it seems sensible to begin an investigation of multi-causal forks with a rather simpler situation. Accordingly, we will initially confine ourselves to multi-causal forks with two prongs, as shown in Figure 7.2.

In our example of heart disease, we will take X = smoking, Y = eating fast food, and Z = heart disease. This is shown in Figure 7.3.

We must now consider another feature of multi-causal forks, which, like the example just given, involve indeterministic causes. These are related to probabilities, as in our earlier

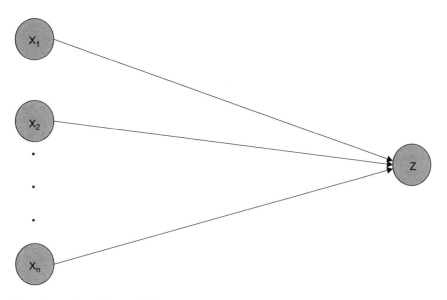

Figure 7.1 n-Pronged multi-causal fork

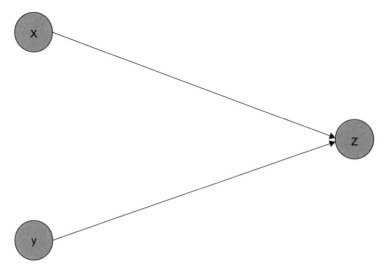

Figure 7.2 Two-pronged multi-causal fork

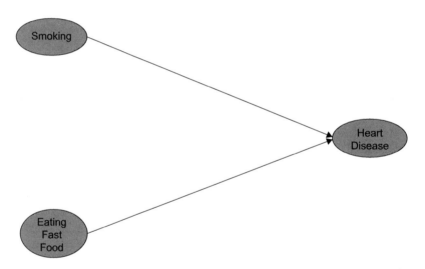

Figure 7.3 Multi-causal fork for heart disease

example of smoking causes lung cancer. Multi-causal forks will therefore often involve not just causes, as represented by the arrows, but also probabilities. This leads us to the concept of *causal model*, which is in the title of the chapter. Causal models are hypotheses about the causes of diseases in medicine that involve a number of different causal factors (represented diagrammatically by arrows) and also probabilities. The hypothesis illustrated in Figure 7.3 is a simple example of a causal model. Causal models belong to theoretical medicine, and so involve generic (or type) causality. The question now arises of how probabilities are to be fitted in with causes in causal models. Let us consider this in connection with our simple model of Figure 7.3.

We need to introduce the probability that someone gets heart disease given that he/she is a smoker [written P(heart disease | smoker)], and contrast this with P(heart disease |

non-smoker). In our earlier example, which concerned smoking and lung cancer rather than smoking and heart disease, the doctors in the table of results were classified into various groups according to the amount that they smoked. However, we will now simplify things by dividing people into smokers and non-smokers. This is obviously a simplification since the probability of heart disease increases with the quantity smoked and the amount of fast food eaten. However, as our project is to examine the relations between causality and probability, it seems best to take a simple, though quite realistic, case first, and then to consider the effects of adding more complexities later.

Mathematically speaking, we can, as in Figure 7.2, represent causes and effects by variables X, Y, Z. If X = smoking, then X has two values 0 and 1. X = 0 corresponds to a non-smoker, and X = 1 to a smoker, and similarly with the other variables. So P(heart disease | smoker) would be written $P(Z = 1 \mid X = 1)$.

Having introduced both causes and probabilities into our model, we are now in a position to consider how they are connected.

3. The Problem of Connecting Causality to Probability

To deal with the connection between indeterministic causality and probability in the context of multi-causal forks, such as that of Figure 7.2, it is helpful to begin by raising the question of how one can test causal claims. Let us consider first a claim of the form A causes B where deterministic causality is used. Here testing A causes B is an easy matter. Since, *ceteris paribus*, B always follows A, we have only to instantiate A. If B fails to follow, then the causal claim is refuted. If B does follow, then the causal claim is confirmed. Testing a claim involving indeterministic causality, such as smoking causes lung cancer is not so easy. Many smokers never get lung cancer, and yet this does not refute the claim that smoking is an indeterministic cause of lung cancer. We need to be able to derive some probabilistic claims from the claim about indeterministic causality. Once these probabilistic claims have been derived, we can then test them using statistical tests. Hence, by testing the consequence of a claim of indeterministic causality, we have tested the causality claim itself.

What we need then is a principle connecting indeterministic causality to probability. There seems at first sight to be an obvious such principle, which I will call the *Causality Probability Connection Principle* (CPCP). A number of principles of the same general type have been proposed by various authors (Gillies and Sudbury 2013: 287–291). The version to be used in this chapter is the following:

If X causes Z, then $P(Z = 1 \mid X = 1) > P(Z = 1 \mid X = 0)$ \hfill (CPCP)

Informally, this states that if X is an indeterministic cause of Z, then the occurrence of X raises the probability of Z occurring. If we take X as smoking and Z as lung cancer, then estimating probabilities from the observed frequencies in the table given earlier, we have that $P(Z = 1 \mid X = 1)$ is about 10 times $P(Z = 1 \mid X = 0)$.

CPCP is simple and intuitive, and works well in a case like smoking causes lung cancer. Unfortunately, however, counter-examples to CPCP have been discovered. These counter-examples are all essentially the same and arise from the fact that several different indeterministic causes may operate at the same time and that these causal factors may not be independent, but rather interrelated. Here I will confine myself to considering the first counter-example to CPCP to be discovered. This was published by Hesslow in 1976 and is still in many ways the best counter-example. A resolution of this counter-example can easily be extended to cover the other counter-examples in the case of generic causality.

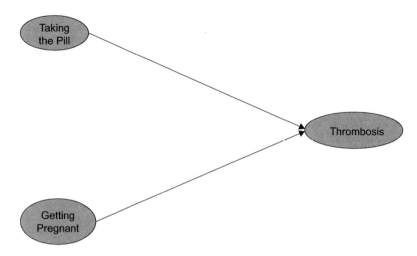

Figure 7.4 Multi-causal fork for Hesslow example

For this counter-example, we consider a population of young women, all living with male partners, where the only contraceptive method available is the contraceptive pill. In our simple causal network of Figure 7.2, we now set X = Taking the Pill, Y = Getting Pregnant, and Z = Thrombosis to get the two-pronged multi-causal fork shown in Figure 7.4.

The clinical background is that both taking the pill and pregnancy are associated with the formation of clots (i.e., thrombosis). Assume, therefore, which is broadly correct, that both the Pill and Pregnancy are indeterministic causes of Thrombosis, but that Pregnancy has a much more powerful effect in producing Thrombosis than does taking the Pill. This means that taking the pill, although it causes thrombosis, actually reduces the probability of thrombosis, since those who don't take the pill will almost certainly get pregnant, and this means in turn that they will have a higher probability of thrombosis. We can reformulate this argument in mathematical symbols to show that it produces a counter-example to CPCP.

$P(Z = 1 \mid X = 1)$ is the probability of getting a thrombosis for those taking the pill. By assumption, this is a finite but quite small probability. Now consider $P(Z = 1 \mid X = 0)$. If $X = 0$, then it is very probable that $Y = 1$, that is to say that those who don't take the pill have a high probability of getting pregnant, since there is no other method of contraception. However, $P(Z = 1 \mid Y = 1)$ is by assumption much greater than $P(Z = 1 \mid X = 1)$. So we are likely to have $P(Z = 1 \mid X = 0) > P(Z = 1 \mid X = 1)$, which contradicts CPCP.

CPCP therefore seems an intuitively reasonable way of connecting causality and probability, and yet it is liable to counter-examples. What should be done about this situation? Should CPCP be abandoned altogether? Or should we modify it to avoid the counter-examples? This problem has been known for some time. In the next section, I will consider how we might get a grip on it by considering the interpretation of the probabilities involved.

4. Interpretations of Probability

Probability can be interpreted in various ways. The most important distinction is between *subjective* and *objective* interpretations. Probability in the subjective interpretation is a measure of the degree of belief that a particular person (Mr. Smith) holds in the occurrence of a particular event, say that it will rain during the course of the day. Mr. Smith's subjective probability will determine whether he decides to take an umbrella with him when he goes out. Pearl has argued that the probabilities in causal models should be interpreted subjectively (see his 1988 and 2000). However, my own view is that it is better to interpret the probabilities in causal models for medicine objectively.

There are two main objective interpretations of probability. The first of these is the frequency interpretation, which was developed by von Mises in his 1928. Von Mises uses the concept of the *collective*. A collective is a large collection of some kind. So, for example, we could consider the collective **C** of English people. We can further consider a set of properties or attributes, which might or might not hold for each member of the collective. So for each English person, we could, for example, consider the attribute ". . . dies of lung cancer." It is obvious that each English person either dies of lung cancer or does not die of lung cancer. Suppose that in the first 1,000 members of the collective considered, 61 die of lung cancer. Then the frequency of cases of lung cancer is 61 divided by 1,000 (i.e., 0.061). As the number of members of the collective considered increases, the frequency of lung cancer cases may tend to a fixed value (say 0.06). This is called the limiting frequency of the attribute in the collective, and the probability of an attribute is defined as its limiting frequency. So, in our example, the probability of dying of lung cancer in **C** is the limiting frequency of those who die of this disease in the whole English population, about 0.06 or 6%.

Now an important point to notice is that the same attribute may appear in a number of different collectives, and it have a different probability in each of these collectives. So we could consider the collective **C and Smoker**, which consists of English people who smoke. P(Dying of Lung Cancer | **C and Smoker**)—i.e., the probability of dying of lung cancer in the collective of English smokers—would be greater than P(Dying of Lung Cancer | **C**)—i.e., the probability of dying of lung cancer in the collective of English people. Similarly, P(Dying of Lung Cancer | **C and Non-Smoker**) < P(Dying of Lung Cancer | **C**). In general then, changing the collective changes the probability, and this led von Mises to propound the principle that probability (in his frequency sense) is relative to a collective. I will refer to this as von Mises' principle.

Although the frequency theory has considerable merits, many philosophers have argued that it links probability too closely to observed frequency. This point of view has led to the propensity interpretation of probability, which was introduced by Popper at the end of the 1950s [(1957) and (1959)]. The idea of such an approach to probability proved to be popular, and several philosophers of science have developed it, though in different ways. As a result, there are now a number of different versions of the propensity theory. I have given a survey and classification of some of these in my article (2000a), "Varieties of Propensity," and in Chapter 6 of my book (2000b: 113–136). For the purposes of this chapter, however, we need consider only a few features of the propensity theory. These features are common to most, though not all, versions of the propensity theory.

In the propensity theory, probabilities are associated with the outcomes of a set of repeatable conditions. Suppose, for example, we are tossing a biased coin. Here the repeatable conditions are those that define the manner in which the coin is tossed. Call the set of them **S**. Then, according to the propensity theory, **S** is endowed with a propensity (or tendency or disposition) to produce a particular result, say heads (H), on any repetition of **S**. So to say that the probability of H given **S** is p, written P(H | **S**) = p, means that the conditions **S** have a propensity equal to p to produce H on any repetition of **S**.

The propensity interpretation is an objective interpretation of probability. What ensures objectivity is the association of propensities with repeatable conditions. This means that it is possible to test whether any value p assigned to a probability agrees with experience or not. In principle all one needs to do is repeat the conditions a large number of times and see whether the observed long-run frequency is close to p or not. Now in formulating causal models for diseases, we are seeking objective knowledge. This makes an objective interpretation of probability, such as the propensity theory, preferable to a subjective one, since an objective interpretation connects probabilities naturally with the statistical frequencies of illnesses in populations. Such statistical frequencies can then be used to test the validity of the claims made by the causal model.

One difference between the propensity theory, as just described, and von Mises' frequency theory is that probabilities are now associated with repeatable conditions rather than collectives. However, von Mises' principle still applies if we alter it to: probability (in the propensity sense) is relative to a set of repeatable conditions. Yet in order to connect the discussion within the propensity theory even more closely to earlier discussion within the frequency theory, I propose the following strategy. It is always possible to consider, instead of the repeatable conditions of the propensity theory, a sequence of repetitions of those conditions. Thus, instead of the condition " . . . is an English person," we can consider "the set of English people"; instead of the condition ". . . is an English smoker," we can consider "the set of English smokers"; and so on. To do this, I propose to use the term "reference class" in the following precise sense. A reference class is associated with a set of repeatable conditions S and consists of a long sequence of repetitions of S. So the key principle now becomes: propensity is relative to a reference class.

Let us now return to our problem about connecting causality to probability. In section 3 we formulated an intuitively plausible Causality Probability Connection Principle (CPCP), but then found that it was liable to counter-examples, notably the Hesslow counter-example. Now CPCP involves probabilities, such as $P(Z = 1 \mid X = 1)$. If we interpret these as propensities, then they must be associated with a reference class. However, CPCP also involves the statement X causes Z, so it would seem that this too must be associated with a reference class. In fact, it is quite reasonable to claim that causality in the generic sense is associated with a reference class. Consider, for example, the case of a medication M, which has no side effects for the average patient but produces vomiting and dizziness in diabetic patients. We then have that, in the reference class of diabetic patients, "M causes vomiting and dizziness" is true, but that in the reference class of non-diabetic patients, it is false. This example can easily be generalized, and so the reference class principle can be extended from propensity to causality. In particular, it applies to both parts of the suggested CPCP.

We can now ask the question: "For which reference class or reference classes does CPCP hold?" In Hesslow's example, the underlying reference class (S say) consists of a set of young women living with male partners in a situation in which taking the pill is the only contraceptive method available. As we showed earlier, CPCP does not hold for this reference class S. However, we now propose resolving the problem by dividing S into two disjoint reference classes, in each of which CPCP holds, though it is not valid in the whole reference class S. The two disjoint reference classes are S and $(Y = 0)$—i.e., the set of those who do not become pregnant in the time period under consideration—and S and $(Y = 1)$—i.e., the set of those who do become pregnant. Now in both of these reference classes, CPCP holds. Consider S and $(Y = 0)$, which is the set of young women who do not become pregnant. In this set, those who take the pill ($X = 1$) have a higher probability of getting a thrombosis than those who do not ($X = 0$) because of the side effects of the pill. Consider next S and $(Y = 1)$, which is the set of young women who do become pregnant. In this case it might be objected that there are no members of this set who take the pill ($X = 1$). However, this objection can be answered by pointing out that the pill is unlikely to be 100% effective, and that, because of the side effects of the pill, someone who both took the pill and became pregnant is likely to have a higher probability of getting a thrombosis than someone who became pregnant without having taken the pill. So CPCP again holds.

5. Simpson's Paradox

I will next observe that the problem generated by the Hesslow counter-example is quite similar to that of Simpson's paradox. To see this, let us consider the most famous example of Simpson's paradox, which concerned the seeming occurrence of gender discrimination in admission to the University of California, Berkeley's graduate school. Let us take as our initial reference class **S**, the set of applicants to this graduate school. These applicants were either male or female, and their application was either successful or unsuccessful. Statistics showed that, in the reference class **S**, P(success | male) > P(success | female). It therefore looked as if gender discrimination in favor of males was at work. However, in practice, applicants applied to particular departments and each department conducted its own admission procedure. So, if there were gender discrimination, it would have to be at the departmental level. Now suppose the departments were $D_1, D_2, \ldots, D_n$. Then we can partition our reference class **S** into n sub-reference classes **S** and D_i where $1 \leq i \leq n$. Statistics showed that, for each of these sub-reference classes, P(success | male) < P(success | female). So gender discrimination in favor of males could not in fact be occurring.

At first sight these results seem impossible— hence the name Simpson's paradox. If in every member of a partition of a reference class, a probabilistic inequality holds, how could it reverse in the reference class as a whole? It is indeed surprising that this occurs, but it is in fact quite possible. The explanation in the case of the Berkeley admissions was that a greater percentage of females than males were applying to departments that had high rejection rates.

The situation as regards CPCP is indeed analogous to Simpson's paradox. Suppose we generalize from our two-pronged multi-causal fork of Figure 7.2 to the n-pronged multi-causal fork of Figure 7.1 in which a number of possibly interrelated indeterministic causes $(X_1, X_2, \ldots, X_n)$ are combining to produce an effect Z. We can only draw conclusions about probabilities from the claim that X_i causes Z, if we assume that the values of the other indeterministic causes X_j for $j \neq i$ are fixed. So these probabilistic conclusions only hold in each member of a partition of the underlying reference class. From this it does not follow that they hold in the reference class as a whole.

6. Conclusions

In contemporary medicine, there are often situations in which a disease does not have a single cause but is brought about by the combination of a number of causal factors. Each causal factor is not sufficient by itself to produce the disease but may well raise the probability of the disease occurring. For example, smoking is not sufficient by itself to produce heart disease, since many smokers never get heart disease. However, smoking does raise the probability of getting heart disease. In these multi-causal situations, the various causal factors have been combined together and also combined with probabilities. Causal models are an attempt to carry out this combination. However, there are many problems in creating satisfactory causal models. Seemingly obvious principles connecting causality with probability turn out to be liable to counter-examples, such as the Hesslow counter-example. One way of resolving these problems is by interpreting probabilities as objective propensities, defined in reference classes. A careful consideration of the relevant reference classes shows that the Hesslow problem can be resolved, and also that it is very similar to what is known as Simpson's paradox. This then is a step in the direction of creating good causal models for medicine.

References

Codell Carter, K. (2003) *The Rise of Causal Concepts of Disease: Case Histories*, Aldershot, England & Burlington, USA: Ashgate.

Doll, Richard, and Richard Peto (1976) "Mortality in Relation to Smoking: 20 Years' Observations on Male British Doctors," *British Medical Journal*, 2(December), 1525–1536.

Gillies, Donald (2000a) "Varieties of Propensity," *British Journal for the Philosophy of Science*, 51, 807–835.

Gillies, Donald (2000b) *Philosophical Theories of Probability*, London and New York: Routledge.

Gillies, Donald, and Aidan Sudbury (2013) "Should Causal Models Always be Markovian? The Case of Multi-Causal Forks in Medicine," *European Journal for Philosophy of Science*, 3(3), 275–308.

Hesslow, Germund (1976) "Two Notes on the Probabilistic Approach to Causality," *Philosophy of Science*, 43(2), 290–292.

Kant, I. (1781/7) *Critique of Pure Reason*. English Translation by Norman Kemp Smith, London: Macmillan, 1958.

Levy, Daniel, and Susan Brink (2005) *A Change of Heart: Unraveling the Mysteries of Cardiovascular Disease*, New York: Vintage Books.

Pearl, Judea (1988) *Probabilistic Reasoning in Expert Systems: Theory and Algorithms*. San Mateo, California: Morgan Kaufmann.

Pearl, Judea (2000) *Causality, Models, Reasoning, and Inference*. Cambridge: Cambridge University Press.

Popper, Karl (1957) "The Propensity Interpretation of the Calculus of Probability, and the Quantum Theory," in S. Körner (ed.), *Observation and Interpretation, Proceedings of the Ninth Symposium of the Colston Research Society, University of Bristol*, London: Butterworth Scientific Publishing, 65–70, 88–89.

Popper, Karl (1959) "The Propensity Interpretation of Probability," *British Journal for the Philosophy of Science*, 10, 25–42.

Von Mises, Richard (1928) *Probability, Statistics and Truth*, 2nd Revised Edition of English Translation by J. Neyman, London: Allen and Unwin, 1961.

Further Reading

Cartwright, Nancy (1983) *How the Laws of Physics Lie*, Chapter 1 Causal Laws and Effective Strategies, 21–43, Oxford: Oxford University Press, gives a good account of the problems of relating causality and probability.

Galavotti, Maria Carla (2005) *Philosophical Introduction to Probability*. Stanford: CSLI Publications, gives clear and detailed accounts of the various philosophical interpretations of probability.

Hitchcock, Chris (2010) "Probabilistic Causation," *Stanford Encyclopedia of Philosophy* (http://plato.stanford.edu) gives an excellent outline of causal models and the philosophical problems they raise.

Neapolitan, Richard (1990) *Probabilistic Reasoning in Expert Systems*. New York: John Wiley, is the best introduction to the mathematics of causal models for those with the requisite mathematical background.

8
REDUCTIONISM IN THE BIOMEDICAL SCIENCES

Holly K. Andersen

1. Introduction

Reductionism has a long history in philosophical and scientific thought, and it is an intuitively appealing view about how the world works. In its simplest form, reduction involves taking something larger, more complex, or more specific, and reducing it to smaller, simpler, or more universal components. Reduction can be understood as a relationship between parts of the world, in which case it involves the view that larger objects, like physical bodies that we can see and hold, are just lots of very small bodies like atoms and molecules, organized by similarly microscopic forces. Reductionism can also be understood as a relationship between theoretical structures like theories, laws, or models. In that context, reductionism involves commitment to the goal of taking multiple distinct and possibly conflicting models of one phenomenon and reducing those to a single, overarching or unifying model of that phenomenon.

In a medical context, reduction often takes three forms. One is the reduction of a system to its component parts and their organization. A second is reduction from a set of models to a single model that incorporates or unifies them. A third form of reduction involves a simplification in the overall complexity required to explain a particular disease, for example, by reducing it to a few primary causal factors with a simplified causal path along which it unfolds and into which interventions can be attempted to alleviate or prevent the disease. Reduction understood as simplification of complexity is an especially useful tool in biology, medicine, and related sciences that study target systems that are characteristically complex.

Reduction is a process of moving downward, to less complexity, smaller sizes, and fewer models. It has a converse process, that of integration, which involves moving upward, to a complex interacting causal structure, to larger size scales or higher levels, and to navigation among multiple overlapping models. Like integration and differentiation in calculus, these directions in which one can move are not in competition—they each have a use and must often be deployed together in the overall process of solving a problem (Mitchell 2002).

Reduction is a useful tool both conceptually and methodologically. Methodologically, it is often helpful to think of a disease in terms of a series of smaller, less complex, interlocked submechanisms, each of which could potentially be intervened on independently. Many diseases stem from dysfunctions at subcellular levels, yet have wide-ranging symptom effects at a variety of size scales in the body. Reduction as a guiding methodological assumption can assist researchers in isolating the earliest causal stages of such diseases and in isolating the main drivers of a disease in a population where there may be a huge variety of causes, but a few causes or causal pathways that are common to most cases. Reduction is also methodologically useful for guiding interventions: targeting dysfunction earlier in a causal chain of cascading problems is, when possible, optimal for preventing or treating disease. Smaller size scales also tend to be (although do note, they are not always) easier to intervene on than are later stages that are higher level or more complicated in terms of how far-reaching the effects become.

Conceptually, reduction can assist us in understanding how different models of the same or closely related phenomena relate to one another, so that we can use these distinct models under appropriate circumstances. Different methods of studying something like protein folding may yield very different models of a target phenomenon like a quaternary protein structure. The differences in the models reflect differences in the kinds of environments in which each method studies the same proteins, and in the ways those environments distort the target to be studied (Mitchell & Gronenborn 2015). The structure might be studied using X-ray crystallography, *ab initio* predictions based on amino acid sequence, or nuclear magnetic resonance spectroscopy. Even though it is the "same" protein studied each way, each method produces different models without being false, since model differences reflect differences in how the experimental method and environment affects the protein being studied. Understanding how different research methods shape the models those methods can produce of their target systems may help us reduce those different models into one overarching or unified model of the target phenomenon. Or, reduction may turn out to not be possible, and instead of reduction, integration of the models is required to guide researchers in knowing which model(s) to deploy under a variety of circumstances. It can still be conceptually fruitful to pursue reduction until we are clear on the details of why reduction fails.

Biomedical sciences are arguably the setting in which we can find the most striking gaps between what we can call the locus of explanation and the locus of control. The locus of explanation is the mechanism or mechanisms, at their characteristic size scale, that are primarily responsible for a given effect. In the case of a misfolded protein, the characteristic mechanisms will involve subcellular machinery, the chemical environment, the base pairs in the protein, and more. These are all at a characteristic and similar size scale, yet the effects of misfolding might cascade up the size scale to produce symptoms at a variety of physical size levels. The locus at which we identify the primary source or origin of a particular causal chain or disease is the locus of explanation for it. This might differ, sometimes dramatically, from the loci available for interventions to prevent or alleviate the disease. The locus of intervention is the mechanism or mechanisms on which we actually intervene in clinical practice. This is constrained by a host of factors beyond the locus of explanation. We might simply not have developed any intervention that could target the locus of explanation; it might not be feasible for logistical or ethical reasons. Instead, knowledge gained through reduction downwards and integration upwards might yield a different target for intervention, a locus for control that is not the main causal driver but nevertheless is pragmatically accessible to us.

The way in which reduction figures in characteristically medical examples will be illustrated by considering Parkinson's disease. This disease is now known to have several variants. One variant is monogenic; another involves the interaction of several genes. One is environmental, involving exposure to known toxins. And one involves the combination of multiple genes that render a patient more susceptible, coupled with subsequent exposure to environmental toxins. Originally defined as an idiopathic disease, with no known origin, researchers have used reductionist methodological tools and conceptual frameworks to isolate the genes and environmental factors that drive this disease and to work towards unification of the various causal etiologies into a common pathway by which to identify and distinguish Parkinson's from other diseases. This yields a variety of potential targets for intervention. A locus of control might involve eliminating use of the pesticides known to trigger it; another locus might involve targeting pharmaceutical interventions that prevent or slow genetically triggered deterioration; or a locus might focus on lessening symptoms directly.

Parkinson's helps illustrate the power of reduction as a guiding assumption and the ways in which reduction can fail in the face of certain kinds of complexities. Ultimately, the success or failure of reduction in the biomedical sciences depends on the particular phenomenon in question. Some causal structures yield great insights when reduction is applied as a methodological and conceptual tool; some causal structures, because of details related to their internal complexity and the difficulty in controlling messy complexity in experimental settings, resist reduction indefinitely.

2. Reduction: Sweeping Physics and Creeping Biomedicine

Physics is a field where reductionism has a long history of leading to useful scientific breakthroughs, and for which most accounts of reduction were originally developed. Many, or arguably all, accounts of reductionism that were developed to apply in physics foundered when they were applied to other sciences such as biology and medicine. The systems being studied in physics versus the biomedical sciences differ in so many relevant ways, and the tools available to study them differ so dramatically, that it is illuminating to first see what reductionism looks like in fields such as physics, and then see how this picture gets complicated in the case of the biomedical sciences.

The most traditional account of reduction can be seen in Oppenheim and Putnam's classic paper, "The Unity of Science as a Working Hypothesis" (1958). They present reduction as key to the unification of different areas of science. The world is divided into levels of physical size and organization, each of which is associated with a characteristic field of study. Physics is the most fundamental. The next, chemistry, is taken to be nothing more than complicated physics that could in principle be eliminated with sufficient knowledge of the relevant physics. Biology is above and reduces to chemistry, and psychology to biology. This hierarchy embeds strong commitments about the "real" objects and the most fundamental fields of study. Molecules just are compounded particles; organisms just are complicated arrangements of chemicals, and so on. Following on these levels relationships in the world, the theories of chemistry should be reducible to the theories of physics, and those of biology to chemistry, etc.

On this way of thinking, reduction is the path by which to unify the disparate sciences together. Oppenheim and Putnam take the unity of science to be, first, an ideal future state of science, in which the vocabulary used in one science can be fully translated into the vocabulary of another. The concept of a genetic disease, for instance, would be entirely translatable into chemical terminology, and then into the terminology of particle physics. Oppenheim and Putnam take the unity of science to be, second, a trend in science, such that unification of apparently disparate phenomena is an ongoing process, and we can continue to unify even if we never reach the ideal end state of total unity. In that regard, reduction as traditionally conceived is somewhat closer to what can be found in medicine in terms of unification of distinct models at distinct levels into a more comprehensive view of the whole, but without the end goal of a single, molecule-by-molecule and moment-by-moment map of everything in the body.

Some systems studied by physics undermine such strong reduction. For instance, in phase transitions for gasses, some behaviors are almost completely independent of the microphysical details of the system in question. That would mean that reduction of such systems would yield less rather than more explanation of the target phenomenon (Batterman 2000). Nevertheless, there is a strong tendency towards successful breakthroughs following the method of reducing complex systems to smaller and simpler parts that are studied separately and then re-assembled, and a track record of making new discoveries by connecting via attempted but failed reduction the theories for higher-level behaviors such as boxes of gas to theories for lower-level behaviors such as energy and momentum of atoms.

Such reductions in physics are examples of what Schaffner (2006, 2011) has called "sweeping reductions." They take a broad swath of phenomena and demonstrate how they are completely explainable with reference to their smaller-scale components, or as specific instances of a more universal phenomenon. Sweeping reductionism is strong: it commits to the idea that, ontologically, there *is* nothing at the higher levels in a metaphysically strong sense. That which is reduced is thereby ontologically dependent on or secondary to that by which it is reduced.

Schaffner contrasts this with "creeping reductions," which are small reductions between specific phenomena at different levels, such that the higher level is not eliminated but rather connected in an explanatory way to the lower level. Creeping reductions lack the synoptic scope and grand unificatory power of sweeping reductions. These are much less dramatic, as least from a philosopher's perspective, than a sweeping reduction. As rare as sweeping reductions are in physics, they are completely absent from fields like medicine, where creeping reductions are considered successes.

Creeping reductions are not thereby less valuable or easy to achieve in biology and related sciences. Schaffner notes that creeping reductions ". . . are fragmentary patchy explanations, and though patchy and fragmentary, they are *very important*, potentially Nobel-prize winning advances" (2011, 139). The systems studied in physics lend themselves to decomposition as a technique for study. The systems in biology and medicine, on the other hand, cannot be so simply decomposed while still retaining the features that make them of interest to us. Reducing a sample organism to component atomic elements might be possible, certainly, but sorting an organism into piles of atomic elements is not a way to figure out what the organism is like when it is intact, what it can do in various environments, and how to intervene on it in medically illuminating ways. Creeping reductions accomplish important explanatory work that lack the universal character of sweeping reductions but that nevertheless yield incredibly useful information when they are found.

Levels and mechanisms are a useful way of understanding the compositional and organizational structure of systems, especially in biology, medicine, and related fields. There are many, many ways to individuate levels in organisms, including humans, but some tend to be robust across populations and across species. Cells are often a useful level on which to focus. Picking out the cell level can be thought of as analogous to focusing a mental microscope at that magnification, or zooming in to a map to the street level but not the house-by-house or city-by-city level. Certain kinds of physical processes, recognizable types of entities (including non-cellular ones), identifiable patterns of activity, and common temporal scales can be characterized at the cellular level. Moving downward, or zooming in on the map, subcellular machinery pops into focus, with a somewhat different set of characteristic processes, on a slightly different time scale, that together constitute the cell at the next level up. Further down we find the genetic level with complex molecular biology, where macromolecules with potentially incredibly complex physical organizations and time scales play key roles. These, taken together in their totality of interactions, constitute the subcellular machinery of another level up.

Going in the other direction from cells, tissues are a frequently used individuation of levels in the human body. Tissues involve cells as components, but not in the simplistic way that a house made of blocks involves those blocks as small pieces. To count as a tissue, there have to be special kinds of interactions between the cells and similarities in terms of their type, or common function, or some other parameter. Many important mechanisms, and many medical issues, can be best described in terms of tissues. Further up, organs are another common level at which to find a distinctive set of processes, entities, activities, and time scales. Systems, such as the endocrine system or the nervous system, are arguably the highest sub-organism level.

Zooming yet further out, though, the human body can also be understood as one distinctive entity, engaging in certain kinds of processes and activities at characteristic time scales. The social environment of a person may be invoked to explain certain kinds of anxiety or stress reactions, and it might be at the larger social or physical environment levels to which more effective interventions might be targeted.

Mechanisms offer another way to keep track of reduction between levels. Mechanisms are causal chains of entities and activities that are connected in such a way that, once triggered, they proceed reliably enough through the causal pathway to a termination state, which may involve a cycle that is maintained (where the termination state is the triggering conditions again for the same mechanism) (Andersen 2014a and b). Mechanisms involve some degree of regularity, in that the same causal pathways retain their stability so that they perform similarly over time and in different instances (Andersen 2012a). For instance, the transcription of DNA by RNA to form various proteins involves mechanisms that work consistently, and about which we can make stable generalizations that might not satisfy a strong reductionist but that do provide weakly reductive characterizations of the relevant upper-level phenomena. Mechanisms in medicine often involve hidden causal relationships, however, to maintain certain kinds of stable parameters in the body (Andersen 2012b). This makes mechanisms a difficult way to construe reduction in medicine, by providing less experimental guidance for effective intervention even when reduction is possible in principle.

As this discussion illustrates, there are some robust ways to individuate levels in the human body. However, there need not be any "one right set" of levels by which to break down such a complex system. Some mechanisms might cross levels, and some phenomena might involves cascades upwards from the molecular level to effects that are macroscopic and phenotypical in size scale. Some phenomena, such as the folding of complex protein structures, might sit awkwardly between the molecular and subcellular levels, and a different way of individuating levels would be required when researching those phenomena. There is no single level-individuation built into the world itself; such divisions always involve an element of pragmatic interest and attention on our part as researchers, clinicians, and patients. Levels should instead be thought of as handy divisions in order to keep track of where the action is, so to speak.

Reduction in the biomedical sciences is thus unique compared to sciences like physics in terms of the complexity of systems to which reduction is applied, the extent to which it is illuminating to reduce systems to component parts, and the extent of the resulting unification from successful reductions—sweeping in physics, creeping in biomedicine.

3. Reduction as a Methodological Approach in Biomedicine

There are many ways in which reduction can manifest in the collection of sciences that are grouped under the heading biomedicine. These sciences share a focus on the human body, at all relevant levels from the subcellular through to the social/environmental, from earliest development through old age. Informally, this goal of understanding is oriented towards *explanation*, in terms of explaining how things work when they work well and what goes wrong when things go wrong; and towards *intervention*, in terms of finding ways to keep things working normally and to restore as much normal function as possible when they are not.

Linus Pauling used the notion of a "molecular disease" to show how mutation involved in the gene for one protein was sufficient to explain sickle cell anemia. This involved a conceptual orientation—namely, commitment to the idea that there was a single or very few primary causal drivers of the disease to which other complexities in the manifestation of the symptoms could be reduced. It also involved a methodological orientation—namely, commitment to such a causal driver being at the molecular level, rather than at a higher level of organization. "Thus, a single molecular change is fundamental to understanding the patient's pathology, symptoms, and prognosis" (Kandel et al. 2000, 867). This encapsulates the potential for reduction in the biomedical sciences—when it works, it works well.

Being able to successfully take a disease as complex as sickle cell anemia or Parkinson's and reduce it to a single genetic error that cascades through various systems, even if that reduction holds for only a proper subset of cases, illustrates that reduction can be a powerful tool for research and explanation in medicine. It cannot be the only tool in the toolkit, since some diseases or dysfunctions may be only partially reducible, or for which only some cases are reducible. But it is a good working assumption, as Oppenheim and Putnam put it, in tackling a problem with an unknown etiology, to look for ways to reduce it to a few or even a single causal driver at a molecular level.

Consider the case of Parkinson's disease. Parkinson's is a disorder affecting the nervous system, resulting in characteristic tremors and slowed and difficult movements. In the monogenic variant, a single allele results in a misfolded version of a key protein in the brain, alpha-synuclein. The triggering condition is the genetic component, which leads to a misshaped protein that then clumps in characteristic ways to form Lewy bodies that accumulate, especially in the substantia nigra. This part of the mid-brain in involved in, for instance, controlling movements, and the accumulation of Lewy bodies inside of the nerve cells in this area lead to the characteristic tremors and other issues with motor control seen at the human level in Parkinson's patients.

The mechanisms by which Parkinson's affects those with the disease involve small molecular changes that cascade up through different level boundaries, per Pauling's description. This holds even though distinct factors such as single genes, gene combinations, and environmental factors can be involved in the onset of Parkinson's. In monogenic cases, though, Parkinson's disease is like sickle cell anemia, in that it is the result of a very localized genetic mutation.

Thus, considering the monogenic case, there are several senses of reduction by which a complex, multi-level phenomenon like Parkinson's might be reduced. The mechanisms involved may be at a lower level of organization than that of downstream causal effects. The locus of causal explanation of the dysfunction might be at a very small size scale, such as a subcellular level, rather than at a higher level such as organs. And there may be a great deal of complexity and noise in the patient population, such that it is very difficult to discern for many patients what caused the onset of the disease, which can be reduced away by identifying a clear genetic component for some variants. There is thus no single way but rather a nexus of ways in which research has reduced Parkinson's: (a) causal complexity must be reduced to a few initial transcription errors; (b) mechanisms for control of movement at the mid-brain level are reduced to mechanisms at the subcellular level; and (c) physical size scale is reduced further along the chain, from organism-level symptoms to mid-brain structures like the substantia nigra to protein aggregates inside nerve cells to allele-sized triggering conditions. Nevertheless, this nexus of ways all point downward in a reductive fashion, such that the locus of explanation for Parkinson's is illuminated by reduction, and options for targeting interventions appear because of the reduction.

There are several examples here of a creeping reduction in Schaffner's sense, a patchy and fragmentary connection between partial explanations. This is highlighted by the fact that this explanation as given only applies to the rare monogenic case; additional factors must be introduced for multigenic cases, and yet more for the environmental cases. Nevertheless, a key insight into Parkinson's came out of the methodological constraints imposed by the reductive assumption that it might be a molecular disease, in Pauling's term, by which possible avenues for intervention were revealed.

4. A Conceptual Guide to Navigating a Plurality of Models

In its traditional understanding from physics, reduction ultimately involves relationships among theories, laws, models, or other representational content being reduced via translation into the terminology of another theory. By extension to medicine, if Parkinson's involves genetic causes, then on this understanding of reduction as a relationship between theoretical structures like models, we should be able to assemble a single overarching or unified model from the concatenated models of various stages of the diseases, of the distorted protein shape through to how this protein clumps into Lewy bodies, through to how these accumulate in the substantia nigra cells, through to the disruption on motor control that is a consequence. This is something of an ideal, like the perfectly unified science that Oppenheim and Putnam envisioned, that we may never reach but towards which researchers do often aim.

There are reasons to think we may never be able to reach such a unification of all the different stages of such a complex mechanism as that involved in the cascading effects of the monogenic variant of Parkinson's. Mitchell and Gronenborn (2015) show how different techniques for modeling the quaternary structure of a protein result in differences in the modeled shape. These different shapes can be incorporated into a fusion model that uses them to triangulate somewhat on the actual shape of the protein. But a reduction in this case would involve being able to eliminate the prior models, either because they are redundant given the new one or because one is found to be more accurate than the other. In the case of protein folding, it would be a mistake to say that one model is more accurate than another: they are different, because they are generated using different techniques that place the protein in different environments and systematically deform it in different ways. These techniques for modeling the same quaternary protein structure have, in an important sense, different objects that they model.

Such protein modeling techniques are involved in studying the protein structures that go awry in genetic variants of Parkinson's (both monogenic and multigenic). Yet in the simplest case, we should not expect a reduction in models in the sense of ending up with fewer of them, but rather an integration of models that must be retained in a pluralistic way. Even the simplest monogenic case appears to resist model reduction to a single overarching and exhaustive representation of the mechanisms by which symptoms are produced. Instead of aiming for an idealized but probably unreachable unification via reduction of the models, we can look for ways to integrate without reducing. This means finding techniques for navigating the pluralism of models with knowledge about how each model was produced that can inform when we should rely on one model or another model based on our pragmatic interests in using it.

Parkinson's disease has yet more complexity in this regard as well, because some variants involve environmental components. There are other ways to construe how Parkinson's can involve a reduction in causal complexity that does not require idealized model simplification. Distinct causal mechanisms lead to four variants of Parkinson's: the monogenic variant, as we saw; a multigenic variant; an environmentally caused variant; and a variant involving both genetic susceptibility and environmental causes. Thus, in the general population of patients with Parkinson's, there are multiple causal etiologies for patients with Parkinson's, at different size scales. Furthermore, some patients may individually have a variety of causal factors involved in their case of Parkinson's, such as the interaction between several gene sites that contribute to susceptibility and environmental factors that trigger Parkinson's in the presence of that genetic susceptibility. Modeling the variants of Parkinson's will thus require the integration of multiple models that may each involve multiple levels of organization. For the variant involving both environmental exposure combined with genetic susceptibility, these models cannot be reduced in the sense of being combined into one overarching model. Rather, they must be integrated so that researchers can navigate the complex causal pathways at each characteristic level of organization and those that cascade up or down levels as well.

Even in the case of Parkinson's with multiple sites for genetic susceptibility in combination with environmental exposure, though, other goals can be served by inter-theoretic or inter-model reduction. For a single patient with a causally complex etiology for Parkinson's, reduction can highlight a few key causal drivers and leave out those that are not the primary causal factors. Even though some of the key causal factors are not available as loci for control (the exposure to chemicals has already occurred for a given individual, for instance), reducing the number of causal factors under consideration in the models to just a few key drivers still may be illuminating as a loci of explanation. Parkinson's disease is an excellent example of a complex disease where several different kinds of reductions have yielded insight into variation across patient populations and in variation across the etiological factors for individual patients.

5. Conclusion

Reduction is a powerful tool for investigating phenomena in the biomedical sciences. It can guide methodological inquiry in useful ways and provide conceptual frameworks that bring clarity to complexity and enable researchers, clinicians, and others to navigate among relevant models and to test and implement various intervention strategies. Reduction can involve zooming in to smaller size scales or levels of organization. It can involve the reduction of many models to one inclusive model for a phenomenon. And, it can involve a reduction in the degree of complexity required to explain or control. Cases like Parkinson's show how many different kinds of reduction applied to the same phenomenon can nevertheless robustly point in the same direction.

Reduction cannot serve alone as such a tool, however, in the biomedical sciences. In physics, the integration of components back into a cohesive system is vastly simpler than it is in biological systems. Here, reduction needs to be paired with integration as a strategy to focus in on the right level: one zooms in and the other zooms out, and with both tools the right level at which to describe and explain, as well as to control and intervene, can be found. Reduction can also yield explanations that are not suitable as direct targets for interventions, but finding the locus on explanation via reduction opens up downstream pathways that might serve as useful loci of control for preventing or alleviating symptoms.

References

Andersen, H. (2012a). "The case for regularity in mechanistic causal explanation." *Synthese*, 189(3), 415–432.
Andersen, H. (2012b). "Mechanisms: What are they evidence for in evidence-based medicine?" *Journal of Evaluation in Clinical Practice*, 18(5), 992–999.
Andersen, H. (2014a). "A field guide to mechanisms: Part I." *Philosophy Compass*, 9(4), 274–283.
Andersen, H. (2014b). "A field guide to mechanisms: Part II." *Philosophy Compass*, 9(4), 274–283.
Batterman, R. W. (2000). "Multiple realizability and universality." *The British Journal for the Philosophy of Science*, 51(1), 115–145.
Kandel, E. R., Schwartz, J. H., & Jessell, T. M., eds. (2000). *Principles of neural science* (Vol. 4). New York: McGraw-Hill.
Mitchell, S. D. (2002). "Integrative pluralism." *Biology and Philosophy*, 17(1), 55–70.

Mitchell, S. D. & Gronenborn, A. (2015, November 27). "After fifty years, why are protein X-ray crystallographers still in business?" *The British Journal for the Philosophy of Science* (online only).

Oppenheim, P. & Putnam, H. (1958). "Unity of science as a working hypothesis." In Herbert Feigl, Michael Scriven & Grover Maxwell (eds.), *Concepts, Theories, and the Mind-Body Problem*. Minneapolis: University of Minnesota Press, 3–36.

Schaffner, K. F. (2006). "Reduction: The Cheshire cat problem and a return to roots." *Synthese*, 151(3), 377–402.

Schaffner, K. F. (2011). "Reduction in biology and medicine." *Philosophy of Medicine*, 16, 137.

Further Reading

Brigandt, Ingo and Love, Alan, "Reductionism in Biology", *The Stanford Encyclopedia of Philosophy* (Winter 2014 Edition), Edward N. Zalta (ed.), URL = <http://plato.stanford.edu/archives/win2014/entries/reduction-biology/>. (An excellent overview of issues particular to reduction in biology)

Oppenheim, P. & Putnam, H. (1958). "Unity of science as a working hypothesis." In Herbert Feigl, Michael Scriven & Grover Maxwell (eds.), *Concepts, Theories, and the Mind-Body Problem*. Minneapolis: University of Minnesota Press, 3–36. (A classic text for discussions of reduction)

Schaffner, K. F. (2011). "Reduction in Biology and Medicine." *Philosophy of Medicine*, 16, 137. (An extended look at the creeping versus sweeping distinction)

Waters, Ken, "Molecular Genetics", *The Stanford Encyclopedia of Philosophy* (Fall 2013 Edition), Edward N. Zalta (ed.), URL = <http://plato.stanford.edu/archives/fall2013/entries/molecular-genetics/>. (A detailed look at reduction and anti-reduction as it relates to molecular genetics)

9
REALISM AND CONSTRUCTIVISM IN MEDICINE

Jeremy R. Simon

Introduction

The central entity of medical science is the disease. The focus of medical research is finding treatments for the various diseases patients suffer from. Indeed, if medicine is the science "of" something, it is the science of diseases, seeking to understand how they are created, how they behave, and how they can be controlled. It would seem, therefore, that medicine assumes that there are such things as diseases and that understanding medicine philosophically will require understanding diseases.

The question for this chapter is, what sort of things are diseases? When we say that a patient has tuberculosis, is that like saying she has a dog, something out there in the world independent of us, in all senses of the word? When a scientist is studying insulin-dependent diabetes, is that similar to a physicist's studying neutrinos? Or perhaps having tuberculosis (or, as was once said, "being tubercular") is more like being American, neither separable from the person nor from the society in which that person lives, with the science of medicine being similarly controlled by social realities, rather than the physical world? These two positions are called realism and anti-realism, or constructivism, respectively, and represent radically different understandings of the nature of diseases.

The question we are addressing here is different from that addressed in Chapter 1. There, the question is, what needs to be true of something for it to be correct to call it a disease? Here, however, we are asking, given that something is a disease, what sort of thing is it, what is its nature? The distinction may be a bit subtle, as both questions can be phrased as "What is disease?" However, as we spend time with our topic, the distinction should become clear. Furthermore, focusing on examples that are uncontroversially diseases, like diabetes, should help keep things clear as well.

Realism and Anti-Realism: The Basics

Let us begin with a somewhat less rough, but still general, understanding of what it is to be a realist or an anti-realist about diseases. First, let me introduce the distinction between types and tokens. Types, roughly, are general sorts of things, and tokens are particular concrete examples of those things. To illustrate, consider a farm that has 12 cows, 7 pigs, and 3 horses.

How many animals are on the farm? One could give two answers. Either one could say there are three animals on the farm (cows, pigs, and horses) or one could say there are 22 animals (12 + 7 + 3). Either would be correct. The difference is that in the first case, we are saying how many animal types there are, and in the second, how many tokens.

Now, consider a patient who comes to a doctor feeling ill, with a list of complaints including cough, fever, runny nose, sore throat, and pain in her ankle, all of which started one week before. The physician first has to decide which of these complaints belong to the patient's primary problem. After ascertaining that the ankle pain began soon after twisting it, and is now mild, the physician will likely set that aside. However, the cough, fever, runny nose, and sore throat will likely be considered part of a single problem. This problem, the particular patient's disease, is a disease token. The next step for the doctor is to decide what particular condition the patient's disease token is—in this case, likely an upper respiratory infection. That is, the doctor needs to decide to what type the disease token belongs.

Realists and anti-realists differ over what occurs (or should be occurring) at both of these steps, that of identifying the disease token and the disease type. For the realist, the physician can be correct in his selection of complaints to combine into a single problem/disease token, and if he is, it is the fact that the patient has some underlying physical entity (of some sort) that accounts for these complaints. The process of identifying the features of a disease in a patient is like deciding whether the tail that appears between two mammals belongs to the dog or the man walking it. There is a correct answer, and this answer is determined simply by the way the physical world is arranged. For the anti-realist, however, the doctor has simply made a choice in selecting complaints. The choice may be good or bad, but it cannot be correct in the way the realist has it, as there is no underlying physical entity against which to judge the correctness of the doctor's choice, nor does the way the world is arranged dictate any particular grouping of patients. It is like asking which side of the road it is proper to drive on—the answer depends not on the way the physical world is arranged but on the convention in the country one is driving in.

When it comes to assigning individual problems to disease types, the disease realist likewise feels that there is a correct answer, underwritten by the way the natural world is organized. That is, disease types form natural kinds—it is a matter of the underlying structure of the world that patients have "the same" disease, just as there is a correct answer, based on the structure of the world, whether two elementary particles are the same—both electrons, say—or not. An anti-realist, again, will deny that there is one naturally correct way to group sick patients into diagnostic groups. Rather, the groupings into which patients are placed are made by people based on various criteria that are, ultimately, chosen, and not driven by the underlying natural world. The process is like deciding which people belong to the same club. Whatever facts of the matter there are, disease types, like clubs, are not determined by the natural world but by human-imposed groupings. One way to put all this is to say that realists believe diseases are discovered and anti-realists believe they are invented.

At this point, many readers may be thinking that the reality of diseases is obvious, and others that their arbitrary nature is obvious. However, far from being obvious, there are strong intuitions behind both positions. On the one hand, medicine seems to deal with entities that are every bit as real as neutrinos and electrons. A scientist who says "I am studying tuberculosis and trying to discover a new cure for it" seems to be saying something very much like one who says "I am studying protons to try to figure out how to take them apart into quarks." If diseases are not out there the way protons are, then how can a scientist study them and learn things about them? In a similar way, having textbooks with chapters entitled "Cystic Fibrosis,"

"Insulin-dependent Diabetes," and "Hepatocellular Carcinoma" makes it seem like these are entities that can be named, studied, and discussed.

On the other hand, anyone who has taken care of patients knows that there are innumerable differences between two patients with the same diagnosis. Indeed, sometimes they seem hardly to be suffering from the same disease. If each case of a disease is different, this seems to belie the notion that all cases of a disease share something, a natural kind to which they all belong. Beyond this, what does it even mean for diseases to be out there in the world? It is difficult to imagine what sort of entity a disease might be. What is there, in front of the doctor, besides the patient?

Realism: The Positions

As we said above, realists believe that in at least some cases, there is a disease present in a patient, and this disease is part of the natural world, not a result of choices the examining doctor makes. The first job of the realist is to explain what sort of entity this disease is. What kind of thing is it that a patient with a disease has?

Realists make claims about the nature of both disease tokens and types. Let us take tokens first. Perhaps the oldest, if least philosophically defensible, form of realism about disease tokens is what we may call concrete realism. This is the position that diseases are non-abstract entities truly separable from the patient. In ancient times, this meant seeing diseases as evil spirits that entered the patient, whereas more recently, this meant indentifying a disease with an infectious agent. Since we no longer believe in evil spirits, and for many reasons, diseases cannot simply be identified with their causes (which may or may not be separable from the patient anyway—consider genetic diseases), this position is difficult to support, though it has certain naïve plausibility.

More acceptable are accounts that see diseases as identifiable parts of the patient in some way. At least three versions of this have been proposed. The first sees diseases as bundles of signs and symptoms (Rather 1959). A second sees diseases as an underlying physical state of the body in question (Virchow 1895/1958). Finally, some see diseases as bodily processes, which, unlike states, take place over time (Whitbeck 1977). Each of these positions can be further modified by insisting that a disease must also have a particular cause as part of its essence.

We can perhaps understand the varieties of realism described in the last paragraph better by applying them to a particular case, insulin-dependent diabetes mellitus (IDDM). This is a form of diabetes where the body cannot process glucose because the pancreas does not produce (adequate) insulin. When untreated, the levels of glucose in the blood rise above normal levels. Symptoms of untreated IDDM include frequent urination, excessive thirst, and weight loss. When untreated for a longer time, the patient develops diabetic ketoacidosis, a life-threatening condition caused by a build-up of byproducts from the metabolism of triglycerides, which is what the body uses as a substitute for the glucose it cannot metabolize. Finally, patients are susceptible to long-term consequences, mostly as a result of damage to blood vessels, including blindness, heart attack, stroke, and loss of limbs. Although the causes of IDDM are unknown, it is thought to be an autoimmune disease with both genetic and environmental risk factors.

According to the signs-and-symptoms view, to have IDDM is to have elevated blood glucose levels, frequent urination, weight loss, etc. For the physical-state view, having IDDM means having the overall physical state of not producing enough insulin, perhaps along with some of the changes this creates in the body at a cellular (or other low) level. Understanding IDDM as a process is a bit harder, but it means focusing not on the state of the body, either at the cellular/molecular or more macro signs/symptoms level, but rather on the process of metabolizing triglycerides rather than glucose, as well as the processes by which this altered metabolism causes damage to the body.

The other part of the realist position, their approach to disease types, is that disease types form natural kinds. That is, the external world provides a correct sorting of these tokens into groups, or kinds. There are many accounts of natural kinds in the philosophical literature, and discussing them is beyond the scope of this chapter. However, any of the major accounts, such as Putnam (1975), Kripke (1980), Boyd (1991), or Ellis (2001; 2002), should work here. In terms of IDDM, the realist claims that the collection of all cases of IDDM form a natural kind. Thus, for a realist, diseases form natural kinds, and the entities that these kinds comprise are of one of the sorts described in this section—a bundle of signs and symptoms, an underlying physical state, or a bodily process.

Realism: Arguments For

What are some arguments that count in favor of realism? Of course, arguments against anti-realism are implicitly arguments for realism, and vice versa, so dividing up the arguments to be presented in this chapter in "for" or "against" one position or the other is somewhat arbitrary. It is, however, necessary in order to keep track of what is happening. Therefore, I have divided the arguments using the following criterion. If an argument could be understood based just on the presentation of realism, it is discussed here. If it makes more sense in the context of anti-realism, it will follow that section.

The first argument we will consider is based on Putnam's (1979) "no-miracles" argument. Generally speaking, this argument for scientific realism says that if the terms of our scientific theories, such as "quarks," do not refer to real entities forming natural kinds, then we cannot explain the fact that our theories successfully describe the world. In the case of medicine, this would mean that the explanation for our success in developing new treatments is that we correctly identify parts of the world, diseases, whose features are fixed by nature and are the same in each particular case. We can then study them, learn how they behave, and develop treatments. Anti-realists, the argument goes, can provide no such story; medicine's success, for them, must be accounted a miracle.

A somewhat related argument, made by Temkin (1961), notes that many advances in medicine, such as the development of the rabies vaccine by Pasteur, were made by people who had no contact with patients (and, indeed, were not even physicians). If diseases are merely conventional groupings of sick individuals, how, the realist asks, could Pasteur have developed a treatment for a disease without any contact with such individuals. Only if rabies exists independent of any patient could Pasteur have cured it from his lab.

There are several responses to Temkin's argument. First, although Pasteur did not treat any patients personally, he was aware of many and could thus have become aware of whatever conventions defined them. Second, in the case of rabies at least, while Pasteur did not work with human patients, he had animal patients. Finally, in arguing that realism means that scientists have something they can study in a lab separate from a patient, we seem to be arguing for concrete realism, which, as we saw, is a difficult position to take.

The next argument for realism is somewhat technical in nature. Reznek (1987) argues that our criteria for distinguishing related diseases from one another make it such that it must be true that diseases are natural kinds. He uses the case of gout and pseudogout as an example. These two diseases both result in inflammation of joints from deposition of crystals. Originally, anyone with inflammation and crystals was considered to have gout. However, it was then noted that two different types of crystals, urate and calcium pyrophosphate, could cause the inflammation. At that point, Reznek argues, medicine had two choices. Either continue to treat gout as a single disease with two possible natures (caused by urate or calcium) or divide it into two diseases. The fact that medicine chose to reserve "gout" for the cases with urate crystals and started calling cases with calcium pyrophosphate crystals "pseudogout" proves that diseases are natural kinds. Otherwise, why would medicine break up a perfectly useful diagnosis just because it was discovered that it could not be considered a single kind?

Again, there multiple responses to this argument, but we will note just one here. Doctors may have felt that it was simply more useful to divide these two "types" of gout, perhaps because different treatments might work for the two groups. That is, the division could just as easily have been driven by pragmatic concerns as by beliefs about the underlying reality.

A final argument for realism looks at the persistence of certain diseases over time (Rather 1959). Despite any number of changes in our culture, our understanding of medicine, and all other factors that might affect how we would choose to group patients, if that were how we identified diseases, certain diseases have continued to be recognized. Thus, we can easily recognize our diseases of epilepsy and mumps in the Hippocratic descriptions of them from more than 2,000 years ago. The best explanation for this, says the realist, is that epilepsy, mumps, and other diseases are natural kinds, out there for us to identify throughout history regardless of our cultures, etc. The problem with this argument is that, as the language of the last sentence makes clear, this is an application of inference to the best explanation (IBE). The realist says that since the existence of diseases is the best explanation for certain observed facts, we can infer that diseases in fact exist. IBE, however, is not an argument for realism so much as a statement of realist methodology. That is, realists think that such inferences provide evidence that entities we cannot see exist, whereas anti-realists generally deny that IBE can provide any reason to believe such entities exist.

Realism: Arguments Against

One set of objections to realism specifically addresses forms of realism that consider causes to be an essential part of diseases. The first of these (Whitbeck 1977) notes that diseases are never the result of a single cause but are always embedded in a complex causal tree. If we cannot talk of *the* cause of a disease, then a disease's cause cannot be part of the nature of that disease; one certainly cannot include the entire causal tree of the disease in the disease itself. Another objection to the causal form of disease realism notes that we often identify and accept diseases (e.g., rheumatoid arthritis) before we know their causes or even whether they have a single cause. The cause, therefore, cannot be a part of the disease (Engelhardt 1985; Severinsen 2001).

Although there are responses to these objections, they are weak, and the best course for the realist is probably to avoid causal realism. There are, however, objections to realism independent of causes. One of these is the problem of "extremal diseases." "Extremal diseases" is Reznek's (1987) term for diseases, like anemia and hypertension, that differ from a state of health only quantitatively. Thus, everyone has a blood pressure. Only once the blood pressure crosses a particular threshold does one "get" hypertension. The problem for the realist is that it is hard to see how the natural kind normotension could transition so smoothly into a different natural kind, hypertension. The response, then, is to admit that while having high blood pressure is a problem for people, it nonetheless does not form a natural kind and is not a disease that can be studied as an underlying part of the natural world.

The next objection to realism is based on the pessimistic meta-induction introduced by Laudan (1981). Far from science's successes pointing to realism, as the no-miracles argument has it, the pessimistic meta-induction takes science's failures as an argument against realism. In brief, the argument is, science has been wrong so much in the past about what there is and what theories describe it (e.g., regarding phlogiston), even while using those theories successfully. Success, therefore, is no reason to believe that a theory is true or its entities real. In the case of medicine, the pessimistic meta-induction would note, for example, that for many years, fever was considered a disease, not a symptom, and that gout included many joint inflammations beyond those associated urate crystals. Although no one would now argue that fever and gout as recognized 500 years ago formed natural kinds, the physicians of the day considered themselves successful at treating them. Why then should our current (perceived) successes be attributed to natural kinds?

Realists have offered several responses to the pessimistic meta-induction. Although this is not the place to rehearse all of them, it is worth noting that the approach of Psillos (1996) may be particularly useful in the case of medicine. Psillos's response to the meta-induction focuses on features of the entities of a science and not its theory or laws (as others do). Since medicine as a science appears not to have a theory in the way that physics does, a response to the meta-induction that, when applied to physics, might be criticized for leaving out theories here could be particularly helpful.

Another objection to medical realism, which can best be brought by a realist about other parts of the world, is that the entities it proposes are just too strange. Unlike more typical real entities (tokens), such as electrons or rabbits, diseases, at least once we discount concrete realism, are not physically contiguous and independently existing. You cannot show a single place in the body where the diabetes is, nor will you ever be able to point to something separate from a body and say "There is a (case of) diabetes." Nor are disease tokens related to each other the way other natural kind tokens are. Elementary particles are each strictly identical (there is no way to distinguish one electron from another), and members of a single species, while not identical to each other, are historically/genetically connected. All rabbits share certain genetic features, and they do so because of a shared historical chain of descent from earlier creatures. Diseases have no such connection to each other. The realist can respond to this objection that it simply prejudges the matter by saying that diseases are not the sorts of things that can be real.

The final argument against realism we will consider here is perhaps the most obvious. If diseases are true parts of the natural world, why do they not act with the regularity of other natural kinds, which follow laws of nature such as quantum mechanics? Even rabbits will reliably die if you remove their heart. Yet for human diseases it is often very difficult to predict how a given patient will respond to a treatment. If they do not behave as real entities, why think that diseases are real?

The realist can respond to this challenge in different ways. One response is to say that perhaps individual disease tokens that respond differently to treatment are not in fact members of the same kind, though we have not yet figured out how to distinguish the two (otherwise) similar kinds. A second response is that just as an electron will move differently depending on what other objects are in its vicinity, diseases are also sensitive to their surroundings. Their surroundings are so complicated, however, that we often cannot predict how a particular case will respond to treatment.

Anti-Realism: The Positions

Anti-realism as such is simply the denial of realism. In the current context, we will take that to mean denying both parts of realism—regarding types and tokens. (It is possible to deny only that diseases form real types, but we will not consider such positions here.) Thus, anti-realists deny that there is an objectively correct way to select features of a patient's current problem/state and say that those features constitute a disease, and also deny that disease tokens form natural kinds. An anti-realist philosopher of medicine, however, must go farther and say what she thinks diseases are. If they are not part of the underlying nature of the world, what are they?

There are many positions in philosophy of science that may be described as anti-realist in one way or another, such as reductionism, instrumentalism, or empiricism. For various reasons beyond our scope here, none of these can provide an account of diseases such as we need here. This leaves us with constructivism as the general form of anti-realism most suited to the case of medicine. For a constructivist about a particular entity, those entities are not part of the furniture of the universe, as the realist has it. Rather, these entities are constructs of the human mind, intentionally created to serve some purpose. Race is generally taken to be a constructed kind. We certainly can divide the majority of Americans into two groups based (primarily) on their skin color, but nothing about the underlying structure of reality drives this division. It was simply created by people for human purposes. When it comes to individual objects, what we have been calling tokens, the constellations are a good example. Although many people look up at the sky and immediately see Orion or the Southern Cross, this is only because their society taught them to see those stars. There is no need to group stars that way, and indeed different cultures have different sets of constellations they recognize (Needham 1974).

Generically speaking, it is easy to see how to apply this to medicine. When a patient presents a doctor with a list of complaints, there is nothing in the outside world that dictates which of the complaints and findings are related to a single disease as opposed to a second one the patient may have. Rather, the physician determines (possibly based on decisions he knows others have made) which of these complaints to explain to the patient as a unified problem. When it comes to disease types, or diagnoses, the constructivist says that the decision to group certain people together as having tuberculosis or diabetes or chronic fatigue syndrome is more or less explicitly made by people.

But what people and what criteria should be used to identify "acceptable" diseases? It is in answering these two questions that different versions of medical constructivism distinguish themselves, and so to which we will now turn. This means that unlike realism, where the primary question was what sort of entities disease tokens are, for constructivism, the question is going to be the nature of disease types. In a functioning medical system, the identification of a given patient's token will not be entirely up to the physician, but will rather be identified with an eye to fitting it into a diagnostic category (i.e., type).

There are several answers to each of the questions we posed above—which people, which criteria—and thus many possible constructivist positions. Not all of these possible positions, however, are particularly plausible, and only a few have been in fact developed. Taking the second question first, there are several answers proposed to the question "What criteria?". The first treats medicine like a science, albeit not a realist one, and says that a disease is legitimate if it furthers our goal of understanding and manipulating the world of diseases. The groupings of patients that best help explain our past observations and make predictions about the future are to be recognized as diseases within medicine. Thus, IDDM is a disease because it helps us understand our observations about glucose in the urine and blood and certain people's severe illnesses. It also allows us to predict that certain people will die if not given insulin, and indeed helped us to develop insulin-based treatments in the first place. Fever, on the other hand, as a clinical entity rather than a symptom, was ultimately found to be useless to medicine and so is not recognized as a disease. A second approach says diseases are recognized not because they help us understand the world, but because they allow us to help people with their complaints. IDDM will still be a disease, but because it allows us to help patients (save their lives with insulin) rather than because it helps in explanation and discovery about the world. Fever again will not be a disease, because we are unable to help patients by using it.

A similar approach looks not to benefiting individual patients but society as a whole. The results of such a "utilitarian" constructivism could be quite different from the patient-centered pragmatism we just considered. A (potential) disease that would be very expensive to treat and leave people unable to contribute to society might not be recognized, even if people so treated would be made happier. It might be more efficient to let them die. IDDM, on the other hand, is a disease, so evidently it is judged (by whom, we have yet to see) that considering it a disease is good for society, perhaps because treatment is relatively inexpensive and untreated diabetics become very (expensively) ill. Finally, one might argue that diseases are recognized with an eye to promoting the power of some particular group.

As for who gets to apply the criteria in question and create diseases, the most frequently identified group is physicians, or medical scientists and professionals more broadly. This is certainly what appears to be happening when, for example, a panel of psychiatrists issues the *Diagnostic and Statistical Manual*, listing those diagnoses that psychiatry will recognize. A second potential group is "the ruling party," those with sociopolitical power. In more concrete terms, this could be the government, which certainly could exert control over diseases by, for example, determining which diagnoses are acceptable to submit for insurance reimbursement. They could also act behind the scenes, for example, by exerting influence on the panels of physicians who otherwise appear to be identifying acceptable diseases.

Let us now consider some of the actual positions staked out on this matter through combinations of the answers to the two questions we just considered. King (1954), although not completely explicit, seems to believe that diseases are recognized by medical professionals according to how they help us understand the world. Gräsbeck likewise sees medical professionals as the arbiters of diseases. However, he does not believe that understanding is their motivation. Rather, they are utilitarians, concerned with "social and legislative arrangements" (Gräsbeck 1984: 57), along with diagnosis and treatment.

Engelhardt (1980; 1984; 1985) and Severinsen (2001) both refer to patient-centered criteria, such as "goods and harms to the lives of patients" (Engelhardt 1980: 45) and attention to the burdens that arriving at diagnoses will place on patients. Although Engelhardt does not say who he thinks applies these criteria, Severinsen is clear that, again, it is medical professionals.

Indeed, the only writer who seems to understand diseases as being determined by a group other than physician-scientists is Illich (1976). Illich says that "disease [is] an instrument of class domination" to "put the worker in his place" (171). Diseases are a tool of power, not of science or benefit to patients or society generally. As to whose power, here Illich is a bit less clear. It may be physicians, but it may also be "the university-trained and the bureaucrats" (171). However, even if it is physicians, these are physicians qua elite, not qua scientists.

Anti-Realism: Arguments For

One type of argument for anti-realism bases itself on the goals and practices of medicine. An example of this is provided by Engelhardt, who says:

> Clinical medicine is not developed in order to catalogue diseases sub specie aeternitatis, but in order for physicians to make more cost-effective decisions . . . so as to achieve various goals of . . . well-being. Thus, clinical categories . . . are at once tied to the likely possibilities of useful treatment and the severity of the conditions suspected.
> (1984: 36)

In other words, since the goals of medicine are pragmatic, its kinds must be too. However, looking at a parallel argument in case of physics will show the weakness here. Engineering is surely a pragmatic practice, but this does not prove that the underlying kinds that physics relies on are not real—so too with the practice of medicine and its underlying science.

Elsewhere, Engelhardt (1985) gives another argument from the practice of medicine. Consider rheumatic fever (RF). This condition is diagnosed based on the "Jones Criteria." If a patient has *either* two "major" criteria (such as rash and joint inflammation) or one "major" and two "minor" criteria (such as fever and cardiogram abnormalities), he is diagnosed with RF. Why precisely this number of criteria? Engelhardt claims that it is because at this point treating such patients becomes pragmatic. More criteria and we will not be able to help as many people as we can. Fewer, and we will spend too much time treating people whom we may not be able to help. Since the determination of a definitive diagnosis relies on criteria that are pragmatically arrived at, and are not driven by scientific necessity, Engelhardt argues, the diagnosis itself cannot be part of the natural world.

The realist has a response here as well. Although she will acknowledge that the Jones Criteria are pragmatically based, she will see this as a function of our ignorance. That is, we still do not clearly understand RF's nature and causes. Therefore, we have no definitive way as yet to diagnose it. In the absence of such a definite means of diagnosis, we use the Jones Criteria as a pragmatic approximate way to pick out patients with RF. However, as we learn more about RF, the realist says, we will abandon this pragmatic tool for one based on the underlying science.

Anti-Realism: Arguments Against

One argument against anti-realism derives, in Temkin's terms, from "the danger of . . . *aeipatheia*, perpetual illness" (1961: 641). Temkin argues that if any physiological state whatsoever can potentially be a disease state if the relevant "authorities" decide to make it so, then there is not such a thing as perfect health. Anyone at anytime may discover that he or she has a disease. Temkin finds this state of affairs problematic. Perfect health, where one is not a decision away from having a disease, should at least be possible.

This argument is open to several responses. First, it is not obvious that Temkin's intuition that perfect health is possible is correct. Second, Temkin seems to believe that anti-realists can identify disease at their whim, which, as we saw, is not true. Other responses require theoretical approaches not developed here, and so will be skipped.

A second argument against anti-realism is based on the indispensability argument used most frequently in philosophy of mathematics against anti-realists. Both Grene (1977) and Nordenfelt (1987) raise versions of this argument. The claim is that medical discourse presupposes, and indeed relies on, diseases being natural kinds. That is, real diseases are indispensable for medicine. Nordenfelt says that communication among medical professionals relies on reference to disease types, thus implying that there are such types. Grene says that "a nominalist would never have world enough or time to produce" many of the assertions we find in medical texts (1977: 79). They would simply take too long to state.

What type of communication and statements are they talking about? It cannot be routine statements, as we refer to constructed types such as American, Nigerian, Freemason, and prime ministers all the time. Rather, it must be a special type of communication. In our context, that is most likely the communication that allows us to make scientific progress. That is, scientific progress in medicine is impossible without real diseases, as the communication among medical scientists needed for them to do science could not otherwise occur. Once we understand this, we can see some constructivist responses to this attack. First, they can deny that progress of the kind realists are talking about occurs in medicine. This is related to the general debate about progress in science between realists and anti-realists. Second, constructivists can respond to

Grene and Nordenfelt by agreeing that progress occurs but denying that it needs natural kinds. What has happened when we learn how to better treat a disease is that we have learned how to better make true generalizations about the people we have grouped together under that diagnosis. Similarly, with more exposure to the enemy, a general can learn more about how to respond to the socially constructed entity "the enemy army."

A final objection to constructivism is that it is not historically plausible. The various potential groups constructivists identify have varied greatly, both across history and across societies, in their actual influence, and thus ability and power to impose diseases, yet medicine has carried on. If medicine's functioning really relied on the ability of a single group to impose diseases, then the rise and fall of the power of this group to do so should significantly affect medicine. Constructivists need to do significant historical work to show either that a given group has had adequate power across history, or show that the power to construct diseases has shifted from group to group across history. It is not clear that this can be done.

References

Boyd, R. (1991). "Realism, Anti-Foundationalism and the Enthusiasm for Natural Kinds," *Philosophical Studies*, vol. 61, pp. 127–148.
Ellis, B. D. (2001). *Scientific Essentialism*, Cambridge, MA: Cambridge University Press.
———. (2002). *The Philosophy of Nature*, Montreal & Kingston: McGill-Queen's University Press.
Engelhardt, H. T., Jr. (1980). "Ethical Issues in Diagnosis," *Metamedicine*, vol. 1, pp. 39–50.
———. (1984). "Clinical Problems and the Concept of Disease," in L. Nordenfelt and B. I. B. Lindhal (eds.), *Health, Disease, and Causal Explanation*, Dordrecht: D. Reidel, pp. 27–41.
———. (1985). "Typologies of Disease: Nosologies Revisited," in K. Schaffner (ed.), *Logic of Discovery and Diagnosis in Medicine*, Berkeley, CA: University of California Press, pp. 56–71.
Gräsbeck, R. (1984). "Health and Disease from the Point of View of the Clinical Laboratory," in L. Nordenfelt and B. I. B. Lindhal (eds.), *Health, Disease, and Causal Explanation*, Dordrecht: D. Reidel, pp. 47–60.
Grene, M. (1977). "Philosophy of Medicine: Prolegomena to a Philosophy of Science," *PSA 1976*, vol.2, pp. 77–93.
Illich, I. (1976). *Medical Nemesis: The Expropriation of Health*, New York: Pantheon.
King, L. S. (1954). "What Is Disease," *Philosophy of Science*, vol. 21, pp. 193–203.
Kripke, S. (1980). *Naming and Necessity*, Cambridge, MA: Harvard University Press.
Laudan, L. (1981). "A Confutation of Convergent Realism," *Philosophy of Science*, vol. 48, pp. 19–48.
Needham, J. (1974). "Astronomy in Ancient and Medieval China," *Philosophical Transactions of the Royal Society of London. Series A, Mathematical and Physical Sciences*, vol. 276, pp. 67–82.
Nordenfelt, L. (1987). *On the Nature of Health*, Dordrecht: Reidel.
Psillos, S. (1996). "Scientific Realism and the 'Pessimistic Induction'," *Philosophy of Science*, vol. 63 (Proceedings), pp. S306–S314.
Putnam, H. (1975). "The Meaning of 'Meaning'," in *Mind, Language and Reality: Philosophical Papers, vol. 2*, Cambridge: Cambridge University Press, pp. 215–271.
——— (1979). "What Is Mathematical Truth," in *Mathematics, Matter and Method: Philosophical Papers, vol. 1*, 2nd ed., Cambridge: Cambridge University Press, pp. 60–78.
Rather, L. J. (1959). "Towards a Philosophical Study of the Idea of Disease," in C. M. Brooks and P. F. Cranefield (eds.), *The Historical Development of Physiological Thought*, New York: Hafner, pp. 353–373.
Reznek, L. (1987). *The Nature of Disease*, London: Routledge & Kegan Paul.
Severinsen, M. (2001). "Principles Behind Definitions of Diseases—A Criticism of the Principle of Disease Mechanism and the Development of a Pragmatic Alternative," *Theoretical Medicine*, vol. 22, pp. 319–336.
Temkin, O. (1961). "The Scientific Approach to Disease," in A. C. Crombie (ed.), *Scientific Change*, London: Heinemann, pp. 629–647.

Virchow, R. (1895/1958). "Hundert Jahre allgemeiner Pathologie," trans. as "One Hundred Years of General Pathology" in Virchow, R. (1958). *Disease, Life, and Man, Selected Essays*. trans. and with an intro. by L. J. Rather, Stanford, CA: Stanford University Press.

Whitbeck, C. (1977). "Causation in Medicine: The Disease Entity Model," *Philosophy of Science*, vol. 44, pp. 619–637.

Further Reading

Foucault, M. (1973). *The Birth of the Clinic: An Archeology of Medical Perception*, trans. A. M. Sheridan Smith, New York: Pantheon, is a fascinating, if somewhat obscure, account of early-19th-century medicine, which at least appears to take a constructivist approach to medicine, although the relevance of Foucault's historical analysis to the ontology of modern medicine is not entirely clear.

"Medical Ontology," in Gifford, F. (ed.) (2011). *Philosophy of Medicine*, Elsevier: Amsterdam, is a more extensive discussion of the material in this chapter by the current author.

Reznek, L. (1987). *The Nature of Disease*, London: Routledge & Kegan Paul, is the only book devoted entirely to the question of disease realism vs. anti-realism, presenting one author's approach to the matter.

Sadegh-Zadeh, K. (2015). *Handbook of Analytic Philosophy of Medicine*, 2nd ed., Springer: Dordrecht, is an advanced, encyclopedic, and idiosyncratic book with a section on medical ontology.

Part II
SPECIFIC CONCEPTS

10
BIRTH

Christina Schües

Birth, puberty, and death are live-changing events that are recognized in culturally and socially different forms. In many cultures, birth is accompanied by special rituals and ceremonies. The technology and medicine-oriented Western culture embraces only few medical and social particular rituals but has largely standardized childbirth, approaching it with a "fear-based" attitude within a discourse of risk (Davis-Floyd 1994: 24). Understanding birth as a risky transition for the mother and the child leads to fear and an urge to put it into a medical practice. Practices of birth vary concerning the decision whether birth takes place in the hospital (98% in the United States), in a birth center (about 10% in the Netherlands), or at home (about 20% in the Netherlands). And these practices differ strongly concerning the process: birth can be "natural," labor-induced, or cesarean, experienced with pain-relief medication or without, in inherently high-risk conditions, and it can end in stillbirth. In the 19th century, the birth practice was taken from the hand of midwives and put into the hands of male physicians. Philosophers have critically debated the "natural" birth movement as well as birth under medication (Kukla 2005, 2011; Lyerly 2006; Purdy 2001).

Despite these differences, we are all born. However, the fact that humans are born has rarely been at the focus in the history of philosophy, which has instead emphasized death and human mortality. The concept of birth had rather been seen as a metaphorical concept in which birth is imagined, like Plato did, as a cognitive act by philosophers or as a divine act of creation by male gods (Plato 2000: 7. book; Plato 1980: Speech by Diotima; Tyler 2000), and it has been considered as the birth of a subject that takes place without women. Hobbes, for instance, writes to "consider men as if even now sprung up out of the earth, and suddenly, like mushrooms, come to full maturity, without all kind of engagement to each other" (Hobbes 1841: 109). Since the 1980s, several feminist works have criticized the neglect of birth as a philosophical concept, the denial that humans are born from women, and the appropriation of birth for masculine action (Irigaray 1985; Young 1990; Cavarero 1998; Schües 2008: Chs. 1–3). Thus, "giving birth" has been transformed metaphorically into the path to knowledge (Plato's way out of the cave) or the technical production of a bomb ("It's a boy!" announced Edward Teller, co-developer of the hydrogen bomb, when it was first successfully tested). Martin Heidegger is one philosopher who has at least referred to "birth," but by a concept of "being thrown" into the world. It is a basic characteristic of being-there (Dasein), whose "there" is already presupposed as being-in-the-world (Heidegger 1996: 127). Yet, in the overall conception, even Heidegger "forgets" the notion of birth and interprets being-there one-sidedly as being-towards-death.

Heidegger's student Hannah Arendt, however, insisted upon recognizing the centrality of natality as opposed to the traditional privileging of mortality. She posits the concept of natality, the capacity to begin because of being born, as a foundational fact of politics and action:

> Philosophically speaking, to act is the human answer to the condition of natality. Since we all come into the world by virtue of birth, as newcomers and beginnings, we are able to start something new; without the fact of birth we would not even know what novelty is, all 'action' would be either mere behavior or preservation.
> (Arendt 1958; Arendt 1970: 82)

Thus, considering birth means recognizing its dual nature as a biological event within nature and as an existential transformation within human society that provides the foundation for creative action. Human beings are thus born in a double sense: "He is new in a world that is strange to him and he is in process of becoming, he is a new human being and he is a becoming human being" (Arendt 1968: 185; also Schües 2011). The first aspect refers to the new relation of the child to the world as a human being, and the second aspect refers to the state of becoming that a child shares with other living beings such as cats and dogs. So we can say that, in respect to life, biological life, that the child is a human being in the process of becoming and developing. In respect to the world, the child is a newcomer to a world that existed before him and will continue to exist after his death. Since human beings are born and introduced to the world, they will spend their life in light of this particular beginning. "Human parents [. . .] have not only summoned their children into life through conception and birth, they have simultaneously introduced them into a world" (Arendt 1968: 185).

Since birth and life have so many different aspects and meanings, they are approached by a variety of different perspectives. Most often, however, they are discussed in the context of medical ethics, but this focus is too narrow to grasp the multilayered dimensions of birth and life. A hermeneutical and phenomenological approach is more appropriate for a philosophy of medicine because it is the task of phenomenology to disclose and unfold the meaning of experiences and the basic structures of the different physical, social, and cultural dimensions of a particular theme such as birth and life. In general, phenomenology takes on the challenge to reveal the "prejudices" and "taken-for-granted presuppositions" that go unnoticed in our unreflective experiences, yet that guide the praxis of human conduct (Husserl 1982; Toombs 2001; Merleau-Ponty 2012: XX–XXXV).

The Beginning

It is not easy to define when a human life begins. Some people believe that it begins with conception, whereas for others it begins with the implantation of the embryo, and still others take life to be a form of continuing development. The moral status of a fetus or a newborn will depend on arguments about individuality, continuity, or potentiality (Guenin 2008). Answers to the question of when human life begins are mostly oriented toward physiological and biological aspects of the neural development of the brain, religious beliefs about the sanctity of human life, as well as political and ethical concepts of the dignity of the human embryo. Ethical debates about the right to live or the question of whether a fetus may be taken for research are closely related to the understanding of whether a fetus is seen as human, on the way of becoming human, or potentially human.

When people are asked about the beginning of their life, they usually think about their life in relation to the world they live in, to the earth and society. They tell about where they come from and who gave birth to them. Every living human being has been born from a woman.

But, concretely, everyone is born in a particular place and time, and from and with at least one particular person—their (biological) mother (Rich 1976). Birth means the beginning of life on earth and in the world. However, it is not the absolute beginning of a human being, because her being is also real prior to her birth. Birth localizes the person who is born within familial, social ,and cultural relations in the world; it puts her in a generative context and a familial biography. Birth marks the biological and social changes between generations and their continuation, because the arrival of a newborn brings a new generation to light. The prenatal existence of the child and the pregnancy of her mother is part of a family biography that can be talked about. When people consider where they come from, the circumstances of the time before the birth is often part of the biography.

We cannot remember our own birth. Epistemically, our own birth is located between withdrawal and apodicticity. With the term "withdrawal" I refer to the fact that one cannot remember his or her birth, because it is excluded from our own memory. The term *apodicticity* refers to a certainty that can be demonstrated, such as the existence of someone. Though birth is withdrawn from inner reflection or memory, I am sure that I was born because I exist. I exist; therefore, I must have been born. I am a natal being. Because of being born, the infant is directed towards the outer world in its three-dimensionality. By this new perspective, intentionality sets in. In Husserlian phenomenology, the term *intentionality* stands for the directedness of consciousness to meaningful things, objects, or events. Being conscious *of* something means to constitute something as an object of sense experience. Hence, intentionality means the structure of being-directed-*toward*-something-*as*-something. Birth, as such a unique event at the beginning of life-in-the-world, posits the possibility of intentionality as "an essential characteristic of each person through which she can initiate the constitution of sense, give birth to sense" (Schües 1997: 245). Hence, birth is the fundamental condition of the possibility of intentionality. It brings the individual in a relation towards the world.

Birth is a social and political fact that is—at least in most countries—recorded in official documents, celebrated at birthday parties, and talked about as an important beginning in life and of life. Birth belongs to the human condition, yet it is contingent because not all human beings who live prenatally will necessarily be born. Some will die before their birth, for instance, because they are aborted. Whether a child is born lies in the hands of others—of physicians, the pregnant woman, or also society. It is also determined by circumstances and conditions like the health of the embryo or the pregnant woman. Hence, birth is a contingent fact of life and society. Even though we do not directly remember our own birth, we know about our own birth because people have told us about it and because we infer our own birth from the birth of others. A person's own birth not only accompanies one as a date in documents but also in the sense of being born from someone and with other people who remain as relatives, if not as the family one lives with. These coordinates of being born *from* and *with* a particular person and *at* a particular place remain parts of an individual's personal history. They make people structurally special. Because of these concrete structural coordinates and constellations, every human being is different. Hence, birth constitutes not only a basic relationship among human beings, but it also denotes a plurality among people in the world. *Plurality* means here that each person is different from every other person alive or who has previously lived; these basic differences are due to the fact of being born in different times and places, and from and with different persons.

Disruption and Transformation

Birth means a disruption and a transformation. It disrupts the life of the woman who gives birth to the child and of the persons for whom a child is born. When a child is born, a mother (of that child) is also born and, at least most of the time, a family is (re-)manifested. With birth, a pregnant woman is transformed into a mother, a man is transformed into a father, and family members are transformed into grandparents, siblings, uncles, aunts, or cousins. Each of these roles is based on societal norms governing what to do or expect. For the child, birth disrupts the continuity of her development. Being born means not living anymore in the inner-bodily connectedness with the pregnant woman, the becoming-mother. Hence, birth for the child means a transition from the intrauterine position to the extrauterine situation in the world, a transformation from living in the womb, with a surrounding wall and in a maternal-fetal relation, to a life in the world and in concrete different relationships and contexts. Thus, transformation can be understood as a fundamental leap from one existential mode of being to another, from prenatal existence to the natal existence of living in the world and of being able to mentally and physically direct oneself towards other objects and beings; that is, having intentionality (Schües 1997). Because of its new spatial location, from the moment of being born, the newborn directly faces other human beings; all objects, humans, voices, and things are in a distance or come closer; they are presented in different perspectives.

The disruption of birth can be understood in different senses and from different subjective perspectives. It can be understood, first, from the perspective of the woman giving birth—that is, birth as birthing, labor, and delivery at the end of pregnancy. Second, there is the perspective of the child who is born—that is, birth as partus or parturition, as the culmination of the development inside the woman's uterus and as the end of the internal bodily connection and the beginning of external bonding with the mother and other persons. And, third, it can be seen from the perspective of an observer, whether the woman who delivers the baby or others—that is, birth as a process in its different phases.

Obstetricians and midwives categorize the process of birth in four phases: (1) the early phase is marked by the onsets of contractions and labor pains; (2) the active or transitional phase involves the dilation of the cervix; (3) the fetal expulsion phase is characterized by the descent and birth of the infant; and (4) the final phase is the placenta delivery. These phases of birthing are talked about in birth preparation classes and are undergone in the routine of a *normal birth* regardless of whether it takes place in a hospital, a birth center, or at home. Yet, these birth phases refer to what is considered "normal." "A normal physiologic labor and birth is one that is powered by the innate human capacity of the woman and fetus" [ACNM (American College of Nurse-Midwives), MANA (Midwives Alliance of North America), and NACPM (National Association of Certified Professional Midwives) 2012]. The term "normal" refers to a birth procedure that does not need medical intervention. The idea of a normal physiologic birth procedure delineates certain factors and benchmarks that define optimal processes, a supportive environment and birth setting, as well several biological, psychological, physiological, or educational conditions of women in labor.

The Relational Structure of Prenatal Existence and of Being Born

There is an ontological distinction between the maternal-fetal relation and the relation after birth, but there are different views about whether or not the pregnant woman and the fetus are separate entities. Even if one is ontologically committed to the idea that a pregnant woman and her fetus are two distinct entities, there remains the question of how these entities relate to each other biologically. The answers to this question indicate the assumptions and values addressed. The "immunological paradox of pregnancy" holds that even though the fetus is foreign to the female body, it is "paradoxically" not rejected (Moffett and Loke 2004). The "foreign fetus model" implies immunologically either that the woman is hostile to the fetus or that she is a passive victim of fetal invasion. Ether way, one is inclined to believe that there is a conflict between the woman and the fetus. The mother's immune system must suppress the foreign body (similarly to organ transplantation), or the fetus resembles a parasite that invaded the mother's body in order to sustain itself and grow. The problem is that the metaphors of "parasite," "foreign," or "victim" ignore the genetic ties and reduce the biological complexity of pregnancy and maternal-fetus relation (Howes 2007). These metaphors are also used in social and psychological discussions about the prenatal relation. Perhaps it can be considered as a symbiotic relation; "they are two, and yet one. They live 'together'" (Fromm 1956: 19).

When the birth takes place, then the prenatal relation is disrupted and replaced by a new relation. A birth means a beginning; the beginning is a relation. This relation takes place between the (biological) mother and the child. It is a concrete, finite relation to another existing person. It is a contingent yet irreplaceable relation; the relation between the pregnant woman and the fetus was conditioned by particular biological and social factors, and it could be ended any time; yet it is unique in kind and necessary for the development of the child. The primacy of the relation in the world is shown with the appearance of the newborn, and it affirms the unity of existence and appearance (Cavarero 1997). When a child is born, it is looked at and—if it is not too exhausted, sick, or sedated—it can look back. This first gaze may ground such a first relation. All human beings who are born have experienced the concrete development and mode of this *primary elemental relation* with at least one person. Regardless of whether this relation has been continued or ended, lived in love or disappointment, with care or neglect, in compassion or superficial interest, it marks the threshold between the inner and the outer, between partus and bonding. If a child is given to someone else after birth—as in the case of adoption or surrogate motherhood—then the first elemental relation is disrupted.

Being born means that a child has undergone the transition from the *prenatal* being-with to the *postnatal, worldly* being-with; that is, the delivery is a release from the symbiotic relation with the pregnant woman to the primary relation with the mother and others in the world. However, this first elemental human relation of "nurturing, clothing, and accommodating" already begins during the pregnancy and with the supportive care of the mother, family, and others. Without this relation and care, a child would be just "'flesh and bones', a piece of reality not invested with any significance or meaning" (Thoné 1998: 121). The Norwegian philosopher Arne Vetlesen formulated an insight in 1995 that has an importance in reference to developmental psychology. He emphasized that the first interpersonal relationship, the mother-child dyad, is characterized by "asymmetry and non-reciprocity." Being the "recipient of unconditional support is [. . .] the first experience made by the human person. And it is thanks to the experience of having been an addressee that we develop the cognitive-emotional capacities required to become a giver" (1995: 379). The times of prenatal being, birth, and postnatal life in the world are times when the child is cared for in many different ways. Traditionally, such care took place in the emotional realm of day-to-day life. With the rise of reproductive technologies, starting with ultrasound screening and continuing with genetic testing of the fetus, the notion of care at the beginning of life has shifted more and more towards medical supervision. If prenatal existence is seen in terms of a symbiotic relation with the becoming-mother, then the pregnancy of a brain-dead woman would mean that the fetus is no longer in such a relation. In a case like this, ontologically speaking, there are two reductions at work: the fetus is reduced to a biological entity of "flesh and bones," and the pregnant woman is a mere fetal "container" (Purdy 1994; also Lindemann Nelson 1994).

The Traces of Birth

Is birth—the prenatal being and the postnatal beginning of existence—forgotten? Certainly, we do not remember our birth explicitly or reflectively. However, since we exist we are certain we were born. Though the birth itself is withdrawn from our reflective memory, human beings have, as the French phenomenologist Maurice Merleau-Ponty and the psychoanalysts Otto Rank and Franz Renggly argue, a "bodily memory." The memory of the body lies in between consciousness and the physical body; it shows itself in our daily bodily movements and rhythms about which we do not need to think or be conscious of (Fuchs 2011; Merleau-Ponty 2012).

One's own birth and pre- and postnatal experiences cannot be explicitly remembered, but they took place as *pre-reflective* experiences that left their traces on the body. Hence, one's own birth is forgotten yet is bodily inscribed as a trace of a particular lived-through pre-reflective experience. This means that our self can only be expressed by our body. Our lived body is always begun before we become conscious of ourselves. Birth in its phases remains as the anonymous past of our presence incorporated into our habituated body. Merleau-Ponty echoes this thought in reference to our relation to the world:

> To be born is to be simultaneously born of the world and to be born into the world. The world is always already constituted, but also never completely constituted: In the first relation we are solicited, in the second we are open to an infinity of possibilities.
> (Merleau-Ponty 2012: 480)

This situation is ambiguous because it includes determination and openness. That is, birth had left its bodily traces in the body. Though they cannot be represented, nevertheless they have their meaning in their silence, a silence that sometimes "speaks" but never with words. Hence, the human conditions are influenced by these traces, and they form the habituated body. Since the habituated body brings the past into the world of presence, the body shapes the "real world" and the actual body. This silence—which can also be called, following Merleau-Ponty, "anonymity"—accompanies all experiences and is even inherent to perception and memory. As Merleau-Ponty describes it: "*One* perceives in me, and not I perceive" (2012: 223). Therefore, being born is located principally in the realm of anonymity as a silent trace of bodily memory. It is a realm of past that, if I search my representational memory, seems to be external to my self-consciousness. In memory, the differentiation of the habituated and the actual body, our own and the other, has always already taken place. Birth is the beginning of difference. And this basic difference, which touches upon the beginning of the beginning, remains silent and incomprehensible, a pure silent experience, an event that might acquire its meaning later in an afterthought (Husserl 1977: 38). I imagine myself, my body, and the world as already preexisting, and I grasp my birth as a "pre-personal horizon" (Merleau-Ponty 2012: 223).

Birth as Trauma?

The psychoanalysts Sigmund Freud and his student Otto Rank believed that each person's birth leaves a trace in his or her subconsciousness that can be understood as bodily memory. Freud supposed that the existential angst experienced during the birth process presents the basic form for all later fear: "The act of birth, moreover, is the first experience with fear, and is thus the source and model of the emotion of fear" (Freud 1921: ch. 5, fn 1). Rank notably put forward the notion of "trauma of birth," that one's own birth is engraved in the subconsciousness and hence has a crucial importance for the psychological development of a human being. The notion of trauma—a notion that lies at the center of trauma and therapy research—describes an intense fear and experience of panic similar to a "near-death experience" (Renggli 2001: 43, my translation).

Rank did not only want to use this concept of birth trauma to ground psychoanalysis "biologically," he also sought to explain with it all neurotic symptoms and creative capacities. In his view, the meaning of the trauma of birth is profound because it is "a real substratum for all psycho-physiological connections and relations" (Rank 1993: 54). In contrast to Freud, Rank believed that the individual can reproduce "the individual intrauterine posture, or peculiarities relating to one's own birth"; consequentially, he predicated a "womb phantasy" (Rank 1993: 84, 83). In this perspective, regardless of whether birth took place naturally or as a cesarean delivery, the perinatal separation from intrauterine being within the womb of the mother is inevitably traumatic. Rank's theory leads to three consequences. First, Freud's father-centered approach is reevaluated and the mother and mother–child relation is given a central place in theory. Rank thereby emancipates himself from the paternalistic and androcentric dogma of the history of ideas (Roazen 1974: 380–395). Second, he postulates the thesis of the trauma of birth, which is brought about by the experience of injury subconsciously inscribed on the body and which can be repeated in particular life circumstances. Third, the pain and fear of separation is the basis for the suppression, or forgetting, of birth.

Birth as Transition

Rank's theory is not without its critics. The pediatrician and psychoanalyst Donald Winnicott describes the "experience of birth" and argues that the child who is not yet born is already able to prepare herself for her birth. Being seen as an active entity, she collects and catalogues nonvolatile bodily experiences (Winnicott 1990: 143ff.). His argument is that the fetus already has the capacity to psychologically structure the influences that it undergoes.

Birth can be experienced as a traumatic experience or as a valuable transition, depending on the different temporal and contextual possibilities and ways in which the birth is prepared for and takes place. A normal birth—that is, the transition from the not-yet-being-born to being-born—is not necessarily traumatizing from the perspective of the neonate. The normal transformation in birth can be characterized in the following ways. First, the infant experiences a radical discontinuity of being regarding different somatic, auditory, or tactile sensations, but it has developed capacities that help it to bridge these breaks in continuity. Second, the infant has bodily memories about subconscious sensations and impulses that are formed in the phase when it lived in unity with itself and the world, the inner and outer (Leboyer 2009: 64). Third, the mechanical and temporal aspects of the birth procedure are important for the question of whether birth is experienced as a trauma or as a valuable transition. It is best if the transition of birth is neither sudden nor delayed. If it is too sudden, then the child is unprepared; if it is delayed, then it is frustrated.

In light of these three aspects, we can see birth as being initiated by the infant itself. It can initiate impulses of movement and the transformation of birth when it has reached the necessary level of biological, physiological, and psychological maturation. These impulses emanate directly from the liveliness of the child (Winnicott 1990: 144). Hence, for Winnicott, the infant "gives" the impulse for being ready to be born. The birth as a transition is a particularly valuable experience for those persons born "normally." If the birth begins too early or too late or if the birth is complicated and problematic, then the child might suffer fear and distress, and hence experience a birth trauma. Winnicott, like Leboyer, therefore advocates a "birth without violence" (2009). The goal of this approach is to arrange for birth to be a smooth transition from prenatal to postnatal being, from an intrauterine world to the extrauterine world. For instance, if the cord is not cut too soon, then the infant can take its first breaths without being shocked by respiratory distress, and then it can experience a smooth transition from the one form of being before and during birth to the form of being in the world as a rich experience for its psychological development. The prenatal world and the postnatal world differ insofar as the latter is characterized by contradictions, such as hard and soft or cold and warm. For Leboyer, birth is understood as a "perinatal leap," which is important and valuable for the infant's development if it is done properly and the baby is welcomed in a room that resembles as much as possible the inner womb, with dimmed lighting, warm temperature, and soft tissues. Both positions—birth as a trauma and birth as valuable natural transition—take it as a psychological fact that birth is continued as bodily memory.

Social and Medical Perspectives

Pregnancy, birth, and other life-changing events are understood from very different perspectives and interests. A child can be delivered by vaginal birth or cesarean section. In the 1970s, the rate of cesarean section was below 5% in the United States; by 2006, the rate had risen to over 30% (Lake 2012; Osterman and Martin 2014). There are several important reasons for the likelihood of a cesarean section. First, medical reasons lead to more labor interventions (for instance, by continuous fetal monitoring or epidural analgesia). Second, for social and psychological reasons, enhancing women's own ability to give birth and supporting a midwifery model of care has a low priority. Third, sometimes the option of a vaginal birth is simply not offered. Fourth, some individuals, society, and institutions have high tolerance for surgical procedures, because of time and organizational constraints. Cesareans are easier to plan, shorter, and more profitable for the hospital (Childbirth Connection 2015). The example of the cesarean rate shows that the question of how a child is born and how a woman gives birth to her child depends upon many different factors that are not only medical but also social and psychological.

This observation leads to two further observations. First, decisions for surgery or for medication are not only made on the basis of medical indication but also out of social or economic interests, organizational structures, or cultural convictions. This observation, however, presupposes already that birth takes place within a medical context. And this thought leads, second, to the observation that birth itself has not always been regarded as an issue for medical praxis. The pregnant woman is (usually) not sick, and birth is not a disease. However, today it is normal that pregnancy and birth are supported, accompanied, and monitored by the medical practice and its standardizations.

Medicalization is a notion that describes the critical determination of a social transformation in the second half of the 20th century. It encompasses the critique that realms of life that had originally been external to or only marginally part of the medical praxis are now seen as medical phenomena or problems (Conrad 1992; Ullrich 2012). Medicalization denotes a process "by which nonmedical problems become defined and treated as medical problems, usually in terms of illness and disorders, and are managed and overseen by medical professionals who are the authorities on that process" (Mullin 2005; Conrad 2007: 4). People are now used to the medicalization of the beginning and end of life, to pregnancy as "medicalized hope" (Duden 1993; Ullrich 2012), as a disruption to health requiring expert medical intervention (Mullin 2005: 54). If pregnancy and birth are understood in terms of health or illness, then they are subject to medical intervention and risk management, which are contingent and depend on social and institutional processes. An example of the contingency of risks and interventions and its dependency on social attitudes was already discussed above in relation to the different rates of cesarean births.

Reproductive medicine in the form of in-vitro fertilization (IVF) or genetic prenatal diagnostics has become part of everyday life. For about 30 years, the philosophy of medicine has discussed the social, psychological, ethical, and political dimensions of medical access to the body of the women and the fetus, and to genetic familial relations and kinship. Most pregnant women easily accept and undergo noninvasive tests (such as DNA tests from the mother's blood or ultrasound); some also accept invasive testing (such as amniocentesis or chorionic villus sampling), although it entails a risk for the fetus. The availability and routinization of noninvasive genetic prenatal testing (NIPD) has the cultural effect that pregnancy is seen increasingly as an inherently risky and perilous process. Accordingly, pregnant women are inclined to disengage from their fetus and keep the relation "tentative" until tests come back negative (Katz Rothman 1993; Mitchell 2001; Kukla 2008). It also has the potential to raise new questions of economic and social impact, of eugenics and of "wrongful birth"—that is, the objection to children being born that will unduly burden their parents and the health care system (de Jong et al. 2010; Greely 2011; Schmitz 2013). NIPD has the potential to change not only the realm of prenatal diagnosis but also procreation and parenting practices in general, because it opens the way for new and cheaper technologies (including next-generation sequencing, microarrays, and exome sequencing) to test human life. It will normalize the idea of the genetic transparency of the future generation for the broader public and not just for high-risk families. Ontologically, in the context of reprogenetics, pregnancy and birth are increasingly becoming an issue about future health.

The consequences are that parenthood becomes an option, while the medical practitioner and expert become part of family planning, including the decision for or against a pregnancy and birth after prenatal tests have been performed. Prenatal and perinatal practices become a discourse of risk and medical options. The medical perspective focuses on the embryo and the fetus, on its health and physiological development. From the perspective of becoming a mother, this medical view is part of her context of being pregnant; yet she also has other perceptions. At the beginning of pregnancy, women would talk about themselves as being pregnant and how they feel about it. Later, often even before women feel the child moving, they refer to it as "my child," or "my son" or "my daughter." For future parents, the fetus is already their "baby." Lynn Morgan observes that often the "social birth" precedes the biological birth into the world (1996: 59).

From the Private to the Public

Unborn children are often already given a name and today can be seen in ultrasound photos, which are then shown around to family and friends. The unborn child is moved into the visual sphere. A "pregnant turn" takes over the becoming-parents. It consists of concern about the results of diagnostic tests, about the aestheticization of the pregnant body as a sign of a new lifestyle, which can be called a "pregnant icon" (Matthew and Wexler 2000: 195). Both trends break the taboo on the visibility of what was once private and hidden. Representations of the pregnant belly could already be found in the early 1990s, and then achieved greater prominence with such pregnant cover girls as Brooke Shields on the cover of *Vogue* in 2003, Christina Aguilera in 2008 on *Marie Claire*, and Claudia Schiffer in 2011, also on the cover of *Vogue*. They prepare the shift from the private to the public of the inside of the womb. And, insofar as the prenatal and postnatal female bodies are supposed to look alike, they support the social idea of strategic control and power over one's body. Heidi Klum, the queen of the "after-baby body," illustrated this by modeling for Victoria's Secret five weeks after the birth of her son.

The medical attempt to make the uterus transparent has its counterpart in the social shift from the private to a public "fetal celebrity" (Barlant 1997: 124; Nilsson 2006). Examples for this are "schools" for prenatal education with music and foreign languages or medically non-indicated four-dimensional ultrasound for public viewing of the fetus. This shift from the private to the public, from the inner to the external realm, can also be seen on YouTube or television, in the two-hour *Birth Night Live* broadcast from a British hospital's maternal unit or the U.S. series *Birth Day* and *Deliver Me!*, which follow women through late pregnancy and childbirth (Tyler 2009). The interested spectator can view countless homemade video clips, especially about natural and unassisted births. These films have thousands of views after only a few months online. The visualization of birth, which was traditionally hidden in the private sphere, has become a public event that can be shared, commented on, and even be clicked with likes and dislikes. Birth and the beginning of life are now part of the media and public world. You are born and your *pictures* are already waiting for you.

Conclusion

Birth marks a beginning and relation not only for the infant but also for the persons to whom it belongs and who care for it. The concept of birth implies a transition from the prenatal to the postnatal existence of the infant in regard to its relations. The understanding of these relations between the pregnant women and the fetus, and the mother, father, siblings (and others), and the child is part of an interdisciplinary and historically changing discourse and its interpretations, experiences, and practices in the private and public, medical, social, and cultural realms.

References

ACNM (American College of Nurse-Midwives), MANA (Midwives Alliance of North America), and NACPM (National Association of Certified Professional Midwives (2012) *Supporting Healthy and Normal Physiologic Childbirth: A Consensus Statement by ACNM, MANA, and NACPM*. http://mana.org/pdfs/Physiological-Birth-Consensus-Statement.pdf (accessed 7/7/2015).
Arendt, H. (1958) *The Human Condition*, Chicago: University of Chicago Press.
Arendt, H. (1968) *Between Past and Future*, New York: Viking Press.
Arendt, H. (1970) *On Violence*, New York: Harcourt Brace Jovanovich.
Barlant, L. (1997) *The Queen of America Goes to Washington City: Essay on Sex and Citizenship*, Durham: Duke University Press.
Cavarero, A. (1997) "Schauplätze der Einzigartigkeit" in S. Stoller and H. Vetter (eds.) *Phänomenologie und Geschlechterdifferenz*, Wien: WUV, 207–226.
Cavarero, A. (1998) "Birth, Love, Politics," *Radical Philosophy: A Journal of Socialist and Feminist Philosophy*, 86, 19–23.
Childbirth Connection (2015) *Cesarean Section: Why Is the National U.S. Cesarean Section Rate So High?* New York: Childbirth Connection. http://www.childbirthconnection.org/article.asp?ck=10456 (accessed 11/14/15).
Conrad, P. (1992) "Medicalization and Social Control", *Annual Review of Sociology*, 18, 209–232.
Conrad, P. (2007) *The Medicalization of Society: On the Transformation of Human Conditions into Treatable Disorders*, Baltimore: Johns Hopkins University Press.
Davis-Floyd, R. (1994) "Culture and Birth: The Technological Imperative," *The Birth Gazette*, 11(1), 24–25.
De Jong, A., Dondorp, W. J., de Die-Smulders, C. E. M., Frints, S. G. M., and de Wert, D. W. R. (2010) "Non-Invasive Prenatal Testing: Ethical Issues Explored," *European Journal of Human Genetics*, 18, 272–277.
Duden, B. (1993) *Disembodying Women: Perspectives on Pregnancy and the Unborn*, translated by Lee Hoinacki, Cambridge/London: Harvard University Press.
Freud, S. (1921) *Dream Psychology: Psychoanalysis for Beginners*, New York: The James A. McCann Company; Bartleby.com, 2010, www.bartleby.com/288/ (accessed 7/3/2015).
Fromm, Eric (1956) *The Art of Loving*, New York: Harper Row.
Fuchs, T. (2011) "Body Memory and the Unconscious" in D. Lohmar, D. and J. Brudzinska (eds.) *Founding Psychoanalysis: Phenomenological Theory of Subjectivity and the Psychoanalytical Experience*, Dordrecht, Netherlands: Kluwer, 69–82.
Greely, H. T. (2011) "Get Ready for the Flood of Fetal Gene Screening," *Nature*, 469, 289–291.

Guenin, L. M. (2008) *The Morality of Embryo Use*, Cambridge: Cambridge University Press.
Heidegger, M. (1996) *Being and Time*, translated by Joan Stambaugh, New York: SUNY.
Hobbes, T. (1841) *Philosophical Rudiments Concerning Government and Society* [1651] (the English version of his *De Cive* [1642]), *The English Works of Thomas Hobbes of Malmesbury*, London: John Bohn.
Howes, M. (2007) "Maternal Agency and the Immunological Paradox of Pregnancy" in H. Kincaid and J. McKitrick (eds.) *Establishing Medical Reality: Essays in the Metaphysics and Epistemology of Biomedical Science*, Dordrecht, Netherlands: Springer, 179–195.
Husserl, E. (1977) *Cartesian Meditations: An Introduction to Phenomenology*, translated by D. Cairns, The Hague: Martinus Nijhoff.
Husserl, E. (1982) *Ideas Pertaining to a Pure Phenomenology and to a Phenomenological Philosophy—First Book: General Introduction to a Pure Phenomenology* (= Ideas) [1913], translated by F. Kersten, The Hague: Martinus Nijhoff.
Irigaray, L. (1985) *Speculum of the Other Woman*, translated by G. C. Gill, Ithaca, NY: Cornell University.
Katz Rothman, B. (1993) *The Tentative Pregnancy: How Amniocentesis Changes the Experience of Motherhood*, New York: Norton.
Kukla, R. (2005) *Mass Hysteria: Medicine, Culture, and Mothers' Bodies*, Lanham, MD: Rowman and Littlefield.
Kukla, R. (2008) "Measuring Mothering," *International Journal of Feminist Approaches to Bioethics*, 1(1): 67–90.
Kukla, R. (2011) Pregnancy, Birth, and Medicine, *Stanford Encyclopedia of Philosophy*. http://plato.stanford.edu/entries/ethics-pregnancy/ (accessed 7/7/2015).
Lake, N. (2012) "Labor, Interrupted Cesareans, "Cascading Interventions," and Finding a Sense of Balance," *Harvard Magazine*, Nov./Dec. http://harvardmagazine.com/2012/11/labor-interrupted (accessed 11/14/15).
Leboyer, F. (2009) *Birth without Violence*, Rochester: Healing Arts Press; 4th Edition.
Lindemann Nelson, H. (1994) "The Architect and the Bee: Some Reflections on Postmortem Pregnancy," *Bioethics*, 8(3), 247–267.
Lyerly, A. D. (2006) "Shame, Gender, Birth," *Hypatia*, 21(1), 101–118.
Matthew, S. and Wexler, L. (2000) *Pregnant Pictures*, London and New York: Routledge.
Merleau-Ponty, M. (2012) *Phenomenology of Perception*, translated by D. A. Landes, London and New York: Routledge.
Mitchell, L. M. (2001) *Baby's First Picture: Ultrasound and the Politics of Fetal Subjects*, Toronto: University of Toronto Press.
Moffett, A. and Loke, Y. W. (2004) "The Immunological Paradox of Pregnancy: A Reappraisal," *Placenta*, 25(1), 1–8.
Morgan, L. M. (1996) "Fetal Relationality in Feminist Philosophy: An Anthropological Critique," *Hypathia*, 11, 47–70.
Mullin, A. (2005) *Reconceiving Pregnancy and Childcare: Ethics, Experience, and Reproductive Labor*, New York: Cambridge.
Nilsson, L. (2006) *Life*, London: Jonathan Cape, Random House.
Osterman, M. J. K. and Martin, J. A. (2014) "Trends in Low-risk Cesarean Delivery in the United States, 1990–2013," *In National Vital Statistics Reports* 63(6). http://www.cdc.gov/nchs/data/nvsr/nvsr63/nvsr63_06.pdf (accessed 11/14/15).
Plato (2000) *The Republic*, translated by Benjamin Jowett, London: Dover.
Plato (1980) *The Symposium*, Greek text with commentary by Kenneth Dover, Cambridge: Cambridge.
Purdy, L. (1994) "Case Study: The Baby in the Body," *Hastings Center Report*, 24(1), 32.
Purdy, L. (2001) "Medicalization, Medical Necessity, and Feminist Medicine," *Bioethics*, 15(3): 248–261.
Rank, O. (1993) *The Trauma of Birth*, translated by E. James Lieberman, New York: Dover.
Renggli, F. (2001) *Der Ursprung der Angst: Antike Mythen und das Trauma der Geburt*, Düsseldorf/Zürich: Patmos.
Rich, A. (1976) *Of Women Born: Motherhood as Experience and Institution*, New York: Norton.
Roazen, P. (1974) *Sigmund Freud und sein Kreis: Eine biographische Geschichte der Psychoanalyse*, Hersching: Manfred Pawlak.

Schmitz, D. (2013) "A New Era in Prenatal Testing: Are We Prepared?" *Medicine, Health Care and Philosophy*, 16(3), 357–364.

Schües, C. (1997) "The Birth of Difference," *Human Studies: A Journal for Philosophy and Social Sciences*, 20(2), 243–252.

Schües, C. (2008, 2nd ed. 2016) *Philosophie des Geborenseins*, München, Freiburg: Alber.

Schües, C. (2011) "Menschenkinder werden geboren. Dackelwelpen geworfen—Die Normativität der leiblichen Ordnung" in P. Delhom and A. Reichhold (eds.) *Normativität und Leiblichkeit*, Freiburg: Alber, 73–95.

Thoné, A. (1998) "A Radical Gift: Ethics and Motherhood in Emmanuel Levinas' Otherwise than Being." *Journal of the British Society for Phenomenology*, 29(2), 116–131.

Toombs, S. K. (2001) "Introduction: Phenomenology and Medicine" in S. K. Toombs (ed.) *Handbook of Phenomenology and Medicine*, Dordrecht / Boston / London: Kluwer, 1–26.

Tyler, I. (2000) "Reframing Pregnant Embodiment" in S. Ahmed, J. Kilby, C. Lury, M. Mcneil, and B. Skeggs (eds.) *Transformations: Thinking Through Feminism*, London: Routledge, 288–302.

Tyler, I. (2009) "Introduction: Birth," in *Feminist Review* 93, 1–7.

Ullrich, C. (2012) *Medikalisierte Hoffnung? Eine ethnographische Studie zur reproduktionsmedizinischen Praxis*, Berlin: Transcript.

Vetlesen, A. J. (1995) "Relations with Others in Sartre and Lévinas: Assessing Some Implications for an Ethics of Proximity2," *Constellations*, 1(3), 358–382.

Winnicott, D. W. (1990) *Human Nature*, London: Routledge.

Young, I. M. (1990) "Pregnant Embodiment: Subjectivity and Alienation," *Throwing like a Girl and Other Essays in Feminist Philosophy and Social Theory*, Bloomington: Indiana University Press, 137–156.

Further Reading

Davis-Floyd, R. E. (2004) *Birth as an American Rite of Passage*, Berkeley / Los Angeles: University of California Press is a lively written analysis about the technocratic model of birth, of its cultural variations and women experiences.

Katz Rothman, B. (1991) *In Labor. Women and Power in the Birthplace*, New York / London: Norton. A social and political analysis how childbirth is perceived and managed in America.

O'Byrne, A. (2010) *Natality and Finitude*, Bloomington: Indiana University Press. (The book discusses the relevancy for a philosophical understanding of human finitude.)

Schües, C. (2008, 2nd ed. 2016) *Philosophie des Geborenseins*, München, Freiburg: Alber. (An insightful written book about role of birth in the history of philosophy and about being-born in a phenomenological and existential perspective.)

Shchyttsova, T. (ed.) (2012) *In statu nascendi. Geborensein und intergenerative Dimension des menschlichen Miteinanderseins*, Nordhausen: Bautz. Collection of essays in philosophy, especially also phenomenology, focusing on birth, generativity and inter-generative relations, such as promises and education.

11
DEATH

Steven Luper

What is death? How can its occurrence be detected? These questions are interrelated, as we can reliably detect death only if we know what it is. The standard view is that dying consists of ceasing to be alive. However, this, the *loss of life* account of death, seems to be open to various challenges. For example, it is sometimes rejected on the grounds that a living thing might cease to exist, losing life thereby, without dying. We will need to determine whether this criticism has merit. Moreover, in order to clarify what death is for you and me, we must consider what type of thing we are, since what it takes to end a creature's existence depends on what that creature is. Many theorists claim that you and I are persons, where being a person entails having various sorts of psychological features, such as the capacity for self-awareness. If these theorists are correct, then the loss of these features would end our existence, and the question will arise as to whether that loss would constitute death.

In this chapter, the loss of life account will be defended and clarified. In order to illustrate its implications, the case of severe dementia will be discussed. Towards the end of the chapter, criteria for detecting death will be considered.

What Life Is

In order to clarify the nature of death, we can begin by discussing what it is for something to be alive. To that end, it will be useful to consider typical examples of things that are alive, such as trees, cats, amoebas, and human animals, and ask what imbues these organisms with life.

The answer, it might seem, is their capacity to develop and maintain themselves using processes such as respiration, photosynthesis or chemosynthesis, cell generation, and waste removal. These may be called vital processes. At least as a good approximation, we might say that being alive consists of having the capacity to deploy processes that are saliently similar to these.

However, this *viability account of life* is often rejected on the grounds that things cease to be alive while in "suspended animation." Consider that infertility clinics often store embryos for extended periods of time by freezing them. While frozen, the embryos are in suspended animation (or simply "suspended"), in the sense that their vital processes are brought to a halt, or at least very nearly so. Other organisms, too, may be suspended. For example, seeds and spores are, for practical purposes, suspended, as are dehydrated water bears. (Water bears, or tardigrades, are tiny animals that have evolved the ability to survive drought for decades by drying up and bringing their metabolism virtually to a stop.) Despite being suspended, these organisms do not lose the *capacity* to deploy vital processes, for their vital processes may be restarted—by wetting the seeds and water bears and by warming the embryos. Hence, these organisms are alive according to the viability account of life, which says that the capacity to deploy vital processes suffices for life. If they are *not* alive, as some theorists have suggested (e.g., Feldman 1992; Belshaw 2009; Gilmore 2013; DeGrazia 2014), then we must reject the viability account. We will also have to countenance the possibility that something can cease to be alive without ceasing to exist, for clearly seeds are still seeds while dormant, an embryo is still an embryo while frozen in an IVF facility, and a water bear continues its existence during its dry hibernation. What is more, on the assumption that suspended organisms are not alive, but also not dead, we will have to reject the loss of life account of death.

Before we decide whether to abandon the viability account, let us consider more carefully how suspension changes an organism. It is one thing to have the capacity to deploy vital processes and another to deploy them. Suspended organisms have this capacity but are not making use of it. To mark this distinction, it is useful to introduce two pairs of contrasting terms. The first pair is *vital* and *nonvital*: when something is employing its capacity for vital processes, let us say that it is vital; when it is not employing its capacity, let us say that it is nonvital. The second pair is *viable* and *unviable*: we can use the former to refer to the condition of something that is capable of engaging vital processes and the latter for the condition of something that has lost this capacity.

With this distinction in hand, we can explain why some theorists are reluctant to say that suspended organisms are alive: such theorists are focusing on the fact that these organisms are not vital. However, their concerns can be accommodated without rejecting the viability account. We need only describe cases of suspended animation by saying that the individuals concerned are nonvital even though still alive, still viable. In this way we can also accommodate the tendency among biologists to characterize things like spores and seeds as "alive" until these are no longer able to grow into plants—that is, they are alive until they are no longer viable.

What Death Is

If life consists of viability, it remains plausible to say that death is the loss of life. However, this view can be challenged, and we will want to see if it can withstand criticism. We will consider two objections. The first is that a living thing could cease to exist without dying. The second is that, unlike death, loss of life may not be irreversible or permanent. We can begin with a more precise statement of the loss of life account itself.

Combined with the view that life is viability (as opposed to vitality), the loss of life account of death can be stated as follows:

> Dying is the loss of a thing's life—the loss of its viability.

Of course, it is one thing to die and another thing to be dead. A thing dies in virtue of losing viability. It dies at the time of that loss. It is dead after that loss. We can add these details to the loss of life account:

> Dying is the loss of a thing's life—the loss of its viability. A thing dies at the time it loses its viability. It is dead afterwards.

Now, this account would be unacceptable if living things could cease to exist without dying, since any living thing that ceases to exist ceases to be alive. But does anything cease to exist deathlessly? According to Jay Rosenberg (1983, p. 22) and Fred Feldman (1992, pp. 68–69), this happens frequently. For example, it occurs when an amoeba reproduces. Dividing into two amoebas ends its existence, but division is a deathless exit. Feldman thinks there are other deathless exits as well; it happens, for example, when two chlamydomonas, which are single-celled algae, fuse to produce a zygote.

However, there is a way to respond to Rosenberg and Feldman: instead of saying that, in ceasing to exist, some things lose life in ways that are deadly and others lose life in ways that are not deadly, it seems reasonable to say that the apparent exceptions, such as division and fusion, are unusual ways of dying, not ways of escaping death altogether. Typically, death involves the destruction of an individual's capacity to engage vital processes. For example, when a person is fatally stabbed, the circulatory system breaks down and its various components, such as red blood cells, are destroyed. By contrast, in division and fusion, the bits of an organism that are constitutive of its vital processes are transferred, as it were, to its children. Although this is an unusual way to lose life, it still entails the loss of life (since the original organism or organisms cease to exist), and, as such, it is reasonably treated as a death.

Now consider the second objection to the loss of life account, which was that loss of life may not be permanent or irreversible, unlike death itself. Perhaps an individual's viability could be restored after it—and not merely vitality—was lost for a time.

To see why such restoration seems theoretically possible, imagine a machine, the *corpse reanimator*, that repairs corpses. It corrects flaws in DNA, restores broken cell membranes, and so forth, making repairs down to the molecular level, and restarting all vital processes. It is so thorough as to return an individual's corpse to much the same state as that individual was in before she lost viability. It seems reasonable to say that the viability of the individual whose corpse is repaired will have been restored. If this is correct, then loss of viability may not be permanent (Luper 2002, 2009, pp. 44–48).

However, what we have just concluded about viability is a problem for the loss of life account only on the assumption that death is irreversible, and that assumption may be convincingly challenged. Consider the corpse reanimator again. If such a device could indeed restore the *viability* of the individual whose corpse is repaired, then it could restore the *life* of that individual, thus reversing her death.

Assuming that it is even theoretically possible to restore life, we would need to refine the last clause of the loss of life account as follows:

> Dying is the loss of a thing's life—the loss of its viability. A thing dies at the time it loses its viability. It is dead at all times after it dies *except while its viability has been regained*.

Moreover, we will want to reject accounts of death that imply that it is necessarily permanent. For example, we should reject the view that dying is the permanent or irreversible loss of viability.

What We Are

The loss of life account implies that you and I will die if we cease to exist, as ceasing to exist entails losing life. Hence, to further clarify what death involves for creatures like us, we will need to clarify what we are.

So, what are we, and under what conditions do creatures like us remain in existence? Few issues are more controversial; it is even difficult to formulate our question in a way that is not question-begging. (For example, if we state it as, What is it to be a person?, we will have presupposed that we are persons, which, as we will see, some theorists deny.) To facilitate the discussion, I will coin the term "persimal" to refer to what you and I are, whatever that turns out to be. So the questions before us are, What are persimals? and What conditions are necessary and sufficient for the continued existence of persimals?

I will discuss three of the most plausible accounts of persimalhood. The first is the simplest: to be a persimal is to be a human animal. This view is known as *animalism* (Snowdon 1990; Olson 2007). The second is that persimalhood consists of being a Lockean person. To be a Lockean person, in turn, is to have the capacity for self-awareness. (According to Derek Parfit, the leading proponent of personism, "to be a person, a being must be self-conscious, aware of its identity and its continued existence over time" [1984, p. 202; cf. 2012, p. 6].) We can call this view *personism*. A third view is that persimals are minds, and being a mind consists of having the capacity for consciousness. Call this view *mindism* (McMahan 2002). All Lockean persons are minds, but mindism differs from personism since some minds, such as those of giraffes and human infants, are not self-aware. Personism and mindism are members of a family of accounts which say that persimalhood consists solely in the possession of some mental feature. Accounts that take this stance may be referred to as *mentalist*. Mentalists say that a persimal is not an animal but instead is "realized in" an animal, in much the same way that a statue is realized in a particular parcel of metal. The statue is not identical to the metal, as it may be destroyed by being melted, yet melting does not destroy the metal.

The three approaches have very different implications concerning our continued existence. Personism says that we continue to exist by virtue of remaining the same Lockean person—the same self-aware being. According to Parfit, we remain the same Lockean person over an interval of time only if we have a substantial degree of psychological continuity over that time (more about this later). Mindism says that our remaining in existence is a matter of being the same mind, and that we remain the same mind as long as the part of the brain that makes consciousness possible remains intact. According to animalism, we remain in existence over time by being the same animal over time. Assuming that this animal need not have psychological features to exist, we could survive the loss of psychological continuity and the capacity for consciousness.

Dementia and Death

Because it is entailed by ceasing to exist, we can illustrate what death might involve by considering how proponents of the three accounts of persimalhood would respond to the question, Is it fatal to incur the sort of damage that severe dementia does to its victim's cognitive abilities? We can begin with a few words about dementia and its effects on awareness and consciousness.

Dementia is a disorder in which a person's cognitive abilities are compromised enough to substantially disrupt normal conduct. The most common cause is Alzheimer's disease, in which tangles or clumps of proteins accumulate in the brain, resulting in a progressive, gradual loss of cognitive capacities. Another form of dementia results when the brain is damaged by obstructions to its blood supply. Dementia may be caused by other things as well, such as thyroid illness and certain vitamin deficiencies.

In damaging cognitive abilities, dementia gradually destroys the capacity for awareness. Awareness is a form of consciousness. An individual is in a state of consciousness if and only if there is something it is like to be in that state (Nagel 1974). By contrast, awareness is a form of consciousness that is directed at some more or less specific object, such as an event or a thought. In particular, *self*-awareness will involve the awareness of some aspect of the self. In degrading awareness, dementia gradually destroys self-awareness.

Eventually, dementia can give way to a persistent vegetative state (Horne 2009). The term *vegetative state* (VS) refers to the condition of individuals who are incapable of voluntary interaction with the environment but who retain, to some degree, a sleep/wake cycle and respiratory or digestive functioning. When individuals remain in a vegetative state for at least a month, they are said to be in a *persistent* vegetative state (PVS). A more recent term for VS is *unresponsive wakefulness syndrome*—"unresponsive" to indicate that movement is limited to reflex behavior and "wakeful" to indicate that eye opening persists. To mark the condition of individuals who display intermittent and minimal voluntary behavior, the term *minimally conscious state* was introduced (Gosseries et al. 2011).

Imaging shows signs of brain function in some patients diagnosed with PVS (Horne 2009), but when their vegetative state persists long enough, they are presumed to have lost the capacity for consciousness. Nevertheless, dementia can progress quite slowly, leaving some forms of awareness intact while ending others, and although it is difficult to determine whether a severely demented individual is aware of anything, some recent research suggests that many retain some minimal forms of awareness well into the final stages of dementia (Clare 2010).

Mentalism implies that persimals who are the victims of lengthy persistent vegetation are dead, as they have lost the capacity for consciousness in any form. Because they distinguish between these persimals and the human animals in which the persimals were once "realized," mentalists will grant that the animals may survive in a vegetative state, languishing in a hospital bed, long after the persimal has ceased to exist. By contrast, animalists would say that the persimal *is* that animal, and since it survived the injury, so did the persimal.

However, some forms of mentalism imply that dementia can be fatal well before the onset of persistent vegetation. It can be fatal even if its victim retains minimal self-awareness. According to Parfit, persimals are Lockean persons, but Lockean persons remain in existence from one time to another only if their mental life displays strong psychological continuity over that time. The requisite continuity consists of "overlapping chains of strong connectedness." Strong connectedness exists just if, over each day, there are "at least half the number of direct connections that hold, over every day, in the lives of nearly every actual person" (1984, p. 206). This formulation is not entirely clear, but it seems reasonable to conclude that if dementia is severe enough, it will eventually disrupt the psychological continuity that is required for the continued existence of a Lockean person. At that point, given Parfit's assumptions, the Lockean person would perish.

Is it plausible to say, with Parfit, that dementia kills people by ending the strong psychological continuity of their mental life? Perhaps, but the consequences of Parfit's view are exceedingly strange. To see why, consider a Lockean person we will call *First*. Suppose that First is "realized in" a human animal we will call *Annie*, who enjoys normal cognitive abilities. Suppose that Annie develops dementia, which eventually drastically diminishes her cognitive abilities. As a result of Annie's decline, First's mental life no longer displays strong continuity. At that point First dies. However, it does not follow that no Lockean person is "realized in" Annie. After dementia takes its toll, Annie's cognitive processes may continue to give rise to some mental life. In fact, that mental life might include the capacity for self-awareness, in which case a Lockean person will be "realized in" Annie after First's existence ends. Call her *Second*. It is hard to believe that dementia might result in the demise of First and the advent of Second, but a further consequence of Parfit's view is even more bizarre, namely that a Lockean person like Second, whose mental life is too impoverished to display strong continuity, cannot remain in existence from one time to another. Hence, severe dementia not only kills a Lockean person, but it may also result in a series of distinct Lockean persons, each replaced by the next, and none lasting beyond the moment.

To avoid these counterintuitive results, mentalists could reject Parfit's assumption that persimals persist only if their mental life displays strong psychological continuity. Mentalists might instead allow that any sort of psychological continuity suffices (McMahan 2002). Alternatively, they could abandon personism in favor of mindism. Mindism says that persimals continue to exist if they retain the capacity for consciousness, so their survival does not require any sort of self-awareness or psychological continuity. Dementia is fatal only if it results in the loss of the capacity for consciousness.

Suppose instead we accept the animalist claim that our existence continues as long as we remain the same animal. Then we can conclude that neither the loss of self-awareness nor consciousness is, by itself, fatal, since neither ends the life of a human animal, assuming that animal retains its capacity to engage its vital processes.

The upshot is that we must resolve the difficult issue of *what we are* before we can answer our question about the fatalness of afflictions, such as dementia or even brain death, that degrade our mental faculties. Mentalism implies that sufficient degradation will indeed be fatal. At the same time, the proponents of mentalism will agree that afflictions of the psyche are not fatal to the human animals in which we are "realized." But given animalism, we *are* those animals, and as such we could survive even the loss of our minds.

Criterion for Death

Our conclusions have powerful practical consequences for the practice of medicine. The loss of life account, if correct, tells us what death is, but as long as we are not sure what we are, we will not always know whether a condition would constitute the loss of life. *Usually* we can know whether death has occurred, for certain features indicate death regardless of whether animalism or a form of mentalism is correct—features such that any of us with such features are dead. Some such features, such as putrefaction and decapitation, are also easy to detect, but they are of little help because they are either rare (decapitation) or appear too late (putrefaction). Another feature that suffices for death is far more useful, namely brain death. Unfortunately, as we will see, we cannot identify a feature that is clearly necessary for death—a feature such that anyone who is dead has that feature. Hence, we cannot formulate adequate criteria for death: conditions that are necessary *and* sufficient for death and that are relatively easily detected.

Brain death occurs just if all of the brain, except perhaps for bits that are peripheral to its functioning, has ceased to be alive. By focusing on this feature, we can state a criterion for (persimal) death, which we may call the *brain death criterion*. It can be formulated quite simply:

An individual is dead if and only if that individual's (entire) brain has died.

This criterion is closely related to the standard for death that is most widely adopted in law in the United States, as set forth by the Uniform Determination of Death Act, according to which "an individual who has sustained either (1) irreversible cessation of circulatory and respiratory functions, or (2) irreversible cessation of all functions of the entire brain, including the brainstem, is dead" (President's Commission 1981, p. 119). There are important differences between our brain death criterion and the standard formulated in the Death Act. For example, our brain death criterion makes no use of the term "irreversible," in view of the theoretical possibility that death could be reversed. Also, the standard set out in the Death Act disjoins two conditions that mark death, saying that either suffices. One is irreversible cessation of brain functions, and the other is the traditional mark of death, namely the irreversible cessation of circulation and respiration. The standard accepted in the UK and in several European nations is similar, but there *brain death* is defined as the "permanent functional death of the brain stem" (Pallis 1982, p. 1488).

Meeting the brain death criterion clearly suffices for death given mentalism, as we lose all of our mental capacities when the brain dies. It is more difficult to show that it suffices given animalism. Still, we can devise a plausible argument by consulting the standard defense of the brain death criterion (offered, e.g., by Pallis 1983; DeGrazia 2014). Putting aside the issue of reversibility, the standard reasoning seems to be that (a) any organism dies when it can no longer function as an integrated whole, since it can *exist* only when able to function as an integrated whole (without that ability it loses its distinctive capacities); and (b) a human animal can function as an integrated whole only if equipped with a functioning brain; so (c) the human animal dies when its brain can no longer function. Given that persimals are human animals, it would follow that we die when the brain loses its capacity to function.

However, this defense faces an obstacle: don't human animals spend a significant amount of time in the womb before they develop a brain? If they are alive in the womb, why can't they be kept alive on life support after the brain dies? To overcome this obstacle, we might develop one of two lines of thought. We might argue that bona fide, full-fledged human animals do not come into existence until they develop functioning brains (perhaps because that is when they are able to function as wholes), and the brain's capacity to function is essential to the continued existence of a (proper) human animal, hence if the brain is lost, the human animal dies. On this view a human animal cannot die before it develops a brain since nothing can die prior to its own existence. (What might die is a mass of tissue that has not yet formed itself into a human animal.) Alternatively, we could claim that what it takes to be a human animal varies with that animal's stage of development. Accordingly, we might argue that (a) a creature is a human animal only if it develops certain specified features according to a certain timeline (unless its development is interrupted); (b) one of these features is the possession of a brain; and (c) once the brain develops, its capacity to function is requisite to the animal's continued existence (even though a functioning brain is not requisite at earlier stages of development); therefore, (d) if the brain's functionality is lost, then the human animal dies. On this latter view, a human animal dies if its brain dies, even though it took time for it to develop a brain.

Although brain death may suffice for death, the claim that it is necessary is disputable, because it is not clear that the entire brain must cease to function for death to occur. Consider the issue from the standpoint of animalism. It would be reasonable for animalists to take the following position: if all of an animal's body except for its brain dies, that animal has ceased to exist, even if the brain itself remains alive for a while, perhaps on some form of artificial life support. This is to say that an animal cannot be pared down to its brain, as a brain is not an animal. Hence, if we are human animals, we could die even though the brain lives on. (Suppose that, contrary to what has just been said, an animal can indeed be pared down to its brain. Presumably, there are substantial components of the brain, such as the cerebrum, that are not animals. So in theory a human animal would die if all of it except for the cerebrum is destroyed. This death would fail to meet the brain death criterion.)

Although animalists can claim that death is consistent with the survival of the entire brain, mentalists will say that death is consistent with the survival of some of the brain. Consider personism. It is reasonably clear that some of the brain can survive the demise of a Lockean person, since the capacity for self-awareness can be lost if various components of the brain die but the entire brain does not. Thus personists are likely to favor what we might call a *self-awareness-centered* criterion for death, designed to detect the loss of the capacity for self-awareness. But this criterion will be difficult to specify, as the facts about which parts of the brain are salient are in dispute. Self-awareness is lost if the cerebrum dies, but it might also be lost if other parts of the brain die or malfunction.

Let us add that different forms of mentalism will consider different parts of the brain to be salient to death. Suppose that some part of the brain that is necessary for self-awareness can no longer function, say because of severe dementia or damage to the cerebrum. Suppose also that the capacity for consciousness remains. Then personists will say that death has occurred, yet mindists will disagree. Mindists will favor what is often called the "higher brain criterion" for death (it might be better to call it the *consciousness-centered* criterion, to bring out the contrast with the self-awareness-centered criterion). According to its proponents, death occurs if and only if the capacity for consciousness is irreversibly lost. (Mindists might drop the requirement of irreversibility, if persuaded that, theoretically, life could be restored.) This criterion is defended by several theorists, including Robert Veatch (1975) and Tristram Engelhardt (1975). Here, too, it will be difficult to formulate the criterion so that it is both accurate and relatively easy to detect, as the meaning of the term "consciousness" is controversial, and so are the facts about how consciousness is related to the various parts of the brain.

So, on the animalist and mentalist views of persimalhood, we might be dead even though the brain is not dead, yet there is no useful necessary condition for death that these approaches will agree on.

Conclusion

To live is to be viable, and to die is to lose life. In theory, death may be reversed, since in theory the loss of viability is reversible. Because death is the loss of life, a living thing dies when it ceases to exist. However, what it is for something to cease to exist depends on what that thing is, and what you and I are is controversial. Animalism says that we are human animals; on this view, we die when the animal to which we are identical ceases to exist. However, most theorists reject animalism. Personism, which says that we are essentially self-aware creatures, is more widely defended. A further alternative is mindism, the position that we are essentially conscious creatures. These views have very different implications. If severe dementia destroyed our capacity for self-awareness yet left behind an animal that was still capable of consciousness, personists would say that we have died, but mindists and animalists would disagree. If the dementia worsened, and destroyed the capacity for consciousness, mindists would now say that we have died, but animalists would say we survive as long as the human animal persists. Because of these disagreements, it is difficult to formulate a useful criterion for death. Animalists, personists, and mindists can accept the view that death occurs if the entire brain ceases to function, but none of them will say that death occurs *only* if the entire brain ceases to function.

References

Belshaw, Christopher, 2009, *Annihilation: The Sense and Significance of Death*, Montreal: McGill-Queen's, 80–100.
Clare, Linda, 2010, "Awareness in People with Severe Dementia: Review and Integration," *Aging and Mental Health* 14.1: 20–32.
DeGrazia, David, 2014, "The Nature of Human Death," in Steven Luper, ed., *The Cambridge Companion to Life and Death*, Cambridge: Cambridge University Press.
Engelhardt, H. T., 1975, "Defining Death: A Philosophical Problem for Medicine and Law," *Annual Review of Respiratory Disease* 112: 312–324.
Feldman, Fred, 1992, *Confrontations with the Reaper*, Oxford: Oxford University Press.
Gilmore, Cody, 2013, "When Do Things Die?" in Ben Bradley, Fred Feldman, and Jens Johansson, eds., *The Oxford Handbook of Philosophy of Death*, Oxford: Oxford University Press, 5–59.
Gosseries, Olivia, Marie-Aurelie Bruno, Camille Chatelle, Audrey Vanhaudenhuyse, Caroline Schnakers, Andrea Soddu, and Steven Laureys, 2011, "Disorders of Consciousness: What's in a Name?" *Neuro-Rehabilitation* 28: 3–14.
Horne, Malcolm, 2009, "Are People in a Persistent Vegetative State Conscious?" *Monash Bioethics Review* 28.2: 12.1–12.12.
Luper, Steven, 2002, http://plato.stanford.edu/archives/sum2002/entries/death/.

Luper, Steven, 2009, *The Philosophy of Death*, Cambridge: Cambridge University Press.
McMahan, Jeff, 2002, *The Ethics of Killing*, Oxford: Oxford University Press.
Nagel, Thomas, 1974, "On What It Is Like to be a Bat," *The Philosophical Review* 83: 435–50.
Olson, Eric T., 2007, *What Are We? A Study in Personal Ontology*, Oxford: Oxford University Press.
Pallis, Christopher, 1982, "ABC of Brain Stem Death," *British Medical Journal* 285: 1487–1490.
Pallis, Christopher, 1983, "Whole-Brain Death Reconsidered—Physiological Facts and Philosophy," *Journal of Medical Ethics* 9: 32–37.
Parfit, Derek, 1984, *Reasons and Persons*, Oxford: Oxford University Press.
Parfit, Derek, 2012, "We are Not Human Beings," *Philosophy* 87: 5–28.
President's Commission for the Study of Ethical Problems in Medicine and Biomedical and Behavioral Research, 1981, *Report*.
Rosenberg, Jay, 1983, *Thinking Clearly About Death*, Englewood Cliffs: Prentice-Hall, Inc.
Snowdon, Paul, 1990, "Persons, Animals, and Ourselves," in Christopher Gill, ed., *The Person and the Human Mind* (pp. 83–107), Oxford: Oxford University Press.
Veatch, Robert, 1975, "The Whole-Brain-Oriented Concept of Death: An Outmoded Philosophical Formulation," *Journal of Thanatology* 3: 13–30.

Further Reading

DeGrazia, D., 2005, Human Identity and Bioethics, Cambridge: Cambridge University Press.
Feldman, Fred, 1992, *Confrontations with the Reaper*, Oxford: Oxford University Press.
Luper, Steven, 2009, *The Philosophy of Death*, Cambridge: Cambridge University Press.
Youngner, S., Arnold, R., and Shapiro, R. eds., 1999, *The Definition of Death: Contemporary Controversies*, Baltimore, MD: Johns Hopkins University Press.

12
PAIN, CHRONIC PAIN, AND SUFFERING

Valerie Gray Hardcastle

The impact of pain on our lives is staggering. Chronic pain affects approximately 116 million adults in the United States annually, and this figure does not include people in long-term care facilities, the military, or prison. It costs up to $635 billion a year in (largely ineffective) health care and lost wages (Committee on Advancing Pain Research, Care and Education 2011). This is more than the costs for cancer, heart disease, and diabetes combined (Pizzo and Clark 2012).

Back pain in particular is pernicious. The recent Global Burden of Disease Study estimates that back pain is one of the top ten conditions that cause disability or early death (Vos et al. 2012). The chance that a citizen in an industrialized nation will suffer low back pain in his or her lifetime is somewhere between 60% and 70% (Kaplan et al. 2013). This pain causes more time to be taken off from work than any other disease or injury, which in turn exacts a high economic cost on the individuals, their dependents, industry, and governments (Andersson 1997, Taimela et al. 1997). In the United States alone, estimates are that 149 million workdays are lost each year due to back pain and its accompanying disability (Katz 2006, Rubin 2007).

But what is pain exactly? It seems to be something with which we all have intimate experience, but at the same time, trying to articulate its fundamental nature is challenging. Is it information about one's body? Is it an imperative to nurse injured tissue? Is it akin to an emotion? The answers are not obvious—and, not surprisingly, more complicated that many might assume.

1. False Things Said about Pain (and What Is True about Them)

Philosophers used to take pain as an easy and obvious example of a simple sensation when reasoning about what mental states are and how we know about them (e.g., Kripke 1980). They did not think too much about what pain was exactly, except as an obvious thing that we have all experienced and know well. Daniel Dennett (1978) put those assumptions to rest, as he adroitly pointed out how complicated pain processing is. He argued that our "folk" conceptions of pain—how we think about it in everyday life—were confused. Our ideas were so confused, in fact, that they were contradictory. For example, we believe that pains are inherently unpleasant, yet there are cases of pain that typically do not bother the holder (e.g., sore muscles in athletes). And we believe that pains are causally connected to tissue damage, yet quite often we can experience severe trauma to our bodies, yet not feel pain; this is a common occurrence in the battlefield. Nothing like what we intuitively think pain is actually exists (see also Hardcastle 1999). [As an aside, Kevin Reuter has done some interesting experimental work that suggests that philosophers are wrong in what they believe the "folk" think pain is (2011). His data demonstrate that people actually have sophisticated and complicated views about pain.]

In the intervening years, many philosophers have tried to analyze the fundamental nature of pain, attempting to identify a simple core underlying all of the complicated details. Most recently, Colin Klein (2015) has advocated that pain is just an internal command to protect a part of the body. Klein compares pain to other homeostatic drives we have, like hunger, thirst, or itch. These are all things that are directly motivating. To be hungry is to be motivated to eat. Similarly, to be in pain is to be moved to protect.

This view has the perhaps paradoxical implication that pains aren't actually painful, at least not at their core. We might have a secondary reaction of "ouch," but we often have pains of which we are not even aware, like those that force us to adjust our posture on an ongoing basis. Just as we can be hungry but not be bothered by it, so too can we be in pain but not be upset by it. (At least for pains that are not severe.) Pains are also uninformative. Having a pain does not tell you why the body part was damaged, how it was damaged, how it might be fixed, or sometimes (in the cases of referred pain) even where it was damaged.

The problem with this view is that not all of our homeostatic systems require a concomitant sensation. Think about the very important systems in us directed toward maintaining a constant temperature. When we get hot, we sweat. Our bodies do that automatically, without needing to run anything through our brains. If pain is a homeostatic system, should it be like being hungry, which does require our brains' cooperation to get us back to equilibrium, or is it like being warm, which does not?

If we consider the pain reflex, which is what jerks our hands off of a hot stove, we can see that that sort of pain processing—the sort that forces you to protect your tissues—happens outside of consciousness awareness. Indeed, the movement does not even require the brain to be enacted. Nociception, the dedicated peripheral response to noxious stimuli, travels to the spinal column, which then can direct the muscles to withdraw from the heat source. Only later do we feel any sensation, after we have moved our hand. Why does that happen? Why would we feel pain after the fact?

Even more confusing is the fact that if we "listen" to our pain, we appear to get into more trouble than if we don't. In a set of experiments conducted 30 years ago, Wilbert Fordyce took patients with back pain and divided them into two groups. One group was told to take some muscle relaxants and rest in bed (the common "cure" from back then, which we now know is very bad advice) until they felt better. The other group was told to take the same medication and rest in bed, but these participants were given a specific length of time to do both. In other words, in the first instance, patients were to use their pain as a guide for protecting themselves; in the second instance, patients were to ignore their pain in favor of following a doctor's orders (Fordyce et al. 1986). Fordyce learned that the second group fared much better, not just over the next few weeks, but even six months later. The second group felt better, was able to move with greater agility, and worked more effectively than the first. Letting your pain tell you what to do might not be a good idea. Indeed, we now know that often moving our bodies during healing, while increasing pain, also promotes recovery.

Klein appears to be right that pains motivate us to protect our bodily integrity (in some sense of "motivate" that we still need to flesh out), but he appears to be wrong that this is all pain is. His story does not explain all facets of pain processing. In particular, it does not explain why pain can and does lead us astray. We need a more nuanced view.

A second, very popular, view of pain in philosophy agrees that there is no single core to pain. Instead, pain is a duality; it has two cores, as it were. David Bain (2011) suggests that pain provides us information about damaged tissues (namely, that it is damaged) and is something that is inherently unpleasant. In contrast to Klein, who thinks that the unpleasantness of pain is a secondary reaction, Bain thinks that being motivated to protect the body is the secondary reaction; it is not part of pain proper. This view of pain lines up nicely with scientific perspectives on pain since the turn of the 21st century.

Historically, pain processing also was divided into two fundamental components: a processing stream that encodes for the emotional response to pain and a processing stream that analyzes location and intensity. This division goes back to Sir Charles Scott Sherrington's and Sir Henry Head's work in the early 1900s on the function of neurons and the somatosensory system (Sherrington 1906, Head and Holmes 1911). The importance of this division was underscored in the 1960s, as neuroscientists discovered two different spinothalamic tracts involved in pain processing: a medial (in the middle) pathway involved in processing affect and a lateral (on the outside) pathway involved in sensory encoding. This two-dimensional structure for pain is ubiquitous in the scientific literature, as well as in philosophical works (e.g., Aydede 2006).

But, important for this discussion, scientists have not made much progress in uncovering the details of the putative duality of pain, even though they have been working on it for over half a century. Indeed, as Vania Apkarian maintains, evidence for division remains "fairly weak" (2012: 6). There are three challenges to the hypothesis. First, there is a lot of crosstalk in our nervous system, starting in the spinal tracts and continuing on up through the cortex (Giesler et al. 1981). Such crosstalk prevents any particular area from being either just affective or just sensory and suggests that various areas in the spine and brain process both types of information together. Second, verbal reports (or other behavioral indicators) of pain intensity (a sensory parameter) and the magnitude of pain's unpleasantness (an affective parameter) are tightly correlated, which makes differentiating the two very difficult. And finally, some brain regions, like the insula, do not match up with either type of processing, for it is seemingly concerned only with magnitudes of responses and not with the emotional or sensory dimensions per se (Apkarian 2012).

The insula integrates magnitude information about a wide range of sensory input, including pain, and then feeds those calculations to the nucleus accumbens (NAc). The NAc calculates the potential reward values for its input and then sends its estimation to other limbic areas involved in analyzing rewards. These areas talk to our cortex, which concerns itself with planning and promoting actions (Cauda et al. 2011). This process, by the way, is not only what happens when we experience pain, but it is also what our brains do when we seek some intuitively pleasurable outcome. So what is going on? Are pains and pleasures two sides of the same coin?

It appears that our brain does in fact process both affective and sensory information when processing pain, as duality theories suggest, but these signals intermingle with one another instead of being divided into two separate circuits. In addition, when we look at which brain areas are involved in pain, we see much more than just these two types of information involved. Perhaps we should stop trying to make pain into something that has a simple nature and take a second look at what we know about the neurobiology of pain processing. As we shall see, doing so will also connect pain with pleasure directly.

Once we stop trying to divine pain's ultimate nature or divide pain into a few simple components, we make room for a new way of looking at nociceptive and pain processing data. For example, Apkarian's lab has found particular temporal sequences in the activation of particular brain areas that correspond to anticipating a painful stimulus, the perception of the pain itself, and relief after the pain is over (Baliki et al. 2010). The NAc and the anterior portion of the insula are most active just at the start of a thermal painful stimulus, but when subjects indicated that they actually felt pain, these regions quieted and the posterior portion of the insula and the anterior cingulate became the most active. Finally, as the stimulus was returning to baseline, the peri-acquiductal gray region became active and the posterior insula and anterior cingulate returned to baseline as well. We can thus identify three distinct and different networks that are sequentially activated during acute pain processing. Moreover, all of these regions are part of our motivational system that calculates and experiences risk and rewards.

This suggests a different approach to understanding pain processing: the pain system as a complex system of motivation. The basic idea is that normal acute pain, which usually indexes tissue damage, first prompts us to escape or avoid our current situation in order to minimize physical harm and then, when the pain stops, provides us with a sense of relief. Together, these two reactions—avoidance and the feeling of relief—not only protect our bodies but also contribute to our being able to predict the utility and costs of competing behavioral goals. That is, we learn from past pain experience how to evaluate the severity of threats, which then informs which goals we pursue and which behaviors we choose.

One consequence of adopting this perspective on pain processing is that it removes, or at least seriously diminishes, the strong distinction we normally assume between pain and pleasure. Intuitively, we believe that pain is unpleasurable and pleasure is, well, pleasurable. But if our pain system works because it ultimately gives us a sense of relief, then it too could be considered, perhaps, part of our pleasure-inducing system. The pain motivates us to do something, not just because we think it will hurt, but also because when we successfully do the something required to stop the pain, we feel more than the absence of pain. We also feel good. Taking this perspective means altering one's views on what a pain is. From this perspective, pain becomes a series of brain states, each of which has a different function in our cognitive economy.

A second consequence of adopting this perspective is that it denies that there is a brain circuit specific to pain processing (Iannetti and Mouraux 2010), contrary to popular theorizing (see, e.g., Brooks and Tracey 2005). Pain processing is simply part of our ongoing risk-reward calculations. Indeed, it might turn out that Melzack's (1989) original idea for a nonspecific, widely distributed network of neurons that crosses many areas of the brain and underlies pain perception might be correct after all. We have one reward calculator, as it were, and some actions of this calculator are what we normally consider to be pain processing. Again, the line between pain and pleasure becomes quite fuzzy, at least from the brain's perspective.

But things get even more complicated, for not all pains are created equal. So far, we have been discussing acute pains, which are triggered by a noxious event and then resolve. But some pains, like some cases of chronic back pain, do not readily (or ever) resolve. These pains, it turns out, are quite different from regular acute pains.

2. Chronic Pain Is Not Acute Pain That Does Not Stop

One might reasonably think that chronic pain is just an acute pain that does not go away, but this is not the case. Acute pains and chronic pains are quite distinct kinds of bodily events, with different impacts on the body and on one's psychology. To take one example of this: chronic pain is represented in different areas in the brain from acute pain, in part because the brain rewires itself when pain becomes chronic. For example, as discussed above, we know that the insula indicates the appearance and magnitude of acute pain. But activity in the medial prefrontal cortex (mPFC), a part of the limbic system, is correlated with chronic back pain. Interestingly, when chronic back pain patients also experience an acute pain, such as a burn on their arm, their insula lights up just as normal subjects' would under similar conditions (Baliki et al. 2006). Hence, people with chronic pain can experience two distinct types of pain—chronic and acute—and these differences are reflected in their patterns of brain activity.

The way the NAc is connected to the rest of the brain is different in chronic pain patients as well. In normal subjects, the NAc and the insula are highly interconnected, but in chronic pain patients, the NAc shifts its functional connectivity to the mPFC. That is, in normal subjects, when the NAc lights up, the insula does as well, but in chronic pain patients, when the NAc is activated, the mPFC responds (Baliki et al. 2010). And, the more chronic pain, the stronger the correlation between activity in the NAc and mPFC. Chronic pain shifts what would be a normal pain reaction to a more emotional one.

In addition, the nearly continuous activation of these areas caused by ongoing chronic pain has specific effects on the brain's firing patterns: the baseline level of activity in the insula and anterior cingulate is much higher in chronic pain patients than in normal controls. Such is not the case for other areas of the brain, not associated with pain, like the sensory cortices (Malinen et al. 2010). One hypothesis is that the near-continuous activation of the limbic areas shifts reward valuation, and these shifts in turn modulate learning and memory (Apkarian 2012). In other words, being in chronic pain fundamentally changes how one thinks, learns, remembers, and feels.

As a result of this rewiring, NAc activity differs between healthy subjects and chronic pain patients for instances of acute pain, especially during the "relief" phase of pain processing. Normal subjects' brain activity signals quite reliably that a reward is coming, but chronic pain patients' brains show activity that reflects a lack of predicted reward and perhaps even disappointment. Normal subjects would be happy and relieved that their pain is ending, but chronic pain patients would still have their chronic pain when the acute pain stimuli ends. Indeed, quite often an acute pain relieves or at least covers over the chronic pain. Under those circumstances, the chronic pain patient would be disappointed that the acute pain is ending.

This change in brain connectivity is a functional rewiring not specific to pain processing, for we see similar effects for monetary rewards in chronic pain patients—their brains show no real response to reward or loss (Apkarian 2012). In other words, chronic pain puts stress on our protective and adaptive motivational systems such that our motivational system fundamentally changes how it operates. And this change in functionality is so large that it distinguishes between normal subjects and chronic pain patients with an accuracy of more than 90% (Baliki et al. 2012). The evidence surrounding chronic pain processing indicates that it is intimately tied to our reward circuitry; chronic pain appears to be a disorder of our motivational/affective system.

Disorder is the operative word, for additional symptoms are associated with chronic pain, beyond the pain itself. Chronic pain patients also experience neuroendocrine dysregulation, fatigue, dysphoria, diminished physical performance, and impaired cognition and executive function (Chapman and Gavrin 1999). Chronic pain clearly is about much more than a simple sensation or bodily imperative.

A clever series of experiments from the 1970s demonstrates this point. Fordyce and his colleagues examined the ability of chronic back pain patients to exercise. At first, the patients were given a repetitive exercise to do and were instructed to do it "until pain, weakness, or fatigue causes you to want to stop" (Fordyce 1979). The researchers discovered that these patients, more often than not, stopped exercising on a number of repetitions that was divisible by five. If the capacity for exercise were really being controlled by pain, then one would expect to find a random distribution of repetitions. Fordyce hypothesized that when movements are easily countable, multiples of five comprise convenient mileposts. They become cues that indicate continuing to exercise risks increasing pain. So patients were stopping not because they could do no more, but because they anticipate (and dread) forthcoming pain.

To test this hypothesis, Fordyce then asked both chronic back pain patients and normal controls to exercise to tolerance in such a way that they could not count how much they were doing. For example, in one trial, subjects were asked to ride a stationary bicycle that had no speed or distance indicators or gear controls in a room with no clock or windows. In these cases, the performance of chronic pain patients and normal subjects was virtually identical (Fordyce 1979). When the pain patients had no way to calculate their effort, they could not anticipate future pain as easily, which led to their being able to do much more physically.

Chronic pain heightens the affective aspects of pain, inducing stress, anxiety, and fear, as it diminishes the ability to think clearly and rationally. These affective dimensions, anticipating increasing pain and having no relief in sight, are key components in the suffering sometimes associated with pain. Although they may often go together, suffering, however, is something over and above pain processing. It is the secondary reaction.

3. Suffering and Chronic Pain

Suffering is not the same thing as pain, and in fact many instances of acute pain do not entail suffering, but suffering often accompanies chronic pain. What is suffering exactly? And how do we know?

Unlike pain, there are no peripheral sensory systems that respond to suffering. We only know that people are suffering when they report that they are. Surveys of advanced cancer patients, whom we often take to be exemplars of suffering, are revealing. The first thing we learn is that we drastically overestimate how much people with advanced cancer do suffer (Schulz et al. 2010). Indeed, only one-quarter of patients with advanced cancer agree that they are suffering at a moderate or extreme level (Wilson et al. 2007). The second thing we learn is that the symptoms associated with suffering resemble those of chronic pain: fatigue, dysphoria, disrupted sleep, hypervigilance, reduced appetite, impaired physical functioning, impaired cognition, and negative ruminations (Chapman and Gavrin 1999). The third thing we learn is that while complaints regarding suffering often revolve around pain, the emotions associated with suffering have a greater effect (Wilson et al. 2007). We might talk about pain, but our feelings impact us the most.

In a survey given to a wide range of people, when asked to indicate the "worst pain you have ever experienced," many different types of answers were given, ranging from accidents and injuries to childbirth to illness to migraines. Only 4% of respondents answered that their greatest pain was emotional (Bendelow 2006). However, when respondents were asked individually to elaborate on their answers, the discussion quickly turned to feelings. One subject claimed a toothache was the worst pain he ever experienced, but talked about his anxiety attacks:

> It makes me totally separate from whatever was going on—it's like an inner terror, it is like a physical pain at times, it's like a vice on my temples and an incredible pressure on my head that does produce a headache but essentially it's just a brooding feeling within the skull.
>
> (2006: 66)

Another subject, who had experienced a horrible motorcycle accident in India while on vacation, talked about a psychological trauma instead:

> The emotional pain goes on longer—I think it's somehow worse than a physical pain because that is usually comprehensible, logical and there's a certain amount of control—you can get your head round it. I'd rather go through all the horror of smashing up my leg and all the being in hospital than two months of what I've just been through.
>
> (2006: 66)

The point is that, while we might assume our greatest suffering is tied to pain, with a little prodding, we can see that often it is connected to emotional duress.

So what is suffering? Not surprisingly, there are a multitude of definitions (Kellehear 2009), but the consensus centers on suffering being a multidimensional experience of physical symptoms, psychological distress, existential concerns, and social-relational worries. It arises from the belief that one's biological, social, or psychological being is being in jeopardy. Eric Cassell (2004) contrasts pain, which he sees as a threat to bodily integrity, with suffering, which he describes as a threat to the intactness of the self. It alters one's sense of self, due to some kind of loss—pain, injury, deprivation, severe illness, some type of significant social or personal change.

Unlike pain, suffering is always a conscious experience. It is about mourning a loss, a sophisticated secondary reaction, dependent on complex interactions of cognition and affect. To suffer requires the ability to project one's self into the future and to anticipate a continued loss. In general, those who are suffering are looking for healing, not for a cure; they are looking for a way to understand themselves in relation to the cause of their suffering (Cassell 1991).

How might one measure suffering, if it is so multifarious? A start would be to look at the case of complicated grief. ("Complicated grief" refers to grief that continues unabated at least six months beyond the time of the loss.) Grief normally activates the anterior cingulate, the insula, and the peri-acquiductal gray areas—all regions associated with pain. In some cases, however, grief does not subside as it should. Like chronic pain, it persists long beyond its expected trajectory. Its principle symptom is a yearning for the missing loved one so intense that it crowds out other wants and needs. And, just like chronic pain, complicated grief activates the NAc, part of our reward system, in addition to the pain areas just mentioned (O'Connor et al. 2008). (The activity of the NAc is depressed in cases of normal grief, just like in cases of acute pain.) Activation of NAc appears to be correlated with a sense of yearning for the lost love, which Mary-Francis O'Connor compares to the craving one finds in addiction (O'Connor et al. 2008).

4. The Reward System and Its Disorders

Interestingly, we see the same changes in NAc and insula activation across chronic pain and complicated grief, as just discussed, but in addiction as well. Could it be that they are all of a piece and that they are all disorders of a single reward system? Many contemporary theories of addiction identify impulse control difficulties as well as compulsive behaviors. Patients with impulse control disorders feel an increasing sense of tension or arousal before committing an impulsive act, and then pleasure, gratification, or a sense of relief at the time of doing the act itself. These types of disorders are generally associated with positive reinforcement mechanisms (American Psychiatric Association 2013). In contrast, patients with compulsive disorders feel anxiety and stress before engaging in some compulsive behavior, and then a sudden release from the stress as they perform the compulsive behavior. These disorders are associated with negative reinforcement mechanisms. Already, we can see interesting comparisons with pain processing, which is associated with positive reinforcement at the cessation of acute pain sensations in normal subjects, but not for the cessation of acute pain in cases of chronic pain.

Impulsivity often dominates early in addiction, and impulsivity combined with compulsivity dominates later in the disease. As addicts move from impulsivity to compulsivity, the driving force motivating their addictive behaviors shifts from pleasure and positive reinforcement over to anxiety, stress, and negative reinforcement (Koob and Le Moal 2001, Edwards and Koob 2010). We see a similar pattern in the shift from acute to chronic pain: the patient shifts from being motivated to seek a pleasurable relief to being unable to experience such relief at all. We also see a similar pattern in complicated grief: the yearning that accompanies the loss is not one of pleasant memories, but of sadness.

The transition from normal consumption to genuine drug or alcohol dependence involves circuitry in the forebrain, including the NAc and prefrontal cortex (Modesto-Lowe and Fritz 2005, Gilpin and Koob 2008, Gianoulakis 2009, Egli et al. 2012). Similar areas are involved in the transition from acute pain processing to a chronic pain syndrome and in comparing normal grief with complicated grief. It does indeed appear that pain, pleasurable consumption, and grieving (and their related disorders) all share the same underlying neural circuitry. All are very complex reactions that stem from our reward circuitry.

Let us return now to where we started: philosophers would like to describe the thing that pain is. We have seen that pain is not a thing, but a multifaceted process. The whole complex of somatosensory information, a representation of badness, and the affective anticipation and motivation comprise pain, of which the experience of "ouch, that hurt" is only a part. And this complex is not specific for pain, but perhaps reflects how the brain processes all incoming stimuli. This suggests that pain, chronic pain, grief, complicated grieving, and addiction (as well pleasurable experiences) are all effects of the basic reward circuitry in the brain. Our reward system normally gives us both pleasures and pains, but with chronic pain processing, unremitting grieving, or extended episodes of intoxication, our brain circuitry and functionality change—and change in very similar ways—such that we can become lost in our disappointment.

Understanding the neurobiological details behind pain can help philosophers analyze its nature. It turns out that their initial intuitions were misguided, and that pain appears to be related to other types of processing that at first blush seem to be categorically different. Moreover, appreciating the complexity and the details of these processes might help all of us to better help those who are in pain and suffering.

References

American Psychiatric Association (2013) *Diagnostic and Statistical Manual of Mental Disorders* (5th ed., DSM-5). Washington, DC: APA

Andersson, G.B.J. (1997) The epidemiology of spinal disorders. In J.W. Frymoyer (Ed.) *The Adult Spine: Principles and Practice*. Philadelphia: Lippincott-Raven, pp. 93–141.

Apkarian, A.V. (2012) Chronic pain and addiction pathways. Society of Neuroscience Annual Meeting, New Orleans.

Aydede, M. (Ed.) (2006) *Pain: New Essays on Its Nature and the Methodology of Its Study*. Cambridge, MA: MIT Press.

Bain, D. (2011) The imperative view of pain. *Journal of Consciousness Studies* 18: 164–185.

Baliki, M.N., Chialvo, D.R., Geha, P.Y., Levy, R.M., Harden, R.N., Parrish, T.B., and Apkarian, A.V. (2006) Chronic pain and the emotional brain: Specific brain activity associated with spontaneous fluctuations of intensity of chronic back pain. *Journal of Neuroscience* 26: 12165–12173.

Baliki, M.N., Geha, P.Y., Fields, H.L., and Apkarian, A.V. (2010) Predicting value of pain and analgesia: Nucleus accumbens response to noxious stimuli changes in the presence of chronic pain. *Neuron* 66: 149–160.

Baliki, M.N., Petre, B., Torbey, S., Herrmann, K.M., Huang, L., Schnitzer, T.J., Fields, H.L., and Apkarian, A.V. (2012) Corticostriatal functional connectivity predicts transition to chronic pain. *Nature Neuroscience* 15: 1117–1119.

Bendelow, G.A. (2006) Pain, suffering, and risk. *Health, Risk, and Society* 8: 59–70.

Brooks, J., and Tracey, I. (2005) From nociception to pain perception: Imaging the spinal and supraspinal pathways. *Journal of Anatomy* 207: 19–33.

Cassell, E. (1991) Recognizing suffering. *Hastings Center Report* 23: 24–31.

Cassell, E. (2004) *The Nature of Suffering and the Goals of Medicine, 2nd Edition*. New York: Oxford University Press.

Cauda, F., Cavanna, A.E., D'agata, F., Sacco, K., Duca, S., and Geminiani, G.C. (2011) Functional connectivity and coactivation of the nucleus accumbens: A combined functional connectivity and structure-based meta-analysis. *Journal of Cognitive Neuroscience* 18: 2864–2877.

Chapman, C.R., and Gavrin, J. (1999) Suffering: The contributions of persistent pain. *The Lancet* 353: 2233–2237.

Committee on Advancing Pain Research, Care, and Education, Institute of Medicine (2011) *Relieving Pain in America: A Blueprint for Transforming Prevention, Care, Education, and Research*. Washington, DC: The National Academies Press.

Dennett, D. (1978) Why you can't make a computer that feels pain. *Synthese* 38: 415–456.

Edwards, S., and Koob, G.F. (2010) Neurobiology of dysregulated motivational systems in drug addiction. *Future Neurology* 5: 393–401.

Egli, M., Koob, G.F., and Edwards, S. (2012) Alcohol dependence as a chronic pain disorder. *Neuroscience and Biobehavioral Reviews* 36: 2179–2192.

Fordyce, W. (1979) Environmental factors in the genesis of low back pain. In J. Bonica, L. Liebeskind, and V. Alba-Fessard (Eds.) *Advances in Pain Research and Therapy, Volume 3*. New York: Raven Press, pp. 659–666.

Fordyce, W., Brockway, J., Bergman, J., and Spengler, D. (1986) A control group comparison of behavioral versus traditional management methods of acute lower back pain. *Journal of Behavioral Medicine* 5: 127–140.

Gianoulakis, C. (2009) Endogenous opioids and addiction to alcohol and other drugs of abuse. *Current Topics in Medicinal Chemistry* 9: 999–1015.

Giesler, G.J., Jr, Yezierski, R.P., Gerhart, K.D., and Willis, W.D. (1981) Spinothalamic tract neurons that project to medial and/or lateral thalamic nuclei: Evidence for a physiologically novel population of spinal cord neurons. *Journal of Neurophysiology* 46: 1285–1308.

Gilpin, N.W., and Koob, G.F. (2008) Neurobiology of alcohol dependence: focus on motivational mechanisms. *Alcohol Research and Health* 31(3): 185–195.

Hardcastle, V.G. (1999) *The Myth of Pain*. Cambridge, MA: MIT Press.

Head, H., and Holmes, G. (1911) Sensory disturbances from cerebral lesions. *Brain* 34: 102–254.

Iannetti, G.D., Lee, M.C., and Mouraux, A. (2010) A multisensory investigation of the functional significance of the "pain matrix". 13th World Congress on Pain, Montreal, Canada.

Kaplan, W., Wirtz, V.J., Mantel-Teeuwissa, A., Stolk, P., Buthey, B., and Laing, R. (2013) *Priority Medicines for Europe and the World—2013 Update*. World Health Organization. http://www.who.int/medicines/areas/priority_medicines/MasterDocJune28_FINAL_Web.pdf?ua=1

Katz, J.N. (2006) Lumbar disc disorders and low-back pain: Socioeconomic factors and consequences. *Journal of Bone and Joint Surgery America* 88(suppl 2): 21–24.

Kellehear, A. (2009) On dying and human suffering. *Palliative Medicine* 23: 388–397.

Klein, C. (2015) *What the Body Commands*. Cambridge, MA: MIT Press.

Koob, G.F., and Le Moal, M. (2001) Drug addiction, dysregulation of reward, and allostasis. *Neuropsychopharmacology* 24: 97–129.

Kripke, S.A. (1980) *Naming and Necessity*. Cambridge, MA: Harvard University Press.

Malinen, S., Vartiainen, N., Hlushchuk, Y., Koskinen, M., Ramkumar, P., Forss, N., Kalso, E., and Hari, R. (2010) Aberrant temporal and spatial brain activity during rest in patients with chronic pain. *Proceedings of the National Academy of Science, U.S.A.* 107: 6493–6497.

Melzack, R. (1989) Phantom limbs, the self and the brain. *Canadian Psychology* 30: 1–16.

Modesto-Lowe, V., and Fritz, E.M. (2005) The opioidergic-alcohol link: Implications for treatment. *CNS Drugs* 19: 693–707.

O'Connor, M.-F., Wellisch, D.K., Stanton, A.L., Eisenberger, N.I., Irwin, M.R., and Lieberman, M.D. (2008) Craving love? Enduring grief activates brain's reward center. *Neuroimage* 42: 969–72.

Pizzo, P.A., and Clark, N.M. (2012) Alleviating suffering 101—Pain relief in the United States. *New England Journal of Medicine* 366: 197–199.

Reuter, K. 2011. Distinguishing the appearance from the reality of pain. *Journal of Consciousness Studies* 18(9/10), 94–109.

Rubin, D.I. (2007) Epidemiology and risk factors for spine pain. *Neurology Clinic* 25: 353–371.

Schulz, R., Monin, J.K., Czaja, S.J., Lingeler, J.H., Beach, S.R., Martire, L.M., Dodds, A., Herbert, R.S., Zdaniuk, B., and Cook, T.B. (2010) Measuring the experience and perception of suffering. *The Gerontologist* 50: 774–784.

Sherrington, C.S. (1906) *The Integrative Action of the Nervous System*. New Haven, CT: Yale University Press.

Taimela, S., Kujala, U.M., Salminen, J.J., and Viljanen, T. (1997) The prevalence of low back pain among children and adolescents: A nationwide, cohort-based questionnaire survey in Finland. *Spine* 22: 1132–1136.

Vos, T., et al. (2012) Years lived with disability (YLDs) for 1160 sequelae of 289 diseases and injuries 1990–2010: A systematic analysis for the Global Burden of Disease Study 2010. *Lancet* 380: 2163–2196.

Wilson, K.G., Chochinov, H.M., McPherson, C.J., LeMay, K., Allard, P., Chary, S., Gagnon, P.R., Macmillan, K., De Luca, M., O'Shea, F., Kuhl, D., and Fainsinger, R.L. (2007) Suffering with advanced cancer. *Journal of Clinical Oncology* 26: 1691–1697.

Further Reading

Aydede, M. (Ed.) (2006) *Pain: New Essays on Its Nature and the Methodology of Its Study*. Cambridge, MA: MIT Press.

Cassell, E. (2004) *The Nature of Suffering and the Goals of Medicine*, 2nd Edition. New York: Oxford University Press.

Egli, M., Koob, G.F., and Edwards, S. (2012) Alcohol dependence as a chronic pain disorder. *Neuroscience and Biobehavioral Reviews* 36: 2179–2192.

Hardcastle, V.G. (1999) *The Myth of Pain*. Cambridge, MA: MIT Press.

Klein, C. (2015) *What the Body Commands*. Cambridge, MA: MIT Press.

13
MEASURING PLACEBO EFFECTS

Jeremy Howick

1. Introduction

The term *placebo* is Latin for "I shall please." In medical research and practice, a placebo is a treatment that is capable of making people believe it is, or could be, the real treatment when in fact it is not that treatment. A sugar pill that is indistinguishable to a patient from real vitamin C, for example, could be a placebo. Within clinical trials, placebo *controls* are often used as a standard against which experimental treatments are compared. For example, when evaluating a new drug, investigators can give some patients a placebo and other patients the experimental drug. If the experimental drug outperforms the placebo by a sufficient margin, it is said to be effective. Patients (as well as, if feasible, doctors, outcome assessors, and data analysts) in these trials are often masked, which means they don't know which treatment is the experimental drug and which is the placebo. Placebo use in clinical practice and clinical trials is both common and controversial (Howick, 2009, Howick et al., 2013a). A great deal of philosophical controversy surrounds what placebos are (Gøtzsche, 1994, Nunn, 2009), as well as whether placebos are ethical (Foddy, 2009, Howick, 2009). There is also a dispute—the one I will concern myself in this chapter—about the correct methodology that should be used to measure placebo response.

Henry Knowles Beecher's early (1955) study claimed that one-third of the benefit of all treatments was due to placebo effects (Beecher, 1955). This is still the most widely cited paper in the field, and the beliefs that placebos are powerful persist because of it. Yet more recent studies conducted by Asbjorn Hróbjartsson and Peter Gøtzsche in Denmark conclude that placebos do not have important effects *at all* (Hróbjartsson and Gøtzsche, 2001, Hróbjartsson and Gøtzsche, 2004a, Hróbjartsson and Gøtzsche, 2010). In this chapter I will review the attempts to quantify placebo effects and list the methodological problems with these attempts. To anticipate, I will argue that while Beecher's 1955 estimate is likely to be an overestimate, the more recent studies underestimate placebo effects. A large body of evidence shows that placebo effects are very powerful for treating some ailments, especially pain.

I will proceed by first describing Beecher's 1955 study and the problems with it. The main objection to Beecher's study is that he incorrectly infers from the fact that someone recovers after taking a placebo to the claim that the recovery was due to the placebo (section 2). I will then proceed to describe the more recent studies conducted by Hróbjartsson and Gøtzsche and list the methodological problems with them. The main issue with Hróbjartsson and Gøtzsche's analysis is that their study includes heterogeneous placebos and conditions, making their average placebo effect estimate compatible with powerful placebo effects for certain conditions such as pain (section 3). Before concluding (in section 5), I describe the evidence and mechanisms that explain why, despite the methodological difficulties in estimating placebo effects, it is reasonable to believe that placebos are effective for treating pain (section 4).

2. Beecher's Powerful Placebos

Although placebos have been used knowingly and unknowingly by physicians for at least several centuries (Kaptchuk, 1998), Henry Knowles Beecher was the first person we know of who rigorously attempted to quantify placebo effects. Beecher was a Harvard Medical School graduate and a doctor in North Africa, France, and Italy during World War II. Morphine was sometimes in short supply, and he noticed that some soldiers did not require any painkillers. Many of these patients had very serious wounds; for example, they had been shot. Beecher thus came to suspect that their mental states and attitudes, as much as their physical wounds, affected how much pain they experienced.

After the war was over, he investigated placebo effects by doing one of the first systematic reviews (a study that includes all available studies) with meta-analysis (statistical pooling of results of the studies within the systematic review). Beecher's review contained 15 placebo-controlled studies measuring the effects of painkillers for treating conditions ranging from postoperative pain to angina pectoris, headaches, and cough. The studies contained a combined total of 1,082 patients. He found that one-third of the patients who had received only placebos got better and concludes: "It is evident that placebos have a high degree of therapeutic effectiveness in treating subjective responses, decided improvement . . . being produced in 35.2 ± 2.2% of cases" (Beecher, 1955, p. 1696). Beecher's study is still the most widely cited in the field, and it was immortalized in his famous article "The Powerful Placebo."

However, in the 1990s, researchers began to formally question Beecher's results. In a provocatively titled article, "The powerful placebo effect: fact or fiction?" Kienle and Kiene (1997) accused Beecher of failing to "consider that many patients with a mild common cold improve spontaneously" (Kienle and Kiene, 1997, p. 1312). In this regard, Kienle and Kiene are correct: Beecher's attribution of change in the placebo control group to the placebo control treatment is a fallacy. Many ailments, including the common cold and postoperative pain, usually go away quite quickly without any treatment at all. The so-called natural history of the disease, and spontaneous remission, are all potential causes of apparent recovery that have nothing to do with placebo effects. Claiming that one thing (giving a patient a placebo) *causes* another thing (reduction in pain) simply because the former comes first is what philosophers call the *post hoc ergo procter hoc* (after, therefore because of) fallacy.

There are at least two additional problems with Beecher's method. First, polite patients taking the placebo could report improvement to please investigators, although no benefit was actually felt (Hróbjartsson et al., 2011). Noting that their doctors were trying to help them, these patients could exaggerate how much benefit they felt after being treated simply to please their friendly doctors. In addition, caregiver and doctor attention given to patients could lead to clinical improvement. Many systematic reviews have demonstrated that empathetic and encouraging caregivers improve outcomes in patients suffering from pain, anxiety, and depression (Crow et al., 1999, Di Blasi et al., 2001, Griffin et al., 2004, Derksen et al., 2013, Kelley et al., 2014, Kelm et al., 2014). Hence, the effects observed in the placebo groups within Beecher's studies could be the effects of caregiver empathy and encouragement rather than the effects of administering, say, a sugar pill. One could, of course, classify practitioner empathy and encouragement as types of placebo effects. At the same time, empathy and encouragement appear to be distinct from the administration of a sugar pill. If the effects of caregiver empathy and encouragement are different from the effects of administering a sugar pill, then the effects in Beecher's studies might not be placebo effects.

Given all of the methodological problems with Beecher's study, Kienle and Kiene thus conclude that "none of the original trials cited by Beecher gave grounds to assume the existence of placebo effects" (Kienle and Kiene, 1997, p. 1316).

3. Hróbjartsson and Gøtzsche's Powerless Placebos

Taking up the challenge of estimating placebo effects more accurately, Peter Gøtzsche and Asbjorn Hróbjartsson conducted a systematic review and meta-analysis of trials that had three groups of patients:

1. patients not treated at all (often people in these groups were placed on waiting lists)
2. patients given a placebo
3. patients given a "real" treatment

They identified 114 trials. Typical pill placebos were lactose pills, typical physical placebos were procedures performed with the machine turned off (e.g., sham transcutaneous electrical nerve stimulation), and typical psychological placebos were theoretically neutral discussion between participant and dispenser. They then compared what happened to patients who received the placebo with patients in the untreated groups. This method avoided the *post hoc ergo procter hoc* fallacy because it included a comparison of patients whose ailment would have followed its natural history: the untreated group. They found that on average placebos only have small effects:

> [There is] little evidence that placebos in general have powerful clinical effects. Placebos had no significant pooled effect on subjective or objective binary or continuous objective outcomes. We found significant effects of placebo on continuous subjective outcomes and for the treatment of pain but also bias related to larger effects in small trials. The use of placebo outside the aegis of a controlled, properly designed clinical trial cannot be recommended.
>
> (Hróbjartsson and Gøtzsche, 2001, p. 1599)

Even in cases where they found a significant placebo effect—notably for treating pain—Hróbjartsson and Gøtzsche question whether the observed effects were actually placebo effects. They raise a similar concern to Kienle and Kiene, noting that the outcomes in the placebo groups could have been due to a form of reporting bias, such as patients politely reporting they had recovered when in fact they did not actually experience any recovery. In their words:

> Patients in an untreated group would know they were not being treated, and patients in a placebo group would think they were being treated. It is difficult to distinguish between reporting bias and a true effect of placebo on subjective outcomes, since a patient may tend to try to please the investigator and report improvement when none has occurred. The fact that placebos had no significant effects on objective continuous outcomes suggests that reporting bias may have been a factor in the trials with subjective outcomes.
>
> (Hróbjartsson and Gøtzsche, 2001, p. 1597)

If we accept their results, Hróbjartsson and Gøtzsche seem to have completely overturned Beecher's estimate. Yet while Hróbjartsson and Gøtzsche's method overcomes the *post hoc ergo procter hoc* fallacy, their conclusions are questionable because of several problems with their review. The main problems are: (1) their review included heterogeneous studies that arguably should not have been lumped together so their average results fail to rule out powerful placebo effects for some conditions; (2) they failed to adequately put strictures on what counts as a placebo; and (3) they failed to recognize that untreated groups within clinical trials are not truly untreated. I will go over each of these problems in turn.

3.1. Problem with the Heterogeneity of the Studies within Hróbjartsson and Gøtzsche's Review

Perhaps the most serious problem with Hróbjartsson and Gøtzsche's review is heterogeneity. It is often acceptable to meta-analyze (pool) results, but the meta-analysis can be misleading if researchers include heterogeneous studies—apples and oranges—within the same analysis. Over 40 clinical conditions were included in the analysis, ranging from hypertension and compulsive nail biting to fecal soiling and marital discord. The placebo interventions were also diverse. Kirsch (2002), for example, notes that Hróbjartsson and Gøtzsche jumble together (along with placebo pills and injections) relaxation (described as a placebo in some studies and a treatment in others), leisure reading, answering questions about hobbies, newspapers, magazines, eating favorite foods and watching sports teams, talking about daily events, family activities, football, vacation activities, pets, hobbies, books, movies, and television shows as placebos.

In this case it is problematic because it means that although they found a very small average placebo effect, some of the placebo treatments within the review had large effects. Just as "real" medicine is not necessarily effective on average for everything, neither are placebos. Instead, like "real" medicine, we might expect placebos to be effective for treating some ailments and not effective for treating others. To illustrate this problem, Howick et al. (2013b) reanalyzed the same three-armed trials from Hróbjartsson and Gøtzsche's review, but instead of just measuring placebo effects (by comparing outcomes in placebo and no treatment groups), they also measured treatment effects (by comparing outcomes in treatment and placebo groups). They found that average treatment effects in that heterogeneous group of studies were also modest and not statistically significantly different from placebo effects. Unless we accept that medicine does not have significant effects, the Howick et al. study is arguably a *reductio ad absurdum* of Hróbjartsson and Gøtzsche's methodology. If it is legitimate to pool heterogeneous placebos for diverse conditions, then it is also legitimate to pool similarly heterogeneous *treatments* for the same diverse conditions. Since pooling treatments in this way reveals minuscule treatment effects, and we know some treatments are highly effective, then it follows that pooling heterogeneous placebos for diverse conditions is not legitimate.

The highly significant heterogeneity of the interventions studied calls into question what conclusions can be drawn from the meta-analysis. An overall finding of insignificant placebo effects does not count against the reasonable view that placebos, like treatments, are common and powerful for certain disorders, but not for others. Similarly, certain placebos—for example, placebo injections—could be more effective than other placebos—for example, placebo pills. In section 4, I will review the evidence supporting the claim that placebo effects are quite large for treating pain.

3.2. Problem with Hróbjartsson and Gøtzsche's Definition of Placebos

The estimate of the placebo effect is based on the average difference between outcome in the placebo and no treatment groups. This estimate depends on whether the placebos in the included trials were "real" placebos and whether the untreated groups were truly untreated. Yet the placebo concept has proven notoriously difficult to define adequately. It is beyond the scope of this chapter to outline the definitional problems—see Howick, 2016 for a review—however, the common definitions of placebos as inactive or nonspecific are clearly mistaken: placebos can be active, at least with respect to some ailments, and their mechanism of action can be as specific as the action of a pharmacological drug (see section 4). To avoid the definitional problem, Hróbjartsson and Gøtzsche decided to adopt a practical approach and characterize placebos "practically as an [any!] intervention labeled as such in the report of a clinical trial." But it hardly needs remarking that this approach is untenable. Suppose, for example, that someone reported using penicillin as a placebo in a trial of some new antibiotic as a treatment for pneumonia. The response will of course be "no one would, and if they did we would not take the trial seriously." But this reaction seems exactly to show that we work with some concept that involves judgments about what can and cannot count as appropriate or legitimate placebos and placebo controls. Since it is unclear whether the placebos within Hróbjartsson and Gøtzsche's study were actual placebos, their estimates of placebo effects must be questioned.

Worse, Hróbjartsson and Gøtzsche go back on their alleged policy of accepting any treatment labeled as a placebo in the report of a clinical trial. For example, they exclude studies where "it was very likely that the alleged placebo had a clinical benefit not associated with the ritual alone (e.g., movement techniques for postoperative pain)" (Hróbjartsson and Gøtzsche, 2001, p. 1595). Here they seem to sneak in a definition of placebos as the effects of "rituals," which is not acceptable: ritual feasting or fasting are not placebos. A similar problem arises because the untreated groups within Hróbjartsson and Gøtzsche's study may actually have benefited from observation by and contact with health care practitioners.

3.3. Hawthorne Effects in the No Treatment Groups

The Hawthorne effect is named not after the scientist discovering the phenomenon but after a series of experiments in the Hawthorne works of the Western Electric Company in Chicago between 1924 and 1933. In one study, the lighting in the factory remained stable for the control group, whereas the lights in the part of the factory where the experimental group worked were made brighter. Productivity increased equally in both groups. In another version of the experiment, the researchers tried the opposite: lighting was kept stable in the part of the factory where the control group worked but was made less bright where the experimental group worked. In the second experiment, productivity also increased steadily and equally in both groups until the lights were so low in the experimental groups that the workers protested that they couldn't see and production fell off (Roethlisberger and Dickson, 1939). Since the productivity increased in both groups whether or not the lights were turned up, turned down, or remained the same, it couldn't be the lighting that caused the increase in productivity. The reason productivity increased was that the workers knew they were being watched. The effect of being watched is called the Hawthorne or observer effect. In the case of medical research, patients in clinical trials could experience some benefit (or change their behavior and responses) because they are being investigated.

Hawthorne effects could have influenced the untreated groups within Hróbjartsson and Gøtzsche's analysis and confounded their results. "Untreated" participants within clinical trials are typically monitored and observed by researchers (Einarson et al., 2001). Knowing that their condition is being monitored, the patients might do various things that are good for their health, such as exercise more, eat healthier, or drink less. These things could all produce a health benefit for many conditions being treated. In addition, by virtue of the fact that they are being monitored, the "untreated" patients have contact with health care practitioners. Contact with practitioners has been shown to have an effect, especially if the practitioners are empathetic (Di Blasi et al., 2001). This presents a problem with Hróbjartsson and Gøtzsche's review that is similar to the objection they level at Beecher. The outcomes in the untreated groups might not be the result of natural history, but instead be the result of Hawthorne effects and the effects of contact with empathetic health care practitioners.

There is good reason to believe that the untreated groups did benefit from Hawthorne effects and the effects of caregiver attention. A recent systematic review found that the untreated groups within Hróbjartsson and Gøtzsche's review experienced a 24% improvement compared with baseline (Krogsboll et al., 2009). There are two plausible explanations for the 24% improvement from baseline. First, it could be natural history of disease. Second, the improvement is partly due to Hawthorne effects and effects of contact with health care practitioners. If it is the second—and this is my preferred explanation since 24% is a large effect that is not likely to be explainable solely in terms of natural history—then Hróbjartsson and Gøtzsche have underestimated the placebo effect. Because the untreated groups were not, in fact, untreated, Hróbjartsson and Gøtzsche's review underestimated placebo effects. This is due to simple arithmetic, noting that the placebo effect is estimated by subtracting outcomes in the placebo effect with outcomes in the untreated groups.

3.4. Other Biases within Hróbjartsson and Gøtzsche's Review

Hróbjartsson and Gøtzsche updated the 2001 meta-analysis in 2004 (Hróbjartsson and Gotzsche, 2004b, Hróbjartsson and Gøtzsche, 2010). The 2010 review included 202 studies—almost twice as many as the initial review. The more studies they included, the greater the placebo effects became, both in terms of statistical significance and (in the case of pain) clinical relevance. Moreover, Hróbjartsson and Gøtzsche recognized the problem that the untreated participants may have been treated and hence led to an underestimation of placebo effects. In their words: "Patients in a no-treatment group also interact with treatment providers, and the patients are therefore only truly untreated with respect to receiving a placebo intervention" (Hróbjartsson and Gøtzsche, 2004a, p. 97). They also acknowledge the problem of heterogeneity: "we cannot exclude the possibility that in the process of pooling heterogeneous trials the existence of such a group [of trials that showed a significant placebo effect] was obscured" (Hróbjartsson and Gøtzsche, 2004a, p. 97). Despite the growing evidence for placebo effects, and recognition of at least a few of the problems with their analysis, their conclusions remained just as skeptical, concluding that they "did not find that placebo interventions have important clinical effects in general" (Hróbjartsson and Gøtzsche, 2010). Yet given that their updated review found greater placebo effects, one may question whether Hróbjartsson and Gøtzsche were justified in maintaining the same conclusion.

Finally, Hróbjartsson and Gøtzsche's study may not be applicable to routine clinical practice because placebos are likely to have a smaller effect in clinical trials than in clinical practice. In a clinical trial, patients believe that they might receive the placebo *or* the experimental treatment. There is therefore some doubt as to whether they will receive the "real" treatment. This could lead to lower expectations about recovery and worse outcomes compared to the case where they believed they were sure to receive the real treatment and thus had higher expectations. If a doctor prescribed a placebo in clinical practice, however—and several studies have shown that most doctors have prescribed a placebo in clinical practice (Fassler, 2009, Howick et al., 2013a)—the patient believes they are receiving a "real" treatment, and hence have higher expectations. The higher expectations, in turn, could lead to greater placebo effects in clinical practice. Since Hróbjartsson and Gøtzsche's review attempts to measure placebo effects within clinical trials, it may not apply to clinical practice where placebo effects could be greater.

3.5. Summary of the Evidence for Placebo Effects

In sum, Beecher's estimate of placebo effects is problematic for failing to take natural history into account and is likely an overestimate. However, Hróbjartsson and Gøtzsche's studies are likely to underestimate placebo effects, most notably because they fail to consider variation in placebo effectiveness and because they fail to acknowledge Hawthorne effects. The area where there is least controversy about the effectiveness of placebos is pain. In the next section I will review the evidence for both the existence of and the mechanism for placebo analgesia.

4. The Evidence and Mechanisms for Placebo Analgesia

Even Hróbjartsson and Gøtzsche's skeptical review acknowledges—albeit with what I argued were partly unjustified skeptical caveats—that placebo analgesia is effective. A large body of evidence has demonstrated the effects of placebo analgesia. For example, a recent systematic review of 198 trials (involving 16,364 patients) of placebo treatments for osteoarthritis pain found that placebos were effective (Zhang et al., 2008). In one of the more innovative studies that points towards a mechanism for placebo analgesia, Benedetti et al. (2004) used four common painkillers on a total of 278 patients who had undergone thoracic surgery for different pathological conditions (Benedetti et al., 2004, p. 680). The patients were then, of course unbeknownst to them, randomized into "overt" and "covert" groups with sex, age, weight, and pain baseline-balanced. The overt group was treated by doctors who "gave the open drug at the bedside, telling the patient that the injection was a powerful analgesic and that the pain was going to subside in a few minutes" (Benedetti et al., 2004, p. 681). Then, one dose of analgesic was administered every 15 minutes until the patients reported a 50% reduction of pain. The covert group, on the other hand, had the analgesic delivered by a pre-programmed infusion without any doctor or nurse in the room. Both the covertly and overtly treated patients already had an intravenous line and machine attached to them because of their operations. The results were that over 30% more analgesic was required by the patients who were treated covertly. The P-values reached statistical significance, ranging from 0.02 to 0.007 depending on the drug.

How come people who expected a positive outcome—those who were treated overtly—felt less pain? The answer lies in the brain's reward mechanism. The expectation of a positive reward (such as decreased pain) excites the body in a way that it produces enough endogenous endorphins to cause a reduction in pain (Benedetti, 2009). Combining the first four letters of the word *endogenous* with the last letters of the word *morphine* gives us the term *endorphin*. From the human cell receptor's point of view, morphine and endorphins are the same thing. To demonstrate that the body's endorphins are responsible for the analgesic effects of placebos, several researchers have done interesting randomized trials where they gave some patients a drug called naloxone, which antagonizes (blocks) the effects of opiates such as morphine and endorphins, and didn't give it to other patients. Now if placebo analgesia works by activating the body's ability to produce endorphins, then blocking the endorphin activity would reduce the placebo effect. In fact, this is just what happened. In the patients who received naloxone, the placebo effect was diminished (ter Riet et al., 1998, Sauro and Greenberg, 2005).

In short, dozens of empirical studies demonstrate an analgesic effect of placebos, and at least one mechanism explaining how placebo analgesia works—endogenous opiates—has also been established in rigorous studies. The existence of a placebo analgesia effect does not fully exonerate Beecher for his methodological flaws, but it does suggest that, at least for killing pain, Hróbjartsson and Gøtzsche are mistaken that placebos have insignificant effects.

5. Conclusion

What can we conclude about the power of placebos from these studies? The first thing is to acknowledge the fact that placebos won't work for everything. If someone has a leg amputated, taking a sugar pill or believing it will grow back is unlikely to have any effect. In addition, many ailments—especially the most common ones people visit their doctors for, such as mild to moderate pain, depression, anxiety, flu, colds, and so on—will go away on their own and without any treatment (placebo or not). Beecher's early and widely cited systematic review of placebo effects made the methodological error of failing to take natural history into account, and it led to an overestimation of placebo effects. However, Hróbjartsson and Gøtzsche's conclusion that placebos do not have much of an effect at all is equally problematic. Just as "real" treatments are exceptionally powerful for treating some conditions (for example, antibiotics for meningitis), so placebo treatments can be effective for treating some conditions, especially mild to moderate pain and depression, anxiety, and smoking cessation, and they can generally improve quality of life. By lumping all types of placebos for all conditions together, Hrobjartsson and Gotzsche obscure important placebo effects for conditions like pain. For example, the effects of placebo analgesia have been demonstrated empirically, and their mechanism of action—endogenous opiates—has been established in numerous studies. Further focused research is required to clarify the placebo concept and provide accurate estimates of placebo effects for specific conditions.

References

Beecher, H. K. 1955. The powerful placebo. *J Am Med Assoc*, 159, 1602–6.
Benedetti, F., Colloca, L., Lopiano, L. & Lanotte, M. 2004. Overt versus covert treatment for pain, anxiety, and Parkinson's disease. *The Lancet Neurology*, 3.11, 679–84.
Benedetti, F. 2009. *Placebo effects: Understanding the mechanisms in health and disease*. Oxford: Oxford University Press.
Crow, R., Gage, H., Hampson, S., Hart, J., Kimber, A. & Thomas, H. 1999. The role of expectancies in the placebo effect and their use in the delivery of health care: A systematic review. *Health Technol Assess*, 3, 1–96.
Derksen, F., Bensing, J. & Lagro-Janssen, A. 2013. Effectiveness of empathy in general practice: A systematic review. *Br J Gen Pract*, 63, e76–84.
Di Blasi, Z., Harkness, E., Ernst, E., Georgiou, A. & Kleijnen, J. 2001. Influence of context effects on health outcomes: A systematic review. *Lancet*, 357, 757–62.
Einarson, T. E., Hemels, M. & Stolk, P. 2001. Is the placebo powerless? *The New England Journal of Medicine*, 345, 1277; author reply 1278–9.
Fassler, M., Gnadinger, M., Rosemann, T. & Biller-Andorno, N. 2009. Use of placebo interventions among Swiss primary care providers. *BMC Health Services Research*, 9, 144.
Foddy, B. 2009. A duty to deceive: Placebos in clinical practice. *The American Journal of Bioethics: AJOB*, 9, 4–12.
Gøtzsche, P. 1994. Is there logic in the placebo? *Lancet*, 344, 925–6.
Griffin, S. J., Kinmonth, A. L., Veltman, M. W., Gillard, S., Grant, J. & Stewart, M. 2004. Effect on health-related outcomes of interventions to alter the interaction between patients and practitioners: A systematic review of trials. *Ann Fam Med*, 2, 595–608.
Howick, J. 2009. Questioning the methodologic superiority of 'placebo' over 'active' controlled trials. *Am J Bioeth*, 9, 34–48.

Howick, J. 2016. The relativity of 'placebos': Defending a modified version of Grünbaum's definition. *Synthese*, 1–34.

Howick, J., Bishop, F. L., Heneghan, C., Wolstenholme, J., Stevens, S., Hobbs, F. D. R. & Lewith, G. 2013a. Placebo use in the United Kingdom: Results from a national survey of primary care practitioners. *Plos One*, 8.

Howick, J., Friedemann, C., Tsakok, M., Watson, R., Tsakok, T., Thomas, J., Perera, R., Fleming, S. & Heneghan, C. 2013b. Are treatments more effective than placebos? A systematic review and meta-analysis. *PLoS One*, 8, e62599.

Hróbjartsson, A. & Gøtzsche, P. 2001. Is the placebo powerless? An analysis of clinical trials comparing placebo with no treatment. *N Engl J Med*, 344, 1594–602.

Hróbjartsson, A. & Gøtzsche, P. 2004a. Is the placebo powerless? Update of a systematic review with 52 new randomized trials comparing placebo with no treatment. *J Intern Med*, 256, 91–100.

Hróbjartsson, A. & Gøtzsche, P. 2004b. Placebo interventions for all clinical conditions. *Cochrane Database of Systematic Reviews*, CD003974.

Hróbjartsson, A. & Gøtzsche, P. 2010. Placebo interventions for all clinical conditions. *Cochrane Database of Systematic Reviews*, CD003974.

Hróbjartsson, A., Kaptchuk, T. & Miller, F. 2011. Placebo effect studies are susceptible to response bias and to other types of biases. *Journal of Clinical Epidemiology*, 64, 1223–9.

Kaptchuk, T. J. 1998. Intentional ignorance: A history of blind assessment and placebo controls in medicine. *Bulletin of the History of Medicine*, 72.3, 389–433.

Kelley, J. M., Gordon, K.-T., Lidia, S., Joe, K. & Helen, R. 2014. The influence of the patient-clinician relationship on healthcare outcomes: A systematic review and meta-analysis of randomized controlled trials. *PLoS One*, 9, e94207.

Kelm, Z., Womer, J., Walter, J. K. & Feudtner, C. 2014. Interventions to cultivate physician empathy: A systematic review. *BMC Med Educ*, 14, 219.

Kienle, G. S. & Kiene, H. 1997. The powerful placebo effect: Fact or fiction? *J Clin Epidemiol*, 50, 1311–8.

Kirsch, I. 2002. Antidepressants and placebos: Secrets, revelations, and unanswered questions. *Prevention & Treatment*, 5.

Krogsboll, L. T., Hrobjartsson, A. & Gotzsche, P. C. 2009. Spontaneous improvement in randomised clinical trials: Meta-analysis of three-armed trials comparing no treatment, placebo and active intervention. *BMC Medical Research Methodology*, 9, 1.

Nunn, R. 2009. It's time to put the placebo out of its misery. *BMJ*, 338, 1015.

Roethlisberger, F. J. & Dickson, W. J. 1939. *Management and the Worker*. Cambridge, MA: Harvard University Press.

Sauro, M. D. & Greenberg, R. P. 2005. Endogenous opiates and the placebo effect: A meta-analytic review. *J Psychosom Res*, 58, 115–20.

Ter Riet, G., De Craen, A. J., De Boer, A. & Kessels, A. G. 1998. Is placebo analgesia mediated by endogenous opioids? A systematic review. *Pain*, 76, 273–5.

Zhang, W., Robertson, J., Jones, A. C., Dieppe, P. A. & Doherty, M. 2008. The placebo effect and its determinants in osteoarthritis: Meta-analysis of randomised controlled trials. *Ann Rheum Dis*, 67, 1716–23.

Further Reading

Beecher, H. K. 1955. The powerful placebo. *J Am Med Assoc*, 159, 1602–6. This is the original study that provides a placebo effect estimate; it is also one of the earliest examples of a systematic review and meta-analysis.

Gotzsche, P. 1994. Is there logic in the placebo? *Lancet*, 344, 925–6. This paper briefly outlines the problems with defining the placebo.

Hrobjartsson, A. 2002. What are the main methodological problems in the estimation of placebo effects? *J Clin Epidemiol*, 55, 430–5. This paper builds on Kienle and Kiene's (1997) critique of Beecher's study and comprehensively outlines the problems with estimating placebo effects.

Hrobjartsson, A. & Gotzsche, P. 2001. Is the placebo powerless? An analysis of clinical trials comparing placebo with no treatment. *N Engl J Med*, 344, 1594–602. This study provides an estimate of placebo effects that addresses the problems with Beecher's study (yet has problems of its own).

Howick, J., Friedemann, C., Tsakok, M., Watson, R., Tsakok, T., Thomas, J., Perera, R., Feliming, S., & Heneghan, C. 2013. Are treatments more effective than placebos? A systematic review and meta-analysis. *PLoS One*, 8.5, e62599. This study provides empirical evidence of the flaw in Hrobjartsson and Gotzsche's estimate of placebo effects.

Kaptchuk, T. J. 1998. Intentional ignorance: A history of blind assessment and placebo controls in medicine. *Bulletin of the History of Medicine*, 72.3, 389–433. This paper provides an interesting history of placebos.

Kienle, G. S. & Kiene, H. 1997. The powerful placebo effect: Fact or fiction? *J Clin Epidemiol*, 50, 1311–8. This study provides a critique of Beecher's estimate.

14
THE CONCEPT OF GENETIC DISEASE

Jonathan Michael Kaplan

1. Introduction: Genetic Diseases

It is generally accepted that there are around 10,000 human diseases associated with variations in single genes ("monogenetic diseases") (WHO 2015; Kingsmore et al. 2011). Many of the genetic diseases known or suspected are rare or "ultra-rare" (by definition, rare diseases occur in fewer than 1 in 2,000 people, ultra-rare diseases in fewer than 1 in 2,000,000) (Hennekam 2011), but the total burden of such genetic diseases is significant, with researchers estimating that around 1% of the population, on average, has at least one condition associated with a single-gene disorder, and perhaps 4–10% of all pediatric hospital admissions are the result of single-gene disorders or chromosomal abnormalities (Dye et al. 2011; Kingsmore et al. 2011).

This description, however, hides the fact that a clear articulation of the "genetic disease" concept remains elusive. Part of the problem is that "disease" itself is a complex and contested concept (Murphy 2015); furthermore, the conceptual difficulties in untangling what it might mean to speak of *any* trait as "genetic" at the very least compounds the difficulties. Descriptions of genetic diseases usually rely on language to the effect that "abnormalities" or "defects" in the DNA (or in the chromosomes, etc.) lead to a disease state, but definitions relying on particular answers to what will count as "normal," or what constitutes a "defect," are also problematic and controversial (Amundson 2005).

It is worth distinguishing between diseases caused by so-called single-gene (monogenetic) diseases, those caused by chromosomal abnormalities, and complex diseases associated with variation in (many different) genes as well as variation in the environments encountered. In addition to single-gene disorders (and the smaller number of chromosomal abnormalities), many much more common medical conditions are said to "have a genetic component." Some researchers estimate that 30% or more of pediatric hospital admissions are the result of diseases with a "strong" genetic component (McCandless et al. 2004; Dye et al. 2011). And many common diseases—e.g., heart disease, diabetes—are thought to be "partially" genetic; this is generally interpreted to mean that one's overall genetic endowment is a risk factor for the disease, along with environmental factors (Craig 2008; McClellan and King 2010; Mitchell 2012).

This essay starts with a brief overview of some of the different ways that genes are referred to in the literature. Next, some of the important distinctions within genetic diseases will be outlined. Complications in the "genetic disease" concept will then be introduced, and phenylketonuria (PKU, or "phenylalanine hydroxylase deficiency") will be used as an example of a "classic" genetic disease and as an opportunity to reflect on the complexities of even relatively well-understood "simple" genetic diseases. Finally, the chapter ends with some further reflections on the genetic disease concept.

2. Genes in Contexts

It is traditional in genetics research to distinguish between an organism's genotype and its phenotype. The genotype of an organism is the complete complement of genetic material—all of its DNA; note that for eukaryotes (organisms whose cells contain a membrane-bounded nucleus), this includes mitochondrial DNA as well. An organism's phenotype, on the other hand, consists of all the measurable traits of the organism except its DNA sequence. So while, for example, the height of a plant is an aspect of its phenotype, so would be the concentration of a particular protein in a particular leaf of that plant. As with most distinctions in biology, there are fuzzy areas—for example, the way a particular chromosome is folded can influence which genes are expressed, what proteins get made, and so forth; is this folding pattern an aspect of the organism's genotype or its phenotype? More generally, *epigenetic* traits (traits that modify gene expression at the molecular level) seem to rest right on the divide between genetic and phenotypic traits (Hanley et al. 2010).

DNA consists of a deoxyribose sugar and phosphate "backbone" linked to nitrogen-based "bases." These bases, adenine (A), guanine (G), cytosine (C), and thymine (T), are the *nucleotides*, and each location on a DNA molecule where one of these bases can occur is a nucleotide site; the "genetic code" is a mapping of *triplets* of base-pairs ("codons") to one of 20 amino acids, plus "stop" codons. When people speak of genes "coding" for proteins, they are referring to the process by which the amino acids that these triplets code for are assembled together to make particular proteins. Particular stretches of DNA are called nucleotide regions; these regions are simply a "mapping" convenience and can be entirely arbitrary. Famously, DNA forms a double helix; these helixes themselves are wrapped tightly, and form chromosomes.

The genome is sometimes referred to as a "blueprint" or a "recipe" (see Condit et al. 2002 for discussion) for producing an organism; both metaphors are misleading. Although both acknowledge that genes are just one of many developmental resources necessary for organismal development, both imply that genes have a unique status as a kind of controller, a unique source of information, or are unique in being instructions of some kind. This implication is problematic; some researchers have argued compellingly that there is no strong sense in which genes can be said to control development or carry information in ways that other developmental resources do not (see e.g., Oyama, Griffiths, and Gray 2001; Moss 2004; see Smith 2007 for this point in the context of thinking explicitly about genetic diseases).

The idea of a "genetic disease," and especially the idea of "single-gene disorders," relies on there being a coherent "gene concept"—something that can be said to be "a gene." But the "gene concept" turns out to be surprisingly contentious (see e.g., Moss 2004; Griffiths and Stotz 2006; Portin 2009). Genes are often thought to be functional nucleotide regions. The most obvious functional regions are those that code for proteins; codons that together specify the sequence of amino acids that get used in forming a protein (note in eukaryotes, *most* DNA is *not* involved directly in specifying sequences for protein synthesis). These proteins, consisting of many amino acids, are used in the cellular processes resulting in growth, reproduction, development, and the like. Other obvious functional regions include regions associated with *gene regulation* (these control how much of a particular protein gets produced); there are a number of different regulatory mechanisms (see e.g., Latchman 2011 for review).

Although these functional conceptions of genes are useful for understanding development generally, the most obviously relevant gene concept for thinking about genetic diseases may be the gene as a *difference maker* (see Waters 2007). Here, the main concern is not with the role played by the nucleotide sequence in development, but rather the locations on the genome where there is variation associated with phenotypic variation—that is, places where the presence of different *alleles* (different versions of a gene) are associated with different developmental outcomes. When people use the language of a "gene for" some trait (including diseases), this is the sense they are usually (if perhaps unwittingly) pointing towards—a place where some version of a gene is associated with some particular trait.

3. Genetic Diseases: A Brief Typology of Kinds

Before interrogating the "genetic disease" concept, it is useful to consider some of the different forms of genetic disease that are usually identified.

Chromosomal abnormalities identify the problematic feature at the level of the chromosome; these can include extra chromosomes (e.g., trisomy 21, which causes Down syndrome), fewer than the normal number of chromosomes (e.g., Turner syndrome, caused by the absence of one of the two sex chromosomes, leaving only a single X chromosome), deletions of part of a chromosome, duplications of part of a chromosome, rearrangements of various kinds (e.g., inversions of parts of a chromosome, or material on different chromosomes swapping places), or structural defects in the chromosome that make it susceptible to breaking (see NIH Fact Sheet: Chromosome Abnormalities).

Complex diseases (aka multifactorial disorders), in which both genetic variation at several different locations and particular environmental variations are thought to play a role in the development of the disease, include essentially all common chronic illnesses, along with most forms of mental illness (see e.g., Craig 2008; McClellan and King 2010; Mitchell 2012). Indeed, while it has proven more difficult to find particular gene variants associated with complex diseases than many researchers had hoped (see Manolio et al. 2009; McClellan and King 2010), it seems clear that genetic variation plays some role in the development and the particular clinical course of *most* diseases, including even classic infectious diseases (see Chapman and Hill 2012). For the reasons explored below, it is difficult to state precisely what makes something a "complex" disease with a genetic component, as opposed to a "genetic" disease that strongly interacts with particular environments. But in general, it seems that if particular alleles increase the disease risk significantly individually, the trend is to think of the disease in question as an ordinary genetic disease, whereas if no individual allele is responsible for any marked increase in the disease risk (but relative risk among genetically similar individuals still suggests an important overall genetic component of some sort), then the disease is usually classed as "complex."

Single-gene diseases, as the name implies, are the result of some variation in a single gene—usually, in a nucleotide region that codes for a protein (or is part of a region that can code for several different proteins) or is part of a regulatory region. Diseases associated with single-gene variations are what are most commonly thought of as genetic diseases. As noted above, over 10,000 diseases in humans are associated with single-gene mutations or variants; most are quite rare. Single genes associated with diseases in humans can be classed according to the general pattern of inheritance that they show.

Autosomal genes are those that are on the *non-sex chromosomes* (i.e., on any chromosome *except* the X or Y). Diseases associated with *autosomal recessive* genes occur only when *two* copies of the mutant gene are inherited (one from each parent); *homozygotes* express the disease, but *heterozygotes* (who have only one copy) do not. Gene variants associated with autosomal recessive disorders are fairly common—indeed, Gao et al. (2015) estimate that on average everyone carries one or two autosomal recessive mutations that, in homozygote form, would lead to complete sterility or death. Rare autosomal recessive genes associated with disease may persist in a population because selection against them is very weak—for an allele that is harmful only in the homozygotic form, it is unlikely that any two individuals will both have that rare allele, and so unlikely that the condition will be expressed (as the vast majority of children will inherit at most one copy of the allele).

More common autosomal recessive genes can be the result either of *heterozygote advantage* or the "founder effect" (drift). Taking the case of heterozygote advantage first, consider sickle-cell anemia. Sickle-cell anemia is the result of being homozygotic for the *HbS* allele, a mutant form of the gene that makes hemoglobin (this is oversimplified: there are several different *HbS* alleles with different consequences for hemoglobin production, and the clinical manifestation of the disease even in homozygotes is variable, etc. See Rees, Williams, and Gladwin 2010). Heterozygotes for the *HbS* allele, however, are resistant to certain forms of malaria. In areas where malaria is common, heterozygotes have a strong fitness advantage. This results in an *equilibrium*; the frequency of the alleles in the population associated with sickle cell will depend on the strength of selection for heterozygotes (how much better heterozygotes do compared to homozygotes for the "normal" form of hemoglobin) and the strength of selection *against* homozygotes (so, for example, the frequency of "modifier" genes that protect homozygotes from the worst consequences of sickle-cell disease may also influence the allele frequencies in these cases; see Rees, Williams, and Gladwin 2010, 2023). The high frequency of cystic fibrosis in some populations has led researchers to suspect that the *CFTR* allele (the mutation associated with cystic fibrosis) might be associated with heterozygote advantage; however, nailing down precisely what advantage heterozygotes might have that could explain the frequency and distribution has been difficult (currently, resistance to tuberculosis is the favored hypothesis; see Mowat and Werpachowska 2015).

The "founder effect" occurs when a population goes through a "bottleneck"—something that reduces effective population size to relatively few individuals. This can occur when, for example, migration leads a small number of people to a new region (or when small populations are cut off from continued migration in other ways), or when a disease or other natural disaster radically reduces effective population size in a particular region. Under these conditions, genetic variation in the population is reduced, and previously rare alleles may increase in frequency via genetic drift. Previously rare alleles associated with autosomal recessive disorders may, under those conditions, become relatively more common, thus increasing the frequency with which the disease occurs (see Chong et al. 2012 for review). The high frequency of the alleles associated with Tay-Sachs disease in the Ashkenazi Jewish population, for example, is generally believed to be due to genetic drift and not to any selective advantage of heterozygotes (Risch et al. 2003).

Autosomal dominant alleles cause the disease in question in heterozygotes as well as homozygotes—only one copy of the allele associated with the gene is required for the disease state. *Huntington's disease* is a relatively common example of an autosomal dominant disease; despite its devastating consequences, the frequency of the allele associated with Huntington's is relatively high, perhaps due in part to Huntington's generally being a *late-onset* disease that, historically, would have struck most people *after* they had already reproduced (the mean age of the onset of symptoms is a bit over 40 years) (see Walker 2007).

Sex-linked genetic diseases involve genes on one of the sex chromosomes (X or Y). X-linked recessive disorders are relatively common among men (chromosomal males), who only inherit one copy of the X chromosome (along with a Y chromosome), and much less common among women (chromosomal females), who inherit two copies (this is complicated somewhat by X chromosome inactivation, the process by which one of the two X chromosomes is inactivated in females). Color blindness, for example, is the result of a mutation on the X chromosome that prevents the formation of certain photopigments; it is far more common in men than in women (color blindness in some form is present in perhaps 8–10% of men and only perhaps half of 1% of women) (see Neitz and Neitz 2000). X-linked dominant disorders are very rare; they are, however, more common among chromosomal females (who, inheriting two X chromosomes, are twice as likely to inherit one with any particular rare allele, though again, this is complicated by X-chromosome inactivation). Rett syndrome, for example, is usually thought of as an X-linked dominant trait. It is worth noting, however, that Rett syndrome occurs almost exclusively in females because it is almost invariably fatal in males—the differential activation of the two copies of the X chromosome in females provides some cells with the functioning *MECP2* genes, whereas males have none (see "Rett Syndrome Fact Sheet"). Part of what the complexities of inheritance patterns in these cases highlight is the inadequacy of the terms "recessive" and "dominant" when dealing with sex-linked genes, as only women will have alternative alleles available at loci on the X chromosome. Y-linked diseases are rare, in part because the Y chromosome is relatively small and contains relatively few genes; most of these diseases involve infertility, and of course only affect chromosomal males (see e.g., Jobling and Tyler-Smith 2000).

Mitochondria are subcellular organelles that are (mostly) maternally inherited through the egg cell (though limited paternal inheritance, from the sperm cell, may be possible; see Duchen 2004). Mitochondrial diseases can be caused by mutations in the mitochondrial DNA (mtDNA); though the mitochondria genome is small, it seems to have a relatively high mutation rate (Duchen 2004; Taylor and Turnbull 2005). Because many physically distinct mitochondria are inherited, mitochondria within a cell can be genetically different from one another; because they replicate within, but in a way partially independent from, the cells that contain them, different cell lineages within an individual can inherit different mitochondrial populations (Collins et al. 2002; Duchen 2004). The inheritance pattern for mitochondrial diseases tends to be complex, in part because, as the above suggests, mitochondria in a cell behave as a "population" with respect to transmission (see Taylor and Turnbull 2005). "Classic" diseases associated with mitochondrial DNA include Kearns–Sayre syndrome—like many mitochondrial diseases, the typical clinical manifestation of this disease includes problems with muscle tissue, heart problems, and eye issues; it is caused by a "large-scale" deletion that knocks out several genes (Taylor and Turnbull 2005, see their Table 2 for a list of mitochondrial diseases). But mitochondria mutations are implicated in some cases of a large number of conditions, only some of which are specific to mitochondria (see Taylor and Turnbull 2005).

4. Genetic Diseases: Concepts and Complexities

As noted above, it is difficult to come up with a definition of "genetic disease" that adequately captures our ordinary understanding of the concept. Part of the problem is that every trait, including disease states, is the result of a complex developmental process that includes any number of different developmental resources. Part of what identifying something as a "genetic disease" does is to point towards a particular developmental resource (a "gene" or "genes") in this process and to identify it as causally special, but explaining why we point towards genes in some cases but not others is tricky.

Consider a simple example: is lactose intolerance (lacking lactase persistence) a "genetic disease" (this example is based on Hesslow 1984; see also Smith 2007)? In the case of populations that have a tradition of raising dairy animals, almost everyone has alleles associated with lactase persistence (the ability to keep producing the enzymes necessary to digest the lactose in milk after infancy), and consuming milk is a regular part of the culture (see, e.g., Ingram and Swallow 2007). Here, someone who lacks a functional version of the allele that, in that population, is associated with lactase persistence, will be lactose intolerant. And given that dairy consumption is a part of the culture's usual diet, he/she will likely be identified as having a (medical) problem that can be tied to a particular variation in the genome—a genetic disease. But now consider a population without a tradition of raising dairy animals; in these populations, almost no one has alleles associated with lactase persistence, and dairy products are not normally consumed after infancy. In such a population, someone who, as an adult, nevertheless consumes a dairy product will get sick, but it would seem odd to blame his or her sickness on the lack of the lactase enzyme (since there was no reason to expect that he or she would have the enzyme after infancy). Rather, the (medical) problem the person has is more naturally thought to be that he or she ate something he or she ought not to have eaten, something that we would expect, in that population, to result in an illness. In the first case, a genetic difference from the relevant population seems explanatory; in the second case, an environmental difference from the relevant population seems to be doing the work. But in both cases, the illness results from *both* lacking the gene to make the relevant enzyme *and* consuming food with lactose.

In a somewhat similar vein, consider one of the class of inherited immunodeficiency diseases; these range from Severe Combined Immunodeficiency Syndrome (SCIDS—there are several different genetic disorders that result in this near-total failure to develop immune function; these are fatal in the absence of treatment) to conditions that result in an increased risk from only certain classes of infectious diseases (see e.g., Janeway et al. 2001, Chapter 11). These conditions are generally regarded as "genetic diseases." But while AIDS is caused by the HIV virus, and so generally regarded as an infectious disease itself, people with the CCR5Δ32 allele are highly resistant to HIV infection; indeed, to date, the only person cured of HIV was treated for leukemia with a bone marrow transplant from a donor with the CCR5Δ32 allele (the patient, at last report, was still virus-free despite no longer taking antivirals) (see Hütter and Thiel 2011). It would seem at best very odd, however, to claim that HIV/AIDS was a genetic disease, caused by the lack of the CCR5Δ32 allele! However, the two cases are, on one description at least, very similar—an infectious agent is only problematic when combined with an immune system lacking certain elements, where those lacks are related to the absence of particular alleles. The difference between the two cases seems to be that *most* people have immune systems able to handle the infectious agents that people with SCIDS cannot, whereas the CCR5Δ32 allele is relatively rare, and so *most* people are susceptible to HIV.

More generally, as McClellan and King point out, genetic diseases are far more heterogeneous than usually thought, in a number of different ways, including at least:

> the same gene may harbor many (hundreds or even thousands) different rare severe mutations in unrelated affected individuals . . . the same mutation may lead to different clinical manifestations (phenotypes) in different individuals; and . . . mutations in different genes in the same or related pathways may lead to the same disorder.
>
> (2010, 210)

McClellan and King are pointing out that the relationship between genes and diseases, even in cases of genetic diseases normally identified as straightforward single-gene disorders, is "many-many" rather than "one-to-one." *Different* mutations in the same region (the same gene) can lead to the same disease (the same phenotype), as can mutations in *different* regions (in different genes); the same illness can be associated with many different genes. Indeed, in some cases a disease with the same clinical features may be associated with either a particular gene or a particular environment. Further, the same mutation—the same change in a particular gene—can lead to *different* phenotypes, from severe forms of a particular disease to apparently normal phenotypes (no clinical manifestations of the disease); the same gene can be associated with many different forms of an illness (or with no illness at all).

PKU (phenylketonuria or "phenylalanine hydroxylase deficiency") provides a good entry into these issues, in part because it has been intensely studied and is relatively well-understood (see, e.g., Williams, Mamotte, and Burnett 2008 and cites therein). PKU is (usually thought of as) a genetic disease caused by having two copies of a mutated human phenylalanine hydroxylase gene (the *PAH* gene), where both copies fail to produce a (sufficient quantity of) properly functional phenylalanine hydroxylase enzyme (the PAH enzyme). People with PKU therefore lack the ability to properly metabolize the amino acid phenylalanine (Phe); untreated, this can lead to the "accumulation of toxic by-products of Phe metabolism," along with shortages of some key metabolites usually produced (primarily Tyrosine, Tyr) (see Williams, Mamotte, and Burnett 2008). Untreated PKU is associated with a range of problematic phenotypic outcomes, including severe deficits in cognitive development and other cognitive problems, and a range of skin and bone problems. Treatment consists mainly of a special diet, very low in phenylalanine, with supplemental tyrosine, although drugs to help metabolize phenylalanine are increasingly part of standard clinical practice (see Vockley et al. 2014). Because treatment is more effective the earlier it is started, most countries test all newborns soon after birth for PKU (along with a number of other disorders). Indeed, PKU screening is generally regarded as the first universal newborn genetic screening program; newborn screening for PKU was first deployed in the 1960s, and by the 1970s newborn screening for PKU (along with some other disorders) had become standard practice in most developed counties (see Brosco and Paul 2013).

The above provides the "sketch" for the "standard" story about PKU. Although there is nothing actually wrong with this story, it is oversimplified in important ways, and perhaps gives a misleading impression of the relationship between both the relevant genetic mutations and the ability (or lack thereof) to metabolize phenylalanine, and the inability to metabolize phenylalanine and phenotypic effects of that inability. Indeed, the relationships here are complex in just the ways that McClellan and King (2010) suggested: many different mutations can give rise to "the same" disease state, the same mutations are associated with different metabolic activity levels, and the same metabolic activity levels are associated with different phenotypic outcomes.

The *PAH* gene codes for the PAH enzyme, a protein made up of 452 amino acids; as might be expected, there are many ways for things to go wrong and many different mutations in the *PAH* gene that are associated with PKU (more than 500 have been identified to date; see Williams, Mamotte, and Burnett 2008, 33). The majority of mutations in PKU individuals seem to be *missense* mutations—places where the "wrong" amino acid was inserted into the protein (see Williams, Mamotte, and Burnett 2008, 33); as suggested above, different missense mutations are associated with different proteins, with different levels of enzymatic activity; many of these missense mutations seem to be problematic largely because they cause the proteins to misfold and (perhaps) to be more rapidly degraded (see Scriver and Waters 1999). Other mutations—nonsense mutations (where a mutation results in a "stop" code and prevents the formation of the full protein), frameshift mutations (where a deletion or insertion of a nucleotide results in a change to all downstream nucleotide triplets, and hence many changes in the amino acids produced), etc.—result in something more like a total lack of enzymatic activity, as even if a protein is produced, it is far too dissimilar to the PAH enzyme to do any of the same work (see Williams, Mamotte, and Burnett 2008, 33).

As Scriver and Waters (1999) note, "there are many instances of discordance between the mutant *PAH* genotype, its predicted effect on enzyme function, and the associated metabolic phenotype" (268). As noted above, different mutations in the *PAH* gene are statistically associated with different levels of enzymatic function, some of which can be predicted from the protein formed, but this association shows significant variation. The mechanisms by which "missense" mutations result in the loss of enzymatic function in the case of PAH enzyme are complex, and differences in the "chaperone" systems (chemical systems that help the protein fold into the correct final form) between individuals may be responsible for the same *PAH* mutations being associated with different levels of enzymatic function (see Scriver and Waters 1999, 268–270). Furthermore, the same level of enzymatic function can be associated with different metabolic outcomes—while two people might share PAH enzymes that are equally unable to metabolize phenylalanine, one might for example be better able to eliminate excess phenylalanine via other metabolic pathways (Scriver and Waters 1999, 269). And even if two individuals have equal metabolic responses to phenylalanine, the *effect* of that metabolic response can be different; differences in the particulars of the blood-brain barrier, for example, can mediate the influence of excess phenylalanine on cognitive development (Scriver and Waters 1999, 268, 270). The result of this is that "patients, even sibs, sharing identical mutant *PAH* genotypes could have greatly different cognitive and metabolic phenotypes" (Scriver and Waters 1999, 268).

Along with the large number of possible errors in the *PAH* gene, in thinking about the many-many relationship between genes and disease states, it is interesting to reflect on tetrahydrobiopterin (BH_4) deficiency, a much rarer genetic disease (PKU is present in perhaps 1:10,000 live births; tetrahydrobiopterin deficiency is present in perhaps 3:1,000,000 or so, though the frequency varies significantly by region; see Blau, Bonafé, and Blaskovics 2005). BH_4 is necessary for the proper metabolism of phenylalanine, and deficiency can result in hyperphenylalaninaemia, the same excess of phenylalanine that characterizes PKU; mutations in one of a number of genes can result in this BH_4 deficiency, and, like mutations in the *PAH* genes, the impacts of these mutations on phenylalanine metabolism can range from mild to severe (see Blau, Bonafé, and Blaskovics 2005). Untreated, the phenotypic impacts of tetrahydrobiopterin deficiency are broadly similar to PKU; indeed, in the 1970s, it was common to refer to tetrahydrobiopterin deficiency as a kind of atypical PKU (see, e.g., Kaufman et al. 1975; Curtius et al. 1979; see Ponzone et al. 2004 for discussion). In recent years, however, it has become standard practice to distinguish PKU (caused by mutations in the *PAH* gene) from tetrahydrobiopterin deficiency, both because they have a different cause and because different treatments are recommended (BH_4 replacement therapy is standard for tetrahydrobiopterin deficiency, sometimes with and sometimes without dietary phenylalanine restriction; see Ponzone et al. 2004; Blau, Bonafé, and Blaskovics 2005; Blau, van Spronsen, and Levy 2010). However, the fact that some forms of "traditional" PKU also respond well to BH_4 therapy (for some patients, increasing BH_4 levels improves PAH enzyme function impaired by errors in the *PAH* gene; see Blau and Erlandsen 2004; Blau, van Spronsen, and Levy 2010) makes this distinction a little less clear-cut than it is sometimes made out to be.

Are PKU and tetrahydrobiopterin deficiency best thought of as two variants of the same disease ("hyperphenylalaninaemia"), or are they best viewed as two different diseases that share some biological pathways and have some broadly similar phenotypic (clinical) outcomes? Is a case of PKU caused by the complete absence of a functional PAH enzyme (say, from a "nonsense" mutation that stops transcription too early), a severe form of the same disease as a case of PKU associated with reduced (but still functional) PAH enzyme activity, or are these different diseases? In individuating "genetic diseases" should we be focused on the gene whose function is being disrupted (all and only disruptions of the *PAH* gene are the same disease, PKU), on the particular form of the disruption (all and only similar failures of PAH enzyme are the same form of PKU), or on the downstream metabolic and phenotypic consequences (all and only conditions that result in the buildup of excess phenylalanine are PKU)? Answering these questions would seem to demand making decisions about the proper use of language, rather than discovering straightforward facts about the world; depending on our particular interests and research agendas, either answer seems defensible. Although one might avoid the conclusion that the same disease can be caused by different mutations in the same gene, or by mutations in different genes, by declaring, as a matter of fiat, that genetic diseases are to be individuated on the basis of the specific mutation associated with them, doing so would merely elide the force of the "many-many" problem.

5. Conclusion: Reflections on the Genetic Disease Concept

Of something purported to be a genetic disease, we can ask at least two questions: How likely is someone to have the disease in question if they have the genetic mutation in question? and How likely is someone to have the disease in question if that person *lacks* the genetic mutation in question? "Ideal" genetic diseases are those in which a particular mutation makes the presence of the disease *very likely* (nearly certain), and in which the clinical manifestation of that disease is *very rare* (all but absent) in the absence of the genetic mutation in question. Smith suggests an "epidemiological" account of "genetic disease," where, "intuitively," (1) "If a disease is genetic, this must mean that those with the gene are more likely than not to develop the disease," and (2) "If a disease is genetic, this must mean that cases of disease more likely than not causally involve the gene in a significant way" (2007, 97). Smith goes on to unpack the work being done by notions of causation in the intuitive account, in order to produce a more precise "minimally epidemiological" account of genetic disease, where (1) gets interpreted in terms of a "*Population Etiologic Fraction* (PEF)"—the fraction of individuals with the disease "whose disease causally involved the gene" and (2) gets cashed out as an "*Attributable Risk* (AR)"—the fraction of people with the gene "whose disease causally involved the gene" (Smith 2007, 100–101; see also Smith 2001). Smith suggests that if both numbers are greater than 50%, the minimal epidemiological account of something's being a genetic disease has been met (Smith 2007, 102).

As Smith notes, the trouble with these accounts is that, in many cases, we simply don't know either the "attributable risk" or the "population etiologic fraction" (2001, 2007, 104). For many "classic" genetic diseases, the belief that everyone, or almost everyone, with the relevant genes would reveal the standard clinical manifestations of the disease was only challenged with the rise of genetic testing; for example, when testing for the most common mutations associated with cystic fibrosis was established, some individuals were discovered with mutations usually associated with the classic forms of the disease, but who were asymptomatic or only mildly symptomatic (see Zielenski 2000; Doull 2001). More generally, as Smith points out, for cystic fibrosis every possible combination of clinical symptoms, sweat chloride level, and genotype has been identified (Smith 2001, 21).

The "population etiologic fraction" is even more likely to be controversial, because, as noted above, for many conditions researchers identify the disease with the cause; when a different causal pathway is uncovered, it is regarded as generating a different disease, even if the clinical manifestations are broadly similar. If we simply declare that nothing is PKU unless it is associated with a mutation in the *PAH* gene that results in functional changes to the PAH enzyme, then of course the "population etiologic fraction" will be very high indeed (100%, by fiat). If we individuate diseases in terms of their most common clinical manifestations, then more diseases will be caused by a number of different factors, where some are genetic and others environmental. But again, which of these approaches is to be preferred would seem to depend on our particular projects (so, for example, the different projects of developing diagnostic tools, developing treatments, and developing classificatory schemes might suggest different ways of individuating diseases). Note well that which projects seem most plausible and worth pursuing will be sensitive to any number of factors—cultural, political, economic, and so forth.

The difficulty in specifying what, precisely, makes something a genetic disease emerges in part from the complexity of the biological world. It is difficult to specify what makes something a disease and what constitutes health. The role played by nucleotide sequences in development are varied and complex, and it is difficult to say what makes some nucleotide sequence a "gene." The genetic disease concept would seem to rely on a distinction between "normal" and "abnormal" genetic sequences that is hard to make precise (and perhaps hard to defend). Arguments about whether to count diseases with partial genetic etiologies as genetic diseases—whether "complex" diseases are a *kind* of genetic disease, or something else entirely—seem to depend more on the researchers' outlooks than anything else.

For all that, while making the distinction precise may prove difficult, the genetic disease concept is firmly entrenched in biomedicine. Although even some of the most "classic" genetic diseases have proven to be far more complicated than anyone would have predicted, those diseases are still regarded as genetic diseases. If we need to adopt a method for distinguishing genetic diseases from other kinds of diseases, Smith's approach (2001, 2007) seems the most adequate attempt so far. It is not perfect, but it does seem to capture at least some of the key intuitions behind the distinction.

References

Amundson, R. (2005). "Disability, ideology, and quality of life: A bias in biomedical ethics." In D. Wasserman, J. Bickenbach, and R. Wachbroit, eds. *Quality of life and human difference*. Cambridge, UK: Cambridge University Press, pp. 101–124.

Blau, N., L. Bonafé, & M. Blaskovics. (2005). "Disorders of phenylalanine and tetrahydrobiopterin metabolism." In N. Blau, M. Duran, M. Blaskovics, K.M. Gibson, eds. *Physicians' guide to the laboratory diagnosis of metabolic disease*. Heidelberg: Springer, pp. 89–106.

Blau, N., & Erlandsen, H. (2004). The metabolic and molecular bases of tetrahydrobiopterin-responsive phenylalanine hydroxylase deficiency. *Molecular Genetics and Metabolism*, 82(2), 101–111.

Blau, N., van Spronsen, F. J., & Levy, H. L. (2010). Phenylketonuria. *The Lancet*, 376(9750), 1417–1427.

Brosco, J.P., & Paul, D.B. (2013). The political history of PKU: Reflections on 50 years of newborn screening. *Pediatrics*, 132(6), 987–989.

Chapman, S. J., & Hill, A. V. (2012). Human genetic susceptibility to infectious disease. *Nature Reviews Genetics*, 13(3), 175–188.

Chong, J. X., Ouwenga, R., Anderson, R. L., Waggoner, D. J., & Ober, C. (2012). A population-based study of autosomal-recessive disease-causing mutations in a founder population. *The American Journal of Human Genetics*, 91(4), 608–620.

Collins, T. J., Berridge, M. J., Lipp, P., & Bootman, M. D. (2002). Mitochondria are morphologically and functionally heterogeneous within cells. *The EMBO Journal*, 21(7), 1616–1627.

Condit, C. M., Bates, B. R., Galloway, R., Givens, S. B., Haynie, C. K., Jordan, J. W., . . . & West, H. M. (2002). Recipes or blueprints for our genes? How contexts selectively activate the multiple meanings of metaphors. *Quarterly Journal of Speech*, 88(3), 303–325.

Craig, J. (2008). Complex diseases: Research and applications. *Nature Education*, 1(1), 184.

Curtius, H. C., Niederwieser, A., Viscontini, M., Otten, A., Schaub, J., Scheibenreiter, S., & Schmidt, H. (1979). Atypical phenylketonuria due to tetrahydrobiopterin deficiency: Diagnosis and treatment with tetrahydrobiopterin, dihydrobiopterin and sepiapterin. *Clinica Chimica Acta*, 93(2), 251–262.

Doull, I. J. (2001). Recent advances in cycstic fibrosis. *Archives of Disease in Childhood*, 85(1), 62–66.

Duchen, M. R. (2004). Mitochondria in health and disease: Perspectives on a new mitochondrial biology. *Molecular Aspects of Medicine, 25*(4), 365–451.

Dye, D. E., Brameld, K. J., Maxwell, S., Goldblatt, J., Bower, C., Leonard, H., . . . & O'Leary, P. (2011). The impact of single gene and chromosomal disorders on hospital admissions of children and adolescents: A population-based study. *Public Health Genomics, 14*, 153–161.

Gao, Z., Waggoner, D., Stephens, M., Ober, C., & Przeworski, M. (2015). An estimate of the average number of recessive lethal mutations carried by humans. *Genetics, 199*(4), 1243–1254.

Griffiths, P. E., & Stotz, K. (2006). Genes in the postgenomic era? *Theoretical Medicine and Bioethics, 27*(6), 499–521.

Hanley, B., Dijane, J., Fewtrell, M., Grynberg, A., Hummel, S., Junien, C., . . . & van Der Beek, E. M. (2010). Metabolic imprinting, programming and epigenetics–a review of present priorities and future opportunities. *British Journal of Nutrition, 104*(S1), S1–S25.

Hennekam, R. C. M. (2011). Care for patients with ultra-rare disorders. *European Journal of Medical Genetics, 54*(3), 220–224.

Hesslow, G. (1984). "What is a genetic disease? On the relative importance of causes." In L. Nordenfelt & B. L. B. Lindahl (eds.), *Health, Disease, and Causal Explanations in Medicine*. Dordrecht, Netherlands: D. Reidel, pp. 183–193.

Hütter, G., & Thiel, E. (2011). Allogenic transplantation of CCR5-deficient progenitor cells in a patient with HIV infection: An update after 3 years and the search for patient no. 2. *AIDS, 25*(2), 273–274.

Ingram, C. J. E., & Swallow, D. M. (2007). "Population genetics of lactase persistence and lactose intolerance." *Encyclopedia of life sciences*. John Wiley & Sons, Ltd. www.els.net.

Janeway, C. A. Jr, Travers, P., Walport, M., & Shlomchik, M. (2001). *Immunobiology: The immune system in health and disease*. 5th edition. New York: Garland Science. Inherited immunodeficiency diseases. http://www.ncbi.nlm.nih.gov/books/NBK27109/

Jobling, M. A., & Tyler-Smith, C. (2000). New uses for new haplotypes the human Y chromosome, disease and selection. *Trends In Genetics: TIG, 16*(8), 356–362.

Kaufman, S., Holtzman, N. A., Milstien, S., Butler, I. J., & Krumholz, A. (1975). Phenylketonuria due to a deficiency of dihydropteridine reductase. *New England Journal of Medicine, 293*(16), 785–790.

Kingsmore, S. F., Dinwiddie, D. L., Miller, N. A., Soden, S. E., Saunders, C.J., & The Children's Mercy Genomic Medicine Team. (2011). Adopting orphans: Comprehensive genetic testing of Mendelian diseases of childhood by next-generation sequencing. *Expert Rev Mol Diagn, 11*(8), 855–868.

Latchman, D.S. 2011. Transcriptional gene regulation in eukaryotes. In: eLS. John Wiley & Sons Ltd, Chichester. http://www.els.net [doi: 10.1002/9780470015902.a0002322.pub2].

Manolio, T. A., Collins, F. S., Cox, N. J., Goldstein, D. B., Hindorff, L. A., Hunter, D. J., . . . & Visscher, P. M. (2009). Finding the missing heritability of complex diseases. *Nature, 461*(7265), 747–753.

McCandless, S. E., Brunger, J. W., & Cassidy, S. B. (2004) The burden of genetic disease on inpatient care in a children's hospital. *Am. J. Hum. Genet., 74*, 121–127.

McClellan, J. & King, M. C. (2010). Genetic heterogeneity in human disease. *Cell, 141*(April 16), 210–217.

Mitchell, K. J. (2012). What is complex about complex disorders? *Genome Biology, 13*(1): 237.

Moss, L. (2004). *What genes can't do*. Cambridge, MA: MIT Press.

Mowat, A. J., & Werpachowska, A. (2015). Why does cystic fibrosis (CF) show the prevalence and distribution observed in human populations? A literature review. *Journal of Cystic Fibrosis, 14*, S56.

Murphy, D. (2015). "Concepts of Disease and Health." Stanford Encyclopedia of Philosophy. http://plato.stanford.edu/archives/spr2015/entries/health-disease/ Accessed 5/8/2015.

Neitz, M., & Neitz, J. (2000). Molecular genetics of color vision and color vision defects. *Archives of Ophthalmology, 118*(5), 691–700.

NIH Fact Sheet. "Chromosome Abnormalities." http://www.genome.gov/11508982

Oyama, S. E., Griffiths, P. E., & Gray, R. D. (2001). *Cycles of contingency: Developmental systems and evolution*. Cambridge, MA: MIT Press.

Ponzone, A., Spada, M., Ferraris, S., Dianzani, I., & de Sanctis, L. (2004). Dihydropteridine reductase deficiency in man: From biology to treatment. *Medicinal Research Reviews, 24*(2), 127–150.

Portin, P. (2009). The elusive concept of the gene. *Hereditas, 146*, 112–117.

Rees, D. C., Williams, T. N., & Gladwin, M. T. (2010). Sickle-cell disease. *The Lancet*, 376(9757), 2018–2031.

"Rett Syndrome Fact Sheet," National Institute of Neurological Disorders and Stroke (NINDS). NIH Publication No. 09–4863. Publication date December 2009. http://www.ninds.nih.gov/disorders/rett/rett_syndrome_brochure_508comp.pdf

Risch, N., Tang, H., Katzenstein, H., & Ekstein, J. (2003). Geographic distribution of disease mutations in the Ashkenazi Jewish population supports genetic drift over selection. *The American Journal of Human Genetics*, 72(4), 812–822.

Scriver, C. R., & Waters, P. J. (1999). Monogenic traits are not simple: Lessons from phenylketonuria. *Trends in Genetics*, 15(7): 267–272.

Smith, K. C. (2001). A disease by any other name: Musings on the concept of a genetic disease. *Medicine, Health Care and Philosophy*, 4(1), 19–30.

Smith, K. C. (2007). "Towards an adequate account of genetic disease." In H. Kincaid & J. McKitrick (eds.), *Establishing Medical Reality: Essays in the Metaphysics and Epistemology of Biomedical Science*. Dordrecht, Netherlands: Springer, pp. 83–110.

Taylor, R. W., & Turnbull, D. M. (2005). Mitochondrial DNA mutations in human disease. *Nature Reviews Genetics*, 6(5), 389–402.

Vockley, J., Andersson, H. C., Antshel, K. M., Braverman, N. E., Burton, B. K., Frazier, D. M., . . . & Berry, S. A. (2014). Phenylalanine hydroxylase deficiency: Diagnosis and management guideline. *Genetics in Medicine*, 16, 188–200.

Walker, F. O. (2007). Huntington's disease. *The Lancet*, 369(9557), 218–228.

Waters, C. K. (2007). Causes that make a difference. *The Journal of Philosophy*, 104(11), 551–579.

Williams, R. A., Mamotte, C. D. S., & Burnett, J. R. (2008). Phenylketonuria: An inborn error of phenylalanine metabolism. *Clinical Biochemistry Review*, 29(February), 31–41.

World Health Organization. (2015). "Genes and Human Disease." Genomic resource centre. http://www.who.int/genomics/public/geneticdiseases/en/ Accessed 5/8/2015.

Zielenski, J. (2000). Genotype and phenotype in cystic fibrosis. *Respiration*, 67(2), 117–133.

Further Reading

The National Center for Biotechnology Information, part of the National Library of Medicine, maintains an excellent collection of materials related to particular genetic diseases ("Genes and Diseases") at http://www.ncbi.nlm.nih.gov/books/NBK22183/.

Along with Smith 2001 and 2007 (cited above), Smith 1992 "The New Problem of Genetics" (*Biology and Philosophy* 7:331–348) and 2001 "Genetic Disease, Genetic Testing, and the Clinician," (*JAMA* 286(1):91–91) provide a fuller picture of Smith's views.

On the problems with thinking of genes as carrying information, constituting "plans," or comprising a special class of "replicators," etc., Susan Oyama's *The Ontogeny of Information* (1985 Duke University Press; new edition 2000) remains a classic treatment, and Griffiths & Gray (1994), "Developmental systems and evolutionary explanation," (*The Journal of Philosophy* 91(6): 277–304) an excellent introduction.

15
DIAGNOSTIC CATEGORIES

Annemarie Jutel

Diagnosis is at the heart of Western medicine and plays an important role in structuring care for the individual patient as well as in creating priorities and measuring outcomes for public health. It is usually in the pursuit of diagnosis that the layperson enters the medical system. Dysfunction or distress, determined by the sufferer to be medical in nature, is brought to the doctor for explication, and that explication is most frequently couched in diagnosis. The diagnosis, both a process and a category (Blaxter, 1978), organizes the symptoms, provides a sense of direction to the patient, points (hopefully) to a treatment, describes the future (prognosis), allocates resources (e.g., sick leave, prescriptions, insurance reimbursement), and even designates a social role to the newly diagnosed patient (Balint, 1964, Leder, 1990, Jutel, 2011).

Diagnosis is a powerful tool as it simultaneously parcels out treatment and other social goods and functions. The diagnosis will determine what sub-specialty should be involved, what resources should be provided, and whether the outlook is bright or dim. It also is the conduit by which the patient assumes the "sick role"(Parsons, 1958), allowing the individual to withdraw from regular social expectations, like going to work or other duties, in exchange for compliance with medical orders. A diagnosis can also transform the identity of the diagnosed person. In the language of diagnosis, one "becomes" diabetic, hypertensive, psychotic. This is the linguistic troping by health professionals that Fleishman has described, where, via synecdoche, the patient *becomes* the ailment (Fleischman, 1999).

It is hard to imagine medicine without diagnostic categories. To develop evidence to determine the best practice, the most effective treatment, the likely outcomes, categories are essential. Without categories, generalization is not possible; without generalization, science cannot fulfill its roles. Statistical analysis is only possible with categories to represent outcomes and interventions.

Outside of medicine as well, we have difficulty speaking of dysfunction without diagnostic categories. As medicine's authority expands, we increasingly use diagnostic language, swapping sadness for depression, distractibility for ADD, and shyness for social anxiety disorder or social phobia. Using the language of medicine adds credibility to suffering, absolves the sick person of many responsibilities, and gives a sense of hope as well: diseases have treatments. Self-diagnosis is also on the rise, with diagnostic apps and online checklists to help shape the experience of dysfunction in medical terms.

Despite the "realness" of disease and dysfunction, the process of categorizing and classifying disease states is one that relies heavily on the social and the political. In the pages that follow, I will describe how diagnostic categories function socially. Considering the social features of diagnostic categories as well as the pathophysiological nature of disease can adumbrate a number of clinical and public health challenges. It punctuates how diagnosis can be a point of tension between patient and physician rather than a point of agreement; how technology can disappoint or trouble a diagnostic picture; and how culture and identity are both tools in, and obstacles to, medicine's success.

Is Nature Really Jointed?

Categories are about cutting nature up into useful packages, what Eviatar Zerubavel (1991) calls "islands of meaning." The question of whether nature comes pre-parceled, with natural categories ("jointed," as Plato suggested), or whether nature is categorizable in unlimited ways, is well known to philosophers. Thomas Arnold espoused the latter. He wrote:

> We are not to suppose that there are only a certain number of divisions in any subject, and that unless we follow these, we shall divide it wrongly and unsuccessfully: on the contrary, every subject is, as it were all joints, it will divide wherever we choose to strike it, and therefore according to our particular object at different times we shall see fit to divide it very differently.
>
> (Arnold, 1839)

When we name things as belonging to one group as opposed to another, we are imposing a structure on our environment that reflects our values, knowledge, and beliefs. Categories and classification thus are at the heart of how we understand the world.

In medicine, how we break up nature into manageable and explainable parts not only reflects the society or group that makes the divisions but also shapes how we will respond to our conditions. How we categorize says as much about who we are as it does about pathophysiology.

Diagnoses are the categories we use to create order, sort through particular symptoms and presentations, place them together or apart, and do the work of medicine. The categories used in medicine reify, serve as heuristic and didactic structures, determine the treatment protocol, predict the outcome, and provide a sense of identity for laypeople and professionals. In the pages that follow, we will review the tradition of medical classification and diagnostic categories, explore how and why we classify disease as we do today, and develop an understanding of the consequences of diagnostic categorization.

Principles of Classification

Classification is based on notions of difference and similarity. It seeks to find items that have more in common with one another than phenomena belonging to another category (Zerubavel, 1996). To classify is to overlook some similarities in favor of others, and similarly to overlook some differences in favor of resemblances. The challenge of classification is to make useful the placing of particular items with one another rather than with something else, without as a result obfuscating important differences that would be better revealed. Occam's Razor, dating from the 14th century, underlines this clearly. He postulated that "entities are not to be multiplied beyond necessity" (Leff, 1997).

We can see how Occam's principle works by thinking about Skip the dog. One could see Skip quite pragmatically as a dog, or possibly as a "mixed-breed, collie-type, mature, male dog." The former is probably all that matters for most cases. But, if we were to classify him as a four-legged, red-furred, long-haired, barking, 8-year-old, male creature named Skip, the reference could probably only be about one dog, as opposed to a class of dogs, which is not useful for categorizing dogs in general, even though it helps to identify Skip the dog from other dogs. By the same token, while classifying dogs, stopping at the level of "creatures" is hardly helpful, as it is so general a term as to include both spiders and Skip.

When discussing disease, the same rules apply. We can see each case of illness as unique, with associated comorbidities, personal circumstances, individual disease expression, and so forth. However, making generalizations about illness is useful. How would we undertake research, examine disease patterns, predict outcomes, without classifying? Classification organizes knowledge (Jacob, 1992); identifies clusterings and similarity (Fleiss et al., 1971); creates stable and predictive classes (Silvestri and Hill, 1964); and perhaps, above all, reduces a disorderly mass to an orderly whole (Richardson, 1901).

So, while categorizing, rather than seeing each case of illness as separate and unique, is certainly helpful in advancing the aims of medicine, as in the case of Skip the dog, categorizing obfuscates difference in the cause of generalization. A person categorized by diagnosis is nonetheless still an individual case, a point the clinician must bear in mind. At the same time, however, the purpose of classification is to generalize, rather than to individualize, with all the benefits that this can bring to understanding a particular situation. Classifying becomes thus a means of summarizing many cases of disease in a way that helps organize human illness in meaningful ways.

Classification Systems

The earliest surviving compilation of diagnostic categories is probably the Ebers Papyrus, which, written circa 1500 BCE, identified remedies for a range of conditions from abdominal obstruction to the "rose," or what we would call herpes zoster today (Veith, 1982). The purpose of this papyrus was to list the remedies for each of these disorders. Treatments were linked to diagnosis in different ways than today, with the proper cure thought to be found in the inverse of the disease (Daly and Brater, 2000).

Another historical example of diagnostic categorization comes from what is often referred to today as the "Galenic tradition" (even though it was in fact initiated by Empedocles some 700 years prior in circa 500 BCE). This categorization saw disease based in the four humors: hot, cold, dry, and moist (Daly and Brater, 2000). Treatment hinged on rebalancing an excess of whichever humor accounted for the diseased state.

Diagnostic categories vary not only over time but also geographically. Traditional Chinese medicine positions the disharmony of yin and yang, qi, meridians, and more at the base of diagnosis, while Tibetan medicine links diagnosis to the three poisons, which give rise to three humors, linked in turn to earth, water, fire, air, and space (Janes, 1995).

The earliest classification of diseases to provide a model for the kinds of records we keep today for public health in Western medicine was likely John Graunt's *Natural and Political Observations*, published in 1662 (Graunt, 1662a), which described the state of the kingdom, its strengths and its vulnerabilities, via the diseases that caused death. In his "Reflections on the Weekly Bills of Mortality" (Graunt, 1662b), Graunt mused about categories of disease, how they were arrived upon, and whether or not they were variable depending upon who used them.

Graunt discussed distinctions among different groups and which categories could be made more explicit. What number of years, he asked, should be called "aged"? Is it, as he suggested it should be, greater than 70? And what is an infant? Is it a child who cannot yet speak, or rather, one who is under two or three years of age? He identified the problems relating to the individuals who undertake classification, their status, and integrity. Can those who record the causes of deaths distinguish fairly between emaciation due to consumption, lung disease, hectic fever, or infection, he wondered, particularly after a glass of ale, or "the bribe of a two-groat fee" (Graunt, 1662b, p. 22)? Disease categories contained in Graunt's analysis included such today-picturesque conditions as "French pox," "purple fever," "rising of the lights," "stopping of the stomach," or "made away themselves."

Belief in an orderly arrangement of nature underpinned other disease classification theories, such as that devised by Thomas Sydenham, a 17th-century English physician (and John Graunts' contemporary). A convinced empiricist, Sydenham wrote that "[a]ll diseases then ought to be reduc'd to certain and determinate kinds, with the same exactness as we see it done by botanic writers in their treatises of plants" and this with the view to "the improvement of physick" (Sydenham, 1742, p. iiv–iv). This classification enables the doctor to "distinguish [a disease] from all other distempers" (p. xvi). He believed that the improvement of medicine depended on creating a collection of disease descriptions and their concordant "methods of cure" (p. ii). Such a collection was based on Sydenham's unwavering belief in a natural world whose existence was available for discovery by the careful observer.

Later classification projects were organized to capture and understand rules of nature, and, as with all things, the presence of divine order. As one 19th-century mother wrote to her children about the study of classification:

> We have reviewed all the classes of beings from insects to man, and we have recognised the hand of God everywhere. We have seen everywhere order and harmony which force our admiration. It would be impossible for us in the face of all these marvels of nature, not to praise he who created all.
> (Anon, 1840, p. 72)

Today

There are dozens of different disease classification systems in use today: general diagnostic systems such as Systematized Nomenclature of Medicine (SNOMED), *International Classification of Diseases* (ICD), *Diagnostic and Statistical Manual of Mental Disorders* (DSM), Read Codes, and so on; specific diagnostic systems such as international classification of headaches, sleep disorders, and cerebral palsy; and administrative diagnostic systems such as Diagnosis Related Groups and Major Diagnostic Groups. Different countries have nation-specific modifications (e.g., the ICD-10-AM/ACHI/ACS, which is the Australian modification of the ICD-10, and the AR-DRG, which is an "Australian Refined Diagnosis Related Groups" classification). Although these diagnostic systems have much in common, they also have many differences that are best understood in terms of social and political factors.

For example, there are differences between how ICD and DSM see particular (mental) disorders and between how different countries apply diagnostic criteria. Diagnoses can be clinical or administrative, and, particularly in countries where medicine is not socialized (and even in those where it is), an administrative disease category may be different than a clinical one. In many American practices, a clinical coder will come after a medical diagnosis is made to determine the most effective diagnostic code for obtaining advantageous reimbursement. Whether a person's mental distress is classified as major depressive disorder, recurrent; major depressive disorder, single episode; or acute anxiety disorder will determine the number of therapy sessions offered, impact on future insurability, and guide the pharmacotherapeutic approach.

Even in countries where the insurance industry plays a smaller role, there are still competing diagnostic categories for many disorders. One example is ankylosing spondylitis, an inflammatory form of arthritis, which causes the vertebra to fuse. In the small country of New Zealand, for example, where limited public resources curtail access to treatment to particular therapeutics, doctors are required to satisfy a checklist of diagnostic criteria administered by the pharmaceutical management agency before they can prescribe some of the more expensive treatments (New Zealand Pharmaceutical Agency, 2010). This checklist includes sacroiliitis and elevated nonspecific clinical markers, which are not necessarily viewed by clinicians as the best means for categorizing the disease and its severity. It is the administrative disease definition that matters, however, because it is a precondition for treatment. Turner has referred to this as "the standardization of illness into phenomena which can be managed by bureaucratic agencies" (quoted in Filc, 2006).

Making Categories

The physical nature of a disorder or dysfunction must be considered when discussing diagnostic categories. However, interlaced with the pathophysiological nature of the problem, and often more easily dismissed, are the important social forces that shape how scientists, physicians, and laypeople talk about diagnosis. So, even while science and medical research play often very important roles in developing disease categories, we cannot overlook how the process of developing disease categories is based in politics, consensus, and pressure, and as we might surmise from the previous paragraph, money. As I wrote in the introduction, no matter how "real" a particular physical dysfunction may be, the process of recognizing it as a disease, deciding that it can be generalized about—so that future incidences of the disorder will be so recognized—creating particular categories into which the condition can be slotted, is done by people in a social setting, involving resources, debates, and consensus, as we will discuss in the paragraphs that follow.

Decisions about what groups of symptoms should receive disease status are often tasked to "expert consensus panels" (Aronowitz, 2001). However, the ultimate decisions of such groups typically conceal some of the factors that have driven the outcomes, and the interests of those who define diseases helps determine how scientific evidence will be used in the development of diagnostic classifications. For example, a consensus panel on Lyme disease ultimately rejected a symptom-based in favor of a laboratory-based diagnostic standard, restricting the diagnosis to the type of patient these doctors would prefer to treat (p. 807).

As this volume goes to press, an interesting debate is taking place about a disease category in-the-making. The fraught problem of chronic fatigue syndrome (CFS), a debilitating, medically unexplained state of extreme exhaustion, is under review. This syndrome is among those for which there is no generally accepted explanation, and Dumit referred to it as an "illness you have to fight to get." By this he means that a patient's receiving a diagnostic label of CFS does not confer legitimacy on the individual's suffering, but instead, causes her suffering to be looked at askance, as if it were less "real" than other diagnosed conditions. It is a condition that lacks biological "facts" (explanations, such as laboratory tests and what might be considered "irrefutable" evidence) and as such, denies its sufferers legitimacy (Dumit, 2006). People with this condition experience extraordinary physical fatigue but often find their condition shrugged off by doctors and explained as psychosomatic in nature, as are so many disorders for which medicine does not have an explanation (Jutel, 2010b). There is thus a hierarchy of diagnoses, and those that rely upon patient description of suffering do not grant the same legitimacy as those that can be measured.

Two recent publications about CFS demonstrate how there are a range of potential avenues for any condition to be categorized and explained. On the one hand, researchers at Harvard have announced they have discovered alterations in the early immune profile of people suffering from CFS, which can be detected by laboratory assay (Hornig et al., 2015). On the other hand, the Institute of Medicine, in response to the frustration about finding root causes or biomarkers for CFS, has proposed symptom-based diagnostic criteria, which include duration of impairment (more than six months), post-exertional malaise, unrefreshing sleep, and either cognitive impairment or intolerance of standing (Institute of Medicine, 2015). Both groups believe in the importance of cementing the disease category in a meaningful way in order to enable research, statistical analysis, and ultimately more effective treatment. Yet, we have here two different ways of creating the diagnostic category: immunological and functional. Which model will win out will be a matter for debate, consensus, politics, and future scientific inquiry. Once established, however, the diagnostic definition will reify the condition for years to come in one way as opposed to the other.

One powerful player in the construction of diagnostic categories is the pharmaceutical industry. The introduction into medical discourse of the diagnosis of female sexual interest/arousal disorder is an excellent example of industry influence in creating diagnoses. In the late 1990s, anticipating the immense commercial success of sildenafil (Viagra) and other drugs developed to treat erectile dysfunction, pharmaceutical companies were eager to bring to market a similar drug to treat sexual dysfunction in women, a "pink Viagra." However, there was no established specific diagnosis for women looking for treatment, parallel to erectile dysfunction in men. Therefore, in order to have a disease for "pink Viagra" to treat, the pharmaceutical industry organized and financed an International Consensus Development Conference on Female Sexual Dysfunction in 1998. Delegates participated by invitation and consisted of a group equally balanced between pharmaceutical representatives and researchers either experienced with or interested in working collaboratively with the industry (Moynihan, 2003).

The consensus they produced lamented the lack of studies investigating female sexual dysfunction and the absence of diagnostic frameworks (Basson et al., 2000). The committee found that urgent investigation was required to develop new classifications and definitions of sexual dysfunction. Their work was supported by "educational grants" from Eli Lily, Pentech, Pfizer, Procter and Gamble, Schering-Lough, Solway Pharmaceuticals, TAP Pharmaceuticals, and Zonagen. Its 19 authors acknowledged financial or other relationships with 24 listed pharmaceutical companies.

Building on this report and its recommendations, a number of screening tools were developed to affirm, define, and reinforce the report's assertions. Not surprisingly, much of the work was both funded and copyrighted by one pharmaceutical company that was at the time working on its own version of "the pink Viagra" (Jutel, 2010a). The need for the category and the means by which it is pinned down and made concrete are, in this case, owned by commercial interests. But, as Ian Hacking (2001) reminds us:

> the idea of nature has served as a way to disguise ideology, to appear to be perfectly neutral. No study of classification can escape the obligation to examine the roots of this idea . . . no study of the word "natural" can fail to touch on that other great ideological word, "real."
>
> (p. 7)

Hacking is pointing out that once a category is cemented, as it is in the case of female sexual dysfunction, the details of its development become opaque, as the category naturalizes the particular disorder as The Way Things Are.

Initially, the U.S. Food and Drug Administration (FDA) did not grant approval for a number of "pink Viagra" substances, but the pharmaceutical industry (Sprout Pharmaceuticals, as one powerful example) worked to create alliances around promoting the disease in order to create a perceived imperative for drug development (see: http://eventhescore.org/). The FDA held a second round of hearings on flibanserin, one of the "Pink viagras" in 2015, with arguments on all fronts, including those who promote the DSM-5 diagnosis and others (notably New View Campaign http://www.newviewcampaign.org) who challenge the premise that sexuality should be medicalized at all. The result of the hearing was that the FDA approved Addyi (flibanserin) to treat acquired, generalized hypoactive sexual desire disorder (HSDD) in premenopausal women (United States Food and Drug Administration, 2015). Thus, the pharmaceutical industry shepherded a disease, female sexual dysfunction, from recognition as a diagnostic entity to treatable condition.

Although sexual dysfunction may seem to be a facile example of social, cultural, and commercial interference in disease categorization, disease categories are all subject to the same influences. In the next section, we turn to another diagnostic category to explore how its "joints" have been defined and how they emerge from a particular set of social circumstances. In this case, however, the diagnosis is physical, observable, and measurable.

Overweight

There is a common tendency to think that social forces and politics only come into play in particular categories of disease, notably those diagnosed by observation of behavior or subjective input. These might include psychiatric illness or other diseases for which laboratory tests or medical imaging technology cannot confirm their "realness." However, even tangible and seemingly concrete conditions that can be measured and observed are subject to social forces and politics. I use the case of overweight and obesity as an example to demonstrate the social contingency of disease categories.

Let us start by thinking about how bodies are categorized. In the case of overweight and obesity, the categorization is based on body mass or size. Organizing people by categories of body weight will necessarily obfuscate some difference, one of the challenges presented by classification. Although people within the category of overweight may be very similar on some grounds (measured body mass index, or weight in kilograms divided by height in meters squared), they may vary greatly on others. They may be tall or short, muscular or flabby, blond or brunette. They may have different dietary intake, exercise patterns, general physical and mental health, blood pressure, blood glucose, and so on. Similarly, a number of other people of lower body mass index (BMI) may be more similar to some members of a group of fat people than is recognized when body weight is the criterion for classification. For example, a slender person may not exercise and may have poor dietary patterns. She will not be classified with heavy non-exercisers on the basis of a mass-based classification system.

Until well into the 20th century, weight was not used to establish the boundaries of a disease category. This is not to say that fatness was not seen as disease, but rather it was seen in qualitative terms. Herrick's dictionary defined it as "increased bulk of the body, beyond what is sightly and healthy" (Herrick, 1889) and Thomas described "excessive development of the adipose tissue" (Thomas, 1891). Scales were not widely available and were too expensive for regular use. The "Reliance Weighing Machine" started marketing scales as a "useful adjunct to the consulting room" in the late part of the 19th century ("Reliance Weighing Machine," p. 940). The importance of weighing undoubtedly grew with their availability.

With this increased availability of scales, the use of weight tables became feasible. In the early 20th century, the insurer Metropolitan Life developed actuarial tables based on weight. These tables were used to evaluate insurance risk. However, weight did not catch on quickly as a tool for diagnosing. In Europe, as in the United Kingdom and United States, doctors argued that weight could not be used to determine who was sick and who was well. Dr. Jean Leray, in France, reduced the various weight tables to "theoretical interest" only and rejected any benefit to their use. He chose rather to accept Leven's definition of safe body weight as "[a person's] average weight, maintained over a number of years, as long as the subject has been well" (in Leray, 1931, p. 7 [translation mine]). Meanwhile, William Christie (1927) warned

English-language readers in 1927 that "no weight table is sufficient by itself to base an estimate of the ideal state." He continued to state that "standard tables which show the average for men and women of our race at any given age and height are fallacious, because no allowance is made for the distinctions of personal physique, nor consideration given to obvious rolls of fat" (p. 23).

However, as we can observe today, weight progressively became a diagnostic criterion. Measurement in general was supplementing, if not replacing, clinical judgment in the assessment of patient health. "Instruments of precision" (like scales):

> . . . promised to provide ways of describing disease that could be built into tight, seemingly objective pictures, useful in diagnosing and monitoring particular cases yet capable of being generalized into larger understandings.
> (Rosenberg, 2002, p. 244)

Progressively, weight tables and calculations became mainstream, and today, weighing oneself and using weight as part of health assessment is commonplace. However, the way in which weight has been used to categorize healthy and non-healthy is not fixed and has been influenced by context and era. Between 1942 and 2000, there were over 17 different methods used for categorizing weight (Kuczmarski and Flegal, 2000). Not only were there different ways of measuring overweight (weight-for-height tables, weight-for-height indexes), but the desirable and undesirable ranges changed repeatedly over the course of the 20th century. When the BMI overweight cut-offs were shifted in 1995, from 27.8 for men and 27.3 for women to 25 for both, 35.4 million adults became overweight overnight (Kuczmarski and Flegal, 2000).

One important factor in the adoption of weight as a measure of illness was the ease of self-diagnosis. Categorizing overweight as a disease was simple in this context and could be undertaken by the individual; all it took was stepping on a scale. Weight seemed an easy proxy for health and disease. However, overweight presents a number of challenges as a disease category as it obfuscates so much salient difference.

Scales do not recognize ethnic difference or muscularity. Although weight *often* co-varies with health, and on this basis is used as a proxy, it doesn't *always* co-vary, and slender people who have health risks get missed, in the same way as fat people without health risks are often presumed to be unwell. Considering overweight as disease creates a situation in which scientists/doctors stop interrogating the data. Rather than ask whether overweight is a disease, they simply ask how to get rid of it. Some data have shown that being overweight carries less risk than being normal weight (Flegal et al., 2005, Flegal et al., 2007, Orpana et al., 2009), and that it is not possible for people to change their weight categories durably anyway. Also, slender people may fail to get health advice on the basis of the assumptions about weight. Although many normal-weight individuals could benefit from weight and exercise advice, Ma et al. (2004) showed that patients were more than five times more likely to get advice about diet and exercise if they were obese than if they were of normal weight.

Numerous industries stand to gain when overweight is categorized as a disease. The pharmaceutical industry is indeed involved, but the gym, fashion, self-help, and diet industries are also actively engaged in the promotion of overweight as a disease. The self-diagnosability of weight that results from the availability of scale, and the public promotion of BMI tables push people towards potentially lucrative weight reduction diets, books, exercise programs, food supplements, and other dietary products.

BMI calculators are available on many public health websites, along with generic information about body weight. The National Heart, Lung and Blood Institute, for example, provides a calculator online and encourages people to calculate their BMI and monitor themselves for symptoms of overweight. They describe clothes feeling tight and needing a larger size, or the scales showing weight gain, as symptoms of this disease (National Heart, 2014). The important point is that overweight has diagnostic tools and is monitored as a matter of national and international public concern (World Health Organization, 2014). Citizens engage in self-monitoring as part of good citizenship and adherence to social expectations.

Although diagnosis can legitimize abnormal behaviors, it can also stigmatize. As has been widely written elsewhere, regarding overweight as a disease results in stigma—making the fat body problematic and the fat person morally deficient. One of the important assumptions supporting fat stigma is the idea that weight is a factor we can independently control and that fatness reflects a failure to engage in healthy activity. This is what Carole Spitzak referred to as an "aesthetic of health," the idea that the physical appearance of the individual speaks to their inner nature. The conception of fatness as a moral deficiency results in the pursuit of slenderness rather than the pursuit of health. Along with this pursuit is the neglect of the health needs of the thin, seeing their body size as virtuously healthy, even though weight loss and disease often go hand in hand.

Fat stigma also leads to discrimination against, and exploitation of, fat people. The discrimination can be enacted on the individual level but also at a population level. For example, in understudied populations, such as Pacific Island peoples, for whom population-specific statistics have not been developed, almost 70% have BMIs of 30 or greater (World Health Organization, 2009). These groups are identified as fat populations and subject to collective stigma associated with being fat. At the same time, an anti-fat industry based in pharmaceutical, diet, exercise, and self-help flourishes, surviving on widely accepted beliefs about, and fear of, being fat, as mentioned above.

If it were necessary to provide further illustration about how weight is socially negotiated as a disease category, the Obesity Society—an American-based association of clinicians focused on obesity advocacy and research—has prepared a white paper entitled "Obesity as Disease," which does so powerfully (Allison et al., 2008). This society, whose stated mission is to "lead the charge in advancing the science-based understanding of the causes, consequences, prevention and treatment of obesity" commissioned a panel of experts to determine whether obesity should be considered a disease or not. The experts came from within its membership and must be considered *a priori* committed to the society's goals. The point here is not to debate whether they were right or wrong in their assessment that the disease label "might have broad effects for a large portion of society for the greater good" (p. 1169). Rather it is to underline the social (and often conflicted) ways in which disease categories come to be recognized as such.

Overweight and obesity are useful examples of how disease categories are socially contingent. They demonstrate how even physical categories are framed by economics, politics, and social norms, with very real consequences for those thus categorized. However, while we note this, we can nonetheless see the important role played by disease categories in understanding health and illness scientifically.

Using Categories

As I wrote above, diagnostic categories are the tools of medicine. They enable classification, which in turn serves as the mechanism by which medicine is enacted as a social process. Diagnosis cements the respective roles of patient and clinician, determines which specialist takes care of which problems and where a patient should turn, provides a language for discussing and reproducing health and disease, allocates resources, shapes research, and much more (Jutel, 2011).

Diagnosis has long served to cement the authority that is vested in the doctor. Hippocrates wrote that

> if [the doctor] is able to tell his patients when he visits them not only about their past and present symptoms, but also to tell them what is going to happen, as well as to fill in the details they have omitted, he will increase his reputation as a medical practitioner and people will have no qualms in putting themselves under his care.
> (Lloyd, 1983, p. 170)

The diagnosis is the starting point for determining a prognosis, variable though it may be, and conveys, as this quotation asserts, authority to the clinician who discloses it.

Diagnosis does provide a sense of direction (prognosis) and an associated treatment, but it also explains and organizes as well as giving access to services, identity, and the sick role. The layperson cannot write her own sick note or prescription, nor can she access disability services or certain therapies without the diagnostic category afforded by the doctor. As Freidson (1972) pointed out, the doctor's ability to label sickness and health, normal and abnormal, is key to medical authority and prestige. Importantly, the labeling also enables a particular social behavior in the diagnosed individual.

The "sick role," as described by Talcott Parsons (1958), is the temporary retirement from social responsibilities afforded to the individual who is diagnosed. Diagnosis allows one to stay in bed, for example, rather than go to work or look after the children. It also often excuses antisocial behavior, as in the case of a legal defense of insanity, or at a lower level, as in the case of attention deficit disorder. The sick role includes not only a shift in social roles, but it also includes an expectation that the diagnosed person will make every effort to submit to medical recommendations to become well.

Not only does diagnosis confirm power relations, it also serves an important didactic and heuristic role. A diagnosis is a kind of short-hand for encapsulating as much information as possible into a word or phrase. To return to the example of the "mixed-breed, collie-type, mature, male dog" above, we can see that within this not-so-short category, we may also be able to make assumptions about fur color and length, activity level, and proclivities. Diagnoses can describe a situation with similar concision. To identify a person with, for example, influenza, is to say that they have a systemic illness probably accompanied by fever and upper respiratory symptoms. Their affliction is contagious, and is generally self-limiting, although presents risks in some cases and in some circumstances. One word, in the case of influenza, captures thirty. But, at the same time, that one word, as with Skip the dog, does not capture the particular case presentation.

The power of diagnosis goes well beyond its use in the doctor's office or the hospital wards. A diagnosis can make a powerful statement about society and its unmaking. For example, to speak of the resurgence of polio in Syria, or of PTSD in post-tsunami Indonesia is to make a statement about social disarray with an intensity that other words cannot convey. Disease categories describes a situation with force and authority.

Diagnosis is also a powerful discourse for politicians, public and corporate figures. To express support of ethnic, national, or gendered groups via diagnostic categories is another powerful statement. By wearing a pink ribbon (symbol of the international breast cancer support charity), a politician positions himself or herself as a supporter of women. In New Zealand, placing prostate cancer screening as a plank in the National Party's campaign platform attempted to demonstrate sensitivity towards male voters. The pharmaceutical industry does not need to peddle its potions as long as it promotes awareness of disease categories for which its therapies are effective. Here, the power of the diagnostic category is in its ability to circumvent drug advertising in countries where direct-to-consumer advertising is not legal.

Conclusion

Diagnostic categories are more influential in daily life than is often acknowledged. On the one hand, they structure the way we practice and consume medicine in Western society. Diagnosis is one of medicine's most important tools: the means by which we talk about, understand, explain, treat, and predict outcomes of disease. It would be hard to imagine a medicine that operated outside of diagnostic frameworks.

On the other hand, diagnoses do far more work than just organize medicine. As described above, from the identification of diagnostic categories to their application in the clinic, and the consequences they entail, diagnoses reflect and reproduce the societies in which they are generated. Recognizing the fluid nature of the diagnosis, challenging the putative perception of firm and naturalized diagnostic categories is important to understanding where power resides and how it operates. Diagnostic categories create social "truths" about the world, by virtue of the fact that they formalize certain parameters, or boundaries, within the continuum of nature. Because of the authority conferred by a diagnosis, the category it creates conceals previous debates, contests, and tensions. We can observe this in real time with the diagnosis-in-the-making chronic fatigue syndrome. If and when there is a shift in the diagnostic configuration of this disorder, one possible model will assimilate the others. This means that certain values (immunological, symptomatic) will be subsumed in favor of others. Each diagnostic category thus promotes certain perspectives and, in a way, puts halt to the debate.

This chapter should have helped to debunk the idea that diagnostic categories simply mark natural boundaries. Diagnostic boundaries are at the same time elusive and necessary (Jutel, 2013). The challenge for those who use diagnostic categories (most of us, as either patients or clinicians) is that while it behooves us to interrogate these categories and step back in order to obtain critical distance, it is difficult to do so from within the medical system in which we are deeply engaged. Mary Douglas makes the point clearly as she both acknowledges the importance of structural constraints and the difficulty of achieving critical distance: "How can an individual [in the grip of iron hard categories] turn round his own thought-process and contemplate its limitations?" (Douglas, 1966, p. 16). Raising our consciousness of the social operation of diagnostic categories is an important first step.

References

"Reliance Weighing Machine." 1898. New Inventions. *Lancet*, 152(3919), 940.
ALLISON, D. B., DOWNEY, M., ATKINSON, R. L., BILLINGTON, C. J., BRAY, G. A., ECKEL, R. H., FINKELSTEIN, E. A., JENSEN, M. D. & TREMBLAY, A. 2008. Obesity as a disease: A white paper on evidence and arguments commissioned by the council of the obesity society. *Obesity*, 16, 1161–1177.
ANONYMOUS. 1840. *Petites études de la nature ou entretiens récréatifs d'une mère avec ses deux filles sur l'histoire naturelle des animaux et des plantes; les phénomenes astronomiques, et les progrès des arts, des sciences et de la civilisation*, Paris, Niogret.
ARNOLD, T. 1839. *On the Divisions and Mutual Relations of Knowledge: A Lecture Read Before the Rugby Literary and Scientific Society, April 7, 1835*, Rugby, Combe and Crossley.
ARONOWITZ, R. 2001. When do symptoms become a disease? *Annals of Internal Medicine*, 134, 803–808.
BALINT, M. 1964. *The Doctor, His Patient and the Illness*, Kent, England, Pitman Medical.

BASSON, R., BERMAN, J., BURNETT, A., DEROGATIS, L., FERGUSON, D., FOURCROY, J., GOLDSTEIN, I., GRAZIOTTIN, A., HEIMAN, J., LAAN, E., LEIBLUM, S., PADMA-NATHAN, H., ROSEN, R., SEGRAVES, K., SEGRAVES, R. T., SHABSIGH, R., SIPSKI, M., WAGNER, G. & WHIPPLE, B. 2000. Report of the international consensus development conference on female sexual dysfunction: Definitions and classifications. *Journal of Urology*, 163, 888–893.

BLAXTER, M. 1978. Diagnosis as category and process: The case of alcoholism. *Social Science and Medicine*, 12, 9–17.

CHRISTIE, W. F. 1927. *Surplus Fat and How to Reduce It*, London, Heinemann/Medical Books.

DALY, W. J. & BRATER, D. C. 2000. Medieval contributions to the search for truth in clinical medicine. *Perspectives in Biology and Medicine*, 43, 530–540.

DOUGLAS, M. 1966. *Purity and Danger*, London, Routledge and Kegan Paul.

DUMIT, J. 2006. Illnesses you have to fight to get: Facts as forces in uncertain, emergent illnesses. *Social Science & Medicine*, 62, 577–590.

FILC, D. 2006. Power in the primary care medical encounter: Domination, resistance and alliances. *Social Theory and Health*, 4, 221–243.

FLEGAL, K. M., GRAUBARD, B. I., WILLIAMSON, D. F. & GAIL, M. H. 2005. Excess deaths associated with underweight, overweight, and obesity. *Journal of the American Medical Association*, 293, 1861–1867.

FLEGAL, K. M., GRAUBARD, B. I., WILLIAMSON, D. F. & GAIL, M. H. 2007. Cause-specific excess deaths associated with underweight, overweight, and obesity. *JAMA*, 298, 2028–2037.

FLEISCHMAN, S. 1999. I am . . . , I have . . . , I suffer from . . . A Linguist Reflects on the language of illness and disease. *Journal of Medical Humanities*, 20, 1–31.

FLEISS, J. L., LAWLOR, W., PLATMAN, S. R. & FIEVE, R. R. 1971. On the use of inverted factor analysis for generating typologies. *Journal of Abnormal Psychology*, 77, 127–132.

FREIDSON, E. 1972. *Profession of Medicine: A Study of the Sociology of Applied Knowledge*, New York, Dodd, Mead & Company.

GRAUNT, J. 1662a. *Natural and Political Observations, Mentioned in a Following Index, and Made Upon the Bills of Mortality*, London, John Martin, James Alleftry and the Dicas.

GRAUNT, J. 1662b. *Reflections on the Weekly Bills of Mortality for the Cities of London and Westminster and the Places Adjacent but More Especially, So Far as It Relates to the Plague and Other Most Moral Diseases that we English-Men Are Most Subject to, and Should be Most Careful Against in this Our Age*, London, Samuel Speed.

HACKING, I. 2001. Inaugural lecture: Chair of philosophy and history of scientific concepts at the Collège de France, 16 January 2001. *Economy and Society*, 31, 1–14.

HERRICK, S. S. 1889. *A Reference Handbook of the Medical Sciences*. BUCK, A. H. (ed.). Edinburgh, UK: Pentland.

HORNIG, M., MONTOYA, J. G., KLIMAS, N. G., LEVINE, S., FELSENSTEIN, D., BATEMAN, L., PETERSON, D. L., GOTTSCHALK, C. G., SCHULTZ, A. F., CHE, X., EDDY, M. L., KOMAROFF, A. L. & LIPKIN, W. I. 2015. Distinct plasma immune signatures in ME/CFS are present early in the course of illness. *Science Advances*, 1(1).

INSTITUTE OF MEDICINE. 2015. *Proposed Diagnostic Criteria for ME/CFS* [Online]. Washington, National Academy of Sciences. Available: http://www.iom.edu/~/media/Files/Report%20Files/2015/MECFS/MECFS_ProposedDiagnosticCriteria [Accessed 3 March 2015].

JACOB, E. K. 1992. Classification and categorization: Drawing the line. *Advances in Classification Research: Proceedings of the ASIS SIG/DR Classification Research Workshop*, 2, 67–83.

JANES, C. R. 1995. The transformations of Tibetan medicine. *Medical Anthropology Quarterly*, 9, 6–39.

JUTEL, A. 2010a. Framing disease: The example of female hypoactive sexual desire disorder. *Social Science and Medicine*, 70, 1084–1090.

JUTEL, A. 2010b. Medically unexplained symptoms and the disease label. *Social Theory and Health*, 8, 229–245.

JUTEL, A. 2011. *Putting a Name to It: Diagnosis in Contemporary Society*, Baltimore, Johns Hopkins University Press.

JUTEL, A. 2013. When pigs could fly: Influenza and the elusive nature of diagnosis. *Perspectives in Biology and Medicine*, 56, 513–529.
KUCZMARSKI, R. J. & FLEGAL, K. M. 2000. Criteria for definition of overweight in transition: Background and recommendations for the United States. *American Journal of Clinical Nutrition*, 72, 1074–1081.
LEDER, D. 1990. Clinical interpretation: The hermeneutics of medicine. *Theoretical Medicine*, 11, 9–24.
LEFF, W. 1997. William of Ockham. *In:* PARRY, M. (ed.) *Chambers Biographical Dictionary*, New York, Chambers, p. 1386.
LERAY, J. 1931. *Embonpoint et obésité: Conceptions et thérapeutiques actuelles*, Paris, Massion et Cie.
LLOYD, G. E. R., ed. 1983. *Hippocratic Writings*, London, Penguin.
MA, J., URIZAR JR, G. G., ALEHEGN, T. & STAFFORD, R. S. 2004. Diet and physical activity counseling during ambulatory care visits in the United States. *Preventive Medicine*, 39, 815–822.
MOYNIHAN, R. 2003. The making of a disease: Female sexual dysfunction. *British Medical Journal*, 326, 45–47.
NATIONAL HEART, L. A. B. I. 2014. *BMI Calculator* [Online]. Bethesda, MD. Available: http://www.nhlbi.nih.gov/health/health-topics/topics/obe/signs.html [Accessed 30 October 2014].
NEW ZEALAND PHARMACEUTICAL AGENCY. 2010. New Zealand Pharmaceutical Schedule. Available: http://www.pharmac.govt.nz/2010/11/18/SU.pdf (online only) [Accessed 15 February 2016].
ORPANA, H. M., BERTHELOT, J. M., KAPLAN, M. S., FEENY, D. H., MCFARLAND, B. & ROSS, N. A. 2009. BMI and mortality: Results from a national longitudinal study of Canadian adults. *Obesity (Silver Spring)*, 18(1), 214–218.
PARSONS, T. 1958. Definitions of health and illness in the light of American values and social structure. *In:* JACO, E. G. (ed.) *Patients, Physicians and Illness: Behavioral Science and Medicine*, Glencoe, IL, The Free Press, pp. 165–187.
RICHARDSON, E. C. 1901. *Classification*, New York, Charles Scribner's Sons.
ROSENBERG, C. E. 2002. The tyranny of diagnosis: Specific entities and individual experience. *The Milbank Quarterly*, 80, 237–260.
SILVESTRI, L. G. & HILL, L. R. 1964. Some problems of the taxometric approach. *In:* HEYWOOD, V. H. & MCNEILL, J. (eds.) *Phenetic and Phylogenetic Classification*, London, UK, Systematics Association Publication, pp. 87–103.
SYDENHAM, T. 1742. *The Entire Works of Dr Thomas Sydenham, Newly Made English from the Originals: . . . To Which Are Added, Explanatory and Practical Notes, from the Best Medicinal Writers, By John Swan, M.D.*, London, printed for Edward Cave.
THOMAS, J. 1891. *A Complete Pronouncing Medical Dictionary*, London, Deacon.
UNITED STATES FOOD AND DRUG ADMINISTRATION. 2015. *FDA Approves First Treatment for Sexual Desire Disorder: Addyi Approved to Treat Premenopausal Women* [Online]. United States Food And Drug Administration. Available: http://www.fda.gov/NewsEvents/Newsroom/PressAnnouncements/ucm458734.htm [Accessed 16 December 2015].
VEITH, I. 1982. The medical world of king Tutenkhamon. *Perspectives in Biology and Medicine*, 26, 98–106.
WORLD HEALTH ORGANIZATION. 2009. *WHO Global InfoBase* [Online]. Available: https://apps.who.int/infobase/report.aspx?rid=114&iso=TON&ind=BMI [Accessed 9 September 2009].
WORLD HEALTH ORGANIZATION. 2014. *Obesity and Overweight* [Online]. Geneva WHO. Available: http://www.who.int/mediacentre/factsheets/fs311/en/ [Accessed 30 October 2014].
ZERUBAVEL, E. 1991. *The Fine Line: Making Distinctions in Everyday Life*, New York, The Free Press.
ZERUBAVEL, E. 1996. Lumping and splitting: Notes on social classification. *Sociological Forum*, 11, 421–433.

Further Reading

Bowker, G. C. & Star, S. L. 1999. *Sorting Things Out: Classification and Its Consequences*, Cambridge, MA, MIT Press (an overview of classification and its consequences, making particular reference to disease classification).

Hacking, I. 1999. *The Social Construction of What?*, Cambridge, MA, Harvard University Press (a clarification of social constructionism in reference to social and natural sciences).

Jutel, A. G. 2011. *Putting a Name to It: Diagnosis in Contemporary Society*, Baltimore, Johns Hopkins University Press (a monograph on the sociology of diagnosis).

Zerubavel, E. 1991. *The Fine Line: Making Distinctions in Everyday Life*, New York, The Free Press (a seminal text on the social aspects of categories and classification).

16
CLASSIFICATORY CHALLENGES IN PSYCHOPATHOLOGY

Harold Kincaid

1. Introduction

Psychiatry, clinical psychology, and other professions that deal with psychopathology, abnormal behavior, and personal psychological problems naturally try to distinguish among the different kinds of phenomena and behavior they encounter. Basic research about the causes and courses of psychopathology and about appropriate treatment depends on sorting the various psychological problems that professionals see into distinct categories. However, there is considerable controversy about how to do that and how those categories are to be understood.

Psychiatrists, who have MD degrees, naturally approach psychopathology with a medical model and think in terms of different diseases or mental illnesses. However, medicine usually picks out diseases by identifying localizable lesions, malfunctions, or infectious agents that cause the symptoms of a disease; yet few if any psychiatric disorders are currently understood in terms of such physical causes.

In North America and much of the world, psychological disorders are classified according to the *Diagnostic and Statistical Manual of Mental Disorders*, produced by the American Psychiatric Association (2013), or other quite similar classification schemes such as the *International Classification of Diseases*, produced by the World Health Organization (2015). The purported disorders in these manuals vary greatly in nature. Some of them do not really look like diseases at all: In the DSM, homosexuality and non-standard gender identities were at one time classified as disorders. Individuals given these labels do not seem to have disorders or diseases in the way that someone who hears voices instructing him or her to murder seems to be mentally ill. Rather, these labels seem to be about behaviors that are socially disapproved.

Social science studies of the processes that produce classification manuals show a variety of interest groups and sociological factors at work in constructing classification schemes that do not seem to promote objectively identifying diseases. Moreover, the classification manuals for psychiatric disorders have very open-ended definitions of disorders, often allowing two individuals to be classified in the same way without having any symptoms in common. These facts have led some critics to doubt that the currently dominant classification systems of psychopathology are identifying diseases and to argue that they are instead a reflection of societal norms being used illegitimately to turn ordinary problems in living into medical problems. Some critics think current systems can be reformed to identify disease; others think the problems are inevitable in any classification systems similar to those currently used.

Because current classification systems are widely used and influence how psychological problems are treated by professionals and how research on them is done, these disputes are of great practical import. The key issues thus revolve around the ability of current and future classification systems to group individuals objectively in ways that inform decisions about treatment and research. In what follows, I first provide some illustrative cases of psychological disorders to use as reference points in our discussion. Then a more detailed outline of the issues at stake follows after that. I look at three approaches to the classification of psychopathology: an approach based on the medical disease model, approaches that emphasize the role of social disapproval of abnormal behavior, and a third position that tries to bridge the first two approaches.

2. Some Illustrative Cases

Describing some cases of typical behavior that can be classified as psychopathology will be useful for the rest of our discussion. The cases that follow are fictional but nonetheless are representative of real-world situations.

Alex was a blue-collar worker in a small Midwestern U.S. city. Alex first started showing symptoms of depression in his early 20s. Aside from a very sad mood, Alex had major difficulties sleeping, very low energy, very low self-esteem, feelings of guilt, and difficulties being around people. Several individuals in his family—a grandfather and uncle—had had similar problems. Alex underwent electroshock treatment and eventually returned to normal. Yet his problems returned at regular intervals, and despite treatment with the latest antidepressant medication, Alex eventually committed suicide at age 44.

Jean, Alex's wife, suffered similar symptoms to those of Alex after his death. She was left with two children to raise and no income; she also experienced very low mood, severe feelings of guilt, difficulty in sleeping, and so on. Yet after six to nine months, her difficulties became manageable and she went on to live a relatively happy and successful life. Her symptoms never returned, though of course like everyone she had her down days.

Shannon lived in the same small Midwestern U.S. city as Alex and Jean. Shannon was born male, but from early on just did not behave like most boys do. His mannerisms were described by those around him as effeminate. In Shannon's early teens, he began to dress on occasion as a woman and had an online identity as a female. By his late teens, Shannon had adopted a female gender. She felt much more at ease with her new identity, yet her behavior met with a fair amount of ridicule and disapproval in the relatively conservative community where Shannon lived. This produced some distress for her and caused her at times to feel conflicted about her new gender identity.

Alex, Jean, and Shannon all saw psychiatrists. Alex and Jean were diagnosed with major depressive disorder and Shannon with gender identity disorder. (Shannon's classification has recently been changed to "gender dysphoria" because the DSM manual changed the title and restricted the condition to individuals who are distressed by their gender identity [this includes Shannon]; however, it is still regarded as a disorder, as are all the entries in the DSM.)

3. The Issues Further Developed

The introduction above sketched some of the issues involved in debates over the status of and the proper methods for classifying psychological disorders. It will be useful to explain those issues further before looking at different positions in detail.

A first thing to note: you may have noticed that I have been switching my terminology when it comes to speaking of what we classify. Sometimes the target is labeled as psychopathology. Other times I have spoken of mental illness, psychological problems, abnormal behavior, or problems in living. I use these different terms because they reflect different viewpoints on how to think about classification systems. Medical approaches want to talk about psychopathology because they believe that disease in the form of dysfunctions in biological or psychological systems is fundamental. Those who reject the medical model will tend to reject the talk of pathology in favor of some less medical description such as "abnormal." There is no completely neutral way to describe the behavior and phenomena we want to classify that does not favor one approach or the other. Thus, I will switch back and forth between these various terminologies.

Even if we approach psychopathology from a medical point of view, it will be useful to make a distinction between disease, illness, and sickness (used in the specific defined senses that are outlined in Chapter 2 and that do not necessarily line up with everyday usage of these terms). The clearest cases of *diseases* are usually conditions where an identifiable physical malfunction exists. AIDS is a paradigm example: a virus causes pathologies in immune system functioning. *Illness* refers to individual suffering (often caused by diminished functioning). For example, chronic fatigue syndrome may be an illness but not a disease at the current state of knowledge, for while there is suffering, there is no identified physical cause. Traditionally, such conditions would be labeled by the medical profession as syndromes until a true disease basis is found. It is also possible to have a disease without an illness. Asymptomatic cancer is an example, but obviously most disease does result in illness. *Sickness* then refers to a socially recognized role or situation that gets special handling from society. For example, individuals labeled as sick may be excused from work and other social obligations. Not every illness gets such treatment. Some societies identify sickness where there may be no obvious disease or illness at all. For example, Germans have the concept of a Gesundheitsurlaub—a health vacation. Individuals in these cases are sometimes entitled to absence from work for the sake of their health, even though they are not ill.

Given these distinctions, the basic paradigm in much of current psychiatry is based on a medical model that treats psychopathology as a disease. This approach asserts that the best-grounded classifications are those based on identifiable aberrations in the proper functioning of bodily systems. Different disease classifications should be based on differences in the kind of malfunctions present. These are objective facts about human bodies and are true independently of what we believe about human maladies, do not depend on human judgments about what is good, bad, or moral, and can be identified by verifiable means by different observers. The different diseases are a part of the world in the same sense that the atomic elements, for example, are. On this view, psychiatric classifications should ultimately be based on identifiable malfunctions.

To refer back to one of our examples, the disease-based view classifies Alex as having a real mental illness. Alex seems to have major depressive disorder that reflects fundamental dysfunction in psychological and probably biochemical systems in ways that neurobiology and cognitive psychology can help identify. Alex has profound depression that is persistent and apparently not simply the result of adverse life events, for example.

This disease-based view can allow for some complications. Current classifications may pick out what we called above syndromes—symptoms of suffering for which we currently have no identifiable disease. Yet, on the medical model approach, classifications of psychiatric syndromes should ultimately be halfway houses on the road to finding the diseases behind them.

To give us a clear map of the possible views on classifying psychopathology, it is useful to contrast the disease view just sketched with its polar opposite, what we can call the social constructivist view. The starkest form of this view claims there is no such thing as mental illness or psychopathology. The view is called social constructivist because it claims that the classifications of behaviors as psychopathological (a term constructivists would reject) are made by society in various ways that are not based on real psychological or biological diseases. Rather, social norms, interest groups such as clinicians and pharmaceutical companies, and various social processes pick out some behaviors to label as mentally ill. In terms of our tripartite distinction above between disease, illness, and sickness, on the social constructivist view what gets called mental illness is not a disease. It may involve suffering and thus might be called an illness, and classified individuals may have different responsibilities than those labeled as normal and thus might fall into the sick category. But the suffering may result entirely due to social stigma, as in the case of homosexuality, and whether the sickness concept applies depends on societal value judgments with no grounding in objective disease processes. Thus, on the social constructivist view, there is no value-free classification of psychopathology; psychopathology does not exist in the world independently of our conceptions of it.

In terms of our examples, Shannon's diagnosis as suffering from "gender dysphoria" is a good example of how the social constructivist thinks of our labeling of behavior as mentally ill. Failing to identify with one's biological sex is a phenomenon that (some) societies choose to treat as a distinct kind of behavior, one that many individuals in such societies find troublesome and disapprove of. There is nothing here, the argument goes, approaching a physical disease like cancer. Society has chosen to relate to these individuals under the description of having a mental disorder. Individuals who do not identify with their biological sex may well suffer, but not from biologically based disease but from social disapproval. Labeling them as having a disorder is likely to contribute to that suffering.

4. The Disease Model of Classification in More Detail

We saw above that the disease model of classification wants to ground psychiatric classification on objective facts about bodily disorders causing psychopathology. Let's look at this approach in more detail, as well as some of its problems. Two of the most developed defenses of this position are those of Murphy (2006) and Wakefield (2007), and I will focus on their accounts.

Wakefield bases his approach on the evolutionary function account of disease. Disease—whether physical or mental—on his view consists in the breakdown of the proper functioning of the systems of the body. The "proper function" of these systems consists in what they were designed to do by evolution. The systems of our body exist and have the functions they do because at some point in our evolutionary history those systems with their functions were selectively advantageous. Wakefield recognizes as noted above that some malfunctions in the systems of our body may not cause us problems, so he adds to his definition that the dysfunctions are harmful—they are conditions we dislike. So there is a value component in Wakefield's definition, but its key component is still value free and supposedly based on objective biological science.

Murphy advances a similar view, except for him disease involves deficits in the normal functioning of the relevant systems of the body. Normal functioning is not understood in terms of functions selected by evolution but instead in terms of the activities of systems typical of our species. This is a notion of disease defended by Boorse (see Chapter 1), which defines disease in terms of deviations below the normal or average functioning of the component systems of human beings. This view does not depend on identifying the selective advantage of our biological or psychological subsystems; it instead focuses on identifying the causal role that various systems play in normal human functioning.

For Murphy, classifications of psychopathology must be grounded in causal knowledge of biological and psychological systems; classifications should be based on the deficits in functioning that explain the symptoms used to classify disorders. Correct classifications are those that identify the causes of psychopathological symptoms in terms of deviations from normal functioning.

Wakefield and Murphy both talk of malfunctioning systems causing symptoms and providing the basis for psychopathological classification. However, psychopathology would seem to be a *mental* disorder. So an obvious question is whether the classification of psychopathology must be based on dysfunctional *psychological* systems. Both Wakefield and Murphy are, reasonably enough, flexible in this regard: the malfunctioning systems may be described in psychological terms or neurobiological terms, or, perhaps best, in terms of integrated psychological-neurobiological terms. So let's return to Alex, who has major depressive disorder. He has clear psychological malfunctions: his memory recall is biased towards negative events—compared to normals he is much more likely to remember negative events in the past. On the neurobiological level, brain imaging shows, for example, that brain regions responsible for fear reactions are much more active in Alex than in individuals without psychopathology.

Given their approaches to classification, which require classifications to be based on identified malfunctions, both Wakefield and Murphy are critical of current mainstream classification systems such as the DSM. Wakefield, with coauthor Horwitz, a sociologist, has argued that the current DSM system significantly overdiagnoses depression, for example. To "overdiagnose" is to mistakenly label people with major depressive disorder who do not have it. In *The Loss of Sadness* (Horwitz and Wakefield, 2007), they argue that the DSM classification system treats individuals with an understandable and temporary depressed mood—for example, that following the death of a close relative or friend—as in the same category as Alex. So they would argue that, contrary to how she would be labeled under the DSM, Jean, who has understandable depressed feelings, should not be classified as psychopathological—there is not an underlying disease behind her problems, but rather the ordinary if regrettable difficulties of life. Alex's problems are recurring symptoms that have no obvious cause in recent major life events; his condition is ongoing or chronic. But the DSM-5 treats the two cases the same. Important revisions in our classification of psychopathology are thus called for by Wakefield and Horwitz.

Murphy is equally or more critical of classification systems such as the DSM. Those systems generally do not have a strong grounding in our scientific understanding of mind-brain dysfunction; to the extent that they do not, they are questionable and cut themselves off from connections needed to make scientific progress.

It is interesting to note that while the current DSM classification system in its introduction advocates the disease view of psychopathology, it vacillates between the evolutionary function account that Wakefield defends and the malfunctioning systems account of Murphy. So the latest version (DSM-5) says that:

> a mental disorder is a syndrome characterized by clinically significant disturbance in an individual's cognition, emotion regulation, or behavior that reflects a dysfunction in the psychological, biological, or developmental processes underlying mental functioning.
>
> (APA, 2013, 20)

But it also says that mental illness is a "psychopathological condition in which physical signs and symptoms exceed normal ranges" (APA, 2013, 19). The deviation from normal ranges is invoked often in the criteria for specific psychopathologies.

These are two different objective definitions of disease; the first is an evolutionary function view and the second focuses on deviations from species-typical functioning. However, this difference is in a way of minor importance, because in practice the DSM classification system openly rejects basing classifications on causes. The first objective of the system is to provide categories that can be consistently applied by clinicians and researchers. It explicitly claims to be "theory neutral"—to provide classifications that make no specific assumptions about the underlying causes of psychopathology. It takes this stance because its primary goal is to allow clinicians and researchers to communicate successfully despite their theoretical differences about the underlying causes of psychopathology. Yet, for Wakefield and Murphy, this stance means that the DSM classification system loses the underpinning it must have to be a scientific and successful medical practice.

There are some significant problems for the disease approaches to classifying psychopathology. As just seen, the disease account really does not do much work in actual psychopathology classification systems. That can, of course, be a basis of criticism of these classification systems, a conclusion that Murphy and Wakefield draw. However, in Wakefield's own work distinguishing real depression from situational depressed mood, he does not actually cite dysfunctions in evolutionary selected mental or biological modules (Kincaid forthcoming a). That is not surprising, since we have no developed way of saying what those are. Talk of capacities selected by evolutionary forces in our evolutionary past as hunter-gatherers when our basic psychology was set is largely speculative sociobiology and not something that can give us detailed guidance on how to classify abnormal behavior. Wakefield's own arguments for separating major depression from understandable grief do not use this evolutionary story but instead rely on much more commonsensical considerations such as that grief is a normal and understandable response to loss.

Psychiatry and psychology can usually provide classifications that allow us to predict with some accuracy the causes and course of behavior without identifying malfunctioning biological or psychological systems. These conditions are thus reasonably seen as *illnesses* or *syndromes* as we defined those terms earlier, but not diseases.

5. Psychopathology Classification as a Social Construction

The polar opposites of a disease-based account of psychopathology classification are social constructivist views. An early and influential statement of this view was made in 1961 by Thomas Szasz in *The Myth of Mental Illness* (anniversary edition, 2010). What gets classified as mental illness is, on his view, the result of medicalizing disturbing or unusual behaviors. These behaviors are symbolic actions in response to specific social situations; Szasz thinks of them as strategies and signs we use in "games" we play with each other all the time. For most behavior that gets classified as mental illness, individuals have their reasons for why they behave the way they do—they are agents with free will whose behavior is not determined in the way that the behavior of a body with a disease is. Our classifications of psychopathology are mostly just the ways society has chosen to deal with certain sorts of behaviors and very rarely reflect real disease classifications (exceptions are things such as neurological diseases). Thus, Szasz thinks mental illness classifications are largely social constructs. Whether we should continue to use our current classifications is a societal decision about how we deal with each other in certain situations and not a problem solvable by the advance of medical knowledge. Szasz doubts the utility of the mentally ill label because he thinks it leads to abuse of individual rights, among other things.

It is important to see how Szasz's view differs from the disease views of Wakefield and Murphy discussed earlier. Defenders of the disease view of classification were willing to grant that some classifications are misapplied, as in the case of depression, and that others have such little scientific backing they probably should be jettisoned. Yet defenders of the disease approach are firmly committed to the idea that classifications of psychopathology can and should be based on identifiable malfunctioning of human systems. For the social constructivist, mental illnesses are just behaviors that society happens to group together and label as dysfunctions. Different societies or the same societies at different times can do this differently, and there is no right or best way to make such groupings; we impose these groupings on the behavior we dislike and label them as diseases.

Szasz has a general conceptual or philosophical argument for his view. His argument has a long history in debates over the scientific understanding of human behavior. The argument is this: we have to understand human behavior as based on reasons for action, as meaningful, and as rule or norm following (these are generally thought to be tightly interrelated). But reasons and meanings are not understandable scientifically; instead they call for the kind of interpretive understanding that we use constantly in our everyday social interactions.

Assessing this argument raises a large debate about what is often called the naturalistic approach to explaining human behavior and social organization. Naturalists claim that the broad methods of the sciences—e.g., observation and testing of hypotheses—are the best routes to explaining human behavior and that arguments like that given by Szasz make humans something special and outside the natural order. Naturalists point to advances in brain science, psychology, cognitive neuroscience, and related fields to support their case. Full consideration of this debate is beyond the scope of this chapter; however, the basic naturalist response to Szasz's argument is that reasons for behavior are psychological states that are subject to scientific explanation.

Another route to constructivism comes from anti-realist positions in debates over what is called scientific realism. One version of scientific realism claims roughly that current science is true, approximately true, or truer than our past science (see Chapter 9). Some anti-realists in those debates argue that results in the natural sciences are best understood as resulting from social processes where what counts as evidence, what counts as sufficient evidence to draw conclusions, how evidence is interpreted, what questions need to be answered, and so on are the product of social negotiation by individuals pursuing their own agendas and not objective statements about the way the world is. There are multiple studies of specific psychopathological classifications or psychopathological systems in general that draw the same conclusions about them. Kirk and Kutchens (1992) argue this about the DSM in general. Kincaid (forthcoming b) argues for this claim about the diagnosis of bipolar disease in children aged two and under, a condition that is identified in a significant number of children in the United States but is nearly unknown outside of the United States.

A further aspect of social construction has been emphasized by Hacking (1995) in *Rewriting the Soul*. When medicine and other professions that deal with psychological problems settle on a classification for a given set of behaviors and that classification becomes widely circulated in society, it may actually come to influence the behavior itself. Individuals with problems naturally enough use what they take to be well-supported professional diagnoses to understand their situation. Individuals may then interpret their problems using the categories of psychopathology advanced by the professions that treat it; they may begin to describe their symptoms in terms of the criteria defining specific psychopathological classifications and then may act in terms of this altered understanding, producing a process Hacking terms "looping." Hacking argues in detail that this is true of the category of multiple personality disorder, a condition that was basically unknown until the mid-1990s but was then quite rapidly claimed to be a problem affecting a number of individuals.

Social constructivism, as described here, like the disease view, has its problems. Is it really reasonable to think that the real difference between ordinary people and Alex, who can only sit in a dark room, is unable to work or function socially, and has identifiable and measurable psychological and biological abnormalities, does not reflect a difference in reality but is just the way we have chosen to label him as a society? It does not seem so. That all behavior classified as psychopathological is only a societal decision to label unwanted behaviors seems an overgeneralization.

6. A Third Pragmatic Route to Thinking about Classifying Psychopathology?

A compromise alternative to thinking of psychopathology as best understood as either disease or as social construction would be good, since both of those views have problems. A third approach is sketched in this section.

As we saw above, the notion of disease is not much help in actual classifications of psychopathology. However, at least for some behaviors that get labeled as mental illnesses, it seems pretty clear that they are not just behaviors that society decides to group together without there being an objective difference in reality. The intuition here is that people like Alex share something important in common that can ground our decision to see those individuals as falling into a specific category and to distinguish them from individuals like Jean, who is having transitory problems. We seemingly can make these distinctions even if we do not know how Alex's behavior results from some evolutionarily selected malfunctioning system or that Jean's behavior is the normal functioning of such systems.

A long philosophical and scientific tradition argues that our categorizations of the world pick out something real if (1) we have multiple independent measures for those categories and (2) they allow us to predict and explain a variety of phenomena. Put this way, the question is whether classifications of behavior into various types of psychopathology seem to be objective and allow the use of those classifications to make predictions, especially causal predictions. Having multiple independent measures—that do not rely on each other—is evidence that what we are classifying is not just something we are making up or subjective impositions. Scientists came to believe in electrons, for example, because they could detect them by many different means. Our evidence gets stronger if we can use classifications to make reliable predictions, especially about causes. So belief in electrons gets stronger if we can predict what will change their behavior (e.g., magnetic fields) and what influences changes in their behavior will have (chemical reactions resulting from "excited" electrons).

So the proposed third way treats psychopathological classifications as real if we have multiple indicators or measures picking out the individuals falling under the category and we can use the categorization to predict the causes and consequences of falling into that category (Kincaid, 2014). So back to Alex: there is good reason to think that the category of major depressive disorder is justified if we can identify various different characteristics that individuals classified like him share and if we can make confirmed predictions about individuals falling into the category.

There is in fact substantial evidence that we can do these things for people like Alex. People classified as having major depressive disorder for example generally:

Score high on questionnaires about symptoms characterizing depression such as suicidal thoughts or feelings of worthlessness
Have greater difficulty than those not categorized as suffering major depressive disorder in recalling positive memories

Have a higher level of stress hormones than those not categorized as suffering major depressive disorder

Have parents who have been classified as having major depressive disorder

Are more likely to attempt suicide

There are numerous other such characteristics that seem to separate those falling into the major depressive disorder category from those that do not. These characteristics are all ones that can be identified or verified by different observers without much need for subjective judgment; they are relatively objective measures, with only the assessment of parental depressive episodes calling sometimes for significant investigator interpretation of patient reports, since they have to rely on subject memories and interpretation of their parents' psychological states.

Contrast Alex with Shannon's gender dysphoria. There are not multiple independent measures of her "condition." We do not have physiological measures or tests of cognitive function that identify those with her condition. Significant value judgments about what is appropriate gender behavior may be a major factor in the distress that individuals who are given this classification report. The case for social construction—for the conclusion that classification represents behavior that is picked out by society for disapproval—seems pretty strong in Shannon's case. It does not in Alex's case.

However, Hacking's insight mentioned above that classification systems imposed by society may nonetheless influence the behavior of those classified suggests that social constructivists need not be as skeptical as Szasz about the classification of psychopathology. Social values may be involved in our classification systems, yet those classification systems may nonetheless still have some ability to objectively pick out and predict behavior. If society classifies according to what it disapproves, it must nonetheless pick out some objective traits of persons to do so. Also, deep-seated social norms of disapproval can have consequences for how individuals come to behave that result in possibly identifiable patterns.

So the third pragmatic route I am suggesting asserts that:

- There are some classifications of psychological problems that seem reasonably grounded in measurable, objective traits of individuals and fit the ideas of illness and syndromes from the medical model of psychopathology, although we may still be some distance from giving a full medical-style account of malfunctioning systems of the body. Alex's depression falls into this category. The extent to which specific individual psychiatric classifications are able to provide objective categorizations like those given for major depressive disorder is likely to be a matter of degree.
- Some behaviors that are really understandable, nonpathological problems in living may be misclassified as more serious illness-like conditions. Wakefield and Murphy make a compelling case for this claim. Jean's temporary depression, if labeled major depressive disorder and treated accordingly, would be a case in point. That does not mean that some better classification—grief-related depression—might not still be a useful way to understand her condition.
- There are some classifications of behavior as psychopathological that reflect social values, that are not reasonably thought of as diseases or illnesses, and that do not have the kind of objective indicators of physiological and psychological abnormality that major depressive disorder does; it may nonetheless still be an open question whether these classifications track behaviors well enough to be of use in dealing with human problems. So, for example, in the case of Shannon and gender dysphoria, there is currently a lively controversy over the usefulness of the category.

References

American Psychiatric Associations. 2013. *DSM 5*. Washington, DC: American Psychiatric Publishing.

Hacking, Ian. 1995. *Rewriting the Soul*. Princeton: Princeton University Press.

Horwitz, A. and Wakefield, J. 2007. *The Loss of Sadness: How Psychiatry Transformed Normal Sorrow Into Depressive Disorder*. Oxford: Oxford University Press.

Kincaid, H. Forthcoming a. "Doing Without 'Disorder' in the Study of Psychopathology." In *Wakefield and His Critics*, eds. L. Faucher and D. Forest. Cambridge, MA: MIT Press.

Kincaid, H. Forthcoming b. "DSM Applications to Young Children: Are There Really Bipolar and Depressed Two Year Olds?" In *Extraordinary Science: Responding to the Crisis in Psychiatric Research*, eds. S. Tekin and J. Poland. Cambridge: MIT Press.

Kincaid, H. 2014. "Defensible Natural Kinds in the Study of Psychopathology." In *Classifying Psychopathology: Mental Illness and Natural Kinds*, eds. H. Kincaid and J Sullivan. 145–173. Cambridge, MA: MIT Press.

Kirk, S. and Kutchens, H. 1992. *The Selling of DSM: The Rhetoric of Science in Psychiatry (Social Problems and Social Issues)*. New York: Aldine.

Murphy, Dominic. 2006. *Psychiatry in the Scientific Image*. Cambridge: MIT Press.

Szasz, Thomas. 2010. *The Myth of Mental Illness*. New York: Harper.

Wakefield, J. 2007. The concept of mental disorder: Diagnostic implications of the harmful dysfunction analysis. *World Psychiatry* 6:149–156.

World Health Organization. *ICD-10-CM Code Reference Guide: Book 5: Mental, Behavioral and Neurodevelopmental Disorders: Codes F01 Through F99*. Tim Dietrich, publisher.

Further Reading

Horwitz, A. and Wakefield, J. 2007. *The Loss of Sadness: How Psychiatry Transformed Normal Sorrow into Depressive Disorder*. Oxford: Oxford University Press.

Murphy, D. 2009. "Psychiatry and the Concept of Disease as Pathology." In *Psychiatry as Cognitive Neuroscience: Philosophical Perspectives*, eds. M.R. Broome and L. Bortolotti, 103–117. Oxford: Oxford University Press.

Szasz, Thomas. 2010. *The Myth of Mental Illness*. New York: Harper.

Wakefield, J. 2007. The concept of mental disorder: Diagnostic implications of the harmful dysfunction analysis. *World Psychiatry* 6: 149–156.

Zachar, P. 2000. Psychiatric disorders are not natural kinds. *Philosophy, Psychiatry and Psychology* 7(3): 167–182.

17
CLASSIFICATORY CHALLENGES IN PHYSICAL DISEASE

Mathias Brochhausen

Introduction

Classifying entities into categories is at the core of scientific practice, especially in the life sciences. In medicine, the discipline of classifying diseases is called *nosology*. In this chapter we will give an overview of the theoretical background and current approaches to the nosology of physical diseases, as well as the aims and contexts of disease classifications and their challenges.

In the first section, we will give a brief overview of the role of classification in science and its ontological background. In the second section, we will introduce a number of best practices in creating classifications. In the third section, we will introduce basic methodologies of how to classify diseases. We will also see that disease classifications are done with specific aims and for usage in specific contexts. This is what creates the challenges in providing both useful and consistent classification in medicine. In the fourth section, we will focus on classification criteria formulated from the requirements created by using computer systems to manage biomedical data. The fifth section provides an overview of the *International Classification of Diseases* against the background of the requirements and best practices. In the last section, we will summarize our discussion regarding the challenges in classifying physical diseases and current approaches used in the area of biomedical informatics and documentation to overcome those issues.

We should note that the distinction between psychiatric disease in general and physical disease is disputed, as psychiatric diseases can indeed have physical causes (Szasz 1960). However, for the purpose of analyzing systems of classification, the distinction is useful, because classification systems exist specifically for psychiatric disease.

That nosology is not merely a theoretical exercise but is highly important to medical practice can be seen from the example of Langerhans Cell Histiocytosis (LCH). LCH, which has been called "Histiocytosis X" in the past, is a rare disease characterized by the proliferation of Langerhans cells. Langerhans cells are immature dendritic cells found in the bone, skin, lymph nodes, and other organs (Morimoto et al. 2014). For some time, there has been uncertainty whether LCH is a neoplastic disease (a disease caused by an abnormal tissue that grows by cellular proliferation more rapidly than normal, as do cancers) or a reactive disease (a disease brought about by an aberrant immune response) (Morimoto et al. 2014; Demellawy et al. 2015). Due to these two conflicting classifications of LCH, it remained unclear what the most promising research agenda or therapeutic regimen for that disease was.

The Role of Classification in Science

How is it that the classification of a disease, such as LCH, is so relevant not only for therapeutic purposes but also for how we conduct research to learn more about a disease? The answer to this question is that classifications (e.g., the classification of mechanisms) are extremely relevant to the development of science. In their influential paper "Thinking about Mechanisms," Machamer, Darden, and Craver analyze mechanisms and break them down into entities and activities. The authors stress the fact that while mechanisms are an important aspect of scientific explanation in all sciences, their role is particularly prominent in the life sciences. According to Machamer et al., specific kinds of mechanisms are interdependent with specific kinds of entities or specific kinds of properties, so in order to explain the phenomena of nature, we refer to mechanisms and kinds of entities that are changed or involved in specific activities (Machamer et al. 2000). One example of this are the pathogenic mechanisms that lead to a disease (e.g., LCH): "The discovery of different kinds of mechanisms with their kinds of entities and different activities is an important part of scientific development" (Machamer et al. 2000: 15). Grouping particulars in kinds that are similar in some ways is a core aspect of the scientific process. When this is done based on the structures of the natural world rather than on human interest, we talk about natural kinds. There are indications that Machamer et al. are talking about natural kinds in the sentence quoted above. In this chapter we will use the term "kind" interchangeably with "type."

Assuming that scientific classifications like the ones mentioned by Machamer et al. reflect the structures of the natural world, the question is on what grounds we create such classifications. In other words, what determines whether individuals are of the same kind or type?

According to Smith, the answer to this question is that systems of classification should be based on universals (Smith 2006: 292). Universals are repeatables that may be instantiated by different individuals at different times and places (Lowe 2007: 10). One example of a universal is *redness*. Imagine standing on a street and watching objects pass by. The first object is a red scooter. If someone asked you what color the scooter was, you would say it was red. Next, a red car passes by. If the same person asked you what color the car was, you would probably say it was also red. So, being red is a repeatable: many things in many locations at many times are red. We might say they share the same property—namely, being red. Phrased in the language of universals, what they share is that they all instantiate the universal *redness*.

If being red is based on a universal, is it possible that being a human being or being a bald eagle is based on universals, too? Or moving to the example of diseases: could a case of LCH be an instance of the universal LCH? Smith thinks so and proposes a classification of diseases that is based on universals (Smith 2006: 292). Smith does not only allow universals, such as redness, which we might call property universals, but also holds that kinds are universals (e.g., being a human or being a case of LCH). Smith, in a paper co-authored with Ceusters, points out that classifying based on universals links classifications closely to the empirical world, since in principle, all universals need to be instantiated (Smith and Ceusters 2010: 178). So, while LCH might be a universal (there have been and are actual cases of LCH), unicorn is not (there are no actual unicorns).

In the history of philosophy, classification based on universals has led to the development of a tree-like representation of the kinds of substances that exist. The intention was to give a unique and comprehensive representation of the types of entities that exist. This representation is called the Porphyrian Tree. One example of such a tree is the Linnaean hierarchy of living beings. Is it possible to build a similar classification of all diseases that is both unique and comprehensive? The answer to that question, at least once we get to the level of classifying particular diseases and their subcategories, seems to be "no."

In *The Disunity of Science*, John Dupré shows why this is so. He argues that for any given domain there might be more than one classification system and that the ways we classify depend on context (e.g., the goals of a particular scientific study). Although he does not deny that there are objective divisions between kinds of things, he holds that the context dependence of classifications allows for what he calls "promiscuous realism" (Dupré 1995). Dupré's main thesis is "that there are countless legitimate, objectively grounded ways of classifying objects in the world. And these may often cross-classify one another in indefinitely complex ways" (Dupré 1995: 18). In discussing biological species, it becomes clear that Dupré rejects the Aristotelian notion of an unambiguous classification based on some essential property, whether it is morphological or genetic (Dupré 1995: 34). Although he argues against essential properties, Dupré agrees that there are properties on which we can build more informative classifications than others. For example, classifying animals based on their having feathers will be more useful than classifying them based on their being gray (Dupré 1995: 64). Dupré advocates that multiple classifications based on meaningful criteria should be allowed to exist beside one another. The call for pluralism in scientific classification has created an echo in philosophy of science and has led to the formulation of scientific pluralism (Longino 2001; Solomon 2001; Kellert et al. 2006). Scientific pluralism holds "that the multiplicity of approaches that presently characterizes many areas of scientific investigation does not necessarily constitute a deficiency" (Kellert et al. 2006: x), but that the question of whether a unified theory to account for all phenomena is still an open, empirical question.

We will see how methodologies created for medical information science can provide systems of classifications that allow the creation of multiple hierarchies along different lines of properties (e.g., diseases classified based on the anatomical site or system effected, diseases classified based on their causes, etc.).

Criteria for Well-Designed Classification Systems

If classifications are such a key element of the scientific endeavor, it raises the question of how we can create better classification systems, especially for scientific purposes. Jansen (Jansen 2009) gives an overview of best practices when it comes to building classifications. He starts by giving a classification that he deems to be a bad example, taken from Jorge Luís Borges's *The Analytical Language of John Wilkins*. Borges claims that, according to an old Chinese dictionary,

> animals can be classified into (a) those belonging to the Emperor, (b) those that are embalmed, (c) those that are tame, (d) pigs, (e) sirens, (f) imaginary animals, (g) wild dogs, (h) those included in this classification, (i) those that are crazy-acting (j), those that are uncountable (k) those painted with the finest brush made of camel hair, (l) others, (m) those which have just broken a vase, and (n) those which, from a distance, look like flies.
>
> (Borges 1975: 103)

Based on the shortcomings of this taxonomy, Jansen proposed a number of criteria that good classification systems should fulfill. Here are the four criteria we will focus on in this chapter:

1. Ontological grounding—In a good classification, classes are based on the properties of their members. This excludes classes like "others" (Jansen 2009: 160). Suppose you are taking it upon yourself to list all birds in a specific area by type, but all you know are silver gulls and bald eagles. You decide to report all birds that are neither a silver gull nor a

bald eagle in a class called "other." The latter class will contain many different birds (like finches, penguins, ostriches, etc.) that all have nothing in common except being a bird and not being a silver gull or a bald eagle. So, this class does not classify birds by their properties, but merely by the fact that they do not have the properties the members of the class "silver gull" or "bald eagle" have.

2. Structure—In building classifications, we need to take into account that types of things can be represented in a hierarchy: types have subtypes and not all classes in a classification are on the same level (Jansen 2009: 160, 162). Considering silver gulls, we can state that they are a species in the family of gulls. Likewise, we can state that bald eagles are one species under the genus sea eagles. In Borges's example, we do not find a hierarchy. All classes are on the same level.

3. Disjointedness—Assuming that we are dealing with a hierarchy, all classes on the same level should be disjoint. That means that if A and B are on the same level, no member of A is also a member of B (Jansen 2009: 160). If we think about the two species, silver gull and bald eagle, as classes, then they would be on the same hierarchical level because they are both species of birds. There are no individual birds that belong to both classes. If a bird is a member of the class "silver gull," then it is not a member of the class "bald eagle." Jansen's principle of disjointedness states that this should be the case for all classes on the same hierarchical level.

4. Uniformity—The criteria used for classification are uniform. They refer to properties that exist throughout the entire domain (Jansen 2009: 160–1). This criterion has two parts: first, the classification should be done based on one type of property. In the example of classifying birds introduced above, we started our classification by creating classes based on species membership. Observing finches, penguins, and ostriches, we might be tempted to also introduce classes based on the bird's preferred mode of locomotion: "flying birds" and "walking birds." However, these two classes should not be part of the classification that also contains "silver gull" and "bald eagle," because these classes are based on a different type of property. Second, the classification should be based on a type of property that exists throughout the entire domain. Looking at Borges's taxonomy, we find those animals that are "painted with the finest brush made of camel hair." However, not all animals in Borges's domain are painted. Hence, classifying animals by the types of brushes they are painted with violates the principle of uniformity. (Borges's example here also contains a category mistake: animals that are painted are not really animals. They are depictions of animals.)

The Aims and Criteria of Classifications of Physical Disease

The classification of diseases is interdependent with the existing system of medicine. Systems of medicine have changed over time, but even at one point in time, multiple systems of medicine can exist in different cultures. In the history of Western medicine, a number of systems of medicine have succeeded each other (e.g., humoral pathology and cellular pathology; Risse 1993). Humoral pathology or humorism is a system of medicine that is based on the assumption that the diseases are the consequence of an imbalance in the proportion of the four bodily fluids: black bile, yellow bile, phlegm, and blood. In his book, *Genesis and Development of a Scientific Fact*, Ludwig Fleck shows how the term *syphilis* was used to cover multiple different diseases over time. Shifts in the extensions and variations of the usage of the term are linked to shifts in the system of medicine as a whole, from humoralism to microbiology and cellular pathology (Fleck 1981). Fleck shows clearly how changes in the system of medicine, while based on the question of what the nature of disease is, affect and alter the classification of specific diseases.

Because of this influence of historical and scientific context on the classification of disease, looking at the historical development of codified and commonly used classification is relevant to understand which factors shaped where we are now. The earliest classifications of diseases were linked to efforts to unify cause-of-death statistics. This effort was partly motivated by an increased attention to preventing epidemics (Moriyama et al. 2011: 1). It is important to note that the classifications did not develop from clinical operations (e.g., diagnosis, therapy). Another historical aspect that has shaped the look of disease classifications used today is that the development of the classification of disease went hand-in-hand with the development of a disease nomenclature. The two are not the same. A nomenclature, such as the *Standard Nomenclature of Diseases and Pathological Conditions, Injuries, and Poisonings for the United States* (U.S. Department of Commerce 1920), merely provides a nonhierarchical list of acceptable terms for diagnosis (e.g., "Abscess, Cornea" or "Acetonemia"). It does not provide any grouping of these terms in more general terms, such as sexually transmitted disease, infectious disease, or histiocytosis. As a result, a nomenclature is not particularly useful for epidemic purposes, like monitoring the spread of infectious disease or STDs in a particular area or during a particular time. From the perspective of the criteria established in the previous section, a nomenclature does not fulfill the criteria of structure since it is not hierarchical.

So, nomenclatures might answer the need for unification of the terms used in diagnoses, but they do not provide the means to classify the diagnoses according to different kinds of pathological mechanisms. The latter is essential to the scientific endeavor. For example, a nomenclature will contain the term *Langerhans cell histiocytosis* and may list symptoms and other diagnostic criteria; however, a nomenclature would not inform us that a whole group of diseases share a proliferation of either tissue macrophages or dendritic cells (called histiocytes) and that these are called *histiocytosis* (Cline 1994). Being able to refer to a group of diseases that share one property is relevant not only for diagnostic purposes (differential diagnosis) but also for scientific purposes. To understand LCH, comparison to other forms of histiocytosis is helpful. However, the example also shows the potential pitfalls when we create classifications: nowadays, *histiocytosis* is still used sometimes, but it is regarded as an umbrella term (Kumar et al. 2015: 621) that captures multiple diseases that, while similar in some respect (proliferation of macrophages or dendritic cells), are different in other clinically and pathologically relevant ways. The more pressing question recently in LCH research has been whether LCH is a neoplastic or immune-reactive disease (Morimoto et al. 2014; Demellawy et al. 2015).

This example also demonstrates that building classifications of physical diseases is not trivial, nor is there one agreed-upon criterion by which to classify. When classifying disease, we could base a classification on a number of traditional criteria: the location of the disease (anatomical), the cause of the disease (etiology), the mechanism that causes the disease (pathogenesis), or the symptoms of the disease. In working toward one classification of physical disease that can be used for all medical activities, more than one criterion is often used (Snider 2003: 678–9).

The fact that the location of the disease is not an optimal criterion for disease classification became apparent fairly early in the history of medicine. Many diseases are not restricted to one body part or one organ system. For example, tuberculosis does not only affect the lungs and the respiratory system but may also affect the spine (Pott's disease) and the lymph nodes (scrofula). Similarly, classifying diseases by symptoms into classes like "diseases presenting with chronic airflow obstruction" is problematic due to the large number of nonspecific symptoms. LCH, for example, shows a multitude of nonspecific symptoms (e.g., of the skin-rashes, etc.).

In the practice of nosology, which criterion is used depends on the particular disease and our knowledge about that disease:

> In a disease without a single etiology but with well-defined pathology, such as mitral stenosis, the pathology of the heart valve would be a defining characteristic. If etiology was not single, and pathology was not specific, well-defined pathophysiology would be used as a defining characteristic.
>
> (Snider 2003: 678)

In the example of LCH, where the cause is yet unknown, some progress regarding the pathogenesis has been made (Demellawy et al. 2015).

Over the last decade, there has also been the vision of linking diseases, even those that are not hereditary, to gene loci. Many small steps have been made toward linking common diseases to genes, but the field still needs to undertake efforts to create more comprehensive genetic mappings for human diseases (Altshuler et al. 2008). If such mappings were created, diseases could also be classified based on genes or gene loci that have been associated with that specific disease. With respect to LCH, recent studies have shown an association between mutations of the *BRAF* gene and the pathogenesis of the disease. This finding led to the re-evaluation of LCH as a neoplastic disease (Demellawy et al. 2015).

In the case of LCH, this shift in classification from autoimmune disease to neoplastic disease has already led to a number of changes that affect patients. First, there now exists a number of clinical trials based on the new understanding of the disease. There have already been successful applications of *BRAF* inhibitors to cure the disease (Charles et al. 2014; Haroche et al. 2015). Second, those seeking funding for research on LCH are experiencing success with institutions that specialize in research on childhood cancers (e.g., St. Baldrick's Foundation). Third, the PDQ database, which is run by the National Cancer Institute and provides cancer information for patients and providers, started to include LCH around 2 years ago. Last, since LCH is now considered a neoplastic disease, families have access to assistance from organizations that were established to help families of children with cancer (P.K. Campbell, personal communication, June 30, 2015).

The classification criteria we have seen so far were without exception criteria based on the biomedical reality of disease. The multiple criteria that can be chosen to classify and the fact that the choice among these criteria is often based on pragmatic considerations reflect Dupré's promiscuous realism: the properties used to classify are actual properties of an objective physical phenomenon. However, some classifications are more based on socio-political assessments and expectations: Johansson und Lynøe (Johansson and Lynøe 2008: 15) give the example of drapetomania. This disease classification was created by surgeon and psychologist Samuel A. Cartwright in the 1850s. Cartwright claimed that slaves who ran away from their owners suffered from a psychiatric disease, drapetomania, which manifested by the afflicted slaves running away from their owners (Johansson and Lynøe 2008: 14–15). It is obvious that sociopolitical assessments of whether slaves ought to be free and the expectation that they ought to suffer their slavery without complaint or resistance shaped this classification of a "disease." There are no physical or biological aspects examined for that diagnosis. The only thing wrong with the afflicted persons is that they do not conform to the social and political norms in their environment. Such disease classifications, which are not based on the physical basis of disease, do not fall under Dupré's call for pluralism, because he stresses the fact that his perspective is a realist one (Dupré 1995: 57–58).

Criteria for Classifications in Biomedical Information Management

We have seen how the need for better statistics about mortality and morbidity drove the development of early disease classifications, showing that the purpose behind creating a disease classification is often rooted in the need for medical documentation. The advent of the computer in assisting with medical documentation changed the practice of how data was captured, accumulated, and interpreted. The use of computer systems and the move towards an Electronic Health Record created new requirements regarding standardization, in particular. Thus, shared vocabularies and terminologies became even more important (Shortliffe and Blois 2013).

In 1998, Cimino published a paper outlining the desiderata for controlled medical vocabularies. Controlled vocabularies are organized lists of terms relevant to a specific domain (e.g., disease) compiled to index publications or other material relevant to the domain. Typically, controlled vocabularies are organized hierarchically and thus can be seen as classifications. Although Cimino's paper is not exclusively about the classification of disease, he is aiming at controlled vocabularies that often contain a classification of disease. Cimino's motivation is rooted in the challenges that arose from biomedical informatics, and thus, his criteria present a proposal for a best practice of modeling medical data. However, some of them are relevant to classifications of disease in general:

1. Polyhierarchy— Although strict hierarchies are more manageable, most users of medical classifications agree that allowing multiple hierarchies is desired. It allows catering to heterogeneous purposes of use (Cimino 1998: 396–7).
2. Formal definitions—Controlled vocabularies should not only provide textual definitions, but formal definitions should link the term to other terms in the vocabulary (e.g., linking *Pneumococcal Pneumonia* to *Streptococcus pneumonia* with a *caused by* relation; Cimino 1998: 397).
3. Rejection of "Not Elsewhere Classified—"Not Elsewhere Classified" (NEC) classes are classes that contain all individuals that did not fit the criteria of any other class that is a subclass to the same superclass as the NEC. Although the use of such classes might seem understandable from the pursuit of comprehensiveness, the members of these classes do not necessarily share any property amongst each other, except the fact that they didn't fit into any other class (Cimino 1998: 397).

We see that Cimino's rejection of NEC classes is in line with Jansen's criterion of "ontological grounding."

It is obvious that Cimino and Jansen disagree greatly regarding polyhierarchy. Jansen rejects the idea that a class can belong to more than one superclass. Allowing for belonging to more than one higher-level group is called multiple inheritance (Jansen 2009: 165) and is likely to exist in polyhierarchies. Cimino, however, considers multiple inheritance to be one of the desiderata embraced by the scientific community in the field of biomedical informatics: "General consensus, seems to favor allowing multiple hierarchies to coexist in a vocabulary. . ." (Cimino 1998: 397).

Smith criticizes the fact that, in this paper, Cimino treats linguistic relations that may hold between linguistic expressions (e.g., terms in writing—*is narrower in meaning than*, *is synonymous*, etc.) exactly like the relations relevant in an ontological view. The latter are relations that hold between the entities that the terms stand for, their referents. The relation between diabetic patient and diabetes mellitus is a relation between the actual patient and one actual instance of the disease; it is not a relation between the terms "diabetic patient" and "diabetes mellitus" (Smith 2006: 289). Linguistic relations, by contrast, hold between the linguistic expressions. The term "diabetic patient" is narrower in meaning than the term "patient."

Failure to differentiate between statements about linguistic expressions and statements about their referents are called use-mention mistakes. Use-mention mistakes are relevant because they lead to incorrect inferences. Consider the following sentences:

(1) Pertussis is a highly contagious bacterial disease.
(2) "Pertussis" has nine letters.

Knowing that pertussis is also known as whooping cough, we might want to replace "pertussis" with "whooping cough." Although sentence (1) is still true if we do so, sentence (2) is not. This is why in sentence (2) we marked that we are talking about the term and not about its referent by using quotation marks. In sentence (1) we use the linguistic expression "pertussis" to refer to a bacterial disease. In sentence (2) we merely mention the linguistic sign, without being interested in what it refers to.

This section shows how the need for more standardized classifications for the medical domain links the practice of providing such classification (including classifications of disease) to the question of what is the best practice in creating classifications. It also shows that while there is some concurrence between the foundational rules for classifications established below, many practical concerns exist, which create some dissent.

The International Classification of Disease and the Classification of Physical Disease

In this section, we will examine the classification of physical disease in *The International Statistical Classification of Diseases and Related Health Problems*, or *International Classification of Diseases* (ICD) for short. It is provided and curated by the World Health Organization (WHO). The ICD is not the only classification of disease commonly used today. There are others, such as the *Systemized Nomenclature of Medicine* (SNOMED), *Medical Subject Headings* (MeSH), and the *National Cancer Institute Thesaurus* (NCIT). Although these all provide slightly different categorizations of diseases, they all follow the basic classificatory criteria outlined in the section above on basic methodologies. They demonstrate once more the multiplicity of uses that motivates the creation of disease classification. MeSH, for example, was created for cataloguing medical libraries (Rogers 1963) and still bears the mark of being a library science–inspired artifact. Today, it is still very useful for annotating medical literature (including journal articles) with keywords.

There are multiple non-identical versions of ICD-10. All examples of classes and codes given in this subsection have been interactively retrieved using the standards browser for ICD-10 (http://apps.who.int/classifications/icd10/browse/2015/en).

Historically, the ICD was developed for classifying causes of death for the purpose of statistical and epidemiological analysis. Since 1948, the ICD has increasingly been aimed to be useful in both morbidity and mortality statistics (Moriyama et al. 2011). Currently, ICD-10 classifies diseases and other relevant phenomena in 22 chapters. Chapters I to IV organize diseases based on etiology, while VI to XIV list diseases anatomically by the organ system that is affected. Chapter V lists mental diseases and behavioral disorders. Chapters XV and XVI deal with diseases related to pregnancy, childbirth, and the perinatal period. Congenital malformations, deformations, and chromosomal abnormalities are listed in Chapter XVII. Chapter XVIII is entitled *Symptoms, signs and abnormal clinical and laboratory finding, not elsewhere classified*. Unlike the preceding chapters of ICD-10, this chapter does not comply with the best practices we have identified, as it contravenes both Jansen's rule of ontological grounding and Cimino's call for rejecting NEC classes. We will see below that NEC classes or terms are not uncommon in ICD-10. The subsequent chapters classify injuries and poisoning (Chapter XIX), occurrences such as accidents, intentional self-harm, operations of war (Chapter XX), and factors like tobacco use, stress, and health care encounters (Chapter XXI). Under the heading *Codes for special purposes*, ICD-10 provides unassigned codes for future use and reassignment (Moriyama et al. 2011: 21). Each chapter is classified in categories (coded with a letter and two numerals). Those categories are further specified into subdivisions (coded with the code of their category plus an additional number). Each subdivision comes with a list of terms called inclusion terms. These identify terms that are to be understood as included in the subdivision. The inclusion terms listed may be synonymous with the subheading itself or may refer to different, more specific disease entities (World Health Organization 2012: 19–20).

Obviously, the ICD-10 is not only a classification of physical diseases, as demonstrated by the inclusion of mental diseases, injuries, and events leading to harm. With respect to the classification of physical disease, it is also obvious that ICD-10 does not classify based on one criterion alone. We find classes based on causes of disease and chapters organized according to the location of the disease with respect to major organ systems. The ICD-10 is understood by its creators to be a variable-axis classification, which combines five main axes (epidemic disease, constitutional/general disease, local disease by site, developmental disease and injuries) into one classification. "It has stood the test of time and, though in some ways arbitrary, is still regarded as a more useful structure for general epidemiological purposes than any of the alternatives tested" (World Health Organization 2012: 14). This statement once more stresses the focus of ICD-10 on statistical usage over clinical usage.

Looking at the ICD-10 and focusing on what kind of hierarchy it provides, it is obvious that the classification methodology used yields a hierarchy where each category is subsumed under one and only one chapter and each subdivision is a subdivision of one and only one category. This is called a monohierarchy. The WHO has identified the fact that ICD-10 provides only one hierarchy as a key shortcoming. Changing this to provide multiple different hierarchies would not only fulfill Cimino's call for polyhierarchy, but it also responds to the need for multiple context-based classifications noted by Dupré. The development of ICD-11, which started in 2007, will provide a disease classification that can be used for Electronic Health Records and is based on a shared logical model that allows for multiple context-specific mono-hierarchical classification (Rodrigues et al. 2014). We will see in the last section that this strategy of using formal logic to provide a multi-hierarchical structure of sets, while not providing a decision regarding how physical diseases ought to be classified, covers much if not all of the actual uses of disease classes.

As an example of the current classifications in ICD-10, let us look at LCH. LCH is classified under C96: *Other and unspecified malignant neoplasms of lymphoid, hematopoietic and related tissue*. (We should briefly note that this subcategory uses "other" as a classification criterion, which we have seen should not be done.) LCH is subdivided into three classes along the distinctions between unisystem vs. multisystem and unifocal vs. multifocal:

- C96.0 Multifocal and multisystemic (disseminated) Langerhans-cell histiocytosis [Letterer-Siwe disease]
- C96.5 Multifocal and unisystemic Langerhans-cell histiocytosis
- C96.6 Unifocal Langerhans-cell histiocytosis

Pulmonary LCH, which is typically classified separately in the clinico-pathological classification, is subsumed under C96.6 as an inclusion term with its alternative name Eosinophilic granuloma. Clinico-pathological classifications are those classifications that are usually found in pathology textbooks (Kumar et al. 2015: 621). They combine pathological considerations with clinically relevant considerations, to facilitate decision making by the health care provider. C96.6 also subsumes the inclusion term *Langerhans cell histiocytosis NOS* ("not otherwise specified"). This is again a clear violation of the criteria of omitting NEC terms. It raises the question of how these inclusion terms can work properly in a classification of entities. All LCH instances for which we do not have enough information to unambiguously categorize them go under 96.6 and are in the same class as instances of unifocal LCH. Such a classification clearly is not helpful for the clinical and pathological assessment of the situation. Although it is obvious how this classification promotes the statistical need to capture everything, it is also clear that clinical interpretation of statistics thus created may be prone to error. Since all cases for which we do not have sufficient information to subsume under 96.0 or 96.5 are counted as 96.6, we cannot be sure that only unifocal LCH cases are captured by that code. There might even be more multifocal cases than unifocal cases.

The Challenges in Classification of Physical Disease and How to Overcome Them

Our brief survey of classifications of physical disease shows that practical requirements seem to collide with best practice of how classifications should be done. Not only does conforming to general guidelines of how to build classifications such as Jansen's appear to be a challenge, but so does conforming to domain-specific criteria such as Cimino's.

The question we need to answer from the perspective of medicine remains: how can we build classifications in such a way that their creation and maintenance follows transparent and commonly acceptable criteria, while still ending up with classifications that allow us to switch perspectives based on the context of use? Multiple authors have stressed the need for multiple hierarchies based on context (Dupré 1995; Cimino 1998).

From the perspective of creating and maintaining a classification, rules like those formulated by Jansen, or Ceuster's and Smith's principle of instantiation, certainly provide advantages. However, these approaches have been criticized for overburdening the field of medical classifications with an uncalled-for commitment to metaphysical realism that can lead to impractical situations (Cimino 2006: 300–1). Facing this criticism, Smith and Ceusters have clarified that the actual ontological commitments of those creating the classifications are not relevant. Although they both are metaphysical realists—they think that universals are real (Smith and Ceusters 2010: 140–1)—in their eyes, the commitment to universals can be treated merely as a part of a methodology that allows building classifications and ontologies for practical use in a more transparent and coordinated manner. Using that methodology does not require that whoever uses it believes in the existence of universals. Smith and Ceusters stress the practical advantages of building and maintaining a single affirmed hierarchy based on "is a" relations, but they concede that multiple inheritance and, thus, multiple hierarchies can be inferred based on other relations and inclusion criteria for classes (Smith and Ceusters 2010: 147).

Although the ability to represent multiple, context-dependent hierarchies corresponding to multiple contexts and purposes is crucial, having a concise, formal methodology to create and maintain disease classifications is also indispensable. This means that instead of sticking to traditional mono-hierarchical classifications like ICD-10, we should strive for classifications that are based on formal logical definitions based on properties of the represented entities. This is certainly possible, and work toward making this a reality is on the way, but it is important to note that scientific progress will continuously make reclassifications necessary, as we have seen from the example of LCH.

References

Altshuler, D., Daly, M.J. & Lander, E.S. (2008) Genetic mapping in human disease. *Science*, **322**, 881–888.

Borges, J.L. (1975) *Other Inquisitions, 1937–1952 (Texas Pan American)*. University of Texas Press, Austin, TX.

Charles, J., Beani, J.-C., Fiandrino, G., & Busser, B. (2014) Major response to vemurafenib in patient with severe cutaneous Langerhans cell histiocytosis harboring BRAF V600E mutation. *J Am Acad Dermatol*, **71**, e97–e99.

Cimino, J.J. (1998) Desiderata for controlled medical vocabularies in the twenty-first century. *Methods Inf Med*, **37**, 394–403.

Cimino, J.J. (2006) In defense of the Desiderata. *J Biomed Inform*, **39**, 299–306.

Cline, M.J. (1994) Histiocytes and histiocytosis. *Blood*, **84**, 2840–2853.

Demellawy, D.E., Young, J.L., de Nanassy, J., Chernetsova, E., & Nasr, A. (2015) Langerhans cell histiocytosis: A comprehensive review. *Pathology*, **47**, 294–301.

Dupré, J. (1995) *The Disorder of Things: Metaphysical Foundations of the Disunity of Science*. Harvard University Press, Cambridge, MA, London, UK.

Fleck, L. (1981) *Genesis and Development of a Scientific Fact*. University of Chicago Press, Chicago, IL.

Haroche, J. et al. (2015) Reproducible and sustained efficacy of targeted therapy with vemurafenib in patients with BRAF(V600E)-mutated Erdheim-Chester disease. *J Clin Oncol*, **33**, 411–418.

Jansen, L. (2009) Classifications. In *Applied Ontology: An Introduction* (Eds., Munn, K. & Smith, B.) Ontos Verlag, Heusenstamm, Germany, pp. 159–172.

Johansson, I. & Lynøe, N. (2008) *Medicine & Philosophy: A Twenty-First Century Introduction*. Walter de Gruyter & Co., Berlin, Germany.

Kellert, S.H., Longino, H.E. & Waters, C.K. (2006) The pluralist stance. In *Scientific Pluralism*, (Eds, Kellert, S.H., Longino, H.E. & Waters, C.K.) University of Minnesota Press, Minneapolis, MS, pp. vii–xxix.

Kumar, V., Abbas, A.K. & Aster, J.C. (2015) *Robbins and Cotran Pathologic Basis of Disease*. Elsevier Saunders, Philadelphia, PA.

Longino, H.E. (2001) What do we measure when we measure aggression. *Studies in History and Philosophy of Science Part A*, **32**, 685–704.

Lowe, E.J. (2007) *The Four-Category Ontology: A Metaphysical Foundation for Natural Science*. Clarendon Press, Oxford, UK.

Machamer, P., Darden, L. & Craver, C.F. (2000) Thinking about mechanisms. *Philosophy of Science*, **67**(1), 1–25.

Morimoto, A., Oh, Y., Shioda, Y., Kudo, K., & Imamura, T. (2014) Recent advances in Langerhans cell histiocytosis. *Pediatr Int*, **56**, 451–461.

Moriyama, I.M., Loy, R.M., & Robb-Smith, A.H.T. (2011) *History of the Statistical Classification of Diseases and Causes of Death*. National Center for Health Statistics, Hyattsville, MD.

Risse, G.B. (1993) Medical care. In *Companion Encyclopedia of the History of Medicine (Routledge Companion Encyclopedias)*, (Eds, Bynum, W.F. & Porter, R.) Routledge, New York, NY, pp. 45–77.

Rodrigues, J.M., Schulz, S., Rectord, A., Spackman, K., Millar, J., Campbell, J., Üstün, B., Chute, C.G., Solbrig, H., Della Mea, V., & Brand Persson, K. (2014) ICD-11 and SNOMED CT Common Ontology: Circulatory system. *Stud Health Technol Inform*, **205**, 1043–1047.

Rogers, F. B. (1963) Medical subject headings. *Bull Med Libr Assoc*, **51**, 114–116.

Shortliffe, E.H. & Blois, M.S. (2013) Biomedical informatics: The science and the pragmatics. In *Biomedical Informatics: Computer Applications in Health Care and Biomedicine (Health Informatics)*, (Eds, Shortliffe, E.H. & Cimino, J.J.) Springer, London, UK, pp. 3–38.

Smith, B. (2006) From concepts to clinical reality: An essay on the benchmarking of biomedical terminologies. *J Biomed Inform*, **39**, 288–298.

Smith, B. & Ceusters, W. (2010) Ontological realism: A methodology for coordinated evolution of scientific ontologies. *Appl Ontol*, **5**, 139–188.

Snider, G.L. (2003) Nosology for our day: Its application to chronic obstructive pulmonary disease. *Am J Respir Crit Care Med*, **167**, 678–683.

Solomon, M. (2001) *Social Empiricism*. MIT Press, Cambridge, MA, London, UK.

Szasz, T.S. (1960) The myth of mental illness. *American Psychologist*, **15**, 113.

U.S. Department of Commerce. (1920) *Standard Nomenclature of Diseases and Pathological Conditions, Injuries and Poisonings for the United States*. Government Printing Office, Washington DC.

World Health Organization. (2012) *International Statistical Classification of Diseases and Related Health Problems, ICD-10*. World Health Organization, Geneva, Switzerland.

Further Reading

Dupré, J. (1995) *The Disorder of Things: Metaphysical Foundations of the Disunity of Science*. Harvard University Press, Cambridge, MA, London, UK. (An extended argument for the need of multiple hierarchies based on examples from biology.)

Fleck, L. (1981) *Genesis and Development of a Scientific Fact*. University of Chicago Press, Chicago, IL. (An inquiry into the development of nosological units against the background of medical history.)

Smith, B. (2006) From concepts to clinical reality: An essay on the benchmarking of biomedical terminologies. *J Biomed Inform*, 39, 288–298. (One of the basic texts pointing toward the shortcomings of traditional approaches of medical classification.)

Smith, B. & Munn, K. (2009) *Applied Ontology: An Introduction* (Eds., Munn, K. & Smith, B.) Ontos Verlag, Heusenstamm, Germany. (A collection of essays about numerous methodological aspects of ontological realism, frequently referring to examples from disease classifications.)

Part III
RESEARCH METHODS
(a) Evidence in Medicine

18
THE RANDOMIZED CONTROLLED TRIAL: INTERNAL AND EXTERNAL VALIDITY

Adam La Caze

Introduction

When discussing the validity of clinical trials, we need to consider both internal validity and external validity. *Internal* versus *external* validity is a distinction made in clinical epidemiology to discuss how well a clinical study answers certain questions. In broad terms, a study has high internal validity if the results can be considered accurate for the sample included in the study. It is an assessment of whether the study's design and conduct is sufficient to reliably answer the research question posed in the study. In clinical drug development, a study with high internal validity is considered to provide a reliable answer to the question: what is the effect of giving drug A to the sample studied in the trial? External validity, by contrast, refers to how well the outcomes observed in the study apply to people outside the sample included in the trial. When most people discuss internal and external validity, they do so in terms of *randomized controlled trials*, also known as *RCTs* or simply *randomized trials*. A common claim is that well-conducted randomized trials have better internal validity than other types of medical studies such as observational studies.

The first objective of this chapter is to examine arguments for the claim that randomized studies are methodologically superior to alternative study designs used in medicine. The section "Internal validity and hormone replacement therapy" introduces internal validity and bias in relation to the challenge of assessing the cardiovascular effects of hormone replacement therapy. More formal arguments for the benefits of random allocation in terms of internal validity are considered in the section "Arguments for randomization." Philosophical criticism of the tendency to claim too much on behalf of randomization is then considered in "Criticisms of randomized trials." The second objective of this chapter is to explore "The challenge of external validity." This section outlines some of the limits of randomized trials and briefly introduces a philosophical approach to improve the assessment of whether the results of a randomized trial will apply to patients in the clinic.

Internal Validity and Hormone Replacement Therapy

Internal validity is defined in similar terms in most clinical epidemiology texts. Fletcher et al. (1996, 12) provide the following:

> *Internal validity* is the degree to which the results of a study are correct for the sample of patients being studied. It is "internal" because it applies to the conditions of the particular group of patients being observed and not necessarily to others. The internal validity of clinical research is determined by how well the design, data collection, and analyses are carried out and is threatened by all the biases and random variation discussed above. For a clinical observation to be useful, internal validity is a necessary but not sufficient condition.

Clinical research—medical studies focusing on patients and the effects of treatments—is conducted using studies of a variety of designs. For each of these study designs, methods are employed to reduce error and improve internal validity. Randomized trials play a prominent role in clinical research because it is argued that they have the capacity to reduce or eliminate more sources of error than the alternative study designs employed in clinical research.

The goal of much clinical research is to estimate the effect of an experimental intervention as opposed to a control. An intervention is considered to "work" if the overall benefits of the intervention are greater than some prespecified clinically important amount (e.g., a new drug for preventing heart attacks might be considered worthwhile if it reduces the absolute risk of heart attack by an additional 5% in an at-risk population who are treated for 5 years). Errors that occur when discerning the effect of an intervention can be subdivided into *random errors* and *systematic errors*. Random errors occur due to chance. To take a simple example, consider 10 tosses of a fair coin. The coin is fair, so the probability of "heads" is 0.5, but in any series of 10 tosses we expect some variation in the number of "heads" observed—this variation is random error. Random error is managed by ensuring the study is large enough to estimate the outcome with sufficient precision.

Systematic errors occur when bias leads to incorrect measures of effect. In epidemiology, the term "bias" refers to any process that systematically promotes erroneous inference. Many sources of bias can affect clinical studies. Rothman et al. (2008) classify biases into three general categories: *confounding*, *selection bias*, and *information bias*. These concepts are illustrated using a real-life example. Estimating the effect of hormone replacement therapy on cardiovascular disease in postmenopausal women provides a classic case study of the effects of different sources of bias. The influence of hormone replacement therapy on cardiovascular disease has been assessed over the past 30–40 years in studies employing a range of different methods. Prior to menopause, women tend to suffer less than men from cardiovascular disease. After menopause the rates of cardiovascular disease in women approaches the rates seen in men. On the basis of this observation, it was thought that long-term use of hormone replacement therapy might protect women from cardiovascular disease. This question was first assessed in the 1970s and 1980s with methods that are collectively labeled "observational studies"—so called because these studies *observe* rather than *intervene* upon participants. Though there were conflicting results, the majority of findings from the observational studies supported the idea that postmenopausal women who take hormone replacement therapy suffer less cardiovascular disease. In a notable review of the observational studies, Stampfer and Colditz (1991, 61) concluded:

> The preponderance of evidence from epidemiologic studies strongly supports the view that postmenopausal estrogen therapy can substantially reduce the risk of coronary

heart disease.... This effect is unlikely to be explained by confounding factors or selection.

Prior to 2002, many women took hormone replacement wholly or partly for the purpose of reducing cardiovascular risk.

Clinical practice regarding the use of long-term hormone replacement therapy to reduce cardiovascular disease changed with the publication of the Women's Health Initiative Study in 2002 (Women's Health Initiative Investigators 2002). This large randomized trial reversed the findings of the earlier observational studies and provides compelling evidence that hormone replacement therapy does *not* reduce cardiovascular disease in older postmenopausal women—indeed, the Women's Health Initiative Study was stopped early due to excess risks in women taking hormone replacement therapy, including excess risks of cardiovascular disease.

What explains the disparity between the observational studies and the Women's Health Initiative Study? How does a single large randomized trial undermine the evidence provided in several large observational studies?

The short answer is that the Women's Health Initiative Study employed methods that rule out or minimize more sources of error than the methods used by the observational studies. Since the key observational studies and the Women's Health Initiative Study were all well conducted, analyzed and reported, the principal difference comes down to the *capacity* of the methods to rule out specific types of error. The key observational studies conducted to assess the association between hormone replacement therapy and cardiovascular disease were *prospective cohort studies*. These studies recruited a sample of postmenopausal women, some of whom were taking hormone replacement therapy and some of whom were not. The participants were observed over time and could start or stop hormone replacement therapy at their discretion. At the end of the study, the investigators compared the rates of cardiovascular disease in the women who were exposed to hormone replacement therapy with those who were not. If women who were exposed to hormone replacement therapy experience less cardiovascular disease, then it may be that hormone replacement therapy reduces the risk of cardiovascular disease.

The inferences licensed by observational studies such as these are valid only to the extent that a rather long list of additional assumptions are met. Sources of bias are legion. If socioeconomic status is positively correlated with exposure to hormone replacement therapy *and* negatively correlated with cardiovascular disease, then unadjusted estimates provided by a cohort study will exaggerate the putative benefits of hormone replacement therapy. If the impact of this confounder is not taken into account in an observational study, then participants exposed to hormone replacement therapy will have higher socioeconomic status and less cardiovascular disease compared to participants not exposed to hormone replacement therapy—and this will be the case *without* there being a causal relation between hormone replacement therapy use and cardiovascular disease. In this example, socioeconomic status is a *confounder*: (i) it is associated with exposure to the factor under investigation; (ii) it is correlated with the outcome; and (iii) the confounder is not an intermediate cause between exposure and outcome (Rothman et al. 2008).

Another type of bias can occur if factors that influence participation in the study also influence outcomes. For instance, the research question of interest is whether long-term use of hormone replacement therapy in postmenopausal women will reduce cardiovascular disease. The challenge here is that hormone replacement therapy is used for a range of indications. Perhaps the most common of these is short-term use in women experiencing symptoms around the time of menopause. If younger women are more likely to participate and their use of hormone replacement therapy poses a different risk of cardiovascular disease, then any estimate of the influence of treatment on cardiovascular disease will differ systematically from the estimate in older women using longer-term hormone replacement therapy. Bias due to the selection of individuals for participation is called *selection bias*. An observational study assessing the cardiovascular effects of hormone replacement therapy will be subject to selection bias to the extent that (i) women using hormone replacement therapy for perimenopausal symptoms are overrepresented in the study and (ii) any difference in effect on cardiovascular events in short-term use among perimenopausal women goes unrecognized. Selection bias seems to have played a part in the observational studies assessing the link between hormone replacement therapy and cardiovascular disease.

The third common type of bias is *information bias*. Information bias occurs from measurement error or misclassification. Studies assessing the effects of hormone replacement therapy need to accurately classify a participant's exposure to the treatment. There are a number of ways this might go awry. Perhaps exposure to hormone replacement therapy is assessed in a database that collects health insurance claims for medicines. If not all types of hormone replacement therapy are covered by the policy, then exposure will be underestimated. Similarly, it may be difficult in the claims database to differentiate women who are taking estrogen for a gynecological indication as opposed to hormone replacement therapy. Any misclassification that occurs may influence the estimate of the effect of hormone replacement therapy on cardiovascular disease. Information bias does not seem to have played an important role in the assessment of the link between hormone replacement therapy and cardiovascular disease.

Well-conducted observational studies are designed to mitigate the influence of known confounders—this is achieved in part through careful selection and adequate control. Further, well-conducted observational studies will employ data-based methods to systematically identify and address additional *potential* confounders. The overall consistency of findings from observational studies and randomized trials regarding the effects of hormone replacement therapy on a wide range of outcomes other than cardiovascular disease is testament to the success of these methods (Lawlor et al. 2004a). Unfortunately, a risk of residual confounding through unsuspected confounding factors or through under- or over-control of known or suspected confounders remains.

The discrepant results observed regarding the cardiovascular risks of hormone replacement therapy from observational studies and the Women's Health Initiative Study have been attributed to residual confounding. Lawlor et al. (2004a) make a case that the specific confounder that was not adequately controlled for in the observational studies was life-course socioeconomic status. *Current* socioeconomic status was controlled for in the observational studies, but socioeconomic status over the participant's life was not. Women with higher life-course socioeconomic status are *both* more likely to take hormone replacement therapy *and* have less physiological risk factors for heart disease—and these relationships are independent of current socioeconomic status (Lawlor et al. 2004b). This hitherto unrecognized confounder plausibly accounts for the apparent benefits of hormone replacement therapy on heart disease in the observational studies. Significantly, the discrepancy between the observational studies and the Women's Health Initiative Study prompted further examination of the association between life-course socioeconomic status, exposure to hormone replacement therapy, and risk factors for heart disease; without the surprising results of the Women's Health Initiative Study, these investigators would not have searched for a possible residual confounding factor.

Observational studies play an important role in epidemiology and clinical medicine, and sophisticated methods can be employed for mitigating the influence of many sources of bias. But, like any methods, the methods employed in observational studies have their limits. One of these limits is that if a confounder is unsuspected and goes unidentified in the variety of checks performed on the data to identify potential confounders, then there is a risk that confounder and its influence on the observed data goes undetected. Random allocation reduces the likelihood of this kind of error. The case of hormone replacement therapy and cardiovascular risk emphasizes the benefits of randomized trials in terms of internal validity. This, and two additional arguments for random allocation, are considered below.

Arguments for Randomization

R. A. Fisher (1966) was influential in promoting randomized allocation in experimental design and for developing frequentist statistical approaches for analyzing randomized studies (see especially, *The Design of Experiments*, which was first published in 1935). Fisher's work remains instructive regarding the key benefits of random allocation in terms of internal validity. Fisher provides three interlinked arguments in favor of random allocation. Random allocation: (i) provides an objective basis for statistical analysis; (ii) allows adequate masking and avoids biases associated with non-random allocation; and (iii) reduces the influence of confounders (especially unsuspected confounders). Each of these arguments is considered below. Although these arguments will be discussed in frequentist terms, it is worth noting that many of the frequentist reasons to randomize are also good reasons to randomize on alternative approaches to statistical inference. Suppes (1982) and Kadane and Seidenfeld (1990), for instance, provide arguments for random allocation in Bayesian terms.

Fisher's (1966, 11–26) tea-drinker experiment provides a helpful illustration of the first two arguments for randomization. Fisher also uses this case to outline and defend the *significance test*—the statistical test he is most famous for developing. The tea-drinker experiment sets out to assess a person's claim to be able to discriminate whether or not milk is added to a cup of tea before the tea infusion. Fisher proposes the following experiment to test the tea-drinker's claim. Present the tea-drinker with eight cups of tea, four of which have been prepared "milk-first" and four "tea-first." The order in which the cups are presented to the tea-drinker is randomized, as opposed to, say, ordered "systematically" where the investigator allocates cups according to some system—e.g., alternate cups or blocks of two "milk-first" cups then two "tea-first" cups, or ordered "haphazardly" where the investigator places the cups in what appears to the investigator to be a "sufficiently random" order (in the sense that there is no apparent order to the sequence). The tea-drinker is considered to have passed the test if she or he correctly identifies all four "milk-first" cups. The tea-drinker is informed of all the details of the experiment, including that the order of the cups will be randomized. Random allocation provides a probability model against which the tea-drinker's claim can be assessed. Fisher uses this probability model as the basis for the significance test. He outlines his approach in the following way:

> In considering the appropriateness of any proposed experimental design, it is always needful to forecast all possible results of the experiment, and to have decided without ambiguity what interpretation shall be placed upon each one of them. Further, we must know by what argument this interpretation is to be sustained.
>
> (Fisher 1966, 12)

The possible outcomes of Fisher's experiment is that the tea-drinker correctly identifies zero, one, two, three, or four "milk-first" cups. Given what matters for the experiment is how many "milk-first" cups the tea-drinker correctly identifies (as opposed to *which* "milk-first" cups the tea-drinker correctly identifies), there are 70 possible permutations of the outcomes of the experiment. In only one of these permutations are all four "milk-first" cups correctly identified (of the remaining 69 permutations, there is one permutation in which no "milk-first" cups are identified; 16 in which one "milk-first" cup is identified; 36 in which two "milk-first" cups are identified; and 16 in which three "milk-first" cups are identified).

Fisher then assigns a pre-experiment probability to each of the experimental outcomes. He does this by hypothesizing that the tea-drinker has no special ability to discriminate "milk-first" cups—this is labeled the *null hypothesis*. If the tea-drinker has no special ability to discriminate "milk-first" cups, then the probability of each outcome of the experiment equals the expected number of successes if the four cups were selected by chance. Fisher thinks of this in terms of the expected outcomes of the experiment if we were to repeat the experiment indefinitely, while assuming the truth of the null hypothesis. In this scenario, the probability of each possible outcome of the experiment is equal to the number of permutations in which the tea-drinker correctly identifies the specified number of cups divided by the total number of permutations. Fisher is particularly interested in the probability of the tea-drinker passing the test if in fact the null hypothesis is true. Despite having no skill in discriminating "milk-first" cups of tea, you would expect the tea-drinker to strike it lucky and correctly identify four "milk-first" cups once in every 70 repetitions of the experiment. Fisher reasons that 1 in 70 is sufficiently unlikely that he is willing to grant the tea-drinker's claim to be able to accurately discriminate "milk-first" cups of tea if she or he passes the test. Notice that the random allocation of the cups provides the objective basis for Fisher's test.

The second argument for random allocation refers to *masking*. "Masking" in clinical studies (also frequently referred to in the medical literature as "blinding") is the attempt to conceal which treatment a participant is allocated to (e.g., the experimental treatment or control; "milk-first" or "tea-first" cups of tea). Allocation concealment can take place at multiple levels; most importantly, allocation may be masked from the participant only ("single-blind"), or allocation may be masked from the participant and investigators ("double-blind"). Employing adequate methods to mask participants and investigators is an important part of trial design because it minimizes a wide range of biases that can occur when participants and/or investigators are aware of treatment allocation. Senn (1994) argues in relation to the tea-drinker experiment that *only* random allocation adequately masks the participant to the order of cups. This is because use of a systematic or haphazard method of allocation leaves the possibility open that the tea-drinker is able to make some educated guesses regarding likely orderings. Even if it is considered unlikely for the tea-drinker to guess the exact order of the cups, if she or he is able to rule some orderings more or less likely, then her or his chances of striking it lucky and correctly identifying all four "milk-first" cups will be considerably greater than 1 in 70, thereby undermining the rigor of the test.

Complete masking throughout a clinical trial is often difficult to achieve. Most experimental treatments cause effects that may unmask treatment allocation. The argument in favor of random allocation is that it is prone to fewer sources of bias than alternative methods of allocation. For instance, haphazard allocation is subject to subtle (perhaps subconscious) biases the investigator might have for allocating participants to experimental treatment or control. And, systematic allocation is subject to biases that may arise from unexpected influences on the order of recruitment. (Perhaps a clinical study sets out to allocate the first four participants enrolled every day to the experimental treatment, then the next four to control, and so on. This system of allocation is likely to result in biased results if, for instance, the first six patients seen in the clinic every day are the sickest.)

Finally, random allocation reduces the influence of extraneous risk factors. "Extraneous risk factors" are confounders: they are associated with both the exposure under investigation *and* the outcome under investigation *without* being an intermediate cause between the exposure and the outcome. Life-course socioeconomic status is an extraneous risk factor between hormone replacement therapy (exposure) and cardiovascular disease (outcome). Random allocation distributes extraneous risk factors in a statistically predictable way—and *it does this independently of whether the extraneous risk factor is recognized*. Exactly what random allocation achieves by distributing extraneous risk factors in a "statistically predictable way" is a key point of contention in the literature debating the merits of random allocation. Appreciating the statistical details of how random allocation influences the distribution of extraneous risk factors is important for understanding what random allocation does (and doesn't) achieve. In what follows, I provide the frequentist account. Consider a single extraneous risk factor, A_1, that is present in 50% of a population. Randomly allocating members of this population to intervention or control in a 1:1 ratio is analogous to tossing a fair coin. The larger the trial, the more likely A_1 is similarly distributed in the treatment and control groups. If the population of prospective participants is infinite, and recruitment is continued *indefinitely*, then we expect the proportion of participants with A_1 in each arm of the trial to approach 50% as the trial size approaches infinity.

We get the same result with a slightly modified scenario. In the modified scenario, the population of prospective participants is finite and the size of each trial is fixed. Consider a trial that recruits, randomizes, and observes 1,000 participants. The size of any single trial is large enough that A_1 is expected to be found in similar proportions in the experimental groups (think of this as something like 1,000 tosses of a fair coin). Now consider replications of this trial. In the indefinite sequence of trials, we expect most trials to have similar proportions of participants with A_1 in each arm, and while we would expect some trials to have an uneven distribution of A_1 simply through the play of chance, this would balance out in the indefinite sequence of trials. The overall effect of A_1 on the trial outcome summed over the indefinite sequence of trials is 0. This simple case can be extended to take into consideration multiple extraneous risk factors, A_i where $I = 1, \ldots, n$. Providing we consider the indefinite sequence of trials, the same can be said: the overall effect of A_i extraneous risk factors on the trial outcome when summed across the indefinite sequence of trials is 0.

Random allocation provides a probabilistic model for how extraneous risk factors will be distributed in the indefinite sequence of trials. This formulation is disputed by Bayesians and other non-frequentists, who question the relevance of what happens in the indefinite sequence of trials. The important point to recognize is that many of the stronger statements made about what random allocation achieves refer to the indefinite sequence of trials, sometimes implicitly. For instance, Devereaux and Yusuf (2003, 107):

> . . . RCTs are superior to observational studies in evaluating treatment because RCTs eliminate bias in the choice of treatment assignments and provide the only means to control for unknown prognostic factors.

Random allocation *eliminates* bias due to known and unknown extraneous risk factors, but only in the indefinite sequence of trials. In any single trial, the most that can be said is that, providing the trial is large, random allocation improves the probability that extraneous risk factors (known and unknown) are roughly balanced. This appears to have been enough for the Women's Health Initiative Study to provide a superior assessment of the effects of hormone replacement therapy on heart disease—assuming that there *are* residual confounders that explain the difference between the cardiovascular outcomes measured in the observational studies and the Women's Health Initiative Study, and that these confounders are sufficiently evenly distributed in the Women's Health Initiative Study. Importantly, *this* more fragile benefit of random allocation is less than what is implied by some proponents of randomized trials.

Criticisms of Randomized Trials

There are many examples in the general medical literature of strong claims made regarding the benefits of randomized trials:

> Without clear confirmatory evidence from large-scale randomized trials or their meta-analyses, reports of moderate treatment effects from observational studies should not be interpreted as providing good evidence of either adverse or protective effects of these agents (and, contrary to other suggestions, the absence of evidence from randomized trials does not in itself provide sufficient justification for relying on observational data).
>
> (Collins and MacMahon 2007, 24)

This is a strong claim in a carefully argued paper on the benefits of randomized trials in medicine. It is easy to find statements that echo this sentiment or make stronger claims. The overarching view of randomized trials provided in this strand of the literature is that they are categorically superior to alternative study designs. The expressed view is that randomized trials are *necessary* for drawing causal inferences about medical therapies. The justification for this view is the idea that random allocation provides some kind of *guarantee* of the observed results of a trial.

This view is targeted by critics, including Peter Urbach (1993) and John Worrall (2002, 2007). Worrall summarizes the problematic view of randomized trials in the following way:

> It is widely believed that RCTs carry special scientific weight—often indeed that they are *essential* for any truly scientific conclusion to be drawn from the trial data about the effectiveness or otherwise of proposed new therapies or treatments.
>
> (Worrall 2007, 452)

Worrall is successful in undermining this view of randomization, and his argument is an important corrective to a common tendency to overstate what randomized trials achieve. Worrall's key argument hinges on what random allocation achieves with regard to the distribution of extraneous risk factors. Worrall suggests that those who hold the view that randomized trials are essential in medicine elide what randomization achieves on any particular allocation with what randomization achieves in the indefinite sequence of trials. If random allocation *did* ensure all possible confounders were *equally* distributed in the intervention and control groups, then randomized trials would have a distinct advantage in terms of internal validity. Other things being equal, any difference observed in such a trial could *only* be due to the effect of the treatment; *this* is the kind of guarantee randomized trials are assumed to provide by those who view randomized trials as essential. But, of course, random allocation does not achieve an equal distribution of all possible confounders in any specific trial (perhaps, for example, in a particular trial more women are allocated to the treatment group than men). Indeed, as Worrall (2002, S324) notes, and as Urbach (1993, 1426) did before him, the more possible confounders there are, the higher the probability that any one of the confounders will be maldistributed in the experimental groups in any particular random allocation. Randomization provides no sure-fire guarantee for the results of the trial.

Most people who write about randomized trials in statistics and medicine know this, and most acknowledge it somewhere, if sometimes only in the small print. Nevertheless, the idea that randomized trials are necessary and sufficient to justify causal inferences about treatments is commonly expressed. And this simple, commonly expressed view is enormously influential in health care and a growing range of disciplines. Worrall takes the view that random allocation in clinical trials provides no additional benefits over those that could be gained from adequate control in an observational study, where "adequate control" for Worrall means comparable based on known confounding factors. But this swings the randomization debate too far in the opposite direction. As discussed above, there are good reasons to randomize even though random allocation does not guarantee the results of a trial. Random allocation provides an objective basis for statistical analysis, permits masking, and reduces the influence of unsuspected confounders. The benefits of randomized studies are, however, relative. Randomized trials employ methods that avoid sources of error that cannot be avoided in observational studies. The capacity to avoid more sources of error means that, other things being equal, the *internal validity* of a well-conducted randomized study will be higher than that of a well-conducted observational study.

The Challenge of External Validity

Being clear on the benefits of randomized trials—and the justifications for these benefits—helps to better identify the limits of randomized trials. Before releasing a new treatment into the market, we want to be sure that it works in a clearly identified group of patients. When thinking about a study testing whether a new treatment works, the focus is rightly on internal validity: can we trust the results? Well-conducted randomized trials provide more reliable results than do alternative methods. In this context the frequently offered advice to focus on randomized trials can be defended, but internal validity is only one aspect to consider. Once the treatment is on the market, clinicians will need to decide whether the treatment is likely to benefit the specific patients under their care; similarly, the patients will need to decide if it is in their best interests to take the treatment. Assessing whether a medicine will work for a specific person is often significantly harder than assessing whether it works, on average, in an experimental sample. These decisions rely on a broader range of evidence than that provided by well-conducted randomized trials. In this context, following the advice provided in many medical resources to focus on evidence from randomized trials leads to poor decisions.

External validity is the "degree to which the results of an observation hold true in other settings" (Fletcher et al. 1996, 12). Assuming an experimental treatment benefits the sample of patients included in a well-conducted randomized trial, the assessment of external validity focuses on the *generalizability* of the observed results to patients outside of the trial. A related term is *effectiveness*—an intervention is "effective" when the intervention benefits patients receiving the treatment undergoing routine care. Effectiveness contrasts with *efficacy*; an intervention is "efficacious" when it benefits patients in an experimental setting (e.g., a randomized trial). There is consensus in medicine as to what needs to be done to achieve internal validity. There is less consensus on the topic of external validity. Indeed, the main point of consensus within medicine is that external validity is difficult to judge (Rothwell 2007; Rawlins 2011).

Well-conducted randomized trials are a more specialized tool than is typically appreciated. The particular function in medicine to which they are well-suited is the rigorous assessment of whether a treatment is efficacious. All of the randomized trials conducted throughout drug development are designed to test a specific hypothesis: does the investigational treatment produce benefits on a specific health measure when compared to control in a particular group of patients? All of the key methodological decisions made in the trial are to ensure the primary hypothesis is assessed in a way that avoids systematic and random error. Many of these decisions limit external validity.

Variability is a threat to internal validity and trial efficiency. Trial participants can vary in many ways that influence their response to treatment; they may vary with respect to their underlying health, the severity of their condition, their adherence to treatments, their ability to absorb, metabolize, and/or eliminate the pharmaceutical treatment, and so on. The more variability, the more likely some of this variability will influence trial outcomes, which means more participants will be needed to demonstrate the purported benefits of the investigational treatment. Variability in how patients are treated *other* than the investigational treatment is also tightly controlled in randomized trials. Many randomized trials conducted for the purposes of regulatory approval are multi-site, and a significant number are multi-national. Considerable effort is taken to ensure that trial participants are treated in similar ways at the various trial sites in order to reduce bias. For instance, non-experimental treatments, treatment progression and withdrawal, and patient monitoring are all standardized. These conditions are important to ensure that the trial results reflect the experimental treatment rather than some other aspect of care. But, at the same time, each one of these conditions shifts a participant's treatment further away from what he or she would have received under routine care. This raises an inevitable question: will the benefit observed under strict trial conditions be observed in the clinic?

Typical patients recruited to clinical trials are often importantly different from typical patients with the same condition seen in the clinic. Patients enrolled in trials tend to be younger and suffer from fewer additional illnesses—this improves trial efficiency and reduces the risk of adverse effects from the investigational treatment. The duration of treatment for many interventions is also substantially shorter in a randomized trial compared to routine care. Furthermore, patients enrolled in trials often have a more *severe* presentation of the primary condition being treated. Patients with moderate-to-severe illness are more likely to experience clinically important outcomes, such as heart attacks or strokes. In this way, the trial will be able to demonstrate the benefits of a new drug in reducing heart attacks or strokes with fewer participants enrolled in the trial. There is no sense in which participants in randomized trials are a random sample of patients in the community with the primary condition being treated. Much progress has been made in improving the reporting of trials in recent years; specifically, inclusion and exclusion criteria, flow of patients through the trial, reasons for trial ineligibility, and dropouts are far better reported now than they were 10 years ago. Nevertheless, it remains difficult to judge the population of which the trial participants are a sample.

The upshot of all of this is that it can be difficult to generalize the results of randomized trials. One response to the challenge of external validity is to conduct *more* randomized trials, especially large pragmatic randomized trials. Large pragmatic trials (also known as "large simple trials") attempt to conduct the trial in a way that is as close to clinical practice as possible (Yusuf et al. 1984). This is achieved by keeping the inclusion and exclusion criteria simple, testing a single intervention, and focusing on a single clinically important outcome (such as mortality). When available, large pragmatic trials provide clinically important evidence on the average outcomes of a treatment and avoid some of the problems of smaller randomized trials. The Women's Health Initiative Study arguably falls into the category of a large pragmatic trial; another positive example is the ISIS-2 study (1988), which enrolled 17,187 patients and demonstrated the benefits of antithrombotic treatments and aspirin in patients suffering from an acute heart attack.

However, large pragmatic trials are not a comprehensive solution to the problem of external validity. The first problem is practical: large trials, no matter how simple, take time and are expensive to complete. If a large pragmatic trial has been conducted for the specific effectiveness question under consideration, it will likely be relevant, but for the vast majority of questions, it is unlikely a large pragmatic trial has been conducted. The second problem is more technical. The main selling point for large pragmatic trials is that by allowing considerably more variability in the patients recruited and in the non-experimental treatments that they receive, the trial provides more insight into the likely effects of the treatment in routine clinical care. This is true to an extent. A well-conducted *successful* large pragmatic trial provides good evidence that the *average* effects of giving the treatment are positive. However, in extending the results of such a trial to a given specific population or individual, the critical assumption is that the positive average effects are consistent across the many subpopulations included in the trial. Sometimes this seems to be a reasonable assumption, but often it is an assumption that is difficult to justify. Subgroup analyses attempt to quantify treatment effects in participant groups *within* a clinical trial, but they are notoriously unreliable (Feinstein 1998; Brookes et al. 2001).

When the assumption of similar effects in different patient groups is brought into question, the results of large pragmatic trials are just as difficult to generalize as those of small trials. The problem in the medical literature is that there is too great a focus on randomized trials. It is often underappreciated how much evidence *external* to a randomized trial needs to be taken into account in order to assess whether the results can be generalized to another setting. Assessing internal validity is largely an assessment of the methods employed in the trial—too often, this focus on trial methods is exported to the assessment of external validity. The important role that evidence external to the randomized trial plays in the assessment of external validity is also often underappreciated. Well-conducted observational studies and the insights provided by basic medical science possess unique strengths in providing evidence that supports judging whether the results observed in an experimental setting are likely to translate to patient benefits in the clinical setting (Black 1996; Vandenbroucke and Psaty 2008; La Caze 2011).

Nancy Cartwright's work highlights what needs to hold *in addition* to a successful randomized trial to be confident that a treatment will be effective. Cartwright and colleagues argue that judgments of external validity and effectiveness need to be *causal* (Cartwright 2010, 2012; Cartwright and Munro 2010; Cartwright and Hardie 2012). Among other causal considerations, assessing the effectiveness of an intervention requires an understanding of how the intervention works and confidence that what is required in the environment to support the intervention working is present in the context in which the intervention is to be employed. Despite some notable exceptions—specifically, Rothman et al. (2008, 129)—explicit causal reasoning is rarely emphasized in the assessment of external validity within medicine. The most common advice provided on assessing external validity in the medical literature is to judge the *similarities* of the randomized trial with clinical practice. For example:

> External validity is matter of judgment and depends on the characteristics of the participants included in the trial, the trial setting, the treatment regimens tested, and the outcomes assessed.
> (Moher et al. 2010, 20–21)

Assessing similarities between the experiment and routine practice will often be important, but it is not enough. What matters is whether or not the causal structure is sufficiently replicated (see Chapter 5, "Mechanisms in medicine," and Chapter 6, "Causality and causal inference in medicine," for further discussion).

Cartwright (2010, 60) argues that the standard way in which external validity is conceptualized risks obscuring the causal judgments that need to made based on an assessment of all the available evidence. Cartwright (2011) and Cartwright and Hardie (2012) focus on *effectiveness arguments*—the argument that supports a claim that an intervention will be effective—and articulate the assumptions present in a strong effectiveness argument.

Argument A:

1. x plays a causal role in the principle that governs y's production [in the experimental setting].
2. x plays a causal role [in the clinical setting] as well as [in the experimental setting].
3. The support factors necessary for x to operate are present for some individuals [in the clinical setting].

Therefore, x plays a causal role [in the clinical setting] and the support factors necessary for it to operate are present for some individuals.

(Cartwright 2011, 222)

Here, x is the intervention and y is the outcome. Argument A must be sound for an intervention to have a causal effect in the clinical setting. A well-conducted randomized trial provides good evidence for premise 1, but more is needed to ground an effectiveness argument. Premises 2 and 3 also need to hold. The assessment of these premises is causal and will need to rely on evidence external to the randomized trial. Is the causal structure in the clinical setting similar enough to the causal structure present in the randomized trial? Are the factors that support x's effect on y present in the clinical setting? Are there factors present in the clinical setting that reduce x's effects or promote y's harms? It is important to note that the conclusion of A is that the intervention will have a causal effect in the clinical setting—postulating that the causal effect will be *positive* on average (or in a particular patient) requires still further conditions to hold (see Cartwright 2011, 2012 for further discussion).

Cartwright's argument schema better explicates the requirements on effectiveness arguments and judgments of external validity. This doesn't make the difficult causal judgments any easier, but it is a step in the right direction for improving the assessment of external validity.

Conclusion

Well-conducted randomized trials reduce or eliminate more sources of error when assessing the efficacy of a treatment compared to alternative study designs. Randomized trials are neither infallible nor the only important source of evidence in medicine. Predicting whether the effects of a treatment observed in a randomized trial will be observed in routine clinical care is often challenging. The assessment of external validity relies on causal knowledge, which is provided by evidence from a range of sources.

References

Black, Nick. 1996. "Why we need observational studies to evaluate the effectiveness of health care." *British Medical Journal* 312 (7040): 1215–8.

Brookes, S T, E Whitely, T J Peters, P A Mulheran, Matthias Egger, and George Davey Smith. 2001. "Subgroup analyses in randomised controlled trials: Quantifying the risks of false-positives and false-negatives." *Health Technology Assessment* 5 (33): 1–58.

Cartwright, Nancy. 2010. "What are randomised controlled trials good for?" *Philosophical Studies* 147 (1): 59–70.

———. 2011. "Predicting what will happen when we act: What counts for warrant?" *Preventive Medicine* 53 (4–5): 221–4.

———. 2012. "Will this policy work for you? Predicting effectiveness better: How philosophy helps." *Philosophy of Science* 79 (5), 973–89.

Cartwright, Nancy, and Jeremy Hardie. 2012. *Evidence-Based Policy: A Practical Guide to Doing It Better.* Oxford, UK: Oxford University Press.

Cartwright, Nancy, and Eileen Munro. 2010. "The limitations of randomized controlled trials in predicting effectiveness." *Journal of Evaluation in Clinical Practice* 16 (2): 260–6.

Collins, Rory, and Stephen MacMahon. 2007. "Reliable Assessment of the Effects of Treatments on Mortality and Major Morbidity." In Peter M. Rothwell, ed., *Treating Individuals: From Randomized Trials to Personalized Medicine.* Edinburgh, UK: Elsevier.

Devereaux, P J, and Salim Yusuf. 2003. "The evolution of the randomized controlled trial and its role in evidence-based decision making." *Journal of Internal Medicine* 254: 105–13.

Feinstein, Alvan R. 1998. "The problem of cogent subgroups: A clinicostatistical tragedy." *Journal of Clinical Epidemiology* 51 (4): 297–9.

Fisher, R A. 1966. *The Design of Experiments.* London: Oliver and Boyd.

Fletcher, Robert H, Suzanne W Fletcher, and Edward H Wagner. 1996. *Clinical Epidemiology: The Essentials.* Philadelphia, PA: Lippincott, Williams & Wilkins.

ISIS-2 Collaborative Group. 1988. "Randomised trial of intravenous streptokinase, oral aspirin, both, or neither among 17 187 cases of suspected acute myocardial infarction: ISIS-2." *The Lancet* 332 (8607): 349–60.

Kadane, Joseph B, and Teddy Seidenfeld. 1990. "Randomization in a Bayesian perspective." *Journal of Statistical Planning and Inference* 25 (3): 329–45.

La Caze, Adam. 2011. "The role of basic science in evidence-based medicine." *Biology and Philosophy* 26 (1): 81–98.

Lawlor, Debbie A, George Davey Smith, and Shah Ebrahim. 2004a. "Commentary: The hormone replacement-coronary heart disease conundrum: Is this the death of observational epidemiology?" *International Journal of Epidemiology* 33 (3): 464–7.

———. 2004b. "Socioeconomic position and hormone replacement therapy use: Explaining the discrepancy in evidence from observational and randomized controlled trials." *American Journal of Public Health* 94 (12): 2149–54.

Moher, David, Sally Hopewell, Kenneth F Schulz, Victor Montori, Peter C Gøtzsche, P J Devereaux, Diana Elbourne, Matthias Egger, and Douglas G Altman. 2010. "CONSORT 2010 explanation and elaboration: Updated guidelines for reporting parallel group randomised trials." *British Medical Journal* 340: c869.

Rawlins, Michael. 2011. *Therapeutics, Evidence and Decision-Making.* London: Hodder Arnold.

Rothman, Kenneth J, Sander Greenland, and Timothy L Lash. 2008. "Validity in epidemiologic studies." In *Modern Epidemiology*, edited by Kenneth J Rothman, Sander Greenland, and Timothy L Lash, 3rd ed., 128–47. Philadelphia: Lippincott Williams & Wilkins.

Rothwell, Peter M., ed. 2007. *Treating Individuals: From Randomised Trials to Personalised Medicine.* Edinburgh, UK: Elsevier.

Senn, S. 1994. "Fisher's game with the devil." *Statistics in Medicine* 13 (3): 217–30.

Stampfer, M J, and G A Colditz. 1991. "Estrogen replacement therapy and coronary heart disease: A quantitative assessment of the epidemiologic evidence." *Preventive Medicine* 20 (1): 47–63.

Suppes, P 1982. "Arguments for randomizing." *Philosophy of Science Association* 2: 464–75.

Urbach, Peter. 1993. "The value of randomization and control in clinical trials." *Statistics in Medicine* 12: 1421–31.

Vandenbroucke, Jan P, and Bruce M Psaty. 2008. "Benefits and risks of drug treatments." *Journal of the American Medical Association* 300 (20): 2417–9.

Women's Health Initiative Investigators. 2002. "Risks and benefits of estrogen plus progestin in healthy postmenopausal women." *Journal of American Medical Association* 288 (3): 321–33.

Worrall, John. 2002. "What evidence in evidence-based medicine?" *Philosophy of Science* 69: S316–30.

———. 2007. "Why there's no cause to randomize." *The British Journal for the Philosophy of Science* 58 (3): 451–88.

Yusuf, Salim, Rory Collins, and Richard Peto. 1984. "Why do we need some large, simple randomized trials?" *Statistics in Medicine* 3 (4): 409–20.

Further Reading

Cartwright, N. 2011. "Predicting what will happen when we act: What counts for warrant?" *Preventive medicine* 53(4–5): 221–4.

Rothman, Kenneth J, Sander Greenland, and Timothy L Lash. 2008. "Validity in epidemiologic studies." In *Modern Epidemiology*, edited by Kenneth J Rothman, Sander Greenland, and Timothy L Lash, 3rd edition, 128–47. Philadelphia: Lippincott Williams & Wilkins.

Senn, S. 2004. "Controversies concerning randomization and additivity in clinical trials." *Statistics in Medicine* 23 (24): 3729–53.

Worrall, J. 2002. "What evidence in evidence-based medicine?" *Philosophy of Science* 69: S316–30.

19
THE HIERARCHY OF EVIDENCE, META-ANALYSIS, AND SYSTEMATIC REVIEW

Robyn Bluhm

Contemporary medical practice is strongly influenced by the idea that clinical decision-making should be based on the results of clinical trials. A number of interrelated trends, stretching back several decades, have converged on a view of what counts as good medical research. This chapter begins with a brief history of three connected movements—clinical epidemiology, evidence-based medicine (EBM), and the Cochrane Collaboration—which aim to promote the use of high-quality evidence from clinical research in patient care. "High-quality evidence" is usually understood as evidence at the top of a "hierarchy of evidence," specifically randomized controlled trials (see Chapter 18) and systematic reviews or meta-analyses of such studies. The remainder of the chapter will discuss these ideas in detail, examining both the arguments for and the criticisms of hierarchical rankings of study methods, and of the combination of research results in a meta-analysis.

Clinical epidemiology is the oldest of the trends toward using the results of clinical trials to inform clinical practice; Jeanne Daly (2005) traces it back to the late 1960s, to the work of Suzanne Fletcher and Robert Fletcher, and, beginning around the same time, to that of Alvan Feinstein. This turn to epidemiological methods as a way of gathering evidence to support clinical decision-making was motivated primarily by the perceived shortcomings of physiological research. In an interview with Daly, Robert Fletcher says:

> We grew up in a very biomedical era. We were trained at Stanford in internship and residency, where everyone was a laboratory scientist, and it seemed to me that the kind of science that they brought to bear on patient care, which was mainly logical argument from laboratory data on mechanisms of disease, just wasn't the best possible way of answering these clinical questions.
>
> (Daly, 2005: 21)

As will become clear below, the contrast between laboratory-based physiological research and clinical trials that use epidemiological methods continued to be prominent in EBM.

According to Daly, "[e]vidence-based medicine was the form in which clinical epidemiology promoted its findings to practicing clinicians, who were its primary target" (2005: 4). The term "evidence-based medicine" was first used in an article in the journal *ACP Journal Club* in 1991 (Guyatt, 1991), but the "manifesto" introducing EBM to a broad clinical audience appeared in *JAMA: The Journal of the American Medical Association* the following year (Evidence-Based Medicine Working Group, 1992). The authors of this second paper were members of the Evidence-Based Medicine Working Group, which was based at McMaster University in Hamilton, Ontario, Canada. In this paper, they contrast "the way of the past" in medicine with "the way of the future" by using a case study of a resident who is trying to find information for a patient who wants to know his risk of experiencing a seizure. The way of the past is based on the knowledge of medical authority figures, as the resident gets an answer to this question by asking her supervisors what to tell the patient. By contrast, the way of the future has her formulating her patient's question in terms that reflect the kind of hypothesis that could be examined in a clinical study, searching the databases of the medical literature for relevant research, appraising the studies she finds to ensure their quality, and then reporting their results to the patient.

Shortly after this, members of the group published a textbook, *Evidence-Based Medicine: How to Practice and Teach EBM* (Sackett et al., 1997), and they began to present a series of articles in *JAMA* on various aspects of EBM; these articles were later published as *The Users' Guides to the Medical Literature* (Guyatt and Rennie, 2002). These publications laid out the main tenets of, and skills required for, EBM. The central theoretical idea underlying EBM is that evidence from clinical research can be ranked hierarchically, with the study methods that are most likely to give good evidence at the top of the hierarchy (see the section below on the hierarchy of evidence). In addition, readers were taught how to critically appraise published research in order to determine its quality. For example, the *Users' Guides* present the techniques of critical appraisal in terms of a series of questions about the methodological features of a study and the details of its results. EBM therefore aimed to teach clinicians how to obtain and assess epidemiological studies that addressed clinical questions relevant to their practice.

By contrast, the third movement that advocated using epidemiological study designs to inform clinical practice focuses on producing summaries of the literature. The Cochrane Collaboration, formed in the UK in 1993, describes itself as "a global independent network of researchers, professionals, patients, carers, and people interested in health" that aims to "gather and summarize the best evidence from research" (www.cochrane.org). One of the problems faced by clinicians who want to practice EBM is the sheer volume of clinical literature. The Cochrane Collaboration aims to solve that problem by gathering, appraising, and systematically reviewing studies relevant to a specific clinical question, and by updating the reviews periodically as new evidence becomes available. From its beginnings, the Cochrane Collaboration was closely allied with the McMaster University group's EBM, as David Sackett (often described as the father of EBM) was also instrumental in the development of the Collaboration.

In summary, looking at clinical epidemiology, EBM, and the Cochrane Collaboration, we can see a convergence on the idea that clinical decision-making should be based on the results of research that examines patient outcomes in large groups, rather than on physiological research or on the advice of medical authorities.

What Kind of Research?: The Hierarchy of Evidence

The central idea underlying EBM is that research can be ranked hierarchically based on study design: studies that use methods ranked higher on the hierarchy are more likely to be high quality and therefore to provide good evidence. Although there are different hierarchies for different kinds of research question (e.g., questions about the efficacy or effectiveness of a treatment, about the harms associated with an intervention, about patient prognosis, or about the accuracy of diagnostic tests), much of the discussion in the EBM literature has focused on research on therapies, and this kind of research will therefore be the primary focus of this chapter.

In addition, there are different versions of the hierarchy of evidence, although all of these have the same basic structure. Here, the hierarchy of evidence presented in the *Users' Guides to the Medical Literature* will be used, as it is among the first and most influential of the proposed hierarchies, and is also therefore representative of the basic structure of hierarchies for treatment evidence. [More recent versions of the hierarchy, notably the one produced by the GRADE Working Group, also incorporate assessments of study quality, although the general ranking of study methods is unchanged (see www.gradeworkinggroup.org).] The hierarchy is as follows:

- Systematic reviews/meta-analyses of RCTs
- Single RCT
- Systematic reviews/meta-analyses of observational studies (cohort, case-control)
- Single observational study
- Physiologic studies
- Unsystematic clinical observations (Guyatt & Rennie, 2002: 7)

There are several important points to note regarding the hierarchy of evidence. First, all of the top levels of evidence use methods adopted from epidemiology; they compare outcomes in large numbers of people, one group of which receives the intervention being tested while the other group does not. Ideally, the groups will be similar (on average), except for differences in their exposure to the intervention being studied. This approach to research is contrasted with studies that aim to understand physiological mechanisms that underlie disease and that are supposed to be affected by therapeutic interventions. This kind of physiological research occupies the lowest level of the hierarchy, even though for most of the 20th century, it was the main kind of medical research. David Sackett, who was strongly influenced by Alvin Feinstein's version of clinical epidemiology, argues for the importance of studies using epidemiological methods, saying that "in sharp contrast to most bench research, the results of clinical-practice research are immediately applicable" (Sackett, 2000: 380). In general, there is more variability toward the bottom of the hierarchy than at the top; some versions include case studies or case series reports, or expert consensus, albeit as relatively poor-quality evidence.

This quotation also draws attention to an additional reason for placing physiological research at the bottom of the evidence hierarchy, which is that EBM emphasizes the importance of using "patient-important," clinically relevant outcomes, such as the occurrence of death, heart attack, or stroke, in clinical research. The alternative approach is to use physiological measurements as "surrogate outcomes" that are related to and predict the clinically relevant events. In many cases, EBM cautions, these surrogate outcomes are only weakly related to the clinical outcomes of real interest, and are therefore poor predictors. This aspect of EBM reflects the concern of the earliest clinical epidemiologists that physiological research is not directly relevant to clinical practice.

Another important thing to notice about the hierarchy of evidence is that it places randomized studies (either a single RCT, a systematic review, or meta-analysis of RCTs) above nonrandomized, "observational" studies. In some places, this ranking is treated as absolute, for example: "[i]f the study wasn't randomized, we'd suggest that you stop reading it and go on to the next article in your search. . . . Only if you can't find any randomized trials should you go back to it" (Straus et al., 2005: 118). In other places, for example, the GRADE Working Group's system, it is acknowledged that a well-designed nonrandomized study can provide better evidence than a less well-designed randomized trial. Furthermore, in some cases, if the evidence (from nonrandomized trials) supporting an intervention is particularly strong, EBM acknowledges that it may not be necessary to conduct RCTs at all. In general, though, the consensus is that randomization is such an important feature of study design that RCTs almost always provide better evidence than studies that do not randomly assign patients to the treatment or the control group.

What does randomization do that makes it so important? There are two main arguments for randomization. First, it is considered to be the best way to ensure that the treatment and the control groups really are similar. Similarity is important because if, for example, one group is significantly older (on average), or contains a greater proportion of women, than the other does, then it is not clear whether any observed outcome differences between the groups are due to the presence or absence of the intervention being studied or to the other (age, sex) differences between the groups. These other factors are said to confound the results of the study. In some cases, specific clinical or demographic factors may be known or suspected to affect the clinical outcomes of interest, but different unknown factors may also have such an influence. For example, an unmeasured genetic or physiological characteristic may be present in one study group more often than in the other, and that may influence either the progression of disease or the action of the treatment. Proponents of randomization say that it is the best way to balance both known and unknown confounders between study groups.

The second major argument for randomization is that if study researchers are deciding who gets assigned to the treatment or the control group, they may (whether consciously or unconsciously) bias the assignment. For example, a clinician may assign sicker patients to receive an experimental drug rather than a placebo because she wants to ensure that they receive an active treatment. Alternatively, a clinician who stands to benefit financially or professionally if a clinical trial shows that a new treatment is efficacious may put patients who are relatively healthy in the experimental group. (Again, this need not be a conscious, deliberate decision.) In both of these cases, the study groups end up being unbalanced with regard to disease status, which is likely to affect the study outcome. This kind of bias is known as allocation bias (because the allocation of patients to study groups biases the trial's results).

Philosophical Criticisms of the Hierarchy of Evidence

It is important to recognize that the hierarchy of evidence is based on and defended by epistemological claims. It is therefore a philosophical theory of medical knowledge. Philosophers have examined and challenged both the implicit assumptions underlying the hierarchy and the explicit defenses supporting it. Most of this philosophical work has addressed the claims made regarding randomization and, in particular, have tended to conclude that the placement of randomized trials above nonrandomized studies is not entirely justified. More recently, there has also been discussion in the philosophical literature of the relative importance of epidemiological and physiological research.

With regard to the first issue, John Worrall's 2002 paper "What Evidence in Evidence-Based Medicine" provides a clear analysis of the claims made in favor of placing randomized studies at the top of the hierarchy and concludes that this placement is not justified. With regard to the balancing of known confounders, he points out that it is just as feasible to deliberately balance the groups, for example, by assigning (nearly) equal numbers of women and of men to each of the treatment and the control groups. With regard to *unknown* confounders, which by definition cannot be deliberately balanced, Worrall says that randomization is not (despite some claims to the contrary) guaranteed to do so; what is actually correct is the much weaker claim that randomization controls for these factors "in some probabilistic sense" (2002: S322).

Worrall also addresses the argument that randomization prevents bias in the allocation of patients to the treatment versus the control groups by ensuring that study personnel do not deliberately assign patients to groups in such a way that confounding factors are not balanced across groups. He concedes the importance of guarding against this bias, but notes that random assignment of patients to study groups is merely a means of achieving this goal: "It is blinding (of the clinician) that does the real methodological work—randomization is simply a means of achieving this end" (S325).

A number of other authors have addressed EBM's claims about the importance of randomization. For example, Grossman and Mackenzie (2005) survey some of the limitations of RCTs and discuss their implications for research. In particular, they worry that the emphasis on randomization will prevent people from conducting, and from appreciating the contributions that can be made by, research that uses other methods. Borgerson (2009) argues that the claim that RCTs are the only type of study that can avoid (or at least the study that best avoids) certain kinds of bias obscures the fact that RCTs are still subject to many other kinds of bias.

More recently, there has been philosophical discussion about the second key feature of the hierarchy, which is the relationship between the epidemiological studies that occupy the top few layers of the hierarchy and laboratory-based research on physiological mechanisms, which is located near the bottom of the hierarchy. One key question in this discussion is whether evidence (or, more generally, reasoning) about mechanisms can ever substitute, as a basis for treatment decisions, for epidemiological research. The EBM position appears to be that, as a last resort, this research can be used: while the *Users' Guides* say that the hierarchy of evidence "implies a clear course of action for physicians addressing patient problems: they should look for the highest available evidence from the hierarchy" (Guyatt and Rennie, 2002: 8), this implies that if no randomized (or nonrandomized) controlled trials are available, then evidence from physiological studies should be used.

A deeper question is whether this kind of research can ever provide *strong enough* evidence to establish the efficacy or effectiveness of a treatment, so that it is not necessary to conduct epidemiological studies. The idea here is that the physiological mechanism by which the treatment works is so well-understood that no further research is needed. The suggestion that RCTs are always necessary to establish whether a treatment works has been spoofed by a study that purported to review the RCT evidence that parachutes reduce mortality associated with "gravitational challenge." The authors noted that they were unable to find high-quality RCT evidence in favor of parachute use (Smith and Pell, 2003). More seriously, people who claim that evidence from physiological mechanisms may be sufficient have observed that nobody claims that it is necessary to do a placebo-controlled RCT to determine whether people with dehydration should be treated with water; our understanding of the relevant physiology is enough to justify using this therapy.

Jeremy Howick (2011) has argued that "mechanistic reasoning" can provide sufficient evidence for the use of a treatment when the following conditions are met: knowledge of the relevant mechanisms must be complete (so that there are no "gaps" between the intervention and the outcome), and both the probabilistic nature and the complexity of mechanisms must be recognized (including the way that their functioning might be influenced by other mechanisms, which may result in "undesirable side-effects or even paradoxical effects" (2011: 144). Howick acknowledges, particularly in his more recent work (Howick et al., 2013), that cases in which knowledge of physiological mechanisms is sufficient to justify the use of a treatment is rare, but he does not want to rule out the possibility that these situations may sometimes occur.

Holly Andersen (2012) has argued that even Howick's guarded optimism about the sufficiency of mechanistic evidence is not justified. She points out that the physiological mechanisms, and the functional relationships among them, are so complex that it is unlikely that we can ever be justified in claiming that Howick's conditions are met. Yet she does (in agreement with the *Users' Guides*) say that there is a second—and legitimate—role that knowledge of mechanisms can play in clinical reasoning. This is the use of knowledge of mechanisms to bridge the gaps between evidence from clinical trials and the circumstances and characteristics of particular patients. Bluhm (2013) has argued, however, that situations in which this kind of reasoning about specific cases is justified are probably almost as rare as ones in which Howick's criteria for reasoning about general claims are met. Similarly, Howick et al. (2013) argue that knowledge of mechanisms cannot justify inferring from clinical trials how a treatment will work in a population that is different from the one in which the study results were obtained, ultimately because of the complexity of physiological mechanisms.

Another important discussion relevant to the relationship between epidemiological and physiological research centers on a position that has come to be known as the Russo-Williamson thesis. Federica Russo and Jon Williamson (2007) have argued that establishing causal claims in medicine (including claims about whether a treatment will have the desired outcomes) requires both population-level, epidemiological research and research on physiological mechanisms. This is because the kind of causal evidence coming from each type of research is different: epidemiological research provides statistical evidence of a correlation between an intervention and an outcome, while physiological research provides mechanistic evidence. According to Russo and Williamson, causal claims in biomedicine rely on both kinds of evidence; they develop the "epistemic theory of causality" in order to do justice to the dual nature of causal claims in medicine. They and their colleagues also note that EBM has failed to do justice to the importance of the evidence provided by mechanisms and that its reliance on correlational, epidemiological research is problematic for several reasons (Clarke et al., 2013). First, correlation on its own cannot establish causation, because of the possibility of confounding variables. Notice that this point can serve as a rejoinder to those who claim that RCTs provide the best method for eliminating the effects of confounding variables by balancing those variables between the treatment and the control group. Second, knowledge of mechanisms is, in a sense, presupposed by epidemiological studies, since they influence the design of clinical trials. Finally, information about mechanisms is necessary to allow researchers to make claims that go beyond the results of a specific RCT, including applying those results outside of the context of the trial. This last point shows that the preceding discussion of Howick's and Andersen's work on generalizing from mechanisms is related to the more general claims about causality in medicine and medical research that are the focus of Russo's and Williamson's work.

Given these problems with the main points raised by the hierarchy of evidence, it has been suggested that it might be worth getting rid of the idea of a hierarchy. Goldenberg (2009), for example, argues that because the hierarchy is problematic, and has therefore been the focus of most criticisms of EBM, attention has been diverted away from the important contributions that EBM *can* make to improving medical care. Bluhm (2005) suggests that a better metaphor for the relationship among different kinds of research evidence is a network; this will better allow us to focus on the relationship between epidemiological and physiological research. Most recently, Stegenga (2014) has examined more recent versions of the hierarchy, such as GRADE, and argued that they fail to overcome the criticisms raised against the earlier hierarchies.

Meta-Analysis and Systematic Reviews

The previous section surveyed issues related to the placement of different kinds of study (randomized, nonrandomized, physiological) on the hierarchy. This section examines questions related to another central feature of the hierarchy, which is the importance of systematic reviews and meta-analyses. A systematic review is "a high-level overview of primary research on a particular research question that tries to identify, select, synthesize and appraise all high quality research evidence relevant to that question" (www.cochrane.org), whereas a meta-analysis is a form of systematic review that combines the results of the included studies in a statistical analysis. Recall that hierarchy is structured so that systematic reviews or meta-analyses of RCTs rank higher than a single RCT, and then similarly for nonrandomized studies. This ranking is based on the (very reasonable) idea that more evidence is better. Note that, even here, things are not straightforward; the hierarchy implies that randomized and nonrandomized trials cannot (or perhaps should not) be combined in a single systematic review or meta-analysis. And beyond this starting point, things rapidly become more complicated; it is not clear how best to amalgamate the available evidence and draw conclusions about what it says. To see why this is the case, we can look at some of the decisions made in the course of conducting a meta-analysis.

Perhaps the most famous example of a meta-analysis in medicine is the one represented graphically in the logo of the Cochrane Collaboration. The study reviewed the results of placebo-controlled RCTs that examined whether giving a short course of corticosteroids to women who were about to give birth early improved the chances of their babies' survival. The first of these RCTs was published in 1972, and ten years later there was sufficient evidence to establish that the odds of infant mortality due to the complications of premature birth were reduced by 30–50%. But, the Cochrane website explains, the first meta-analysis on this topic was not published until 1989. Prior to this, because the evidence was located in a number of studies, many of which were too small to have the power to detect any effect, the studies did not affect practice and, as a result, "tens of thousands of premature babies have probably suffered and died unnecessarily (and needed more expensive treatment than was necessary)" (www.cochrane.org).

This story illustrates powerfully the important contribution that can be made by meta-analysis; however, it is not clear that this particular meta-analysis is representative. The intervention in these studies (corticosteroid therapy) was simple and short term. The outcome measured (neonatal death) is clear and easy to measure, and was not tracked over the long term. The results (when combined) are also dramatic: the improvement in survival rates with therapy is large.

By contrast, in many other cases, the studies that have been conducted on a particular intervention may be highly variable, both in the details of their design and in their quality. RCTs may differ in a number of ways, including the details of the intervention (e.g., dosage, duration of treatment, use of concomitant therapies), the outcomes measured (and when and how they are measured), and the population included in the study (e.g., age, presence of comorbid conditions). Those who conduct a meta-analysis must decide how similar studies must be with regard to these features before they can be combined. They may even decide, if the studies are different enough, to conduct a systematic review without a meta-analysis and present information for all of the studies separately. In cases where there is a large number of studies, this approach will soon become unwieldy; moreover, the fact that meta-analyses are generally conducted if possible suggests that they are the preferred method (though the hierarchy of evidence does not explicitly rate them above systematic reviews that do not conduct a meta-analysis).

Similarly, decisions must be made about when to include studies on the basis of their quality. One option is to set a threshold, with studies falling below this quality cut-off being excluded from the meta-analysis. Another is to assign different weights to the study results being included, so that higher-quality studies make a bigger contribution to the summarized results than do lower-quality studies. Because of the variability of studies pertaining to a clinical question, part of the process of conducting a meta-analysis is deciding which studies to include. Different groups conducting meta-analyses on the same treatment may make these decisions differently. Stegenga (2011) has pointed out that one of the advantages claimed for meta-analysis is that it is an objective way to assess evidence, but that because these different choices are possible, this claim is false.

A review by Deshauer et al. (2008), on the use of selective serotonin reuptake inhibitors (SSRIs) to treat depression, illustrates the kinds of decisions researchers must make in conducting a review. This particular paper both reports the methods and results of individual studies and conducts several analyses that combine the results of these studies for different treatment outcomes (e.g., response to treatment, overall acceptability of the treatment). Only six RCTs met the criteria for inclusion in the review. When the results of all of the studies were combined, they showed a statistically significant response to treatment, but a closer look at the individual studies suggests that combining them may obscure important information. Half of the trials included only patients who had major depression with no comorbid conditions; combining only these studies showed a statistically significant treatment response (and a larger effect size than the analysis that included all six of the studies). The other half did include patients with comorbidities, and when these studies were combined, there was no statistically significant improvement. This means, though, that clinicians making decisions about the treatment of patients with comorbidities on the basis of the meta-analysis of all six studies may overestimate the effect of SSRIs for their patients.

The presentation of the details of each of the six studies also shows clearly that they varied with regard to the treatment dose and duration, and even the drug being tested; although most of the studies examined sertraline, others tested paroxetine or citalopram. The studies also differed in how they defined "response to treatment" and in the eligibility criteria for the study. These differences among the studies do not mean that the authors of the review were wrong to combine them, but different choices—ones that might have resulted in rather different results—could also be defended. Deshauer et al. also chose to exclude a number of studies, most notably any that lasted less than 6 months and any that used a "discontinuation" study design (which begins by treating *all* participants with the study drug and then assigning them to either continued treatment or placebo). Again, had these studies been included, the conclusion regarding the effectiveness of SSRIs may well have been different.

Conclusion

The trend toward basing clinical decision-making on the results of clinical research raises interesting philosophical questions. Although much has been written by philosophers on EBM and, in particular, the hierarchy of evidence, most of the discussion of the merits of meta-analyses has taken place in the medical literature.

References

Andersen, H. (2012) "Mechanisms: What Are They Evidence for in Evidence-Based Medicine?," *Journal of Evaluation in Clinical Practice* 18(5): 992–999.

Bluhm, R. (2005) "From Hierarchy to Network: A Richer View of Evidence for Evidence-Based Medicine," *Perspectives in Biology and Medicine* 48: 535–547.

———. (2013) "Physiological Mechanisms and Epidemiological Research," *Journal of Evaluation in Clinical Practice* 19(3): 422–426.

Borgerson, K. (2009) "Valuing Evidence: Bias and the Evidence Hierarchy of Evidence-Based Medicine," *Perspectives in Biology and Medicine* 52: 218–233.

Clarke, B., Gillies, D., Illari, P., Russo, F., and Williamson, J. (2013) "The Evidence that Evidence-Based Medicine Omits," *Preventive Medicine* 57(6): 745–747.

Cochrane collaboration website: www.cochrane.org
Daly, J. (2005) *Evidence-Based Medicine and the Search for a Science of Clinical Care*. Berkeley: University of California Press.
Deshauer, D., Moher, D., Fergusson, D., Moher, E., Sampson, M., and Grimshaw, J. (2008) "Selective Serotonin Reuptake Inhibitors for Unipolar Depression: A Systematic Review of Classic Long-Term Randomized Controlled Trials," *CMAJ* 178(10): 1293–1301.
The Evidence-Based Medicine Working Group. (1992) "Evidence-Based Medicine: A New Approach to Teaching the Practice of Medicine," *The Journal of the American Medical Association* 268: 2420–2425.
Goldenberg, M. (2009) "Iconoclast or Creed? Objectivism and Pragmatism in Evidence-Based Medicine's Hierarchy of Evidence," *Perspectives in Biology and Medicine* 52(2): 168–187.
GRADE Working Group: www.gradeworkinggroup.org
Grossman, J., and MacKenzie, F.J. (2005) "The Randomized Controlled Trial: Gold Standard, or Merely Standard?" *Perspectives in Biology and Medicine* 48: 516–534.
Guyatt, G., and Rennie D. (2002) (eds.) *Users' Guide to the Medical Literature*. Chicago: AMA Press.
Guyatt, G.H. (1991) "Evidence-Based Medicine," *American College of Physicians Journal Club* 114(suppl 2): A16.
Howick, J., Glasziou, P., and Aronson, J.K. (2013) "Problems with Using Mechanisms to Solve the Problem of Extrapolation," *Theoretical Medicine and Bioethics* 34(4): 275–291.
Howick, J.H. (2011) *The Philosophy of Evidence-Based Medicine*. Chichester, West Sussex: BMJ Books, Wiley-Blackwell.
Russo, F., and Williamson, J. (2007) "Interpreting Causality in the Health Sciences," *International Studies in the Philosophy of Science* 21(2): 157–170.
Sackett, D.L. (2000) "The Fall of 'Clinical Research' and the Rise of 'Clinical-Practice Research'," *Clinical and Investigative Medicine* 23:379–81.
Sackett, D.L., Richardson, W.S., Rosenberg, W., and Haynes, R.B. (1997) *Evidence Based Medicine: How to Practice and Teach EBM*. New York: Churchill Livingstone.
Smith, G.C., and Pell, J.P. (2003) "Parachute Use to Prevent Death and Major Trauma Related to Gravitational Challenge: Systematic Review of Randomised Controlled Trials," *British Medical Journal* 327(7429): 1459–1461.
Stegenga, J. (2011) "Is Meta-Analysis the Platinum Standard of Medicine?," *Studies in history and Philosophy of Biological and Biomedical Sciences* 42: 497–507.
———. (2014) "Down with the Hierarchies," *Topoi* 33(2): 313–322.
Straus, S.E., Richardson, W.S., Glasziou, P., and Haynes, R.B. (2005) *Evidence-Based Medicine: How to Practice and Teach EBM*. Toronto: Elsevier.
Worrall J. (2002) "What Evidence in Evidence-Based Medicine?," *Philosophy of Science* 69(S) (Supplement): S316–S330.

Suggested Readings

Bluhm, R. and Borgerson, K. (2011) "Evidence-based medicine" In: Gifford, F, editor. *Philosophy of Medicine* Volume 16 of the *Handbook of the Philosophy of Science* (Series Editors: Dov Gabbay, Paul Thagard, John Woods). Oxford, North Holland: Elsevier. pp. 203–238.
The Evidence-Based Medicine Working Group. (1992) "Evidence-Based Medicine: A New Approach to Teaching the Practice of Medicine," *The Journal of the American Medical Association* 268: 2420–2425.
Grossman, J. and Mackenzie, F.J. (2005) "The Randomized Controlled Trial: Gold Standard, or Merely Standard?," *Perspectives in Biology and Medicine* 48(4): 516–534.
Howick, J.H. (2011) *The Philosophy of Evidence-Based Medicine*. Chichester, West Sussex: BMJ Books, Wiley-Blackwell.
Stegenga, J. (2011) "Is Meta-Analysis the Platinum Standard of Medicine?," *Studies in History and Philosophy of Biological and Biomedical Sciences* 42: 497–507.

20
STATISTICAL EVIDENCE AND THE RELIABILITY OF MEDICAL RESEARCH

Mattia Andreoletti and David Teira

Statistical evidence is pervasive in medicine. In this chapter we will focus on the reliability of randomized clinical trials (RCTs) conducted to test the safety and efficacy of medical treatments. RCTs are scientific experiments and, as such, we expect them to be *replicable*: if we repeat the same experiment time and again, we should obtain the same outcome (Norton 2015). The statistical design of the test should guarantee that the observed outcome is not a random event, but rather a real effect of the treatments administered. However, for more than a decade now, we have been discussing a *replicability crisis* across different experimental disciplines including medicine: the outcomes of trials published in very prestigious journals often disappear when the experiment is repeated (see, e.g., Lehrer 2010, Begley and Ellis 2012, Horton 2015).

There are different accounts of the reason for this replicability crisis, ranging from scientific fraud to lack of institutional incentives to double-check someone else's results. In this chapter we will use the replicability crisis as a thread to introduce some central issues in the design of scientific experiments in medicine. First, in section 1, we will see how replicability and statistical significance are connected: we can only make sense of the p-value of a trial outcome within a series of replications of the test. But in order to conduct these replications properly, we need to agree on the proper design of the experiment we are going to repeat. In particular, we need to prevent the preferences of the experimenters from biasing the outcome of the experiment. If there is such a bias, when the experiment is replicated by a third party, the observed outcome will vanish. In section 2, we will argue that trialists need to agree on the debiasing procedures and the statistical quality controls that feature in the trial protocol, if they want the outcome to be replicable. In section 3, we will make two complementary points. On the one hand, replicability per se is not everything: we need trial outcomes that are not only statistically significant but also clinically relevant. On the other hand, trials are not everything: the experts analyzing the evidence can improve the reliability of statistical evidence, although they sometimes fail; we need to further study how they make their decisions. In section 4, we will use a controversy about the overprescription of statins to show how non-replicable effects are obtained in trials and how experts may fail at detecting such flaws, if the commercial interests are big enough.

1. What Sort of Statistical Evidence Is the p-Value of a Trial?

Mathematical statistics, with different degrees of sophistication, has been used for different purposes in medicine since the 19th century (Matthews 1995). One major purpose has been the assessment of the efficacy of treatments, and a significant step forward in our ability to assess this efficacy was the implementation of the RCT as a testing standard in the 1940s (Marks 1997). The RCT is an experimental design articulated by the statistician Ronald A. Fisher in the 1930s, endowing a comparative method for causal inference with statistical foundations that allowed an interpretation of the outcome. In its simplest form, an RCT assesses the effect of a treatment on a given population comparing it to a standard alternative or a placebo—see Hackshaw (2009) for a quick overview. The treatments are randomly allocated to the individuals in the test, usually an equal number in each treatment group. After the administration is complete, we measure the variable of interest to assess whether there is any significant difference between the two groups of patients.

In order to quantify the significance of the difference, Fisher arranged the experiment as a test of the hypothesis that there is no difference between the two treatments (Teira 2011). This is known as the null hypothesis. Under this assumption, you can calculate the probability distribution of all potential outcomes of the experiment. In other words, a statistically significant difference is an outcome for which the probability, under the null hypothesis, is very low. Fisher introduced as an index of significance the p-value, the probability of obtaining a result as extreme as the observed trial outcome or more if there is indeed no difference between treatments. A p-value of 0.05 means that, assuming that the null hypothesis is true, if you repeat the trial time and again, only in 5% of the repetitions will you observe such an extreme outcome or an even more extreme one.

If you obtain a statistically significant result, with a p-value below the conventional threshold of 0.05, there are two possible ways to interpret this outcome: either the initial hypothesis is true (there is no difference between treatments) and you have observed a rare event, or the event is actually not rare at all, and the hypothesis is just false. There is no way to tell which is the case other than replicating the experiment and seeing whether further outcomes confirm or disconfirm the hypothesis that there is no difference between the effects of both treatments. If repeated trials of the experiment continue to give "unexpected" results, then the therapy probably works, and the null hypothesis is probably false. If most trials give no significant difference, then the trial that did so was probably just a fluke, and the null hypothesis is probably true. Thus, ultimately, drawing conclusions from clinical trials is in an inductive inference: you are trying to prove the truth of a general proposition (or its negation) on the basis of a finite number of instantiations. There is no surefire method to decide whether the hypothesis is actually true or not. As Ronald Fisher put it, one has a real phenomenon when one knows how to conduct an experiment that will rarely fail to give a statistically significant result: we can show time and again that there is a real difference between the effects of the tested treatments (Spanos and Mayo 2015).

We should notice a crucial point in this argument. The p-value estimates how often an outcome will appear in a series of replications of the experiment. Thus, Fisher's interpretation of the trial outcome requires a *frequentist* understanding of probabilities as opposed to a Bayesian approach where probabilities measure our degree of belief in the truth of a given statement (see Chapter 21 by Nardini, in this volume). A Bayesian trial would measure how strong our belief in the safety and efficacy of a treatment is. In a frequentist trial we measure instead how often we will observe a given outcome if we repeat the same experiment time and again. Our p-values are tied to an experimental design. If we conduct a somewhat different trial of the same therapy, the probability distributions of outcomes will be different, and thus an outcome that was statistically surprising in the original experiment may not be in the new one. Thus, paradoxically, identical outcomes in two differently designed experiments may not confirm each other. Confidence intervals, alpha values, and other frequentist concepts for hypothesis testing are equally tied to an experimental plan.

As a general epistemic principle, scientific experiments should be replicable: if we implement the same design properly, we should obtain the same outcome independently of any subjective feature of the experimenter or the contingent circumstances of the experimental setup. The more replicable an outcome, the more reliable it is. In clinical trials, as in other fields in science, p-values provide an implicit index of the replicability of an outcome: if we reject a hypothesis about both treatment effects being equal, then we should expect the new treatment to perform better than the alternative whenever we administer it according to the trial protocol (patients, dosage, etc.). However, as we will see in the next section, the p-value may be a misleading index of replicability.

2. The Sources of Non-replicability

In 1962, the U.S. Food and Drug Administration (FDA) received the mandate to test the safety and efficacy of new treatments with "well controlled investigations," later specified as two RCTs plus one further confirmatory trial (Carpenter 2010). This new regulatory standard created the contemporary trial industry, with pharmaceutical companies heavily investing in the design and conduct of RCTs in order to gain market access for their compounds. The FDA experts are supposed to assess these trials and infer whether the outcome observed in the sample of patients participating in the trials will obtain when the treatment is used on the general population. In other words, the FDA experts should assess the *external validity* of the trial (see Chapter 18 by La Caze, in this volume), that is, whether the causal connection established in the trial between the treatment, on the one hand, and improved patient outcomes, on the other, will hold in non-experimental clinical settings. If the drug is approved and then turns out not to be safe and efficacious—e.g., if unexpected adverse effects are observed once the treatment is released commercially—we would have accepted the wrong hypothesis in the trial: the experimental treatment would actually be inferior to the standard alternative.

A correct decision should be grounded on reliable trial outcomes and in order to obtain these latter, the experimenters testing a drug should agree, at least, on the proper controls to be implemented in the trial and on the adequate statistical design of the experiment. Otherwise, the p-values of their trials may be pointing to different experimental designs, providing non-comparable evidence. Ideally, a good trial should be *internally valid* (see Chapter 18 by La Caze, in this volume): the experimental protocol should properly capture the causal connection between the administered treatment and the observed effect. A correct causal inference should be grounded in a *like with like* comparison. The different arms of the trial should be entirely alike except for the treatment each group of patients receives. Othewise, we would be unable to tell whether the observed difference between treatments originates in the causal effect of each treatment or in a non-controlled factor that creates a difference between groups. For several centuries, physicians have been debating the proper experimental *controls* that a *fair* test should implement in order to fend off confounding factors. The reader should bear in mind that this debate is endless (e.g., Franklin 1990): every experimental setup is different, and so are the potential confounding factors and the corresponding controls. Experimenters in all disciplines have their checklists updated according to the progress in their fields.

In medicine, researchers have paid particular attention to the biases originating in the preferences of either the experimenters or the experimental subjects and how to control for them.

Non-replicable outcomes are usually blamed on these sort of biases: the interests of the pharmaceutical industry spoils the design of their sponsored trials, so that their outcome disappears once these tests are conducted in an unbiased manner. There are a large number of biases (e.g., Bero and Rennie 1996), so we can only illustrate here some that are particularly relevant for the replicability crisis. We will focus on two stages of the experiment: the conduct of the test and its statistical interpretation.

As to the former, there is a clear consensus on some of the biases that may spoil a trial outcome. Selection bias occurs when the allocation of subjects to study groups is contaminated by the preferences of the experimenter (e.g., the healthiest patients receive the experimental treatment). Randomization controls for selection bias and is therefore considered a prerequisite for a methodologically sound trial. So is the masking of treatments, so that the physicians and patients in the trial cannot ascertain what they are giving or getting, guaranteeing that their preferences do not bias the treatment effect. However, there is still no consensus on the full list of controls that should be implemented in a trial in order to consider it unbiased.

Peter Gøtzsche (2013) illustrates the risk of unmasked trials as follows. In trials of antidepressant drugs, we usually assess subjective outcomes, even if the assessor is often a third party and not the patient. There is evidence from a meta-analysis (Hrobjartsson et al. 2013) that when the assessor is not masked to the treatment patients receive (i.e., the assessor knows whether the patients got the experimental drug or a placebo), the assessor overestimates the effect on average by 36%.

Reaching statistical significance is often a matter of getting a few more positive outcomes. Following Gøtzsche (2013), if you are testing an antidepressant versus a placebo on 400 patients, the p-value of observing 19 more patients improve with the experimental drug than with the control is 0.07.

	Improved	Not improved	Total
Drug	119	81	200
Placebo	100	100	200

However, if you observe two more patients improve with the active treatment (121 instead of 119), then your trial will reach statistical significance ($p = 0.04$). A non-masked assessment of outcomes increases thus the chances of getting a positive result. We may suspect that failing to mask the assessor could have been intentional if the sponsor of the trial was seeking such a favorable outcome. Here we see what is at stake with the *internal validity* of the trial: the design of the experiment may fail to grasp the causal connection (or rather, in this case, lack thereof) between the treatment and its study's outcome, with the p-value providing misleading evidence about the treatment efficacy.

Biases, which by their nature do not (necessarily) repeat each time a trial is redone, can thus be a cause of non-replicability. If we wish to eliminate bias, we need to agree on the list of controls that would guarantee an unbiased outcome and incorporate them into the trial protocols, in order to maximize our chances to observe the same outcome whenever we repeat the experiment. How far are we from this ideal list of debiasing controls? In principle, we should aim at controlling for every source of human intervention, but this is difficult to achieve. For instance, Helgesson (2010) has illustrated practices of out-of-protocol data cleaning in large Swedish RCTs. Helgesson tracks the ways in which data are informally recorded and corrected without leaving a trace in the trial's logbook, from sticky notes to guesses about the misspelling of an entry. In his view, those who make such corrections do so in good faith, in order to increase the credibility of their results. Would these corrections threaten the internal validity of the outcome? After all, if the experiment was replicated elsewhere, the corrections might be different, and the test would yield a different outcome. But if we tried to explicitly control for these cleaning practices, the experimental protocol would become extremely cumbersome. This is why it is so difficult to agree on a full list of controls: experimenters have different standards as to what constitutes an unbiased experiment, and we need to reach a compromise between absolutely unbiased (but unfeasible) protocol and protocols that are too open to interested manipulations.

As we noted above, the statistical analysis of trial results, as well as the study design itself, can lead to problems in replicability, as statistical analyses can also be biased (e.g., according to the preferences of the sponsor), most notoriously when the sample size is not chosen according to statistically justified principles. In biomedical research, a particularly vocal critic of this statistical flaw is the epidemiologist John Ioannidis. Although some of his claims are controversial ("most published research findings are false": see, e.g., Soric 1989, Ioannidis 2005a, 2005b, 2014a), his contributions are worth considering as a focal point in the replicability debate. Take, for instance, his empirical evaluation of very large treatment effects (VLEs) of medical interventions (Pereira et al. 2012). A standard complaint about industry-sponsored research is that trials are designed to detect small treatment effects that would guarantee regulatory approval without any clinical innovation (e.g., "me too" drugs): in principle, VLEs would sort out this problem. Ioannidis and his coauthors define a statistical threshold for VLEs, and they used data from the Cochrane Database of Systematic Reviews to identify studies that showed such effects and track further studies on such outstanding outcomes. They found that VLEs usually arise in small trials with few events, and their results typically become smaller or even lose their statistical significance as additional evidence is obtained. According to Ioannidis (2008), this is a problem of statistical literacy: biomedical researchers tend to claim discoveries based exclusively on p-values, focusing on significance while ignoring statistical power, which is a measure of whether a study is large enough to detect what it is looking for. Without a proper sample size, it is impossible to tell a random spike in the data from a true treatment effect. If the sample is small, we may observe a large difference by chance, but if the experiment were repeated and the sample size grew, chance would gradually give way to the true treatment effect (see, e.g., Button et al. 2013). Replicability fails to obtain the effect because there might have been no effect to grasp—even if the trial protocol itself was unbiased. Although adequate sample size is usually included in lists of requirements for well-designed studies, it is still often not met, as not all medical journals require it for publication. As before, part of the problem is lack of agreement as to which tools for bias control to require of researchers.

Summing up, biases can contaminate the trial and spoil the statistical reliability of the outcome, both while the experiment is being conducted and when the data are interpreted. The replicability of a trial will depend on which debiasing procedures and statistical quality controls that experimenters adopt in their experimental protocols. The more replicable the trial, the more reliable the information it yields.

3. Is the Problem Truly a Crisis?

Although we have discussed some of the sources of the replicability crisis, the question remains whether it is reasonable to refer to the problems we have with replicability as a crisis. On the one hand, a trial may be replicable and yet it may not deliver the information we actually need: we want clinical, not just statistical, reliability. Replicability is no guarantee of clinical benefit. On the other hand, despite the problem with the replicability of trials, regulators seem to have coped with them reasonably well until recently, according to the available data. In other words, even without replicability, expert judgment has allowed us to make proper decisions about the safety and efficacy of drugs.

Let us argue for the first point: Pereira et al. (2012) note that VLEs usually appear with treatments whose efficacy is defined by a laboratory test (e.g., hematologic response), as opposed to a clinically defined efficacy (e.g., symptomatic improvement) or a fatal outcome (e.g., death). There were only three reliably documented VLEs that used mortality as an endpoint (out of 2,791). This is another contentious point in contemporary debates on biomedical research: sometimes there are good reasons to adopt *soft* endpoints instead of *hard* trial outcomes (death); sometimes not. According to the industry critics, *soft* endpoints are chosen in order to get a statistically significant effect of a treatment, even if it is clinically not very interesting. This positive effect is just enough for the manufacturing company to request regulatory approval. Such trials may be unbiased, statistically well-grounded, and perfectly replicable, but the research question they are addressing may just concern the commercial interest of the manufacturer sponsoring the trial rather than the clinical interests of patients and physicians alike—as we will see in the case study below. This point suggests that some of the issues at stake in the replicability crisis go beyond the methodological quality of trials as scientific experiments and rather pertain to their clinical goals: what trial outcomes should we look for and who should decide about them?

Let us argue for the second point now: expert judgment can improve the reliability of the information provided by trials. If trials were systematically unreliable, then the decisions of regulatory agencies such as the FDA would be systematically misguided. Critics like Gøtzsche (2013), for instance, think that this is actually the case: 70% of FDA scientists are not confident that the drugs they approve are safe. If the internal or external validity of a trial fails, we will observe outcomes in the population that were not anticipated in the trial.

Dan Carpenter has tracked such unanticipated outcomes through label changes: adverse effects observed in the commercial use of a drug are often incorporated into its brochure. From 1980 to 2000, the average drug received five labeling revisions, about one for every three years of marketing after approval (Carpenter 2010, p. 623). Clearly, there is much about the full range of effects of a drug that we discover only after it reaches the market. Regulatory trials are testing the safety and efficacy of a compound, so these new findings do not necessarily call the original studies and their evaluation into question. Indeed, if we judge the reliability of trials by the number of market withdrawals due to serious adverse effects, the figures seem more promising: between 1993 and 2004, only 4 out of the 211 authorized drugs (1.9%) were withdrawn (Carpenter et al. 2008). In other words, the external validity of trials might be far from perfect (they don't track the full range of effects), but when it matters (serious adverse effects), the FDA seems to have been making the right decision. How is this possible?

The FDA combines the statistical evidence of clinical trials with expert deliberation: decisions about drugs are not made on the basis of RCTs alone, but in committees with adversarial confrontation of experts (Urfalino 2012). These committees seem to be able to make correct decisions as to the safety and efficacy of drugs and ponder the reliability of the evidence provided by trials—for a critical discussion, see Chapter 31 by Stegenga, in this volume. At least, under certain conditions: a 1.9% error rate (drug withdrawal) in a decade seems a reasonable standard. But when the FDA committee was given a shorter deadline, still in the same period (1993–2004), 7% of the drugs approved were later withdrawn (Carpenter et al. 2008). In other words, under certain conditions, expert judgment can improve the reliability of the information from RCTs when it comes to making decisions about medical treatments. Further investigation is needed as to how these expert judgments work, but the effect cannot be discounted.

4. Case Study: A Controversy Over Statins

Let us illustrate with a case study about two of the previous points: not large enough trials and the relevance of expert judgment. The treatment under discussion will be statins, a class of drugs that inhibits the cholesterol synthesis associated with cardiovascular diseases (CVDs). Statins have been widely used over the last 30 years to prevent CVD, with excellent success in many different trials—and an equally successful record in sales. However, there is a growing concern that statins are being overprescribed on the basis of trials that verify their ability to decrease cholesterol in many groups of patients without evaluating whether they prevent these patients' death (see, e.g., Goldacre 2012, González-Moreno et al. 2015). The reader should bear in mind that this is a controversial issue, and the question is far from settled.

This concern about overprescription was highlighted by the controversy that followed the publication, in November 2013, of The American College of Cardiology/American Heart Association guidelines on the topic. These new guidelines recommend the use of statins for primary prevention of CVD (prevention of CVD in patients who do not yet have it) in patients with a 10-year predicted risk of CVD of 7.5% or greater; statin therapy was suggested as an option in patients with a predicted risk between 5% and 7.4%. These are very low thresholds and, consequently, more than 45 million (about one in every three) middle-aged, asymptomatic Americans qualified for treatment with statins. If we consider that the U.S. population is about one-twentieth of the global population in the same age range, and assuming that the distribution of risk profiles is similar, this would suggest that approximately one billion people should take statins. In Ioannidis's (2014b) words, this would amount to a "statinization" of the planet.

Taking statins is not completely harmless, because there are side effects (Macedo et al. 2014). So what were the grounds for such a massive public health intervention? According to Ioannidis (2014b), the guidelines were based on trials that tracked the cholesterol reduction in patients but did not follow them for long enough to see whether such reductions also lowered their mortality rate. This was the case of JUPITER, one of the biggest trials testing a statin in patients who had not yet shown evidence of CVD (primary prevention) (Ridker 2009). It showed that the treatment significantly reduced the risk of myocardial infarction, stroke, and vascular events, but, because it showed strong evidence of benefit early, the trial stopped following patients after 1.9 years instead of the planned 4 years, and thus was unable to detect an effect on mortality in the participants (de Lorgeril et al. 2010).

Trials are statistically designed to reveal a treatment effect of a given size with a minimal error rate. We need a certain amount of data (a designated number of patients: the sample size) to minimize error. If we interrupt the trial, we are losing data, and we can only be certain of identifying the true effect of a treatment under a number of statistical assumptions. JUPITER was interrupted because the preventive effect of statins was judged big enough to make the remaining two years of data accumulation unnecessary. In other words, the implication was that if other researchers tried to replicate JUPITER in full, they would observe the same effect, as the effect JUPITER observed was so large that, even before it was completed, it could not reasonably be supposed to be due to chance.

But, in fact, when other researchers tried to reproduce the same effect, they were unsuccessful. For instance, CORONA (2007) aimed to test the efficacy of statins in secondary prevention, treating patients who already have had a cardiac event, with a view to reducing the probability of a second one. The conclusion was that "there were no significant differences between the two groups in the coronary outcome or death from cardiovascular cause." This was an unexpected outcome, since the trial population should clearly benefit from the preventive effects of statins. Indeed, if the physio-pathological mechanism of stroke or myocardial infarction is always the same, statins should be at least as efficacious in the secondary prevention as in the primary, and we do not have any scientific reason to think the opposite. In fact, the only difference between the two populations is the probability of observing an infarction, which is obviously higher in patients who already had one than in healthy people.

This has an important consequence in designing and performing trials. As we have just mentioned in primary prevention, if the population is at lower risk, this means that the probability of observing a myocardial event is low; therefore, the detection of the outcomes needs both a bigger sample size and a longer follow-up of patients. Whereas in secondary prevention, we need fewer people and a shorter follow-up to show an effect of statins since the probability of observing a cardiac event is high. Therefore, from a statistical point of view, it should be easier to demonstrate the efficacy of statins in secondary prevention than in primary, yet this did not happen. The outcome of CORONA was also reached by two more trials: GISSI-HF (2008) and AURORA (2009). In patients undergoing hemodialysis with high cardiovascular risk, rosuvastatin lowered the low-density lipoprotein (LDL) cholesterol level but had no significant effect on a *hard* composite endpoint (i.e., death, myocardial infarction, and stroke). CORONA, GISSI-HF, and AURORA appear to be trying to reproduce the effect observed in JUPITER in conditions where it should be even easier to detect. Why did these replications fail? Perhaps because the decision to interrupt JUPITER for evidence of early benefit was mistaken. (Although it was not exceptional. A systematic review showed that the number of trials that are being stopped early for apparent benefit is gradually increasing [Bassler et al. 2010].) The decision to stop is often not well justified in the ensuing reports: the treatment effects are often too large to be plausible, given the number of events recorded. Thus, the observed effects are not replicable because researchers ground their conclusions too optimistically on sample sizes that are not large enough (insufficient power).

Unlike the FDA experts discussed in the previous section, The American College of Cardiology/American Heart Association did not correct for the flaws in JUPITER, and we suspect that they may have been somehow biased by the huge commercial interests at stake. Hence, we need to pay attention not just to the replicability of trials, but also to the way in which experts judge their conclusions.

Conclusion

We have only covered (partially) the methodological side of the replicability crisis. We have shown how a proper epistemic interpretation of *p*-values requires replicability. This latter depends, on the one hand, on the controls we impose on the experiment to secure that it is not biased by any particular preference or skill of the experimenter (or any other participant in the trial) and, on the other hand, on a proper statistical design for the trial, in which the sample size plays a crucial role. Without prior agreement on the list of controls and statistical features that characterize a fair trial, we may be missing replicability due to ambiguity in our experimental plan. And yet, not only statistical replicability matters. As John Norton (2015) has recently argued, the epistemic value of a replication is domain-specific: it depends on what we already knew about a given condition and the goals we seek to reach with a treatment. On the one hand, we need clinically (and not just statistically) significant outcomes. On the other, we need to investigate how expert judgment can properly assess the statistical evidence provided by trials.

References

AURORA STUDY GROUP. 2009. Rosuvastatin and cardiovascular events in patients undergoing hemodialysis. *New England Journal of Medicine*, 360, 1395–1407.

BASSLER, D., BRIEL, M., MONTORI, V. M., LANE, M., GLASZIOU, P., ZHOU, Q., HEELS-ANSDELL, D., WALTER, S. D., GUYATT, G. H., STOPIT-2 STUDY GROUP, FLYNN, D. N., ELAMIN, M. B., MURAD, M. H., ABU ELNOUR, N. O., LAMPROPULOS, J. F., SOOD, A., MULLAN, R. J., ERWIN, P. J., BANKHEAD, C. R., PERERA, R., RUIZ CULEBRO, C., YOU, J. J., MULLA, S. M., KAUR, J., NERENBERG, K. A., SCHÜNEMANN, H., COOK, D. J., LUTZ, K., RIBIC, C. M., VALE, N., MALAGA, G., AKL, E. A., FERREIRA-GONZALEZ, I., ALONSO-COELLO, P., URRUTIA, G., KUNZ, R., BUCHER, H. C., NORDMANN, A. J., RAATZ, H., DA SILVA, S. A., TUCHE, F., STRAHM, B., DJULBEGOVIC, B., ADHIKARI, N. K., MILLS, E. J., GWADRY-SRIDHAR, F., KIRPALANI, H., SOARES, H. P., KARANICOLAS, P. J., BURNS, K. E., VANDVIK, P. O., COTO-YGLESIAS, F., CHRISPIM, P. P., RAMSAY, T. 2010. Stopping randomized trials early for benefit and estimation of treatment effects: Systematic review and meta-regression analysis. *JAMA*, 303(12), 1180–7.

BEGLEY, C. G., & ELLIS, L. M. 2012. Drug development: Raise standards for preclinical cancer research. *Nature*, 483, 531–3.

BERO, L., & RENNIE, D. 1996. Influences on the quality of published drug studies. *International Journal of Technology Assessment in Health Care*, 12(2), 209–37.

BUTTON, K. S., IOANNIDIS, J. P., MOKRYSZ, C., NOSEK, B. A., FLINT, J., ROBINSON, E. S., & MUNAFÒ, M. R. 2013. Power failure: Why small sample size undermines the reliability of neuroscience. *Nature Reviews Neuroscience*, 14(5), 365–76.

CARPENTER, D. P. 2010. *Reputation and Power: Organizational Image and Pharmaceutical Regulation at the FDA*, Princeton, Princeton University Press.

CARPENTER, D., ZUCKER, E. J., & AVORN, J. 2008. Drug-review deadlines and safety problems. *New England Journal of Medicine*, 358, 1354–61.

CORONA GROUP. 2007. Rosuvastatin in older patients with systolic heart failure. *New England Journal of Medicine*, 357(22), 2248–61.

DE LORGERIL, M., SALEN, M. P., ABRAMSON, J., DODIN, S., HAMAZAKI, T., KOSTUCKI, W., OKUYAMA, H., PAVY, B., & RABAEUS, M. 2010. Cholesterol lowering, cardiovascular diseases, and the rosuvastatin-JUPITER controversy: A critical reappraisal. *Arch Intern Med*, 170(12), 1032–6.

FRANKLIN, A. 1990. *Experiment, Right or Wrong*, Cambridge, Cambridge University Press.

GISSI-HF INVESTIGATORS. 2008. Effect of rosuvastatin in patients with chronic heart failure (the GISSI-HF trial): A randomised, double-blind, placebo-controlled trial. *The Lancet*, 372(9645), 1231–9.

GOLDACRE, B. 2012. *Bad Pharma*, London, Fourth State.

GONZÁLEZ-MORENO, M., SABORIDO, C., & TEIRA, D. 2015. Disease-mongering through clinical trials, *Studies in History and Philosophy of Biological and Biomedical Sciences*, 51, 11–8.

GØTZSCHE, P. 2013. *Deadly Medicines and Organised Crime: How Big Pharma Has Corrupted Healthcare*, London, Radcliffe Health.

HACKSHAW, A. K. 2009. *A Concise Guide to Clinical Trials*, Chichester, UK and Hoboken, NJ, Wiley-Blackwell/BMJ Books.

HELGESSON, C.-F. 2010. From dirty data to credible scientific evidence: Some practices used to clean data in large randomised clinical trials. In: WILL, C., & MOREIRA, T. (eds.), *Medical Proofs, Social Experiments*, Farnham, Ashgate, 2010, pp. 49–64.

HORTON, R. 2015. Offline: What is medicine's 5-sigma? *The Lancet*, 385(9976): 1380.

HROBJARTSSON, A., THOMSEN, A. S., EMANUELSSON, F., TENDAL, B., HILDEN, J., BOUTRON, I., RAVAUD, P., & BRORSON, S. 2013. Observer bias in randomized clinical trials with measurement scale outcomes: A systematic review of trials with both blinded and nonblinded assessors. *CMAJ*, 185, E201–11.

IOANNIDIS, J. P. 2005a. Why most published research findings are false. *PLoS Med*, 2(8), e124. doi:10.1371/journal.pmed.0020124.

IOANNIDIS, J. P. 2005b. Contradicted and initially stronger effects in highly cited clinical research. *JAMA*, 294.2, 218–28.
IOANNIDIS, J. P. 2008. Why most discovered true associations are inflated. *Epidemiology*, 19, 640–8.
IOANNIDIS, J. P. 2014a. Clinical trials: What a waste. *BMJ*, 349, g7089.
IOANNIDIS, J. P. 2014b. More than a billion people taking statins?: Potential implications of the new cardiovascular guidelines. *JAMA*, 311, 463–4.
LEHRER, J. 2010. The truth wears off. *The New Yorker*, http://www.newyorker.com/magazine/2010/12/13/the-truth-wears-off. Accessed June 15, 2015.
MACEDO, A. F., TAYLOR, F. C., CASAS, J. P., ADLER, A., PRIETO-MERINO, D., & EBRAHIM, S. 2014, Unintended effects of statins from observational studies in the general population: Systematic review and meta-analysis. *BMC Medicine*, 12, 51.
MARKS, H. M. 1997. *The Progress of Experiment: Science and Therapeutic Reform in the United States, 1900–1990*, New York, Cambridge University Press.
MATTHEWS, J. R. 1995. *Quantification and the Quest for Medical Certainty*, Princeton, NJ: Princeton University Press.
NORTON, J. 2015. Replicability of experiment. *Theoria*, 30/2, 229–48.
PEREIRA, T. V., HORWITZ, R. I., & IOANNIDIS, J. P. 2012. Empirical evaluation of very large treatment effects of medical interventions. *JAMA*, 308, 1676–84.
RIDKER, P. M., & COOK, N. R. 2013. Statins: New American guidelines for prevention of cardiovascular disease. *The Lancet*, 382(9907), 1762–5.
SORIC, B. 1989. Statistical "discoveries" and effect-size estimation. *Journal of the American Statistical Association*, 86(406), 608–10.
SPANOS, A., & MAYO, D. 2015. Error statistical modeling and inference: Where methodology meets ontology. *Synthese*, doi:10.1007/s11229-015-0744-y.
TEIRA, D. 2011. Frequentist versus Bayesian clinical trials. In: GIFFORD, F. (ed.), *Philosophy of Medicine*, Amsterdam, Elsevier, pp. 255–97.
URFALINO, P. 2012. Reasons and preferences in medicine evaluation committees. *In*: ELSTER, J., & LANDEMORE, H. (eds.), *Collective Wisdom: Principles and Mechanisms*, Cambridge, Cambridge University Press, pp. 173–203.

Further Reading

Because the replicability crisis is still unfolding, it is probably better to use the Internet for updated references.

For a general overview, in open access, you will find *Nature*'s special issue on reproducibility at http://www.nature.com/nature/focus/reproducibility/index.html.

For updates on withdrawn papers from scientific journals, often (but not only) for replicability issues, see http://retractionwatch.com/.

Deborah Mayo's blog is a rich source of (statistically informed) philosophical discussion on the replicability crisis: http://errorstatistics.com/category/reproducibility/.

An extensive historical source about controls implemented medical experiments for grounding like with like comparisons is the James Lind library: http://www.jameslindlibrary.org/.

Related Chapters

Chapter 7: Frequency and propensity: the interpretation of probability in causal models for medicine
Chapter 18: The randomized controlled trial: internal and external validity
Chapter 19: The hierarchy of evidence, meta-analysis, and systematic review
Chapter 21: Bayesian versus frequentist interpretations of clinical trials
Chapter 30: Outcome measures in medicine

21
BAYESIAN VERSUS FREQUENTIST CLINICAL TRIALS

Cecilia Nardini

1. Introduction

Randomized controlled trials (RCTs) currently represent the gold standard within evidence-based medicine for the evaluation and approval of new therapeutics. RCTs are experiments in which we test the efficacy of an intervention with respect to a control by comparing their performance in two separate groups of patients. In the typical procedure, patients are assigned to either the experimental or the control group. Patients in the first group receive the new treatment that is being tested, while patients in the control group receive a treatment that constitutes the benchmark for the evaluation of the newly proposed alternative—this can be either no therapy or the standard therapy for the condition under study. In order to avoid bias, allocation to the two groups is random, and the investigators and treating physicians are generally masked to group allocation.

The experimental data of an RCT generally consist in the record of the outcomes for all patients involved in the trial, with respect to one or more endpoints of interest—for instance, the average time-to-recurrence for patients taking one or the other drug. The trial is positive for the new treatment if the analysis reveals a difference that favors the new treatment group. Clearly, not any difference will do. Events in one group may be higher than in the other just due to chance and not to an actual difference in effectiveness. The role of statistics in clinical trials is precisely that of separating results that could depend on random fluctuations from the conspicuous trends that point to genuine differences in patient-relevant outcomes.

Far from being a mere mathematical tool, statistics represents a fundamental element in the architecture of clinical trials. The statistical methodology adopted influences such aspects as which conclusions are licensed on the basis of data or what degree of support is granted to a hypothesis. This chapter will explore how the use of two different statistical frameworks, which diverge at the level of foundational principles, impacts the interpretation of results of clinical trials, and how it can eventually lead to different approaches to designing and conducting them. The two frameworks are the frequentist (also called classical) and the Bayesian. These are two long-established and competing paradigms, diverging in the stance they take on the interpretation of probability.

Frequentist statistics is so called because it identifies probabilities with the frequency of occurrence of events, whereas the Bayesian school derives its name from a 17th-century clergyman,

Rev. Thomas Bayes, who laid the theoretical foundation for the Bayesian interpretation of probability as degree of confidence in a hypothesis. From this contrast in probability interpretation, which is fundamentally philosophical, the two frameworks derive a very pragmatic difference in that they license the use of distinct techniques, they contemplate a different kind of warrant upon the conclusion reached through such techniques, and they can eventually point to different conclusions in the face of the same experimental data.

The discussion of the two frameworks that will take place in this chapter does not imply that one of the two frameworks is wrong while the other is right, even though both sides can claim cases where the other leads to error. Rather, the factors that will be highlighted reveal that each framework is more suited to analyzing and interpreting trials designed according to certain principles. Thus, the bottom line is that the choice of the statistical framework to use in clinical trials is not neutral, but it should be made with an eye to the consequences that the choice can have upon the conduct and interpretation of clinical trials.

2. The Frequentist Framework

Frequentists subscribe to an "objective" interpretation of probability. For the frequentist, probabilities are "in the world" in the sense that they correspond to frequencies characterizing observable events. For instance, the expectation that a fair coin has 50% probability of landing heads is grounded on the fact that, tossed a large number of times, the coin would land heads on average in half of them. Therefore, frequentist analytic techniques focus on deriving conclusions that can be valid over a long series of attempts. In order to see how this principle finds practical application in clinical trials, we will examine a concrete example of analysis of a trial. In the second part of the section, we will derive some general considerations about the use of frequentist statistics.

As an example, we consider a trial analyzed by Spiegelhalter et al. (2004). The original reference is GREAT Group (1992). The study was set up to assess the efficacy of anistreplase, a thrombolytic drug, in the setting of myocardial infarction. The trial was comparing anistreplase, administered at home by a general practitioner as soon as possible after myocardial infarction, against conventional thrombolytic therapy administered in hospital. The endpoint of the study was 30-day mortality rate under each treatment measured by the *odds ratio (OR)*— i.e., the ratio of the odds of death under the new treatment to the odds of death under the conventional one. The odds ratio is a measure of difference in performance between the two treatments: the OR is smaller than 1 when the new treatment is more effective than the control and greater than 1 if the control is more effective.

The evidence from this study was as follows:

	Anistreplase	Control	Total
Death	13	23	36
No death	150	125	275
Total	163	148	311

The 30-day mortality was 13/163 on new treatment and 23/148 on control, yielding an estimated OR of 0.48, approximately corresponding to a reduction of the risk of death by 52% using anistreplase.

The statistical analysis proceeds to determine whether this apparent reduction in risk of death is due to the effectiveness of the new treatment or to chance. The situation is analogous to that of determining whether a coin is fair: we toss a coin 10 times and record on which side it lands. Suppose it lands heads seven times and tails three times. The imbalanced result, seven over three, could be the result of chance alone, or it could be a sign that the coin is not fair. How do we decide?

The frequentist response to this question consists in setting a threshold on how unlikely a result needs to be for us to decide it is not due to chance. We start from the *null hypothesis* that there is no difference in effect between anistreplase and control: that is, the OR equals 1. Under this hypothesis, it would be very likely to observe small differences in death rates between the two treatment groups, while it would be fairly unlikely to observe large differences. This intuition is translated in numerical terms by the p-value. The p-value measures the probability of observing this particular value of the difference between the two groups in the case that there is actually no difference in effect between treatments (null hypothesis). The closer the p-value is to zero, the larger the deviation of the result that was observed from the null hypothesis, and the more unlikely that the observed result is due to chance alone. In our case, OR = 0.48 is fairly distant from 1, and it generates a p-value of 0.04. Is this value unlikely enough that we may reject the null hypothesis, and thus claim the new treatment to be more effective? Or is the deviation from an OR of 1 small enough that it could still be due to chance?

In order to decide, we need a threshold for how unlikely a result must be in order to convince us that the null hypothesis is false and the new treatment is more effective. This threshold is generally referred to as *statistical significance*: when the p-value calculated from the trial data is lower than the statistical significance threshold, we agree that the result is so unlikely that we can conclude that the new treatment is more effective than the control. It is conventional to use a significance threshold of 0.05; since the p-value in the anistreplase trial is 0.04, lower than 0.05, we can conclude that anistreplase is more effective than conventional thrombolytic therapy.

A further frequentist measure that is worth considering is the *confidence interval* (CI), which provides a range of values (in our case, of the OR) that are compatible with the observed one. A confidence interval is a range of values within which we can be confident the true value lies, given the trial result. A CI is always associated with a certain level of *confidence* (e.g., 95%) that it contains the true value. The interval is centered on the observed value, and its width depends on the confidence level we want to express and on some study parameters such as the sample size (the number of patients that were enrolled). In the anistreplase trial, the CI at the 95% confidence level is [OR = 0.22, OR = 0.97], corresponding to a range of risk reduction from 3% to 88%. The fact that the interval does not include 1 means that we can say with 95% confidence that the equivalence of anistreplase with conventional treatment is not compatible with the data—the same thing we learned by comparing the p-value with the statistical significance threshold. However, the interval is extremely wide, and it also includes insubstantial values of risk reduction. This is because the information provided by the trial is relatively limited: only 13 + 23 = 36 deaths were observed. Small studies like this produce CIs that are wide and not very precise; by contrast, large studies provide CIs that are narrow and yield a precise estimate for the plausible value of the new treatment effect. Confidence intervals may appear to be more useful than p-values, since they provide a range of values for the OR (and thus for the new treatment effectiveness); however, both p-values and confidence intervals ought to be interpreted carefully.

In clinical trials we measure a certain quantity that expresses the effectiveness of the new treatment versus the control—such as the OR in our anistreplase example. This quantity is subject to statistical fluctuation in any given trial, so a statistical analysis is necessary in order to determine from the observed value (e.g., OR = 0.48) whether or not there truly is a benefit to the new treatment. As we saw above, the frequentist procedures fulfill this task in an indirect manner: they start from a hypothesis about the extent of treatment benefit, and they measure it against the observed value, to see if they are compatible. In the Discussion section below, we will see the advantages and disadvantages of this approach, especially as compared to the other approach, the Bayesian. There is, however, one consequence of the frequentist approach that is worth noticing already at this point. Frequentist measures, such as the p-value and the confidence interval, are not related in any intuitive manner to the clinical importance of the observed result. The p-value measures a *statistical discrepancy* between the observed result and the null hypothesis of no difference. However, a large effect in a small group will produce the same deviation from the null (and thus the same p-value) as a small effect in a large sample. Hence, a statistically significant p-value does not necessarily correspond to a result that is also clinically significant. For instance, Ocana and Tannock (2011) analyzed several recent oncological trials that have sample sizes of between 500 and 800 patients: they found that the low p-value observed in such studies often corresponded to a performance of the new treatment that was only marginally superior to the control, and thus failed to be relevant from the patients' perspective.

Furthermore, the p-value expresses strength of evidence against the null hypothesis that the new treatment is equally effective as the old; it does not and it cannot express strength of evidence for or against a particular hypothesis, such as a particular value for the new treatment's effectiveness. Confidence intervals are more informative in this respect, since they provide a range of values for the new treatment effectiveness that are acceptable within a given level of confidence. However, this fixed confidence level leaves no room for considerations about how strongly the trial result speaks for or against a particular hypothesis. Confidence in frequentist terms does not refer to confidence in the result of a single, particular application of the methodology, such as our anistreplase trial; it refers to the correctness *on average* of a large number of trials that have the same characteristics. Frequentist measures were conceived with certain analytical aims: namely, with a focus on the long-run properties of the procedure. However, in medical application of statistical procedures, it is generally the result of the one trial that was performed which is of interest, and not the property of a hypothetical infinite series of trials with the same characteristics. As we shall see in the upcoming section, this focus is better captured by the concurrent statistical methodology, the Bayesian.

3. The Bayesian Framework

Bayesians subscribe to a "subjective" interpretation of probability. For Bayesians, probabilities reflect the degree of uncertainty in our knowledge about the world. The Bayesian interpretation explicitly links probability to what it is rational to believe in light of the data. In the coin toss case, a Bayesian would ground the statement that a fair coin has 50% probability of landing heads upon the knowledge that the possible outcomes are only two, and that none is more likely than the other.

Most Bayesian techniques proceed from a fundamental mathematical result, Bayes' Theorem. This theorem applies to events that are related, such as someone having an appendicitis (event a) and someone having a fever (event b). Since these two events are not independent (people with appendicitis tend to have fever), knowledge about the occurrence of one will make a difference about the probability we assign to the occurrence of the other—e.g., when visiting patients with abdominal pain and fever, the expectation that they have appendicitis is higher than if we had not observed fever. Bayes' Theorem is a formula relating such probabilities, and it can be used to tell us what degree of confidence we should have in a hypothesis: for instance, it can be used to determine how likely it is that a patient showing fever (b) has appendicitis (a). In order to calculate this probability, the theorem uses two main inputs: the probability of event a, $P(a)$, prior to knowing about b—in our case, this would be the general probability of someone having appendicitis, i.e., the prevalence of appendicitis in the population—and the conditional probability $P(b|a)$ relating b with a—in our case, $P(b|a)$ (read "the probability of b given a") is the probability of having fever in case of appendicitis. Bayes' Theorem is then used to *update* the prior probability $P(a)$ with the evidence, contained in the conditional probability $P(b|a)$, about occurrence of event b. This gives us the probability $P(a|b)$ that our patient showing up with fever has appendicitis.

The full Bayesian analysis of a clinical trial proceeds in a similar fashion, by employing the evidence coming from the trial to update a prior probability estimate about the effectiveness of the new treatment. In the remainder of this section, we will see how this works by examining our example anistreplase trial from a Bayesian perspective; this will enable us to conclude the section with a review of the main highlights of Bayesian methods.

A Bayesian analysis of the anistreplase trial (GREAT Group 1992) was performed by Pocock and Spiegelhalter (1992), and it is described in detail in Spiegelhalter et al. (2004, 69–72). As the starting point of a Bayesian analysis, it is necessary to specify the prior distribution for the possible values of the effect size—i.e., the distribution of how probable the various values are (believed to be), in view of the knowledge available prior to performing the trial. Spiegelhalter et al. elaborated the prior distribution based on the subjective judgment of a senior cardiologist, informed by empirical evidence—derived from three previous trials of the same treatment—who expressed belief that "an expectation of 15–20% reduction in mortality is highly plausible, while the extremes of no benefit and a 40% relative reduction are both unlikely" (p. 69). This information was used to construct the prior probability distribution—i.e., to assign a probability to each possible value for the effect of anistreplase, according to what we know before performing the trial.

The Bayesian analysis consists in using Bayes' Theorem for combining this prior distribution with the trial data. This is done by combining the prior probability of effectiveness of the new treatment with the probability of obtaining the value of the difference that was observed. The resulting *posterior distribution* has a sample mean of OR = 0.73, i.e., a relative risk reduction around 27%. The posterior distribution is the final result of the Bayesian analysis, and it can be interpreted in a direct manner: its mean value (OR = 0.73) is the most likely value for the effectiveness of anistreplase in light of the trial result. From the posterior distribution we can also derive credible intervals, the Bayesian counterpart to confidence intervals. The 95% credible interval for the OR is (0.58, 0.93)—corresponding to a relative risk reduction of approximately 7% to 42%—and this interval can be interpreted as having 95% probability of containing the true value.

As we see, the conclusion about the effectiveness of anistreplase in the Bayesian analysis, though still positive, is far less impressive than the estimate coming from the frequentist analysis of the trial. The Bayesian interval is narrower than its frequentist counterpart (which was 0.22, 0.97), and it is centered upon more modest values for the effect of anistreplase. In combining the prior probability distribution with the trial data, the prior had a dampening effect. The in-trial performance of anistreplase did surpass expectations, but the previous evidence, expressed through the senior cardiologist's judgment, moderated the impact of this performance. The trial data were overwhelmed to a degree by prior information also because the trial recruited a small sample and hence yielded weak evidence.

As this example of Bayesian analysis shows, the state of knowledge about a newly proposed treatment option prior to conducting a trial can be formulated in terms of a hypothesis H about the effectiveness of the new treatment that has a certain prior plausibility. When the trial is conducted, the value e of effect that is observed can be used to update our belief about the plausibility of H via Bayes' Theorem. The resulting posterior probability distribution can be exploited along two main routes:

- We can evaluate the plausibility of a hypothesis that the treatment effect has a specific value.
- We can derive a range of values for treatment effectiveness that has a certain probability of being true.

Notice the difference from frequentist procedures. In frequentist analysis we are only able to decide whether certain values can or cannot be ruled out based on evidence provided by the trial: we cannot establish a range of values that are more likely to be true, nor can we assign to a certain value a probability that it is correct. This is because frequentist procedures focus on the long-run properties, whereas Bayesian techniques focus on making the most out of the available evidence, both coming from the trial and external to it. We will return to this distinction in the upcoming Discussion, as we conclude this section with a discussion of the main problem with the Bayesian approach.

On the one hand, Bayesian techniques capture the spirit with which trials are conducted: results of a Bayesian analysis are readily interpretable as statements about our knowledge of treatment effectiveness, in the light of trial results. On the other hand, when performing a Bayesian analysis, it is necessary to start from a prior hypothesis about treatment effectiveness, which will then be updated according to the observed trial results. This prior hypothesis, unlike the null hypothesis in a frequentist analysis, must be chosen by the investigators from among various possibilities, and this choice may bias the analysis.

One possible solution to this problem is to use "reference" priors that are intended to provide a kind of default Bayesian analysis free from subjectivity; essentially this means using "flat" distributions that do not mark any of the possible value as more likely—all possibilities are assigned equal probability. Using a uniform prior implies that the posterior distribution will reflect exactly the results of the trial, as the prior hypothesis does not lend any weight to any particular result. Clearly, this kind of analysis might be of interest, but it does not exploit the full power of the Bayesian methodology.

Several solutions have been proposed for introducing priors that are both reasonable and unbiased. These solutions can be grouped under two main strategies for selecting prior probabilities: the opinion-based and the evidence-based. Opinion-based priors are derived from experts' opinions and expectations about treatment effectiveness, as done by Pocock and Spiegelhalter in the anistreplase trial. Evidence-based priors instead can be constructed by collecting available data from previous studies. In the case of the anistreplase trial, taking an evidence-based approach would have implied using the previous trials—the trials upon which the opinion of the senior cardiologist was based—to construct priors. Notwithstanding the possibility for sensible prior construction using the strategies just highlighted, though, the fact remains that whenever the prior is based on a large evidence basis or a confident subjective judgment, it will have a strong influence on the final result of the analysis, just as happened in the anistreplase trial. Therefore, it is legitimate to worry about the prior's possibly introducing bias and the risk of contaminating the conclusion. In the final section, we will consider this worry alongside the advantages and disadvantages of the two frameworks in a general evaluation that will conclude this chapter.

4. Discussion

The main characteristic of the frequentist methodology, as we saw, is that each trial result is analyzed in isolation and is evaluated against a fixed yardstick, the significance level. Bayesian analysis features a diverging approach to inference, whereby trial results are cast against a background of available knowledge or of prior opinion. Each of these approaches bears strengths as well as limitations, and in this section we will analyze them in turn.

The frequentist strategy for drawing conclusions from data is based on confronting the trial result with the null hypothesis of no difference at a fixed confidence level. This approach requires that the design features of the study, such as its size and duration, be fixed in advance, since they are in every respect a part of the analytic procedure. The number of patients to enroll, the duration of the study, and the number of events that have to be observed must be planned beforehand and cannot be modified once the trial has started. This is useful under some respects but disadvantageous in others. The fact that the trial design has to be planned beforehand and then remain immutable is an advantage for regulatory bodies, which can thus maintain complete oversight of the trial. A methodology that is set up prospectively, before data collection begins, is an equally important advantage in the eye of regulators (Teira & Reiss 2013): since the analytical procedure is constructed before data even begin to accrue, the trial progress will not be allowed to influence the result of the analysis. The fact that frequentist methods are easy to supervise effectively is an asset in the present context, given the importance of clinical trials in the current regulatory system: they represent the gold standard for the generation of medical evidence and, at the same time, they constitute the main gateway to market approval for medical interventions in the Western world.

On the other hand, the rigidity of frequentist trial design can be difficult to reconcile with the dynamic setting of ongoing trials. When conducting a trial, investigators may often be confronted with unexpected observations—such as very large treatment effect, whether positive or negative—or new external evidence—such as concurrent trials of the same treatment. Such occurrences may pose ethical problems to trial continuation (Joffe & Truog 2008). This situation is deeply problematic for frequentist methodology. From the frequentist standpoint, it is hard to say anything about the results of a trial that was stopped earlier than planned because, if a trial did not reach its planned termination, it cannot be regarded as part of a sequence of trials of its specific preplanned design.

By contrast, the Bayesian approach is able to tackle the same situation effectively, since the data that have accrued before the trial was stopped can be analyzed by Bayesian methods like any other data. It is simply evidence with which to update our beliefs based on Bayes' Theorem. An early-stopped trial will simply contain less evidence, and thus arrive at a conclusion that has (measurably) less credibility than the full trial would have. Bayesian techniques can make the most efficient use of available information because they interpret it in context. Since a Bayesian analysis consists in a direct update of the state of knowledge, trials can be analyzed at any point, even as data are still accruing (Parmar et al. 2001), and new sources of external evidence can be incorporated simply by employing a different prior in the analysis. This possibility offered by Bayesian methodology is of great interest in recent years, as dissatisfaction is growing with the traditional design of large-scale trials (Sharma & Schilsky 2012). It has been suggested that the proper testing of novel therapeutic agents—such as molecularly targeted oncological drugs—requires the possibility to adapt the study protocol to information emerging during the trial (Berry 2012).

However, the dynamic nature of Bayesian methods also makes them less amenable to effective supervision by regulatory bodies. Concerns about the possibility of effectively supervising trials run under a Bayesian methodology represent a long-standing issue, mostly in the light of reliance on priors, discussed in Section 3. It is feared that prior distributions could represent a possible source of bias and external influence on the evaluation of a new treatment.

This worry is at least in part misplaced. Reaching sensible conclusions from any statistical analysis inevitably requires some subjective input (Berger & Berry 1988). In Bayesian analysis this is incorporated into the prior; in frequentist analysis, more subtly, the investigators' knowledge as well as their expectations about treatment effectiveness shape the design parameters of the study. Indeed, since with frequentist methods the size of the trial has to be fixed from the onset and may not be modified once the trial has started, it is necessary to ensure that a sufficient number of patients is enrolled in order to observe enough events to reach a definite conclusion. Therefore, expectations and information about the effectiveness of the experimental treatment are actively used in the design phase in order to determine how large the trial will need to be (Lachlin 1981). It could even be argued that Bayesian methods in this regard provide regulators with *better* ways for supervising the trial process, since all assumptions and external evidence that enter Bayesian inference have a single point of entry, the prior distribution. Therefore, it may be possible for regulators to keep potential sources of bias in Bayesian analysis in check simply through controlling and evaluating the prior (Berry 2006), whereas the review of the assumptions used in frequentist analysis for sizing the trial is arguably less direct.

5. Conclusion

So far, the attitude of considering clinical trials purely as scientific experiments has been dominant and is a cornerstone of evidence-based medicine (Howick 2010). However, it is increasingly being recognized that trials are embedded in a complex web of social and ethical values (Ashcroft 1999). Clinical trials are a particular form of experiment, given that the choice and assessment of medical therapies is not exclusively a matter of scientific interest. Almost invariably, a new treatment option has to be evaluated within the context of its clinical application, taking into account alternative available treatments, accessibility of care and therapy administration in clinical context, cost-effectiveness considerations, and last but not least ethical aspects. Each of these factors, all external to the trial, should be allowed to influence our evaluation of a new treatment under test. It is apparent then that what is asked of the statistical framework guiding the interpretation of trial results is not to rule out every possible source of bias, but rather to maintain the ability to discriminate between legitimate and illegitimate ones. In other words, what is needed for keeping bias in check is not a framework that is totally "epistemically pure," also given that no framework is capable of that, but rather a framework that allows for an open appraisal of influence.

The discussion undertaken in this chapter should have clarified that the two frameworks, the frequentist and the Bayesian, are equally legitimate but differ in that they give precedence to contrasting considerations in conducting statistical inference. Bayesian methods give precedence to exploiting efficiently the information yielded by the trial and by valid external sources; frequentist methods give precedence to a procedure that can be easily standardized and supervised. As highlighted in discussion, these different priorities reflect into a different set of characteristics for trials designed according to one or the other framework. As a consequence, the choice between the two frameworks is not neutral, and it should depend on what epistemic objectives are set for treatment evaluation through clinical trials. While the objective of controlling the trial process remains an important goal, especially from the perspective of regulators, the drive toward improving trial efficiency opens the way to alternative considerations. Nonetheless, implementing Bayesian methods within the current regulatory context would require a shift in focus, from considering the appropriateness of the design choices for the trial to assessing the appropriateness of the prior used in the analysis. This problem, more than the classical objections to the use of priors in medical research, may hinder the spread of Bayesian methods in clinical trials in the near future.

References

Ashcroft, R (1999) 'Equipoise, knowledge and ethics in clinical research and practice', *Bioethics*, 13(3/4), 314–326.

Berger, JO & Berry, DA (1988) 'Statistical analysis and the illusion of objectivity', *American Scientist*, 76, 159–165.

Berry, DA (2006) 'Bayesian clinical trials', *Nature Reviews Drug Discovery*, 5, 27–36.

———— (2012) 'Adaptive clinical trials in oncology', *Nature Reviews Clinical Oncology*, 9, 199–207.

GREAT Group (1992) 'Feasibility, safety, and efficacy of domiciliary thrombolysis by general practitioners: Grampian region early anistreplase trial', *British Medical Journal*, 305, 548–553.

Howick, J (2010) *The Philosophy of Evidence-Based Medicine*. Oxford: Wiley Blackwell.

Joffe, S & Truog, R (2008) 'Equipoise and randomization', In EJ Emanuel, CC Grady, RA Crouch, RK Lie, FG Miller, & DD Wendler, eds., *The Oxford Textbook of Clinical Research Ethics*, New York: Oxford University Press, pp. 245–260.

Lachin, JM (1981) 'Introduction to sample size determination and power analysis for clinical trials', *Controlled Clinical Trials*, 2(2), 93–113.

Ocana, A & Tannock, I (2011) 'When are "positive" clinical trials in oncology truly positive?', *Journal of the National Cancer Institute*, 103(1), 16.

Parmar, MKB, Griffiths, GO, Spiegelhalter, DJ, Souhami, RL, Altman, DG & van der Scheuren, E (2001) 'Monitoring of large randomised clinical trials: A new approach with Bayesian methods', *The Lancet*, 358, 375–381.

Pocock, S & Spiegelhalter, D (1992) 'Domiciliary thrombolysis by general practitioners', *British Medical Journal*, 305, 1015.

Sharma, MR & Schilsky, RL (2012) 'Role of randomized phase III trials in an era of effective targeted therapies,' *Nature Reviews Clinical Oncology*, 9(4), 208–214.

Spiegelhalter, DJ, Abrams, KR & Myles, JP (2004) *Bayesian Approaches to Clinical Trials and Health-Care Evaluation*. Hoboken, NJ: Wiley.

Teira, D & Reiss, J (2013) 'Causality, impartiality and evidence-based policy', in Chao H-K, Cheng S-T, and Millstein RL (eds.), *Mechanism and Causality in Biology and Economics*, Dordrecht, Netherlands: Springer, pp. 207–224.

Suggested Readings

Goodman, SN (1999) 'Toward evidence-based medical statistics. 1: The P value fallacy', *Annals of Internal Medicine*, 130(12), 995–1004. (A methodological piece warning against easy misinterpretations of frequentist measures)

Hansson, SO (2006) 'Uncertainty and the ethics of clinical trials', *Theoretical Medicine and Bioethics*, 27(2), 149–167. (An ethically informed view on the concept of statistical uncertainty in trials)

Moyé, LA (2008) 'Bayesians in clinical trials: Asleep at the switch', *Statistics in Medicine*, 27, 469–482. (A provocative take on the use of Bayesian methods in clinical trials)

Sprenger, J (2009) 'Evidence and experimental design in sequential trials', *Philosophy of Science*, 76, 637–49. (A philosophical discussion of the problem of interim analyses and data-dependent designs)

Teira, D (2011) 'Frequentist versus Bayesian clinical trials', in Gifford F (ed.), *Philosophy of Medicine* (pp. 255–298). Amsterdam: Elsevier/North Holland. (An extended philosophical comparison of the two frameworks)

22
OBSERVATIONAL RESEARCH
Olaf M. Dekkers and Jan P. Vandenbroucke

What Is Observational Research?

In medical research, a fundamental distinction can be made between *observational* and *experimental* studies. In experimental studies, investigators perform an action intended to interfere with the course of the disease or condition under study. The prototype is the randomized experiment, in which patients are randomly allocated to an experimental intervention and a comparator or a placebo. For example, patients with subclinical hypothyroidism (a condition in which thyroid function is slightly diminished) might be randomized between treatment with thyroid hormone substitute or an identical looking but inert placebo. With such an experiment, investigators could aim to answer the question of whether treatment improves cardiovascular risk associated with this condition. The patients would be treated for the purpose of the study, and arguably, many patients would not have been treated in the same way had they not been study participants (e.g., because their doctors do not think that their subclinical hypothyroidism importantly affected their health). The outcome of such an experimental study is a priori uncertain, and besides examples in which the treatment helped or made no discernible difference, there are examples of experimental treatments that turned out to worsen the disease course (Besselink et al. 2008).

In what is commonly called "observational research," the investigator does not intervene to allot the treatment (more generally, the exposure of interest). Some say that a better term for observational research would be "non-randomized research." However, we will stick with the term observational research as it is commonly used, as not all experiments are randomized. Observational research may take several forms, some of which the patients will be aware of (when questionnaires are to be filled in or additional blood samples are taken). In other instances, patients will not notice that their disease course is studied (e.g., when routinely gathered data from electronic medical records or large administrative health care databases are analyzed). An observational study can even be conducted several years after disease has occurred. For example, investigators may want to know whether hypothyroidism is associated with an increased risk for mortality. For this purpose, they could select all patients with a hypothyroidism diagnosis from a medical database, link these data to data regarding vital status, and compare mortality risk between patients with hypothyroidism and a control group without the disease.

Although in observational studies investigators do not intervene in the treatment of patients, they are not passive recorders of data. In almost all observational studies, substantial preparation is required, and decisions have to be made on what to observe and how to observe it. These decisions fundamentally influence what can be found and how those findings are interpreted (Popper 1957).

Observational studies can be primarily *descriptive* or *comparative*. In descriptive studies the main goal is to determine risks or prognosis in specified groups, without further analytical exploration of underlying causes. The proportion of individuals with a certain chronic condition in a specified population can be estimated (e.g., the proportion of people with diabetes in the Netherlands), or the risk of an outcome can be assessed in a population that has a (disease) characteristic in common (e.g., mortality risk after a specific type of surgery). Such descriptive studies are important for informing policy makers, patients and doctors; they are usually not comparative but give a broad idea about the likely outcome of a disease in a particular patient.

In comparative studies, groups are compared to determine the effect of an "exposure"; such exposures can be external agents (e.g., smoking, air pollution) or internal agents (e.g., a gene variant). Medical interventions can also be compared observationally. The crucial difference from experiments is that the type of medical intervention is not allotted by the investigator to study the effect of the intervention; the investigator observes what happens in the course of usual medical care, where different doctors make different decisions. The purpose of these comparative studies is generally to disentangle causal effects. A comparative observational study could compare mortality risks between two types of surgery for the same indication, aiming to determine which is the best procedure. Although this is not an experiment, the aim is the same as that of an experiment: to learn from a comparison. A comparative study could also compare cardiovascular risk between patients with hypothyroidism and a population without this condition, thereby aiming to identify a potential causal role of the condition in disease occurrence. Of note, descriptive studies can be presented as an apparent comparison, if risks are presented for several populations, such as when diabetes risks are presented for different countries. As long as the studies do not aim to disentangle underlying causes for the differences, such studies are still descriptive—even if they can give clues to potential causes (Pearce 2011). The distinction between descriptive and comparative studies is important, as the considerations regarding validity of the study's conclusions differ between these types of studies.

Threats to Validity in Observational Studies: Bias and Confounding

In this section, we describe the two most common threats to the validity of comparative observational research. The results of a study are considered to be valid if, under ideal circumstances with an infinite number of participants and infinite resources for assessment, the result of the study would yield "the truth." A study is not valid if, even with infinite resources, its results will still be wrong. The two notions under which the possibilities for being wrong are described are *bias* and *confounding*.

All studies in medicine rely on measurement and classification of the disease (e.g., breast cancer) or disease-related states (e.g., quality of life in cancer patients, survival without metastases). However, diseases and other study endpoints are prone to inadequate measurement or misclassification, potentially leading to bias in study results. If such a misclassification depends on whether or not the person has a particular exposure, comparing exposed and non-exposed persons can give biased results. An example of potential misclassification might be a study where the risk of pneumonia is assessed and compared between patients treated with chemotherapy and a control group from the general population. In case of clinical suspicion, chemotherapy patients are more likely to be admitted to the hospital, where X-rays will be made, whereas in the control group the diagnosis is more often based on the clinical evaluation by a general practitioner, and diagnosis is not confirmed with an X-ray. Bias may occur if the X-ray is more sensitive for the diagnosis compared to clinical evaluation only. Misclassification bias can occur in descriptive as well as comparative observational studies. It is important to note that misclassification occurs in different degrees, wherein some diseases, diagnostic tools, or classification systems are more prone to misclassification than others. For example, it is hardly possible to misclassify death. Of note, misclassification is also an issue in randomized trials, even though blinding aims to prevent the misclassification being different between the compared groups.

As mentioned, in comparative observational studies the aim of the study extends beyond the mere description of data and investigators want to study the effect of a treatment on a disease course, or the effect of a potential risk factor on the occurrence of a disease (so-called etiologic study). Potential causes of disease are often called *risk factors*. The comparison among two or more groups is central to comparative studies, and the validity of this comparison needs close attention when assessing validity in observational studies. The basic idea of a comparative study is that the group without the risk factor under study is used to determine the disease course in the absence of the risk factor under study. By comparing an outcome between the two groups (with and without the risk factors), investigators aim to ascribe any difference in outcome to a difference in the risk factor (or treatment). If there is a difference, the conclusion would then be that the risk factor under study is indeed a cause of the outcome under study—moreover, the study will also describe how strong the risk factor is.

This leads to the second major threat to validity of comparative observational studies: *confounding*. Causal inference by making the comparison between exposed and non-exposed is only valid under the assumption that all other risk factors (other than the risk factor under study) are balanced—i.e., on average similar, between the groups compared—the underlying idea is the notion of *ceteris paribus* (everything else being equal). *Confounding* is the term used to describe groups being different at baseline with regard to prognostic factors. This problem can occur by two mechanisms: in etiologic studies, risk factors tend to cluster (e.g., alcohol drinking is related to smoking), whereas in observational studies on therapeutic effects, treatment prescription is related to disease prognosis (this is called *confounding by indication*). For example, patients with newly diagnosed diabetes are more likely to be treated with drugs if their glucose levels are very high, but only with general lifestyle advice if their glucose levels are low. The opposite can also occur, such as when patients with advanced-stage cancer are no longer treated with chemotherapy. Of note, doctors are trained to treat patients according to the perceived prognosis of the patient. This is crucial for patients' care, but it hampers the comparison of therapeutic interventions by observational studies. As such, confounding (by indication) is often held to be the central problem in observational studies of therapeutics.

However, the problem of confounding is also problematic in studies of general risk factors. Suppose that investigators want to study the effect of a vegetarian diet on mortality risk, and they compare a group of vegetarians to a group of non-vegetarians. In this context, no confounding would mean that all factors (other than vegetarianism) that influence mortality risk are similar in these two groups. How likely is that to be the case? Not very, as vegetarians will probably have a healthier lifestyle in general, being physically more active, are smoking less, possibly being higher educated, etc. So, numerous prognostic differences will exist between these groups, and it will be difficult to ascribe a difference in mortality risk between the groups to a difference in vegetarian diet only.

In the analysis of comparative studies, statistical methods are often used to tackle the problem of confounding, to try to statistically restore the "balance" between the groups. The basic principle of such statistical models is that any comparison is done between groups of patients with similar confounder characteristics (e.g., the comparison between vegetarians and non-vegetarians is made separately for smokers and non-smokers—as there will be more non-smokers among the vegetarians—and afterwards the two comparisons are averaged over the smokers and the non-smokers to yield an overall comparison). Statistical models can indeed compensate for baseline differences between the groups, *but only under the assumption* that all differences are known and perfectly measured. This condition is rarely met. How likely is it that we have measured all prognostically important differences between vegetarians and non-vegetarians? The investigators might use statistical methods to account for differences in blood pressure, but more subtle and difficult-to-measure differences (such as differences in driving style, physical activity, or mood) will probably not be measured and therefore not be incorporated into the statistical model; these factors will remain unbalanced when comparing these two groups. The final verdict regarding the causality of the association remains uncertain if unmeasured confounding is an issue in a study. The same goes for confounding by indication in an observational study of therapeutic effects; some baseline prognostic factors might be known and measured (such as age or cancer staging), but other risk factors might be known only very crudely (e.g., if the presence of diabetes is known for all patients, but information about duration and severity of the diabetes is lacking).

It should be emphasized that confounding is not a yes/no phenomenon but a matter of degree depending on the association under study (see Figure 22.1). For some associations, such as the association between a vegetarian diet and mortality, the confounding might be very large. At the other end of the spectrum, when studying whether a specific genetic polymorphism is a risk factor for myocardial infarction, the association will likely not be confounded as the segregation of genes is a random process (there are exceptions to this general rule, for example when two genes are in linkage disequilibrium) (Smith & Ebrahim 2004).

In the above discussion, we have dealt with the typical situation in which two (or more) groups are compared. However, verdicts about causality are not always made in exactly the same way. In some instances a study compellingly shows a causal effect, even if the study does not involve a control group. An example is the use of pancreatic islet transplantation in patients with type 1 diabetes mellitus. Type 1 diabetes is characterized by destruction of pancreatic islets, and therefore the inability to produce insulin. In an experimental study, eight

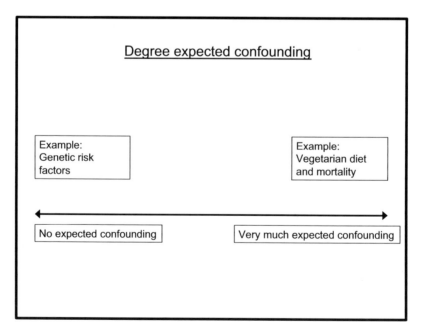

Figure 22.1 Degree of expected confounding in etiologic studies

patients received pancreatic islet transplantation, and after a follow-up period of one year, five patients were still not in need of insulin treatment (Hering et al. 2005). The study showed a therapy to be effective, even though no control group was included.

Why is this a convincing study? The reason is that we know for sure what would have happened to the diabetes patients in the absence of an islet transplantation: all diabetes patients would with certainty have continued their insulin therapy (assuming no deaths), as insulin is by no means capable of curing the disease, and spontaneous recovery is not reported. More generally speaking, a study without a control group can be convincing if two conditions are met: (1) we have a perfectly clear and certain picture of what would have happened in the absence of the experimental intervention, and (2) the difference between the known course without the intervention and the actual course with the experimental therapy (such as islet transplantation) is clear-cut and large. This also means that the known disease course should not have large fluctuations, as it is otherwise difficult to distinguish a treatment effect from fluctuation in the disease course. Both conditions are met in the islet transplantation study: the difference in effect is clear-cut (insulin dependence versus insulin independence), and the natural course of type 1 diabetes is characterized by a continuous need for insulin, so the difference is also large. Moreover, we know about the pathophysiology of diabetes, and the intervention counters the mechanism of the disease.

Causal Inference in Observational Research—Counterfactual Reasoning

Over the past decade, the above-mentioned problems in causal inference in observational research have led to attempts by epidemiologists to structure observational research according to what is called "counterfactual reasoning." There is at present an extensive literature about this counterfactual view on epidemiology (Hernan 2004). In fact, from a philosophical perspective, this newer counterfactual view is likely a mixture of different philosophical views on causation, such as counterfactual thinking, interventionism, and contrastivism (Vandenbroucke et al. 2016). We will introduce this counterfactual reasoning by examples.

Suppose there is a heavy storm. We see a tree falling down, and we witness how the tree hits and ruins a parked car. It is not risky to argue that the falling tree ruined the car, as the whole causal chain was observed: storm, tree falling, car ruined. We can then reason counterfactually that "if there had not been a storm, the tree would not have fallen, and the car would not have been damaged." Unfortunately, such obvious causal inference is hardly possible in the study of risk factors for diseases, for the simple reason that we are not in a position to directly observe the causal chain that leads to a disease. There are several reasons for this (e.g., some risk factors may take years to lead to a disease), but the fundamental problem is that we have few tools to directly observe medical causes in action. We cannot see high cholesterol levels actually causing heart disease; we have no observations that show high glucose levels in patients with diabetes causing neuropathy—in marked contrast to the tree that we see falling and crashing onto the car. Causal inference (i.e., inferring that an association is causal) is therefore necessarily also based on judgment. So the central question is how to infer a causal relation from data when we cannot directly observe the causal process and how such inference can be justified. Here philosophy enters the picture.

According to the idea of counterfactual reasoning in epidemiology, if we study the effect of a risk factor or treatment on a disease in a specific population, we need to know what would have happened in the absence of the risk factor/treatment, the so-called *counterfactual situation*. Think about the use of a control group in medical research (the non-exposed group as mentioned above). The control group is used to quantify what would have happened to a specific study population in the absence of the risk factor or experimental treatment under consideration. Inferring causality would indicate that a disease occurs or at least is more likely to occur, in the presence of the risk factor, whereas it would not have occurred, or would have been less likely to occur, without the risk factor. Using control groups to define what would have happened in the absence of the risk factor or treatment under study resembles a counterfactual way of defining causality: *"X causes Y because the counterfactual 'if not X then not Y' is true"* (Menzies 2001). Intuitively, such a counterfactual theory has some appeal, as it resembles reasoning in ordinary life. A simple statement with causal connotations, such as "he was late (Y) because it was snowing (X)," implicitly assumes knowledge of what would have happened "if it had not been snowing"; it assumes that the counterfactual "if it had not been snowing, then he would not have been late" is true.

Despite its intuitive appeal, applying a counterfactual framework to observational studies has some drawbacks. First, a risk factor can be a cause of a disease even if the counterfactual "if not X then not Y" is not true. In other words, a risk factor can be a disease cause even if in the absence of the risk factor (the counterfactual situation) the disease still occurs. This is the case when, in the absence of the risk factor under study, a "back-up cause" makes the disease occur. For example, if in a patient a stroke is caused by smoking, then it might be that in the absence of smoking the stroke still would have occurred but be caused by the patient's diabetes (as a back-up cause). This so-called problem of *preemption* shows that counterfactual dependence is not a necessary condition for causation. A second problem is that the counterfactual framework does not distinguish between meaningful causes and causes that fit the counterfactual definition but that are obviously irrelevant (Broadbent 2013). Think about causes as the presence of oxygen or the existence of a temperature on Earth that is compatible with life: these factors would be causes of a myocardial infarction within a counterfactual framework. It would make more sense to call these conditions "feasibility conditions" rather than causes, as they set the stage for all causes to act. A third problem is that the type of counterfactual thinking that is advocated in epidemiology is often explicitly limited to "humanly possible actions" and limited to external interventions. In contrast, "states," such as being obese or having a certain genetic trait, fall outside of this type of counterfactual thinking in epidemiology. For a more detailed critique, see Glymour and Glymour 2014.

Fundamentally, a counterfactual theory provides no rules or tools for justifying causation in observational studies. More formally, the counterfactual framework is semantically appealing as it seems to give a well-described meaning to causation that fits with some types of observational studies, but it is hardly of epistemic value, as it does not add to justification of causal claims. It might be true that if we know the counterfactual outcome for certain, we can infer causality, but this only shifts the problem, as the new question arises how to know whether the counterfactual situation is in fact truly known. By definition, we cannot know the counterfactual outcome for certain, as this counterfactual situation does not exist.

Think again about the impact of a vegetarian diet on mortality. We do not know what the outcome in the vegetarians would have been had they been carnivores, and the closest we can get is to perform a study including a comparison group. How can we know with certainty that the non-vegetarians represent the outcome the vegetarians would have had they not been vegetarians? We only have some certainty if there is no confounding or if confounding is adequately dealt with statistically. This means that a counterfactual reasoning cannot in itself justify causality in a specific observational study. Claiming that X is a cause of Y because the counterfactual statement "if not X then not Y" is true adds nothing to a claim that something is the cause of a disease. Such a claim can only be put forward if the central assumptions of epidemiologic assumptions are met—i.e., the assumptions of no bias and no (unmeasured) confounding.

In randomized trials the situation is slightly different, as the randomization produces a control group that can be seen as a tool to determine the counterfactual outcome. Why? Because the randomization procedure is designed to equally distribute prognostic factors between groups. As treatment decisions in a randomized trial are not influenced by prognosis, confounding by indication is thereby circumvented. However, randomization is relying on a chance procedure, and there is no guarantee that for a particular trial the groups still will be balanced. Randomization gives only an "asymptotic" reassurance (i.e., that in the long run, if an infinite number of trials is done, on average they will show truth). The difference between the actual outcome and the expected outcome can be explained by a game of throwing two dice. The prior probability of throwing any number below 12 is 35/36. Now suppose the dice are thrown and the outcome is 12. Referring to the prior probability and complaining that this was not expected is no longer meaningful. Unlikely events can occur by chance. This is similarly true for a randomized experiment: the randomization provides an *expected balance* but not a guarantee of balance in a particular study (Senn 2013).

Intended and Unintended Effects

Confounding by indication is a central problem in observational studies of *intended* effects of interventions (i.e., effects that one hopes to see). The question is whether confounding plays a similar role in studies of *unintended* effects, such as side effects of treatments. Suppose two antibiotics are available for a specific disease (e.g., urinary tract infection). When studying treatment effects (e.g., whether symptoms of the infection recede more quickly or not), confounding by indication is an issue, as the one antibiotic might more often be given to patients with more severe symptoms. But what about side effects such as allergic reactions? As doctors cannot predict such side effects, for a study that compares the two drugs with regard to side effects, confounding is not an issue (Vandenbroucke 2004). This is graphically illustrated in Figure 22.2 (Schneeweiss 2007): an intended effect of Coxibs (fewer gastrointestinal adverse effects compared to standard painkillers) leads to selective prescribing to people with previous gastrointestinal problems and is strongly confounded in contrast to the unintended effect of causing cardiovascular disease, which was unknown at the time of marketing, so that cardiovascular risk factors were not taken into account on prescribing. To further guarantee the absence of confounding, patients with a known history of side effects (e.g., a known history of allergic reactions to drugs) can be excluded from the analysis (Schneeweiss et al. 2007). These theoretical considerations that confounding by indication is usually not an issue in the study of side effects are supported empirically by several studies showing that observational studies on side effects show results similar to results from randomized trials (Golder et al. 2011).

This idea of no confounding for unexpected events does not only apply to studies of drugs. In fact, most causes of disease that are investigated successfully in epidemiology could be called unintended effects: lung cancer is an unintended effect of smoking. One starts smoking, say at age 11, without any knowledge as to one's probability to develop lung cancer, and for reasons that have nothing to do with one's risk of lung cancer. In the same way, exposure to asbestos and its subsequent risk of mesotheliomas is due to a choice of work at a particular factory, at the time (in the 1950s) when the risk of exposure to asbestos was not known at all. Similar reasoning applies to leukemia in children being caused by prenatal X-ray exposure, which a long time ago was routinely applied to assess the width of the mother's pelvis before birth (Vandenbroucke 2004). These are examples where not much confounding is expected, as the outcomes studied are unintended and unpredictable. Mind that confounding still needs consideration, as risk factors might cluster.

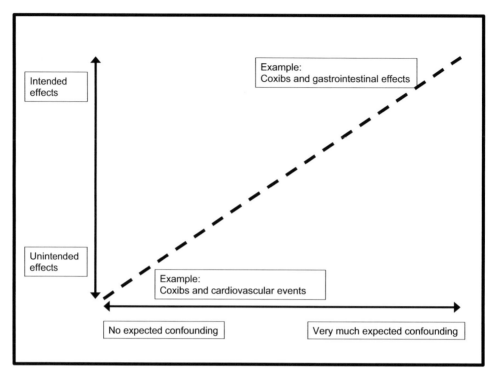

Figure 22.2 The potential for confounding in relation to intended and unintended effects (adapted from Schneeweiss 2007)

Should Only Interventions Be Studied?

Experimental studies are characterized by studying interventions (in a broad sense), whereas observational studies can also study risk factors. Some risk factors are characterized by an unclear onset (obesity, hypercholesterolemia) or represent a lifelong state (gender, genetic make-up). It should be emphasized that studying risk factors and not interventions poses one important restriction on the interpretation of study results: one might very well come to the verdict that some risk factor is causal, but that does not yet guide the type of intervention that is needed to reduce or normalize the risk, nor does it predict what will happen if the risk factor would be wiped out totally. Say that researchers have performed the perfect study (no confounding), with a valid estimate, of the effect of obesity on mortality, and they conclude that mortality risk is doubled in the obese compared to the non-obese. This does not automatically mean that intervening on obesity (e.g., by a highly effective diet) normalizes mortality risk among the obese. It might well be that, despite our spectacular diet being truly effective, obesity has induced irreversible organ damage, making it impossible to normalize mortality risk at all. Moreover, dieting might be less effective than exercise, or either might be less effective than a combination. Even more difficult to foresee, people might change their behavior when on a diet (e.g., by starting to smoke). Others might become depressed by the intervention and turn to alcohol. This means that, even assuming etiologic causality can be inferred from observations, such causal inference cannot directly be translated into an estimate of the effect of specific interventions on the risk factor (Greenland 2005).

Thus, from a health policy perspective, studying risk factors has disadvantages compared to studying interventions, as the latter can inform treatment policy directly. Randomized studies on interventions have the additional advantage that the justification of causal claims is more straightforward compared to observational studies. Does this mean that we should only study interventions and abstain from studying hypertension, hypercholesterolemia, obesity, gender, genes, and socio-economic status because these are not interventions? A view stating that only interventions can be studied neglects the fact that the well-conducted study of risk factors is an important step in identifying targets for interventions (Glymour & Glymour 2014). If we had no clue that obesity increases mortality risk, we would not have thought to perform studies that intervene on obesity. In a world where no observational studies of risk factors were performed, nor their potential causality assessed, researchers likely would have no idea what conditions to intervene on.

Causal Inference: The Role of Integrating Evidence

As discussed, arguments about the comparability of groups are important for causal inference. We can, however, never be certain that all measured and unmeasured prognostic factors are equally distributed—not even in a randomized trial, because chance baseline differences remain possible. This is to a certain extent a limitation of current knowledge, as we have no guarantee that we know all factors that are prognostically relevant to a specific outcome. Moreover, chance cannot be ruled out as a possible explanation for results in a single study. One could argue that given this inherent uncertainty, causality should never be judged based on a single observational study. One way forward is to place the results from studies in a much broader context. Until now, whether talking about randomized studies or about observational studies, we have mainly considered numerical results from individual studies. However, these data need integration for causal inference. In general, empirical studies are interpreted in the light of other knowledge. For example, it was found that certain compounds of tobacco smoke induce mutations at a p53 mutational hotspot in epithelial lung cells. Without epidemiological knowledge about smoking and lung cancer, searching for such carcinogenic mechanisms in tobacco smoking would be nonsensical. Even more to the point, if this particular mechanism would not have "worked," the basic scientist would not have concluded that smoking is not a cause of lung cancer, but she would conclude that she has been looking at the wrong mechanism. Thus, basic science depends on epidemiology for interpretation of studies (Vandenbroucke 1998). In exactly the same way, epidemiology needs input from the outside: numbers and risks can only be collected and interpreted meaningfully in the light of reasoned arguments about explanations and mechanisms. One cannot set up a randomized trial and decide on a dosage and a particular type of patient without strong outside knowledge (e.g., from basic science, animal research, general physiology, or consistent observations on humans) that can guide our understanding about the likely mechanism of the disease and the likely mechanism of drug action. The same is true for observational research on risk factors; the epochal studies on cardiovascular risk factors like serum cholesterol were in part based in mice experiments about producing atherosclerosis. Thus, deciding how and what to study, and rendering the ultimate verdict that something is a cause, requires a judgment based on results from various types of research.

Ultimately, there are no rules for combining the results from different studies into one final judgment. Ruling out alternative explanations is one key way to help this judgment. For example, findings about the influence of smoking habits on pregnancy outcomes could be questioned, because it might be argued that perhaps women who smoke during pregnancy will also be different in a host of other factors (confounding by lifestyle, nutrition, or socio-economic class). However, when pregnancy outcomes were related to smoking by fathers, almost no effect was found. This is compatible with the idea of a direct effect of smoking in mothers (Howe et al. 2012). Such a way of reasoning is called *triangulation* of evidence. In other instances, a pathophysiological mechanism tilts the scales: a contentious debate on the question of whether a new type of oral contraceptives led to more venous thrombosis than older types of contraceptives was strengthened by a finding of a stronger effect on coagulation by the newer contraceptives, showing the epidemiologic findings to be in line with basic science (Vandenbroucke et al. 2001).

This means that in observational research, and actually in all research, different types of reasoning play a role when coming to a final verdict: quantitative contrast-thinking, counterfactual thinking, thinking about mechanisms, and reasoning about the conditions under which an exposure might cause an outcome to occur. In observational studies it seems apparent that its conclusions rely on all of these types of reasoning. Randomized trials give the impression of standing on their own, given the single intervention that is studied experimentally. However, also in randomized trials, in their set-up and interpretation, ancillary knowledge from other science is crucially included. Lastly, all causal inference is ultimately "inference to the best explanation" and remains an inference under uncertainty. In the end, the judgment that something is a cause always remains a risk, as judgment is fallible. People try to escape this situation with diverse methodological strategies, making the rules for establishing causal inference ever stricter. Still, regardless of strategy, judgments about causality will, for the foreseeable future, remain grounded in the integration and weighting of different types of relevant research and theory.

References

Besselink, M.G.H., van Santvoort, H.C., Buskens, E., Boermeester, M.A., van Goor, H., Timmerman, H.M., Nieuwenhuijs, V.B., Bollen, T.L., van Ramshorst, B., Witteman, B.J., & Rosman, C., 2008. Probiotic prophylaxis in predicted severe acute pancreatitis: A randomised, double-blind, placebo-controlled trial. *Lancet*, 371(9613), 651–659.

Broadbent, A., 2013. *Philosophy of Epidemiology*. London: Palgrave and Macmillan.

Glymour, C. & Glymour, M.R., 2014. Commentary: Race and sex are causes. *Epidemiology*, 25(4), 488–490.

Golder, S., Loke, Y.K., & Bland, M., 2011. Meta-analyses of adverse effects data derived from randomised controlled trials as compared to observational studies: Methodological overview. *PLoS Med*, 8(5), e1001026.

Greenland, S., 2005. Epidemiologic measures and policy formulation: Lessons from potential outcomes. *Emerging Themes in Epidemiology*, 2, 5.

Hering, B.J., Kandaswamy, R., Ansite, J.D., Eckman, P.M., Nakano, M., Sawada, T., Matsumoto, I., Ihm, S.H., Zhang, H.J., Parkey, J., & Hunter, D.W., 2005. Single-donor, marginal-dose islet transplantation in patients with type 1 diabetes. *JAMA*, 293(7), 830–835.

Hernan, M.A., 2004. A definition of causal effect for epidemiological research. *Journal of Epidemiology & Community Health*, 58(4), 265–271.

Howe, L.D., Matijasevich, A., Tilling, K., Brion, M.J., Leary, S.D., Smith, G.D., & Lawlor, D.A., 2012. Maternal smoking during pregnancy and offspring trajectories of height and adiposity: Comparing maternal and paternal associations. *International Journal of Epidemiology*, 41(3), 722–732.

Menzies, P., 2001. Counterfactual theories of causation. In: *Stanford Encyclopedia of Philosophy*. Available online at: http://plato.stanford.edu/entries/causation-counterfactual/

Pearce, N., 2011. Epidemiology in a changing world: Variation, causation and ubiquitous risk factors. *International Journal of Epidemiology*, 40(2), 503–512.

Popper, K.R., 1957. Philosophy of science: A personal report. In: C.A. Mace, ed., *British Philosophy in Mid-Century*. London: George Allen & Unwin, pp. 182–183.

Schneeweiss, S., 2007. Developments in post-marketing comparative effectiveness research. *Clinical Pharmacology and Therapeutics*, 82(2), 143–156.

Schneeweiss, S., Patrick, A.R., Sturmer, T., Brookhart, M.A., Avorn, J., Maclure, M., Rothman, K.J., & Glynn, R.J., 2007. Increasing levels of restriction in pharmacoepidemiologic database studies of elderly and comparison with randomized trial results. *Medical Care*, 45(10 Supl 2), S131–S142.

Senn, S., 2013. Seven myths of randomisation in clinical trials. *Statistics in Medicine*, 32(9), 1439–1450.

Smith, G.D., & Ebrahim, S., 2004. Mendelian randomization: Prospects, potentials, and limitations. *International Journal of Epidemiology*, 33(1), 30–42.

Vandenbroucke, J.P., 1998. Medical journals and the shaping of medical knowledge. *The Lancet*, 352(9145), 2001–2006.

Vandenbroucke, J.P., 2004. When are observational studies as credible as randomised trials? *Lancet*, 363(9422), 1728–1731.

Vandenbroucke J.P., Broadbent, A., & Pearce, N., 2016. Causality and causal inference in epidemiology: The need for a pluralistic approach. *International Journal of Epidemiology*. Advance online publication only.

Vandenbroucke, J.P., Rosing, J., Bloemenkamp, K.W., Middeldorp, S., Helmerhorst, F.M., Bouma, B.N., & Rosendaal, F.R., 2001. Oral contraceptives and the risk of venous thrombosis. *New England Journal of Medicine*, 344, 1527–1535.

Further Reading

Broadbent, A., 2013. *Philosophy of Epidemiology*. London: Palgrave and Macmillan.

Glymour, C., & Glymour, M.R., 2014. Commentary: Race and sex are causes. *Epidemiology*, 25(4), 488–490.

Greenland, S., 2005. Epidemiologic measures and policy formulation: Lessons from potential outcomes. *Emerging Themes in Epidemiology*, 2, 5.

Senn, S., 2013. Seven myths of randomisation in clinical trials. *Statistics in Medicine*, 32(9), 1439–1450.

Vandenbroucke, J.P., 2004. When are observational studies as credible as randomised trials? *Lancet*, 363(9422), 1728–1731.

23
PHILOSOPHY OF EPIDEMIOLOGY

Alex Broadbent

The Nature of Epidemiology

Epidemiology is the study of the distribution and determinants of disease and other health states in human populations by means of group comparisons for the purpose of improving population health (Broadbent 2013, 1). But this definition does not make obvious the distinctive character of epidemiology, nor its philosophically interesting aspects.

Epidemiology is a detective-like science, seeking to identify exposures or traits that cause ill health, or trying to find out what the effects of certain exposures or traits might be. It relies on a mix of evidence types, depending on the topic under study. In clinical epidemiology, where one is testing the effect of a medication, it is possible to construct large and methodologically strong studies, randomized controlled trials (RCTs), where individuals are randomly assigned between treatment and control group. In theory, this allows the probability of various kinds of bias being responsible for the finding to be calculated, although in practice, RCTs often fail to live up to their on-paper promise. For example, they remain open to financial bias (where the funder's financial interests influence the finding) and to publication bias (where findings of no effect or negative effect are less likely to be published than positive findings).

However, often, RCTs are impossible, and studies are "observational," meaning that the investigator does not assign subjects to exposure/treatment or control group, but passively observes the groups as s/he finds them. For example, the effects of smoking are studied entirely through observational studies, because of the ethical and practical difficulty of forcing people to take up or stop smoking. This makes epidemiologists methodologically creative in coming up with ways to get the most out of apparently unpromising investigative conditions, and it also forces them to take different kinds of evidence from a wide range of sources.

The unifying principle of epidemiology is normative. Epidemiologists seek to find out things that will improve population health. Not all epidemiologists are happy with this value-laden goal, but it is hard to make sense of epidemiology if that is not the goal. If improving population health were not the goal, then the discipline would not cohere. Because the unifying principle comes from a goal and not from the field of study itself, the scope of epidemiology is unlimited: anything that might give rise to differences in health status between identifiable populations is a potential topic of study, including environment, occupation, nutrition, lifestyle, genes, and medication. Likewise, the outcomes that epidemiologists measure are not necessarily lined up with existing medical disease classifications. The combined effect is an expansive pressure on the scope of medicine. This is most evident in calls of some social epidemiologists for physicians to take an active interest in reducing socioeconomic inequality, on the basis that socioeconomic inequality is detrimental to population health (Marmot 2006; Venkatapuram and Marmot 2009).

Epidemiology is quite unlike a typical philosopher's picture of science. Besides often proceeding without experiment, as already discussed, it also lacks theory, in the sense that philosophers use the term. Epidemiologists do not seek to fit their findings into an overarching mathematical framework, like physicists or economists (surprisingly, since on paper economics and epidemiology might look similar to a visiting alien). Nor do they work within the auspices of a grand non-mathematical theoretical framework, as biologists do with the theory of evolution. Epidemiologists do not aim to offer a comprehensive description or understanding of their domain—indeed, their domain is not clearly delimited, as already mentioned, but is rather defined with reference to the goal of improving population health. Thus a truth-apt epidemiological claim about the empirical world, such as "Smoking causes lung cancer," is not helpfully seen as a piece of epidemiological theory. It is a piece of causal knowledge, but it is not a theory in the sense that physics or even biology has theories—grand, overarching, often mathematical structures explaining or describing large chunks of the observable world. Epidemiologists do not devise theories in this grand sense. Asked to identify a piece of epidemiological theory, an epidemiologist would be more likely to come up with something like the fact that you can estimate a relative risk from an odds ratio given the assumption that the disease is rare. Thus epidemiological theory consists, not in what philosophers would call theoretical claims about the nature of the world, but of a bundle of methods for finding things out.

There are many different philosophical approaches to science, and it is perhaps dangerous to generalize. However, even those philosophers who place method in the foreground tend to see science as a matter of developing overarching theoretical frameworks having, ideally, universal application. Indeed, those philosophers who have emphasized method (such as Carl Hempel and Karl Popper) have tended to be the most enamored of the "physics model" of science. The idea that science is distinguished by its methods is certainly familiar to philosophers, but the idea that these methods might not be directed at developing theory, rather than solving piecemeal problems like a detective, is not widespread.

This disconnect between the philosopher's received image of science and the actual nature of epidemiology gives familiar philosophical topics new aspects as they arise in epidemiology. It also makes the study of those topics within epidemiology fruitful for the discipline of epidemiology, since there really is room for conceptual and methodological advances in this young and developing science. The detective-like nature of epidemiology gives rise to questions about the inferences it seeks to make. These inferences concern causation, prediction, and explanation—all central philosophical topics. These topics will be prominent themes in the remainder of this chapter, which is divided into three sections, as follows. First, we will explore the significance of epidemiological thinking for *disease classification*. Second, we will consider recent thinking about *causal inference* in epidemiology. Third, we will look at *prediction* in epidemiology, and the philosophical questions it raises.

Epidemiology and Disease Classification

One striking effect, or perhaps cause, of the growth of epidemiology in the latter part of the 20th century was the rise of the view that many or perhaps all diseases are *multifactorial*. Prior to the rise of multifactorial thinking, it seemed natural to classify infectious diseases by their causes, since their causes come ready-classified as distinct species of bacteria or virus. It is much harder to classify chronic, non-communicable diseases (CNCDs), such as cancer or coronary heart disease, by their causes, since the various cases of cancer or coronary heart disease do not display any obvious pattern of causes. Some factors are more commonly found—perhaps much more commonly—among those with these diseases than among those without, but there will be cases of these diseases among those without these exposures and cases where the disease fails to occur even with them.

Lung cancer provides a case in point. Before the widespread uptake of smoking, it was virtually unknown and was not identified as a distinct form of cancer. Nonetheless, it does occur in non-smokers. And among smokers, it is not an especially common disease—the majority of smokers do not get lung cancer. Thus there is little appetite for defining lung cancer as "smoking-itis" or "smoking disease," even though lung cancer is very rare in the absence of smoking. In particular, we do not know what makes the difference between those smokers who succumb to lung cancer and those who do not. So even though smoking is far and away the most significant cause of lung cancer, we do not go so far as to define lung cancer with reference to this cause. Contrast cholera, where the definition of the disease does mention a particular cause—namely, *vibrio cholerae* in the small intestine. This is part of the definition of cholera: it distinguishes cholera from non-choleric fever and diarrhea.

Much of the classic epidemiological writing on the multifactorial nature of disease focuses on the multiple causes of various diseases (e.g., Rothman 1976). However, this is misleading: everything is "multifactorial" in the sense of arising from many causal factors and not one alone. The real question—the question that would make sense of the literature on multifactorialism in epidemiology—is not "How many causes do diseases have?" but "Can diseases be defined or classified causally?"

This issue arose because of a prevailing view in the first part of the 20th century that diseases should be classified by their causes, and by just one cause. This view was pushed hard by the originators of germ theory, notably Robert Koch: ". . . each disease is caused by one particular microbe—and by one alone. Only an anthrax microbe causes anthrax; only a typhoid microbe can cause typhoid fever" (Koch 1876; quoted in Evans 1993, 20).

This principle may be appealing for infectious diseases (though even there, it does not make room for diseases arising from symbiosis between two or more infectious agents, such as swine fever, which arises from the symbiotic action of a bacterium and a virus), but it is clearly not appealing for something like cancer, for which we simply cannot identify anything like a universal general cause that is present when cancer is present and absent when cancer is absent. As industrialized countries passed through the epidemiologic transition (where the main causes of mortality shift from infectious diseases to CNCDs), epidemiologists found this view of disease unhelpfully restrictive. Hence the trend toward multifactorialism.

This current trend toward multifactorialism amounts to a view that diseases need not be classified by reference to causes in any strict way. Interestingly, this is in many ways an echo of the way diseases were thought about before the advent of germ theory in the late 19th century. Jacob Henle (who gave his name to a loop in the kidneys) issued this wonderful complaint:

> Only in medicine are there causes that have hundreds of consequences or that can, on arbitrary occasions, remain entirely without effect. Only in medicine can the same effect flow from the most varied possible sources. One need only glance at the chapters on etiology in handbooks or monographs. For almost every disease, after a specific cause or the admission that such a cause is not yet known, one finds the same horde of harmful influences—poor housing and clothing, liquor and sex, hunger and anxiety. This is just as scientific as if physicists were to teach that bodies fall because boards or beams are removed, because ropes or cables break, or because of openings, and so forth.
>
> (Henle 1844; quoted in Carter 2003, 24)

This sort of complaint might equally be directed at some areas of contemporary epidemiology, which identify risk factors for disease that, though clearly part of the causal story in some cases of disease, are absent in other cases, and often present where the disease is not.

The fact that Henle was complaining about it in 1844 shows that multifactorialism is not new, and this ought to raise a question mark over the modern trend toward it. To be specific, it raises the question: what is at stake in classificatory decisions about disease entities? Clearly, Henle and Koch felt that it *mattered* whether diseases were classified by their causes or not. Note that Henle is not criticizing the medical science of his day for poor causal inference. In the analogy, he is not denying that breaking ropes and beams, openings, and so forth cause falling. He is rather pointing out that merely cataloguing the causes of falling is not a good explanatory theory (he says not "scientific") of the phenomenon.

Just as one might hope that a theory of falling will unify all the various cases of falling and not merely provide a catalogue of causes (e.g., boards breaking, ropes snapping, etc.), one might hope that a disease classification system expresses or is underwritten by an explanatory theory. In this context, an explanation would not be a law like the theory of gravity, but would be a causal difference between cases with the disease and cases without. Hence the feeling of dissatisfaction that some have expressed about contemporary risk factor epidemiology (Vandenbroucke 1988), which catalogues causes fairly enough, but does not attempt to offer explanations of the difference between cases of disease and cases of health. Multifactorialism about disease licenses the development of a catalogue of causes, but does not put pressure on medical science to systematize this catalogue so as to arrive at something more explanatory (Broadbent 2009). It also licenses the development of medications that are not "magic bullets" but that reduce the risk at the population level, without necessarily helping all or even most patients who take them (Stegenga forthcoming). (Statins and selective serotonin reuptake inhibitors are two examples of such medications.) If we settle for an understanding of etiology consisting of risk factors only, then we must also settle for interventions on risk factors only, and the best these will do is reduce the probability of a disease. Yet we know there have been "magic bullets" in the history of medicine (such as penicillin). Whether it is reasonable to insist upon magic bullets or not, the present point is that the effects of our conceptual framework for thinking about disease affect medical practice, can encourage or discourage medical breakthroughs, and can shape the development of medications and the assessment of their effectiveness.

Causal Inference

The paradigm case of causal inference in modern epidemiology is the discovery that smoking causes lung cancer. This episode involved significant methodological developments (in study design, e.g., in relation to case-control and cohort study methodologies) and conceptual developments (insights into how to use given data to draw inferences, e.g., the argument that a potential confounder could not be entirely responsible for an association if its relative prevalence is lower than the observed risk ratio [Cornfield et al. 1959]). It also saw the authoring of some famous commentaries on causal inference by epidemiologists and statisticians. The two most famous are those of Austin Bradford Hill (Hill 1965) and the principles of causal inference included in the first Surgeon General's report on the effects of smoking (Advisory Committee to the Surgeon General of the Public Health Service 1964). Thus, for example, Hill lists strength, consistency, specificity, temporality, biological gradient, plausibility, coherence, experiment, and analogy as properties of an association that one can look for when seeking to make a causal inference.

However, it is important to note that Hill insists that the presence or absence of none of these properties is decisive on its own. Hill's nine viewpoints have often been misunderstood as "criteria," but this is not how Hill intended them:

> None of my nine viewpoints can bring indisputable evidence for or against the cause-and-effect hypothesis and none can be required as a sine qua non. What they can do, with greater or less strength, is to help us to make up our minds on the fundamental question—is there any other way of explaining the set of facts before us, is there any other answer equally, or more, likely than cause and effect?
>
> (Hill 1965, 11)

Hill is effectively implementing *inference to the best explanation (IBE)*, some years before that topic was paid comprehensive attention by philosophers of science. The core idea of IBE is that, in some circumstances, we can infer the truth of an explanation on the basis that it is good. For example, if you hear a roaring noise outside the window, you can reasonably infer that a bus or car is revving its engine. The noise *could* be coming from a recording, or from a helicopter, but these are probably less good explanations of the sound because they are more complex or raise further questions (e.g., why would someone play a recording of an engine outside the window?). In Peter Lipton's terminology (Lipton 2004), Hill was in effect listing a number of criteria that would need to be considered in order to decide whether an explanation was "lovely"—whether it was a *good* explanation. If so, then one can make a causal inference, albeit tentatively. But there is not a formal balancing of factors, no strict checklist; sometimes very strong evidence of one kind (say, an overridingly *strong* association) might outweigh the lack of evidence of another kind (say, the lack of *biological plausibility*, relative to our current biological knowledge). Instead, there is a focus on qualitative judgment taking into account as much evidence as possible, from as many sources as possible—and, crucially, considering the question of whether there is a *better* possible explanation.

This reflects the character of the evidence that was available concerning smoking and lung cancer. The direct epidemiological evidence was entirely based on observational studies. Many of these were case-control studies, and it is difficult to base causal inferences on case-control studies alone. (In a case-control study, cases of the disease are identified, and the prevalence of the exposure among cases is compared to the prevalence of the exposure among controls, who do not have the outcome.) Large-scale cohort studies were started, but since lung cancer takes a long time to develop, the results of these studies were not decisive at that time. (In a cohort study, a "cohort" of people are followed, and their exposures and outcomes are recorded. In a famous cohort study on smoking running from 1951 to 2001, Richard Doll and Austin Bradford Hill wrote to registered British doctors requesting information on their smoking habits.) Laboratory evidence from animal experiments showed carcinogenic effects of tar painted onto the ears of rats, but it is a long way from there to the conclusion that tar carried in smoke inhaled into the lungs of a human will cause cancer in the lung. Time-trend data showed that lung cancer incidence, previously virtually unknown, had increased about 20 years after the widespread uptake of smoking, but time-trend data invites the slogan "correlation is not causation": the fact that two factors trend in the same way does not mean that one is causing the other; it could equally be explained by some other factor causing both trends.

In this situation, the only way to make a causal inference is to triangulate from the various kinds of evidence available. Each piece of evidence alone is weak, but the combination, handled correctly, amounts to a strong case. An example of such "triangulation" concerns the use of time-trend data, which, as mentioned, is not impressive evidence for causality on its own. The main alternative explanation for the association between smoking and lung cancer was the "constitutional hypothesis," which suggested that smoking and lung cancer might share a genetic cause of some kind. This hypothesis does not, however, accommodate time-trend data, since it is not plausible that a genetic variant would have spread fast enough to explain the increase in lung cancer incidence. Thus a piece of evidence that is not compelling on its own becomes an important part of the larger case, by refuting a competitor hypothesis.

Against this background, it is not surprising that a qualitative, informal approach to causal inference should have become dominant. However, contemporary epidemiology is witnessing a trend in the opposite direction, toward formal methods for causal inference. This trend emphasizes the importance of study design. Because of this emphasis, it implicitly locates causal inference within studies, or with data that can be amalgamated into one data set. The implication is that a causal inference is something that you do with a data set, using formal tools. This is not how causal inference was done in all historical cases: for example, there is no formal way of amalgamating the results of experiments involving painting tar on the ears of rats with observational data about smoking and lung cancer.

The core of the contemporary approach is the idea that, in order to make a meaningful causal claim, one must clearly specify the intervention one has in mind to bring about the difference between the exposed group and the unexposed group. It is not enough, according to this approach, merely to seek to investigate the effects of obesity on mortality, for example. There are many ways to intervene so as to reduce (or increase) body mass index (BMI), and these may have different effects on mortality (Hernán and Taubman 2008). Exercise and diet may have different effects; smoking or amputation will reduce BMI (the measure we are supposing is used in this case to detect obesity), but are likely to have quite different effects on mortality from either exercise or diet. This means that if one conducts a study of the association between obesity and mortality, one cannot make reliable claims about the *potential outcome* of reducing obesity in the study population. It would depend on how obesity was reduced. Proponents of this line of reasoning argue that this inability to identify the potential outcome—the inability to assert a counterfactual—means that such a study is not well-placed to provide evidence for a causal inference.

This "potential outcomes approach" (POA) insists that investigators clearly specify the counterfactuals whose truth they are investigating. The POA is also often thought to imply a kind of ranking among study designs, with experimental studies coming out well, because in such studies the investigator actually makes an intervention to create the exposed and control groups. A well-specified intervention is causally uniform: it has the same effect on the outcome of interest in different cases. The intervention "reduce obesity" does not satisfy this criterion with respect to the outcome "mortality" because the different ways that you might reduce obesity have different effects on mortality.

One of the confusions that arises in connection with the POA is the idea that experimental studies, in which you "do" the intervention rather than just observe an exposure of interest, guarantee that interventions are well-specified. That is not correct, because one may under-specify an intervention, even if one actually "does" it. For example, "one hour of strenuous exercise a day" could include boxing, powerlifting, running, and so forth, and these exercises might have different effects on mortality. So well-specified interventions are not guaranteed in experimental trials.

Nonetheless, the emphasis on clearly specified alternatives lends itself better to implementation in experimental studies than in observational ones. And among observational studies, it lends itself much better to cohort studies than case-control studies, which—despite their enormous influence in the history of epidemiology—are barely discussed at all in the POA literature, receiving just one page in the authoritative POA textbook (Hernán and Robins 2015, 98).

Thus, the nature of causal inference in epidemiology is undergoing active and energetic rethinking. There are real philosophical issues at stake in this rethinking. These can be divided into two main kinds. One concerns *the nature of causation*. Do causal claims only make sense if they are relativized to a clearly specified intervention? Must the intervention be humanly possible or only physically possible—either to be admissible in a causal question or to make the answer useful? Must we be able to actually move tectonic plates in order to understand or use the claim "Movements of tectonic plates cause earthquakes"? The other set of issues concerns the *methodological implications of the POA*. Is the focus on precise formulation of causal questions incompatible with the kind of triangulation that is so central to the history of epidemiology? Does it make space for study designs that are admittedly weak on their own but may provide important pieces of a larger puzzle, as with time-trend data on smoking and lung cancer? Is causal inference an informal, qualitative judgment, located "between" studies, or is it a formal, quantitative finding, located "within" studies?

There is not space here to enter into these debates (but see Further Reading). However, it is worth noting that the POA is partly motivated by the idea that unless one clearly specifies one's intervention, one is unable to make good predictions about the effects of a proposed intervention. This connection is both welcome and misguided. It is welcome because, as the POA has noticed, there is a lack of clear thinking about prediction and about the way causal knowledge can be used to predict, as we will see in the next section—both within epidemiology and much more broadly. It is misguided because the POA sets the generation of useful predictions as a necessary condition for a causal claim to be well-formulated. This is too strong, as we will see in the next section as well.

Prediction

Prediction is clearly central in epidemiology, and epidemiologists often use predictive language and concepts. For example, they will describe a risk factor as *predicting* a certain outcome. However, the exact nature of epidemiological prediction has received less attention than one might expect within the theoretical branches of epidemiology. This may be because the paradigmatic problem of epidemiology, the question of whether smoking causes lung cancer, was about causal inference. Once that causal inference was made, good predictions appeared to follow fairly automatically. Before smoking, lung cancer was rare; smoking was responsible for the increase in the incidence of lung cancer; thus it seems barely to require discussion that reducing smoking prevalence will result in a decrease in lung cancer incidence.

However, the apparent ease of moving from a piece of causal knowledge to a prediction in this instance is misleading. There is no denying that causal knowledge is useful for predicting and, in particular, for predicting what will happen if one intervenes to bring a certain situation about. However, it is still a further step to make a good prediction using causal knowledge. This can even be seen in the context of smoking and lung cancer, where some false predictions were made. Reducing the quantity of tar in cigarettes appeared not to help much, since smokers adapted their habits to get the tar anyway (e.g., puffing more, covering ventilation holes with their fingers). And advising smokers to take shallower puffs appeared to make matters worse, since—it turns out—the tissue at the top of the lungs is more susceptible to carcinogens in tobacco smoke. (Both cases are discussed in Parascandola 2011.)

The dangers of assuming that causal knowledge is sufficient, or almost sufficient, for prediction are part of what drives the POA, as already noted. Merely knowing that obesity causes a certain excess mortality does not allow you to predict how your obesity-reduction program will affect mortality. To take another example, establishing that acetaminophen can cause asthma (for which there is inconclusive evidence) does not license a recommendation not to give acetaminophen to children who are at risk of asthma, as one eminent pediatrician suggested (McBride 2011). It may be that fever and alternative fever-reducing medication would be just as likely to cause asthma, or more so (Broadbent 2012). This difficulty also connects with the difficulties previously identified for risk factor epidemiology: a causal risk factor is typically difficult to use for prediction in a public health context, precisely because the knowledge that a certain risk factor is causal does not tell you what the effect will be of an intervention on that risk factor.

Even if the causal knowledge has been obtained through asking a causal question that corresponds to a well-specified intervention, the most that licenses you to predict is what would have happened had the unexposed group been exposed or vice versa. It does not allow you to predict with confidence beyond your study group. This is sometimes called the problem of external validity (where internal validity is the soundness of a causal inference within a study, and external validity is the transportability of that conclusion to other populations). This is not an entirely happy formulation because it implies that a study can be rendered externally valid, in some general way. A better way to put the question is Nancy Cartwright's question: "Will this work for me?" A prediction in relation to a particular population must always take account of the properties of that population, and must take an all-things-considered, total-evidence view as to whether the same outcome will be found as was found in the studies that are used as an evidence base. Otherwise, the possibility remains open that some difference between the study and target populations matters for the prediction (Cartwright 2011; Broadbent 2011a)

It does not help matters that the philosophy of science is very poor in literature on prediction as a topic in its own right. Prediction receives far less attention from philosophers than do explanation, causation, and laws of nature (Douglas 2009; Broadbent 2013, 84–89). If there is one area ripe for new work in the applied philosophy of science, it is prediction, and epidemiology is a prime example of a science where philosophical attention to prediction could be really useful.

References

Advisory Committee to the Surgeon General of the Public Health Service. 1964. *Smoking and Health*. PHS Publication No. 1103. Washington DC: Department of Health, Education and Welfare.

Broadbent, Alex. 2009. "Causation and Models of Disease in Epidemiology." *Studies in History and Philosophy of Biological and Biomedical Sciences* 40: 302–311.

———. 2011a. "What Could Possibly Go Wrong?—A Heuristic for Predicting Population Health Outcomes of Interventions." *Preventive Medicine* 53 (4–5): 256–259.

———. 2012. "McBride's Recommendation on Acetaminophen Is Not Warranted by His Argument or Evidence." *Replies to The Association of Acetaminophen and Asthma Prevalence and Severity*. http://pediatrics.aappublications.org/content/128/6/1181.abstract/reply#pediatrics_el_53669.

———. 2013. *Philosophy of Epidemiology*. New Directions in the Philosophy of Science. London and New York: Palgrave Macmillan.

Carter, K. Codell. 2003. *The Rise of Causal Concepts of Disease*. Aldershot: Ashgate.

Cartwright, Nancy. 2011. "Predicting What Will Happen When We Act: What Counts for Warrant?" *Preventive Medicine* 53 (4–5): 221–224.

Cornfield, Jerome, William Haenszel, E. Cuyler Hammond, Abraham M. Lilienfeld, Michael B. Shimkin, and Ernst L. Wynder. 1959. "Smoking and Lung Cancer: Recent Evidence and a Discussion of Some Questions." *Journal of the National. Cancer Institute* 22: 173–203.

Douglas, Heather. 2009. "Reintroducing Prediction to Explanation." *Philosophy of Science* 76: 444–463.

Evans, Alfred S. 1993. *Causation and Disease: A Chronological Journey*. New York, NY: Plenum Publishing Corporation.

Henle, Jacob. 1844. "Medicinische Wissenschaft Und Empirie." *Zeitschrift Fur Rationelle Medizin* 1: 1–35.

Hernán, Miguel A., and James M. Robins. 2015. *Causal Inference*. http://www.hsph.harvard.edu/miguel-hernan/causal-inference-book/.

Hernán, Miguel A., and Sarah L. Taubman. 2008. "Does Obesity Shorten Life? The Importance of Well-Defined Interventions to Answer Causal Questions." *International Journal of Obesity* 32: S8–S14.

Hill, Austin Bradford. 1965. "The Environment and Disease: Association or Causation?" *Proceedings of the Royal Society of Medicine* 58: 259–300.

Koch, R. 1876. *Verfrahen Sur Untersuchung Zur Conserviren Und Photographie Der Bakterien, Beitrag Der Pflanzen*. Breslow: Cohn's Bicr.

Lipton, Peter. 2004. *Inference to the Best Explanation*. Second edition. London and New York: Routledge.

Marmot, Michael. 2006. "Health in an Unequal World: Social Circumstances, Biology, and Disease." *Clinical Medicine* 6 (6): 559–572.

McBride, John T. 2011. "The Association of Acetaminophen and Asthma Prevalence and Severity." *Pediatrics* 128: 1181–1185.

Parascandola, Mark. 2011. "Tobacco Harm Reduction and the Evolution of Nicotine Dependence." *Public Health Then and Now* 101 (4): 632–641.

Rothman, Kenneth J. 1976. "Causes." *American Journal of Epidemiology* 104 (6): 587–592.

Stegenga, Jacob. forthcoming. *Strange Pill: Evidence, Value, and Medical Nihilism*. Oxford: Oxford University Press.

Vandenbroucke, J.P. 1988. "Is 'The Causes of Cancer' a Miasma Theory for the End of the Twentieth Century?" *International Journal of Epidemiology* 17 (4): 708–709.

Venkatapuram, Sridhar, and Michael Marmot. 2009. "Epidemiology and Social Justice in Light of Social Determinants of Health Research." *Bioethics* 23 (2): 79–89.

Further Reading

Broadbent, Alex. 2013. *Philosophy of Epidemiology*. New Directions in the Philosophy of Science. London and New York: Palgrave Macmillan.

Broadbent, Alex. 2015. "Causation and Prediction in Epidemiology: A Guide to the Methodological Revolution." *Studies in History and Philosophy of Biological and Biomedical Sciences*, 54, 72–80.

Hernán, Miguel A., and Sarah L. Taubman. 2008. "Does Obesity Shorten Life? The Importance of Well-defined Interventions to Answer Causal Questions." *International Journal of Obesity* 32: S8–S14.

Hill, Austin Bradford. 1965. "The Environment and Disease: Association or Causation?" *Proceedings of the Royal Society of Medicine* 58: 259–300.

Greenland, Sander. 2005. "Epidemiologic Measures and Policy Formulation: Lessons from Potential Outcomes." *Emerging Themes in Epidemiology* 2: 1–7.

24
COMPLEMENTARY/ ALTERNATIVE MEDICINE AND THE EVIDENCE REQUIREMENT

Kirsten Hansen and Klemens Kappel

1. Introduction

Homeopathy is a paradigmatic example of a complementary/alternative therapy (CAM). Homeopathy relies on the premise that "like cures like," so a homeopath uses the same substances, though extremely diluted, in the treatment of ailments, which are thought to produce the symptoms of the ailment in question. The basic idea in homeopathy is that a putatively active substance is diluted in water and administered to a patient. The distinctive feature of homeopathic medicine, however, is that the dilution is so extreme that at the end of the process it is unlikely that any molecules of the putatively active substance remain. A common remedy has strength of 30C. This means that the original substance has been diluted 30 times by a factor of 100 each time (called a 30C remedy). This implies that the original substance has been diluted by a factor of no less than 1,000,000,000,000,000,000,000,000,000,000,000,000,000,000,000,000,000,000,000,000. Homeopaths argue, however, that the water in which the remedy is diluted (or the remedy itself) has a memory of the original substance, which is why the homeopathic remedy that may consist of nothing but pure water can nonetheless treat an ailment. There are readily available homeopathic remedies for a wide array of conditions such as anxiety, asthma attacks, broken bones, chicken pox, rubella, and many more ailments.

Clearly, our general physical, biological, and chemical theories provide no reason to believe that homeopathy has any effect whatsoever. Conversely, if we were to accept the theories behind homeopathy, we would need to reject or at least revise a large number of established physical and chemical theories. In addition, homeopathy has been evaluated rigorously by randomized controlled trials, and the research has been examined in meta-analyses. Some trials have shown that there might actually be some effect of homeopathy, but it has been widely discussed what to conclude from this. The conducted studies have overall been of a poor quality, and it is likely that some will produce misleading results. The overall conclusion in the established research community has been that there are no clinical effects of homeopathy apart from placebo effects, if any such are found (Shang et al., 2005).

Although nowhere admitted as a part of established health care, homeopathy is nonetheless widely used in most of the Western world. Homeopathy is particularly popular in France, where it is the leading alternative therapy; it is advocated strongly by the royal family in England, and according to the 2012 National Health Interview Survey, which included a comprehensive survey on the use of complementary health approaches by Americans, an estimated 5 million adults and 1 million children used homeopathy in the previous year (National Institutes of Health, 2015).

Something similar is true of other modes of CAM. Though prima facie unlikely to have any effects, and with no systematic evidence of efficacy, they are widely, and perhaps increasingly, used outside of the established health care systems. Here we consider some philosophical questions that this information raises.

2. What Is CAM?

The terms complementary/alternative medicine or CAM cover many different types of therapy. The expression "alternative" indicates that a therapy is used as an alternative to conventional therapy, whereas "complementary" indicates that CAM therapies are provided merely to supplement conventional medicine. Apart from homeopathy, typical forms of CAM include acupuncture, reflexology, herbal medicine, osteopathy, and meditation.

There are three typical characteristics of CAM of interest here, and they are illustrated by the case of homeopathy. First, although there might be certain exceptions, such as meditation for anxiety and acupuncture for pain, CAM therapies are generally not supported by evidence derived from scientific methods or findings. Second, CAM theories are even in some cases inconsistent with our best and most corroborated theories of the natural world. Third, CAM is not provided in the established health care system, whether this is publicly funded or not. Of course, there are exceptions to these general rules, yet they hold in many instances, and they are important for what we want to discuss below.

Conventional medicine is conventional in the three senses that CAM is not. Conventional medicine is based on evidence generally provided by the use of methods that are congruent with the methods of inquiry used in science, whereas the efficacy of CAM has generally proved difficult to establish with the same methods. The efficacy of conventional medicine is usually explained by, or at least consistent with, broadly accepted scientific theories, and conventional medicine is broadly implemented in the established health care system.

Some CAM therapies have been tested by randomized controlled trials (RCTs), showing a possible effect, and not all conventional therapies have been tested by RCTs. This does not, however, change the characteristics of CAM and conventional medicine. We will return to these aspects later.

In the literature, one finds various attempts to offer relatively broad definitions of CAM (e.g., National Institutes of Health, 2015; World Health Organization, 2015). For example, the definition offered by WHO refers to "a broad set of health care practices that are not part of that country's own tradition and are not integrated into the dominant health care system." No consensus on a satisfactory definition has emerged in the literature, but we need not go into this dispute here. It is enough that we have the paradigm cases of CAM in mind and the characteristic features that distinguish CAM from conventional medicine.

3. CAM and the Evidence Requirement

The use of CAM in the Western world appears to be increasing (Eisenberg et al., 1998; World Health Organization, 2013), though the exact extent is difficult to estimate due to the heterogeneity of data, as well as the use of a variety of different definitions of CAM (Eardly et al., 2012).

Below we will focus on some of the wider philosophical questions that this increasing usage raises. Maybe CAM should be admitted as part of established health care? If not, then what are the exact reasons that justify the exclusion of CAM?

It is easiest to present the various positions in this debate if we start by focusing on the concerns that members of the medical profession and the research community have often felt about admitting CAM into the established health care system (e.g., Goldacre, 2008; Singh and Ernst, 2008). We can represent the main concern in the form of the following argument:

1. The Evidence Requirement. Treatments offered in established health care/public health care should undergo testing by RCTs to ensure evidence of efficacy or effectiveness.
2. Most CAM interventions have not been evaluated rigorously by RCTs, and those that have show little or no effect. So, either there is no evidence suggesting that various modes of CAM are effective, or there is evidence showing that they are not effective.
3. So, CAM should not be provided as part of an established health care/public health care system.

Let us grant that this is a valid argument, and consider the plausibility of the premises of the argument. Many proponents of CAM have not been persuaded by this argument, as they have disputed one or more of the premises. Thus, a number of replies to the argument can be discerned in the literature:

1. Some argue that CAM should be exempted from the Evidence Requirement, as evaluation by RCT is impossible even in principle. In section 4, we consider and reject an argument to this effect.
2. Some proponents of CAM have insisted that there are other ways (apart from RCTs) by which one can gather the evidence necessary for evaluating CAM, or that the notion of evidence as presupposed in RCTs is irrelevant for CAM. We discuss this objection in section 5.
3. Finally, it has regularly been suggested that the Evidence Requirement should be rejected. After all, it might be said, conventional medicine is far from always tested by RCTs, and this shows that we should lower the standards of evidence in general, admitting more and less rigorously tested treatments. We argue in section 6 that this is not the case. However, the Evidence Requirement needs to be modified, though not in a way that accommodates CAM, or so we argue.

Finally, in section 7 we discuss the claim that choices of CAM need not be made on the basis of evidence of the sort assumed in the evidence requirement above. Rather, CAM should be seen as akin to a lifestyle or value choice. Such choices may be entirely reasonable, we argue, even when made in the absence of evidence of efficacy, or when evidence of lack of efficacy is available. We consider whether this would constitute a reason to reject the Evidence Requirement (and argue that it is not) and consider certain other normative implications.

4. Randomized Clinical Trials and CAM

Today's medical practice generally seeks to adhere to the Evidence Requirement (i.e., that medical practices should be backed by evidence). Although there are many admissible types of evidence and procedures by which evidence is collected, the gold standard for evidence in medicine is often considered to be the randomized controlled trial, or RCT. This is the main idea in what is known as evidence-based medicine.

Let us briefly outline the main idea of an RCT. The aim of an RCT is to assess the effect or efficacy of some sort of clinical treatment or other form of medical intervention, typically a specific medication aimed at a particular health condition, but in principle it could be any form of intervention offered with the aim of improving health. The strategy in RCT is to assess the outcome of the intervention by using it on human subjects, typically subjects drawn from a larger group of patients with a particular disorder. The subjects are divided into two groups. One group receives the intervention, and the other group, the control group, does not. The control group instead receives either a placebo or some well-known treatment. This permits a comparison of the outcome of the intervention. It is considered crucial for the RCT that subjects are divided into an intervention group and a control group by a randomization procedure. Moreover, at least the most rigorous trials are masked, which means that subjects, administrators, and/or researchers are kept ignorant about whether a particular subject belongs to the intervention group or the control group.

The first question is whether the use of RCT is possible or appropriate with respect to CAM. Some CAM practitioners reject this idea and hence reject the Evidence Requirement. One argument focuses on the claim that CAM essentially is holistic and relies on a unique and individual relationship between the patient and the practitioner (Frank, 2002; Walach, 2003). Thus, the argument goes, the RCT procedure itself will destroy the beneficial property of the treatment. For instance, Walach (2003) writes:

> Suppose that the "active" principle of homeopathy resides in a complex mix of the homeopathic situation between patient, practitioner, remedy, history of medicinal substances and their use as codified in the homeopathic material medica, with some mental interaction between the doctor and patient—such as a flash of security, a spark of trust and hope. In other words, suppose homeopathy is a kind of field effect with no single element that can be isolated and attributed to the remedy alone. If that were the true picture, then testing the remedy alone would be like taking one transistor out of a radio set and testing it for its capacity to play music.
>
> (9–10)

The suggestion is that the effect of CAM cannot be assigned to any single causal factor of a specific component in the intervention. CAM is not like a pill whose chemical properties account for the entire effect. Rather, the beneficial effects of CAM reside holistically in the whole of the interaction between patient and practitioner.

This worry about the use of RTC to evaluate CAM fails to acknowledge the distinction between what is known as *explanatory (or causal) trials* and *pragmatic trials* (White, 2002). Explanatory (or causal) trials generally measure *efficacy*; that is, they seek to measure the specific effects of a causally active component in an intervention. Typically, explanatory trials do so by assessing the treatment effects of a treatment produced under ideal, controlled conditions in a research clinic by carefully isolating the treatment effects from other effects. So, typically, explanatory trials require substantial deviations from the usual clinical practice.

Pragmatic trials, by contrast, measure what is known as *effectiveness*. They measure the benefit of a treatment produced in routine, real-world clinical practice, no matter what specific causal factors may contribute to that benefit. So, pragmatic trials typically do not provide conclusive information about causally active components in a particular treatment (Roland and Torgerson, 1998; Cardini et al., 2006).

Explanatory trials as applied to CAM, then, would seek to investigate the causal efficacy of a specific CAM component, such as the homeopathic medicine as such, or the prick of an acupuncture needle. Pragmatic trials, on the other hand, would aim to evaluate the clinical effectiveness of a CAM practice as a whole. For example, a pragmatic trial of a homeopathic medical intervention would aim to assess the homeopathic consultation as such, rather than to seek to determine the causal effects of the specific homeopathic component involved in that practice. Pragmatic trials of homeopathic practices can be conducted by providing the treatment group with the whole consultation, including the homeopathic medicine, the meeting with the practitioner, conversation, time spent on the individual, and so on. The control group will receive no treatment, a sham treatment, or the prevailing treatment. The clinical effectiveness of these two regimens can then be compared. So, there is no reason in principle why CAM cannot be assessed in this way. Clearly, however, masking is bound to present practical difficulties in pragmatic trials, and this may affect the quality of the trials. If a pragmatic trial is not masked, there is a risk that if it shows an effect, it will be unclear whether this effect is due to the intervention as such or results from bias due to lack of masking (due to experimenter error or placebo effects).

A somewhat similar objection to the Evidence Requirement highlights what is felt by some to be a principal obstacle to using conventional research methods to evaluate treatments with a perspective on illness and disease other than that of conventional medicine (Hammerschlag, 1998; MacPherson et al., 2002). The objection is that since CAM and conventional medicine rely on fundamentally different assumptions about the nature of disease and human biology, RCTs cannot be used to measure the effect of CAM. The following quote exemplifies this view:

> The whole process [of evaluating CAM with RCTs] can be equated to asking a sculptor to sculpt with a paintbrush to prove he is an "artist." The need to conform to an existing tool can undermine the very process we are trying to evaluate. In the case of the sculptor, the need to use the paintbrush undermines his or her ability to demonstrate his or her artistic skills, and in the case of the acupuncturist, the need to use standardized interventions (as in most RCTs) may undermine his or her ability to effectively treat the patient.
>
> (Ahn and Kaptchuk, 2005, 41)

Again, it seems that this objection (even if otherwise sound) applies only to RCT in the form of explanatory trials, not to pragmatic trials. So, even if it is true that the very process of conducting a RCT will "undermine the very process we are trying to evaluate," this would seem to apply only to explanatory trials, not to pragmatic trials. Nothing in the objection shows that it is not possible to consider a CAM intervention as a whole and compare it to a conventional intervention.

In saying this, we have not, of course, considered the various logistical and practical problems that are bound to arise in setting up pragmatic trials. Surely, masking, and in particular double masking, may pose practical difficulties. It is, for example, not easy to mask an acupuncture trial so that the patient and/or the practitioner does not know whether they actually receive acupuncture. Several attempts have been made to solve this practical problem, however. There have, for example, been RCTs in which the control group received sham acupuncture, conducted by gluing needles to the patients' skin in order to mask the patients (Filshie and Cummings, 1999). Other types of sham acupuncture include shallow needling and needling at non-acupoint sites. The masking requirement might also be met by masking either the assessor of the results or the statisticians involved.

There seems to be no reason in principle why CAM could not be evaluated by RCT. Note, however, that even if it were correct that CAM could not be evaluated using RCTs (including pragmatic trials), this by itself would *not* imply that CAM should be *exempted* from the Evidence Requirement. All this would show would be that CAM could not even in principle *meet* the Evidence Requirement.

5. Are There Other Kinds of Evidence for CAM?

So far no reason has emerged why CAM cannot be evaluated by using pragmatic trials. Indeed, a considerable number of RCTs assessing the effectiveness of CAM have been conducted, though the general quality is debatable (a search in the Cochrane Library in June 2015 returned more than 12,000 trials, more than 300 Cochrane Reviews, and more than 1,400 other reviews of complementary therapies). Generally, there are many poor studies allowing for much bias and risk of misleading evidence. So, even though there exist RCTs indicating a possible effect of CAM (mostly in relation to acupuncture and meditation), the quality of the studies makes the results very uncertain. Furthermore, systematic reviews performed on these studies generally suggest that CAM has no discernible effects or that there is insufficient evidence to judge whether a therapy is effective, due to the poor quality of the studies included (e.g., McCarney, Linde and Lasserson, 2008; Paley, Johnson, Tashani and Bagnall, 2011).

At this point some proponents of CAM insist that we might draw upon other sources of evidence. Sometimes it is suggested that the fact that most CAM treatments have existed for many years provides evidence of their effectiveness. For example, Walach (2003) argues:

> Homeopathy has some clinical effectiveness. If it did not, it would have died out. Indeed it is more sought after now by patients at a time when modern medicine prides itself in being more powerful than ever.

(7)

We might refer to these lines of reasoning as the *evolutionary argument*. Although the evolutionary argument may at first seem intuitively compelling, it is far from clear that evolutionary arguments are sound when applied to treatments offered in CAM or medicine in general. For the evolutionary argument to carry conviction, people need to be able to adapt their purchasing choices to what is clinically effective, but it is hardly reasonable to believe that this condition is met. It may seem plausible that people can adapt their purchasing choices to treatments that immediately and significantly decrease pain or improve well-being. However, in general, it seems hard to believe that consumers can adapt their purchasing choices to treatments that are clinically effective or more effective than alternatives short of systematic independent information about this choice. The reason, again, is that it seems impossible for the individual consumer (patient) to predict whether one purchase would be better than another in terms of effect, or whether a purchase was unnecessary because the patient would have recovered spontaneously.

Thus, it is doubtful that the continued existence of a variety of CAMs on the market is best explained by their clinical effectiveness. Consumers and patients simply lack the information necessary to develop preferences ensuring that only clinically effective treatments survive, and no other mechanism to select effective modes is in place. It is worth reminding that many treatments now known to be ineffective were used in conventional medicine up to the 1800s and had at that time existed for hundreds of years. It should also be noted that providers of CAM have a direct financial incentive in maintaining the presumption that the therapies they offer are effective. Survival on the market is no guarantee for effectiveness.

But there might be other forms of evidence suggesting the effects of CAM, a view suggested in the following quote:

> In therapeutics as well, there are numerous examples where the causal relationship of treatment and effect is convincing without appeal to anything other than simple observation of a single case.
>
> (Tonelli and Callahan, 2001, 1215)

Both providers and users of CAM could appeal to this idea: at least in some instances, we simply know from our experience of individual patients that CAM is effective. There is surely a certain intuitive appeal to cases where a patient's long-standing symptoms disappear shortly after an intervention. Yet one needs to stress the fundamental problem in this line of reasoning: the practitioner, relying on her or his sense that individual clients benefit from the treatment provided, simply lacks information about whether the clients would *also* have improved by another treatment, for example, a treatment provided by another practitioner or by no treatment at all (i.e., whether the patient would have recovered spontaneously).

In short, the main problem with appeals to individual experience of clinical effect is the lack of a control group. Control groups are essential to demonstrations of the effectiveness of treatment. No reason why comparison with a control group when considering the effectiveness of CAM is *not* needed has yet been put forward. It seems very unlikely, therefore, that evidence for the effectiveness of CAM can reasonably come from any other source than RCTs.

6. Lowering the Standards of Evidence?

As is well known, conventional medicine is far from being fully evidence based in the sense of being thoroughly based on evidence of effectiveness determined by RCTs. Most newly introduced conventional interventions are tested rigorously, but this is not the case for many established interventions that are used routinely. Proponents of CAM may object that the Evidence Requirement cannot be consistently applied to CAM, while parts of conventional medicine is exempted. The question is whether a case can be made that the standards of evidence should be lowered for CAM, given that many conventional treatments have not been evaluated by RCTs.

There are two related responses to this objection. First, this typically concerns treatments that are used routinely and where there are good reasons—deriving from observational studies and our background knowledge—to believe that if the treatment is not provided, the patient will die, or suffer great harm, and there are compelling reasons why these conventional therapies have not been tested by RCTs—as it would in many cases be ethically objectionable to do so. An example could be surgery on children. Few parents are willing to let their child participate in an RCT if the standard treatment is used routinely and there are convincing reasons to believe in the effect of the treatment. This suggests that one should accept a modified version of the Evidence Requirement:

> The Modified Evidence Requirement. Treatments offered in established health care/public health care should undergo testing by RCT to ensure evidence of efficacy/effectiveness, except when ethical constraints prevent this, which can be the case when observational evidence or background knowledge sufficiently strongly suggests foregoing a treatment will have serious adverse consequences.

So, the Modified Evidence Requirement acknowledges that decisions about whether or not to include a treatment can be based on types of evidence other than RCT. Yet, the Modified Evidence Requirement does *not* justify exempting CAM from RCT. The reason is that we do not have background knowledge or observational evidence that suggests that *foregoing* any particular mode of CAM will have serious adverse effects. Similarly, the Modified Evidence Requirement does not imply the existence or even the possibility of strong observational evidence *for* the efficacy of any particular mode of CAM. So, there is no ethical obstacle to evaluating CAM by pragmatic trials. Note that the same applies to many of the conventional therapies that have not been tested rigorously.

The second related reason for not exempting CAM from a requirement of providing evidence for efficacy is more general. CAM and conventional medicine differ significantly with respect to plausibility when viewed against our widely accepted background theories. Consider again homeopathy. The simple fact is that when viewed against the backdrop of the vast accumulated body of relevant scientific theory (in particular physics, biology, molecular biology, and chemistry), it is just extremely implausible that homeopathic interventions can have any effects whatsoever. If we accept standard theories in physics and chemistry, we are bound to be very skeptical that homeopathy works (and if we accept homeopathic theory, we are committed to rejecting a large number of physical and chemical theories).

Conventional treatments, by contrast, are fully compatible with established scientific theories. It is generally not the case that acceptance of standard theory commits us to thinking that any particular mode of conventional medicine is very unlikely to have any effects, although the fact that one conventional therapy is compatible with scientific theories is typically not a reason to favor this therapy over *other* equally compatible conventional therapies. In part, this is why RCTs are needed. However, any CAM therapy whose effects appears highly unlikely given accepted scientific theory, thereby incurs an *extra* burden of proof or more genuine evidence of effectiveness. Hence, it seems reasonable to insist that there is a heavier burden of proof resting on CAMs than on conventional therapies. Extraordinary claims require extraordinary evidence (Vickers, 2000).

7. Should Decisions About CAM Be Made on the Basis of Evidence of Efficacy?

At this point one might take a step back and query the basis of the Evidence Requirement. The most straightforward justification of this demand is a conjunction of two claims. First, a rational individual would want to choose between optional treatments on the basis of reliable information about which treatment is likely to be more effective (or plain effective). Second, this information can be provided only by RCTs, and this holds for both CAM and conventional medicine.

As we have just argued, the Evidence Requirement is implausible—in some cases treatments should be admitted though they have not been shown to be effective in RCT, and this is why the Evidence Requirement should be replaced with the Modified Evidence Requirement. We also saw that this does not justify an exemption for CAM. However, some commentators suggest a very different reason that CAM should not be subject to the (Modified) Evidence Requirement.

Some commentators suggest that a choice of CAM need not be based on evidence about clinical effects. For example, Borgerson (2005) writes that:

> . . . . while certain acupuncture points and procedures might be proven effective in RCTs and adopted into mainstream medicine, the underlying philosophy of traditional Chinese medicine, including the existence of the chi or vital force and the

commitment to health as the balance of chi will be lost [. . .] the naturopathic approach to health (including a commitment to holism, highly individualized care, and a principle of self-healing) will likely be left behind. For the millions of people choosing to spend out-of-pocket for alternative health care to- day, these elements of healing philosophy are of critical importance, and their loss would be substantial.

(506)

Borgerson says that, if subjected to RCT, a crucial aspect of CAM is likely to be lost. This must imply that Borgerson thinks that one can choose CAM knowing that no evidence from RCT is available, and she suggests that is because of the crucial importance of one's commitment to a wider healing philosophy, as this is the essence of the treatment. Typically, of course, those accepting a healing philosophy will assume that there are beneficial effects to doing so and that the particular modes of treatment recommended by the healing philosophy are effective. However, it is natural to interpret Borgerson as implying that the commitment to a wider healing philosophy can be made without possessing robust evidence to the effects of that commitment or the particular modes of treatment that it involves.

Whatever the details of Borgerson's view, we think that it is attractive to view a choice of CAM in this light. The choice of CAM could be considered a lifestyle choice or a value choice, a decision to join or sustain a community of shared values and beliefs, or perhaps something that is similar to undertaking a religious commitment. This has several implications.

First, this sort of value choice need not be based on reliable information about clinical effectiveness, and it therefore makes sense that an individual might choose CAM despite knowing that there is no reliable information about clinical effectiveness.

Second, CAM communities and those purchasing CAM treatments might be compared to religious and quasi-religious communities. By implication, we might say that CAM communities and those using the services should be accorded the same freedoms and protections as religious communities. There are many ways of spelling out the implications of this. One appeals to what is known as Mill's harm principle (after British philosopher J.S. Mill [1806–1873]), according to which the state is justified in limiting a person's actions or to interfere in a person's way of living only in order to prevent harm to third parties (Gray and Smith, 1991). Accordingly, the state should not interfere if people wish to join a CAM community, as long as their doing so harms no one and limits nobody's freedom of action. This is so even if there is no evidence showing an effect of the treatment in question or even evidence showing that the treatment does not have any effect. Hence, when CAM use is viewed as a lifestyle choice, the natural implication is that neither the state nor the individual rational agents need require rigorous evidence of the clinical effectiveness of CAM. However, the state might impose restrictions in special or extreme cases: for example, in connection with harmful or dangerous varieties of CAM—say, in the sense that some users are mislead into foregoing more beneficial conventional treatments—or when CAM is marketed under what are clearly false pretenses, it being claimed, for example, that there is a well-documented clinical effect.

Arguably a third implication (again, if we accept the basic values of liberal democracy) is that the state or government authorities and public officials should not support CAM, or offer CAM in publicly funded health care, unless reliable evidence testifies to its clinical effectiveness. CAM can be offered privately without making any claim to be backed by evidence on equal terms with, for example, various services offered by religious communities. However, if a CAM treatment is offered by the publicly funded health services, we ought to demand that it is evidence based. State health services should not offer treatments that can be accepted only by those with religious or quasi-religious convictions. The state should only offer treatments that survive rational scrutiny, and evidence is crucial to fulfill this purpose. A further implication might be that insofar as the state has a role in guaranteeing the quality of the established health care (even if not funding it), the state should adopt a similar stance and back the Evidence Requirement (in its modified form). Consequently, the state should be reluctant to admit CAM in established health care without sufficient evidence of efficacy.

8. Conclusion

We have argued that a main controversy regarding CAM concerns the Evidence Requirement. We have argued not only that RCTs can in principle be conducted on CAMs, but also that there is no other way to gather evidence about effectiveness in medicine. The fact that many conventional treatments have not been subjected to RCTs does not justify exempting CAM from rigorous testing. We propose that CAMs could be treated in the same way that the practices of religious and quasi-religious communities are. This implies that the state should not interfere if people wish to join a CAM community (i.e., use CAM or practice CAM), as long as this does not harm other people and does not limit other people's freedom of action, regardless of lack of evidence of the medical effectiveness of the CAM in question. It also implies that the state should not offer CAM in public health care or admit CAM in established health care. The state may legitimately impose restrictions in cases (if there are such) where a CAM is harmful in some way.

References

Ahn, A. C., and T. J. Kaptchuk. 2005. Advancing acupuncture research. *Alternative Therapies* 11:40.
Borgerson, K. 2005. Evidence-based alternative medicine? *Perspectives in Biology and Medicine* 48:502–15.
Cardini, F., C. Wade, L. A. Regalia, S. Gui, W. Li, R. Raschetti, and F. Kronenberg. 2006. Clinical research in traditional medicine: Priorities and methods. *Complementary Therapeutic Medicine* 14:282.
Filshie, J., and M. Cummings. 1999. Western medical acupuncture. In *Acupuncture—A scientific appraisal* (pp. 31–59), eds. E. Ernst, and A. White. Oxford: Butterworth-Heinemann.
Frank, R. 2002. Integrating homeopathy and biomedicine: Medical practice and knowledge production among German homeopathic physicians. *Sociology of Health and Illness* 24:796–819.
Eardley, S., F. Bishop, P. Prescott, F. Cardini, B. Brinkhaus, K. Santos-Rey, J. Vas, K. von Ammon, G. Hegyi, S. Dragan, B. Uehleke, V. Fønnebø, and G. Lewith. 2012. CAM use in Europe—The patient's perspective. Part I: A systematic literature review of CAM prevalence in the EU. Final Report of CAMbrella, Work Package, 4.
Eisenberg, D. M., R. B. Davis, S. L. Ettner, S. Appel, S. Wilkey, M. V. Rompay, and R. C. Kessler. 1998. Trends in alternative medicine use in the United States, 1990–1997: Results of a follow-up national survey. *JAMA* 280:1569–75.
Goldacre, B. 2008. *Bad science*. London: HarperCollins.
Gray, J., and G. W. Smith (eds.). 1991. *J.S. Mill—On liberty*. London: Routledge.
Hammerschlag, R. 1998. Methodological and ethical issues in clinical trials of acupuncture. *Journal of Alternative and Complementary Medicine* 4:159–71.
MacPherson, H., K. Sherman, R. Hammerschlag, S. Birch, L. Lao, and C. Zaslawski. 2002. The clinical evaluation of traditional East Asian systems of medicine. *Clinical Acupuncture Oriental Medicine* 3:16–9.
McCarney, R. W., K. Linde, and T. J. Lasserson. 2008. Homeopathy for chronic asthma (Review). *The Cochrane Collaboration*. Hokoken, NJ: John Wiley & Sons.
National Institutes of Health. 2015. Complementary, alternative, or integrative health: What's in a name?. Available: https://nccih.nih.gov/health/integrative-health (Accessed May 2015).
Paley, C. A., M. I. Johnson, O. A. Tashani, and A. M. Bagnall. 2011. Acupuncture for cancer pains in adults (review). *The Cochrane Collaboration*. Hokoken, NJ: John Wiley & Sons.

Roland, M., and D. J. Torgerson. 1998. Understanding controlled trials: What are pragmatic trials? *BMJ* 316:285.

Sackett, D., W. M. Rosenberg, J. A. Gray, R. B. Haynes, and W. S. Richardson. 1996. Evidence based medicine: What it is and what it isn't. *BMJ* 312:71–2.

Shang, A., K. Huwiler-Müntener, L. Nartey, P. Jüni, S. Dörig, J. A. Sterne, D. Pewsner, and M. Egger. 2005. Are the clinical effects of homoeopathy placebo effects? Comparative study of placebo-controlled trials of homoeopathy and allopathy. *Lancet* 366:726–32.

Singh, S., and E. Ernst. 2008. *Trick or treatment*. London: Bantam Press.

Tonelli, M. R., and T. C. Callahan. 2001. Why alternative medicine cannot be evidence-based. *Academic Medicine* 76:1213.

Vickers, A. J. 2000. Clinical trials of homeopathy and placebo: Analysis of a scientific debate. *Journal of Alternative and Complementary Medicine* 6:49–56.

Walach, H. 2003. Reinventing the wheel will not make it rounder: Controlled trials of homeopathy reconsidered. *Journal of Alternate and Complementary Medicine* 9:7–13.

White, A. 2002. Acupuncture research methodology. In *Clinical research in complementary therapies* (pp. 307–23), eds. G. Lewith, W. Jonas, and H. Walach. New York: Churchill Livingstone.

World Health Organization. 2013. *WHO traditional medicine strategy 2014–2023 Report*. Geneva: WHO.

World Health Organization. 2015. Traditional Medicine: Definitions. Available: http://www.who.int/medicines/areas/traditional/definitions/en/ (Accessed May 2015).

Further Reading

Ernst, E. (ed.). 2008. *Healing, hype or harm: A critical analysis of complementary and alternative medicine*. Exeter, UK: Societas.

Ernst, E., M. H. Pittler, and B. Wider. 2001. *The desktop guide to complementary and alternative medicine*. Second edition. Maryland Heights, MO: Mosby Elsevier.

Hróbjartsson, A., and S. Brorson. 2002. Interpreting results from randomized clinical trials of complementary/alternative interventions: The role of trial quality and pre-trial beliefs. In *The role of complementary and alternative medicine* (pp. 107–121), ed. D. Callahan. Washington, DC: Georgetown University Press.

Linde, K., Scholz, M., Ramirez, G., Clausius, N., Melchart, D., and Jonas, W. B. 1999. Impact of study quality on outcome in placebo-controlled trials of homeopathy. *Journal of Clinical Epidemiology* 52: 631–6.

Singh, S., and E. Ernst. 2008. *Trick or treatment? Alternative medicine on trial*. London: Bantam Press.

Part III
RESEARCH METHODS
(b) Other Research Methods

25
MODELS IN MEDICINE
Michael Wilde and Jon Williamson

1. Introduction

The major goals of medicine include predicting disease, controlling disease, and explaining disease. The main way of achieving these goals proceeds by modelling. In this chapter, we provide an introduction to the use of models in achieving the goals of medicine. To begin with, we introduce the notion of a model in medicine and distinguish experimental models from theoretical models. Then we provide an overview of the extensive array of these models by giving an account of animal models, which are a kind of experimental model, as well as association models, causal models, and mechanistic models, which are kinds of theoretical models. Next, we argue that in order to achieve the goals of medicine we need all of the latter three kinds of theoretical model—none are redundant. We go on to provide a framework for systematizing the production of theoretical models. Lastly, we present an example involving benzene and leukemia to illustrate the approach of this chapter.

2. Models in Medicine

The use of models is an important feature of scientific practice. Accordingly, scientific models have received a good deal of attention from philosophers of science. This attention has tended to focus on general problems such as the nature of models, how models are related to scientific theories, and how scientists can learn about the world by using models (Frigg and Hartmann, 2005). The scope of this chapter is much narrower, concerning only models in medicine. Given this, we shall set aside a number of important general issues in order to focus on those matters most pertinent to medicine.

Broadly speaking, we take a model to be a structure that represents some target system and that is used as a means of drawing conclusions about that target system. It is difficult to draw conclusions directly about a target system where that target system is inaccessible or very complicated. Usually, in such cases, it is more straightforward to reason instead about a model, since the model may involve considerable simplifications. The utility of models lies in the fact that conclusions drawn from the model can carry over to the target system, as long as the model is a sufficiently good representation of that target system.

As far as models in medicine are concerned, the target system is typically an individual or population of biomedical interest. The aim is to draw conclusions about such an individual or population. The conclusions that are relevant to medicine include claims about associations and causal relationships between exposures and diseases, as well as claims about biological mechanisms. On the one hand, it is important to establish associations and causal relationships in medicine, since the goals of medicine include predicting and controlling disease, and it is not possible to achieve these goals without information about associations or causal relationships (Russo and Williamson, 2007). On the other hand, it is also important to find out about biological mechanisms, since another of the goals of medicine is to explain disease, and it seems that explanations are best given by appealing to mechanisms (Williamson, 2013).

There have been a number of attempts to classify the different types of model in science, but none of them seems entirely satisfactory (Mäki, 2001). However, for our purposes, models in medicine can be usefully classified into two types, experimental models and theoretical models.

An experimental model is typically a concrete object that is experimented upon in order to draw conclusions about the model. This experimentation also licenses conclusions about a target system, insofar as the experimental model adequately represents the target system. The experimentation is intended to gather brand-new information about the target system. A theoretical model is a more abstract construct that systematizes information that has already been gathered about the target system. This systematization allows further conclusions to be drawn from that information more straightforwardly than would be possible if the information were not systematized in a model.

We shall shortly fill in the details of this classification by presenting animal models as examples of experimental models, and then presenting association models, causal models, and mechanistic models as examples of theoretical models. The aim is to show exactly how all of these models help achieve the major goals of medicine, i.e., predicting and explaining the occurrence of diseases, as well as providing recommendations about the control of such diseases.

3. Animal Models

One kind of model is an *animal model*, e.g., an *experimental organism*. An experimental organism, at least as far as medicine is concerned, is a non-human organism that is experimented upon in order to gather information relevant to the prediction, explanation, and control of disease in humans. A particular experimental organism is typically chosen on the grounds of its tractability to experimentation and its suitability to the biomedical phenomenon under investigation. Often, practical considerations, such as the availability of the organism for investigation, also inform the selection of experimental organism (Kohler, 1994).

An experimental organism is a means of gathering biomedical information about humans in cases where it is not possible to gather this information by directly conducting experiments involving human subjects, e.g., where such experiments would be unethical or difficult to carry out. As long as the experimental organism is representative of humans in the appropriate respects, conclusions arrived at by experimenting upon the organism also license further conclusions about humans. The conclusions may include claims about associations between exposures and disease, as well as claims about biological mechanisms.

A famous example of a model from the history of experimental physiology is the frog, which was studied in order to learn about the biological mechanisms of muscle contraction in humans and mammals more generally (Holmes, 1993). In this case, it was difficult to learn about muscle contraction in humans directly, since it was not possible to carry out the relevant experiments on humans. Instead, the frog was experimented upon as a representative model of the mechanisms of human muscle contraction. The conclusions drawn about muscle contraction in frogs were taken to apply also to muscle contraction in humans, insofar as the frog was appropriately representative of human muscle contraction.

A further example is the use of experimental organisms in preclinical animal trials for toxicology testing. Such tests are often conducted using mice or rats in order to assess the safety or efficacy of a drug before comparative clinical trials in humans are attempted. A comparative clinical trial in humans is carried out only if it has been established that the drug is not associated with adverse outcomes in the experimental organism. This is because establishing the safety of the drug in the experimental organism is taken to support the conclusion that the drug is safe also in humans, to a sufficient extent that it is deemed safe for comparative clinical trials in humans to proceed.

As these examples make clear, the use of experimental organisms can help with all of the main goals of medicine. First, experimental organism research can help with explanation because it allow claims about biological mechanisms in humans to be established (as long as the experimental organisms are appropriately representative). Second, it can help with predicting and controlling disease since it enables claims about associations and causal relationships between exposures and disease to be established (again, as long as the organisms are representative).

There have been a number of debates about experimental organism research. Claude Bernard (1865) believed that the results of animal experiments were straightforwardly applicable to humans, since the differences between animals and humans were only a matter of degree. However, Hugh LaFollette and Niall Shanks (1996) have argued that evolutionary theory casts doubt on the claim that it is justified to extrapolate from experimental organisms to humans, and that this makes experimental organism research morally questionable. Recently, it has been argued that significant findings in preclinical animal trials rarely lead to successful treatments in humans (Djulbegovic et al., 2014). Some have suggested that this may be because many animal trials are poorly conducted (Hirst et al., 2014).

Rachel Ankeny and Sabina Leonelli (2011) have argued that *model organisms* should be distinguished from the broader class of experimental organisms. Some examples of model organisms include the fruit fly, the nematode worm, and certain strains of mouse. Among other things, Ankeny and Leonelli argue that model organism research is unlike experimental organism research in that it aims to provide a detailed account of the model as a whole organism, in terms of its genetics, physiology, evolution, and so on. Arnon Levy and Adrian Currie (2015) have argued that model organisms are not models in the traditional sense. In traditional modelling, conclusions about the target system are supported by assessing whether the model is sufficiently analogous to the target system. In model organism research, they argue, the models are not analogues of some target system but instead are samples from a broader class of organisms. They maintain that the conclusions drawn from model organisms are the result of empirical extrapolation mediated by indirect evidence concerning the similarity of members of a broader class, where this broader class includes both the model organism and its target system. This indirect evidence might be that the broader class of organisms have a shared evolutionary ancestry or shared phylogeny. Marcel Weber (2005) argues that extrapolations from model organisms to humans can be reasonably sound, as long as they are based on known phylogenetic relationships.

This concludes our discussion of animal models. Now we survey the principal kinds of theoretical model: association models, causal models, and mechanistic models.

4. Association Models

One simple kind of theoretical model employed in medicine is an association model. This charts the main correlations among variables measured in a data set, so that one can use the observed values of certain measured variables to predict the value of an unmeasured variable in a new patient.

When applied to diagnosis, for example, an association model might be used to determine the probabilities of a range of possible diseases, given a particular combination of symptoms observed in a particular patient. These probabilities can then be used to motivate a particular treatment decision. An association model for prognosis, on the other hand, will usually be used to predict severity of disease, given observed clinical features of the patient and any observed biomarkers of the disease in question. Either way, the main use of the association model is *prediction*.

By way of example, we shall present two kinds of association model: a Markov network model and a Bayesian network model.

From a qualitative point of view, an association model can typically be represented by an undirected graph, sometimes called a *Markov network*, with nodes corresponding to variables and edges corresponding to the significant associations:

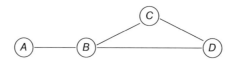

Separation in the graph can be used to denote probabilistic independence. In the above graph, B separates A from C and D, in the sense that all paths from A to C or D proceed via B. This separation relationship can be used to signify that A is probabilistically independent of C and D, conditional on B. (A is probabilistically independent of C and D, conditional on B, written $A \perp\!\!\!\perp C, D \mid B$, just when $P(a \mid bcd) = P(a \mid b)$ for all values a, b, c, d, of A, B, C, D respectively.) Thus, if one wants to predict A and one can observe B, it would make no sense to also observe C and D, because these would provide no further information about A. For example, suppose that a blockage in the main coronary artery (A) raises the probability of a heart attack (B), which in turn raises the probability of particular electrocardiogram results (C and D), in such a way that can be charted by the above association model. Then, to predict that the patient has had a blockage in the main coronary artery, one need only observe that the patient has had a heart attack, since learning in addition that the patient had certain electrocardiogram results provides no more information about the blockage.

From a quantitative point of view, in order to determine the probability of any variable conditional on any given combination of values of the other variables, one needs to specify the joint probability distribution, defined over all the variables of interest. In the above example, one would need to specify $P(abcd)$ for each combination of values a, b, c, d, of A, B, C, D, respectively. In a Markov network this is achieved by specifying the probability distribution over variables in each clique of the graph. A clique is a maximal subset of nodes of the graph such that each pair of variables in the subset is connected by an edge. The cliques in the above graph are $\{A, B\}, \{B, C, D\}$, so one would need to specify $P(ab)$ and $P(bcd)$ for all combinations of values a, b, c, d, of A, B, C, D, respectively.

Alternatively, one can use a *Bayesian network* model to represent the joint probability distribution. A Bayesian network has a qualitative and a quantitative component. The qualitative component of a Bayesian network consists of a directed acyclic graph, i.e., a graph with arrows such that there is no path in the direction of those arrows from a node to itself. In our example, one possible directed acyclic graph would be:

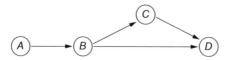

The directed acyclic graph needs to be constructed in such a way that each variable is probabilistically independent of its non-descendants, conditional on its parents. (A non-descendant of a variable is any node that cannot be reached by a directed path from the variable in question. A parent of a variable is any node from which there is an arrow to the variable in question. For example, in the graph below, A is a parent of B, and B is a parent of C and D. This means that C and D are descendants of B, and so B is a non-descendent of both C and D.) The quantitative component of a Bayesian network consists of the probability distribution of each variable conditional on its parents. The probability of a particular combination of values of variables is then a product of specified conditional probabilities:

$P(abcd) = P(a)\,P(b\,|\,a)\,P(c\,|\,b)\,P(d\,|\,cb)$.

There are a wide range of algorithms for producing a Bayesian network that represents the observed probability distribution of a set of variables measured in a data set (see, e.g., Neapolitan, 2004). There are also many algorithms for drawing predictions from a Bayesian network (see, e.g., Darwiche, 2009). Note that the directions of the arrows in the Bayesian network do not represent causal relationships in this sort of association model—the arrows are merely a technical device for representing certain probabilistic independencies.

Another kind of association model, called a classifier, is often used when it is only necessary to predict the value of a single variable, such as severity of disease. Many such models have been devised in the fields of machine learning and statistics (see, e.g., Alpaydın, 2010).

5. Causal Models

A second kind of theoretical model widely used in medicine is a causal model. A causal model seeks to chart the causal connections between the variables of interest. Such a model has three uses: prediction, explanation, and control. Like an association model, a causal model can be used for prediction, since it can be used to infer the probability of one variable conditional on others taking certain observed values. It can also be used to construct rudimentary explanations, since one can explain the fact that a particular variable takes the value that it does in terms of the causes of the variable in question taking certain values. Most importantly, perhaps, it can be used for control: it can be used to predict the effects of interventions and so can be used to decide how best to intervene in order to control the disease or symptoms of a particular patient.

A causal model can represent causal connections qualitatively by means of a directed acyclic graph. For example:

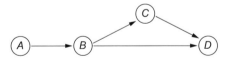

In contrast to the arrows of the directed acyclic graph presented in the previous section, which featured in an association model, in a causal model the arrows have significance in that they represent direct causal connections. For example, the above graph says that A is a cause of C, but only via a single pathway that proceeds through B.

A *causal Bayesian network* or *causal network* is a Bayesian network model built around a causal graph such as the above. Formally, it is a Bayesian network, but now the arrows in the graph have causal significance. Because it is a Bayesian network, it can be used to define a joint probability distribution over the variables in the graph, and thus can be used for prediction. But because the arrows have causal significance, it can also be used to predict the effects of interventions, as follows (see Pearl, 2000). When an intervention is performed to fix a variable to a certain specific value, one modifies the Bayesian network by deleting all arrows into this variable in the graph and updating the conditional probability distribution of each variable conditional on its parents in the graph to take into account the new value of the intervened-upon variable. Then this modified network can be used to infer changes to the probabilities of variables of interest, given the intervention. For example, intervening to fix the variable C to a specific value c will lead to a modified network in which the arrow from B to C is deleted:

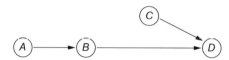

There are other sorts of causal model besides causal Bayesian networks (Illari et al., 2011); such models tend to portray causal processes in a similarly schematic way, representable by means of a directed acyclic graph. The explanations offered by such models can be superficial in that they only pick out key variables—milestones on the causal pathways to the effect in question—rather than the detailed structure of the underlying mechanisms that are responsible for the phenomena to be explained.

6. Mechanistic Models

Mechanistic models are used to generate explanations that are less superficial than the explanations yielded by causal models, in that they tend to include a richer set of explanatory features. There are two principal sorts of mechanism. A *complex-systems mechanism* consists of entities and activities organized in such a way that they are responsible for some phenomenon of interest (Machamer et al., 2000; Illari and Williamson, 2012). Examples include the mechanism for the circulation of the blood (which includes the features responsible for operation of the heart as well as the organization of the other components of the cardiovascular system) and the mechanism in an artificial pacemaker for producing electrical impulses to stimulate the heart (which includes its power source, clock, sensors and pulse generator, and the features of their arrangement that ensure its correct operation). On the other hand, what one might call a *mechanistic process* is a spatio-temporally contiguous process along which a signal is propagated (Reichenbach, 1956; Salmon, 1998). Examples include the process of an artificial pacemaker's electrical signal being transmitted along a lead from the pacemaker itself to the appropriate part of the heart, and the process by which an airborne environmental pollutant reaches the lining of the lung. While complex-systems mechanisms are often multi-level—e.g., involving coordinated activity at the levels of the organism, the organ, the cell, and the gene—mechanistic processes usually take place at a single level. Furthermore, whereas complex-systems mechanisms typically operate in a regular way, repeatedly producing the phenomenon of interest, mechanistic processes are often one-off, transmitting a single signal on a single occasion. In either case, however, the mechanism's structure and its organization—particularly its spatio-temporal organization—tends to be crucial to its operation.

A mechanistic explanation will often appeal to both sorts of mechanism. An explanation of the circulation of the blood in a particular individual may appeal to the complex-systems mechanism by which the heart pumps the blood, as well as the complex-systems mechanism of the individual's pacemaker and the mechanistic process linking the two. An explanation of a failure of blood to circulate may appeal to the same mechanisms, any faults of these mechanisms, and any pathophysiological mechanistic processes that these faults give rise to.

Mechanistic models are used to model the salient features of mechanisms in order to explain phenomena of interest. They differ from causal models in that they appeal to a richer set of features—entities, activities, organization, hierarchical structure, processes, etc.—instead of simply variables or events. Some of these features cannot be easily incorporated into a causal model: spatio-temporal organization and hierarchical structure, for example, are not naturally represented using the nodes and arrows that typically characterize causal models. However, these features are often essential components of an adequate explanation. Only in cases where these features are not essential to the explanation will an explanation generated from a causal model be adequate, in the sense that it picks out all the main features of an adequate mechanistic explanation (Williamson, 2013).

We noted above that a single mechanistic model may seek to represent two kinds of mechanism: complex-systems mechanisms and mechanistic processes. In addition, mechanistic models can be classified into two kinds: qualitative and quantitative.

Qualitative mechanistic models fill textbooks and research papers in medicine. They usually take the form of diagrams that highlight the main features of the mechanism. For example, Figure 25.1 portrays a part of the mechanism for apoptosis (cell death). Increasingly, animations are employed as qualitative mechanistic models, in order to portray activities and processes developing over time. Agent-based models are another kind of qualitative mechanistic model. Such a model represents a target system, e.g., a human population, in terms of a large numbers of similar individuals that interact in a restricted set of ways, e.g., colored cells in a grid that

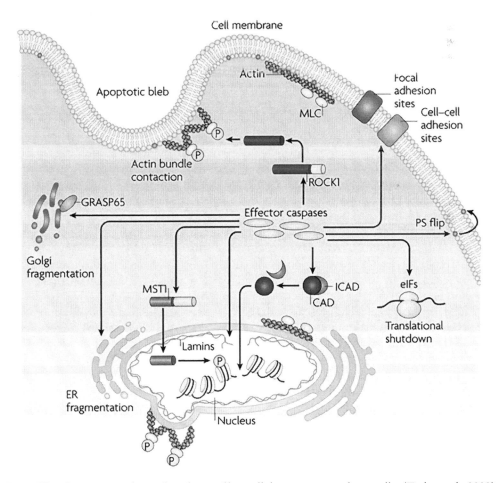

Figure 25.1 Caspases coordinate demolition of key cellular structures and organelles (Taylor et al., 2008)

influence the colors of their neighbours. Computer simulations are used to determine the typical behaviour of such a system. To the extent that this simulated behavior tallies with some observed phenomenon, such as the spread of a contagious disease, the agent-based model can be used to explain the occurrence of phenomenon.

Qualitative mechanistic models can be used to point to the underlying structure of reality that is responsible for producing the phenomenon to be explained, but normally cannot, on their own, explain why certain quantities within the mechanism take the values that they do, or explain the probability of a certain phenomenon. For this sort of explanation, a quantitative mechanistic model is required. A quantitative mechanistic model might, for example, consist of a diagram that portrays the qualitative structure of the mechanism, together with differential equations that can be used to model the changes in certain quantities over time. Another example of a quantitative mechanistic model is a *recursive Bayesian network*, which can represent a hierarchically structured mechanism by means of a collection of causal Bayesian networks, and which can be used to infer the probability of variables in the mechanism, given the observed values of other variables (Casini et al., 2011; Clarke et al., 2014b).

7. Combinations of Theoretical Models

Given this extensive array of theoretical models—association, causal, and mechanistic—two questions arise. Do we really need all these kinds of model in medicine? If so, is there any way of systematizing and unifying the production of these models? We will argue in this section that both of these questions should be answered affirmatively.

Do we need all these models? As we hope to have made clear in the above discussion, different kinds of theoretical model are put to different uses. Association models are for prediction; causal models are for prediction, explanation, and control; mechanistic models are primarily for explanation. One might think, then, that in medicine we should strive to produce good causal models, which can be put to the widest variety of uses, and we should avoid association and mechanistic models. There are three main reasons why this is not a sensible suggestion.

First, as we have mentioned, causal models generate more impoverished explanations than do mechanistic models. Causal models abstract away from the details of mechanistic structure, generating explanations that invoke only variables and the "thin" causing relation, i.e., explanations that invoke only claims of the form X causes Y. Mechanistic models, on the other hand, invoke entities, "thick" activities such as dilating and osmosing (i.e., a rich variety of kinds of causing), organizational features such as the structure and location of the cell wall, constitutive relations between components at different levels of a hierarchical mechanism, and spatio-temporally contiguous processes. Therefore, mechanistic models are far from redundant in situations in which a detailed explanation is required.

Second, causal and mechanistic models tend to be less reliable than association models. It is relatively easy to build an association model from some given data: one merely needs to model the joint probability distribution that generates the data. Often, in cases in which there are ample, good-quality data and no reason to suspect bias in the way the data were sampled, one simply models the joint distribution of the data and treats that data distribution as an approximation to the data-generating distribution. In a causal model, however, one needs to model not only the associations in the data but also the causal relationships among all the measured variables. Determining causal relationships is a harder problem than determining associations, so a causal model will normally be more speculative than an association model. Harder still is the task of establishing the details of the underlying mechanisms. This has long been the primary goal of biomedical science, and while great progress has been made, many mechanistic models are either very speculative or "gappy," with important features missing. This is less the case with a causal model: one needs to establish causal connections between those variables that are in the model, but there is no requirement to include in the model every variable that represents a component of one of the pertinent mechanisms. A causal model that omits some salient variables can still generate useful inferences for prediction and control, and capture some explanatory factors. In sum, there is a sense in which association models are normally less speculative than causal models, which are in turn normally less speculative than mechanistic models. Association models, in particular, retain an important place in medical research.

Third, association and mechanistic models are epistemically prior to causal models. This is a consequence of the following epistemological thesis, put forward by Russo and Williamson (2007). In order to establish a causal claim in medicine, one normally needs to establish two things: (1) that the putative cause and putative effect are appropriately correlated; and (2) that there is some underlying mechanism that links the cause to the effect in an appropriate way and that explains this correlation.

Some points of clarification are needed. First, the latter two claims are existence claims: in order to establish causality, one normally needs to establish the existence of a correlation and the existence of a mechanism, not the precise extent of the correlation nor all the details of the mechanism (Darby and Williamson, 2011, 2). Second, the mechanism involved might be a complex-systems mechanism, or a mechanistic process, or a combination of the two—whatever connects the putative cause to the putative effect in such a way that can *explain* occurrences of the latter. Third, this thesis concerns the evidence required to *establish* a causal claim, i.e., to settle the question according to the standards of the community, in such a way that warrants a high degree of confidence that the causal claim will not be overturned by any new evidence.

This epistemological thesis is plausible for the following reason. Recall that in medicine, causal claims are used for prediction, explanation, and control. If the putative cause and putative effect were not appropriately correlated, one would not be predictive of the other and one would not be able to intervene upon the cause to control the effect. Moreover, if there were not some mechanism that links the cause to the effect in an appropriate way, one would not be able to invoke the cause to explain the effect. Why is establishing a correlation not normally sufficient on its own for establishing causality? This is because many correlations are best explained by relationships other than causal connection—such as semantic, logical, or mathematical relationships—or by confounding, bias, or chance. Mechanistic evidence steers the causal discovery process toward those connections that are genuinely causal (Clarke et al., 2014a). Investigations of cases of causal discovery that provide further evidence in favor of the epistemological thesis include Clarke (2011), Darby and Williamson (2011), Gillies (2011), and Russo and Williamson (2011, 2012).

This epistemological thesis applies to each causal claim in a causal model. One therefore normally needs to establish the pattern of correlations, as represented by an association model, as well as the pattern of mechanistic connections, as represented by a mechanistic model, in order to establish the qualitative causal connections posited by the causal model. Association and mechanistic models are epistemically prior to causal models, since one needs to establish features of the former two kinds of model in order to establish the claims made by the latter kind. (Since the epistemological thesis merely requires one to establish existence of a correlation and a mechanism for each causal connection, in order to determine the pattern of causal relationships one only needs to establish the *pattern* of correlations and the *pattern* of mechanisms, not other features of association and mechanistic models.)

For these three reasons, one should not seek to abandon association and mechanistic models in favor of causal models.

How can the production of models be systematized? The third of the above three reasons suggests one way of systematizing and unifying the production of these models. First, consider an idealized case in which the available evidence is so extensive and of such high quality that it allows one to establish the full pattern of associations posited by an association model and the full pattern of mechanisms posited by a qualitative mechanistic model. Then one is in a position to establish the full pattern of causal claims made by a causal model, as well as the quantitative component of the causal model, which determines the joint probability distribution over the variables in the model. Having specified the quantitative component of a causal model, one is then in a better position to augment a qualitative mechanistic model by adding quantitative information.

Of course, in practice it is almost never the case that evidence is so plentiful and of such high quality as to establish every association and every mechanistic connection. In practice, evidence is often inconsistent, some data sets are extensive, others not, and items of evidence are of very varying quality. Thus some intermediate steps are needed to evaluate the relative merits of the items of evidence, and to determine which claims of association and mechanism can be considered established and which others are merely plausible or conjectural. It is possible, then, to establish some causal claims on the basis of what can be established in an association model and a mechanistic model. Other causal claims in the causal model will be more tentative, in proportion to the uncertainty of the corresponding association and mechanism claims.

This epistemological picture is depicted in Figure 25.2. Evidence of correlation (of which data sets are key) needs to be evaluated and graded with regard to the support it provides for associations. For example, data sets arising from larger numbers of observations will normally be more highly graded, and experimental studies will normally be favored over observational studies. Evidence of mechanisms informs this evaluation process because such evidence is crucial to determining whether trials are well-designed and their results correctly interpreted (Clarke et al., 2014a). This will lead to an association model. On the other hand, evidence

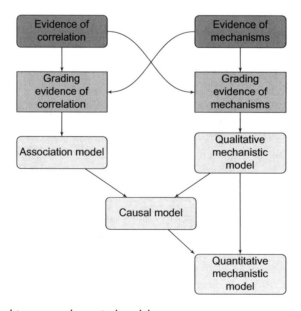

Figure 25.2 Relationships among theoretical models

of mechanisms (which can also be gained from basic lab research, imaging, autopsy, etc.) will need to be evaluated in order to construct a qualitative mechanistic model. Here, evidence of correlation is important in identifying the most salient components in the mechanisms and in determining the net effect of several interacting mechanisms or components of a mechanism. With an association model and a qualitative mechanistic model in place, one is in a position to construct a causal model and to determine whether each claim made by the model can be considered established or more conjectural. The quantitative causal model will go on to inform a quantitative mechanistic model.

The Bayesian network family of models can be used as a unifying formal framework that dovetails with this epistemological picture. As discussed above, a standard Bayesian network can be used as an association model. One way of constructing a Bayesian network from a range of data sets is provided by the *objective Bayesian network* approach: an objective Bayesian network represents the probability distribution that best fits the range of data available, where this optimal distribution is determined by the principles of objective Bayesianism (Williamson, 2005; Nagl et al., 2008). Next, a causal Bayesian network can be constructed from the Bayesian network association model and a qualitative mechanistic model. Finally, a recursive Bayesian network might be employed as a quantitative mechanistic model.

We should conclude this section by noting that the unified account presented above is not the standard way to approach the problem. Where causal Bayesian networks are advocated, it is usually in the context of a data mining approach: the idea is to learn a causal Bayesian network directly from data, in a similar way to the way in which association models are often constructed directly from data (e.g., Spirtes et al., 1993). In contrast, we advocate developing a qualitative mechanistic model on the way to producing a causal model. This is because we hold that causal relationships track mechanistic connections as well as associations, and because one needs to establish that a posited causal connection does indeed track these two things before one can consider the causal claim to be established.

8. An Example: Benzene and Leukemia

One example that illustrates the approach of this chapter concerns benzene and leukemia. Benzene is a clear and highly flammable liquid. Among other things, benzene is added to gasoline in order to reduce engine knocking, and it is also used in the manufacture of organic chemicals. A number of studies established a relationship between benzene exposure and leukemia in humans (Infante et al., 1977; Rinsky et al., 1981). These results are corroborated by studies in mice and other rodents (Goldstein et al., 1982; Cronkite et al., 1984). The association between benzene exposure and leukemia can be charted in an association model, which can be used to predict the disease given the environmental exposure. The relationship may even be charted in a causal model, e.g., a causal Bayesian network, insofar as the relationship is causal. But such a causal model could not yet provide anything other than an impoverished explanation, since it claims only that there exists a mechanism linking benzene exposure and leukemia, rather than providing the details of that mechanism.

This is an example of a more general problem in epidemiology, which is the study of health and disease in defined populations. A key working hypothesis in epidemiology is that diseases are often the result of environmental exposures. However, despite much epidemiological research, the biological processes linking many environmental exposures and diseases remain unknown. This is the case despite the technological advances in measuring certain biomarkers, i.e., biological markers of events at the molecular and physiological levels. Unfortunately, the details of these biological processes are required in order to provide less-impoverished explanations of the occurrence of disease.

Molecular epidemiology is a response to this state of affairs (Schulte and Perera, 1993). Molecular epidemiology is a branch of epidemiology that uses advances in biomarker technology in order to elucidate the biological mechanisms between environmental exposures and diseases. An important methodology in molecular epidemiology involves utilizing complementary studies in order to validate biomarkers that mediate between environmental exposures and disease outcomes (Vineis and Perera, 2007). For example, some studies may provide information associating a certain biomarker to a particular environmental exposure. Other studies may provide information relating a disease outcome to the same biomarker. By bringing together the results of these studies, the disease may be associated with the environmental exposure while providing some insight into the biological processes responsible for this association by highlighting the intermediate biomarkers (Russo and Williamson, 2012).

In the case of benzene and leukemia, studies revealed that certain chromosome aberrations were predictive of cancer in humans (Bonassi et al., 2000). In other case-control studies, those same chromosome aberrations were seen to be more frequently present in leukemia patients who had been exposed to benzene (Zhang et al., 2007). These results are corroborated by animal models (see, e.g., Eastmond et al., 2001). Not only, then, was a chain of associations established between benzene and leukemia, but also some insight was provided into the biological mechanism underlying this chain, i.e., the role of chromosomal aberrations; Vineis and Perera, 2007. These insights can be represented in a mechanistic model, and the model may be used to provide a less-impoverished explanation of leukemia in terms of exposure to benzene. Furthermore, details of the mechanism underlying the association between benzene exposure and leukemia, along with the details of the association, can all be charted in a quantitative mechanistic model.

References

Alpaydın, E. (2010). *Introduction to machine learning*, second edition. MIT Press, Cambridge, MA.

Ankeny, R., and Leonelli, S. (2011). What's so special about model organisms? *Studies in History and Philosophy of Science*, 42:313–323.

Bernard, C. (1865). *An Introduction to the study of experimental medicine (1949)*. Henry Schuman, New York.

Bonassi, S., Hagmar, L., Strömberg, U., Montagud, A. H., Tinnerberg, H., Forni, A., Heikkilä, P., Wanders, S., Wilhardt, P., Hansteen, I.-L., Knudsen, L. E., and Norppa, H. (2000). Chromosomal aberrations in lymphocytes predict human cancer independently of exposure to carcinogens. *Cancer Research*, 60(6):1619–1625.

Casini, L., Illari, P. M., Russo, F., and Williamson, J. (2011). Models for prediction, explanation and control: Recursive Bayesian networks. *Theoria*, 26(1):5–33.

Clarke, B. (2011). *Causality in medicine with particular reference to the viral causation of cancers*. PhD thesis, Department of Science and Technology Studies, University College London, London.

Clarke, B., Gillies, D., Illari, P., Russo, F., and Williamson, J. (2014a). Mechanisms and the evidence hierarchy. *Topoi*, 33(2):339–360.

Clarke, B., Leuridan, B., and Williamson, J. (2014b). Modelling mechanisms with causal cycles. *Synthese*, 191(8):1651–1681.

Cronkite, E., Bullis, J., Inoue, T., and Drew, R. (1984). Benzene inhalation produces leukemia in mice. *Toxicology and Applied Pharmacology*, 75(2):358–361.

Darby, G., and Williamson, J. (2011). Imaging technology and the philosophy of causality. *Philosophy & Technology*, 24(2):115–136.

Darwiche, A. (2009). *Modeling and reasoning with Bayesian networks*. Cambridge University Press, New York.

Djulbegovic, B., Hozo, I., and Ioannidis, J. P. A. (2014). Improving the drug development process: More not less randomized trials. *Journal of the American Medical Association*, 311(4):355–356.

Eastmond, D., Schuler, M., Frantz, C., Chen, H., Parks, R., Wang, L., and Hasegawa, L. (2001). Characterization and mechanisms of chromosomal alterations induced by benzene in mice and humans. *Research Report, Health Effects Institute*, 103:1–68.

Frigg, R., and Hartmann, S. (2005). Scientific models. In Sarkar, S., editor, *The philosophy of science: An encyclopedia*, pages 740–749. Taylor & Francis, New York.

Gillies, D. A. (2011). The Russo-Williamson thesis and the question of whether smoking causes heart disease. In Illari, P. M., Russo, F., and Williamson, J., editors, *Causality in the sciences*, pages 110–125. Oxford University Press, Oxford.

Goldstein, B., Snyder, C., Laskin, S., Bromberg, I., Albert, R., and Nelson, N. (1982). Myelogenous leukemia in rodents inhaling benzene. *Toxicology Letters*, 13(3–4):169–173.

Hirst, J. A., Howick, J., Aronson, J. K., Roberts, N., Perera, R., Koshiaris, C., and Heneghan, C. (2014). The need for randomization in animal trials: An overview of systematic reviews. *PLoS ONE*, 9(6):1–11.

Holmes, F. (1993). The old martyr of science: The frog in experimental physiology. *Journal of the History of Biology*, 26:311–328.

Illari, P. M., Russo, F., and Williamson, J., editors. (2011). *Causality in the sciences*. Oxford University Press, Oxford.

Illari, P. M., and Williamson, J. (2012). What is a mechanism? Thinking about mechanisms *across* the sciences. *European Journal for Philosophy of Science*, 2:119–135.

Infante, P. F., Wagoner, J. K., Rinsky, R. A., and Young, R. J. (1977). Leukaemia in benzene workers. *The Lancet*, 310(8028):76–78.

Kohler, R. (1994). *Lords of the fly: Drosophila genetics and the experimental life*. University of Chicago Press, Chicago, IL.

LaFollette, H., and Shanks, N. (1996). *Brute science. Dilemmas of animal experimentation*. Routledge, London.

Levy, A., and Currie, A. (2015). Model organisms are not (theoretical) models. *British Journal for the Philosophy of Science*, 66:327–348.

Machamer, P., Darden, L., and Craver, C. (2000). Thinking about mechanisms. *Philosophy of Science*, 67:1–25.

Mäki, U. (2001). Models. In *International encyclopedia of the social and behavioural sciences*, volume 15, pages 9931–9937. Elsevier, Oxford, UK.

Nagl, S., Williams, M., and Williamson, J. (2008). Objective Bayesian nets for systems modelling and prognosis in breast cancer. In Holmes, D., and Jain, L., editors, *Innovations in Bayesian networks: Theory and applications*, pages 131–167. Springer, Berlin.

Neapolitan, R. E. (2004). *Learning Bayesian networks*. Pearson / Prentice Hall, Upper Saddle River, NJ.

Pearl, J. (2000). *Causality: Models, reasoning, and inference*. Cambridge University Press, Cambridge.

Reichenbach, H. (1956). *The direction of time*. University of California Press, Berkeley and Los Angeles, 1971 edition.

Rinsky, R., Young, R., and Smith, A. (1981). Leukemia in benzene workers. *American Journal of Industrial Medicine*, 2(3):217–245.

Russo, F., and Williamson, J. (2007). Interpreting causality in the health sciences. *International Studies in the Philosophy of Science*, 21(2):157–170.

Russo, F., and Williamson, J. (2011). Generic versus single-case causality: The case of autopsy. *European Journal for Philosophy of Science*, 1(1):47–69.

Russo, F., and Williamson, J. (2012). Envirogenomarkers: The interplay between mechanisms and difference making in establishing causal claims. *Medicine Studies*, 3:249–262.

Salmon, W. C. (1998). *Causality and explanation*. Oxford University Press, Oxford.

Schulte, P., and Perera, F., editors. (1993). *Molecular epidemiology: Principles and practices*. Academic Press, Cambridge, MA.

Spirtes, P., Glymour, C., and Scheines, R. (1993). *Causation, prediction, and search*. MIT Press, Cambridge MA, second 2000 edition.

Taylor, R. C., Cullen, S. P., and Martin, S. J. (2008). Apoptosis: Controlled demolition at the cellular level. *Nature Reviews Molecular Cell Biology*, 9(3):231–241.

Vineis, P., and Perera, F. (2007). Molecular epidemiology and biomarkers in etiologic cancer research. *Cancer Epidemiology, Biomarkers and Prevention*, 16:1954–1965.

Weber, M. (2005). *Philosophy of experimental biology*. Cambridge University Press, Cambridge, UK.

Williamson, J. (2005). Objective Bayesian nets. In Artemov, S., Barringer, H., d'Avila Garcez, A. S., Lamb, L. C., and Woods, J., editors, *We will show them! Essays in honour of Dov Gabbay*. volume 2, pages 713–730. College Publications, London.

Williamson, J. (2013). How can causal explanations explain? *Erkenntnis*, 78:257–275.

Zhang, L., Rothman, N., Li, G., Guo, W., Yang, W., Hubbard, A. E., Hayes, R. B., Yin, S., Lu, W., and Smith, M. T. (2007). Aberrations in chromosomes associated with lymphoma and therapy-related leukemia in benzene-exposed workers. *Environmental and Molecular Mutagenesis*, 48(6):467–474.

Further Reading

A good introduction to scientific models more generally is given by Frigg, R., and Hartmann, S. (2005). Scientific models. In Sarkar, S., editor, *The philosophy of science: An encyclopedia*, pages 740–749. Taylor & Francis, New York.

More detail on Bayesian networks and causal models is given by Pearl, J. (2000). *Causality: Models, reasoning, and inference*. Cambridge University Press, Cambridge.

Russo and Williamson put forward the epistemological thesis that establishing a causal claim in medicine typically requires establishing both an appropriate correlation and an underlying mechanism. Russo, F., and Williamson, J. (2007). Interpreting causality in the health sciences. *International Studies in the Philosophy of Science*, 21(2):157–170.

Some more information on this thesis is given by the following: Clarke, B. (2011). *Causality in medicine with particular reference to the viral causation of cancers*. PhD thesis, Department of Science and Technology Studies, University College London, London; Clarke, B., Gillies, D., Illari, P., Russo, F., and Williamson, J. (2014a). Mechanisms and the evidence hierarchy. *Topoi*, 33(2):339–360; Gillies, D. A. (2011). The Russo-Williamson thesis and the question of whether smoking causes heart disease. In Illari, P. M., Russo, F., and Williamson, J., editors, *Causality in the sciences*, pages 110–125. Oxford University Press, Oxford; Russo, F., and Williamson, J. (2011). Generic versus single-case causality: The case of autopsy. *European Journal for Philosophy of Science*, 1(1):47–69; Russo, F., and Williamson, J. (2012). Envirogenomarkers: The interplay between mechanisms and difference making in establishing causal claims. *Medicine Studies*, 3:249–262.

26
DISCOVERY IN MEDICINE
Brendan Clarke

Introduction

Medical textbooks often begin with a concise account of key discoveries in the field. In the first chapter of the textbook *Medical Microbiology* (Greenwood, Slack and Peutherer 2000), for example, several of the important milestone discoveries in microbiology are set out. I quote one of these below:

> Among notable events were: the discovery by James Paget (while a first-year medical student at St Bartholomew's Hospital, London) of the larvae of *Trichinella spiralis* in muscle during an autopsy (1835).
>
> (Greenwood et al. 2000: 4)

Excerpts such as these illustrate the importance of discovery to medical practice. We can find other signifiers, too, of the esteem in which discoveries (and discoverers) are held, such as the Nobel Prize in physiology or medicine. We might therefore expect discovery to be of the foremost importance for philosophers of medicine. So it might come as a surprise that, far from being a central topic, discovery has been often neglected by philosophers. In the first section of this chapter, I will discuss when and why this was the case. In order to do this, I will give a brief account of the early history of philosophy of science, concentrating in particular on the positivist roots of much contemporary philosophical work. In section two, I will then interrupt this historical story about the place of discovery in the philosophy of science to introduce an extended case study of discovery in medicine. This is the case of the discovery of McArdle disease during the 1950s. Finally, in section three, I will use the details of the McArdle case to introduce two substantive accounts of discovery that offer complementary understandings of discovery (and its role) in the sciences.

1. The (pre-) History of the Philosophy of Discovery

In the Introduction, I noted that the topic of discovery has been neglected by many philosophers of science. In this section, I will substantiate and clarify this claim by emphasizing that the status of discovery, as a topic of philosophical study, has ebbed and flowed across the 20th century. I begin by suggesting that the neglect of discovery can be effectively localized to philosophical work influenced by logical positivism during the early 20th century. I will discuss the reasons that discovery was cast out by these authors by concentrating on the distinction that they drew between discovery and other kinds of scientific work. I will then conclude this section by noting what happened once this distinction was challenged.

Much in contemporary philosophy of science is rooted in the influence of the Vienna Circle (Reisch 2005 gives an accessible introduction). A detailed discussion is beyond the scope of this chapter, but the Vienna Circle was a group of scientists and philosophers who met in Vienna in the period between the first and second world wars. Skeptical of the traditional metaphysical interests of philosophers, they sought to develop a philosophical method compatible with the rigorous and mathematical techniques used in the sciences. Drawing on Whitehead and Russell's attempt to axiomatize mathematics using symbolic logic (Whitehead and Russell 1910), the philosophers of the Vienna Circle sought to bring similar methods of analysis to bear on the structure of scientific reasoning. This importance placed on symbolic logic led to the neglect of discovery as a philosophical question. Hempel, for example, describes the way that new scientific ideas arise. New scientific ideas are first invented by "guesses" (Hempel 1966: 15), "free invention," or the "imagination" (Hempel 1966: 16) of scientists. These guesses do not come about from "any process of systematic inference" (Hempel 1966: 15). By this, Hempel means to reject the "narrow inductivist account of scientific inquiry" (Hempel 1966: 11) that might claim the existence of "generally applicable 'rules of induction' by which hypotheses or theories can be mechanically derived or inferred from empirical data" (Hempel 1966: 15). These "guesses" are then strenuously tested by a process of "critical scrutiny" via "careful observation or experiment" (Hempel 1966: 16). This two-step process—free invention of a hypothesis and then its strict testing—Hempel terms "the method of hypothesis" (Hempel 1966: 17). According to Hempel (and other followers of the Vienna Circle), as the creation of theories is untrammelled and creative, the method of their acceptance bears the onus of ensuring that they are correct. Hempel summarizes:

> . . . the scientist may give free rein to his imagination . . . yet scientific objectivity is safe-guarded by the principle that while hypotheses and theories may be freely invented and *proposed* in science, they can be *accepted* into the body of scientific knowledge only if they pass critical scrutiny, which includes in particular the checking of suitable test implications by careful observation or experiment.
>
> (16)

Hempel illustrates this process by reference to the process by which Semmelweis discovered the cause of childbed fever (Hempel 1966, chapter 2, see also discussion in Gillies 2005). Here, much more of the chapter is concerned with the manner in which various hypotheses were tested, rather than how they arose.

It is worth emphasizing at this point that the "guesses," "free invention," or "imagination" required to invent new scientific ideas made their genesis not at all compatible with the methods of logical analysis used by followers of the Vienna Circle. This difficulty was avoided by drawing a distinction between discovery and justification, and by confining the use of the methods of logical analysis to the context of justification. According to Brown (1994: 1), "In this view, discoveries come from chance or intuition; what counts as science is their logic and validity—not their discovery, but their justification."

This distinction between discovery and justification was first expressed by Hans Reichenbach (1938), but was most forcefully stated by another philosopher associated with philosophical work in Mitteleuropa during the 1930s: Karl Popper. Brown (1994: 1) suggests that Popper was responsible for the way that discovery was "banished . . . from the kingdom of philosophy to the netherworld of 'empirical psychology.'"

With a turn towards historically influenced philosophy of science during the early 1960s, several accounts were written that were critical of this strict distinction between contexts. One such author was the philosopher Norwood Russell Hanson. His complaint was that the dominance of the context of justification, at the cost of the context of discovery, was mistaken. Hanson argued that philosophers working in the positivist tradition that arose from the Vienna Circle unfairly neglected the process of scientific discovery in favor of analyzing the logic of completed scientific work (Hanson 1958a, 1958b, 1960a, 1960b). Of this kind of philosophical work, he wrote that it "reads less like a Logic of Discovery than like a Logic of the Finished Research Report" (Hanson 1958a: 1073). Far from being an inexplicable creative act, fit only for the attentions of psychologists, the study of discovery in the broad sense was an essential task for philosophy of science and a necessary complement to work in the context of justification:

> More philosophers must venture into these unexplored regions in which the logical issues are often hidden by the specialist work of historians, psychologists, and the scientists themselves. We must attend as much to how scientific hypotheses are caught, as to how they are cooked.
>
> (Hanson 1958a: 1089)

We can see the emphasis on "cooking" (justification) at the expense of "catching" (discovery) in Hempel's account (discussed above). Discovery, and more specifically the project of accounting for discovery in terms of inductive logic, became a key issue for historically motivated philosophers of science like Kuhn and Hanson. In terms of finding an inductive logic of discovery, capable of algorithmically arriving at novel discoveries from observation, this project was unsuccessful (Downes 1990; Laudan 1980; Lugg 1985; Shah 2007). But in popularizing new approaches to philosophy of science, unlike those wielded by the logical positivists, it was a great success, which set the scene for the flourishing of historically informed philosophy of science that would occur from the late 1960s. We can see this flourishing in the many philosophical accounts of discovery that followed these pioneers (Brown 1994; Darden 1976; Maxwell 1974; Nersessian 1984; Siegel 1980; Simon 1973; Thagard 1982, 2003).

2. The Discovery of McArdle Disease

Let's interrupt this philosophical story to discuss an individual discovery in some detail. This is the discovery of McArdle disease, which occurred in the United Kingdom and the United States during the 1950s. In contrast to the short example of discovery given in the introduction, the recurring motif here is vagueness: of what is discovered, and by whom, and when. We will examine the implications of this vagueness in section three.

McArdle disease is a rare disorder of glycogen metabolism characterized by fatigue and cramps on exertion. Skeletal muscle contains large quantities of glycogen, a polymer of glucose, which provides a reserve of energy to power the muscle during exertion. Two enzymes maintain this energy store. At rest, glycogen synthase adds glucose molecules to glycogen, while a second enzyme called myophosphorylase removes them during exertion. McArdle disease is caused by a functional lack of this myophosphorylase enzyme. Thus, individuals with the disorder can form glycogen normally, but they are unable to break glycogen back to glucose. This means that their skeletal muscles function normally until called to perform strenuous activity, upon which exertion their muscles rapidly run short of energy, leading to fatigue and pain.

McArdle disease is named after Brian McArdle (1911–2002), whose 1951 paper (McArdle 1951) was the first case report of this (apparently novel) condition. This case report recounts the extensive clinical investigation of a 30-year-old male patient, George W., at Guy's Hospital in London. He had presented to the hospital in September 1947 complaining of lifelong fatigue, stiffness, and pain on exertion:

> For as long as the patient could remember, light exercise of any muscle had always led to pain in the muscle and, if the exercise were continued, to weakness and stiffness. For example walking a few hundred yards, particularly if fast or uphill, provoked pain in the calves and thighs, and lifting heavy weights resulted in pain in the arms. Even chewing sometimes gave rise to pain in the masseters. The pain, at first dull and aching, increased with continued exercise, while the muscles became progressively stiffer and weaker. Usually all the symptoms rapidly disappeared on resting, but when he continued the exercise not only did the symptoms increase in severity, but they persisted longer when he was finally forced to rest.
>
> (McArdle 1951: 13)

Despite an approximately normal physical examination, McArdle suspected that this individual was suffering from a disorder of muscle metabolism. This suspicion was compatible with George W.'s biochemical investigations. First, the level of lactate (a breakdown product of glucose) found in the blood during exercise was significantly less than expected, indicating that the patient was utilizing less glucose than a normal control subject. Second, the patient experienced electrically silent muscle cramps during exercise, indicating that his muscles were failing to contract normally.

On the basis of these findings, McArdle made three claims about George W.'s illness. First, he claimed that these symptoms were both real—rather than spurious—and arose from the operations of a physical—rather than mental—disease. Second, he claimed that this disease process was novel and not, say, an unusual presentation of a known disease. Third, he claimed that the etiological process responsible was some kind of defect in the glucolytic pathway. More specifically, McArdle suggested that (for reasons detailed in the next paragraph) a deficiency of an enzyme known as glyceraldehyde phosphate dehydrogenase (GPD), which is involved in producing energy from glucose, was the cause of the disease.

Although the first two of these claims—that the symptoms were caused by a novel, physical disease—became accepted without serious controversy, McArdle's claim that the disease was caused by GPD deficiency was more problematic. As I noted above, GPD metabolizes glucose, rather than glycogen. This puts McArdle's third claim strangely at odds with the title of McArdle's paper ("Myopathy Due to a Defect in Muscle Glycogen Breakdown"). It is worth therefore briefly quoting McArdle's argument in favor of this claim. In contrast to the empirical evidence provided to support the other claims, the evidence in favor of GPD deficiency was largely analogical. George W.'s symptoms were noted to be similar to the effects of iodoacetate poisoning. As this agent was known to affect GPD function, McArdle rather tentatively concluded:

> Theoretically, the phenomena following iodoacetate poisoning of this enzyme [GPD] could also be caused by interference with the other components of the enzyme system. It is suggested therefore that it is the glyceraldehyde phosphate dehydrogenase system that is the site of the biochemical lesion in the muscles of G.W.
>
> (McArdle 1951: 32)

It was not until the late 1950s that subsequent cases of McArdle disease were reported. Toward the end of the 1950s, two other groups of researchers reported cases of McArdle disease. One was patient D.G., a 19-year-old man with a lifelong reduction in his tolerance for physical exercise, who came to the attention of clinicians at the University of California, Los Angeles (Mommaerts et al. 1959; Pearson et al. 1959; Pearson, Rimer and Mommaerts 1961). The second patient, A.D., was a 54-year-old man investigated at Harvard Medical School in Boston

(Larner and Villar-Palasi 1959; Schmid and Hammaker 1961; Schmid and Mahler 1959a, 1959b; Schmid et al. 1959).

Many of these two patients' symptoms were broadly similar to those described by McArdle, although with minor variations in degree and kind. For example, the severity of the condition appeared to increase with age. Understanding this variability was important for the understanding of McArdle disease in general terms, rather than on the basis of isolated, individual cases. The most important point of difference between this general account and McArdle's account concerns the etiology of the syndrome. Rather than abnormal GPD, it was argued that the characteristic symptoms were due to a lack of myophosphorylase (Mommaerts et al. 1959: 792). This claim was based on the following evidence. First, in skeletal muscle samples from patients, very little phosphorylase activity was detected (Mommaerts et al. 1959: 793–5). Second, normal function was restored if myophosphorylase was added, indicating that this lack of activity was due to an absence of functional myophosphorylase, rather than any sort of regulatory problem (Mommaerts et al. 1959: 793–5). Third, the defect could similarly be bypassed if downstream products of glycogen metabolism were added (Mommaerts et al. 1959: 793–5), indicating that the metabolic problem was specific to the glycogenolysis pathway, rather than the glucolytic pathway. Finally, it was noted that muscle glycogen stores were very much higher than normal (Mommaerts et al. 1959: 793–5), suggesting that the pathology resulted from a primary problem with the way that glycogen was broken down to glucose (catabolism), rather than a problem from the way that glycogen was made (glycogen synthesis).

In conclusion, from a total of nine papers, detailing the investigation of three individuals with McArdle disease, the following was claimed about the condition. First, that it was a real (rather than spurious), physical (rather than mental), and etiologically distinctive disease with a variety of clinical features. Second, that the etiology responsible was an isolated absence of myophosphorylase and that the disorder could be characterized in such terms. Third, that various consequences of this abnormality could be identified, including clinical features (fatigue, muscle weakness, myoglobinuria), heritability, and possible avenues for treatment. These diverse features could all be explained by reference to the mechanistic consequences of myophosphorylase deficiency (see Chapter 5). Finally, a range of further disease features were identified for further investigation, including the progressive nature of the disease, muscle wasting, and so on.

3. Philosophers of Discovery

In this section, I will sketch out two accounts of discovery and use them to discuss the McArdle case. As these accounts aim their responses against Hempel-like ways of understanding discovery (which we might term the "creative" view), it will be helpful to distinguish two slightly different positions that have already been sketched out in the introduction and in section one. These are:

Point conception of discovery: a sudden moment of inspiration in which a new idea is revealed to a researcher

Positivist conception of discovery: discovery is a psychological process and, as such, is not amenable to logical analysis

First comes the account of discovery produced by Thomas Kuhn (1962a). Kuhn argues against the point conception of discovery. Far from being sudden moments of individual inspiration, Kuhn argues that discoveries are typically evolving processes that are necessarily extended in time. Because of this duration, they are therefore subject to historical (and philosophical) analysis.

Hanson (1960a), on the other hand, takes aim primarily at the positivist conception of discovery. Hanson suggests that, although there is no simple inductive logic of discovery, it does not follow that there is no logic capable of accounting for discoveries. In fact, he suggests that a kind of logical inference called "retroduction" might be a viable candidate for such a logic of discovery.

In both cases, I will illustrate by means of the McArdle case. Finally, I make some concluding remarks linking these two accounts of discovery to the developing philosophy of scientific practices approach. It is worth emphasizing that Kuhn and Hanson's accounts are different and (strictly speaking) incompatible. However, they share the intention of showing that discovery, far from being just an act of creative imagination, could be an object of philosophical scrutiny.

Even a glance at the first page of Kuhn's paper shows that he disagrees with the model of point discovery: "To the historian discovery is seldom a unit event attributable to some particular man, time, and place" (Kuhn 1962a: 760). Kuhn begins with a puzzle. Knowing who discovered something is important. Historians (and others, like the Nobel Prize committee) spend a great deal of effort trying to find out about discoveries. Despite this attention, though, often it has not proven possible to pinpoint "the time and place at which a given discovery could properly be said to have 'been made'" (Kuhn 1962a: 761). Why is this?

Kuhn's response to this puzzle is as follows: this apparent failure of finding out arises because of a category mistake about discovery. Far from being point events, occurring in a particular place at a particular time, most discoveries occur gradually and involve more than one researcher:

> ...there is no single moment or day which the historian, however complete his data, can identify as the point at which the discovery was made. Often, when several individuals are involved, it is even impossible unequivocally to identify any one of them as the discoverer.
>
> (Kuhn 1962a: 763)

The key to understanding Kuhn's point here is the phrase "most discoveries," because it indicates a classification of kinds of discovery. There are **troublesome** discoveries and **expected** discoveries. Kuhn elucidates:

> The troublesome class consists of those discoveries . . . which could not have been predicted from accepted theory in advance and which therefore caught the assembled profession by surprise.
>
> (Kuhn 1962a: 761)

Expected discoveries, on the other hand, are those that are predicted by theory. Kuhn's examples here are rather schematic, but a good example might be the discovery of a new element that fills a blank spot on the periodic table or the discovery of a planet via some mathematical prediction. The point is that the discoverer in these cases both (a) "knew from the start what to look for" (Kuhn 1962a: 761) and (b) had "criteria which told them when their goal had been reached" (Kuhn 1962a: 761). As a result, this kind of discovery does not suffer from the problems that characterize troublesome discoveries. A consequence of some interest to the historian is that resolving questions about, say, the date of an expected discovery, or of the identity of its discoverer, are only limited by the available evidence.

Things are rather different for the troublesome discovery. Kuhn illustrates by giving three examples—the discovery of oxygen, the discovery of Uranus, and the discovery of X-rays. In these cases, neither of Kuhn's two characteristics of expected discovery (a and b above) were the case. Instead, even when there is plenty of historical evidence available, the kinds of questions that we like to answer about discoveries cannot be answered in the unequivocal manner that we associate with expected discoveries. As Kuhn suggests, this is largely a matter of recognizing that discovery is usually a more complex process than stories of point discovery would suggest. Discoveries—far from being points or "Eureka" moments—have a proper internal history. Kuhn suggests that this history of discovery occurs in three phases:

1. Awareness of anomaly (Kuhn 1962a: 762–3)
2. Making the anomaly behave (Kuhn 1962a: 763)
3. Adjustment, adaptation, and assimilation (Kuhn 1962a: 763)

Anomalies are "nature's failure to conform entirely to expectations" (Kuhn 1962a: 762). For a discovery to occur, a scientist must first notice something novel or unexpected. This unexpected finding is an anomaly. Kuhn gives more detail at this point about exactly what is involved in this process. First, a scientist must have available the necessary tools—conceptual or tangible—to make some anomaly occur. Examples might include developing the necessary astronomical theories to be able to recognize that an orbital perturbation is really a perturbation, or having available photographic film that might become anomalously clouded (see Kuhn's examples). Next, once the anomaly has been produced, the scientist must have the "individual skill, wit, or genius to recognize that something has gone wrong in ways that may prove consequential" (Kuhn 1962a: 763).

In relation to the McArdle case, George W.'s clinical features were the anomalies. Here, though, a detailed historical story about how each clinical finding was noted is important because it illustrates Kuhn's argument about the difference between the recognition, and the production, of anomalies. Although George W. had some features that could have been recognized by almost anyone likely to have met him (tiredness, pain on exertion, and so on), most of the telling anomalies required special equipment to produce and to detect. For example, the finding that George W. was suffering electrically silent muscle cramps depended on techniques capable of measuring tiny electrical currents in muscles, a technology intimately related to the development of the electrocardiogram (ECG) during the first decade of the 20th century (Porter 1997: 582). Similarly, discovering that George W. had abnormal serum levels of lactate after exertion required laboratory techniques capable of accurately measuring the concentration of small molecules in blood. Again, this was a development of the early 20th century (Porter 1997: 582). Thus, this discovery was highly dependent on the available technology.

Once an anomaly has been produced and recognized, the process of discovery moves into a second phase: that of making the anomaly behave. This involves a complicated period of negotiation in order to "make the anomaly behave" in a regular fashion. For example, once a researcher has made and recognized an anomaly, other researchers might seek to replicate it under a range of different conditions, try and understand it using their theories, compare it to other similar phenomena, and so on. Kuhn's account of the way in which anomalies are made to behave is rather schematic, but a similar process can be found in the McArdle case. To recap, McArdle differed from later researchers as to the etiology of the clinical syndrome that he described:

McArdle: the syndrome is caused by abnormal functioning of GPD, largely by analogy with the known effects of iodoacetate poisoning
(Others): the syndrome is caused by abnormal functioning of myophosphorylase, largely from *in vitro* investigations of the behavior of muscle enzymes

The final stage of Kuhn's model of discovery is to fit behaving anomalies into existing knowledge. The working of this phase is best illustrated by thinking about the differences between expected and troublesome discoveries. Once expected discoveries have been discovered, they become an addition to the store of scientific knowledge. By contrast, troublesome discoveries may also transform other parts of scientific knowledge in more revolutionary ways. Again, there are resonances here with Kuhn's work on scientific revolutions (Kuhn 1962b):

> In a sense that I can now develop only in part, they also react back upon what has previously been known, providing a new view of some previously familiar objects and simultaneously changing the way in which even some traditional parts of science are practiced. Those in whose area of special competence the new phenomenon falls often see both the world and their work differently as they emerge from the extended struggle with anomaly which constitutes that phenomenon's discovery.
>
> (Kuhn 1962a: 763)

That concludes this rather skeletal summary of Kuhn's article. As the above quote suggests, this work on discovery was part of a larger project for Kuhn. Discovery was important for Kuhn because new discoveries contributed to grand changes in scientific theories—scientific revolutions. As Bîgu summarizes:

> One of Kuhn's well-known theses is that a distinction can be drawn between periods of normal science and periods of extraordinary science. The first periods are characterized by the existence of a substantial set of shared commitments, on the basis of which the scientific activity is carried out. Still, after a period of research, anomalies, i.e. difficulties met by scientists when trying to solve the problems of normal research, become more numerous and serious. This will lead to a scientific crisis, and, possibly, to a scientific revolution.
>
> (Bîgu 2013: 331)

Perhaps it is overstating the case to say that the discovery of McArdle disease led to a revolution. Yet it was in the vanguard of diseases known to be caused by single gene abnormalities. And the consequences of trying to account for disease in genetic terms are not yet clear to us. Perhaps, to medical historians of the far future, the case of McArdle disease will have played some small role in leading to a revolution in the way that genetics might influence health.

If Kuhn's account of discovery is mainly concerned with showing that discoveries are usually not point events, Hanson aims at describing a logical structure of these processes. To do this, he uses a form of inference that he terms "retroductive reasoning." This form of inference is as follows:

1. Some surprising, astonishing phenomena $p_1, p_2, p_3 \ldots$ are encountered.
2. But $p_1, p_2, p_3 \ldots$ would not be surprising were an hypothesis of H's type to obtain. They would follow as a matter of course from something like H and would be explained by it.
3. Therefore, there is good reason for elaborating an hypothesis of type H—for proposing it as a possible hypothesis from whose assumption $p_1, p_2, p_3 \ldots$ might be explained.

(Hanson 1960a: 104)

Let's illustrate using the McArdle case, before moving on to discuss why this might be an effective rejoinder to the positivist philosophers with whom Hanson was arguing.

First, we need some surprising phenomena. As with the discussion of Kuhn's account of discovery above, I will take these to be the various clinical features belonging to George W. Next, we need some kind of hypothesis that is capable of explaining these clinical features. Happily for us as philosophers (although maybe not as medical practitioners)m we have at least two of these H's:

H_1: the syndrome is caused by abnormal functioning of GPD (McArdle).
H_2: the syndrome is caused by abnormal functioning of myophosphorylase (Mommaerts and colleagues).

Having two competing hypotheses is useful to see how Hanson intended this logical framework to operate. As he admits in the final paragraph of his paper, the aim here is to provide a way of discarding hypotheses in a logically definite way: "With such a rich profusion of data and technique as we have, the arguments necessary for eliminating hypotheses of the wrong type become a central research inquiry" (Hanson 1960a: 106). And we can see, by a careful study of the McArdle disease literature, that H_2 does a superior job to H_1 in explaining how our surprising phenomena come about. Thus, we can give a logical account of how part of the discovery of McArdle disease came about.

Is this an effective rejoinder to the positivist? Well, on the one hand, it does seem to provide a way of understanding how part of the process of discovery might happen in logical terms. On the other hand, this is clearly not the algorithmic kind of discovery that Hempel characterized as the "narrow inductivist view" (Hempel 1966: 15–1K). Retroductive reasoning also does not offer us a comprehensive logical account of discovery, because we still have to create some hypotheses before we can examine them against our surprising phenomena. Perhaps this just moves the context distinction further forward, such that part (albeit a smaller part) of discovery remains a creative, psychological, non-logical process.

Conclusion

In this chapter, I have tried to sketch out the beginnings of discovery as a topic for philosophers of science. This work has continued, with an increasing number of philosophers treating discovery as an object of philosophical inquiry. With reference to the philosophy of medicine, one important facet of this work has been the various attempts to produce algorithmic methods by which discovery from data could be (somewhat) automated. Herbert Simon, for example, became interested in logical ways of producing discoveries from patterns in data (Simon 1973). Note, though, that Simon does not claim to produce these discoveries in a completely certain way (as proponents of what Hempel called the "narrow inductivist account of scientific inquiry" [Hempel 1966: 11]). Instead, Simon speaks of a "recommended strategy" that alone "does not guarantee the achievement of the goal" (Simon 1973: 474). Simon's work is also intimately connected to Kuhn and Hanson's rejection of the purely psychological view of discovery popularized by Popper and others:

> ... we see that we must reject Popper's assertion that the "question how it happens that a new idea occurs to a man ... may be of great interest to empirical psychology; but it is irrelevant to the logical analysis of scientific knowledge."
> (Simon 1973: 474)

Simon's (1973) paper is a modest start to a grand project about digital automation of discovery from data. Moving through various technologies, this work has been the subject of much interest in the philosophical community. Just to pick one example of this work on the interface of philosophy and computer science, Paul Thagard and David Croft note that the retroductive inferential pattern that Hanson described for discovery is also to be found in the way in which technological innovations in computer science occur:

> Although abductive inference to explanatory hypotheses is much more central to scientific discovery that to technological innovation, inference to possible solutions to technological problems seems to involve very similar representations and processes. We can therefore conclude that scientific discovery and technological innovation are cognitively very much alike.
>
> (Thagard and Croft 1999: 137)

This consilience between fields is, I think, another good reason to think that there is more to discovery in medicine than a sudden flash of inspiration.

References

Bîgu, D. (2013) "A similarity-based approach of Kuhn's no-overlap principle and anomalies," *Studies in History and Philosophy of Science Part A*, **44**(3): 330–8.

Brown, R.H. (1994) "Logics of discovery as narratives of conversion: Rhetorics of invention in ethnography, philosophy, and astronomy," *Philosophy & Rhetoric*, **27**: 1–34.

Darden, L. (1976) "Reasoning in scientific change: Charles Darwin, Hugo de Vries, and the discovery of segregation," *Studies in History and Philosophy of Science Part A*, **7**(2): 127–69, ISSN 0039-3681, http://dx.doi.org/10.1016/0039-3681(76)90014-5.

Downes, S. (1990) "Herbert Simon's computational models of scientific discovery," *Philosophy of Science Association*, **1990**(1): 97–108.

Gillies, D. (2005) "Hempelian and Kuhnian approaches in the philosophy of medicine: The Semmelweis case," *Studies in History and Philosophy of Science Part C: Studies in History and Philosophy of Biological and Biomedical Sciences*, **36**(1): 159–81.

Greenwood, D., Slack, R., and Peutherer, J. eds. (2000) *Medical Microbiology*. 15th edition. London: Churchill Livingstone.

Hanson, N.R. (1958a) "The logic of discovery," *The Journal of Philosophy*, **55**: 1073–89.

———. (1958b) *Patterns of Discovery: An Inquiry into the Conceptual Foundations of Science*. Cambridge: Cambridge University Press.

———. (1960a) "Is there a logic of scientific discovery?" *Australasian Journal of Philosophy*, **38**: 91–106.

———. (1960b) "More on 'the logic of discovery'," *The Journal of Philosophy*, **57**: 182–8.

Hempel, C.G. (1966) *Philosophy of Natural Science*. Englewood Cliffs, NJ: Prentice-Hall.

Hempel, Carl G. and Fetzer, J.H. (ed.) (2001) *The Philosophy of Carl G. Hempel: Studies in Science, Explanation, and Rationality*. New York: Oxford University Press.

Kuhn, T.S. (1962a) "Historical structure of scientific discovery," *Science*, **136**(3518): 760–4.

———. (1962b) *The Structure of Scientific Revolutions*. Chicago: University of Chicago Press.

Larner, J. and Villar-Palasi, C. (1959) "Enzymes in a glycogen storage myopathy," *Proceedings of the National Academy of Sciences of the United States of America*, **45**: 1234–5.

Laudan, L. (1980) "Why was the Logic of Discovery Abandoned?" In T. Nickles (ed.) *Scientific Discovery, Logic, and Rationality*. Dordrect, Netherlands: Boston Studies in the Philosophy of Science, **56**: 173–83.

Lugg, A. (1985) "The process of discovery," *Philosophy of Science*, **52**(2): 207–20.

Maxwell, N. (1974) "The rationality of scientific discovery Part I: The traditional rationality problem," *Philosophy of Science*, **41**: 123–53.

McArdle, B. (1951) "Myopathy due to a defect in muscle glycogen breakdown," *Clinical Science*, **10**(1): 13–33.

Mommaerts, W.F.N.M., Illingworth, B., Pearson, C. M., Guillory, R. J., and Seraydarian, K. (1959) "A functional disorder of muscle associated with the absence of phosphorylase," *Proceedings of the National Academy of Sciences of the United States of America*, **45**: 791–7.
Nersessian, N.J. (1984) "Aether/or: The creation of scientific concepts," *Studies in History and Philosophy of Science Part A*, **15**: 175–212.
Pearson, C.M., Rimer, D.G. and Mommaerts, W.F.N.M. (1959) "Defect in muscle phosphorylase: A newly defined human disease," *Clinical Research*, **7**: 298.
———. (1961) "A metabolic myopathy due to absence of muscle phosphorylase," *The American Journal of Medicine*, **30**: 502–17.
Porter, R. (1997) *The Greatest Benefit to Mankind: A Medical History of Humanity from Antiquity to the Present*. London: HarperCollins.
Reichenbach, H. (1938) *Experience and Prediction*. Chicago: University of Chicago Press.
Reisch, G.A. (2005) *How the Cold War Transformed Philosophy of Science: To the Icy Slopes of Logic*. Cambridge: Cambridge University Press.
Schmid, R. and Hammaker, L. (1961) "Hereditary absence of muscle phosphorylase (McArdle's syndrome)," *The New England Journal of Medicine*, **264**: 223–5.
Schmid, R. and Mahler, R. (1959a) "Syndrome of muscular dystrophy with myoglobinuria: Demonstration of a glycogenolytic defect in muscle," *The Journal of Clinical Investigation*, **38**: 1040.
———. (1959b) "Chronic progressive myopathy with myoglobinuria: Demonstration of a glycogenolytic defect in the muscle," *The Journal of Clinical Investigation*, **38**: 2044–58.
Schmid, R., Robbins, P.W. and Traut, R.R. (1959) "Glycogen synthesis in muscle lacking phosphorylase," *Proceedings of the National Academy of Sciences of the United States of America*, **45**: 1236–40.
Shah, M. (2007) "Is it justifiable to abandon all search for a logic of discovery?" *International Studies in the Philosophy of Science*, **21**: 253–69.
Siegel, H. (1980) "Justification, discovery and the naturalizing of epistemology," *Philosophy of Science*, **47**: 297–321.
Simon, H.A. (1973) "Does scientific discovery have a logic?" *Philosophy of Science*. **40**: 471–80.
Thagard, P. (1982) "Artificial intelligence, psychology, and the philosophy of discovery," *PSA: Proceedings of the Biennial Meeting of the Philosophy of Science Association 1982*, 166–75.
———. (2003) "Pathways to biomedical discovery," *Philosophy of Science*, **70**(2): 235–54.
Thagard, P. and Croft, D. (1999) "Scientific Discovery and Technological Innovation," In Magnani, L., Nersessian, N.J. and Thagard, P. (eds.) *Model-Based Reasoning in Scientific Discovery*. New York: Kluwer, pp. 125–37.
Whitehead, A.N. and Russell, B. (1910) *Principia Mathematica*. Volume 1. Cambridge: Cambridge University Press.

Further Reading

For a general introduction to the Vienna Circle and the subsequent development of philosophy of science in the 20th century, see the extremely accessible Gillies, D. (1993) *Philosophy of Science in the Twentieth Century*. Oxford: Blackwell.
For a more detailed historical treatment of the Vienna Circle, their followers, and their subsequent influence on philosophy of science in the later 20th century, see Reisch, G.A. (2005) *How the Cold War Transformed Philosophy of Science: To the Icy Slopes of Logic*. Cambridge: Cambridge University Press.
For a set of neatly drawn case studies of important biomedical discoveries, see Norrby, E. (2010) *Nobel Prizes and Life Sciences*. Singapore: World Scientific Publishing.
For far more on the role of discovery in the philosophy of Norwood Hanson, see Lund, M. (2010) *N. R. Hanson: Observation, Discovery, and Scientific Change*. Amherst, NY: Humanity Books.
For more examples of the way that discovery has become part of the mainstream of philosophy of science, see the papers in Magnani, L., Nersessian, N.J. and Thagard, P. (eds.), (1999) *Model-Based Reasoning in Scientific Discovery*. New York: Kluwer pp. 125–137.

27
EXPLANATION IN MEDICINE
Maël Lemoine

Introduction

The scientific part of medicine seeks explanations. Some think that nowadays evidence-based medicine embodies scientific medicine. Yet evidence-based medicine does not focus on explanation. Moreover, as Thompson puts it, "the results of randomized controlled trials cannot underwrite explanations, notwithstanding the current emphasis on them in clinical medicine" (Thompson 2011).

This chapter aims to determine the nature, the function, and the importance of explanation in medical science, from basic biological studies of diseases to epidemiology. Medicine uses diverse explanatory strategies. After a review of the basics of scientific explanation according to general philosophy of science, as applied to medical cases, the chapter will focus on the specifics of medical explanations, as more recent developments in the philosophy of medicine have established and discussed them. A satisfactory account of explanations in medicine should:

1. encompass all types of medical explanation (molecular, epidemiological, psychiatric, pathophysiological, social. . .) or explicitly define a more limited target;
2. state what is specific to medical explanation, if anything, as opposed to non-medical explanation;
3. account for the function of explanation as opposed to other activities such as providing evidence.

We will successively examine several accounts of explanation in medicine, none of which is likely to be entirely satisfactory.

1. Basics of Scientific Explanation

1.1. The Deductive-Nomological (D-N) Model of Explanation

In 1844–1848, the Austrian physician Ignaz Semmelweis was looking for an explanation of the epidemic of puerperal fever that killed significantly more women in one of the two Maternity Divisions of the Vienna General Hospital than in the other. The ward with the higher mortality was staffed by doctors and medical students, and the other one with midwives. After ruling out many possible explanations that neither matched the facts nor prompted actions that succeeded in controlling the infection, Semmelweis hypothesized that students and doctors, who often arrived in the ward directly from conducting autopsies, transported "cadaveric matter" from the dissection tables into the women's bodies. Midwives, on the other hand, did not perform autopsies. This hypothesis suggested thorough hand washing prior to delivery, which led to a sharp decline in mortality in the ward staffed by doctors and students (Hempel 1966; Gillies 2005).

We can see this short case study as, among other things, an explanation. Semmelweis explained the higher rate of infection among one set of women by the transferring of cadaveric material to them by one set of caregivers but not another. Hempel took this as an example of how scientists seek explanations. His conclusion is that they look for a hypothesis that may "explain" the observed facts and deduce from it further observations that can be checked in turn. This so-called hypothetico-deductive view has been much discussed. For instance, Lipton claims that Semmelweis's choice of the cadaveric matter hypothesis over other hypotheses was not logical but aesthetic ("loveliness"), as none of Semmelweis's various hypotheses were, strictly speaking, incompatible with the facts. Thus, Lipton claims, Semmelweis's reasoning is an example of *inference to the best explanation* (i.e., of the conclusion that the *best* explanation is *the* explanation; Lipton 2003), with best, in this case, being loveliest according to Lipton. Adopting a less demanding view of evidence than Lipton, Bird argues that Semmelweis's successive contrastive hypotheses were indeed refuted, so that just one remained in the end, which makes this a case of inference to the *only* explanation (Bird 2007), or eliminative abduction (Bird 2010), a specific form of the inference to the best explanation.

Should an explanation be the loveliest hypothesis or the last one remaining? This may be an important characteristic of what should count as an explanation, but it hardly counts as a definition of explanation. What all three authors above seem to agree on is the fact that hypothesis confirmation implies deduction, because hypotheses are explanations, and explanations imply deduction. In fact, the main classic description of scientific explanation is called the Deductive-Nomological (D-N) view. First put forward precisely by the same Hempel, along with Oppenheim (1948), the D-N view claims that a hypothesis explains an observation by logically implying it. In particular, scientific hypotheses explain by being general statements of the form "All X are Y." If the phenomenon of interest is indeed X, then it follows that it is Y as well. Moreover, if the general statement is verified, it is considered a law of nature. In that case, the phenomenon of interest is considered explained by this law of nature: we know why the phenomenon of interest is Y; i.e., because it is X (observation) and "all X are Y" (law of nature). In the case of childbed fever, for the hypothesis to be a possible explanation, it is necessary (and sufficient) that (1) the general statement or hypothesis "childbed fever is blood poisoning produced by cadaveric matter" be true, and that (2) observations (such as "its prevalence is reduced by antiseptic measures") logically follow from the general statement. However, the observations do not prove the general statement, as Hempel emphasizes.

Since its introduction, the D-N account has been subject to many objections. A few of these are particularly relevant to its (potential) use in philosophy of medicine. One such objection is that the D-N view should, but does not, require that the *explanans* (what explains) be a *cause* of the *explanandum* (what is explained). Thus, for example, from mercury rising in a thermometer, fever can be deduced, but the thermometer does not explain the fever. On the other hand, fever explains the rising of mercury because it also causes it. *Causality*, not only correlation, should be a further requirement of proper explanations, and a version of the D-N view thus amended would account for most explanations in biology and medicine, according to Schaffner (1993: 267).

Yet this does not appear to be an adequate solution to the problem of applying the D-N view to medicine, as it is often the case, in medicine, that we observe the correlation between facts, but do not understand how one causes the other. Many signs of diseases have been known long before we understood what causes them, and some are yet to be properly explained. Schaffner suggests that such non-causal "explanations" are considered "provisional surrogates for deeper causal explanations" (Schaffner 1993: 274). This raises as many questions as it solves: are they necessary, to whom and why? How long can they remain provisional? In what sense is a surrogate explanation, an explanation?

A second reason to doubt the relevance of the D-N model in medicine is that most medical explanations do not refer to laws of nature. Examining short-term potentiation in neurobiology, inheritance, and AIDS, Schaffner concludes that the general premise in the *explanans* is not a law of nature, but "rather a set of causal sentences of varying degrees of generality" (Schaffner 1993: 286). Some of these sentences are very general and resemble laws of nature. One can think of often implicit principles such as "a phenotypic trait of a heterozygotic individual is determined by the dominant allele." But others are very specific. For instance, "Gaucher disease is prevalent among Ashkenazi Jews" cannot be thought of as a law of nature. Schaffner attempts to solve this problem for applying the D-N view to medicine by proposing a second amendment to the D-N view: theories in biology and medicine are what he calls "middle range" theories—i.e., not encompassing the whole realm of the living but limited to a species (as *Aplysia californicum*), a particular strain in a species (a mutant *E. Coli*), or even a subpopulation (Ashkenazi Jews). Yet it is questionable whether this should still be considered a nomological view of explanations. Often, explanatory general claims come in numerous versions, depending on the species, or even a particular strain in a species. And many generalizations are required in most medical explanations, not just one law.

1.2. Statistical Models of Explanation

A modified version of the D-N view may fit some explanations in medicine, but it certainly does not account for *all* of them. Hempel provided other models for scientific explanation in general (1965), some of which are of obvious importance to medical explanations in particular. First, deductive-statistical (D-S) explanations rely on a statistical general premise. An example is:

> 25% of children with heterozygous parents have a disease associated with a recessive allele.
> Sickle-cell anemia is associated with a recessive allele.
> Population X has heterozygous parents carrying the recessive allele of sickle-cell anemia.
> *25% of population X has sickle-cell anemia.*

Note that the 25% figure is not an observation but logically follows from the assumption that humans have two alleles of the same gene and that meiosis (i.e., the selection of alleles from parents) is random. Note also that the *explanans* is a general statistical proposition and that the *explanandum* is a less general statistical proposition, derived from the former. Such explanations obviously define a different sort of medical explanations, not all of them.

A further sort of explanation according to Hempel, Inductive-Statistical (I-S) explanation, derives a particular fact, the *explanandum*, from a general statistical fact, the *explanans*. The derivation is not deductive (i.e., logically included in the general claim) but inductive (i.e., consists in an extension of what is observed to a more general claim). Hempel's (1965) famous example is:

> The statistical probability of recovery of patients with streptococcus treated with penicillin is close to 1.

Jones was infected with streptococcus.
Jones is being treated with penicillin which
(*made very likely that*) he would recover from his infection.

As noted by Schaffner, many explanations in medicine are similar to this one (Schaffner 1993: 268), not only those involving individuals, but populations as well (e.g., replace "Jones" with "n patients in Ohio in 1959" in the previous derivation). This argument is inductive rather than deductive because no particular case follows necessarily from an *almost general* statement—Jones may have been infected with a streptococcal strain resistant to penicillin.

As Hempel points out, I-S explanations are ambiguous in an essential sense. If it is highly likely, but not necessary, that people like Jones, who are infected with streptococcus and treated with penicillin, recover, then there is also a subset of these same people for whom it is highly likely that they will not recover, including people infected with penicillin-resistant streptococcus and octogenarians with weak hearts. There are therefore two equally sound I-S explanations, with true premises, leading to opposite conclusions about people infected with streptococcus and treated with penicillin. We can thus explain both Jones' recovery and his death equally well. Additionally, I-S explanations are "epistemically relative" (Hempel 1965): they are not true or untrue, but relevant or irrelevant, given the state of knowledge. Indeed, we know that we should not indifferently apply the same probability of recovery to all populations.

But what does "relevance" mean? Since Hempel and Carnap, many have defined it as the choice of the proper reference class to which an individual belong. In this context, a reference class is a group of individuals with causally associated features: for instance, having a weak heart (or being old) *and* being more likely to die from streptococcus infection. If an individual belongs to several reference classes for which varying I-S explanations can be given, the question of which reference class should be used in the explanation is known as "the problem of relevance." Consider the following example:

Jones had a cold.
Jones took vitamin C.
Almost all people taking vitamin C recover from a cold within one week.
(*makes it very likely that*) Jones would recover from his cold within one week.

Salmon objects that this inference conforms to the I-S model of explanation but is not an explanation. The reason it is not is that most people recover from a cold within one week, with or without vitamin C. "People treated with vitamin C" is not a relevant reference class in this situation because this apparent *explanans* does not make any difference to the outcome. Salmon thus proposes the Statistical Relevance (S-R) model of explanation (Salmon 2006). The basic idea is that the *explanans* should contain all (and only) statistical information *relevant* to the *explanandum*. Jones' recovery is explained by his being treated by penicillin, because the probability of his recovery given that he has strep infection *and* has been treated with penicillin is higher than the probability of his recovery given that he has strep infection. On the other hand, vitamin C is not relevant because it does not raise the probability of recovery from common cold. It is also relevant that Jones is host to a penicillin-sensitive bacterium, but irrelevant that Jones has been contaminated by his mother rather than his sister.

Yet Salmon's proposal still leaves unsolved a problem as to the choice of alternative relevant reference classes. Of course, the probability of recovery given both strep infection and penicillin treatment can be more accurately predicted by referring to narrower subclasses (e.g., including whether the strain of strep is resistant to penicillin). But say Jones, who is 35 years old and has a weak heart, is infected with a non-resistant strain, is given penicillin, and recovers from strep infection. What is the relevant reference class for the explanation to hold? As a first attempt to resolve this problem, Carnap introduced the "requirement of total evidence," which demands that all the knowledge we have about a specific situation be used in the assessment of the probability of an outcome (Carnap 1962). Does this mean that probabilities add up, or that the right probability to use is attached to the intersection of all classes? Hempel proposed a different criterion—that we should use the *narrowest reference class* the individual belongs to—that is, the narrowest class for which we have information relevant in the situation (Hempel 1965). However, these requirements may conflict. The reason is that no matter how narrow the reference class for which evidence of probability exists, there still is further information relevant to the probability of the outcome for a particular patient—for instance, that Jones may not be observant because he is afraid of antibiotics, changes the probability of recovery. Even if the probability that he does not take the treatment given that he is afraid was known, the narrowest reference class requirement states that we use the conditional probability given everything else as well, which is different (Fuller and Flores 2015). In addition to the questions just raised, other problems with identifying reference classes in medical explanations have been raised (Clarke 2011).

In the end, just as Schaffner does in the case of the D-N model, Salmon argues that causation should be involved between the *explanans* and the *explanandum* in the S-R model. In the explanation of GI Joe's death by leukemia, Salmon says, many facts are relevant, such as the atomic blast he was exposed to, his position relative to it, his shelter, etc. All of them are causally relevant to his death, and that is why they are explanatory. Because of the apparent universality of the requirement of causality for explanations, and because of the difficulties of the D-N, D-S, I-S, and S-R models of explanations, many now prefer to start all over again from the simpler idea that explanations are causal, and more specifically in the life sciences, that they are *mechanistic*.

1.3. Mechanistic Explanations in Medicine

In accounting for scientific explanations in general, philosophers of science have increasingly moved their attention away from *laws* and toward *models* of phenomena. In medicine, mechanistic models in particular have been the focus of much attention when discussing explanations. Mechanistic models are the kind of schemas you find in biology or physiology textbooks: they represent entities (e.g., cell receptors and neurotransmitters), the interaction of which account for the general outcome, that is, regular behaviors of the system (Machamer, Darden, and Craver 2000). In medicine, mechanisms are deeply connected to physiology. Physiology has been a basic medical science since Hippocratic medicine, and one could say that it is the science of the mechanisms of organisms. Another way to put it is that physiology is the science of biological functions. This means that our understanding of mechanism and function within medicine are closely connected. Therefore, an important debate in philosophy of biology about the definition of "function" is important to understanding mechanistic explanations in medicine. One approach to functions in biology is to understand them as a causal role in a system (see Ariew, Cummins, and Perlman 2002). A causal role is the contribution a part of the system is expected to provide for a general effect to obtain: for instance, the heart pumps blood for circulation, so pumping blood is the heart's causal role in the body. In a different understanding, a function is an effect that is currently present in a population because it has been selected during evolution (Millikan 1989): at a certain time, a certain contractile structure improved circulation by pumping fluids, and this ancestor of hearts was selected because it improved fitness. This general debate leads to two complementary understandings of what a mechanistic explanation in medicine should look like: "functional explanations" and "evolutionary explanations." Whereas the former focuses on what parts do and what their functions are—*explaining based on functional effects*—the latter focuses on why the parts are here in the first place (i.e., because their effects were selected)—*explaining why effects such and such, rather than anything else, are functional effects*. There may, for instance, be vomiting in the course of a bacterial infection. One can either explain the vomiting as a dysfunctional mechanism, some part of the system not fulfilling its function, or explain that vomiting may have been selected during evolution as a defense mechanism (Nesse, Williams, and Brown 1996). Let us examine each of these approaches.

1.3.1. Functional Explanations in Medicine

Explaining an observed effect by the causal contributions of biological functions is the essential task of physiology, thereby playing a major role in medical explanations. Organisms can be thought of as systems within which interacting parts tend to maintain or attain certain states, such as keeping the level of blood glucose fairly constant or minimizing pain (Sommerhoff 1950; Nagel 1961: 411–9). These phenomena can be explained more specifically through "functional analysis."

The most famous description of functional analysis is Cummins' (1975, 1985). According to Cummins, functional analysis is a strategy used to explain how a system can exert a capacity. The functional analysis consists in decomposing this capacity into an organized series of sub-capacities: for instance, blood circulates (capacity), because heart pumps (sub-capacity 1), artery walls keep the system close (sub-capacity 2), etc. Each of these capacities may in turn be explained in the same way, until it bottoms out of biological science at the molecular level. *Functional analysis* is the analysis of one capacity of a system into a more or less complex concatenation of functions.

There is widespread acknowledgment that functional analysis accounts for a very large set of medical explanations, roughly, *physiological* explanations, such as the example of blood circulation above. However, this view cannot account for a very important, related sort of medical explanations. Take the following example:

> The function of the adrenal glands is to produce cortisol.
> Jones' adrenal glands do not produce cortisol.
> Jones' adrenal glands are dysfunctional.

Explanations of this sort, which are important in medicine, say why an observed fact should be judged normal or abnormal. The mechanistic approach, however, which does not make any difference between mechanisms of diseases and mechanisms of healthy processes (Craver 2001: 67), does not account for such explanations: the outcome of the mechanism of Jones' adrenal glands is what they do, not that they are dysfunctional. Yet, doctors do consider functional statements such as the one above about the adrenal gland to be explanatory. Moreover, they seem to consider them to be laws of nature from which judgments are deduced.

The D-N model of explanation, thus, seems to give a better account than the mechanistic approach for this specific class of medical explanations. This is a superficial understanding of these types of explanations, however. For D-N to apply here, we would need to understand "the function of adrenal glands is to produce cortisol" as a law of nature. But surely, "the function of adrenal glands is to produce cortisol" does not mean "they always produce cortisol" (Jones' don't). It is not a statistical law of nature either (i.e., it does not mean "the frequency of adrenal glands producing cortisol is close to 1"). In fact, something being functional or dysfunctional cannot be derived from frequency or rarity: the frequency of people dying before they turn 100 is close to 1, which does not entail that Jones surviving over 100 is dysfunctional; being left-handed or red-headed is not dysfunctional because it is rare; dental caries are not functional because they are frequent, etc. Functional statements are important as *explanans*, yet they are not reducible to general or statistical statements about reference classes.

Thus, D-N explanation does not in fact work well here. On the other hand, the mechanistic approach is blind to the functional/dysfunctional difference. Some have proposed that this normative nature of functional statements, determining what a part's contribution ought to be, is not statistical, but rather theoretical (Wachbroit 1994). What is meant is that functional analysis results in an explanatory and predictive model, and that what is called dysfunction is simply departure from this model. In this approach, the question remains of determining how to choose the factual basis of the model.

The upshot is that functional analysis accounts for *how*-explanations of physiological processes, either healthy or pathological, but does not account for *which*-explanations (i.e., explanation of which processes are healthy and which are pathological). For this type of explanation, some turn to evolutionary explanations.

1.3.2. Evolutionary Explanations

Just as the theory of evolution has translated the notion of an organ's "goal" into an explanation of how it came to being, some have hoped that it would account for which observed effects are functions and which are not, and maybe, which are healthy and which are pathological. In their view, ascribing functions can only be justified in light of evolution. Jones' adrenal glands are dysfunctional because they ought to do what adrenal glands were selected for. In this way, the evolutionary account makes function relevant to medical explanations of dysfunction without entailing that functional statements are laws of nature (Millikan 1989; Neander 1991).

Whether evolution alone can provide which-explanations or not, the evolutionary sense of "function" is also relevant to medical explanations in a different sense. Philosophers of biology have long emphasized the difference between *how*- and *why*-questions. Medical explanations have been more concerned with how-questions than with why-questions about diseases. Explaining how diseases act is indeed more important, but explaining why they exist in the first place is valuable as well. The latter question has been addressed through evolutionary means by Darwinian medicine, whose agenda has been expressed in Nesse and Williams (1995). In this book, a case is made for the explanation of many diseases as resulting from a mismatch between the optimal, evolution-selected hunter-gatherer of the Pleistocene era we genetically still are and modern life. Moreover, some symptoms, such as vomiting in bacterial infection, are actually naturally selected defense mechanisms. Finally, they explain why there still are diseases—why total resistance to disease has not been the result of evolution. This paradigm has also been popular in evolutionary psychiatry, as is illustrated by Cosmides and Tooby's work mainly since the 1990s (Barkow, Cosmides, and Tooby 1995), and discussed by philosophers of psychiatry (Wakefield 2001; Faucher 2012).

Méthot (2011) has proposed a distinction between Darwinian medicine and evolutionary medicine. Darwinian medicine is "backward-looking" and seeks for an explanation of why we get sick in the way we do at present, given what our evolutionary history (presumably) is. It is thus purely hypothetical. Evolutionary medicine, on the contrary, does not provide any theoretical framework for medicine but uses evolutionary thinking as a tool for solving specific problems. Méthot proposes the example of bacterial resistance to antibiotics, one of the few observable processes of evolution through selection, to illustrate his distinction. The theory of evolution explains and even predicts why and how resistance occurs, making resistance an example of Darwinian medicine. But evolution also suggests how resistance could be minimized and prevented (e.g., by alternating antibiotics), making this case amenable to problem solving by evolutionary medicine.

So far, there does not seem to be any general picture of explanations in medicine. Yet most of the accounts we have discussed are general accounts of explanation that were applied to medicine rather than accounts designed for medicine. Given the deficiencies in these general accounts when applied to medicine, philosophers of medicine have developed approaches that deal with problems of explanation specific to medicine.

2. Specifics of Medical Explanation

Specific problems in explanation for philosophy of medicine are the role, variety, and structure of explanations in medical science. First, a focus on explanation in medicine, as opposed to many other less action-oriented sciences, may lead to the conclusion that they may be of a lesser importance, leading to the view that medicine is an engineering knowledge rather than a science. Second, some have doubted that all medical explanations are or should be mechanistic: the originality of medicine might be to resort to types of explanation as different as mechanistic and statistical explanations. Eventually, the variety of medical sciences has led some to develop the view that what is explanatory in medicine is what unifies knowledge, a view here labeled as "coherentism."

2.1. Does Medical Science Model Mechanisms or Establish Evidence?

One possible response to the problem of providing an account of explanation in medicine is to see medical science as an engineering practice, to which explanations are of minor importance. In this view, evidence for action would be of primary interest for medicine. There has indeed been a trend in recent years to think that medical science should focus on establishing evidence for effective treatments and diagnostic and prognostic tools. The so-called evidence-based medicine movement is largely responsible for this trend. To an engineering practice, explanation is deemed less essential or even largely dispensable. Establishing *that* something works does not involve establishing *how* it works. Whereas the former is done through establishing difference-making, the latter is done through the investigation of mechanisms, according to a distinction now common among philosophers of medicine. Indeed, some studies just establish *that* a factor makes a difference to an outcome; others establish *how* a mechanism links this factor to the outcome. There does not seem to be any alternative. Knowledge of difference-makers is necessary and sufficient for prediction and control.

However, even if one admits that prediction and control, not explanation, is all medicine looks for, it remains that causal claims must be established to that end: there is no prediction, nor control, of an effect E by a cause C, without knowing that C causes E. In particular, the Russo-Williamson Thesis claims that establishing evidence of causality between C and E implies establishing both *that* C makes a difference to E and *how* a mechanism may link C to E (Russo and Williamson 2007). Establishing a mechanism is *ipso facto* explaining. Some even admit that it also is the main, if not the only, way to explain in medicine:

> [C]ausal explanations are only explanatory to the extent that they can be viewed as providing a glimpse of the structure of a corresponding mechanistic explanation.
> (Clarke et al. 2014)

So even if explanation is not the primary goal of medicine, the Russo-Williamson Thesis shows that it is central to medical science.

A further argument against ignoring explanation in medicine goes that causal explanation, in the form of models of mechanisms, provides a stronger and broader basis to clinical practice. Paul Thompson, who also strongly opposes the view of medicine as mere engineering, acknowledges the difference between "clinical medicine" and "basic medical science" (Thompson 2011: 115). Medical science, he insists, requires explanations, in the form of quantitative models of both basic physiological processes and epidemics, and not just evidence gathered from randomized controlled trials (RCTs). Moreover, the view that medical science would only consist in the knowledge that environmental factor X causes disease Z or that treatment Y improves it weakens medical science: isolated causal claims are stronger when they are based on integrated, multicausal models. Such explanatory models are precisely what basic medical science is interested in. Only *clinical* medicine is to medical science what engineering is to physical science (Thompson 2010: 273). Medicine itself, however, is not merely engineering for Thompson.

2.2. Alternatives to Mechanistic Explanations in Medicine

Whatever their position on the importance of explanation in medical science, some philosophers of medicine have raised doubts about whether mechanisms matter much in medical explanation. The great complexity of organism-environment interactions often makes a satisfactorily exhaustive mechanistic model of diseases unattainable (Howick 2011). Fortunately, mechanisms are not all there is to explanations of causal relationships, as recent work in the philosophy of epidemiology demonstrates.

A traditional view of epidemiology is that it is both a descriptive science of population-related aspects of disease and an empirical inquiry into their potential causes, among which is the environment, that does not need to establish mechanisms. "Epidemiology is the study of the distribution and determinants of disease frequency in human populations" (Rothman, Lash, and Greenland 2012). It contains no theoretical framework, only methods. Therefore, the discipline is often rather seen as heuristic, descriptive, and predictive at best, rather than as explanatory.

Recently, however, Broadbent has developed an approach to understanding epidemiological explanations as truly explanatory (Broadbent 2013). Epidemiology is an inquiry into what associations signal causation. According to Broadbent, in epidemiology, a cause should be interpreted as an explanatory factor. In particular, it explains a *difference* in observed effects in two contrasted groups of people, but it does not consist in the disclosure of the complete causal network that led to the presence of an effect. Exposure may thus explain the difference between exposed and unexposed samples, say, the presence or absence of an infectious syndrome when exposed, or not exposed, to an infectious agent. It does not, however, explain why exposure, rather than non-exposure, produces the effect. This latter task would be much more demanding. To prove that exposure, rather than non-exposure, produces the syndrome, one would have to know what else could possibly cause it, whereas establishing that exposure produced the syndrome rather than the absence of the syndrome, one only has to show a difference in outcome when exposed and unexposed. Epidemiological explanation is also less demanding than mechanistic explanation, as epidemiological explanations admit of so-called black-box approaches to causation (i.e., the claim that to establish causality, one does not have to identify the mechanisms).

This suggests that epidemiological explanations are an alternative to mechanistic explanations. The same sort of alternative to simple mechanistic explanation is provided by quantitative genetics and genome-wide association studies, that is, systematic investigation of parts of the genome that may be associated with a given disease. Infectious diseases again provide interesting examples: for instance, only one out of ten people infected with influenza virus develops the flu. Darrason (2013) points out that multiple factors play a role in explaining such differences, one of which being genetic variability. Genetic variability is expressed by the notion of heritability. Heritability is a ratio of phenotypic variance to genotypic variance: it expresses how much of the former is explained by the latter, but as a statistical concept, it bypasses the individual genes and the various molecular pathways by which the variation of a trait such as susceptibility to flu explains. Yet, heritability provides an explanation of which patients are susceptible to the flu. It is a non-mechanistic explanation, as both Darrason (2013) and Schaffner (2016) have emphasized. Indeed, explanation by heritability is only contrastive, of the sort Broadbent emphasized in epidemiology: it expresses a proportion of *differences* in populations. The same could be said of recent approaches based on Big Data Biology: they seem to provide a statistical sort of explanations as compared to mechanistic explanations. However, it remains an open discussion whether epidemiology and data-driven approaches really provide explanations, and not just descriptions or predictions (Ratti 2015).

2.3. Coherence Approaches to Explanation in Medicine

Contemporary analyses of medical explanations have led to the observation that explanations are somewhat subordinate to therapeutic action and that there may be at least two sorts of explanations in medicine, statistical and mechanistic. This suggests that medical knowledge might be more pragmatic and eclectic than other sciences such as physics. In fact, just as unification of knowledge has been proposed in the general philosophy of science as defining of what explanation consists in (Kitcher 1981), coherence of apparently pluralistic knowledge may reveal that what is explanatory, is a unificatory schema that provides coherence to various beliefs held by medical scientists.

Thagard articulates such a view about the example of the science of peptic ulcer (1999). Until the 1980s, peptic ulcer was understood as the result of high acid concentration in the stomach or duodenum, which in turn was explained by stress. These hypotheses were supposed to be robustly established by converging observations and meshed well with traditional representations of the "bilious type" in medicine and their consequences in popular cultures. This was coherent and held as explanatory. In the early 1980s, Warren and Marshall discovered the existence of a bacterium in the stomach, *Helicobacter Pylori*, that survives the low pH of this environment. They then observed that almost all patients with peptic ulcer are hosts to this bacterium, although some of those who host it are not affected. Ultimately, they showed that antibiotic treatment was efficient on almost all cases of peptic ulcer. Their conclusion was that *H. Pylori* infection *explains* peptic ulcer.

To be explanatory, the coherentist view goes, this belief had to provide a coherent picture of all other beliefs held about peptic ulcer; it could not be explanatory if it remained isolated from all other widely held beliefs about peptic ulcer. According to Thagard's view, "coherence of beliefs" is defined as the "positive constraint" (Thagard 1999: 67) that two hypotheses exert on each other, so that adopting either one both forces one to adopt the other and to reinforce it. For example, before Warren and Marshall's discoveries, the accepted hypothesis that bacteria could not survive in environments as acidic as the stomach was not coherent with the hypothesis that *H. Pylori* colonizes the stomach; after their discoveries, the coherent picture is that infection to *H. Pylori* causes most peptic ulcers, that treatment against *H. Pylori* cures most cases, and that stress plays a minor part in the disease, if any (Thagard 1999: 135–47). Progress in explanation has thus been made, in that this wider set of consistent beliefs can now be held.

Thagard also notes that "unified understanding does not come from the availability of a general overarching theory but from the availability of a system of explanation schemas" (Thagard

1999: 34–5). For Thagard, explanations are based on a *disease explanation schema* consisting of an *explanation target* (i.e., an answer to the question "why does a patient have a disease with associated symptoms?") and an *explanatory pattern* (i.e., the demonstration that the patient "is or has been subject to causal factors" and that "the causal factors produce the disease and the symptoms") (Thagard 1999: 20). This explanation schema may be fulfilled by many different approaches. One example is the *Germ Theory Explanation Schema*:

> "*Explanation target*
>
> Why does a **patient** have a **disease** with **symptoms** such as fever?
>
> *Explanatory pattern*
>
> The **patient** has been infected by a **microbe**.
> The **microbe** produces the **disease** and the **symptoms**."
>
> (Thagard 1999: 25)

These statements are not interpreted by Thagard as stating (mechanistically related) facts, but as expressing beliefs. Other medical explanations may rely on different entities than microbes, for instance, presence or absence of a nutrient, genes-RNA-protein and mutations, etc. The form of the explanation schema remains the same; that is, it always relies on the same sort of cognitive arrangement of beliefs. These beliefs, however, may evolve with time and the succession of dominant theories; that is, they are also under the influence of social processes that spread them. In other words, the fact that beliefs are widely held is also important to their being considered as explanatory.

Thagard's approach to explanations is *cognitive* —it accounts for why arguments are accepted as explanations, rather than accounting for their logical structures. It is also *social*, in that what makes a causal reasoning an explanation are social beliefs shared by a community. Of particular interest here is the case of the adoption of a new theory, as illustrated by the explanation of peptic ulcer by infection. Adoption of any theory is explained following the same explanation schema:

> "*Integrated Cognitive-Social Explanation Schema*
>
> *Explanation target:*
>
> Why did a group of **scientists** adopt a particular set of **beliefs**?
>
> *Explanatory pattern:*
>
> The **scientists** had a set of **mental representations** that include a set of **previous beliefs** and a set of **interests**.
> The **scientists'** cognitive mechanisms included a set of **mental procedures**.
> The **scientists** had **social connections** and **power relations**.
> When applied to the **mental representations** and **previous beliefs** in the context of **social connections** and **power relations**, the **procedures** produce a set of **acquired beliefs**.
> The **scientists** adopted the **acquired beliefs**."
>
> (Thagard 1999: 9)

Thagard's cognitive view also emphasizes that part of the difficulty of medical explanation comes from the existence of multiple types of explanations to medicine. Heterogeneous beliefs about various entities involved are generally held about the same disease: recall the various beliefs about peptic ulcer before the *H. Pylori* revolution.

Lemoine (2011) also endorses a cognitive, coherentist view of explanations. However, he emphasizes the fact that *all* beliefs about the same disease do not have to, or even cannot, be coherent. Lemoine suggests that only subsets of beliefs about the same disease have to be coherent with one another. They are defined by different, incommensurable scientific views of explanation, which can converge or diverge about a specific disease. These scientific views of explanation, Lemoine calls "explanatory values." They state what counts as a sufficient *explanans* (i.e., set of facts) for the explanation to be satisfactory in their perspective. For instance, the clinical explanation that "patient p presents signs x, y, and z because he suffers from disease entity D" is satisfactory only if all relevant signs in the exhaustive list semiology provides, have been observed, and if diagnostics has been correctly conducted. According to a different explanatory value, the explanation is considered satisfactory only if a continuous mechanism from etiology to symptomatology has been established. Lemoine distinguishes several other types of explanation—pharmacological, evolutionary, infectious, etc.—and shows that each states the specific set of conditions accepted as sufficient for a satisfactory explanation of one sort.

Lemoine also explores the various possibilities that explanations of various types converge, diverge, or just coexist, which he dubs the "disunity of medicine." He considers the plurality of medical explanations to play an important role in medicine, as strengthening or weakening evidence.

Conclusion

A satisfactory account of medical explanations should be explicit about what is specifically medical (if anything), whether all medical explanations are of the same sort, and what the role of explanations is in medical science. Philosophical accounts of scientific explanations in general cast light on the logical structure of medical explanations in particular, and also reveal something of their diversity: D-N, D-S, I-S, and S-R explanations are resorted to in medicine, albeit in a modified version. Mechanistic explanations focus on causal relations and come in two forms: functional and evolutionary. These investigate respectively how a process works and why it is there in the first place. Despite the relevance of these accounts of explanations in medicine, they gloss over the specific problems medical explanations raise. These pertain to the role of explanation in medicine—is it prominent or not, is medicine a search for knowledge, or is it imbued with pragmatism?—and to the question whether all explanations should ultimately be mechanistic in nature. In the end, this chapter suggests that from a cognitive point of view, how distinct types of explanations relate to each other is critical to understanding the specificity of medical science when it comes to explaining. Thagard and Lemoine both think that there is no unitary theory of diseases from which explanations are derived. However, Thagard emphasizes the need for *coherence* of different beliefs, whereas Lemoine emphasizes the usefulness of *convergence* of differently grounded explanations.

References

Ariew, André, Robert Cummins, and Mark Perlman. 2002. *Functions: New Essays in the Philosophy of Psychology and Biology*. Oxford, UK: Oxford University Press.

Barkow, J. H., Cosmides, L., and Tooby, J. (Eds.). 1995. *The adapted mind: Evolutionary psychology and the generation of culture*. London: Oxford University Press.

Bird, Alexander. 2007. "Inference to the Only Explanation". *Philosophy and Phenomenological Research* 74 (2): 424–32.

———. 2010. "Eliminative Abduction: Examples from Medicine". *Studies in History and Philosophy of Science Part A, Explanation, Inference, Testimony, and Truth: Essays Dedicated to the Memory of Peter Lipton* 41 (4): 345–52.

Broadbent, Alex. 2013. *Philosophy of Epidemiology*. London: Palgrave Macmillan.

Carnap, Rudolf. 1962. *Logical Foundations of Probability*. London: University of Chicago Press.

Craver, Carl F. 2001. "Role Functions, Mechanisms, and Hierarchy". *Philosophy of Science* 68 (1): 53–74.

Cummins, Robert. 1985. *The Nature of Psychological Explanation*. New edition. Cambridge, MA: MIT Press.

Cummins, Robert C. 1975. "Functional Analysis". *Journal of Philosophy* 72 (November): 741–64.

Darrason, Marie. 2013. "Unifying Diseases from a Genetic Point of View: The Example of the Genetic Theory of Infectious Diseases". *Theoretical Medicine and Bioethics* 34: 327–44.

Faucher, Luc. 2012. "Evolutionary Psychiatry and Nosology: Prospects and Limitations," *Baltic International Yearbook of Cognition, Logic and Communication*, Vol 7, http://newprairiepress.org/biyclc/vol7/iss1/5/.

Fuller, J., & Flores, L. J. 2015. "The Risk GP Model: The standard model of prediction in medicine". *Studies in History and Philosophy of Science Part C: Studies in History and Philosophy of Biological and Biomedical Sciences*, 54, 49–61.

Gillies, Donald. 2005. "Hempelian and Kuhnian Approaches in the Philosophy of Medicine: The Semmelweis Case". *Studies in History and Philosophy of Biological and Biomedical Sciences* 36 (1): 159–81.

Hempel, Carl. 1965. *Aspects of Scientific Explanation and Other Essays in the Philosophy of Science*. New York: The Free Press.

———. 1966. *Philosophy of Natural Science*. 1 edition. Englewood Cliffs, NJ: Pearson.

Hempel, Carl G., and Paul Oppenheim. 1948. "Studies in the Logic of Explanation". *Philosophy of Science* 15 (2): 135–75.

Kitcher, Philip. 1981. "Explanatory Unification". *Philosophy of Science* 48: 507–31.

Lemoine, Maël. 2011. *La désunité de la médecine: Essai sur les valeurs explicatives de la science médicale*. Paris: Hermann.

Lipton, Peter. 2003. *Inference to the Best Explanation*. London: Routledge.

Machamer, Peter K., Lindley Darden, and Carl F. Craver. 2000. "Thinking about Mechanisms". *Philosophy Of Science* 67 (1): 1–25.

Méthot, Pierre-Olivier. 2011. "Research Traditions and Evolutionary Explanations in Medicine". *Theoretical Medicine and Bioethics* 32 (1): 75–90.

Millikan, Ruth G. 1989. "In Defense of Proper Functions". *Philosophy of Science* 56 (June): 288–302.

Nagel, Ernest. 1961. *The Structure of Science: Problems in the Logic of Scientific Explanation*. New York: Harcourt, Brace & World.

Nesse, Randolph M., and George C. Williams. 1995. *Evolution and Healing: New Science of Darwinian Medicine*. Second Edition. London: J.M. Dent & Sons Ltd.

Nesse, Randolph M., George C. Williams, and Jared M. Brown. 1996. *Why We Get Sick: The New Science of Darwinian Medicine*. New York: Vintage Books.

Ratti, Emanuele. 2015. "Big Data Biology: Between Eliminative Inferences and Exploratory Experiments". *Philosophy of Science* 82(2): 198–218.

Rothman, Kenneth J., Timothy L. Lash, and Sander Greenland. 2012. *Modern Epidemiology*. Philadelphia, PA: Lippincott, Williams & Wilkins.

Salmon, Wesley C. 2006. *Four Decades of Scientific Explanation*. Pittsburgh, PA: University of Pittsburgh Press.

Schaffner, Kenneth F. 2016. *Behaving: What's Genetic and What's Not, and Why Should We Care?* New York: Oxford University Press.

———. 1993. *Discovery and Explanation in Biology and Medicine*. Chicago, IL: University of Chicago Press.

Sommerhoff, G. 1950. *Analytical Biology*. London, Royaume-Uni: Oxford University Press.

Thagard, Paul. 1999. *How Scientists Explain Disease*. Princeton, NJ: Princeton University Press.

Thompson, R. Paul. 2010. "Causality, Mathematical Models and Statistical Association: Dismantling Evidence-Based Medicine". *Journal of Evaluation in Clinical Practice* 16 (2): 267–75.
———. 2011. "Models and theories in medicine". In Fred Gifford, ed., *Philosophy of Medicine*, vol 16: *Handbook of the Philosophy of Science*. Oxford, UK: Elsevier, pp. 115–136.
Wachbroit, R. 1994. "Normality as a biological concept". *Philosophy of Science* 61(4): 579–591.

Further Reading

Bird, Alexander. 2010. "Eliminative Abduction: Examples from Medicine." *Studies in History and Philosophy of Science Part A, Explanation, Inference, Testimony, and Truth: Essays Dedicated to the Memory of Peter Lipton* 41 (4): 345–52.
Craver, Carl F. 2007. *Explaining the Brain: Mechanisms and the Mosaic Unity of Neuroscience*. London: Oxford University Press, Clarendon Press
Schaffner, Kenneth F. 1993. *Discovery and Explanation in Biology and Medicine*. Chicago, IL: University of Chicago Press.
Thagard, Paul. 1999. *How Scientists Explain Disease*. Princeton, NJ: Princeton University Press.
Thompson, R. Paul. 2011. "Models and theories in medicine". In Fred Gifford, ed., *Philosophy of Medicine*, vol 16: *Handbook of the Philosophy of Science*. Oxford, UK: Elsevier, pp. 115–136.

28
THE CASE STUDY IN MEDICINE

Rachel A. Ankeny

Introduction

The case report continues to be exceedingly popular within medicine. A survey in the early 2000s documented publication of approximately 40,000 new case reports per year, representing 13.5% of all publications in core medical journals (Rosselli and Otero 2002). With the recent founding of several peer-reviewed, open-access journals dedicated to cases (Kidd and Hubbard 2007) and improvements in the communication and dissemination of cases via internet-based databases, rapid and easy exchange of information via cases has increased significantly. Hence, although the preferred venues may have changed for this genre of publication, it is indisputable that cases remain an important form of medical literature.

In short, a case report describes a medical problem experienced by one or more patients, usually involving the presentation of an illness or similar that is puzzling or otherwise difficult to explain or categorize based on existing understandings of disease or understandings of physiology and pathology. Typical cases outline the presentation of the disease, diagnosis, treatment, and outcomes of the patient, with a focus on practice-based observations and clinical care (rather than the results of randomized controlled trials [RCTs] or other experimental methodologies). Although cases are not equivalent to narratives, many researchers have commented on their highly standardized narrative structure (Hunter 1991; Hurwitz 2006). The goal is to capture information about particular cases, including many details that may not be immediately relevant, but that could prove to be, so that the information contained in the case and the case itself can be useful over the long term and so it can be systematically combined with other cases into larger data sets.

The published case report can be used for a variety of purposes (see Gagnier et al. 2013): for instance, it can be utilized in clinical settings should other instances of a similar condition or disease arise, or it may serve as the basis for clinical research based on the hypotheses suggested in the account of the case. In addition, cases often prove pedagogically useful, particularly when they recount the processes through which clinicians came to a better understanding of the patient's condition, prognosis, or outcomes, even where negative outcomes resulted. Cases can assist with identification of adverse reactions to drugs or other stimuli, documentation of beneficial effects of certain therapies particularly in new contexts, or recognition of novel or emerging diseases as well as unusual presentations of recognized diseases. An oft-cited example of the success of case reporting is the recognition of the relationship between use of the drug thalidomide by pregnant women and severe limb and other deformities in newborns (even though the underlying causal mechanism only came to be understood over time). In this chapter, I use the term "case" to refer collectively to published case reports as well as case series (examining multiple patients with similar symptoms or conditions) presented as a published report, as these are the most easily accessible forms of cases and are not dissimilar from those typically presented in pedagogical contexts (as will be discussed later).

One of the most famous examples of a case was a series of five cases published in June 1981, which presented information about young homosexual men with a specific form of pneumonia, viral infections, and severe oral thrush, two of whom had died (CDC 1981a). Once a second report appeared in early July detailing a spike in cases of Kaposi sarcoma (an aggressive, cancerous tumor caused by infection with a form of herpes virus) and pneumonia among homosexual men (CDC 1981b) and the popular media began its coverage (Kinsella 1989: 115), the public as well as the medical community began to take notice. By December 1981, there was enough evidence to warrant publication of several articles related to this disease condition (which had yet to be named, but came to be called acquired immune deficiency syndrome or AIDS), which appeared as research articles but the content of which still was largely case-based (Gottlieb et al. 1981; Masur et al. 1981; Siegal et al. 1981). I will return to this example throughout the paper to illustrate various features of cases and the types of reasoning done using them in the field of medicine.

What are some of the common characteristics for what makes a case? The overwhelming majority depict complaints arising in specialty or subspecialty settings (with some differences in format and style across subspecialties, which I do not discuss in this chapter), often where diagnosis would be difficult or particularly tricky, and describe uncommon or even "unique" clinical occurrences (McCarthy and Reilly 2000) observed under uncontrolled conditions (Simpson and Griggs 1985). In its instructions to authors, *The Lancet* stresses that it aims to publish those cases that have a "striking message" (Bignall and Horton 1995). Many report rare conditions for which trials of various types of therapies (particularly RCTs) are not feasible due to low patient numbers or ethical issues (Albrecht et al. 2005).

Single cases are seen by some as problematic as a form of evidence, inasmuch as they capture exceptions or highly unusual manifestations of illness and disease, rather than typical ones that might support generalizable rules. Because of this, they remind practitioners and others that the field of medicine is in fact a "science of particulars: (Gorovitz and MacIntyre 1976), or even as often claimed, an art rather than a science. They highlight the importance of clinicians having diagnostic "puzzle-solving skills" and not only scientific knowledge, a fact that has been well recognized in popular culture through Oliver Sacks's books and more recently in the television series *House, MD*. Although publication of case reports has a long history in prominent journals, systematic guidelines for case reports have only recently been developed through the articulation of a consensus-driven international framework for case reporting that the guidelines' authors argue will promote more consistency, transparency, completeness, and impact (Gagnier et al. 2013).

How Do Practitioners Use Cases as a Form of Knowledge?

Use of cases involves complex processes of pattern recognition among other epistemic and practical skills. Suppose a patient presents to a physician with several symptoms that are individually identifiable. Perhaps these symptoms have not previously been seen in this combination and have not been recognized as together constituting a discrete disease condition (or an adverse reaction). Often a particular cluster of symptoms does not "make sense" for the patient in question, for instance due to her baseline health condition or personal history. In this type of case, the pattern being sought is a diagnostic category describing a syndrome or disease (or details of an adverse reaction), which can be used to make decisions about the provision of therapies and prognoses. Otherwise, if no pattern is detectable, the symptoms may need to be treated individually, or else multiple diagnoses will need to be made and treated as appropriate.

Practitioners then can engage in a process of gathering additional information using a variety of mechanisms such as case reporting or comparison to other cases in the published literature. By organizing (and reorganizing) the various details about a particular patient, they can propose a diagnosis that establishes which facts are relevant and how they are interrelated. Published cases thus allow practitioners to recognize similar patterns as new patients present themselves, to make decisions about the most appropriate clinical care, and to expand their background knowledge beyond their experiences of the typical or the usual in the clinic.

To return to the example of what came to be known as AIDS, the disease was first noted in late 1979 by clinicians in California who saw a number of young, homosexual men presenting with symptoms of a mononucleosis-like syndrome, including fever, weight loss, and swollen lymph nodes, as well as thrush and diarrhea, which are not symptoms of mononucleosis (Grmek 1990; see also Shilts 1987; for a longer discussion of this example, see Ankeny 2011). Meanwhile in New York City, Kaposi sarcoma was observed in a number of young, homosexual men with friends in common; this cancer had previously only been seen in elderly men of Mediterranean descent. Thus, the mismatch between the observed symptoms and the patients' background conditions triggered increased scrutiny about what the shared patterns in the patients might be. The original published case proved useful because its epistemological structure reflected selectivity about what information was included and emphasis on particular facts about the patients, which signaled its authors' suspicions about the nature and mode of transmission of the disease: for instance, although the total number of patients described was relatively small (five), the authors clearly thought it could be no coincidence that these were homosexual men who were sexually active, which in turn pointed to the hypothesis that the underlying disease condition was likely to be infectious in nature. Hence, in this example, it is clear that the information contained in the case is not disputed; in other words, the truth status of the various claims is not in question, but more importantly their relevance and whether they are essential (or just incidental) correlations becomes a key matter for debate and clarification. Ongoing documentation of what amounted to nearly 200 patients, apparently based in only three geographic locations (New York City, Los Angeles, and San Francisco) (Grmek 1990: 11), solidified theories about the disease's mode of transmission (and that in fact it was a distinct and new disease condition).

By 1982, the literature on what came to be known as AIDS began to grow, and within about one year, cases with similar symptoms emerged in different populations, including Haitian migrants, blood transfusion recipients with and without known risk factors, and women, which helped to further refine what philosophers would term the necessary and sufficient conditions for the diagnostic category. These new examples triggered re-evaluation of the relationships and assumptions about the existing facts associated with the case, particularly the association of the disease solely with homosexual men.

This case also allows us to understand that much of the information contained in cases may well prove irrelevant or even misleading: as mentioned above, the opening paragraph of the published case noted that all of the patients also were infected with a particular virus (cytomegalovirus or CMV). As one of the authors, the immunologist Michael Gottlieb, later wrote, "we rushed to judgment, overlooking the fact that CMV is a common opportunistic infection in immune compromised organ transplant recipients" (1998: 367), and hence likely to be common in any immunocompromised group. This example underscores the potential fruitfulness of even those cases that contain information that later proves to be irrelevant (or even incorrect), so long as the balance of the information within a case proves to be relevant.

Misleading information can be identified and weeded out with the addition of more data or information, which can reveal mere correlations or coincidences (as in the case of CMV infection, which proved not to be directly relevant to the etiology of AIDS) or that a particular attribute does not apply beyond the small population described in the case at hand and hence is inessential to it. In a similar way, the fact that many who had the disease were homosexuals was later determined to be relevant but not essential, which in part lead to the abandoning of one of the original names for the disease, GRID (gay-related immunodeficiency disease).

How Are Cases Utilized in Medical Education?

It is claimed that a good case study "begets awareness, jogs the memory and aids understanding" (Morgan 1985: 353), a description that indicates the mixture of educative and epistemological goals inherent in cases. For generations, medical training depended on cases, arguably back to ancient times and the Hippocratic corpus, which was highly narrative and contained considerable material that resembles contemporary published cases (for more on the Hippocratic corpus and other relevant historical examples, see Hurwitz 2006). As the physician-philosopher Howard Brody puts it, "without a storehouse of case exemplars to draw upon medicine could be neither taught nor practiced" (2003: 9).

Medical education also can produce cases: for instance, the generation of the AIDS case discussed above in fact occurred in a teaching setting. When Gottlieb asked one of his immunology fellows to look for interesting teaching cases, he was told about a young homosexual man with what seemed to be a severely damaged immune system (Fee and Brown 2006); this man then became the index patient (the first case of a condition to be described in the medical literature, otherwise known as patient zero) for the original case report.

As used in medical training, the case comes to have a sort of "generic" existence as it is used as a way of thinking through differential diagnoses (Montgomery 1992; Ankeny 2011); no longer is the published case about an individual patient, but instead it becomes the prototype of a particular disease or condition. However, the use of cases in medical education also is rather paradoxical, given that published cases often recount unusual, rare, or even unique instances; at worst, cases might be considered to have the status of anecdotes (Hunter 1986), particularly as compared to evidence generated by RCTs, systematic reviews, and so on.

However, cases continue to be used and to have considerable value in medical pedagogy, in part because of their standardized and highly stylized formats, which provide a way of organizing the type of detailed information typically present in clinical settings. When used comparatively, cases can allow the tracing of evolving practices and theories of disease over time, including changes in what factors are given the most weight in the clinical encounter and in turn changes in the doctor-patient relationship (Monroe, Holleman, and Holleman 1992; Hurwitz 2006). In addition, cases (as accounts of single instances) can serve as reminders to pay attention to the details of any particular patient, and that in some cases, such patients will be exceptions to the "rules" that dominate much of medical training and practice (Hunter 1996). As Kathryn Montgomery Hunter notes (1996), the oft-heard clinical maxim "When you hear hoof beats, don't think zebras" underscores that despite uncertainties inherent in clinical care, many cases are in fact typical, and thus there is a need to cultivate both scientific and practical knowledge through medical training. Cases are a fundamental means of capturing these trade-offs, inasmuch as they model the dissection of the complexities of real-life instances of illness or other clinical conditions and sorting through what is relevant and essential and what is incidental or mere correlation.

So to return to the case of AIDS, a certain process was followed that reflects the logical elimination strategies that guide much of clinical care. First, the clinician ruled out the known causes of this sort of immunosuppressed condition, including chemotherapy for cancer, immune-suppressing drugs used post-transplantation, and an inborn autoimmune disorder (Gottlieb 1998), none of which applied to the patient in question. Candidiasis (the particular form of thrush in evidence) was known to be typically associated with a deficiency in T-lymphocytes, and thus blood tests were done that revealed a marked decline in the number and function of this type of lymphocyte. A case report discussion among the teaching hospital's immunology postdoctoral fellows and internal medicine residents then resulted in a decision to use an experimental technique to identify subclasses of lymphocytes, which led to the identification of a severe depletion of CD4+ helper cells, which initially could not be explained. Only after the patient was discharged without a diagnosis, then readmitted a week later with a rare form of pneumonia usually only found in immunocompromised patients, and additional patients who also were homosexual men were referred with the same symptoms, the same form of pneumonia, CMV infection, and similar blood results, did the medical team narrow in on the new disease condition and its key characteristics, which became known as AIDS.

Many commentators have noted not only the benefits but the limitations of using cases in medical education, perhaps most pointedly Rita Charon: "The genre, in the end, is the distillation of many medical lessons, and by teaching our students how to tell this type of story, we teach them deep lessons about the realms of living that are included and excluded from patient care" (1986: 10). In other words, although cases do allow patients' stories to be told, in their contemporary form they also are highly abridged, standardized, and formalized, and thus typically lack the voice and point of view of the patient. This silence has not always been the case if we compare the conversational style of cases in the 17th and 18th centuries (Hurwitz 2006). Some journals now actively promote the inclusion of patient perspectives in case reporting, although the uptake of this approach remains somewhat limited (Ankeny forthcoming).

What Kind of Evidence Do Cases Provide?

As noted previously, cases have sometimes been viewed pejoratively as too speculative and lacking in any real evidence, particularly because they often provide information about singular instances (i.e., they are n's of 1) without any controls or underlying experimental methods, hence raising questions about their validity and generalizability. This criticism has become heightened in the current era, which is dominated by evidence-based medicine (EBM) and which places RCTs at the top of its hierarchy of evidence. Cases are typically and essentially observational, and the circumstances under which the observations and even the documentation of the observations are performed are uncontrolled as they occur within real-life clinical settings. Some might even claim that the narrative form common to many cases in fact obscures what should be taken to count as evidence, as the information contained in cases is not filtered or prioritized; this claim taken to an extreme is, of course, debatable as the example of the processes associated with the formulation of the AIDS diagnosis recounted above reveals.

However, it is important to note that even standard accounts of EBM include as a type of evidence a form of research that aggregates individual cases known as the case series, where patients with similar attributes (e.g., who received the same treatment or who were exposed to the same chemical or other toxic substances) are tracked over time using descriptive data and without utilizing particular hypotheses to look for evidence of cause and effect. Admittedly, EBM places the case series rather low in its hierarchy of evidence but nonetheless acknowledges its potential usefulness in instances where forms of evidence that rate more highly are not available, as may often be the case where human patients are concerned due to practical or ethical reasons, or where the available evidence at higher levels has been produced in a manner that is methodologically flawed.

In the simplest sense, even singular cases clearly can be used as hypotheses to refute existing generalizations, at least if we accept a simplistic form of confirmation theory. So, for instance, once AIDS was described and came to be understood as caused by HIV infection, it was initially thought that HIV was universally fatal. However, with the development of a range of highly successful treatments for HIV, people with it who are treated are living much longer than those who are not treated and sometimes do not develop AIDS, a phenomenon that was recognized in part out of individual cases. However, this type of use of cases as a form of evidence is fairly limited in medicine, which rarely relies on simple generalizations such as the one described here.

A more common use of published cases to provide evidence is associated with the considerable detail often contained in them, which allow them to serve as springboards for future investigations. Cases can be reused as evidence of something other than that which they were intended to show or be reanalyzed with attention to different details than in their initial usage, particularly because of the rich information that they typically contain. In the case of AIDS, we can see that cases often include information without making the logic of the inclusion of particular claims explicit, but which opens up avenues for additional reporting and research. Thus, the closing paragraph of the published case listed a series of facts without additional comment, for instance, that the patients did not know each other and had no known common contacts, that they were not aware of sexual partners with similar illnesses, and that all five patients reported using nitrite inhalant drugs ("poppers"). It is likely that this information was included to hint at possible hypotheses about the etiology of the disease (namely, that the disease likely was infectious and transmissible through blood and/or sexual contact) but for which not enough evidence (in conventional terms) was available. The specific mention of lack of known common contacts or of ill sexual partners points toward what would later emerge when tracing the disease's epidemiology, namely that transmission was occurring in part through movement of individuals with numerous and often casual sexual partners and that the disease was not affecting all people who were carrying HIV in a similar manner, conclusions that were drawn from this case in retrospect once it was reanalyzed in light of additional epidemiological and other information.

Can Cases Help Us to Identify Causes?

Although cases often are considered to be highly limited in terms of their abilities to allow causes of disease and illness or other clinical phenomena to be identified, whether this claim is in fact valid depends on what account of causality is utilized. If we take the search for causes to be a matter of identifying both necessary and sufficient conditions, individual published cases typically will fail the test, as they are not documented under controlled conditions that would allow sufficient conditions to be detailed. An alternative account might utilize a causal network model, which would be highly compatible with the narrative form taken by most published cases as well as with clinicians' actual practices of working through the details of a disease condition and its potential causes, particularly in discussions with patients, as Paul Thagard (2000) has noted. As detailed in the initial reporting of the immunocompromised patient seen by Gottlieb and his team, a detailed narrative was developed that outlined various putative causes and systematically excluded certain ones as irrelevant based on additional information from the patient's medical history, bloodwork results, and so on. Even the content and format of the case supports this type of view: as more information is gathered, certain hypotheses are narratively "tested" against what appear to be putative necessary and/or sufficient causal conditions. The processes of ruling in (and out) various possible causal factors also reflect aspects of clinical practice relating to what often yet misleadingly is termed the "art" of the case. Although reasoning via cases does involve particular skills that may be difficult to axiomatize and teach, experienced practitioners also engage in a detailed, systematic sorting and weighing process of various factors in ways that are highly specific, and derive from both medical education and training, especially in differential diagnosis (Hunter 1991), as well as the long history associated with the particular format, language, and contents in which cases appear in the published literature. And given that clinical practice is about actually caring for patients and not just coming up with generalizations and laws, as might be the case in a purely theoretical science, such processes are an essential part of medicine.

Cases and their analyses also can be viewed as preliminary steps in a longer process of medical research by providing working hypotheses about causal attribution that can ground further tests of causal relations (for additional examples relating to a series of cases associated with ingestion of high-dose caffeine, see Ankeny 2014). Individual cases can play this sort of role, but these attributions are most common when cases are brought together and compared, which allows similarities and differences to be articulated and new information to be added, until a clear hypothesis about causality can be identified and tested. Hence, the attribution of a putative cause in clinical settings via cases can provide information that allows medical researchers to engage in more "conventional" or traditional methodologies such as RCTs, cohort studies, and so on to explore various causal hypotheses. Without the information provided via cases, the range of potential causal hypotheses would be much less narrow, and thus not allow as specific of a focus on key likely causal pathways.

However, with regard to clinical care, cases tend to play a much narrower role: they assist practitioners to identify a cause that can be manipulated to cure (or prevent) the condition in question, in order to treat patients who might present with similar patterns of symptoms until better evidence can be produced (if ever). This approach to causation is known as a manipulability approach, and it has been defended most forcefully in the context of medicine by Carolyn Whitbeck, who notes that clinicians focus on the (causally related) sufficient conditions for cure and "speak of it as 'the' factor which cured the disease" (1977: 628). This attribution is similar where prevention is the main goal: practitioners focus on the (causally related) necessary conditions, avoidance of which would have prevented the development of the disease condition. A manipulability approach is highly instrumental and as such is greatly dependent on the current state of medical knowledge, as well as on the particular details available in relation to any one patient or case.

Thus, as argued elsewhere (Ankeny 2014), the approach to causality that appears to be embedded in case studies due to their narrative structures, the context within which they are written and disseminated, and the type of evidence typically presented in them relies on manipulability. The key idea of such an approach is that causal relations are best analyzed in terms of the production or prevention of some state of affair by manipulation of another (see also Collingwood 1940, Gasking 1955, and for a more technical examination, Woodward 2003). Given that cases are aimed in large part at documenting instances of clinical practice, and assisting with future patients who might present with similar conditions, what is key in the determination of causes is what can be manipulated in practice (based on current knowledge and technologies), and not in principle, which in turn also guides the hypotheses that are likely to emerge for testing within a research context.

Conclusion

In summary, although cases are perennially criticized for their lack of rigor and consistency, and as low-standard substitutes written by those who are not able to publish "real" medical research, this chapter has attempted to explain why cases continue to retain their critical place within the medical literature as an important means of communication among practitioners, as well as for promoting improved clinical practice and for educating health care professionals. Further philosophical research is warranted into the use of cases within diverse clinical specialties in medicine, with attention to epistemological and other types of differences in their status, value, and use, as well as about the evolution of the use of cases as reporting guidelines and mechanisms for dissemination continue to emerge.

References

Albrecht, J., A. Meves, and M. Bigby (2005) "Case Reports and Case Series from *Lancet* Had Significant Impact on Medical Literature," *Journal of Clinical Epidemiology* 58: 1227–32.

Ankeny, R.A. (2011) "Using Cases to Establish Novel Diagnoses: Creating Generic Facts by Making Particular Facts Travel Together," in P. Howlett and M.S. Morgan (eds.), *How Well Do Facts Travel? The Dissemination of Reliable Knowledge*, Cambridge: Cambridge University Press, pp. 252–72.

Ankeny, R.A. (2014) "The Overlooked Role of Cases in Causal Attribution in Medicine," *Philosophy of Science* 81: 999–1011.

Ankeny, R.A. (forthcoming) "The Role of Patient Perspectives in Clinical Case Reporting," in R. Bluhm (ed.), *Knowing and Acting in Medicine*, New York: Rowman & Littlefield.

Bignall, J., and R. Horton (1995). "Learning from Stories—The *Lancet's* Case Reports," *The Lancet* 346: 1246.

Brody, H. (2003). *Stories of Sickness*, 2nd edition, Oxford: Oxford University Press

Centers for Disease Control (CDC) [M.S. Gottlieb, H.M. Schanker, P.T. Fan, A. Saxon, J.D. Weisman, and I. Pozalski] (1981a) "Pneumocystis Pneumonia—Los Angeles," *Mortality & Morbidity Weekly Report* 30: 250.

Centers for Disease Control (CDC) (1981b) "Kaposi's Sarcoma and Pneumocystis Among Homosexual Men—New York City and California," *Mortality & Morbidity Weekly Report* 30: 305.

Charon, R. (1986) "To Listen, to Recognize," *Pharos* 49: 10–13.

Collingwood, R. (1940) *An Essay on Metaphysics*, Oxford: Clarendon Press.

Fee, E., and T.M. Brown (2006) "Michael S. Gottlieb and the Identification of AIDS," *American Journal of Public Health* 96: 982–3.

Gagnier, J. J., Kienle, G., Altman, D. G., Moher, D., Sox, H., Riley, D., and CARE Group (2014) "The CARE guidelines: consensus-based clinical case report guideline development," *Journal of Clinical Epidemiology*, 67(1), 46–51.

Gasking, D. (1955) "Causation and Recipes," *Mind* 64: 479–87.

Gorovitz, S., and A. MacIntyre (1976) "Toward a Theory of Medical Fallibility," *Journal of Medicine and Philosophy* 1: 51–71.

Gottlieb, M.S. (1998) "Discovering AIDS," *Epidemiology* 9: 365–7.

Gottlieb, M.S., Schroff, R., Schanker, H.M., Weisman, J.D., Fan, P.T., Wolf, R.A., and Saxon, A. (1981) "Pneumocystis Carinii Pneumonia and Mucosal Candidiasis in Previously Healthy Homosexual Men: Evidence of a New Acquired Cellular Immunodeficiency," *New England Journal of Medicine* 305: 1425–31.

Grmek, M.D. (1990) *History of AIDS: Emergence and Origin of a Modern Pandemic*, trans. R.C. Maulitz and J. Duffin, Princeton: Princeton University Press.

Hunter, K.M. (1986) "'There Was This One Guy. . .': The Uses of Anecdotes in Medicine," *Perspectives in Biology and Medicine* 29: 619–30.

Hunter, K.M. (1991) *Doctors' Stories: The Narrative Structure of Medical Knowledge*, Princeton, NJ: Princeton University Press.

Hunter, K.M. (1996) "'Don't Think Zebras': Uncertainty, Interpretation, and the Place of Paradox in Clinical Education," *Theoretical Medicine* 17: 225–41.

Hurwitz, B. (2006) "Form and Representation in Clinical Case Reports," *Literature and Medicine* 25: 216–40.

Kidd, M., and Hubbard, C. (2007) "Introducing the Journal of Medical Case Reports," *Journal of Medical Case Reports*, 1(1), 1.

Kinsella, J. (1989) *Covering the Plague: AIDS and the American Media*, New Brunswick: Rutgers University Press.

Masur, H., M.A. Michelis, J.B. Greene, Onorato, I., Vande Stouwe, R. A., Holzman, R. S., . . . and Cunningham-Rundles, S. (1981) "An Outbreak of Community-Acquired *Pneumocystis Carinii* Pneumonia: Initial Manifestation of Cellular Immune Dysfunction," *New England Journal of Medicine* 305: 1431–8.

McCarthy, L.H., and K.E. Reilly (2000) "How to Write a Case Report," *Family Medicine* 32: 190–95.

Monroe, W.F., W.L. Holleman, and M.C. Holleman (1992) "Is There a Person in This Case?" *Literature and Medicine* 11: 45–63.

Montgomery, K. (1992) "Remaking the Case," *Literature and Medicine* 11: 163–79.

Morgan, P.P. (1985) "Why Case Reports? (editorial)," *Canadian Medical Association Journal* 133: 353.

Rosselli D., and A. Otero (2002) "The Case Report is Far from Dead (letter)," *The Lancet* 359: 84.

Shilts, R. (1987) *And the Band Played On: Politics, People, and the AIDS Epidemic*, New York: St. Martin's Press.

Siegal, F.P., C.M. Lopez, G.S. Hammer, Brown, A.E., Kornfeld, S.J., Gold, J., Hassett, J., Hirschman, S.Z., Cunningham-Rundles, C., Adelsberg, B.R., and Parham, D.M. (1981) "Severe Acquired Immunodeficiency in Male Homosexuals, Manifested by Chronic Perianal Ulcerative Herpes Simplex Lesions," *New England Journal of Medicine* 305: 1439–44.

Simpson, R.J., and T.R. Griggs (1985) "Case Reports and Medical Progress," *Perspectives in Biology and Medicine* 28: 402–6.

Thagard, P. (2000). *How Scientists Explain Disease*, Princeton, NJ: Princeton University Press.

Woodward, J. (2003) *Making Things Happen*, Oxford: Oxford University Press.

Further Reading

K. M. Hunter's *Doctors' Stories: The Narrative Structure of Medical Knowledge* (Princeton: Princeton University Press, 1991) synthesizes her publications on the topic of narrative in medicine and remains a key resource.

Though not directly about cases, P. Thagard's *How Scientists Explain Disease* (Princeton: Princeton University Press, 2000) provides an excellent account of related issues associated with diagnostic reasoning and the causality of disease.

A. Kleinman's *The Illness Narratives: Suffering, Healing, and the Human Condition* (New York: Basic Books, 1988) and H. Brody's *Stories of Sickness* (Oxford: Oxford University Press, 2nd edition, 2003) explore stories and narratives in clinical medicine from philosophically and sociologically informed perspectives.

29
VALUES IN MEDICAL RESEARCH

Kirstin Borgerson

1. Introduction

Medical research is thought to involve the objective or unbiased pursuit of knowledge. This is often expressed by saying that research is, or at least ought to be, value-free. In this chapter, I aim to draw attention to some of the problems with this ideal. I begin by developing a thought experiment aimed at demonstrating the significant challenges of controlling for values in the design, execution, and reporting of medical research. Following this, I outline the implications of pervasive unacknowledged values in medical research, from direct harm to research subjects to the indirect harms that arise as a result of loss of public trust in the research enterprise. Given the undesirability of these potential harms, I clarify what is meant by values in this context and then turn to recent work in philosophy of science, on two questions: "Is value-free medical research possible?" and "Is value-free medical research desirable?" In both cases, the answer is no. In light of these answers, I present a different way of thinking about the role of values in medical research, and trace the practical implications of this approach. Rather than pretending that medical research is more objective than it really is, we ought to improve the social mechanisms that allow for the exposure and critical evaluation of the values that necessarily persist in medical research.

2. Effectivex

Imagine there are two medical researchers: one scrupulous and honest, and the other quite unscrupulous. Each of them is given the same task: design a research trial on a new drug for cholesterol: *Effectivex*. The good, scrupulous researcher does so diligently, with the sole aim of producing knowledge. The unscrupulous researcher, on the other hand, has a strong desire to obtain positive results in this trial because she stands to benefit, financially and through enhanced reputation, from the success of *Effectivex*. However, while unscrupulous, she is not willing to go so far as fabricating the results of a study or plagiarizing a study done by someone else, nor is she crude enough to want to falsify the results by directly altering the data once it has been collected. These strategies would be too risky, since she wants to publish the research in a high-impact journal, and the reviewers might notice these more-familiar types of research misconduct. Like the scrupulous researcher, she wants to obtain regulatory approval for the drug, if it works, and this added level of scrutiny provides extra reason for caution. As a more sophisticated breed of unscrupulous researcher, she is keen to rig the trial, through design, to produce the results she wants.

Here's the challenge, then: none of the tactics the unscrupulous researcher employs can affect the quality of the trial as judged by reviewers and regulators. It will have to be the same design used by the scrupulous researcher, a double-masked, randomized controlled trial (RCT): the gold standard of medical research. This means that selected study participants will be allocated at random to a treatment and control group, and neither the researchers nor the participants will know which group they are in until the trial is complete. There will have to be appropriate follow-up of subjects in the trial and a rigorous statistical analysis. Within these parameters, what tactics might the unscrupulous researcher employ? How might she accomplish her goal?

When I pose this scenario in university classrooms, I am astonished at how many strategies the students are able to come up with. Occasionally the proposed tactics won't work, because they violate one of the quality conditions, such as by requiring that the researcher knows which group participants have been assigned to during the trial so that she can tamper with the experience of that group (which violates the double-masked condition), but in most cases the tactics abide by the rules and seem like they would work. One such tactic my students often come up with involves giving the treatment to everyone in a set of potential research subjects, before the trial begins, and then eliminating any subjects who are not responding well. Then the real trial (and randomization) can take place, using only the subjects already known to be good responders. (In case this tactic seems too outlandish, I should note here that this strategy is employed regularly in clinical research: it is called a pre-trial run-in period.) I suspect, like my students, by this point you have already thought of one or two strategies the unscrupulous researcher might try.

There is a deeper point to be made here, however: even the scrupulous researcher, in the design of a trial of *Effectivex*, will have to make decisions during the course of research that make it more or less likely the drug will be shown to be effective. So, for instance, if the scrupulous researcher does *not* include a pre-trial run-in, this choice has implications for how generalizable the results are and how highly regarded the study will be by other researchers. So even the scrupulous researcher will have to consider the use of such tactics, even if they are ultimately rejected. As we will explore later in this chapter, values guide this assessment all the way through, whether a particular researcher is scrupulous, unscrupulous, or somewhere in between. I will turn now to the empirical evidence on the use of such tactics in actual medical research.

3. Evidence of Values in Medical Research

Many of the tactics known to have been employed by unscrupulous researchers for the purposes of obtaining positive results have been exposed in recent years (cf. Borgerson 2009). One of the most common ways to meddle with results is to be selective about what ends up in the official record. So, for instance, if the results of a trial were quite positive, researchers have been known to publish them many times, in many different journals (Tramer et al. 1997). This creates the (false) appearance of a preponderance of evidence in favor of the treatment. Conversely, if the results are not what the investigators had hoped for, they might decide not to publish them at all or to publish them in obscure journals in other languages so they won't be found (Lexchin et al. 2003). Taking this strategy a step further, researchers might publish only the early results of a trial, ignoring the final results, in cases where the trial was looking good for some period of time, but then took a turn for the worse by the end of the study period (the Celebrex scandal arose because of the use of this sort of tactic).

Variations also exist in ways to make the comparison treatment worse (Psaty, Weiss, and Furberg 2006). Specifically, there are studies in which researchers employed what is called "sub-optimal dosing," where this involves using too low or too high a dose of the competitor's drug, in order to make one's treatment look better by comparison (Rochon et al. 1994). Researchers have also been known to administer comparison treatments in less absorbable forms (e.g., pills rather than intravenous [IV]) for similar reasons (Johansen and Gotzsche 1999).

Another common tactic involves testing treatments for very short periods of time, for instance just a few weeks, even when the researchers know the treatments will be used long term by patients. This makes it less likely they will discover (and then have to reveal) adverse side effects or long-term effects. And there are others: inappropriate surrogate endpoints (Psaty et al. 1999); stopping trials early when the results look favorable (Montori and Guyatt 2007); and perhaps the most common strategy, simply refusing to allow many potential research subjects to enroll in the trial by setting strict inclusion/exclusion criteria (Humphreys et al. 2013). My students' idea of a pre-trial run-in period is a variation on this last strategy. (Note that some of these tactics have justifiable variations. I discuss this challenge further in section 7.)

In addition to these methods, there are dozens more such tactics already in use and no doubt many more that haven't yet been dreamed up. The authors of one satirical take on this problem, in an article on "how to achieve positive results without actually lying to overcome the truth" identified a handful of very common practices and proposed others (Sackett and Oxman 2003). Drawing on decades of experience in the design and assessment of clinical trials, clinical researchers Alex Jadad and Murray Enkin identified dozens of tactics in their 2007 guide to randomized controlled trials (RCTs) (Jadad and Enkin 2007). Updates to these lists would certainly be required today. For instance, there has been increasing awareness very recently of the tactic of "p-hacking," which involves selecting and reselecting data or methods of analysis until nonsignificant results appear to be significant; this tactic has been employed not only in medical research but also in many other fields of scientific research (Head et al. 2015).

As this exercise and the supplementary empirical evidence demonstrates, it is possible to introduce various biases into even the highest-quality research designs. That is, even our gold standard highest-quality research trials (RCTs) may not be as unbiased as we might have believed them to be. And because tactics to manipulate trials are extensive, widespread, and always evolving, this is a problem we can't ignore.

4. Implications of the Widespread Use of These Tactics

The widespread use of these tactics constitutes what Ben Goldacre has referred to as a "cancer at the core of evidence-based medicine" (2012). Physicians who rely upon research evidence to determine how best to treat patients are being misled and misinformed, whether intentionally or not. As a result, they are prescribing treatments to patients that, in some cases, do more harm than good. Many drugs have been withdrawn from the market after these harms came to light: rofecoxib (Vioxx) is a classic example here, since as a result of its popularity it is estimated to have seriously harmed more than 100,000 people in the United States alone. In other cases, the uncertainty about the evidence stalls the process of decision-making and this delays proper treatment, also resulting in harm to patients who suffer without treatment for some period of time.

There are also direct harms to research subjects, who believe they are participating in a socially valuable enterprise, when they are not because the results are too biased to contribute anything meaningful to knowledge. Public awareness of the tactics identified may lead to a loss of trust in the research enterprise. Over time, this would exacerbate the difficulties already encountered by researchers hoping to recruit new research subjects. In addition to these direct and indirect harms, the widespread use of these tactics corrupts our knowledge base, and because researchers "stand on the shoulders of giants" (i.e., use previous research as the basis for proposed new research), this may have a corrupting effect on future research as well.

A common reaction to this dismal situation is to call for stricter adherence to a value-free ideal of medical research. In what follows, I will outline some of the limitations of this type of solution. Before doing so, I will clarify some key terms and review recent developments in philosophy of science as they relate to this issue.

5. What Are Values?

Before going any further, it is important to clarify what is meant by *values* in this discussion, since we have been using the term very loosely thus far. Values are often contrasted with facts or empirical evidence. It can be useful to keep this sort of contrast in mind when we are talking about values in the sciences, since scientific inquiry may aim to produce facts but draw on values along the way: "Values are statements of norms, goals, and desires; evidence consists of descriptive statements about the world" (Douglas 2007: 126). We are all familiar with facts, or descriptive statements, such as "the earth is round" and "the sky is blue." Value statements may contain terms like "ought" and "should" or otherwise signify that they are meant as norms or rules. So, when we say that simplicity is a value in the sciences, we mean that scientists operate with a norm, or rule, such as "the simplest explanation is often the best one" or, put more directly, "when all else is equal, choose the simplest explanation." (This may or may not be a good rule.)

A distinction is sometimes drawn between epistemic values, such as truth or accuracy, and non-epistemic (moral or political) values, such as equality or freedom. This distinction is also contested; we will stick with it for now because it will help to clarify a key position in the debate over the role of values in science.

The unscrupulous researcher described above aimed to design research studies in such a way that they would produce only positive results. In doing so, she might be said to have valued her own financial interests more than accuracy or truth. Sometimes people talk about negative values as those that compete with the standard positive epistemic values like truth. These values lead us away from knowledge. Because one of the core scientific values is accuracy in our descriptions of the world, one of the most common forms of negative value are biases. Biases are any factors that interfere with our ability to obtain accurate results; in other words, they are the negative value that contrasts with the positive value of accuracy. A bathroom scale may be biased if it systematically displays a higher or lower weight than is accurate, and dice at a gambling table can be biased if they have been weighted to turn up lucky 7's more, or less, frequently than balanced dice. Determining whether a value is positive or negative can sometimes be tricky and requires close attention to the particular context. For instance, there is nothing inherently wrong with valuing financial gain. But it can become a negative value, or disvalue, in the context of medical research.

6. What Does It Mean for Someone to Claim That "Science Is Value-Free"?

Philosophers and historians of science have written extensively on value-freedom as an ideal in the sciences. A systematic overview of this literature can be found in a collection of essays entitled *Value-Free Science? Ideals and Illusions* (Kincaid et al. 2007). Many of the case studies and examples used in the collection stem from areas such as biology, archaeology, evolutionary psychology, and economics. The lessons, though, are readily applicable—perhaps even more immediately applicable—in the practically oriented domain of medicine, where the stakes, and the interests, are very high. The value-free ideal maintains that whereas moral and political decisions often depend on interests, values, and preferences, scientific claims are dependent only on empirical evidence. Kincaid and colleagues characterize this view as follows: "The job of science is to tell us the facts" (2007: 4).

The first thing to realize about the claim that "ideal science is value-free" is that, in its strongest form, it is obviously false. This is not simply because there are so many cases of biased science (as in the examples of medical research described above). It is for two deeper reasons. First, epistemic values seem to be a necessary component of all elements of scientific research. Insofar as scientists value truth or accuracy (and we hope they do), these epistemic values (and others) ought to permeate and guide all elements of scientific research. Second, non-epistemic values are a necessary component of scientific research at key stages. In the earliest stages of scientific research, scientists must decide what topic or area to study, what specific question to ask or hypothesis to test, what study population to use, and so on. These choices will be interest-driven and thus value-laden. They will respond to things like pressing health needs in the world or a desire for financial gain. For instance, choosing a homogenous and healthy study population limits the generalizability of results but arguably gives the treatment the "best shot" at success because it limits confounding factors. Which matters more, in a particular case, comes down to values: for instance, the value of increased knowledge for its own sake versus the value of results directly relevant to clinical care. And when the research is complete, the scientists may have to decide whether to make the results public, whether to patent the discovery, and so on. Again, these decisions depend on commitments to different values.

So when people claim that science is value-free, they really mean something much less dramatic than their statement suggests (at least, if they have thought through the position). They allow for epistemic values throughout the process of scientific inquiry, and they allow that non-epistemic values affect science "before and after" the crucial experimental phase of research (where that includes the interpretation of results). What they mean to deny is that non-epistemic values play a role in the middle stage: during hypothesis testing. The reason they might want to maintain some "pure" domain within science is because they are concerned that the intrusion of non-epistemic values into the core of science will compromise the focus on obtaining knowledge. The worry is that researchers might simply manipulate (or "torture") any data they gather in order to obtain the results they'd like, and there will be a free-for-all that loses all purchase on the truth. It is important to keep this worry in mind, and to ask whether proposed middle-ground positions, which allow a role for non-epistemic values throughout scientific inquiry, avoid the dangers just described.

7. Is Value-Free Medical Research Possible?

I will assume for the sake of this discussion that even if medical *practice* is a blend of art and science, as is often claimed, medical *research* is, or aims to be, scientific. We can then approach the question of whether the core of scientific inquiry is value-free (in the limited sense clarified above) through the case of medical research.

Someone faced with evidence of the widespread use of the tactics used to bias medical research outlined above may well call for stricter adherence to a value-free ideal in the hopes that it will provide resources for eliminating bias in the sciences. To someone with this position, stricter adherence to, for instance, methodological rules in medical research, is needed to solve the problems outlined above. I will show that this approach, while sometimes helpful, cannot fully resolve the problem, and I will do so by returning to the case of research on *Effectivex*.

Some of the biasing tactics employed by medical researchers seem like they might be prevented through clearer rules for researchers. So, for instance, researchers who refuse to publish unfavorable results could be made to do so by funding agencies or could be tracked through registration systems designed to publicly track all clinical trials. In fact, these solutions have been implemented in recent years, though they are not wildly successful as yet. However, other biasing tactics seem to be not only challenging to eliminate through rules imposed from outside, but actually so like the decisions necessarily made in the course of research that they could not be imposed from outside without serious harm to inquiry.

Take, for example, inclusion and exclusion criteria for trial participation. In the design of a trial, researchers will have to decide which potential research subjects to include and which to exclude. There is no option to just not decide this or to have an independent committee decide it, though we could require that a burden of justification be placed on researchers to defend any exclusions if we were worried about heavy-handed use of exclusion criteria (as has been common). However, if researchers are committed to a certain type of causal or explanatory trial, with the fewest number of participants possible, and this is key to their research question, then a strict rule prohibiting this choice seems misguided. What rule, then, would prevent the sorts of abuses outlined above? However many safeguards we put in place, researchers will always have some degree of autonomy, and to that extent values will continue to shape research. This effect can be limited, and constrained, but not eliminated.

Consider also the practice of stopping a trial early due to dramatic effects. Surely there are some cases where this is appropriate, even if there are others where researchers are stopping trials early because they are afraid that the dramatically positive effects will fade with time (as is common). These sorts of judgment calls will have to be made, and while rules can help guide these decisions, and some regulatory mechanisms exist for checking on these decisions (such as Data and Safety Monitoring Boards), there will always be room for individual judgment.

A third example concerns the choice of comparison treatment in a controlled trial. Again, some rules can be put in place to ensure roughly appropriate comparison treatments (for instance, ethical prohibitions on the use of placebo comparisons when other treatments have been approved for the condition, with oversight by Research Ethics Committees). But there will still be room for judgment about what the best comparison is, and what the appropriate dose of that control treatment ought to be. Where judgment calls are needed, biased intervention is possible, yet we cannot eliminate these moments of judgment.

Hopefully these examples from recent medical research help to illustrate why philosophers of science have come to believe that the value-free ideal is not achievable. Philosophers such as Helen Longino have arrived at similar conclusions using case studies from biology and the behavioral sciences, and investigations into other domains have turned up similar conclusions (Longino 1990, 2002, 2013; Kincaid et al. 2007). But philosophers, including Longino, have also gone beyond the claim that value-free science is unachievable to consider whether the ideal might actually be undesirable. In the next section, I will review Heather Douglas's argument that we ought to reject the value-free ideal entirely.

8. Is Value-Free Medical Research Desirable?

Douglas argues that, in many areas of science, the value-free ideal is actually detrimental to good scientific practice (2007). The areas of science she focuses on include all those in which the knowledge produced is used to inform decisions, guide actions, and solve problems in the real world. Medical research is a paradigm case of this sort of science. Douglas focuses on the choices made by scientists during the course of research, and in particular on the internal or core stages of hypothesis testing where proponents of the value-free ideal would hope to eliminate all non-epistemic values. Throughout scientific inquiry, including this internal stage, Douglas reminds us that scientists must make choices. Any choice they make has the possibility of being wrong: they may choose the wrong methodology, the wrong study sample or sample size, or the wrong form of statistical analysis. There will be some uncertainty about what the right thing to do is in any particular case. If the scientists know that the results of their research will be used to inform decisions in the world, they will have to choose what level of error they are prepared to accept in their own decisions throughout the study. These choices will be affected by things like how much harm would result if the wrong results were applied in practice. It seems, then, that these are value-laden choices (Douglas 2007).

Douglas illustrates this point using a study that involved the liver tissue of rats that had been exposed to dioxin, a type of organic compound thought to be harmful to humans, at three different dose levels, plus one control group (2007: 124). The slides of the liver tissue were analyzed initially and classified as either free from tumors or containing tumors, whether benign or malignant. Over the next 14 years, "those slides were reevaluated by three different groups, producing different conclusions about the liver cancer rates in those rats" (2007: 124). This exposes the uncertainty around what counts as liver cancer in rats. Douglas stresses this point: all scientific inquiry contains some degree of uncertainty and some possibility of error in decisions made to manage that uncertainty.

> Which errors should be more carefully avoided? Too many false negatives are likely to make dioxin appear to be a less potent carcinogen, leading to weaker regulations. This is precisely what resulted from the 1990s industry-sponsored reevaluation that was used to weaken Maine water-quality standards. Too many false positives, on the other hand, are likely to make dioxin appear to be more potent and dangerous, leading to burdensome and unnecessary overregulation. Which consequence is worse? Which error should be more scrupulously avoided?
>
> (2007: 125)

There is no way for scientists doing this kind of research to avoid making decisions about the trade-off between, for instance, human health and economic efficiency. They will have to decide which type of error is appropriate. These value-laden decisions are, as Douglas points out, "*at the heart* of doing science," not at the periphery as advocates of the value-free ideal might hope (p. 125).

Douglas responds to an obvious objection to this argument: why not require that scientists estimate the uncertainty involved in any judgment and report it rather than take a position? There are two problems with this position. First, it is extremely difficult to estimate, with any precision, the effect of any and all methodological choices on the outcome of a study. This is particularly clear in cases where methodological choices made very early in the study have an effect on the outcome. Douglas offers another medical example to make this point. Medical researchers sometimes rely on the cause of death reported on death certificates to track outcome measures in an experiment, but death certificates are sometimes wrong about the cause of death, for instance in cases involving soft-tissue sarcoma. Nobody knows what the error rates are, but they exist. Medical researchers would be hard-pressed to produce a precise estimate of the effect of this possible error on their results. Now consider that many other such decisions will be made during the course of research, each with some imprecise level of uncertainty. Because research has to proceed on the basis of some decision, these uncertainties will come along for the ride.

Second, and even more fundamentally, even the estimate of error has its own chance of error. That is, it too is imprecise and might be wrong, and so can be assigned some probability of error. It seems researchers are stuck in a regress, in which the task required (e.g., estimating error) itself requires an estimate of error, and that estimate also requires a further estimate of its error, and so on. (A version of this argument was first made by Richard Rudner in 1953, in an admirably clear paper: "The scientist qua scientists makes value judgments.") While the value judgment might, in some cases, be pushed back one level, or set by convention, it can never be eliminated (Douglas 2007).

9. What Are We Left with, Once We Reject the Value-Free Ideal in Medicine?

Recall how entrenched the commitment to a value-free ideal is in the sciences: "when the value freedom of science is questioned, a fundamental institution in our lives is being challenged" (Kincaid et al. 2007: 4). It is likely that many of you reading this chapter will have been taught, whether in elementary school, through the media, or even at university, that the sciences produce objective knowledge through a trustworthy and reliable scientific method, which has been carefully designed to eliminate biases. The alternative, it seems, is quite frightening: if we reject the view that the sciences are value-free, we seem to be left only with the view that scientific inquiry is so thoroughly value-laden that it is merely a reflection of our own prejudices (whether individual or social). This is clearly an undesirable position!

The choice of positions is, happily, not as stark as this. Philosophers of science have, since the 1980s, explored the conceptual territory between these extremes, and many middle-ground positions have been constructed and defended. This is good news, since many important practical decisions are made on the basis of scientific research, including decisions about the care of individual patients, and these decisions can have tremendous importance for the lives of individuals. Health insurance organizations, whether public or private, also need evidence to decide which treatments to cover, and governments (at least ideally) use evidence to guide allocation decisions at the highest level, so having trustworthy evidence is vitally important.

The most promising class of proposals for how to proceed with our new value-laden understanding of scientific inquiry emphasize the importance of transparency and critical engagement within scientific communities (Longino 1990, 2002, 2013). The central idea is this: since values are ineliminable in the sciences, we must identify and acknowledge the values and background assumptions guiding decisions and subject them to critical scrutiny. That critical scrutiny should come from as many diverse perspectives as possible. Elizabeth Anderson nicely draws out the contrast between this approach and the one currently preferred under the value-free ideal:

> Insistence on the value-neutrality of scientists is self-deceptive and unrealistic. . . . Indeed, it is self-defeating: when scientists represent themselves as neutral, this blocks their recognition of the ways their values have shaped their inquiry, and thereby prevents the exposure of these values to critical scrutiny.
>
> (2012, np)

Instead of allowing scientists to hide value-laden choices under the guise of neutrality, they ought to be required to acknowledge and critically discuss those choices with other community members. Douglas sums this up nicely:

Giving up on the ideal of value-free science allows a clearer discussion of scientific disagreements that already exist and may lead to a speedier and more transparent resolution of these ongoing disputes [over public policy issues].

(2007: 136)

The "open science" movement of recent years provides specific recommendations for ensuring the transparency and publicity of scientific research. In other words, it provides recommendations for ensuring recognized avenues for criticism and for ensuring that community members use these avenues. These include: disclosure of competing interests in publications, prohibitions on ghost authorships, a mandatory clinical trials registry, open peer-review, open-access journals, and more comprehensive analytic training for health professionals.

Of these, the latter is perhaps the least developed but most important. All the transparency in the world won't solve anything unless educated and informed community members engage critically with the identified value-laden assumptions. The analytic training could, for instance, take the form of education in philosophy, which tends to be focused on developing critical thinking skills through the examination of taken-for-granted assumptions. It is also vitally important that critics represent different perspectives and viewpoints, since these diverse perspectives are most likely to expose common assumptions to scrutiny and provide new ideas on how to proceed. Finally, it is worth pointing out that certain values may turn out, through this critical evaluation, to be bad for medical research. (Financial interests, for instance, may well interfere with a commitment to knowledge. We saw this in the *Effectivex* case, and the problem may be widespread.) If we have reason to believe this is true, then the medical research community will need to organize itself, or advocate for policy makers, to limit or prohibit the influence of such values. Although some values might require a lot of debate, and policies might thus be subject to revision over time, others, such as the value of truth or accuracy, will be more settled.

10. Conclusion

Consider a medical researcher who is making a decision about which disease to investigate in her research. She is not value-free if she chooses to investigate ovarian cancer rather than pancreatic cancer because that disease affected, or even killed, a friend or loved one. She is not value-free if she chooses to investigate a disease affecting impoverished people in developing nations, such as malaria, on the basis of concerns about social justice and maximum benefit. And she is not value-free when she chooses to conduct research on prostate cancer because there is funding dedicated specifically for research on prostate cancer by the national research institute in her home country.

The difficulty in characterizing a process of value-free reasoning persists in cases where we examine choices of research method, of outcomes, of which journal to publish in (or whether to publish), and so on. That is, it persists at every stage from the very first to the very last choice made in the research process. It is no less a value-laden choice when the researcher aims to do the right thing and choose the most judicious topic, method, outcome, or journal, than when she aims to act self-interestedly and chooses the topic, method, outcome, or journal based on personal gain. Some of these choices are laudable and others are not, and we have good reasons for making these assessments, but they are all underpinned by values. It is not clear that we can make any sense out of the claim that scientists can learn to reason in a wholly value-free manner. Values pervade all aspects of knowledge production. If we are to avoid losing sight of the truth, these values should be carefully and critically evaluated by a diverse set of people, but in order to do that, they will first have to be acknowledged. However counterintuitive it might seem, giving up the value-free ideal is a good place to start.

References

Anderson, E. (2012) "Feminist Epistemology and Philosophy of Science," in Edward N. Zalta (ed.) *The Stanford Encyclopedia of Philosophy*. URL = http://plato.stanford.edu/archives/fall2012/entries/feminism-epistemology/

Borgerson, K. (2009) "Why reading the title isn't good enough: An evaluation of the 4S approach to evidence-based medicine," *International Journal of Feminist Approaches to Bioethics* 2 (2): 152–75.

Douglas, H. (2007) "Rejecting the Ideal of Value-Free Science," in Kincaid, H., Dupré, J., Wylie, A. (eds.) *Value-Free Science? Ideals and Illusions*. Oxford: OUP.

Goldacre, B. (2012) "What doctors don't know about the drugs they prescribe," TED Talk, URL= http://www.ted.com/talks/ben_goldacre_what_doctors_don_t_know_about_the_drugs_they_prescribe?language=en

Head, M.L., L, Holman, R. Lanfear, A.T. Kahn, and M.D. Jennions. (2015) "The extent and consequences of P-hacking in science," *PLOS Biology*, np. DOI: 10.1371/journal.pbio.1002106

Humphreys, K., N.C. Maisel, J.C. Blodgett, I.L. Fuh, and J.W. Finney. (2013) "Extent and reporting of patient nonenrollment in influential randomized clinical trials, 2002 to 2010," *JAMA Internal Med* 173 (11): 1029–31.

Jadad, A., and M. Enkin. (2007) *Randomized Controlled Trials: Questions, Answers and Musings*. Second Edition. Oxford: Blackwell Publishing.

Johansen H.K., and P.C. Gotzsche. (1999) "Problems in the design and reporting of trials of antifungal agents encountered during meta-analysis." *JAMA* 282 (18): 1752–1759.

Kincaid, H., J. Dupré, and A. Wylie (eds). (2007) *Value-Free Science? Ideals and Illusions*. Oxford: OUP.

Lexchin, J., Bero, L. A., Djulbegovic, B., and Clark, O. (2003) "Pharmaceutical industry sponsorship and research outcome and quality: Systematic review." *BMJ* 326: 1167–77.

Longino, H. (1990) *Science as Social Knowledge: Values and Objectivity in Scientific Inquiry*. Princeton: Princeton University Press.

Longino, H. (2002) *The Fate of Knowledge*. Princeton: Princeton University Press.

Longino, H. (2013) *Studying Human Behavior: How Scientists Investigate Aggression and Sexuality*. Chicago: University of Chicago Press.

Montori, V.M., and G. Guyatt. (2007) Corruption of the evidence as threat and opportunity for evidence-based medicine." *Harvard Health Policy Review* 8 (1):145–55.

Psaty, B.M., N.S. Weiss, and C.D. Furberg. (2006) Recent trials in hypertension: Compelling science or commercial speech? *JAMA* 295 (14): 1704–6.

Psaty, B.M., Weiss, N.S., Furberg, C.D., Koepsell, T.D., Siscovick, D.S., Rosendaal, F.R., Smith, N.L., Heckbert, S.R., Kaplan, R.C., Lin, D., and Fleming, T.R. (1999) Surrogate end points, health outcomes, and the drug-approval process for the treatment of risk factors for cardiovascular disease. *JAMA* 282 (8): 786–90.

Rochon, P.A., Gurwitz, J.H., Simms, R.W., Fortin, P.R., Felson, D.T., Minaker, K.L., and Chalmers, T.C. (1994) A study of manufacturer-supported trials of nonsteroidal anti-inflammatory drugs in the treatment of arthritis. *Archives of Internal Medicine* 154: 157–63.

Rudner R. (1953) The scientist qua scientist makes value judgments. *Philosophy of Science* 20 (1): 1–6.

Sackett, D.L., A.D. Oxman. (2003) Harlot plc: An amalgamation of the world's two oldest professions. *BMJ* 327: 1442.

Tramer, M.R., Reynolds, D.J.M., Moore, R.A., and McQuay, H.J. (1997) Impact of covert duplicate publication on meta-analysis: A case-study. *BMJ* 315 (7109): 635–40.

Further Reading

Brown, J.R. (2001) *Who Rules in Science: An Opinionated Guide to the Wars*. Cambridge: Harvard University Press.

Borgerson, K. (2014) Redundant, secretive, and isolated: When are clinical trials scientifically valid? *Kennedy Institute of Ethics Journal* 24 (4): 385–411.

Elliot, K. (2011) *Is a Little Pollution Good for You? Incorporating Societal Values in Environmental Research*. Oxford: Oxford University Press.

Goldacre, B. (2012) *Bad Pharma: How Medicine Is Broken, and How We Can Fix It*. London: Fourth Estate.

Michaels, D. (2008) *Doubt Is Their Product: How Industry's Assault on Science Threatens Your Health*. Oxford: Oxford University Press.

30
OUTCOME MEASURES IN MEDICINE

Leah McClimans

Outcome measures assess the effects of health care on the health status of patients (Donabedian, 1988). Although the documentation of outcome measures such as mortality has long history, modern outcome assessment was not established until the 1970s as the result of at least three factors (Epstein, 1990; O'Connor and Neumann, 2006). First and perhaps most significant was the perception of variation in clinical processes from doctor to doctor, hospital to hospital, and region to region. This variation raised questions about the quality of health care, and measuring health outcomes was seen as one way to answer these questions (Epstein, 1990; Kassirer, 1993). A second contributing factor was the influence of cost containment on quality. The growth of managed care and the emphasis on efficiency led to concerns about the negative effect this might have on health care quality. Outcomes measures were seen as one way to monitor possible deterioration in the system (Epstein, 1990). Third, heightened competition among insurers also led to the perceived need for outcome measures that provide a metric of value for different health care plans. Although price plays a key role in this competition, buyers were and still are interested in the relative value of the health care they purchase.

It is difficult to exaggerate the impact that measuring health outcomes has had on contemporary health care. In terms of recent funding and impact, consider that in 2009 Congress allocated $1.1 billion of the American Recovery and Reinvestment Act for comparative effectiveness research—the goal of which is to improve health outcomes. To give a sense of the scale of this investment, the U.S. Agency for Healthcare Research and Policy (AHRQ) received $300 million of this funding, which represented a 500% increase in the Agency's expected 2009 budget for comparative effectiveness research. Moreover, in 2010, Congress authorized the establishment of the Patient-Centered Outcomes Research Institute (PCORI) as part of the Patient Protection and Affordable Care Act. Furthermore, the U.S. is not alone in its focus on health outcomes. In the 2008 review of the UK's National Health Service (NHS), *High Quality Care for All*, Lord Darzi writes, "...we can only be sure to improve what we can actually measure." (Department of Health, 2008). The UK has since put into place a number of reforms focused on measuring health outcomes, including the NHS's Quality and Outcomes Framework and the Patient-Reported Outcomes Measures (PROMs) Programme.

As the measurement of health outcomes plays an increasing role in the management and evaluation of contemporary health care, it has also begun to receive attention within philosophy. To be sure, concepts of health and well-being have long had a place in philosophy, but only more recently have philosophers started to engage with the measurement practices and epistemic consequences of contemporary outcome assessment. In beginning such an engagement, one of the first things to note is that terms such as "outcome," "outcome measure," and "measurement outcome" are used frequently—and sometimes ambiguously—in the medical and philosophical literature. It is thus useful to be clear about what these different terms mean.

When authors refer to "outcomes" or "health outcomes," they are usually referring to what Eran Tal calls the object of interest (Tal, 2015). The primary outcomes or objects of interest in medicine and surgery are mortality (how many people die after or during receipt of health care), morbidity (a population's disease load as the result of health care), and quality of life (patients' perceptions of their health-related quality of life). These outcomes are measured using a variety of instruments; for instance, quality of life is typically measured using self-reported questionnaires, and morbidity may be inferred from cholesterol assays or clinician-reported questionnaires. In the medical literature these instruments are sometimes referred to as "outcome measures" (e.g., quality of life is often measured using questionnaires collectively referred to as Patient-Reported Outcomes Measures, PROMs). Outcome measures are then put to use within a study (i.e., descriptive or analytic), the aim of which is to answer a question about a health outcome(s) in a particular context. Within the context of a particular study design, the results gathered via these instruments are reported using as a statistic, e.g., relative risk (comparison of risk of, e.g., mortality, for different groups of people), difference between means (subtraction of two group averages, e.g., the mean quality of life score of two populations), or Quality Adjusted Life Year (QALY, a measure of disease burden over time). These results, or what are sometimes called "measurement outcomes," are the knowledge claims about the values of one or more quantities attributed to the outcome of interest (Tal, 2015).

In what follows I begin with a look at how measures of mortality and morbidity are used in clinical research. To illustrate the role that outcomes, outcome measures, and measurement outcomes play, as well as some of the philosophical questions that they raise, I turn to the UK's Place of Birth study (Birthplace in England Collaborative Group, 2011). After my consideration of mortality and morbidity, I turn to quality of life. As will become clear, the kinds of philosophical questions that concern quality of life measures are somewhat different from those that frame philosophical interest in mortality and morbidity. In part this difference is because quality of life measures are a more recent addition to outcome measurement and in part it is because quality of life is a more complex outcome.

Mortality and Morbidity

In the U.S. and UK, births in obstetric units vastly outnumber births that take place outside of obstetric units. Still, non-obstetric births are increasing in both countries. For example, in 2004 only 0.87% of U.S. births occurred in non-obstetric units (home or midwifery units), but by 2012 1.36% occurred in a non-obstetric unit. In England and Wales, they have seen an even steeper increase, with only .9% of births occurring at home between 1985–1988, rising to 2.4% in 2011 (Office for National Statistics, 2013). Is it professionally responsible to support a non-obstetric birth? Should clinical guidelines support choice in place of birth? These kinds of questions shape the debate over place of birth. These questions tend to be answered—and debated—via empirical studies investigating the relative riskiness in terms of mortality and morbidity of different birthplace choices.

Take for example the Birthplace in England Research Programme, which in 2011 published the largest prospective cohort study of its kind, looking at 64,500 births to women at low risk for complications, including almost 17,000 women who planned to give birth at home (Birthplace in England Collaborative Group, 2011; Hollowell, 2014). The Birthplace in England Research Programme was a multidisciplinary research program. The main publication coming out of this research, "Perinatal and maternal outcomes for planned place of birth for healthy women with low risk pregnancies: the Birthplace in England national prospective cohort study," compared three non-obstetric unit locations (i.e., home, free-standing midwifery units, and midwifery units alongside obstetric units) with obstetric units to determine whether outcomes in the non-obstetric units differed from those in the obstetric group. Based on the study's results, the authors conclude that both healthy low-risk women who have never before given birth to a live baby (nulliparous women) and healthy low-risk women who have given birth to at least one live baby (multiparous women) should be offered a choice of birth setting (Birthplace in England Collaborative Group, 2011)

Primary Outcome

The primary outcome or object of interest in this study was a mixture of neonatal morbidity and mortality. Specifically, researchers were interested in measuring the following morbidity and mortality events: stillbirth after start of care in labor, early neonatal death (within seven days of birth), disturbed neurological functioning (neonatal encephalopathy), meconium aspiration syndrome (when a baby breathes in a mixture of amniotic fluid and feces near the time of delivery), brachial plexus injury (damaged shoulder nerves resulting in loss of movement or weakness of the arm), and fractured humerus or clavicle (Hollowell, 2014). A mixed outcome of mortality and morbidity events was used instead of focusing solely on a single mortality or morbidity event in order to give the study more statistical power.

Statistical power refers to a study's ability to detect a difference between two groups if a difference between them really exists. In this study, power refers to the study's ability to detect differences among planned places of birth. When the differences you are searching for are relatively rare, as they are in the case of the above adverse events, then it is difficult for a study to detect genuine differences with regard to the rare event. One way to increase a study's power is to combine rare events and thus increase the probability of their occurrence in the study. Nonetheless, the choice to use a mixed outcome in this study has been controversial.

Lachlan de Crespigny and Julian Savulescu (2014) write that, as a result of this choice, the study is "unhelpful" in determining the relative risk involved in choosing to give birth at home. They argue that it is unhelpful because the outcome includes disparate conditions (e.g., perinatal mortality, encephalopathy, humeral and clavicle fractures), which vary in the seriousness of their prognosis. For instance, while hypoxic-ischemic encephalopathy (HIE) contributes to long-term disability, fractures are typically less debilitating. Their argument is that without evidence of the relative risk of individualized morbidity outcomes (e.g., HIE), we cannot properly assess the safety of home births.

De Crespigny and Savulescu go on to argue that because children born at home can expect a delay in diagnosis, delivery, and/or transfer following an acute intrapartum event, it is more likely that homebirths lead to a higher rate of HIE events and subsequent disability (de Crespigny and Savulescu, 2014). Jennifer Hollowell and Peter Brocklehurt, authors of the Birthplace study, respond to this claim arguing that while mixed outcomes can obscure differences in outcomes among birth sites, it is speculation to say that the risk of long-term disability is higher at home. Indeed, although the confidence intervals are wide, the Birthplace study does not suggest an excess of neonatal encephalopathy in multiparous women planning a homebirth: neonatal encephalopathy occurs in 1.2 events per 1,000 planned homebirths (99% CI .6 to 2.2) vs. 1.8 per 1,000 planned obstetric unit births (99% CI .8 to 3.7) (Hollowell, 2014). Confidence intervals (e.g., .99% CI .6–2.2) provide an estimate of uncertainty associated with the statistical estimate of a population parameter (e.g., mean, odds ratio, weighted incidence).

De Crespigny and Savulescu respond to this point arguing that the Birthplace study does not provide us with information regarding the differences in long-term avoidable disability according to place of birth. This is both because the study was not powered to individuate the primary outcomes and because it did not include a long-term follow-up. Consequently, we cannot conclude anything about the relative riskiness or safety of homebirth. Another way to put de Crespigny and Savulescu's argument is that the logic of the Birthplace study is invalid: the conclusion does not follow from the premises.

Measuring Instrument

In the above paragraphs I discussed the outcomes that the researchers in the Birthplace Study were interested in measuring; now I move on to discuss the instrument through which they acquired information about these outcomes. In this study information regarding the primary outcome was collected via forms completed by the midwife who started intrapartum care and in some cases with the help of a member of the clinical team on the admitting neonatal unit. Epistemologically speaking, these data collection forms are complex and in part serve as the measuring instrument for the primary outcomes (in part they also serve to provide other kinds of information, e.g., demographics of the study). Consider the fifth and sixth questions in Section E of the Home Birth Data collection form (Birthplace in England Collaborative Group, 2011):

"Was this baby admitted to a neonatal unit within 48 hours of birth?"   yes □   no □
"If yes, to where was the baby admitted?"
□ Special Care Baby Unit   □ High Dependency Unit   □ Neonatal Intensive Care

Date baby was discharged from neonatal unit: □□/□□/□□
Not yet discharged   □

Were any of the following identified in the baby within 48 hours after birth?

□ Meconium aspiration syndrome   □ Cephalohaematoma
□ Neonatal encephalopathy   □ Cerebral haemorrhage
□ Brachial plexus injury   □ Kernicterus
□ Fractured humerus   □ Seizures
□ Fractured clavicle   □ Admission to neonatal unit within 48 hrs
□ Fractured skull   of birth for at least 48 hrs with evidence of
□ Neonatal sepsis   feeding difficulties or respiratory distress
□ No morbidity identified   □ Other morbidity Please specify_____

Tal makes a useful distinction between measurement indications and measurement outcomes, which help mark the transition between answers to questions such as those above and the results of the study, which I discuss below (Tal, 2013). Measurement indications describe states of the measuring instruments (e.g., answers to questions), and measurement outcomes refer to knowledge claims about quantities attributed to the object being measured (e.g., morbidity).

The measurement indications in this example are the positively marked tick boxes that a midwife checks to indicate answers to the questions posed on the data collection forms. Measurement outcomes are acquired by statistical analysis of multiple indications, often taking into account missing data and systematic error. With regard to the latter in the Birthplace study, neonatal encephalopathy required clinical review and coding to ensure accurate and consistent treatment. For example, a clinician reviewed each data form in which there was no confirmed diagnosis of neonatal encephalopathy, but there was a record of isolated seizures (which might indicate encephalopathy). If there was no cause of seizure other than presumed oxygen deprivation, then neonatal encephalopathy was coded as the outcome (Birthplace in England Collaborative Group, 2011).

Measurement Outcome

Measurement outcomes of mortality and morbidity are typically given as a statistic. It is not uncommon for these statistics to represent some kind of risk. Risk can be expressed as absolute or relative risk. Absolute risk refers to an individual's chances of an adverse outcome over a period of time (e.g., intrapartum period). In the Birthplace study, absolute risk was expressed as the weighted incidence of the primary outcome per 1,000 for women planning to give birth at difference locations. Relative risk compares the risk between two groups of people. In the Birthplace study, relative risk was expressed as an odds ratio between, for example, nulliparous women planning a homebirth and nulliparous women planning an obstetric birth.

As both Jacob Stegenga and Alex Broadbent have noted, it is common practice for clinical studies to report their results in terms of relative risk at the expense of absolute risk (Broadbent, 2013; Stegenga, 2015). This practice can be misleading because relative risk does not take into account the baseline rates of the outcome in question, and as a result doctors and patients tend to overestimate the risk of the intervention in question. Stegenga argues that it is less misleading for studies to present both kinds of statistics (Stegenga, 2015). But while the Birthplace study does present both absolute and relative risk, controversy regarding which statistic to emphasize continues.

Consider the results of the Birthplace study for all of the women planning obstetric and homebirths. There were 4.4/1,000 primary events in the planned obstetric unit group and 4.2/1,000 in the planned homebirth group. Looking at absolute risk, planned homebirths appear safer for healthy women with low-risk pregnancies than planned obstetric unit births, but there are different ways of dividing up the cohort of women planning an obstetric and homebirth, which can alter the interpretation of the results.

For example, part of the way through collecting data for the study, it was discovered that almost 20% of women in the obstetric unit group, compared with ≤ 7% in each of the other settings, had at least one complicating condition at the start of labor (e.g., high blood pressure or abnormal fetal heart rate). This difference suggested to the researchers that the risk profile was different among low-risk women in the study. The study's steering group decided to analyze a restricted population of women who did not experience complication at the start of labor. In this cohort we see a weighted incidence per 1,000 of 3.1 for the obstetric group and 4.0 for the homebirth group. The odds of a primary event was 59% higher for those planning a homebirth.

What should be made of the findings from this study? In their conclusions the Birthplace study authors repeatedly emphasize that, "the incidence of adverse perinatal outcomes was low in all settings" (Birthplace in England Collaborative Group, 2011). In other words, the authors emphasize the statistical findings expressed in terms of absolute risk and as a result see individual choice as a reasonable policy. Some, including the UK's National Institute for Clinical Excellence (NICE), have agreed with these conclusions, and in April 2015 NICE decided to update its intrapartum care guidelines. These guidelines now recommend that all healthy low-risk women should be given a choice of birthplace (NICE, 2014).

But others referring to the same study have interpreted the evidence differently. For example, Frank Chervenak and colleagues argue that offering women a choice of birth setting "is irrational and cannot be supported in light of the reported adverse outcomes for birth outside of an obstetric service" (Chervenak et al., 2013). In this article Chervenak et al. focus exclusively on outcomes reported in the Birthplace study, referencing the population of women who did not experience complications at the start of labor. Moreover, they report solely the odds ratio of 1.59 to argue that the choice of a homebirth is "irrational," and they never mention absolute risk.

What should we take from this disagreement over whether the measurement outcomes from the Birthplace study support a reasonable or irrational choice regarding place of birth? First, while it is certainly better to include absolute and relative risk in a study, doing so does not forestall disagreement over the interpretation of the results. Second, this disagreement points to philosophical issues beyond the scope of the present chapter. One of those issues is a question regarding what population cohort best represents the general population of women with low-risk pregnancies: the restricted population or the study population? Independent of absolute or relative risk, the answer to this question could help direct our attention to one set of statistics.

Another issue concerns non-epistemic values (e.g., ethical, aesthetic, prudential values) that are at play in interpreting these measurement outcomes. For instance, while the authors of the Birthplace study consider low-risk pregnancies to be a low-risk activity, de Crespigny and Savulescu look at the same mortality and morbidity data and conclude that birth is "inherently risky" (de Crespigny and Savulescu, 2014). Inmaculada de Melo-Martin and Kristen Intemann (2012) have argued that this difference in interpretation is at least partly due to the non-epistemic values that the different authors hold. Although we can only speculate, the position that Holowell and colleagues take on the riskiness of childbirth seems to be motivated by values of respect for nature, individuality, expertise (midwifery), trust, and what we might call clinical frugality. The positions that Crespigny and Savulescu, and Chervenak and colleagues take on this same issue seem to be influenced by values of respect for the unknown, duty, science, control, and restraint (McClimans, 2015). It is in virtue of holding these different values that these researchers can read the same data and understand its significance differently.

Quality of Life

Although "hard" outcome measures such as mortality and morbidity originally led the outcome assessment movement, "soft" outcome measures of quality of life were quickly taken up as part of it. The popularity of these measures—often referred to collectively as Patient-Reported Outcome Measures (PROMs)—is due to at least two factors. First, for many contemporary health interventions, particularly elective ones such as hip and knee replacements, the point is less one of mortality and morbidity and more a matter of improvements in quality of life. Indeed, even in cases such as place of birth, if the mortality and morbidity is roughly equivalent among birthplaces, then quality of life becomes a deciding factor—indeed it *is* a deciding factor for many low-risk women. For instance, women may choose to give birth in a non-obstetric unit because they want to avoid some of the risk factors associated with urinary incontinence. Non-obstetric units are associated with decreased probability of having an episiotomy (a cut made in the pelvic floor muscle during delivery) (Birthplace in England Collaborative Group, 2011). Episiotomies are associated with an increased probability of urinary incontinence. The quality of life associated with urinary incontinence is measured with a PROM (e.g., Urinary Incontinence Quality of Life Scale, I-QOL).

Second, patients play an essential role in the assessment of quality of life, since the measuring instruments used (i.e., questionnaires) seek answers directly from patients about their quality of life. One reason to ask patients directly about their quality of life is that third parties such as clinicians and caregivers tend to underestimate patients' quality of life (Sprangers and

Aaronson, 1992). Another reason is because patient-centered measures promise to ameliorate some of the hermeneutic marginalization that occurs when patients enter the technology and measurement-heavy setting of contemporary health care (Institute of Medicine, 2001). Once again, consider the Birthplace study. Although there is controversy over how to understand the significance of the measurement outcomes associated with neonatal mortality and morbidity, there is almost zero controversy regarding the maternal morbidity measurement outcomes. Maternal morbidity is significantly lower in non-obstetric units (Birthplace in England Collaborative Group, 2011; Chervenak et al., 2013). Although studies have not yet done so, using quality of life measures to compare place of birth might help to move women and their health from the periphery of this debate.

But for all their popularity, PROMs have received significant criticism. One line of criticism focuses on the under-theorized nature of quality of life and the problems this creates when trying to develop a valid and interpretable measuring instrument. Another line of criticism involves one form that the measurement outcome can take: the quality adjusted life year (QALY). In what follows I will consider both criticisms, but before doing so it is helpful to keep in mind one basic distinction: as measuring instruments, PROMs can be either utility or non-utility measures.

Philosophers tend to be more interested in utility measures than non-utility measures. This is because quality of life measures that are associated with utility values enable the generation of QALYs. QALYs aid in priority setting for scarce health care resources and thus pose philosophical questions regarding, for example, justice. For instance, we might wonder how place of birth impacts one's quality of life over time. Indeed, one way to read de Crespigney and Salvulescu's argument is that planned homebirths lead to decreased QALYs since they lead to greater disability. To be sure, this point is speculation as there is currently an absence of QALY data for these outcomes (Schroeder et al., 2012).

Utility measures are always index measures; this means that a patient's quality of life score (e.g., the sum of the instruments indications) can be added together to produce a single number. In the context of a utility measure, this score is correlated with utility values that are derived from the public's ranking of different health states to yield a measurement outcome. The public ranks health states using thought experiments such as the time trade-off technique. The time trade-off technique provides individuals with two scenarios, and they are asked which scenario they prefer. The length of time in the second scenario is varied until the individual is indifferent between the two. Traditionally, the first scenario offers a medical intervention, which results in a chronic condition, i, that lasts for 10 years followed by death. The second scenario offers a medical intervention, which results in full health for $x < 10$ years followed by death. When the value of x is such that an individual is indifferent between the two scenarios, then the utility value for i is given by $x/10$. The EQ-5D (EuroQol) is an example of a utility measure of quality of life that uses the time trade-off technique (Brooks, 1996).

There are more non-utility measures than utility measures (Measuring Health, 2006). The main difference between them is that non-utility measures cannot generate QALYs because they are not associated with utility values. As a result, non-utility measures are not used in conjunction with priority setting in health care, and they do not raise the same philosophical questions about justice. Instead, non-utility measures are used in other contexts (e.g., as endpoints in trials to determine the clinical effectiveness of different medical interventions, to support the claims made on the labels of pharmaceuticals, and to help determine the quality of the health care patients receive).

Non-utility measures can be either profile or index measures (i.e., producing a single score or multiple scores for each questionnaire). A profile measure is developed when the construct under measurement (i.e., quality of life of those with cancer) is deemed to consist of multiple independent measurement scales (e.g., physical functioning, emotional role functioning, mental health). Two examples of non-utility measuring instruments are the European Organization for Research and Treatment of Cancer Quality of Life Questionnaire (EORTC QLQ) and the aforementioned Urinary Incontinence Quality of Life Scale (I-QOL).

Problems with Quality of Life Measurement: Interpretability

The first line of criticism regarding the under-theorized nature of quality of life typically refers to non-utility measures. This criticism can be summarized by a single deficiency that has multiple consequences—namely, quality of life measurement generally lacks the theoretical resources to provide a representation of the measurement interaction (i.e., the relationship between the primary outcome or construct, e.g., quality of life and its instrument, i.e., a PROM) (Van Fraassen, 2010). This criticism is not uncommon amongst psychologists and others working within the field. Consider Donna Lamping's (2008) Presidential address to the International Society of Quality of Life Research, in which she identified the need for a theoretical framework as one of three challenges facing the future of PROMs. Or take Jeremy Hobart et al.'s (2007) *Lancet Neurology* article, in which they lament the lack of explicit construct theories in their article criticizing the current state of PROMs (Hobart et al., 2007). Sonja Hunt, in her 1997 editorial for *Quality of Life Research*, argues that the surfeit of poorly designed measures suggest that we do not know what quality of life *is* (Hunt, 1997). In what follows, I provide a brief overview of one consequence that results from this lack of theory: problems with interpretability.

Over the last 15 years, the discussion of how to interpret change in patient-reported outcomes has received considerable attention. Interpretability refers to the clinical significance of increases or decreases on a particular measure over time. For instance, if I score a 30 on the Urinary Incontinence Quality of Life Scale (IQOL) directly after having an episiotomy, then we know that I have scored toward the bottom end of the scale—the IQOL has a 100-point scale, and higher scores indicate better quality of life. But imagine two months later I score 42. What does this 12-point increase mean from a clinical point of view? Should my drug regime change? If so, how should it change? Most PROMs only provide ordinal-level information; i.e., we know that someone who scores 42 is more depressed than someone who scores 30, but we do not know the degree of that difference. PROMs are thus difficult to interpret.

This difficulty has led to the development of methods to enhance their interpretability. One popular method is the identification of a minimal important difference (MID). A MID is the smallest change in respondent scores that represent clinical, as opposed to merely statistical, significance and that would *ceteris paribus* warrant a change in a patient's care (Jaeschke et al., 1989). One method for determining a measure's MID is to map changes in respondent answers onto some kind of control. The idea is to determine the minimal amount of change that is noticeable to patients and to use this unit of change as the MID. This method asks the control group of patients to rate the extent of their symptom change over the course of an illness or intervention on a transition-rating index (TRI). TRIs are questionnaires that ask patients questions, such as "Do you have more or less pain since your first radiotherapy treatment?" Typically, patients are given seven possible answers ranging from "no change" to "a great deal better" (Fayers and Machin, 2015). Those who indicate minimal change (i.e., those who rate themselves as just "a little better" than before the intervention) become the patient control group. The mean-change score of this group is used as the MID for the PROM.

This approach of acquiring a MID via a patient control group assumes that respondents who rate their symptom change as "a little better" on a transition question should *ceteris paribus* also have comparable change scores on the PROM. Put differently, similarities in respondent answers to transition questions ought to underwrite similarities in respondents' magnitude of change over the course of an intervention or illness, but qualitative data from interviews with patients suggests that this assumption is ill-founded (Taminiau-Bloem et al., 2011, Wyrwich and Tardino, 2006). Whether one understands the magnitude of change over the course of an illness as large or small is a matter of interpretation.

Consider Cynthia Chauhan, a patient advocate during the deliberations on the FDA guidelines for the use of PROMs in labeling claims. During her testimony, she discussed the side effects of a drug called bimatoprost, which she uses to forestall blindness from glaucoma. One of the side effects of bimatoprost is to turn blue eyes brown. Chauhan has "sapphire blue" eyes, in which, she says, she has taken some pride. As she speaks of her decision to take the drug despite its consequences, she notes that doing so will affect her identity in that she will soon no longer be the sort of person she has always enjoyed being—i.e., she will no longer have blue eyes (Chauhan, 2007).

We can imagine that, even if the bimatoprost is only minimally successful and Chauhan's resulting change score from the PROM is low, she will nonetheless have experienced a significant change—she will not be the same person she was before. But this significance is tied to the place that her blue eyes had in her understanding of herself and what she took to be a good life; *ceteris paribus* we would not expect a brown-eyed person to summarize their experience in the same way. Thus, it would not be surprising if Chauhan's answer to the transition question was "quite a bit," while the magnitude of her change score was minimal.

I suggest that what examples such as this illustrate is that our understanding of clinical significance ought to be closely linked to our understanding of quality of life given the cohort of respondents for whom the outcome measure is targeted (e.g., octogenarians). To put this point slightly differently, understanding change in PROMs requires that researchers have a grip on what quality of life means in the context of a particular PROM and the population it serves.

Quality Adjusted Life Years

Thus far I have been discussing quality of life in terms of difficulties with theorizing the construct and the consequences this has for developing an interpretable outcome measure. The second line of criticism regarding quality of life refers to utility measures and one statistic commonly used to express their measurement outcomes: the quality-adjusted life-year (QALY).

QALYs are a way of valuing and comparing health outcomes. They combine the length of survival with a measure of the quality of that survival and assume that given a choice a person would choose a shorter life of high quality over a longer life of poor quality. A year of life in perfect health is associated with 1 QALY and death is associated with 0. To determine the values between a PROM, such as the EQ-5D (EuroQol), provides indications that are correlated with utility values derived from the public's ranking of different health states. Earlier I gave the example of the time trade-off technique as one way to value the health states provided by the EQ-5D (EuroQol). These respective utility values are multiplied by the number of years that patients are expected to live in that particular state of health. This multiplication provides a QALY between 1 and 0. QALYs can then be used to compare the quality of life of different individuals in a population or the same individual given different treatment regimes.

Consider the following example. Since we do not have QALY data for place of birth, the following QALY gains are fictitious but not inconceivable. Imagine a group of women giving birth in an obstetric unit and a group giving birth at home. For the obstetric units, further imagine that the QALY gain (before and after birth) is .0 and the QALY gain (before and after birth) for the homebirth group is -0.01. According to these QALY scores, obstetric unit births are associated with better quality of life than homebirths, but where should women give birth? The answer depends at least in part (but as we saw earlier when we discussed the neonatal mortality and morbidity measurement outcomes, only in part) on whether obstetric unit births are cost effective. The average cost of a homebirth in the UK is approximately $1,180, and the average cost of an obstetric unit birth is approximately $1,730 (Schroeder et al., 2012). To determine cost effectiveness we divide the difference of the cost of the respective locations by the difference of the QALY gains.. The cost per QALY in this case is $55,000. Institutions such as the UK's National Institute for Clinical Excellence (NICE) often set cost per QALY thresholds over which new treatments are not considered cost effective. NICE's current threshold is at the £20,000–£30,000 mark (Appleby et al., 2007).

This example illustrates an interesting point. Even if quality of life over time is slightly worse when giving birth at home, it still may not be cost effective to encourage women to give birth in obstetric units (as we currently do in the United States). To be sure, this point begs the question of using cost per QALYs to determine healthcare priorities. One common criticism of QALYs is that they use a societal point of view instead of an individual point of view to value health states. With regard to the first point, it is not the women giving birth who determine the utility value associated with their health state. Recall from earlier that the general population determines utility values when they participate in thought experiments such as the time trade-off exercise. Thus, women giving birth provide information about their health state via a PROM such as the EQ-5D, but their score from this measuring instrument is associated with a utility value determined by the general population. This utility value is used, along with time, to generate a QALY. Is it appropriate for the general population, some of which are individuals who will never give birth, to value the different places where one can give birth?

QALYs have also been criticized on the grounds that they are ageist (Harris, 1987). Because part of the QALY algorithm requires the number of years lived in a particular health state, older members of a population are at a disadvantage simply because their natural life expectancy is less than younger members of the same population. Other criticisms of discrimination have also been leveled at QALYs. If low cost per QALY interventions translate into health care priorities (at the expense of high cost per QALY interventions), then those individuals with conditions that are relatively cheap to treat will be prioritized. This way of priority setting could introduce systematic bias if, as is in fact the case, certain people in a population require a greater investment of resources to obtain a decent quality of life.

Despite these criticisms, QALYs are the measurement outcome of choice in most resource allocation contexts. They are the cornerstone of economic analysis for governments, managed care, and other health care payers (Kind et al., 2009). Few are overly enthusiastic over QALYs, but many believe that while they are not perfect, they are useful (see, e.g., Hausman, 2006). In 2006 the International Society for Pharmacoeconomics and Outcomes Research (ISPOR) held an invited panel entitled "Will the QALY Survive?" The panelists acknowledged various degrees of difficulties, but they agreed that it would survive. Dennis Fryback usefully summarized the reasons: (1) the QALY provides a metric for the allocation of resources that combines capacity and functioning; (2) it does a decent job indicating the average impact of an intervention on a population; and (3) we do not have an alternative.

References

Appleby, J., Devlin, N., and Parkin, D. (2007) "NICE's cost effectiveness threshold," *British Medical Journal* 335 358–359.
Birthplace in England Collaborative Group. (2011) "Perinatal and maternal outcomes by planned place of birth for healthy women with low risk pregnancies: The Birthplace in England national prospective cohort study," *British Medical Journal* 343 d7400–d7400.

Broadbent, A. (2013) *Philosophy of Epidemiology*, Basingtoke: Palgrave Macmillan.
Brooks, R. (1996) "EuroQol: The current state of play," *Health Policy* 37 53–72.
Chauhan, C. (2007) "Denouement: A patient-reported observation," *Value Health* 10 S146–S147.
Chervenak, F.A., McCullough, L.B., Brent, R.L., Levene, M.I., and Arabin, B. (2013) "Planned home birth: The professional responsibility response," *American Journal of Obstetrics and Gynecology* 208 31–38.
de Crespigny, L., and Savulescu, J. (2014) "Homebirth and the future child," *Journal of Medical Ethics* 40(12), 807–812.
de Melo-Martín, I., and Intemann, K. (2012) "Interpreting evidence: why values can matter as much as science," *Perspectives in Biology and Medicine* 55(1), 59.
Department of Health. (2008) *High Quality Care for All: NHS Next Stage Review Final Report*, London: The Stationary Office.
Donabedian, A. (1988) "The quality of care: How can it be assessed?," *Journal of the American Medical Association* 260 1743–1748.
Epstein, A.M. (1990) "The outcomes movement—will it get us where we want to go?," *New England Journal of Medicine* 323 266–270.
Fayers, P. M., and Machin, D. (2015) *Quality of life: The assessment, analysis and reporting of patient-reported outcomes*, Hoboken, NJ: John Wiley & Sons.
Harris, J. (1987) "QALYfying the value of life," *Journal of Medical Ethics* 13(3), 117–123.
Hausman, D.M. (2006) "Valuing health," *Philosophy and Public Affairs* 34 246–274.
Hobart, J.C., Cano, S.J., Zajicek, J.P., and Thompson, A.J. (2007) "Rating scales as outcome measures for clinical trials in neurology: Problems, solutions, and recommendations," *Lancet Neurology* 6 1094–1105.
Hollowell, J. (2014) "Homebirth and the future child: Factual inaccuracies in commentary on the Birthplace study," *Journal of Medical Ethics, Letter* http://jme.bmj.com.pallas2.tcl.sc.edu/content/40/12/807.abstract/reply.
Hunt, S.M. (1997) "The problem of quality of life," *Quality of Life Research* 6 205–212.
Institute of Medicine (IOM). (2001) *Crossing the Quality Chasm: A New Health System for the 21st Century*, Washington, DC: National Academy Press.
Jaeschke, R., Singer, J., and Guyatt, G.H. (1989) "Measurement of health status: Ascertaining the minimal clinically important difference," *Controlled Clinical Trials* 10 407–415.
Kassirer, J.P. (1993) "The quality of care and the quality of measuring it," *New England Journal of Medicine* 329 1263–1265.
Kind, P., Lafata, J.E., Matuszewski, K., and Raisch, D. (2009) "The use of QALYs in clinical and patient decision-making: Issues and prospects," *Value Health* 12 Suppl 1 S27–S30.
McClimans, L., 2015. "Place of birth: Ethics and evidence," *Topoi*. doi:10.1007/s11245-015-9353-0.
McDowell, I. (2006) (3rd ed.) *Measuring Health: A Guide to Rating Scales and Questionnaires*, Oxford: Oxford University Press.
National Institute for Health and Clinical Excellence (NICE). (2014) *Intrapartum Care for Health Women and Babies*, London: National Institute for Health and Clinical Excellence.
O'Connor, R.J., and Neumann, V.C. (2006) "Payment by results or payment by outcome? The history of measuring medicine," *Journal of the Royal Society of Medicine* 99 226–231.
Office for National Statistics. (2013) *Births in England and Wales by Characteristics of Birth 2, 2012*, London: Office for National Statistics
Schroeder, E., Petrou, S., Patel, N., Hollowell, J., Puddicombe, D., Redshaw, M., and Brocklehurst, P. (2012) "Cost effectiveness of alternative planned places of birth in woman at low risk of complications: Evidence from the Birthplace in England national prospective cohort study," *British Medical Journal* 344 e2292.
Sprangers, M.A.G., and Aaronson, N.K. (1992) "The role of health care providers and significant others in evaluating the quality of life of patients with chronic disease: A review," *Journal of Clinical Epidemiology* 45 743–760.
Stegenga, J. (2015) "Measuring effectiveness," *Studies in the History and Philosophy of Biological and Biomedical Sciences* 54 62–71.

Tal, E. (2013) "Old and new problems in philosophy of measurement," *Philosophy Compass* 8 1159–1173.
———. (2015) "Measurement in Science", in E. N. Zalta (ed.), *The Encyclopedia of Philosophy*. Available online at: http://plato.stanford.edu/entries/measurement-science/
Taminiau-Bloem, E.F., Van Zuuren, F.J., Visser, M.R.M., Tishelman, C., Schwartz, C.E., Koeneman, M.A., Koning, C.C.E., and Sprangers, M.A.G. (2011) "Opening the black box of cancer patients' quality-of-life change assessments: A think-aloud study examining the cognitive processes underlying responses to transition items," *Psychology and Health* 26 1414–1428.
Van Fraassen, B.C. (2010) *Scientific Representation: Paradoxes of Perspective*, Oxford: Clarendon Press.
Wyrwich, K.W., and Tardino, V.M. (2006) "Understanding global transition assessments," *Quality of Life Research* 15 995–1004.

Further Reading

McClimans, L. (2015) "Place of Birth: Ethics and Evidence", in M. C. Amoretti & N. Vassallo (eds.), *Topoi: An International Review of Philosophy*, 1–8. Special issue on "Philosophy of sex and gender in gender-medicine." (Extended discussion of role of non-epistemic values in the place of birth debate)
McClimans, L. (2010) "A theoretical framework for patient-reported outcome measures", *Theoretical Medicine and Bioethics*, 31: 225–40. (Extended discussion of need for theoretical framework for PROMs)
McClimans, L. (2011) "Interpretability, validity and the minimum important difference", *Theoretical Medicine and Bioethics*, 32: 389–401. (Detailed discussion of Interpretability in the context of PROMs)
Weinstein, M., Torrance, G., and McGuire, A. (2009) "QALYs: The Basics", *Value in Health*, 12: S5–9. (An introduction to QALYs)

31
MEASURING HARMS
Jacob Stegenga

Introduction

The benefits and harms of pharmaceuticals are principal subjects of investigation in clinical science. In this chapter I discuss how harms are measured and note the challenges that clinical research faces in detecting harms of pharmaceuticals. This chapter provides an introduction to the structure of clinical research with a focus on harm detection. As it is usually performed today, clinical research does not reliably measure the harms of pharmaceuticals. There are at least three categories of problems with clinical research that lead to the underestimation of the harm profile of pharmaceuticals: subtle features of research methodology, secrecy surrounding the evidence from clinical research, and inadequate regulation.

Clinical research is organized into four phases. Phase 1 trials evaluate an experimental drug in a small number of subjects to see if any very serious harms result from the drug. Phase 2 trials are usually randomized controlled trials in several hundred subjects and are preliminary evaluations of the benefits of an experimental drug. Phase 3 trials are larger randomized controlled trials, with several hundred to several thousand subjects, in which the potential benefits of the experimental drug are precisely measured. Approval of the experimental drug for general use occurs after Phase 3 trials. Once the drug is being used in general clinical settings, Phase 4 trials are sometimes done to gather more evidence about the benefits and harms of the drug.

Several important problems impede the detection of harms of experimental drugs in clinical research. Even before any evidence is gathered, the ways that harms are defined and planned to be measured in clinical research contributes to their underestimation. Although harms of experimental drugs are initially evaluated in Phase 1 trials, evidence from these trials is rarely published, although such trials provide the foundation for assessing the harm profile of pharmaceuticals. Phase 2 and 3 randomized controlled trials (RCTs) are usually designed to be as sensitive as possible to detecting potential beneficial effects of pharmaceuticals, but not sensitive enough to detecting rarer potential harmful effects of pharmaceuticals. This is especially troublesome given that RCTs are usually thought to produce the best evidence for causal hypotheses in medicine (although the epistemological status of RCTs is a matter of debate; see Chapters 18, 19, and 20). Phase 4 surveillance of harms contributes to underestimating the harm profile of pharmaceuticals, because harms are under-reported in general clinical settings. At every phase of medical research, the evidence regarding harms of pharmaceuticals is shrouded in secrecy (as is the evidence regarding benefits), which further contributes to the underestimation of the harm profiles of new drugs (and the overestimation of benefits). I use several examples to illustrate the problems with detecting harms of pharmaceuticals, and one example in particular—rosiglitazone (trade name: Avandia) and the research about its harm profile—runs throughout the chapter.

The net effect of these conceptual, methodological, and social factors is that our available pharmaceuticals appear to be safer than they truly are. Were these factors mitigated in medical research, the harm profile of pharmaceuticals would be more faithfully represented, and pharmaceuticals would be deemed more harmful than they now are.

Defining and Measuring Harms

A harm of a pharmaceutical is, of course, an effect of the intervention, just as a benefit is an effect of the intervention. The interpretation of an effect *as* a harm (or conversely, as a benefit) is a normative judgment, and such is influenced by social values. Such judgments are not always straightforward. A compelling illustration is provided by the drug methylphenidate (Ritalin), which is frequently prescribed to treat attention deficit hyperactivity disorder (ADHD). The alleged benefits of methylphenidate depend on a particular social nexus and are conceptually intertwined with its potential harms. Empirical tests of methylphenidate suggest that it mitigates bodily motions of children diagnosed with ADHD and mitigates frequency of social interactions, which might be seen as a benefit by a school teacher. But this effect could be seen as harmful by someone who thinks that children moving around, playing, and socializing are generally positive behaviors, with a large range of normal or healthy levels. Thus, the same effect of a pharmaceutical may be considered a benefit or a harm depending on one's broader normative commitments and sociocultural position. There are, though, many effects of pharmaceuticals that can be considered harms with little ambiguity in typical cases, from insubstantial effects such as a minor headache to more severe effects such as death.

Harmful effects of drugs are often thought of as discrete outcomes, referred to as adverse events, or if they are extremely harmful, serious adverse events. Harms, however, can be changes of continuous parameters (and not just changes of discrete parameters); many harms should not be thought of as events, since discrete events constitute only a subset of the potential harms of a pharmaceutical. As I discuss below, much data about harms of drugs come from passive surveillance after a drug has been approved for general use. Passive surveillance can only identify a token event as a harm if a patient or a physician observes and interprets the event as a harm and reports it as such. Small effects and common effects are often not reported. If a drug causes a patient to gain several pounds, this effect might go unnoticed by the patient and the physician, and even if it were noticed, there is no way that the patient or the physician could reliably assess the drug as a cause of the weight gain. In other words, such an effect might not be attributed as an effect of the drug, and if it were, such an attribution would not be reliable.

Terminological choices also contribute to obscuring the harm profiles of pharmaceuticals. Concern about harms of pharmaceuticals are often referred to with terms such as "drug safety" (indeed, the discipline sometimes referred to as "pharmacovigilance" is also referred to as drug safety). A report of a new kind of harm of a pharmaceutical is a "signal" of a "safety finding" (!), which is documented via "safety reporting." For example, the FDA, when talking about serious harms such as death and strokes caused by peroxisome proliferator activated receptor (PPAR) modulators (described below), referred to these events as "clinical safety signals," and some drugs in this class were removed from the market because of "clinical safety." We should beware of use of the term "safety" to refer to harm.

Beyond mere terminological choices, the way harms are defined and measured in trials contributes to their underestimation. For example, before it was established that some antidepressants cause suicidal ideation and behavior, meta-analyses of clinical trial data suggested that these drugs do not in fact cause such harms. The data in these trials came from patient outcomes measured with the Hamilton Rating Scale for Depression (HAMD). The HAMD is a poor instrument for measuring benefits of antidepressants, because interventions with effects that are irrelevant to the core elements of depression (such as mitigation of fidgeting or a slight change in sleeping patterns) can contribute to a large change in HAMD score (Stegenga, 2015). But the HAMD is an even worse instrument for measuring harms of antidepressants, including suicidality. There is one question on the HAMD regarding suicidality, as follows: "Suicide. 0 = Absent. 1 = Feels life is not worth living. 2 = Wishes he were dead or any thoughts of possible death to self. 3 = Suicidal ideas or gesture. 4 = Attempts at suicide (any serious attempt rates 4)." The numbers refer to the points contributed to the overall HAMD score (the higher the score, the more severely depressed a patient is said to be). The problem is that an antidepressant could cause a patient who has occasional but passing suicidal thoughts to develop severe and frequent suicidal ideation and self-mutilation, yet the patient's HAMD score would not change, since both before and after the antidepressant the patient would receive a score of 3 on suicidality. The HAMD is insensitive to such harms of antidepressants. Since the HAMD is the measuring instrument employed in research on antidepressants, such research systematically underestimates the harms of antidepressants. This example illustrates the general point that the way harms are defined and measured in research can contribute to their underestimation.

Another example of how defining and measuring harms contributes to their underestimation is from the drug rosiglitazone (Avandia), once the world's best-selling drug for type 2 diabetes. By 2007, evidence was mounting that rosiglitazone causes cardiovascular disease and death. GlaxoSmithKline, the manufacturer of rosiglitazone, funded a large trial (the RECORD trial) in an attempt to show that rosiglitazone was safe. The primary outcome measured in the RECORD trial was a composite outcome that included all hospitalizations and deaths from any cardiovascular causes. Although this outcome measure appears reasonable given the goal of the study, it in fact is quite problematic. The problem is that it included hospitalizations that were very likely not related to the randomized interventions. Since we can presume that hospitalizations and deaths that were not caused by the interventions occurred at roughly the same rate between the different groups, the overall larger number of this outcome in both groups minimized the relative difference in outcomes observed between the groups (in other words, the important outcome of interest—cardiovascular disease and death *caused by rosiglitazone*—was watered down, making it less likely that the study would detect a statistically significant difference between the groups). Moreover, the outcome hospitalization depends, obviously, on a patient being hospitalized, but this is a socio-economic decision as much as a health-related outcome; the trial included patients in dozens of countries, including many in eastern Europe, with diverse hospitalization practices. This diversity of practice could have introduced variability into the data, which would make it more likely that a statistically significant difference between experimental groups would not be detected. In short, the way that the potential harm was defined and measured in this trial artificially lowered the chance of detecting the harm of the drug in question.

The broader point illustrated by the examples above is that harms of pharmaceuticals will only be found if the harms are properly defined and measured. Defining and measuring a harm in the wrong ways—such as by employing a measuring instrument or an outcome that is insensitive to the harm in question—amounts to not looking for the harm.

The definition and measurement of harms influences (and constrains) the evidence that is gathered at each phase of clinical research. I now turn to the particular phases of clinical research.

Phase 1: First in Human

A first-in-human study is an experiment in which a pharmaceutical is first tested in humans. Experimental pharmaceuticals are often initially evaluated with cell and animal experiments, and if such experiments provide evidence that suggests that the pharmaceutical is relatively safe and potentially effective for humans, then a first-in-human study is performed, usually in healthy volunteers. These are also referred to as Phase 1 trials.

Such trials are obviously risky for the research subjects. Despite the risk, Phase 1 trials are important because they provide the foundation for assessing harms of pharmaceuticals. Given that first-in-human studies are the first time a potential new pharmaceutical is tested in humans, they provide crucial evidence regarding harms. Such evidence is relevant, obviously, to the harm profile of the particular molecule under investigation, but it is also relevant to the harm profile of the class of molecules to which the particular molecule belongs and is more broadly relevant to the harm profile of drugs, generally. Evidence from first-in-human studies is, therefore, extremely important.

Unfortunately, such evidence is rarely shared publicly. The vast majority of Phase 1 trials are not published (Decullier, Chan, & Chapuis, 2009). Publication bias in clinical research refers to the phenomenon in which studies that suggest that a drug is beneficial and/or safe are more likely to be published than those studies that do not. Those molecules that appear to be relatively safe in a Phase 1 trial often go on to be tested in Phase 2 and Phase 3 trials, and thus the broader scientific community can infer that such molecules have passed a Phase 1 trial and so are at least somewhat safe. Those molecules that appear to be relatively harmful in Phase 1 trials hardly ever go on to be tested in later-phase trials, and results from such Phase 1 trials are rarely published. This failure to publish Phase 1 trials is wasteful, because future scientists who become interested in the same compound may needlessly repeat Phase 1 trials they were unaware of (trial registries have been proposed to address publication bias of Phase 3 trials, but thus far have not been very helpful, and I am not aware of trial registries for Phase 1 trials). Beyond being wasteful, this needless repetition also has the potential for causing needless harm to subjects in these subsequent trials. There is, though, a further consequence of not publishing Phase 1 trials that is even more widespread and sinister.

The problem is that when assessing the harm profile of an experimental pharmaceutical, if one is unaware of past evidence regarding harms, then one's prior assessment that the molecule is harmful will be lower than it should be (that is, lower than it would be if one was aware of such evidence). Since molecules that appear safe in Phase 1 trials go on to be evaluated in larger and more public Phase 2 and 3 trials, and molecules that appear harmful do not, and since most Phase 1 trials are not published, it follows that, of all drugs that are approved for clinical use, the proportion that *appears* harmful is lower, perhaps much lower, than is truly the case. In other words, publication bias of Phase 1 trials is systematically skewed: publicly available evidence from Phase 1 trials tends to suggest that experimental drugs are safe, but unavailable evidence from Phase 1 trials tends to suggest that experimental drugs are harmful. Since we have access to the former but not the latter, our general assessment of the harms of pharmaceuticals is grossly underestimated.

Returning to our example, rosiglitazone is a modulator of proteins called peroxisome proliferator-activated receptors (PPARs), which regulate the expression of genes. In recent years, more than 50 PPAR modulators have failed clinical tests, and many of these failures have been due to harms caused by the PPAR modulators (Nissen, 2010). Indeed, evidence of such harms was available even prior to first-in-human studies: for example, PPAR modulators were found to cause numerous types of tumors and cardiac toxicity in rodent studies. Unfortunately, "few publications have detailed the precise toxicity encountered" (Nissen, 2010).

In addition to past evidence regarding harms of experimental drugs from Phase 1 trials, another factor that ought to influence one's assessment of the probability that an experimental drug is harmful is background knowledge of the way the experimental drug intervenes in normal and pathological physiological mechanisms. PPAR modulators are again a good example: any given PPAR modulator can influence the expression of many dozens of genes, and thus "the effects of these agents are unpredictable and can result in unusual toxicities" (Nissen, 2010). Unfortunately, given the emphasis on RCTs in clinical research, this kind of knowledge is often downplayed. Indeed, mechanistic reasoning is typically denigrated by contemporary evidence-based medicine. Taking such knowledge into account, even if such knowledge were incomplete, would render our estimations of the harm profiles of pharmaceuticals more accurate.

Given this argument, we should expect examples of drugs that appear to be relatively safe based on evidence from Phase 1 trials, but then come to be viewed as relatively harmful based on evidence from clinical trials and post-market surveillance (Phases 2–4). This is precisely what we observe. Just among the class of PPAR modulators, troglitazone has been withdrawn in some jurisdictions because it appears to cause liver damage; tesaglitazar has been withdrawn in some jurisdictions because it appears to cause elevated serum creatinine; pioglitazone has been withdrawn in some jurisdictions because it appears to cause bladder cancer; and muraglitazar has been withdrawn in some jurisdictions because it appears to cause heart attacks, strokes, and death.

In short, the lack of availability of evidence from Phase 1 trials contributes to the systematic underestimation of harms of pharmaceuticals.

Phases 2 and 3: Randomized Controlled Trials

The next steps in testing an experimental pharmaceutical are Phase 2 and Phase 3 RCTs. Although these are larger than Phase 1 trials, as typically employed in clinical research, RCTs are not good methods for detecting harms of pharmaceuticals. In principle, RCTs could be more reliably employed to hunt for harms, though such trials would have to be larger and longer than most trials performed today and incorporate other more fine-grained methodological changes. In practice, RCTs are designed to be sensitive to the detection of benefits of pharmaceuticals instead of being sensitive to the detection of harms.

Trial designers try to maximize the ability of a trial to detect the potential benefits of an experimental pharmaceutical; they employ a number of design strategies to do so. One is to maximize the observed effect size in an RCT by including only subjects who are most likely to show the most benefit of the intervention in question. For example, many trials testing antidepressants include only the most severely depressed patients, and antidepressants have only been shown to work for such patients. Another trial design strategy is to minimize the potential variability of data generated by trials—for example, RCTs usually include a relatively homogeneous group of subjects. The greater the similarity among subjects in a trial with respect to parameters that can influence outcomes of a trial (such as age, sex, or the presence of other diseases), the less variable the data. Moreover, RCTs usually exclude subjects who have diseases other than the one being intervened on in the trial, who are on other pharmaceuticals, and who are elderly. The most egregious examples of inclusion and exclusion criteria are called "enrichment strategies." Enrichment strategies involve excluding subjects—after subject recruitment but prior to the collection of hard data—who respond especially well to placebo or who respond especially poorly to the experimental pharmaceutical. Such strategies are widely employed in clinical research.

The net effect of these strategies is that subjects of trials are different in many important respects from the patients who might use new pharmaceuticals once they are approved for use in a clinical setting. Some of these features that often differ between research subjects and clinical patients are known to influence the harm profiles of pharmaceuticals. For example, older people, pregnant women, and patients on other drugs are more likely to be harmed by pharmaceuticals, but they are also precisely the kinds of people who are excluded from RCTs. For example, the most common harm of statins is myopathy, which ranges from simply myalgia (muscle pain) and muscle weakness, to rhabdomyolysis, a severe condition in which muscle tissue dies and releases proteins into the blood, which can cause kidney failure and death. This risk is higher among women, elderly people, and people with other conditions like infections, seizures, and kidney disease—precisely the kinds of people that are excluded from clinical trials.

In short, the exclusion of certain kinds of patients and inclusion of other kinds of patients entails that the sensitivity of RCTs to detect harms is typically much less than the sensitivity of RCTs to detect benefits. The subjects who are included in a trial are less likely to be harmed by an experimental pharmaceutical than an average person is, and more likely to benefit. In another striking example, Worrall (2010) notes that in the large ASSENT-2 trial, one exclusion criterion was "any other disorder that the investigator judged would place the patient at increased risk." Of course, there is a basis for such an exclusion criterion—namely, the protection of patients who are more liable to be harmed by experimental pharmaceuticals. However, this exclusion criterion directly reduced the ability of the trial to detect the harms of the experimental pharmaceutical that would result when it would be employed in a real-world clinical setting (since it is precisely in the clinical setting in which patients have other disorders that put them at increased risk of harm).

To return to rosiglitazone, a meta-analysis showed that rosiglitazone causes an increased risk of heart attack and death (Nissen & Wolski, 2007), although the RCTs in the meta-analysis were too small individually to have adequate power to detect this rare but severe harm. GlaxoSmithKline funded the RECORD trial in an attempt to refute this finding. RECORD employed seven inclusion criteria and 16 exclusion criteria, and 99% of the subjects were Caucasian. One effect of these criteria was that subjects in the trial were, on average, healthier than the broader population; for example, subjects in both the experimental group and control group of the trial had a heart attack rate of close to half that in the equivalent demographic group in the broader population. Thus, the results from this trial were less applicable to the target population than it would have been had the trial not employed such restrictive exclusion and inclusion criteria.

In addition to restrictive exclusion and inclusion criteria, a further reason why RCTs are typically not very good at finding harms is that there is often insufficient measurement of a wide range of physiological parameters that could indicate that the pharmaceutical causes harm. In principle, a pharmaceutical could negatively impact any of the countless physiological parameters of a subject, but these harms will not be detected if they are not measured. Moreover, even if such parameters are measured, they are often not reported. A survey of 142 randomly selected publications of RCTs of psychiatric interventions found that only a fraction bothered to address harms, and on average, reports of RCTs used a small fraction of a page in the results section to discuss harms (Papanikolaou, Churchill, Wahlbeck, & Ioannidis, 2004). In other words, harms that are not looked for will not be found.

Two other limitations of RCTs contribute to the underestimation of harms of pharmaceuticals: their size and their duration. RCTs normally enroll enough subjects to detect the potential benefit of the pharmaceutical for the disease in question. Any further subjects add expense. However, this number of subjects is often not enough to detect harms that are rare. The duration of a trial is normally also just long enough to detect the potential benefit of the pharmaceutical for the disease in question. Some RCTs only evaluate pharmaceuticals for a period of weeks, as did many of the trials of rosiglitazone. However, some harms of drugs manifest only after years of taking the drug. Methylphenidate (Ritalin), for example, has been shown to cause stunted growth in children, but this is only found three years after the initiation of treatment with the drug (Swanson et al., 2007). For these reasons, larger and longer Phase 4 observational studies are usually relied on to detect harms (but as I note below, Phase 4 studies have their own practical and epistemic shortcomings).

Since RCTs underestimate harms, we should expect to observe examples of drugs that appear relatively safe after RCTs but come later to appear more harmful once used in a clinical setting. This phenomenon is widespread. In the worst-known cases, pharmaceuticals are pulled from the market by manufacturers or regulators, such as valdecoxib (Bextra), fenfluramine (Pondimin), gatifloxacin (Gatiflo), and rofecoxib (Vioxx) over just the last few years. In other cases, the harm profile in the clinical setting appears greater than RCTs suggested, but the drugs have been left on the market for whatever reason (often, regulators consider the benefit-harm profile of the drug to remain favorable despite the increased estimation of its harm profile). A few examples include celecoxib (Celebrex), alendronic acid (Fosamax), risperidone (Risperdal), and olanzapine (Zyprexa).

Phase 4: Post-Approval

The majority of data on harms of pharmaceuticals come from observational studies and passive surveillance conducted after a pharmaceutical has been approved for clinical use. These studies are sometimes called Phase 4 studies. The fact that the majority of data regarding harms of pharmaceuticals comes from post-market studies has an important practical consequence, and the fact that such studies are usually observational designs has an important epistemic consequence.

The bar for approval of a new pharmaceutical is low. The FDA, for example, requires only that two Phase 3 RCTs show some benefit, regardless of how many RCTs were performed, and although such RCTs are usually not very sensitive to the harm profile of the pharmaceutical. After the new pharmaceutical has been approved for clinical use, the potential harms of the pharmaceutical are assessed by passive surveillance systems and observational studies. Although the harm profile of new pharmaceuticals are not yet known at this point, they are nonetheless prescribed to and consumed by patients, often numbering in the millions. Only when the new pharmaceuticals are used in a clinical setting (rather than an experimental setting) are most data on harms gathered. These data come from patients who are prescribed the drug by their physician and thus unwittingly become subjects in a study of the harm profile of the drug.

There is strong reason to think that, like the earlier phases of drug evaluation, post-market passive surveillance also significantly underestimates harms of pharmaceuticals. One empirical evaluation puts the underestimation rate at 94% (Hazell & Shakir, 2006). Part of the reason for this is that there is widespread under-reporting of possible harmful effects (as noted above, patients have to identify harmful effects, and patients and physicians have to infer that the harmful effect is indeed a result of a drug, and the physician must then take the time to report the harmful effect).

A few large-scale randomized trials have included a thorough hunt for harms, and Papanikolaou, Christidi, and Ioannidis (2006) compared estimates of harms from these trials to equivalent non-randomized trials. They found that non-randomized studies, on average, have conservative estimates of harms of pharmaceuticals relative to randomized trials of comparable sizes on the same intervention. One reason cited for such a finding is that patients who reliably take their prescribed medications on schedule tend to be healthier than non-compliant patients, and thus there is a confounding factor when comparing the outcomes of those patients who consume more of a particular pharmaceutical with those patients who consume less of the pharmaceutical. That is, those who consume more of a medication (those who are faithful to their medication schedule) also tend to be healthier than those who take less, so observational studies tend to overestimate the benefits of pharmaceuticals and underestimate their harms.

The kind of evidence most often available in post-approval research can influence our estimations of the harm profiles of drugs. Most evidence regarding harms of pharmaceuticals comes from non-randomized studies, and the dominant view of the evidence-based medicine (EBM) movement is that non-randomized studies are unreliable. Therefore, the EBM movement denigrates the majority of evidence regarding harms of pharmaceuticals. This is important because EBM is now a very influential movement—for instance, EBM proponents' views about RCTs has influenced regulators such as the FDA. According to this line of thinking, only RCTs can provide compelling evidence regarding harms of pharmaceuticals. Since the majority of data regarding harms comes from non-randomized studies, and since the data regarding harms that does come from RCTs is fundamentally limited for the reasons noted above, regulators are liable to make unreliable judgments regarding the harms of pharmaceuticals.

Vandenbroucke (2008) and Osimani (2014) have argued against the view that RCTs are better than non-randomized (e.g., observational) studies at detecting harms. Because harms of drugs are unintended and often unknown effects, physicians cannot bias treatment allocation with respect to such effects. Thus, so-called selection bias is less of an epistemological worry for unintended harmful effects as it is for intended beneficial effects, and so one of the central advantages of RCTs over observational studies is mitigated in the context of the hunt for harms.

Secrecy

Evidence regarding harms of pharmaceuticals is shrouded in secrecy. Companies that pay for RCTs claim that they own the data from the trials, and clinical researchers who participate in RCTs are bound by gag clauses in their contracts that constrain their ability to share data, even if they suspect that the pharmaceutical under investigation causes harm. One form of secrecy is the ubiquity of publication bias (discussed above about Phase 1 trials). Just like first-in-human studies, Phase 3 trials suffer from substantial publication bias, resulting in the benefits of novel pharmaceuticals being exaggerated and the harms of novel pharmaceuticals being underestimated. Another form of secrecy involves the withholding of data from independent researchers by both corporate manufacturers and government agencies. As with first-in-human studies, if regulators do not have access to all of the available data, then they cannot make a reliable inference regarding the harms of drugs.

As one example, reboxetine is an antidepressant (SSRI) sold in Europe during the past decade. Recently, researchers with access to both published and unpublished data performed a meta-analysis of the drug's effects (Eyding et al., 2010). Of the 13 trials that had been performed on reboxetine, data from 74% of patients remained unpublished. Seven of the trials compared the drug to placebo: one had positive results and only this one was published; the other six trials (with almost 10 times as many patients as the positive trial) showed no benefit for reboxetine, and none of these were published. The trials that compared reboxetine to competitor drugs were worse. Three small trials suggested that reboxetine was superior to its competitors, but the other trials, with three times as many patients, showed that reboxetine was worse than its competitors on the primary outcome and had worse side effects. (For more details on this episode, see Goldacre, 2012.) Thus, just as with Phase 1 trials, Phase 3 trials suffer from publication bias.

Once one starts looking for examples of the secrecy surrounding evidence of harms of pharmaceuticals, they become easy to find. Olanzapine (Zyprexa) is now known to cause extreme weight gain and concomitant diabetes, but the manufacturer, Eli Lilly, "engaged in a decade-long effort to play down the health risks of Zyprexa according to hundreds of internal Lilly documents and e-mail messages among top company managers" (Berenson, 2006). Oseltamivir (Tamiflu) provides yet another example: evidence on the harms of oseltamivir largely remain unpublished, despite the massive stockpiling of the drug by Western countries in recent years (Doshi, 2009).

Rosiglitazone, again, provides a striking example. After several trials suggested that rosiglitazone may cause cardiovascular harms, Nissen and colleagues requested this trial data from GlaxoSmithKline, which refused to share it. However, due to a lawsuit regarding secrecy about another of its drugs (Paxil), the company had agreed to develop a registry of data from its clinical trials, and Nissen was able to access this data. These investigators identified 42 RCTs of rosiglitazone, of which only seven had been published. The resulting meta-analysis, which included the unpublished studies, showed that rosiglitazone increases cardiovascular harms by 43%. Within 24 hours of the meta-analysis being submitted to the *New England Journal of Medicine*, one of the peer reviewers faxed a copy of the manuscript to GlaxoSmithKline. Internal company emails discussed the similarity of Nissen's findings to GlaxoSmithKline's own analysis, which they had performed years earlier but had not published. Moreover, the FDA had performed their own analysis, which reached similar conclusions, but also did not publicize its findings. The director of research at the company wrote: "FDA, Nissen, and GSK all come to comparable conclusions regarding increased risk for ischemic events, ranging from 30% to 43%!" (Harris, 2010). In other words, the FDA and GlaxoSmithKline already knew of the cardiovascular harm caused by rosiglitazone, but neither the regulator nor the company had publicized this finding.

When the secrecy of evidence about harms of pharmaceuticals is threatened by vigilant researchers, manufacturers occasionally respond belligerently. Rosiglitazone, again, provides a good illustration. John Buse, one of the world's foremost diabetes researchers, gave two talks in 1999 arguing that rosiglitazone may have cardiovascular risks. GlaxoSmithKline then executed an orchestrated campaign to silence Dr. Buse. This plan appears to have been initiated by the company's head of research, and even the chief executive officer was aware of it. The company referred to Dr. Buse as the "Avandia Renegade," and in contact with Buse and Buse's department chair there were "implied threats of lawsuits." The company's head of research wrote in an internal email:

> I plan to speak to Fred Sparling, his former chairman as soon as possible. I think there are two courses of action. One is to sue him for knowingly defaming our product even after we have set him straight as to the facts—the other is to launch a well planned offensive on behalf of Avandia. . . .

Buse responded to the company with a letter that ended by capitulating and asking them to "please call off the dogs." (This episode was detailed in a U.S. Senate Finance Committee Report, Anon, 2007.) Later, Buse expressed embarrassment that he caved to the pressure of GlaxoSmithKline. Only with the publication of Nissen's meta-analysis in 2007 was the potential of rosiglitazone to cause ischemic events revealed. By that time, the FDA estimated, rosiglitazone had caused about 83,000 heart attacks since coming on the market in 1999.

Conclusion

Various solutions have been proposed to address some of the problems of detecting harms of pharmaceuticals. There are some obvious candidates, including increasing the quality of evidence in the hunt for harms, improving the accessibility of such evidence, and improving regulation of clinical research. The harm profile of a pharmaceutical is, obviously, necessary in order to evaluate the benefit-harm balance of the pharmaceutical. Because harms of pharmaceuticals are systematically underdetected at all stages of clinical research, policy makers and physicians generally cannot adequately assess the benefit-harm balance of pharmaceuticals.

References

Anonymous. (2007). *The Intimidation of Dr. John Buse and the Diabetes Drug Avandia*. Committee staff report to the chairman and ranking member, Committee on Finance, United States Senate, November 2007. Available at: http://www.finance.senate.gov/imo/media/doc/prg111507b.pdf

Berenson, A. (2006, December 17). Eli Lilly said to play down risk of top pill. *The New York Times*. Available at: http://www.nytimes.com/2006/12/17/business/17drug.html?_r=0

Decullier, E., Chan, A.-W., & Chapuis, F. (2009). Inadequate dissemination of phase I trials: A retrospective cohort study. *PLoS Med, 6*(2), e1000034. doi:10.1371/journal.pmed.1000034

Doshi, P. (2009). Neuraminidase inhibitors—the story behind the Cochrane review. *BMJ, 339*, b5164. doi:10.1136/bmj.b5164

Eyding, D., Lelgemann, M., Grouven, U., Härter, M., Kromp, M., Kaiser, T., . . . & Wieseler, B. (2010). Reboxetine for acute treatment of major depression: Systematic review and meta-analysis of published and unpublished placebo and selective serotonin reuptake inhibitor controlled trials. *BMJ, 341*. doi:10.1136/bmj.c4737

Goldacre, B. (2012). *Bad pharma: How drug companies mislead doctors and harm patients*. New York: Farrar Straus & Giroux.

Harris, G. (2010, February 22). A face-off on the safety of a drug for diabetes. *The New York Times*. Available at: http://www.nytimes.com/2010/02/23/health/23niss.html

Hazell, L., & Shakir, S. A. (2006). Under-reporting of adverse drug reactions: A systematic review. *Drug Saf, 29*(5), 385–396.

Nissen, S. E. (2010). The rise and fall of rosiglitazone. *Eur Heart J, 31*(7), 773–776. doi:10.1093/eurheartj/ehq016

Nissen, S. E., & Wolski, K. (2007). Effect of rosiglitazone on the risk of myocardial infarction and death from cardiovascular causes. *New England Journal of Medicine, 356*(24), 2457–2471. doi:10.1056/NEJMoa072761

Osimani, B. (2014). Hunting side effects and explaining them: Should we reverse evidence hierarchies upside down? *Topoi, 33*(2), 295–312.

Papanikolaou, P. N., Christidi, G. D., & Ioannidis, J. P. A. (2006). Comparison of evidence on harms of medical interventions in randomized and nonrandomized studies. *Can Med Assoc J, 174*(5), 635–641. doi:10.1503/cmaj.050873

Papanikolaou, P. N., Churchill, R., Wahlbeck, K., & Ioannidis, J. P. (2004). Safety reporting in randomized trials of mental health interventions. *Am J Psychiatry, 161*(9), 1692–1697. doi:10.1176/appi.ajp.161.9.1692

Stegenga, J. (2015). Measuring effectiveness. *Studies in the History and Philosophy of Biological and Biomedical Sciences, 54*, 62–71.

Swanson, J. M., Elliott, G. R., Greenhill, L. L., Wigal, T., Arnold, L. E., Vitiello, B., . . . & Volkow, N. D. (2007). Effects of stimulant medication on growth rates across 3 years in the MTA follow-up. *J Am Acad Child Adolesc Psychiatry, 46*(8), 1015–1027. doi:10.1097/chi.0b013e3180686d7e

Vandenbroucke, J. P. (2008). Observational research, randomised trials, and two views of medical science. *PLoS Med, 5*(3), e67. doi:10.1371/journal.pmed.0050067

Worrall, J. (2010). Do we need some large, simple randomized trials in medicine? In M. Suarez, M. Dorato, & M. Redei (Eds.), *EPSA Philosophical Issues in the Sciences*. Dordrecht, Netherlands: Springer, pp. 289–301.

Further Reading

Biddle, J. (2007). Lessons from the Vioxx debacle: What the privatization of science can teach us about social epistemology. *Social Epistemology, 21*(1), 21–39. (A detailed examination of a particular case in which clinical research underestimated the harm profile of a pharmaceutical)

Fuller, J. (2013). Rationality and the generalization of randomized controlled trial evidence. *J Eval Clin Pract, 19*, 644–647. (An examination of the limitations on extrapolation from RCTs to clinical practice)

Osimani, B. (2014). Hunting side effects and explaining them: Should we reverse evidence hierarchies upside down? *Topoi, 33*(2), 295–312. (An argument that the evidence-based medicine evidence hierarchies should be turned "upside down" for assessing the harms of pharmaceuticals)

Worrall, J. (2002). What evidence in evidence-based medicine? *Philosophy of Science, 69*, S316-S330. (By now a classic article that invigorated the study of the epistemology of medical research)

32
EXPERT CONSENSUS
Miriam Solomon

Introduction

"Expert consensus" is a shorthand way of referring to the agreement of experts on a matter about which they have expertise. The practice of medicine involves frequent reliance—by clinicians, patients, and others—on expert consensus. Sometimes that consensus is informal (achieved through the usual processes of research and clinical practice) and sometimes formal (developed at a medical consensus conference). An example of an informal consensus is, "The experts (pediatric pulmonologists) agree that early aggressive treatment of cystic fibrosis is best." An example of a formal consensus is, "The Cystic Fibrosis Foundation recommends that airway clearance therapy be increased as part of the treatment of an acute exacerbation of pulmonary disease." The goal of this chapter is to investigate the social epistemology of expert consensus in medical contexts. Why do we rely on expert consensus and under what circumstances (if any) is it appropriate to challenge it?

Trust in the knowledge of others is a foundation for all knowledge (Hardwig 1991, Scheman 2001). Our knowledge—whether of auto mechanics, culinary techniques, legal processes, or medical practice—rests on trust in the work of epistemic peers and epistemic experts in various domains. Solipsism (reliance only on one's own mind) is a route to ignorance, especially in scientific and technical fields where individuals are dependent on their education by teachers and mentors, on prior research, and on their collaborators and critics. In medicine specifically, the often-cited reliance "on the literature" is reliance (of researchers and clinicians) on the work of other researchers, and reliance "on medical advice" is reliance (of patients) on the knowledge of medical experts. Expert consensus produces the most trust, since it combines both the epistemic advantages of expertise and the reassurance that several apparently independent individuals have come to the same conclusion. When the experts agree, it is typically thought, claims to knowledge have the most chance of being true, or at least true enough to be useful. Trusting expert consensus seems like a reasonable—although not infallible—strategy.

I find it helpful to think of expert consensus as a modern version of the ancient and medieval "argument from authority." Instead of reliance on the writing of some individual giant or genius (typically white, male, and deceased), the reliance is on the (hopefully) democratic deliberations of a free (unbiased) research community composed of experts in their fields. For a layperson who knows little about the matter at hand, reliance on expert consensus is an excellent epistemic strategy.

Of course, no epistemic trust, even in the consensus of experts, should be absolute. There was a consensus among astronomers from antiquity until the 16th century that the Earth was the center of the universe, and that turned out to be wrong. There was consensus among surgeons from the late 18th century until at least 1970 that shaving the skin before surgery prevents infection, and that turned out to be wrong. Experts are not infallible. Nevertheless, the agreement of experts deserves to be taken seriously. It is also the most concise way of appealing to the complex body of evidence and reasoning that experts are assumed to have mastered and based their decisions upon.

Here is a well-known non-medical example of the use of both informal and formal expert consensus in communicating a scientific result. Activist-scholar Naomi Oreskes uses expert consensus to argue for the reality of climate change. She points out that the publications acknowledging climate change vastly outnumber the publications denying it (Oreskes 2004), reflecting an informal consensus. In addition, there has been a formal scientific consensus on the reality of climate change, in the statements of the Intergovernmental Panel on Climate Change (IPCC). These are carefully composed joint statements agreed to by thousands of scientists working on climate change. In medicine, it is similarly common to appeal to expert consensus, either informal (when individual scientists do not attempt to make a joint statement) or formal (as in consensus statements, typically produced at medical consensus conferences).

It is appropriate to interrogate the reliance on expert consensus in some cases by asking questions such as: do any of the experts dissent? If so, what do they say and what do they claim is the evidence for their position(s)? How difficult is it to maintain a dissenting position in this scientific community? Or, to put it another way, what is the strength of peer pressure? Are there assumptions that this scientific community takes for granted that should be challenged? If the consensus is formal, who is included and who is excluded from the process used to make a consensus statement? Leading researchers should be included in some way, but care should be taken to avoid simplistic reliance on markers of prestige such as seniority, institutional affiliation, etc. to identify the positions to be taken seriously. For a sophisticated account of how to ask such questions of an apparent expert consensus, see the work of philosopher of science Helen Longino (1990).

The common medical practice of asking for a second opinion is an example of the reliance on informal expert consensus—albeit a consensus of just two experts. When the second opinion agrees with the first, clinicians and patients feel reassured about the correctness of the recommendation. Sometimes it takes a third or fourth consistent expert opinion—especially when the matter is consequential—to settle any questions. When the expert opinions differ from one another, however, there is much uncertainty about how to proceed. In such cases, decisions are often made based on considerations other than the preference for good health outcomes. For example, they may be made based on cost, and the cheapest intervention selected. In effect, medical experts can lose their voice in practical decision making when they disagree with each other.

Disagreement is not in itself problematic for science—it is a valuable source of constructive criticism and division of epistemic labor (Solomon 2001)—but it is a problem for the role of science in matters of public concern and practical decision making. As cigarette manufacturers and climate change deniers know well, if they can create even the appearance of dissent in science, they can produce public distrust of science (Oreskes & Conway 2010). Similarly, if there is expert disagreement in a medical context, decisions can by hijacked by other interested parties, such as health insurance companies, professional interests, or government policies. Expert consensus is thus important for maintaining the authority of science in the medical context.

The Medical Consensus Conference Movement

While scientific medicine dates back at least to the 19th century, the pace, complexity, and sophistication of medical research increased greatly in the post–World War II era. In the early stages of testing, the effectiveness of most health interventions is uncertain. This is because the effectiveness tends to be modest and non-uniform: some are helped, whereas others may be not helped or even harmed. The overall effectiveness of health care interventions is reliably discerned through the aggregated results of high-quality clinical trials. By the 1970s, the need arose for authoritative and unbiased evaluations of new medical interventions such as drugs, surgery, and tests. In response, the U.S. National Institutes of Health (NIH) developed an assessment process, based on expert consensus, called the NIH Consensus Development Conference Program. This was the beginning of an international medical consensus conference movement. Begun in 1977, the NIH conferences had a formal structure, taking place over three days, with a written statement presented at a press conference on the third day. Each conference was run like a court of law, with a "judge" (the panel chair), "jury" (10–20 panelists with medical expertise in the area under discussion), and "witnesses" (prominent researchers in the area under discussion). Details of the conduct of the conferences show the effort to give both the appearance and the reality of an objective process, with freedom from bias (whether financial or intellectual), openness to the public for most of the proceedings, and instructions to the scientific "witnesses" to present as they would at a scientific conference rather than as legal advocates. (See Solomon 2015, Chapter 2, for more details about this.)

The expectation was that the medical consensus conference would be a social epistemic process in which experts who might at first disagree with one another come to consensus on the basis of sharing their knowledge of relevant evidence and arguments. It was anticipated that medical consensus conferences would produce authoritative, speedy, and well-publicized statements communicating new and reliable knowledge to clinicians, policy makers, and all interested parties.

Examples of topics of early NIH Consensus Conferences were Breast Cancer Screening (1977), Intraocular Lens Implantation (1979), and Drugs and Insomnia (1983). The immediate reception of the conferences was very positive, both in the U.S. and in other (First World) countries. The NIH model was widely adopted and adapted, so that it became the dominant method of assessment for new medical interventions throughout the 1980s and early 1990s. Examples of U.S. consensus conference programs are run by the Medicare Coverage Advisory Committee and the U.S. Preventative Services Task Force. Examples of non-U.S. programs are run by the Canadian Preventative Services Task Force and the Danish Medical Research Council. A 1990 Institute of Medicine report on the international use of consensus conferences began by stating that "Group judgment methods are perhaps the most widely used means of assessment of medical technologies in many countries" (Baratz, Goodman & Council on Health Care Technology 1990).

There were important differences among the various consensus conference programs. The NIH represented one extreme, focusing on what they called "technical" (meaning "scientific") rather than "interface" (meaning "broader" issues such as economics, ethics, etc. as well as "scientific") consensus. The NIH wanted to steer clear of policy issues, whereas other consensus conference programs (especially those in Canada and Europe) eagerly took on the policy questions, even seeing these as their main charge. In Europe, the focus on expertise was modified to allow an emphasis on public participation: the well-known "Danish" consensus conference model, developed at the Danish Board of Technology (now the Danish Board of Technology Foundation) used a panel of selected lay volunteers, along with the guidance of experts. The sense in Europe was that policy deliberations should have—at least in microcosm—the semblance of a public democratic process. In the U.S. and Canada, any public participation was limited to one or two participants on a panel. The European, particularly Scandinavian, method of consensus conferences with public participation coheres with the style of participatory democracy that is politically established in those countries (Horst & Irwin 2010). In North America there has been more deference to expertise.

Although the early reception of medical consensus conferences was largely positive, a minority expressed skepticism about the proposed use of a three-day deliberative process to move from scientific dissent to scientific consensus. For example, Arthur Holleb wrote an editorial deploring the idea of "medicine by committee" and argued that the conclusions reached would be susceptible to bias (1980). Others noted that consensus conferences are not generally used by scientists for their own research purposes. When physicists, or chemists, or biologists disagree about a scientific matter, they do not convene a consensus conference to settle their disagreements; instead, they engage in further research designed to gather new evidence and arguments to support (and/or modify) their positions. This minority expressing skepticism went largely unheard, so great was the early enthusiasm about consensus conferences.

(The case of the Intergovernmental Panel on Climate Change may seem like an exception to the rule that scientists do not settle disagreement among themselves with consensus processes. However, it is not: the IPCC consensus statement is not intended for the scientists, but for the public and governments. It is needed only because of the influence of climate change deniers.)

Over time, experience with medical consensus conferences yielded more nuanced assessments than the early enthusiasm showed. NIH consensus conferences tended to go particularly well when there was *already* a consensus in the research community, an irony since these conferences were originally designed to *develop* a consensus that did not yet exist in the research community. When consensus in the research community did not exist, NIH consensus conferences tended to go worse, even on occasion ending with a split panel (Institute of Medicine 1985). Because NIH consensus conferences took more than a year to plan and organizers wanted to wait until there was sufficient evidence to draw plausible scientific conclusions, they usually took place after a consensus was reached in the research community. The NIH argued, plausibly, that they still served a purpose, namely to communicate authoritatively the results of that consensus. (Of course, that purpose might be achieved more easily without such a choreographed public event.)

Medical consensus conferences also had differential impact. In countries with centralized health care provision (Canada and Europe), the consensuses informed health care policy, and thus had greater impact than in the U.S., where health care is more privatized and decentralized. In general, the impact of medical consensus conferences was less than was hoped for. We now know that written dissemination of expert consensus changes clinical practice slowly, if at all. However, in the 1970s and 1980s, this was not appreciated. Instead, efforts went into improving access to the consensus statements (making statements available in pamphlets, in medical journals, and eventually on the internet) rather than into creating more personal networks or incentives for change (strategies that we use now to engineer change in practice).

The Challenge of Evidence-Based Medicine

By 1990, it was generally appreciated that medical consensus conferences can produce an authoritative expert voice, especially when there is already an informal consensus in the research community. Consensus conferences were widely used. The development of evidence-based medicine in the early 1990s, however, destabilized the role of consensus conferences. Evidence-based medicine championed techniques for evaluating complex evidence that were explicit rather than implicit and dependent on expertise in epidemiology and statistics rather than seat-of-the pants judgment and group process. (See Chapters 18 and 19 on The Randomized Controlled Trial: Internal and External Validity and The Hierarchy of Evidence, Meta-Analysis, and Systematic Review, respectively, for details about evidence-based medicine.)

EXPERT CONSENSUS

From the point of view of evidence-based medicine, expert consensus, without an accompanying explicit evaluation of the evidence, is the *lowest* form of evidence on the evidence hierarchy used in systematic reviews (if it is counted as evidence at all). This is because consensus is susceptible to bias. It can be brought about by non-rational processes such as peer pressure, time pressure, and aggressive leadership. In such cases, active dissent is silenced rather than incorporated into the group deliberative process. Consensus can also be brought about through non-evidential, seemingly rational but actually quite fallible considerations shared by experts, such as pathophysiological reasoning. For example, in the 1990s and early 2000s, there was expert consensus that vertebroplasty was effective for treating pain from osteoporotic spinal fractures. This consensus was reached on the basis of anecdotal reports of pain relief after the surgery, and also on the pathophysiological reasoning that stabilizing the fractures would relieve the pain. In 2009, two randomized controlled trials showed that vertebroplasty is ineffective for treating osteoporotic spinal fractures. Pathophysiological reasoning is quite fallible, and therefore low on the evidence hierarchy. However, it can be persuasive, and is sometimes all that lies behind a consensus of experts.

Thus, it might be expected that with the rise of evidence-based medicine would come the fall of consensus conferences. But that is not exactly what happened, although to be sure the rise of evidence-based medicine forced a reassessment of consensus conferences, a decline in some of the programs, and changes in those programs that continued. Yet expert consensus, and even expert consensus conference programs (sometimes renamed to include the words "evidence-based" along with the term "consensus conference") have continued to be important. For example, the Medicare Coverage Advisory Committee was renamed the Medicare Evidence Development Coverage Advisory Committee in 2007. The current situation is that expert consensus and medical consensus conferences have a continuing role, integrated with evidence-based medicine, in producing authoritative knowledge. The consensus work typically follows the work of systematic evidence review. There are at least three reasons for this continuation of reliance on expert consensus and consensus conferences.

The first reason is that consensus conferences are often thought of as a check on the prior work of formal evidence review. When well-known experts participate in the consensus conferences, they show their agreement with the evidence synthesis, which may be done by less well-known statisticians. Expert consensus thus adds a layer of assurance about the results of the evidence review. Analyses of the evidence from statisticians and epidemiologists do not stand alone, but are supplemented by a joint statement of credible domain experts in the medical field.

The second reason for the continuing role of medical consensus conferences is that decisions about whether to recommend a medical intervention may rest on *policy* as well as scientific considerations. Cost considerations, comparisons of harms and benefits, and specific challenges with implementation may need to be considered, often in nationally and culturally specific contexts. Consensus conferences are particularly well suited for considering policy issues in a manner that is perceived as democratic as well as authoritative. The addition of ethics and politics experts to the panel, as well as laypersons (more common in Scandinavian countries), improves the perceived value of the consensus.

Finally, the practice of evidence-based medicine itself rests on a foundation of expert consensus. Evidence-based medicine is really "evidence-hierarchy medicine," and the exact hierarchy to use is a matter of controversy. Many rating systems are in use. They all rank randomized controlled trials high and anecdotal evidence low, but they disagree on matters such as the status of high-quality observational trials, some regarding them as highest-quality evidence and some not. The GRADE Working Group, established in 2000, is attempting to reach expert consensus on one system of rating the quality and strength of the evidence in clinical trials. It is ironic that expert consensus may supply the foundations for evidence-based medicine.

The Current Role of Consensus and Consensus Conferences in Medicine

The NIH Consensus Development Conference Program, which began the consensus conference movement, was "retired" in 2013 (Solomon 2015). Many of the 1980s and 1990s consensus conference programs, such as the King's Fund Forum in the UK and the Norwegian Institute for Hospital Research, no longer exist. With the dominance of evidence-based medicine, appeals to "expert consensus" rarely stand alone. But, unlike some other items on the lowest tier of the evidence hierarchy—like anecdotal evidence—expert consensus and even consensus conferences are still very much a part of making knowledge in medicine.

An example of a medical knowledge resource that shows the importance of both evidence and expert consensus is the National Guidelines Clearinghouse. Supported by the U.S. governmental Agency for Health Care Research and Quality, it describes itself as "a public resource for evidence-based clinical practice guidelines." It has two main requirements for listing clinical guidelines: that they come from a professional group (and in particular, not from individuals) and that they are evidence-based. Professional groups typically devise clinical guidelines in workgroups comprising respected experts who meet face-to-face. The importance of such meetings has not changed, even though the "golden age" of medical consensus conferences is over. Evidence review typically takes place before the consensus meeting, and the results anchor the conference discussions, which focus on clinical practice guidelines.

The evidence review together with an analysis of benefits and harms is, really, all that is needed to evaluate the current state of knowledge. For effective dissemination and uptake of new knowledge by clinicians, the public, and policy makers, the assurance of expert consensus on guidelines based on the evidence review is also needed. The requirement for professional group authorship of the guidelines ensures that experts in the field jointly agree on the guidelines, and thereby increases confidence in them. The guidelines are supported by the reputation of the professional group. Moreover, the participants in the consensus conference begin the process of practice change by taking these results back to their communities, in which they typically play the role of thought leaders. These days, it is understood that publication of guidelines—no matter how trustable the source—is not sufficient for bringing about change in medical practice. Dissemination of recommendations for change is more effective when it is done at the local level, through the active influence of thought leaders and even the creation of local incentives.

Expert Disagreement

So far, I have focused on the rhetorical force created by the dissemination of expert consensus. In some fields—such as cystic fibrosis treatment—there is a high degree of consensus and a correspondingly smooth delivery of effective health care. Atul Gawande (2004) praises the medicine in this field, writing that "cystic fibrosis care works the way we want all of medicine to work." The flip side is that lack of expert consensus—expert disagreement—leaves a rhetorical vacuum. As I said at the beginning of this chapter, medical experts can lose their voice in practical decision making when they disagree with each other. This creates an opportunity for other interested parties to become more influential, sometimes to the detriment of medical care. The medical profession has resisted this by minimizing their disagreements, as well as by doing "damage control" when disagreement is inescapable.

It is worth looking at an example of such disagreement: the controversy over recommendations for screening mammography, especially for women aged 40–49. I choose this example for two reasons. First, it shows several kinds of effort to minimize dissent. Second, there are important differences between screening mammography and cases when there is consensus on clinical guidelines.

Recommendations for screening mammography have been controversial ever since the technology was developed in the late 1960s. It was certainly expected, in the early days of the technology, that screening mammography would reduce breast cancer deaths, because the prevailing understanding of cancer was that early detection saves lives. The early days of the technology also coincided with the women's health care movement, and not surprisingly breast cancer screening became both a symbol of women's self-empowerment and a measure of adequate health care coverage for women's health. Many professional organizations made early consensus statements about the importance of mammographic screening, yet clinical trials have shown that the benefits of breast cancer screening are at most fairly modest. As the quality of clinical trials has improved, the estimated benefit of breast cancer screening has decreased. Some professional organizations (e.g., the U.S. Preventative Services Task Force and the American College of Physicians) now recommend against routine mammography for women aged 40–49. Others (e.g., the National Cancer Institute and the American College of Radiologists) have stayed with the earlier recommendations of annual mammograms for women in this age group. As this chapter goes to press, the American Cancer Society is in the process of changing its recommendations and taking an intermediate position, making different recommendations for women aged 40–44 and women aged 45–54.

The medical profession has "managed" the dissent over screening mammography (and other controversial interventions) in three different ways—reframing, promoting trust, and producing guideline syntheses. Reframing involves regarding areas of dissent positively, as areas for clinical judgment and patient preferences, rather than as a reason to distrust the experts. Even though the scientific uncertainty is not equal to either patient biological variability or variability in patient preferences, the strategy is a convenient compromise that makes positive use of expert disagreement.

Disagreements about medical efficacy are frequently accompanied by suspicions about costs and health care coverage. It is not unreasonable for someone to suspect that a reduction in the recommended frequency of mammography screening is motivated by a desire to save costs. The medical profession has done much to allay such distrust in the case of screening mammography by insisting that individuals who want more frequent mammographic screening than is recommended should still be covered for this by health care insurance.

Finally, so-called guideline syntheses, produced by The National Guidelines Clearinghouse, are attempts to reconcile different guidelines for the same condition. They are reviews of those of the guidelines that are of high enough quality to be listed (i.e., written in the past five years, evidence-based, and from a professional organization) and a discussion of the areas of agreement and disagreement. In general, such guideline syntheses emphasize the areas of agreement and acknowledge the rationales of the areas of disagreement. Overall, they support medical authority.

Controversy about clinical recommendations is the exception rather than the rule in health care. The case of screening mammography for women aged 40–49 is unlike most areas of clinical medicine, in two ways. First, the evidence is in the "gray area," with no clear evidence of benefit. For many recommended clinical interventions, the evidence is clearly positive, even if the intervention is less effective than we would like it to be. Second, consensus recommendations for screening mammography were produced *before* the era of evidence-based medicine. Since the end of the 1990s, it has been general practice for there to be a systematic evidence review *before* a consensus conference on a recommendation, making it less likely that the consensus conference approach will end up disagreeing with the evidence-based recommendations. This is a structural means of generally avoiding conflict between evidence-based medicine and consensus conference approaches by giving priority to evidence-based medicine.

Conclusions

Expert consensus is an essential part of making medical knowledge. It is vital for establishing the trust needed to disseminate new research findings and clinical recommendations. Medical consensus conferences at first used expert consensus to both evaluate evidence and disseminate conclusions. Evidence-based medicine has taken over much of the former task, but expert consensus is still indispensible for communicating results with authority.

References

Baratz, S.R., Goodman, C. & Council on Health Care Technology. 1990, *Improving consensus development for health technology assessment: An international perspective*, National Academy Press, Washington, DC.
Gawande, A. 2004, "The Bell Curve", *The New Yorker*, December 6, p. 82.
Hardwig, J. 1991, "The Role of Trust in Knowledge", *The Journal of Philosophy*, vol. 88, no. 12, pp. 693–708.
Holleb, A.I. 1980, "Medicine by Proxy?", CA: *A Cancer Journal for Clinicians*, vol. 30, no. 3, pp. 191–192.
Horst, M. & Irwin, A. 2010, "Nations at Ease with Radical Knowledge: On Consensus, Consensusing and False Consensusness", *Social Studies of Science*, vol. 40, no. 1, pp. 105–126.
Institute of Medicine. 1985, *Assessing medical technologies*, National Academies Press, Washington, DC.
Longino, H.E. 1990, *Science as social knowledge: Values and objectivity in scientific inquiry*, Princeton University Press, Princeton, NJ.
Oreskes, N. 2004, "Beyond the Ivory Tower: The Scientific Consensus on Climate Change", *Science (New York, N.Y.)*, vol. 306, no. 5702, p. 1686.
Oreskes, N. & Conway, E.M. 2010, *Merchants of doubt: How a handful of scientists obscured the truth on issues from tobacco smoke to global warming*, Bloomsbury Press, New York.
Scheman, N. 2001, "Epistemology Resuscitated: Objectivity as Trustworthiness," in *Engendering rationalities*, eds. N. Tuana & S. Morgan, SUNY Press, Albany, NY, pp. 23–52.
Solomon, M. 2001, *Social empiricism*, MIT Press, Cambridge, MA.
Solomon, M. 2015, *Making medical knowledge*, Oxford University Press, Oxford.

Further Reading

Hardwig, J. 1991, "The Role of Trust in Knowledge", *The Journal of Philosophy*, vol. 88, no. 12, pp. 693–708.
Horst, M. & Irwin, A. 2010, "Nations at Ease with Radical Knowledge: On Consensus, Consensusing and False Consensusness", *Social Studies of Science*, vol. 40, no. 1, pp. 105–126.
Solomon, M. 2015, *Making medical knowledge*, Oxford University Press, Oxford.

Part IV
CLINICAL METHODS

33
CLINICAL JUDGMENT
Ross Upshur and Benjamin Chin-Yee

> Life is short, the art long, opportunity fleeting, experiment treacherous and judgment difficult.
> (Hippocrates)

Consider the following two cases, on a typical busy day of a primary care physician.

Case 1

A 3-year-old girl is brought to the clinic. Her father tells you that he has had a temperature of 38.2 C for 24 hours. She is slightly listless and has a cough and a runny nose. Physical examination reveals no worrisome evidence of serious disease. The physician reassures the father that the child has a common virus that is circulating in the community, counsels the use of antipyretic medication to reduce the fever, encourages rest and hydration, and clearly states that he should bring the child back for reassessment if there is any worsening of symptoms or failure to improve.

Case 2

An 84-year-old woman attends the clinic with her daughter. She has a long history of high blood pressure, type 2 diabetes, and suffers from osteoarthritis. She lost her husband 15 months ago. The woman denies any problems, but her daughter mentions that her mother is not eating well, is increasingly forgetful, and is not keeping her household as tidy as usual. The daughter is worried that her mother has dementia and is keen to get a specialist assessment. When asked directly, the woman admits to ongoing sadness at the loss of her husband, and jokes that everyone her age has problems with their memory.

We shall return to these cases below.

Clinical judgment refers to the range of complex reasoning tasks and actions performed by clinicians in the context of offering diagnosis, therapeutic options, and prognosis to patients regarding their health and illness. The philosophically relevant aspects of clinical judgment relate to the status of the reasoning and/or logic that inform clinical judgment.

It is evident that the concept of clinical judgment covers a very wide range of considerations in the philosophy of medicine. It is not possible to cover all of these in one chapter. In recent years the focus has moved away from consideration of overall clinical judgment, and research has concentrated on the study of dimensions of clinical reasoning through the lens of statistics, decision science, and cognitive science. Philosophers with an epistemological orientation have similarly moved away, being more concerned with analyzing and critiquing the claims of evidence-based medicine. Therefore, we will seek to revive the philosophical discussion around clinical judgment. We will revisit accounts of clinical judgment and then situate the discussion within recent scholarship in virtue theory in medicine.

Clinical Versus Statistical Judgment

In 1954, the psychologist Paul Meehl published a book summarizing the evidence comparing clinical versus actuarial judgment (Meehl 1954). Evidence that reasoning aided by some form of mathematical model, usually based in statistics or probability, was superior to unaided reasoning indicated a need for investing resources into what are now termed decision support systems. The general conceptual strategy would be to suggest that clinical judgment is "subjective" and actuarial judgment "objective." Unaided reasoning is notoriously subject to numerous biases. These have been extensively documented in the psychology literature as well as in studies examining clinical reasoning. These biases increase risk and harm to patients and therefore are a concern for patient safety (Patel et al. 2015). Concerns about such biases have been influential in stimulating research into medical reasoning in order to make it more reliable and objective.

Much of the thrust of evidence-based medicine (EBM) has been to support more "actuarial" approaches to clinical judgment. Systematic reviews have shown how computer-based decision supports or simplified standardized clinical assessment rules result in improvements in diagnosis and management. Examples of the former are seen in interventions to reduce medication errors and of the latter in the "Ottawa Ankle Rules," which are simple reproducible steps in clinical examination that have been shown to be accurate in the diagnosis of ankle fractures and effective in reducing the unnecessary use of radiographs (Stiell et al. 1994).

However, EBM proponents recognized the need for clinical expertise to inform clinical reasoning. Sackett et al. state: "External clinical evidence can inform, but can never replace, individual clinical expertise, and it is this expertise that decides whether the external evidence applies to the individual patient at all and, if so, how it should be integrated into a clinical decision" (Sackett et al. 1996: 72). The exact nature of this expertise is not well articulated, but it seems clear that expertise and judgment can be regarded as synonymous.

Meehl's distinction between clinical and statistical judgment may represent an oversimplification in the case of medical reasoning. In what follows we will argue that a robust account of clinical judgment does not see as clear a distinction such as that posited by Meehl. Both clinical and statistical judgment are required. We will argue that there is much to gain from research into methods to make clinical reasoning more reliable by integrating decision aids and statistical approaches. However, we will also argue that given the extent to which uncertainty shapes the landscape in which clinical reasoning occurs, clinical judgment will always require more than simply actuarial approaches.

What Is Clinical Judgment?

It is important at the outset to get some clarity around what we are referring to when we discuss clinical judgment. What sorts of clinicians and what kinds of judgments?

In this chapter we will refer primarily to physicians, but other health care professionals, such as nurses, physiotherapists, occupational therapists, pharmacists, dentists, etc., exercise judgment and work as clinicians. "Clinical" also relates to the world of practice that is based in encounters with patients, families, and caregivers related to the health of patients or clients.

This will differentiate clinical judgment from the sorts of reasoning employed in research or in policy contexts. Although there is good reason to believe that judgment is exercised in these spheres of practice and they are no doubt relevant to health care and medical practice, they shall not be the focus of our analysis.

By judgment we refer to the collection of heterogeneous reasoning tasks involved in the formulation of a differential diagnosis, initiating and communicating a management plan, or communicating a prognosis. Judgment relates to the ability to take into account multiple sources of information as well as multiple constraints on a decision in order to formulate a course of action relevant to the problem at hand. In simple terms, physicians with good clinical judgment reliably demonstrate this quality in daily practice. The goal of educating health care professionals is to foster and develop good clinical judgment. These elements of clinical judgment are therefore core to medical practice and are constitutive of the expertise expected of health care professionals.

How Has Clinical Judgment Been Defined?

H. Tristam Engelhardt was one of the first to apply the insights of contemporary philosophy of science to medicine. In the introduction to a volume of essays devoted to the varied dimensions of clinical judgment, he observed:

> . . . an adequate assessment of the significance of clinical judgment is a complex endeavor in the epistemology of medicine. It involves . . . a fundamental critique of medical knowledge and the methods of clinical knowing. It is, moreover, much more than simply an enterprise in epistemology. Clinical judgments in their rich and full sense are freighted with values, including ethical and moral values. Evaluation and explanation properly and inextricably are bound together in medicine in general and in clinical judgment in particular. . . . Ideas, concepts, and notional presuppositions and structures, including value judgments, fashion the actual practice of medicine. Medicine more than most endeavors of knowledge and technology is involved in the entire range of human values and the whole gamut of levels of reality (i.e., from subcellular processes to psychological and sociological interactions). In studying medical knowledge, in analyzing clinical judgment, one thus addresses a first instance of knowing and doing, the better comprehension of which is likely to illuminate our understanding of science and technology in general. That is, understanding medicine may shed light on areas of science and technology where the interplay of facts and values may not be as salient, or the consequences of different views of science and technology as immediately or as intimately intrusive.
>
> (Engelhardt 1979: xxii–xxiii)

Engelhardt argues that the study of clinical judgment is part of the long tradition of making medicine more rational and of the evolution of medical reasoning. Medical reasoning since antiquity has abjured purely "metaphysical" accounts of disease and disease management and held closely to the requirement that physicians devote considerable time to develop their skills in physical examination through close observation of their patients.

Edmund Pellegrino, another renowned philosopher of medicine, notes that clinical reasoning and clinical judgment are goal oriented, aimed at clinical action, and often concerned with application of generalities to particular cases. Such judgments are often made with clinicians working with imperfect information (inaccurate history, equivocal physical or laboratory findings) and uncertainty of prognosis and therapeutic effectiveness. He distills the process of clinical judgment to three questions:

1. What can be wrong?
2. What can be done?
3. What should be done? (Pellegrino 1979)

The difference between question two and question three is important in demonstrating the inescapable ethical dimensions of clinical judgment. This is an area where the actuarial approach has been much less successful in providing guidance to clinicians, as the particularities of each patient require tailored strategies for management.

Feinstein and Clinical Judgment

As discussed above, recent scholarship has shifted attention away from understanding the nature of clinical judgment in favor of developing more "objective," algorithmic approaches to clinical decision-making, exemplified by EBM epistemologies. Nonetheless, earlier work on clinical judgment remains a valuable source for contemporary medicine. To this end, Alvan Feinstein's work remains especially salient, particularly his two seminal books *Clinical Judgment* (1967) and *Clinimetrics* (1987) (Feinstein 1967, 1987). Feinstein, hailed as the "Father of Clinical Epidemiology," played an important role in shaping the EBM movement, although he later became an outspoken critic (Feinstein & Horwitz 1997; Fletcher 2001). Returning to Feinstein's original arguments provides a useful starting point for understanding evidence-based approaches to clinical reasoning, and allows us to appreciate how current EBM epistemologies depart from this earlier work on clinical judgment.

For Feinstein, clinical reasoning is based firmly in clinical observation and measurement. Feinstein's argument is both descriptive and prescriptive—not only does he view clinical judgment fundamentally as a process of logical inference based on available evidence, but further, he claims that the practice can be made more "scientific" through renewed attention to and explication of its methods (Feinstein 1967: 26). Feinstein hopes to dispel the "mystique" surrounding clinical judgment to make it transparent and teachable. His overall aim is to develop a scientific approach to clinical reasoning, which utilizes the tools of mathematics (i.e., set theory, symbolic logic, and Boolean algebra) to underwrite more reliable clinical decision-making. A key metaphor throughout Feinstein's work is the identification of the clinician as an apparatus of scientific measurement. Feinstein argues that clinical medicine ought not pursue the scientific ideal by turning away from the clinic, seeking validity from paraclinical domains such as the laboratory, but rather should focus inward on the epistemology of everyday clinical practice to develop more robust standardized clinical measures and generate improved clinical taxonomies that allow for reliable and reproducible modes of inference for diagnosis, prognosis, and treatment (Feinstein 1983a, 1983b). This approach emphasizes the importance of data from the patient history and physical examination, in contrast to the more recent privileging of "objective" paraclinical data in EBM epistemologies (Henderson et al. 2012). Feinstein contends, "clinicians can bring science to clinical judgment by better exercise of the very human capacities that appear to impair it, and by giving increased attention not to laboratory substances and inanimate technology, but to sick people and the human methods of evaluating sick people" (Feinstein 1967: 29).

Feinstein begins his exposition of clinical judgment by discussing the nature of clinical evidence and types of data inputs into clinical decision-making. He identifies three types of data: (1) the description of disease in impersonal terms (i.e., morphologic, chemical, microbiologic, physiologic etc.); (2) the description of the host, in environmental, personal terms (i.e., age, sex, education, social status etc.); and (3) the description of the illness, the interaction between disease and environment (i.e., signs and symptoms) (Feinstein 1967: 24–25). All three types of data contribute to clinical reasoning, which he further categorizes into "therapeutic" and "environmental" reasoning. Therapeutic reasoning, which treats the patient as a representative case of disease, occurs alongside environmental reasoning, which individualizes management of an illness for a particular patient. Environmental reasoning, however, goes beyond simply individualizing therapy for a specific case and further encompasses the humanistic aspects of clinical practice, determining the "choice of methods of communication, accommodation, and human interchange that will best enable the sick host to bear the burdens of both ailment and treatment." According to Feinstein,

> A clinician's privilege and power in clinical therapy is his ability to make both the therapeutic and the environmental decisions concomitantly. The clinician combines treatment for the patient as a personal case of disease, with concern for the patient, as a personal instance of mankind, into the unified mixture that is clinical care.
> (Feinstein 1967: 26)

By breaking down clinical judgment into distinct yet concomitant processes of environmental and therapeutic reasoning and identifying their respective data inputs, Feinstein hopes to begin penetrating the complexity of clinical practice. This discussion provides a useful heuristic for clinicians in thinking about the domain to which Feinstein intended his "scientific" method of clinical judgment to apply. Feinstein's "clinimetrics" is an approach to therapeutic reasoning (i.e., the treatment of patients as instances of disease), which he recognizes leaves out important domains of clinical evidence, namely the "environmental" and "host" data, and must always be performed in light of these contextual factors (Feinstein 1967: 25).

We believe that this basic insight, emphasized at the outset of Feinstein's work, is often overlooked in contemporary treatments of clinical judgment, particularly those promoted by the EBM movement since the 1990s. Although Feinstein endeavors to make clinical judgment "scientific," he is cognizant of the limitations of this approach and of the specific types of data to which such methods apply. Therefore, although he was a major influence in early EBM, the contemporary movement shows a two-fold departure from Feinstein's clinical epistemology. Not only did it neglect to develop robust clinical measures based in the patient history and symptomatology, favoring instead "objective" paraclinical data, but it also failed to acknowledge the parts of clinical reasoning not included in Feinstein's "scientific" approach, the "environmental" and "host" factors, which contribute to the value-laden, humanistic aspects of clinical practice (Feinstein 1967). As Feinstein argued, clinical judgment is a complex process that necessarily considers the totality of the evidence to arrive at clinical decisions. Virtue-based clinical epistemologies, which we turn to now, recognize the plurality of clinical evidence and give a more prominent role to the humanistic data often left out of quantitative, evidence-based methods.

Uncertainty and Clinical Judgment

Thus far we have established that clinical judgment integrates the totality of considerations in clinical practice relevant to the care of an individual patient. In agreement with the original description by Engelhardt, we view clinical judgment as a complex process. However, we will resist arguments that try to pit the strengths of clinical judgment against "statistical" or algorithmic forms of reasoning. Consistent with Feinstein, we believe that clinical judgment is better conceived as an amalgam of both. Resisting the obvious advantages of reasoning aided by either computational devices or statistical tools would seem irrational. Similarly, arguing that all clinical reasoning can and should be reduced to such processes seems farfetched and overly reductionistic. A balanced account of clinical judgment would then have room for both styles of reasoning. There are good grounds for striving to value both.

One argument for valuing both clinical and statistical judgment comes from the extensive uncertainty that arises from the nature of the clinical encounter combined with the limitations of medical knowledge. Most clinical encounters result in a fragmentary understanding of the patient's problems. Diagnoses are often provisional, and even if definitive, seldom accompanied by certainty that treatments will be effective. Prognoses are always hedged with statements of uncertainty. This situation likely prompted Osler to call medicine the art of probability and the science of uncertainty. Djulbegovic, Ozo, and Greenland (2011) provide a comprehensive taxonomy of uncertainty in clinical medicine. Uncertainty arises from lack of evidence, conflicting interpretations of evidence, inability to access evidence in a timely manner, concerns about the application of aggregate statistical data to individual cases, and lack of clarity regarding patient preferences and values. In any clinical case, therefore, uncertainty plays a role. As they state: "Uncertainty is inherent in medicine. It occurs at every level of the clinical practice and research. It has multiple causes, with important implications for decision making, quality of care, and patient management. It will remain a central feature of medical practice" (Djulbegovic, Ozo & Greenland 2011: 347).

The inherent uncertainty in medicine means that it is unlikely that statistical judgments will completely determine decision-making. There will always be the need to integrate values into decisions, and medical decisions have an inherent value dimension to them.

A Virtue-Based Framework

Some believe that a future state awaits wherein complete and perfect knowledge will occur that will render all clinical reasoning amenable to mathematical characterization. But as we have shown, this is not possible or achievable. An adequate philosophy of clinical judgment must rest on a theoretical basis that can accommodate both factual and normative considerations. Virtue theory seems to have the requisite components of both ethics and epistemology that could provide both a descriptive and normative account of clinical judgment.

As Rosalind Hursthouse notes, virtues are more than simple traits, but "well entrenched" dispositions that are consistently manifested:

> A virtue such as honesty or generosity is not just a tendency to do what is honest or generous, nor is it to be helpfully specified as a "desirable" or "morally valuable" character trait. It is, indeed a character trait—that is, a disposition which is well entrenched in its possessor, something that, as we say "goes all the way down," unlike a habit such as being a tea-drinker—but the disposition in question, far from being a single track disposition to do honest actions, or even honest actions for certain reasons, is multi-track. It is concerned with many other actions as well, with emotions and emotional reactions, choices, values, desires, perceptions, attitudes, interests, expectations and sensibilities. To possess a virtue is to be a certain sort of person with a certain complex mindset. (Hence the extreme recklessness of attributing a virtue on the basis of a single action.)
>
> (Hursthouse 2003)

Clinical judgment is no doubt part of a "complex mindset" of particular dispositions with respect to the pursuit, analysis, interpretation, application, and responsibility for knowledge and information as it pertains to the practice of medicine. James Marcum, in a paper entitled "The epistemically virtuous clinician," explicates the features of this complex mindset. He argues that epistemic virtue in clinical practice requires both reliabilist and responsibilist virtues (without arguing for the primacy of one over the other). Reliabilist virtues include "sight or hearing for sensory or perceptual faculties, and memory, intuition, inferential reasoning, insight, or introspection for cognitive or conceptual faculties" (Marcum 2009: 250). Responsibilist virtues include "honesty, courage, open mindedness, humility, fairness, curiosity, tenacity and integrity" (Marcum 2009: 251). In addition to these, Marcum argues that two additional considerations are required for epistemic virtue: love of knowledge and theoretical and practical wisdom. Clinical judgment would then be the exemplification of virtue in practice.

It has been argued that EBM is compatible with virtue theory. Zarkovich and Upshur (2002) argued that the definition of EBM lends itself to a virtue theory account in that the most influential definition of EBM is "the conscientious, explicit, and judicious use of current best evidence in making decisions about the care of individual patients." They further argue virtue theory may provide an organizing account of clinical judgment as it is able to incorporate both considerations of intellectual and moral capabilities of clinicians. Intellectual virtues relate to the clinician's motivation for knowledge and the quest for truth through knowledge. These are typically demonstrated by the ability to access and assess information relative to the patient problem at hand. Moral virtues aid in the pursuit of patient welfare. Applying both the intellectual virtues and moral virtues to medicine, it can be argued that the definition of clinical judgment requires a combination of intellectual and moral virtues. Zarkovich and Upshur (2002) argue that the term "conscientiousness" is an intellectual virtue and "judiciousness" a moral virtue. The call for explicitness would indicate an attempt, as much as it is possible, to have clinical decisions rest on clearly stated reasons.

Back to the Cases

How does the above discussion relate to the cases stated at the beginning of the chapter?

Case 1 represents one of the most typical clinical encounters in medicine. Children with fevers are among the most frequent reasons for consulting physicians, and this child has a fever. Viral illnesses are very common in this age group, causing listlessness, cough, and runny noses. For the most part, the diagnosis is straightforward, management simple, and prognosis uniformly good. In this case, one may wonder whether any sort of clinical judgment is required. Some would wonder why a physician or other health care provider need be involved at all. If the condition is common and self-limiting, it could be argued that it is preferable to have this managed by the parents themselves (and indeed many parents do so).

Yet, good clinical judgment is required to manage even the simplest case, and such judgment must be exercised with every patient encounter regardless of its nature. In this case, once the father and child have engaged the physician, the physician has an obligation to exercise all of the relevant skills he or she possesses in determining the nature of the illness, the best course of action, and the likely prognosis. The physician may avail him or herself of a structured scoring system or algorithm to determine the likeliest diagnosis and best management plan. However, few such tools are used in routine clinical practice, particularly in North America. Context matters in clinical diagnosis, and this will very much influence clinical judgment. If this case were in a malarial zone, one would take a very different course of action, and the consideration of cause would also be adjusted to explain the fever.

Clinical judgment is required in this case for the clinician to be assured that there is no more serious underlying cause that may explain the fever. In the case of children, the concern may be for a more serious underlying bacterial infection such as pneumonia or meningitis. These conditions can be somewhat reliably excluded on the basis of the physical examination, but experienced clinicians know that the observations they make on any particular patient are limited to the time they are examining the patient. As all disease states are dynamic, it is prudent, as in this case, to communicate the diagnosis and explain the conditions that would warrant reconsideration of the initial diagnosis. In the case of illness in children, considerable parental anxiety often needs to be managed.

A simple case thus demonstrates the complex interplay of facts and values, and the requisite exemplification of reliabilist virtues (physicians must be competent and confident in their history and physical examination skills) and responsibilist virtues (physicians must demonstrate the requisite concern for the welfare of the patient, be attentive to the anxiety of the parent, and demonstrate a willingness to revisit the diagnosis in the event that either the initial diagnosis was incorrect or the illness worsens).

Case 2 is more complex as it involves more interacting issues and a more subtle set of judgments. In this case, the use of more actuarial or aided reasoning may play a role in improving upon the clinician's reasoning. As in the first case, context matters, but regardless of how much we vary the case or add detail, it simply reinforces the point of how complex clinical reasoning and the exercise of clinical judgment is in practice.

In this case, the patient denies any concern, but the collateral information given by the daughter requires revisiting the situation in more depth. The woman has many risk factors for conditions that can impair her memory, as well as recent life events that are well known to impair well-being and manifest with self-neglect and sadness (grief). In this case, there are many factors to consider, but to simplify the case we can center the concerns on two conditions: depression and dementia. Both of these are common conditions in older adults that can often be mistaken for or overlap with grief. In this situation, a physician may choose to employ standardized assessment tools for both dementia and depression. There are many such valid instruments to use, all with suitable limitations, but it would be good judgment to employ them in this case to assess cognitive function and depression. Relying on their knowledge of the patient or their intuition in this case would likely not be sufficient. Depending on the results, a management plan could be instituted at that particular visit, or a referral made to an appropriate specialist. Again, both reliabilist and responsibilist virtues are exemplified in the management of the case.

In both cases, it is unlikely that any actuarial tool or model could be employed that would answer all of the questions raised in very typical cases. They no doubt would aid in certain elements of decision-making and contribute to more accurate, less error-prone decisions. However, given the multitude of uncertainties in both cases (not fully catalogued in the analysis to be sure, but easily elicited), judgment will be required in many of the elements. The analysis also shows how the humanistic elements of clinical reasoning that Feinstein believed to be important can be integrated into more structured reasoning.

Conclusion

In this chapter we have discussed the complex nature of clinical judgment. We have argued that clinical judgment should not be conceived in opposition to more structured and guided forms of reasoning (actuarial judgment), but both are required. The ubiquity of uncertainty in clinical medicine necessitates that some form of "unaided" reasoning will be required in many cases. We have further argued that the work of Alvan Feinstein, coupled with virtue theory, provides a basis for a theory of clinical judgment that is more aligned with actual clinical practice than one that is reliant on either clinical or statistical judgments on their own.

References

Djulbegovic, B., Ozo, I. & Greenland, S. (2011) "Uncertainty in Clinical Medicine" in Gifford, F. (ed.), *Handbook of the Philosophy of Science. Volume 16: Philosophy of Medicine*, Amsterdam: Elsevier, pp. 299–356.

Engelhardt Jr., H.T. (1979) "Introduction" in Engelhardt Jr., H.T., Spicker, S., & Towers, B. (eds.), *Clinical Judgment*, Dordrecht: D. Reidel, pp. xi–xxiv.

Feinstein, A.R. (1967) *Clinical Judgment*, Baltimore: The Williams & Wilkins Company.

Feinstein, A.R. (1987) *Clinimetrics*, New Haven: Yale University Press.

Feinstein, A.R. (1983a) "An additional basic science for clinical medicine: II. The limitations of randomized trials," *Annals of Internal Medicine* 99: 544–50.

Feinstein, A.R. (1983b) "An additional basic science for clinical medicine: IV. The development of clinimetrics," *Annals of Internal Medicine* 99: 843–48.

Feinstein, A.R. & Horwitz, R.I. (1997) "Problems in the 'evidence' of 'evidence-based medicine'," *American Journal of Medicine* 103: 529–35.

Fletcher, R.H. (2001) "Alvan Feinstein, the father of clinical epidemiology, 1925–2001," *Journal of Clinical Epidemiology* 54: 1188–90.

Henderson, M., Tierney, L. & Smetana, G. (2012) *The Patient History: Evidence-Based Approach*, New York: McGraw-Hill Professional.

Hursthouse R. (2003) "Virtue Ethics" (Stanford Encyclopedia of Philosophy) [Internet]. Stanford Encyclopedia of Philosophy. [cited June 2, 2015] Available from: http://plato.stanford.edu/entries/ethics-virtue/

Marcum, J.A. (2009) "The epistemically virtuous clinician," *Theoretical Medicine and Bioethics* 30: 249–65.

Meehl, P. (1954) *Clinical versus Statistical Prediction: A Theoretical Analysis and a Review of the Evidence*, Minneapolis: University of Minnesota Press.

Patel, V.L., Kannampallil, T.G. & Shortliffe, E.H. (2015) "Role of cognition in generating and mitigating clinical errors," *BMJ Quality and Safety* 24: 468–74.

Pellegrino, E.D. (1979) "The Anatomy of Clinical judgments: Some Notes on Right Reason and Right Action" in Engelhardt Jr., H.T., Spicker S., & Towers B. (eds.), *Clinical Judgment*, Dordrecht: D. Reidel, pp. 169–194.

Sackett, D.L., Rosenberg, W.M., Gray, J.A, Haynes, R.B. & Richardson, W.S. (1996) "Evidence based medicine: What it is and what it isn't," *British Medical Journal* 312: 71–2.

Stiell, I.G., McKnight, R.D., Greenberg, G.H., McDowell, I., Nair, R.C., Wells, G.A., Johns, C. & Worthington, J.R. (1994) "Implementation of the Ottawa ankle rules," *Journal of the American Medical Association* 271: 827–32.

Zarkovich, E. & Upshur, R.E. (2002) "The virtues of evidence," *Theoretical Medicine and Bioethics* 23: 403–12.

Further Reading

The essays in the volume *Clinical Judgment* (Engelhardt Jr., H.T., Spicker S., & Towers B. (eds.), *Clinical Judgment*, Dordrecht: D. Reidel Publishing Company, 1979) make for indispensable reading.

Alvan Feinstein's book *Clinical Judgment* (Baltimore: The Williams & Wilkins Company, 1967) still contains much that is of great value.

Edmond Murphy's *The Logic of Medicine* (Baltimore: Johns Hopkins Press, 1997) contains a section on epistemology that covers diagnostic reasoning and focuses on issues related to measurement.

The second edition of *Clinical Epidemiology: A Basic Science for Clinical Medicine* by David Sackett et al. (Boston: Little Brown, 1991), published before the advent of evidence-based medicine, contains an excellent account of how quantitative reasoning aids in clinical judgment.

34
NARRATIVE MEDICINE
Danielle Spencer

> Listening is a rare happening among human beings . . . [it] is a primitive act of love, in which a person gives self to another's word, making self accessible and vulnerable to that word.
> — William Stringfellow, A *Keeper of the Word* (1994: 169)

Introduction

As physician and literary scholar Rita Charon describes, narrative medicine is medicine practiced with the "narrative skills of recognizing, absorbing, interpreting, and being moved by the stories of illness" (Charon 2006: 4). One of its fundamental tenets is the importance of *narrative competence*, of attending to the form and structure of storytelling. A clinician taking a patient history, like the skilled reader of a literary work, must recognize and understand not simply data as it fits into checkboxes of the electronic medical record but also the *way* the patient relates his experience, including context, narrative voice, tone, figurative language, and temporality. Where does the account begin? What is left out? Which perspectives are represented, and how? Further, how do both teller and receiver construct meaning reciprocally—and what is the form and ethics of this exchange?

Such focus on clinical care enriched and strengthened by an understanding of narrative is integral to the field, which also includes varied areas of enquiry and modes of practice. Intrinsically interdisciplinary, narrative medicine draws upon literary theory and criticism, philosophy, creative writing, disability studies, psychoanalytic theory, critical race theory, qualitative research, sociology, and oral history, among other discourses. Key themes include:

- Humanistic healthcare education and training, with a particular emphasis on close reading
- Progressive pedagogy and a commitment to social justice
- Scholarship and creative expression about experiences of illness, disability, and caregiving

Through such practices the field explores the operation of storytelling in the clinical encounter, in conceptions of the self and intersubjectivity, in broader narratives informing health policy and social justice, and in the relationships among these different contexts.

Genealogy

Narrative medicine affirms and explores the intrinsically narrative and humanistic nature of medical practice, offering a corrective to an accelerating trend toward bureaucratization and specialization, which has proven increasingly alienating for clinicians and patients alike.

Given the field's emphasis on narrative, it is especially important to note some of the circumstances in health care, society, and academic discourse contributing to its genesis and evolution—and also to acknowledge the contingency and selectivity of any particular account.

In the United States, Abraham Flexner's (1910) highly influential Carnegie Foundation report instantiated a robust emphasis on biomedical science in medical education, and until recent years his recommendations for admissions requirements and curricular focus have remained largely intact in medical education. Inspired by the German system of that era and concerned primarily with standardization and scientific rigor, Flexner also acknowledged that medicine requires "requisite insight and sympathy on a varied and enlarging cultural experience" as well as ethical and social responsibility; however, critics point toward the omission of such humanistic values in his model, which describes the human body as "belonging to the animal world." Lamenting what he viewed as the growing impoverishment of the doctor-patient relationship, in 1926 Flexner critic Francis Peabody memorably reminded medical students that "one of the essential qualities of the clinician is interest in humanity, for the secret of the care of the patient is in caring for the patient."

Throughout the 20th century many voices joined Peabody in addressing the scope of care and warning against dehumanization in medical practice. Eric Cassell, decrying the alienating legacy of Cartesian mind-body dualism, explained that "Suffering is experienced by persons, not merely by bodies, and has its source in challenges that threaten the intactness of the person as a complex social and psychological entity" (1982: 639). Similarly, Charles Odegaard (1986: 16), critiquing Flexner's mechanistic description of the human body, cautioned that

> The physician who is educated to see his patient as only a collection of interrelated tissues and organs is not seeing his patient whole; and, except as he may be aided fortuitously by untutored intuition, he will not be able to deal with his patient's health in all its aspects.

Over time such calls to tutor the clinician's intuition became increasingly acute as awareness grew of the alienating effects of health care education and practice; studies demonstrated a dramatic drop in students' levels of empathy and compassion throughout medical school, and writings by patients and clinicians reflected growing cynicism and estrangement.

Such a rising tide of discontent in health care joined with social trends and political movements of the 1960s and 70s—the civil rights movement, the growth of women's health awareness, community health centers, patient advocacy, and disability rights, among many others—combining to produce a range of innovative and interdisciplinary responses. The biopsychosocial model promulgated by George Engel (see Chapter 40) emphasizes the multifaceted psychological and social context of health and illness, while patient-centered care, relationship-centered care, and the patient's rights movement privilege patient/family values as well as collaboration and transparency in medical decision-making. The hospice movement, influenced by the work of Elizabeth Kübler-Ross and Cicely Saunders, offers a more humanistic approach to the processes of dying and grieving. The field of bioethics arose in the wake of the Nuremburg Trials concerning human experimentation by Nazi researchers and clinicians as well as the abuses of research projects such as the Tuskegee syphilis study and the Willowbrook hepatitis experiments, with Henry K. Beecher (1966) acting as a particular catalyst, pointing toward "ethical errors" and "troubling practices" in human experimentation. Bringing philosophy into conversation with health care, law, and sociology, bioethics addresses questions in research ethics, medical decision-making, and social justice, with particular emphasis on issues raised by new developments in biotechnology. In addition, medical humanities as well as "Literature and Medicine" entered the curricula of medical schools in the U.S. beginning in the early 1970s, accentuating the study of literature—typically topical works oriented toward clinical care and illness—as a means of improving clinicians' understanding and enriching ethics education. This era also saw a tremendous growth in writing and activism about experiences of illness, disability, and caregiving, which accelerated dramatically during the AIDS epidemic of the 1980s and '90s.

Many of these trends coincided with a "narrative turn" in the humanities, social sciences, and popular culture beginning in the later decades of the 20th century. Fields as diverse as history, sociology, cognitive science, law, business, psychology, literature, and cinema experienced a resurgent recognition of the prevalence and relevance of narrative —a revival of interest in storytelling taking different forms in various disciplines. Each of these discourses possesses its own framework and terminology, and so a brief précis can, at best, gesture toward these contexts: in literary criticism and theory, the narrative turn is comparable in magnitude to the earlier "linguistic turn," indicating a move away from deconstructionism while retaining a keen attentiveness to the particularity and contingency of narratives and their role in structuring knowledge and power. Meanwhile, Hayden White elaborates an influential formal analysis of historiography—its modes of emplotment, argument, and ideology—exploring the narrative tropes structuring our understanding of the past. Similarly, in areas such as post-colonial and feminist studies the identification of hegemonic narratives offers insight into their ideological effect on historical and scientific orthodoxy, as in Homi Bhabha's study of narrative and nation-states, or Donna Haraway's exploration of "allowable stories" in biology.

In philosophy Richard Rorty offers an influential narrativist methodology, turning from scientific definition and argument toward a literary style, analyzing philosophical works in comparison with novelistic fiction and stressing the importance of "telling a new story." Literature and the arts, too, reflect a resurgent concern with storytelling after the interest in "anti-narrative" exemplified by the French *nouveau roman*, and (in some quarters) moving beyond what Paul Ricoeur described as a "hermeneutics of suspicion"—a deep-rooted distrust of narrative's seemingly illusive legerdemain and power to ideologically conscript the reader. The field of narratology—the study of narrative—itself shifts from a structuralist conception of narrative coherence grounded in literary forms towards "postclassical" narrative theory, embracing a broader range of modes of expression and media types: human exchange, performance, dance, visual art, spoken word, interactive online communication, corporeal narratives, and hybrid permutations. Thus, each of these turns is particular to its field, yet such common interest in narrative emphasizes its importance in expressing and structuring discourse, identity, and experience.

Narrative medicine arises in this context, responding to the crisis in health care and drawing upon scholarship in the humanities and social sciences. There is no single starting point to the field, but physician-scholars Trisha Greenhalgh and Brian Hurwitz's (1998) formative *Narrative Based Medicine: Dialogue and Discourse in Clinical Practice* notably united clinicians, historians, psychotherapists, literary scholars, computer scientists, creative writers, biologists, epidemiologists, anthropologists, ethicists, and educators bridging biomedicine and the humanities. With writings on illness narratives, pedagogy, and ethics, the volume explored the myriad ways in which experiences of illness and care unfold in narrative terms and the implications for clinical practice, reflecting a rich history of such work. The field reached a critical point of articulation and development at Columbia University in New York with Rita Charon and colleagues' scholarship and implementation in a range of contexts, including the medical school curriculum. Charon, whose 2006 *Narrative Medicine: Honoring the Stories of Illness* is a canonical work in the field, was joined at Columbia by literature and film scholar Maura Spiegel, philosopher Craig Irvine, physician-scholar Sayantani DasGupta, psychoanalyst Eric Marcus, fiction-writer David Plante, and scholars Marsha Hurst and Rebecca Garden, among others. Sharing an interest in healthcare-themed literature with "Literature and Medicine", the scope of enquiry broadens to study the narrative structure of works of art addressing a wide spectrum of topics, with "texts" reflecting the range of postclassical narrative theory: stories expressed in the body, in silence, in dialogue, graphic novels, oral history, and technology, alongside more traditional literary genres. Utilizing a variety of techniques, narrative medicine emphasizes the importance of developing narrative competence and explores its relevance to the health care setting. One of the distinctive practices developed by the Columbia practitioners is a workshop method of close reading, prompted writing, and dialogue, which is described and discussed below.

Close Reading in Practice

Close reading is central to narrative medicine: skilled attentiveness to a text, with particular interest in the form of an account. The act of reading in this sense parallels the act of listening, and by learning and practicing narrative competence we may offer better care and understanding. As Charon (2006: 107) explains, a physician

> must be prepared to comprehend *all* that is contained in the patient's words, silences, metaphors, genres, and allusions. Listening and watchful clinicians must become fluent in the tongues of the body and the tongues of the self, aware that the body and the self keep secrets from one another, can misread one another, and can be incomprehensible to one another without a skilled and deft translator

A narrative medicine workshop dedicated to honing narrative competence may take various forms and occur in a range of contexts; it may include health care students, nurses, physicians, persons experiencing illness or disability, or another type of group.

Taking the example of clinicians reading and discussing a poem: the group may first read the poem aloud together, perhaps twice, listening for and discussing its syntax, details, and nuance, its cadence and silences, its narrative voice, its arc. A trained facilitator often guides the reading and discussion, encouraging participants to heed their own affective responses while simultaneously engaging and honing their analytic faculties to better understand the work. The piece under discussion may or may not explicitly address themes of illness. A group of clinicians often finds that non-medical topics prevent them from slipping into a familiar mode of analysis and diagnosis; the effect can be both unsettling and generative, and inevitably the discussion will return to health care, but with a different valence. Through close attention to detail, participants become aware of the ways in which the work elicits emotional and intellectual reactions, and the group typically experiences a variety of different responses. Such diversity can be surprising and revelatory, and readers must learn to sit comfortably with multiple perspectives and to tolerate ambiguity.

Although the tools of narratology and literary theory are operative in such discussions, technical terminology is often de-emphasized, particularly in an introductory workshop. A facilitator trained in literary criticism and theory may prompt a group of physicians to consider concepts arising from critical sources—such as French narratologist Gerard Genette's distinction between story, narrative, and narration, or Russian literary critic Mikhail Bhaktin's conception of chronotope, describing the interwoven aspect of time and space in literature— all without necessarily using such terms nor referencing the theorists. This practice certainly varies, and many narrative medicine workshops and seminars will explicitly invoke and study the critical tradition, but some groups may be unlikely to embrace a corpus of unfamiliar technical terms. Broadly speaking, the pedagogical approach is more progressive than didactic, nourishing a rigorous collaborative interpretive process. Some facilitators may de-emphasize the author's biographical circumstances in order to avoid the presumption of a singular interpretation (what New Critics termed the "intentional fallacy"), while other discussions will take as their starting point a work's social, cultural, racial, and political context, the ways that it reflects and addresses specific historical conditions. In this sense the range of practices reflects the heterogeneity of contemporary critical discourse, which we will explore further below.

Following upon a discussion of the literary work, the group may be offered a short writing prompt and a brief amount of time in which to write —perhaps five minutes. The prompt often bridges the themes of the piece under discussion with individual reflections and experiences. For example, after discussing Elizabeth Alexander's (1997: 60) prose poem "Haircut," which describes a visit to a Harlem hair salon and includes the line, "What am I always listening for in Harlem?"—the group may be asked to write, beginning with the phrase: "What am I always listening for . . ." Typically many participants share what they have written, and then the group discusses the responses with close attention to form, structure, detail, and so forth, practicing the tools of narrative competence. The text and prompt must of course be crafted with sensitivity to context and with respect for participants' privacy and the operation of hierarchy and privilege within the group. Such an exercise calls upon interpretive and creative capacities that in many cases have long remained dormant.

How does this practice affect medical decision-making? The combination of close reading, writing, and discussion echoes the "different movements within clinical telling and listening" articulated by Charon (2005) as attention, representation, and affiliation: *attention*, as a form of listening and understanding, is reciprocally nourished by the receiver's attempt to formulate and *represent* what she has heard through a process such as writing, and both contribute to greater *affiliation* between teller and receiver. In the clinical context attention often takes the form of a physician's receptive and focused listening to a patient's account; representation may consist of the clinician's act of reflective writing about the experience, which then nourishes the quality of attention she is able to offer, thus enriching the affiliation between doctor and patient and improving diagnosis and treatment.

In the example of a group of clinicians in a narrative medicine workshop, the poem and discussion may offer specific insights informing medical decision-making and treatment, such as greater awareness of a patient's particular cultural context or of the clinician's own presumptions and projections. Through the act of inhabiting another point of view—both in the literary work and in the discussion of participants' different responses—a clinician may become more adept at imagining various perspectives. And by honing narrative competence through the triad of attention, representation, and affiliation, clinicians approach the care of the sick with greater awareness of stories' structure and import. The practice, too, offers an opportunity for clinicians to reflect upon their own experiences and to nourish—or in some cases revive—the humanistic basis of health care, and it notably contributes to greater collegiality and inter-professional collaboration as well.

Charon (2006: 177) tells the story of a new patient in her internal medicine practice and the ways in which she employs narrative medicine techniques to improve clinical care:

> A 46-year-old Dominican man visits me for the first time, having been assigned to my patient panel by his Medicaid Managed Care plan. He has been suffering from shortness of breath and chest pain, and he fears for his heart. I say to him at the start of our first visit, "I will be your doctor, and so I have to learn a great deal about your body

and your health and your life. Please tell me what you think I should know about your situation." And then I do my best to not say a word, to not write in his medical chart, but to absorb all that he emits about himself—about his health concerns, his family, his work, his fears, and his hopes. I listen not only for the content of his narrative but also for its form—its temporal course, its images, its associated subplots, its silences, where he chooses to begin in telling of himself, how he sequences symptoms with other life events. After a few minutes, the patient stops talking and begins to weep. I ask him why he cries. He says, "No one ever let me do this before."

In this example, Charon exhibits a form of *attention* that her patient—who she calls Mr. Ignatio Ortiz—had not experienced in a clinical context. She then writes about the experience and shares it with Mr. Ortiz (who also grants permission for its publication), thus *representing* it, resulting in heightened *affiliation* between the two. Charon describes the complexity of Mr. Ortiz's health concerns, including depression, joint pain, cardiac disease, and difficulty working, as well as her commitment to listen for his physical symptoms as well as his "story of himself." When she refers him for specialized cardiac care, his willingness to trust and accept treatment is strengthened by the manifest affiliation between the two. When she attempts to discern the complex relationship between his inability to work and support his family and his coronary artery disease, she sees them as part of an integral whole and deploys a keen attentiveness to the narrative form his experience takes. As she describes, such an approach toward receiving a patient's story does not necessarily take longer than a standard interview, but the clinician must be trained and willing to hear it with a nuanced appreciation for the import of its narrative structure alongside the factual data: "I try my best to register the diction, the form, the images, the pace of speech. I pay attention—as I sit there at the edge of my seat, absorbing what is being given—to metaphors, idioms, accompanying gestures, as well as plot and characters represented for me by the patient" (2006: 187–188). These narrative skills have a direct bearing on the clinical care Charon is able to provide as a physician.

There are many and varied forms of narrative medicine practice that engage clinicians and non-clinicians alike. Creativity, in various forms, is integral—clearing a space for a participants to exercise and express the imagination. The "Parallel Chart," another formulation of Charon's, encourages health care students to write reflective notes about their clinical experiences, complementing the rubric of a standard H&P (history and physical examination) or SOAP (subjective, objective, assessment, and plan) note. A graduate-level seminar may combine the workshop experience with readings in literature and philosophy, while a medical school seminar or patient group may study thanatology or visual art. Film is a particularly rich medium for exploring the effects of narrative; as literary and film scholar Maura Spiegel and psychologist Arthur Heiserman (2006: 464) explain, "movies can light up parts of ourselves that have been dark for awhile," and a film prompts the viewer to engage in a way that evokes Arthur Frank's description of "thinking with" stories.

Illness and disability narratives, too, are an important area of study, from published memoirs to oral history to spoken word, with attentiveness to the operation of privilege and disenfranchisement: how might we understand "contested narratives," and how to interpret the role of silence? Does a category such as illness or disability narratives risk perpetuating a reductive pathologizing lens? In this sense the field is continually engaged in a dynamic process of interrogating the terms of health care, their implicit ideology and effects, as well as its own premises. It is, then, particularly important to situate the field, to explore its critical roots and context.

Critical Topography

True to the breadth of its practices and its interdisciplinary character, narrative medicine draws upon a variety of literary and philosophical schools of thought. Just as our account of its genealogy is necessarily selective, any map of its critical topography will represent a particular (and necessarily partial) narrative, privileging certain perspectives and eliding others. We will endeavor to navigate portions of its terrain, traversing some paths and—given the constraints of an introductory essay—merely nodding toward many essential areas.

The practice of close reading, integral to the field, has roots extending to exegesis of sacred texts, and enters into the tradition of literary theory and criticism with varied interpretations of what it entails, how it is situated, and what constitutes a text. Notably, the mid-century literary movement of New Criticism elaborated a focus on close reading of literary works, particularly poetry. Drawing upon and responding to British thinkers such as I. A. Richards, American writers and critics including John Crowe Ransom, T. S. Eliot, Robert Penn Warren, and Cleanth Brooks encouraged close study of a poem or work of prose as an entity expressing a type of organic harmony—Brooks likened the beauty of a poem to the unity of a flowering plant, including roots, stalk, and leaves. The methodology of New Criticism offers tools for formal explication of the structure and operation of a work, including its paradoxes and tensions, the relationship between the literal and the figurative, and so forth. However, following Brooks's metaphor, meaning does not quite extend to the *soil* of the plant, as interpretation is largely detached from historical context, the intention of the author, and the response of any given reader. While such a-historicism is, now, itself an historical chapter in mid-20th-centry academic discourse, the practice of close reading has remained essential to literary study through the passage of many different models of criticism and theory.

As Rita Felski (2008: 52) describes, "Academic fashions may come and go, but a sharply honed attentiveness to nuances of language and form is still held, by most scholars and teachers of literature, as an indispensable sign of competence in their field." Narrative medicine draws upon this tradition, asking how we may attend closely to the form of a text, slowing down to see what is at work in its operation. When a group lingers over three lines in a poem, discussing the resonance of particular words or a subtle shift in narrative voice, we become attuned to the operation of language in a different register. Such awareness informs our capacities as readers and as listeners. We may emerge with a different understanding of the metaphors employed in speaking about illness experiences, their evocations and significance. We may also hone our sensitivity to the way a narrative voice is figured, noting with fresh awareness the *lack* of a first-person narrator in a standard clinical or scientific account and the import of such rhetorical choices in framing medical and ethical decisions.

Fortunately, many modes of literary criticism offer direction in practicing close reading with attentiveness to different dynamics of a creative work and its context. For example, Derridean Deconstructionism analyzes a text (nearly infinitely broadly understood) as a series of unending signifiers; language is also, in this view, laden with hierarchical terms that can be exposed through close reading and synthesized in a new form encompassing unending interplay, eternally resisting closure. New Historicism considers the exchange between culture and literary works, emphasizing historical context and the ideological import of any form of discourse—in this sense, exploring the soil of the plant, the setting that is discounted in New Criticism's model of formal "unity." And, admitting another critical element, reader-response theory stresses the importance of the reader—to belabor our analogy, the *beholder* of the beautiful flowering plant—and her role in co-creating meaning, bringing her own memories, associations, thoughts, and feelings to bear on a particular text. Scholars such as Norman Holland, Louise Rosenblatt, Stanley Fish, and Roland Barthes advance different models of reader-response theory, and the school of criticism contains a range of perspectives on the locus of meaning-making, from underscoring the way a given text provokes certain reactions to accentuating the singular response of each particular reader.

A narrative medicine workshop or class setting draws upon a breadth of critical perspectives, an approach that highlights the contingency of any given school of thought. In particular, readers are encouraged to consider the means by which their reactions to a complex artistic work are activated, including the role of socio-historical context and individual identifications and projections. Through close reading and discussion of a literary work and of the group's prompted writing, readers learn to acknowledge the particularity of any given response and to recognize the range of different understandings. Such revelations are experiential and quite potent; they cannot be conveyed with the same effect through a lesson in literary criticism or cognitive science. A workshop participant comes to perceive her own role in a different and more nuanced fashion and—critically—applies that understanding to *listening*. For just as the reader of a text participates in meaning-making, the listener, too, co-constructs meaning with the teller of an account. And just as reader-response critics vary in their view of the relationship between text and reader, we may also ponder a range of possible relationships between teller and listener. When a person gives an account of a bodily experience to another—such as a patient describing his pain to a doctor—how is meaning constructed? To what degree does the listener bring her own associations to bear on what she hears? As a clinician becomes aware of the ways she is *reading* the stories told in language and in the body—how she is part of the meaning-making process—she attends with humility and care, strengthening both efficacy and affiliation.

Considering the physician who listens closely to her patient's account of pain, we may ask if she can ever *fully* understand another's experience, or if such a presumption may be incorrect and even appropriative. As philosopher Elaine Scarry (1985: 4) describes, physical pain is characterized by its intrinsic "unshareability" and "resistance to language." Beyond the experience of pain, we may also consider whether we can ever empathize (a contested term in the health humanities fields) with one another's experiences, or if we remain in some sense opaque to one another. Although this perspective may seem defeatist, such opacity to oneself and among persons can serve as the very basis for ethics, as Judith Butler (2005) describes in *Giving an Account of Oneself*, drawing upon philosophers such as Nietzsche, Hegel, Adorno, Foucault, Levinas, Hegel, and Cavarero. In particular, philosopher Emmanuel Levinas emphasizes the radical unknowability and primacy of the Other—an otherness that calls me to care for him—as the foundation for ethical action. In contrast to this alterity, Levinas describes the way in which I also possess a "totalizing" impulse to understand and assimilate the world in my own terms. As philosopher Craig Irvine (2005) describes, health care embodies this Levinasian paradox: it is a response to the fundamental call to care for the Other, yet medical science enacts the drive to circumscribe the world into rational classification. In Irvine's analysis, literature—in its capacity to represent experience—mirrors the way medicine inevitably reduces and systematizes. Thus it *exemplifies* this objectifying stance while also nourishing critical thinking, which enables it to play a vital role in health care education.

In considering such a view of alterity we bring a particular humility to the act of listening—see Sayantani DasGupta's 2008 discussion of "narrative humility"— and this awareness underscores the importance of attending to one another with subtlety and skill. Such competency proves particularly invaluable in health care, for despite its dogged positivism, medicine is steeped in stories. Comparing the diagnostic process to a detective's reconstruction of a crime, Kathryn Montgomery Hunter (1991: 25) likens it to a Sherlock Holmes mystery: "Like the master, the physician uses narrative first as a means of organizing the details that with luck and careful thought will flower into a testable generalization and then to demonstrate the accuracy of that generalization in the chronological chain of its details." Critical to the diagnostic process, too, is a recognition of the importance of a patient's account. As Greenhalgh and Hurwitz (1998: 6) describe, "The narrative provides meaning, context and perspective for the patient's predicament. It defines how, why, and in what way he or she is ill. It offers, in short, a possibility of *understanding* which cannot be arrived at by any other means."

A growing interest in illness and disability narratives honors such experiences and accounts well beyond the sphere of diagnosis and treatment, and narrative medicine joins with scholars in literature and the arts, sociology, and ethnography to explore the resonance of these stories. For example, psychiatrist and anthropologist Arthur Kleinman conducts an ethnographic investigation of narratives and signs of illness, with a particular emphasis on stigma and its role in healthcare. Making the distinction between *disease*, understood as the medicalized disorder, and *illness*, elaborated as the "innately human experience of symptoms and suffering" embedded within culturally specific normative standards of bodily experiences, Kleinman (1998: 3, 10) also suggests guidance for clinicians in offering "empathic witnessing of the existential experience of suffering." And drawing upon narratology and sociology, scholars such as Anne Hunsaker Hawkins (*Reconstructing Illness*) and Arthur Frank (*The Wounded Storyteller*) offer analysis and structural typologies of "pathographies," exploring the effect of prevailing cultural narratives and objectifying healthcare practices on illness experiences.

Scholarship about illness and disability narratives often bridges analytic, affective, and corporeal understandings—Frank, for example, describes a distinction between thinking about a narrative at a critical remove and "thinking with" a story, allowing it to touch and resonate with one's own experience. Along with scholars such as Havi Carel, Kathlyn Conway, S. Kay Toombs, and others, Frank also includes personal accounts of illness in his scholarship, breaching the passive or impersonal academic voice to explore the relationship between lived experience and the various means we use to understand it. At the same time, scholars such as John Hardwig question the putative authority of memoir and autobiography, and—broadly speaking—late-20th-century literary criticism expresses a deep-seated wariness of the effects of such stories' purported access to the unmediated experience of their authors. Recent scholarship by thinkers such as Rita Felski and Ann Jurecic seeks to resuscitate the importance of such works while heeding and responding to the critical tradition.

Positioned at this intersection of storytelling, identity, embodiment, and critical enquiry, narrative medicine explores and interprets the ways in which stories both reflect and structure experiences. Such a project of narrative hermeneutics (a formulation emphasized by Jens Brockmeier, among others) draws upon the philosophical tradition of Heidegger, Gadamer, and Ricoeur to explore the notion of "narrative identity," the ways in which we both construct and are formed by narratives. Narrativization is here understood as an integral component of experience—we are always already embedded within a range of stories—rather than as a descriptive postscript. Moreover, the process of forming and understanding narrative meaning is an ongoing and sometimes disruptive process. For example, invoking the work of canonical symbolic anthropologist Clifford Geertz, Anna Donald (1998: 22) describes the ways in which illness experiences are often characterized by conflict over contested narrative understandings and social constructions: a particular disease has a different etiology and resonance if one is informed by traditional Chinese medicine, Western medicine, or various specialties within these disciplines; furthermore, it carries different metaphorical and historical associations in different cultures (notably critiqued by Susan Sontag in *Illness as Metaphor*). Jerome Bruner (1987: 708) develops this perspective as well, exploring the ways in which narratives "become so habitual that they finally become recipes for structuring experience itself, for laying down routes into memory, for not only guiding the life narrative up to the present but directing it into the future."

Drawing upon the phenomenological tradition, corporeal narratives become vital to the interpretive project, offering an antidote to Cartesian mind-body dualism. Indeed, understanding the stories told by the body has long been integral to medical practice, from Galen to Madhavakara, from Hildegard of Bingen to Elizabeth Blackwell to Atul Gawande—and reaffirming and expanding the scope of such narratives becomes an integral part of the field. (See Chapter 42 for an elaboration of phenomenology and narrative hermeneutics in medicine.) Also critical to our understanding is continued attentiveness to the role of stories in cultural discourse, reflecting and affecting the operation of power and privilege. Informed by and participating in scholarship and activism in disability studies, LGBTQ studies, critical race theory, and other disciplines, narrative medicine explores the means by which certain narratives silence and disenfranchise individuals and groups as well as the possibility for stories to operate as a powerful tool of advocacy for social change. As Hilde Lindemann Nelson (2001) describes, "master narratives" may perpetuate detrimental identities—often internalized by oppressed individuals and groups—against which counterstories offer the possibility of "narrative repair." For example, according to Nelson, the "Clinically Correct" master narrative of trans* identity is one that understands a trans* person as passing or not passing as a particular gender (both before and after "transitioning"), thus perpetuating a gender binary; furthermore, it is reliant upon a pathologizing medical model. Alternative stories reframing such gender categories and affording greater personal determination without prejudice are, as Nelson argues, critical to improving the moral agency of and respect toward affected individuals and groups (2001: 125–135). Thus narrative becomes vitally important in understanding and addressing marginalization of particular groups within society.

Conclusion

Narrative medicine occupies a dynamic and evolving terrain, drawing upon and informing many disciplines. Its scope is global, with clinicians and scholars from many regions around the world working at this nexus of clinical care and the humanities. The field has a particular influence on health care education—at Columbia University College of Physicians and Surgeons in New York, for example, it is an established part of the required curriculum—and many programs turn to the field in order to strengthen humanistic education and address required competencies in inter-professional collaboration, emotional intelligence, cultural sensitivity, and listening skills. A growing body of scientific literature demonstrates that reading literary fiction improves theory of mind and that narrative medicine interventions advance clinical efficacy and humanistic care. And as we continue to address pervasive issues in public health and disparities in access to care, the field's focus on the operation of power and privilege in narratives offers an important perspective and set of analytic tools.

In addition, the field offers a significant corrective to the rule-based principlism of bioethics articulated by Beauchamp and Childress; emphasizing the particularity of any given case/story, it joins casuistry and narrative ethics—with notable contributions from Richard Zaner, Tod Chambers, Anne Hudson Jones, Hilde Lindemann Nelson, and many others—to approach ethical questions and medical decision-making with a critical *literary* understanding rather than a normative approach. In its emphasis on relationality and context, too, the field challenges the pervasive emphasis on "autonomy" in bioethics. It also offers a meaningful interdisciplinary complement to "evidence-based" medicine; in this context it is sometimes termed "narrative-evidence-based medicine"—honoring different forms of evidence such as illness narratives, and demonstrating the ways in which medical science is itself narrative in nature, bound by its own set of rhetorical and epistemological models.

Positioned at the nexus of medicine, literature, philosophy, and many other discourses, narrative medicine seeks to bring such trans-disciplinary understanding to bear on clinical care, and to continue the rich and complex enquiry into our understanding of the stories we tell ourselves and one another.

References

Alexander, E. (1997) *Body of Life: Poems*. Chicago: Tia Chucha Press.
Beecher, H.K. (1966) Ethics and clinical research. *The New England Journal of Medicine*, 274, 1354–1360.
Bruner, J. (1987) Life as narrative. *Social Research*, 54(1), 11–32.
Butler, J. (2005) *Giving an Account of Oneself*. New York: Fordham University Press.
Cassel, E.J. (1982) The nature of suffering and the goals of medicine. *N Engl J Med*, 306(11), 639–645.
Charon, R. (2005) Narrative medicine: Attention, representation, affiliation. *Narrative*, 13, 261–270.
———. (2006) *Narrative Medicine: Honoring the Stories of Illness*. New York: Oxford University Press.
DasGupta, S. (2008) Narrative humility. *The Lancet*, 371, 980–981.
Donald, A. (1998) The words we live in. In *Narrative Based Medicine* (Eds, Greenhalgh, T. & Hurwitz, B.) London: BMJ Books, pp. 17–26.
Felski, R. (2008) *The Uses of Literature*. New York: Wiley-Blackwell.
Flexner, A. (1910) *Medical Education in the United States and Canada, Bulletin Number Four*. New York: The Carnegie Foundation for the Advancement of Teaching.
Greenhalgh, T. & Hurwitz, B. (Eds.) (1998) *Narrative Based Medicine*. London: BMJ Books.
Heiserman, A. & Spiegel, M. (2006) Narrative permeability: Crossing the dissociative barrier in and out of films. *Literature and Medicine*, 25, 463–474.
Hunter, K.M. (1991) *Doctors' Stories: The Narrative Structure of Medical Knowledge*. Princeton: Princeton University Press.
Irvine, C. (2005) The other side of silence: Levinas, medicine, and literature. *Literature and Medicine*, 24, 8–18.
Kleinman, A. (1988). *Suffering, healing & the human condition*. New York: Basic Books.
Nelson, H.L. (2001) *Damaged Identities, Narrative Repair*. Ithaca: Cornell University Press.
Odegaard, C.E. (1986) *Dear Doctor: A Personal Letter to a Physician*. Menlo Park, CA: H.J. Kaiser Family Foundation.
Scarry, E. (1985) *The Body in Pain: The Making and Unmaking of the World*. New York: Oxford University Press.
Stringfellow, W. (1994) *A Keeper of the Word: Selected Writings of William Stringfellow*. Grand Rapids: Wm. B. Eerdmans Publishing Co.

Further Reading

Brockmeier, J. & Meretoja, H. (2014) Understanding narrative hermeneutics. *StoryWorlds: A Journal of Narrative Studies*, 6, 1–27. (For context and philosophical references in the area of narrative hermeneutics)
Charon, R., DasGupta, S., Hermann, N., Irvine, C., Marcus, E., Rivera Colón, E., Spencer, D., & Spiegel, M. (2017) *The Principles and Practice of Narrative Medicine*. New York: Oxford University Press. (Elaboration of theoretical foundation as well as practical methodology in narrative medicine, co-authored by faculty members in the Program in Narrative Medicine at Columbia University)
Jones, E.M. & Tansey, E.M. (Eds.) (2015) *The Development of Narrative Practices in Medicine, c. 1960–c.2000*. Queen Mary, University of London, London. Available from: <http://wellcomelibrary.org/player/b22674871#> [7 June 2015] (Transcript of "Witness Seminar" including many key figures in the field discussing the origins, development, principles, and practice of narrative medicine)
Jurecic, A. (2012) *Illness as Narrative*. Pittsburgh: University of Pittsburgh Press. (A very thoughtful discussion of the genre of illness narratives, including its treatment in academic discourse and detailed study of several examples)
Lewis, B. (2011) *Narrative Psychiatry: How Stories Can Shape Clinical Practice*. Baltimore: The Johns Hopkins University Press. (A valuable application of narrative medicine principles in a psychiatric context, including an astute review of the field's history and scope)

35
MEDICAL DECISION MAKING: DIAGNOSIS, TREATMENT, AND PROGNOSIS

Ashley Graham Kennedy

Introduction: What Is the Science of Medical Decision Making?

Medical decision making (MDM) refers to the process of clinical reasoning that is used to make estimations of, and decisions regarding, patient diagnosis and treatment. Although as we will see in this chapter, this process is often complex, it is nevertheless used throughout clinical practice from routine primary care cases to complex cases encountered in specialty clinics and emergency departments. Because all clinicians engage, either formally or informally, in the process of medical decision making, the study of this process is important both for those who practice medicine as well as for those, such as philosophers of medicine and decision analysts, who aim to understand and improve it. Thus, the goal of this chapter is to provide an introduction to medical decision making that is relevant to both medical practitioners and philosophers of medicine.

The process of MDM generally begins by working toward a diagnosis of the patient's problem or illness. This means that MDM involves the consideration of how to understand and characterize information gained from the medical interview and physical examination of the patient, as well as decisions regarding whether and which diagnostic tests or procedures to use when considering the possible causes of a patient's problem, and how to best interpret these tests. In some instances, a clinician might make a diagnosis and suggest treatment based on the interview and exam alone, whereas in many others, diagnostic tests or procedures may be required during the process of differential diagnosis, which is a probabilistic method of diagnostic reasoning used to rule in or rule out diagnostic options. When considering whether or not to perform diagnostic testing, the first concern is with test or procedure accuracy. This is a function of the specificity (how well the test excludes patients without disease) and sensitivity (how well the test identifies patients with disease) of the diagnostic. In some instances, data from randomized controlled trials (RCTs) might be used to determine test or procedure accuracy, whereas in other cases, observational studies are used.

Even once a diagnostic test is found to be accurate—that is, once it is found to provide reliable epistemic information about the patient's condition—the question of whether or not it is worth performing remains. Some medical textbooks simply state that, "test selection should be restricted to those diagnostic tests whose results could change the physician's mind as to what should be done for a patient" (Sox et al. 2007). However, the situation is often not as simple as this. For instance, in many cases, patients want to know what is wrong with them (Bossuyt et al. 2009), even when a treatment for their illness or condition is not available. Further, clinicians are divided over whether an accurate test for an untreatable disease should ever be performed. Whether or not it should depends on how we understand value in the medical context—for example, whether we believe that medical value is tied inextricably to patient outcomes, or whether we believe that a test that provides knowledge is valuable in some epistemic sense even when it does not lead to improved patient health.

The most generally accepted position in current medical practice is that, unless performing a test will lead, via treatment, to an improved patient outcome, it should not be performed. However, the information provided by diagnostic testing does not, on its own, have a direct effect on patient outcomes (although it can have positive or negative indirect effects on the patient's mental state; see, e.g., Cournoyea and Kennedy 2014). Rather, it is what we do with this information that has an impact (positive or negative) upon the patient's health. Thus, in order to determine whether or not performing a test will in turn lead to improved patient health, one needs to know whether or not a given diagnostic test is a good predictor, not only of the condition it is intended to diagnose, but also of treatment outcomes for this condition, and thus this consideration too must enter into the medical decision-making process: one might decide to perform a test only if it has been shown to not only accurately diagnose a given condition, but also to lead to improved patient health in those with the condition it diagnoses.

Once a patient's condition has been diagnosed, medical decisions regarding treatment must then be made. Treatment decisions can be quite complex and involve evaluating (generally on the basis of data from RCTs or observational studies) whether or not a treatment is therapeutically effective, whether or not it is affordable, whether or not it will interact with other patient treatments and/or comorbid conditions, and whether or not the long-term benefits and/or effects of the treatment are in line with patient expectations and values. Because RCT or other clinical study data is not always available, or because, in some cases, it is not clear whether or not the available data is applicable to a particular patient, in practice, clinicians must often decide whether or not to begin treatment for a condition despite being unsure of the diagnosis or of the way the treatment will ultimately affect the particular patient.

Before making the decision to begin treatment, the clinician must consider both the risks and the benefits of the proposed treatment, as well as the patient's attitudes toward these risks and benefits. Then, on the basis of this information, the clinician must decide whether to continue gathering information in order to make a more certain diagnosis, to observe the patient without treatment, or to begin treatment of the patient's (potential) condition.

The long-term prognosis of the patient's condition must also be taken into consideration when deciding whether or not to begin treatment. The term "prognosis" describes the probability that a patient with a specific clinical profile will develop a certain health outcome over a period of time. Prognostic information is helpful for deciding when to start and when to stop treatment and for monitoring disease progression or remission, and symptomatic improvement or worsening. Further, an accurate prognosis can also be invaluable for selecting the most effective treatment for an individual patient. However, the reliability of currently available prognostic tests is limited, and thus, in many instances, prognostic estimations must be made on the basis of very limited data.

As we will see in the example that follows, many decisions need to be made regarding diagnosis and treatment in each patient case. In order to facilitate these decisions, the science of MDM draws on insights from decision science, economics, probability theory, and theoretical models of the clinician-patient relationship and shared decision-making. Thus MDM incorporates both quantitative approaches, such as using Bayes' Theorem to determine pre- and post-test probabilities, and qualitative approaches such as discussing with the patient his or her values when estimating the expected utility of a given diagnostic test or treatment plan. Further, the process of MDM involves both the concerns of the clinician (such as the harms vs. benefits of a test or an intervention) as well as those of the patient (such as how the patient feels about these harms or benefits). As such, MDM is a science that can be understood only from a multidisciplinary perspective.

Example

In order to better understand both the process of MDM in clinical practice and the philosophical issues that this process raises, it is helpful to consider an example. Let us consider the following plausible scenario:

> A 32-year-old male presents to the ENT clinic with facial pain, fatigue, low-grade fever, and a history of chronic/recurrent sinusitis. Further, he reports that he experiences six to ten such infections per year and has so continuously for the past 5–6 years. These infections have been diagnosed via computerized tomography (CT) scan and treated by his primary care physician with 10–14 days of antibiotic therapy. The infections resolve successfully (temporarily), only to return after a few weeks or months. The patient has also undergone allergy testing, with no significant environmental allergies documented. The patient is frustrated and expresses a desire to find the "cause" of his ailment so that the recurrences of sinusitis can be stopped.

Encounter to Differential

In this section we will see how a clinician might use the process of MDM to initially evaluate a patient and create a list of possible diagnoses. In our example, the patient has expressed a clear desire to know why he has recurrent infections, so that they can be stopped. Thus, this clinical encounter will begin, as most do, with a medical interview in order to gather evidence toward making a diagnosis of the cause of the patient's recurrent illness. The questions asked of a patient during this interview will differ depending upon the presentation and the specific patient circumstances. In some cases, such as in the emergency setting, there will be little time for questioning at all, and the clinician will proceed directly to physical exam. In our specific example, however, the situation is not emergent. Instead, the pressing matter is to determine why the patient's infections are recurring so they can be stopped. If the clinician in this case works on the assumption that the patient's infections were properly diagnosed, and did in fact respond to antibiotic therapy, then he or she would next need to perform a physical exam to see if further diagnostically relevant information can be obtained. However, with sinusitis, *as with many conditions*, the ailment cannot be diagnosed on the basis of an exam alone. A patient who presents with recurrent nasal swelling, sore throat, fatigue, and lymphadenopathy, for example, could be suffering from a number of illnesses, including viral upper respiratory tract infection, allergies, or bacterial sinusitis. However, a patient who has a history of documented bacterial sinusitis, and does not have environmental allergies, is more likely to be experiencing a recurrence of this condition rather than a viral infection. Absent the history of bacterial sinusitis, the opposite would be true (a viral infection would be more likely). Thus, one can see how in this instance, as is again the case with many others, the diagnostician must use the information gathered from the patient history *in conjunction* with clinical data to make a diagnosis—clinical data alone would not be enough to ensure accuracy in diagnosis.

In this particular example, the clinician, based on patient history and a physical exam, could reasonably rule out causes for the patient's condition other than recurrent bacterial sinusitis and then diagnose the patient with this condition. Once the patient has been diagnosed with recurrent bacterial sinusitis, the clinician will not only need to treat the current infection but will also need to investigate the underlying cause of the recurrent infections, with the goal of making a causal diagnosis that will facilitate resolution of these infections.

Causal diagnoses (as opposed to label or syndromic diagnoses), particularly those that posit pathophysiological mechanisms, are considered to be the gold-standard for medical diagnosis (Smith and Francesca 2007), because it is thought that understanding the underlying pathophysiology of an illness best allows for effective treatment. In this particular case, uncovering the underlying reason for the recurrent bacterial sinus infections is important, because unless the cause of the issue is identified and treated, the infections will most likely continue to recur. This case, in that aspect, is again representative of others, because often the initial reason that a patient presents in the clinic does not automatically reveal the underlying cause of the ailment in question.

In order to try to find the underlying pathophysiological cause of a patient's complaint, the clinician will often begin by making a list (known as the "differential") of the most likely predisposing conditions. For chronic or recurrent sinusitis, these include:

- Allergic reactions (which cause the nasal passages to swell)
- Nasal passage abnormalities such as nasal polyps or a deviated septum
- An immune system disorder (either acquired or primary)
- Asthma

Although there are certainly other possible underlying conditions, the initial differential list is composed of only the most likely ones. If all of the conditions in the initial differential are eventually ruled out, then the clinician will create a new differential that includes other, less likely, conditions.

In order to eliminate conditions in the differential and eventually arrive at an accurate diagnosis, the clinician needs to evaluate, either formally or informally, the probability of each of the conditions individually. This is generally done via an appeal both to personal experience and to published data. Personal experience in the clinical setting provides a wealth of knowledge that can assist in making a diagnosis. Published data can also provide diagnostic evidence. However, while both sources of evidence are important, both are subject to biases. In the former, the biases include physician overconfidence (which can lead to a disinterest in investigating supplemental decision support), a tendency to assume that the current patient has a condition that the physician has previously seen, and premature closure, which is the narrowing of the diagnostic options too early in the process of differential diagnosis so that the correct diagnosis is never seriously considered (Berner and Graber 2008). In the latter case, the biases most often result from faulty clinical trial design (Emmanuel, Wendler, and Grady 2000) or from an assumption that the available trial data is applicable to a particular patient when it is not.

Further, there is often disagreement among clinicians about the relative probabilities of the conditions in the same differential. This disagreement is due both to differing personal experiences and to the fact that different clinicians read different journals. For instance, the primary care physician is more likely to read articles in primary care journals, which report the prevalence of disease in unselected patients, while specialists, on the other hand, more often read journals that report studies of patients who have been referred to specialists. Understanding the prevalence of each of the conditions in the differential is pivotal in facilitating a correct and timely diagnosis.

Once probabilities (either formally or informally) have been assigned to the various conditions in the differential, the clinician must then decide what diagnostic testing to perform in order to decide between them. Of the conditions on the list in our example, allergies and structural abnormalities are far more common than primary immunodeficiency diseases. Because of this, they should be investigated first and, in this case, can be ruled out on the basis of previous allergy testing and CT scan evaluation, respectively. Asthma, which is also more common in the general population than immunodeficiency disease, can in turn be ruled out via a careful patient history and/or simple testing. This leaves only immunodeficiency disease in the differential. Although it is the last to be ruled out, this does not mean that the patient definitely has such a condition, but it does raise the probability that he does, especially given the frequency of bacterial infection reported (which would be uncommon in someone with a well-functioning immune system).

At this point in the diagnostic process the clinician will then form a new differential that includes both HIV/AIDS (the most common form of secondary immunodeficiency) and common variable immune deficiency (the most common form of primary immune deficiency). Given this new differential, the physician will then proceed to diagnostic testing in order to gather further evidence toward a diagnosis.

Obtaining and Interpreting Diagnostic Information

Since the clinician in this case suspects an immune disorder, she or he will want to use diagnostic testing to determine the likelihood that the patient has one of these conditions. Importantly, this likelihood, or probability estimate, depends upon both the pre- and post- test probability that the patient has the disease in question. The pre-test probability of a disease is the probability that disease is present before any testing is conducted. This probability will depend upon whether or not, given the patient history, the patient is in a high- or low-risk group for the condition in question. The post-test probability of a disease depends upon the pre-test probability as well as the accuracy (sensitivity and specificity) of the test that is being performed. Thus, there is no such thing as being absolutely certain of what a test result means; it varies from one patient to another depending on his or her pre-test probability. Having a positive test result does not necessarily mean that a patient has the condition, nor does having a negative test result mean that the patient does not have the condition. However, clearly, these results do affect the likelihood of whether or not the patient has the condition that the test is intended to diagnose.

Suppose that the patient in our example is in a low-risk group for secondary immunodeficiency due to infection with the HIV virus and is also seronegative for the virus. In that case, the clinician can reliably remove this condition from the differential list. Of course, as we have noted, even with a negative test, one cannot be certain that the patient isn't infected with HIV, but when the pre-test probability of the condition is low, and the test is negative, the likelihood of the patient having the disease is very low.

Once secondary immune deficiency has been ruled out, the possibility of a primary immunodeficiency in a patient who has frequently recurring infections of the skin or respiratory tract should be investigated. There are more than 200 primary immunodeficiency diseases recognized by the World Health Organization. Common Variable Immune Deficiency (CVID), so named because it is the most common one (but still very rare, according to the National Institutes of Health the incidence in the population is approximately 1 in 25,000 to 1 in 50,000 people worldwide, although the prevalence can vary across different populations) can be tested for with simple and relatively inexpensive serum tests, as it is characterized by low levels of serum immunoglobulins. The low level of these antibodies in CVID is the underlying cause of the increased susceptibility to infection in patients with the condition.

Further, patients with CVID also have an increased incidence of autoimmune or inflammatory manifestations and susceptibility to cancer when compared to the general population.

The diagnosis of CVID is usually made in the third or fourth decade of life, so this condition should be suspected in patients of this age who present with any of the above conditions.

In our particular example, the patient has at least two markers that increase the pre-test probability that he has this condition: frequent respiratory infections (Wood et al. 2007) and age in the mid-thirties. Thus, diagnostic testing for CVID, given the above markers, seems reasonable in this specific case, particularly since it is minimally invasive and relatively inexpensive, even though it would likely not be warranted in most people who present with acute bacterial sinusitis.

In some cases, a physician might elect to conduct an empirical trial of a suspected condition instead of opting for diagnostic testing. However, in the case of CVID, the testing for the condition is relatively inexpensive, while the treatment is not. Thus, in this particular case, it is better to have more decision support for treatment (in terms of a positive serum test) than to elect for an empirical trial without further testing.

Testing for CVID involves measuring the levels of IgG (and sometimes IgM and IgA) in the serum. If the levels are low, then the patient is treated with replacement IgG, generally either monthly intravenously or weekly subcutaneously at a cost, in most cases (depending upon dose), of more than $20,000 per year. Because of the difficulty and expense in treating CVID, the clinician will want to be as confident as possible in the diagnosis. Although all medical diagnoses are at some level uncertain, a clinician must be confident that she or he is at least "approaching a solution" before initiating treatment (Barrows 1991), particularly if, as in this case, the solution involves significant expense.

Because of the inherent uncertainty of diagnosis, there is always the potential for missed or mistaken diagnosis in clinical practice, and clinicians must be aware of this. Some have argued that we might have more confidence in our diagnoses if clinicians more regularly used computer-based diagnostic algorithms. There are currently many diagnostic programs either available or in development, and dozens of studies have compared computer-aided diagnosis with expert diagnosis. Although some of these studies show that computer-aided diagnoses outperform expert diagnoses in certain settings (Grove 2000), at the current time, the use of such diagnostic aids in clinical practice is not widespread, and until it is, clinicians must continue to rely on more traditional methods for diagnosing their patients.

Let us suppose that the clinician in our example, given the patient history and presentation, decides to test for total IgG and finds that the IgG level is 530 mg/dL (where the normal value is 700–1600 mg/dL). What does this mean for the diagnosis? In order to answer this question, the test information needs to be used to estimate the post-test probability of CVID. In other words, in our particular example, the clinician must ask "what is the probability of CVID given a total IgG level of less than 700 and a history of frequent bacterial sinusitis?" The answer to this question in turn depends upon an evaluation of both pre- and post-test probability of the disease. This can be done informally (which is what is usually done in busy clinical practice) or formally, using Bayes' Theorem. Bayes' Theorem is a mathematical formula that is used to calculate post-test probabilities given pre-test probability and new evidence (such as a test result). For instance, Bayes' Theorem can be used to determine the probability that a person who tests positive for CVID actually has the disease. Specifically, in our example, we are interested in the probability of a person having CVID (X), if he has a positive test result (+). According to Bayes' Theorem, this probability is:

$$\Pr(X \mid +) = [\Pr(X) \times \Pr(+ \mid X)] / \Pr(+)$$

(This is read: The probability of X given a positive result is . . .) Let us assume that a positive test result for CVID is indicated by an IgG level of 700 or less, and that the occurrence of the disease in the general population is .004%. Further, for the sake of illustration, let us also assume the following values, which represent features of the test itself:

Pr (positive test result | subject has CVID) = 96%
Pr (negative test result | subject has CVID) = 4%
Pr (negative test result | subject does not have CVID) = 91%
Pr (positive test result | subject does not have CVID) = 9%

First, we need to find Pr(+). This is given by:

$$Pr(+) = [Pr(+ \mid X) \times Pr(X)] + [Pr(+ \mid \sim X) \times Pr(\sim X)]$$

Where
X is the occurrence of the disease in the general population,
+ is a positive test result, and
~X is the probability that someone in the general population does not have CVID. Substituting in our numbers from above, we get:

$$Pr(+) = [.96 \times .00004] + [.09 \times .99996] = .0900348$$

Now we can substitute the calculated value of Pr(+) into our formulation of Bayes' Theorem:

$$Pr(X \mid +) = (.00004 \times .96) / .0900348 = .0004265$$

This result shows us that, even with a positive test result for CVID, it is very unlikely that a person randomly chosen from the general population has the disease. However, the probability that the patient in our example (given a positive test result) has CVID is not the same as the probability (calculated above) of a person in the general population (given a positive test result) having it. The diagnosis of CVID in this particular patient depends largely upon his history of frequent, recurrent bacterial infections. This information makes it far more likely that he, rather than a randomly chosen individual, has CVID. This reliance of diagnosis on history, even given positive test results, highlights the clinical and diagnostic importance of taking a careful patient history.

Formulating a Treatment Plan

Once a diagnosis is made, the clinician can turn her attention to formulating, together with the patient, a plan for treating the diagnosed condition. This plan must consider the cost and risks on the one hand, as well as the potential benefits of the proposed treatment on the other. In our example, the treatment, as we will see, is very high in cost but carries little risk and a high potential for patient benefit.

When it comes to treatment decisions, it is generally understood that the patient has the final decision-making authority and can choose whether to accept or reject a proposed treatment plan. However, decisional *priority* might lie with the clinician, the patient, or both (Whitney 2003). What this means is that, in some cases, a patient might wish to defer a treatment decision to his or her clinician. However, in many cases, patients want to participate with their clinicians in the decision-making process.

This concept of shared decision making is much discussed in current medical literature. While it is agreed that sharing the burden of decision making is best for both patients and clinicians, exactly how this should be done in practice is debated. For instance, one might wonder whether or not patients should be given *all* the relevant information about various treatment options or whether instead clinicians ought to narrow down treatment choices according to their own expertise and preferences before presenting them to their patients. On the one hand, it is argued that patient autonomy depends upon completeness of information. On the other hand, it has been pointed out that patient satisfaction "has been shown to be higher when people choose from a smaller set of options" (Botti and Iyengar 2004), perhaps because too many options can seem overwhelming to a patient when he feels that he doesn't have enough education or information to make an intelligent choice from among the options. Although patient satisfaction is not the sole goal of clinical medicine, it is nevertheless something that needs to be taken into account in medical decision making, especially in the U.S. context, and thus cannot be ignored.

Further, although accurate information must be given to a patient before he or she can make an informed decision regarding a treatment option, "precise information (e.g., about risk) is frequently ineffective in changing decisions and behaviors" because patients and professionals rely on the personally relevant *meaning* of the information, rather than on the information itself (Reyna 2008). In other words, it is important that patients are told what their options are, but clinicians must also allow them the opportunity to interpret these options according to their own values and preferences. Once this interpretation has taken place and been discussed, the clinician and the patient can then decide together upon a mutually agreeable course of action.

In addition to clinician and patient values and preferences, in many cases extra-patient considerations such as health justice issues might inform or influence treatment decisions. If a treatment under consideration is a scarce resource, or in some other way affects the population at large, then this factor must be taken into account before deciding to use the treatment on an individual patient.

In our example, if the patient is diagnosed with CVID, then the patient and the clinician will have to decide together how to best treat the condition. In some cases, if the infections are mild enough (that is, if they do not require, for instance, intravenous antibiotics or extended hospital visits) the clinician might suggest trying prophylactic antibiotic therapy before moving to IgG replacement. The benefit of continuous antibiotic treatment is that it is less expensive than either intravenous or subcutaneous IgG; however, there are downsides as well. First, while antibiotics might work to prevent bacterial infections such as the sinusitis this patient has been experiencing, they do not prevent viral or parasitic infections—infections that CVID patients are also predisposed to getting. Second, long-term antibiotic therapy can lead to reduced and/or changed intestinal flora, with resulting decreased immunity to certain infections: clearly not a benefit to someone who is already immune deficient. Third, long-term treatment with antibiotics can result in antibiotic-resistant organisms, which might be passed to others in the community, raising an issue of health justice.

On the other hand, IgG therapy, while not entailing the downsides of continuous antibiotic therapy, has its own risks. The most significant risks of IgG therapy are blood clots and complications due to hyperviscosity (increased blood thickness) (Katz 2007). Further, IgG treatment is both expensive and inconvenient. Intravenous IgG must be given either in an infusion center or by a home health care practitioner and takes 4 to 6 hours per infusion, one day per month. For the patient, this means a potential loss of one full workday per month, if an infusion center or home health care practitioner with weekend hours is not available. In that case, subcutaneous IgG therapy might be a better option. Although subcutaneous IgG is generally more expensive than intravenous IgG, it can be administered by the patient at home (without assistance) on a weekly basis. Infusion time for this method of delivery ranges from

1–3 hours per infusion. Some patients, however, are not willing or able to self-infuse, so this option cannot be utilized in all cases.

How Prognostic Considerations and Patient Interests Affect Treatment Decisions

After the clinician arrives at a diagnosis in a patient case, then decisions regarding treatment must be made. These decisions must take into account both prognostic considerations and patient interests. After diagnosis, the first question a patient often asks is, "What's going to happen to me?" (Mebius, Kennedy, and Howick forthcoming). A prognosis helps to answer this question. However, accurate prognoses are notoriously difficult to make, and the reliability of available prognostic tests is currently very limited. Recently, it has been suggested that, as with diagnosis, computer-aided prognosis might be able to remedy this situation by improving prognostic reasoning and data acquisition. There are a growing number of trials comparing clinical reasoning aided by real-time, computer-based decision and action algorithms with expert decision-making alone. The results from these trials so far suggest that computer aids may improve treatment decisions based on prognostic considerations in certain health care settings, such as trauma care (Matheny et al. 2005, Ahmed et al. 2009, Fitzgerald et al. 2011), and these aids might be more widely available for clinical use in the near future.

Until such time, however, clinicians must rely on published data of prognostic studies in order to improve treatment decisions. While the gold standard for testing the efficacy of medical treatments is the randomized controlled trial, in the case of prognosis, observational (particularly cohort) studies are considered to be superior for providing prognostic evidence and answering prognostic questions (Greenhalgh 2014). Observational studies allow for a longer follow-up period, which is useful for many types of prognostic information. However, many worry that observational studies do not provide internally consistent evidence (Howick 2011) and are more prone to bias than randomized controlled trials. Because of this, some clinicians are skeptical about the reliability of information gained from observational prognostic studies and tend to make treatment decisions without taking these studies into account.

Let us turn again to our example to see how the consideration of prognosis affects treatment decision in clinical practice. Suppose that the clinician and patient are considering whether to start treatment with weekly subcutaneous infusions of IgG as a treatment for the patient's CVID. They might want to know what the prognosis for a 32-year-old male patient with CVID would be, given this treatment, before deciding to start the infusions.

However, this prognostic question is difficult to answer for several reasons. First, "high-quality controlled trial data on the therapy [. . .] is generally lacking" (Wood et al. 2007). While we do know that higher doses of immunoglobulin are associated with reduced infection frequency and a decreased risk of a variety of organ-specific complications and malignant cancers that can occur with un- or under-treated CVID, no long-term clinical data is available on exactly how IgG dose affects these risks. Further, while some patients find that their rate of infection decreases dramatically with IgG therapy, others find that the number of their infections remains the same, but that they recover more quickly from each infection. Also, patients with CVID are more likely than the general population to have autoimmune and endocrine disorders, and these conditions can complicate, or in some cases even hinder, the treatment of the concomitant diseases. For example, IgG therapy, since it is composed mainly of proteins, can bind certain protein-bound medications in the blood, such as levothyroxine, which is often given for hypothyroidism, a condition that commonly occurs in conjunction with CVID.

Despite the lack of high-quality prognostic evidence, it is likely, based upon observational and case studies, that the patient will have a better prognosis when given the treatment, than he would have if he remained untreated. In our example, then, the clinician can be confident in advising her patient that his health will improve with the proposed treatment for his condition, even in the absence of sufficient clinical trial data that supports this prognosis.

Conclusion

As can be seen from the analysis of the preceding example, even in patient cases that might initially appear to be relatively uncomplicated, the process of medical decision making can be very complex, and in all instances, involves considerations of epistemology (such as how to understand diagnostic information), ethics (such as consideration of patient values and preferences, or of how much information to give to a patient), probability (such as, for instance, when interpreting diagnostic test results or making prognostic estimations), economics, and health justice. Thus, learning to make medical decisions requires at least a basic understanding of concepts in each of these fields and careful consideration on the part of both the clinician and the patient.

Note

Ashley Graham Kennedy would like to acknowledge the research assistance of Lauren Lowy.

References

Ahmed, B., Matheny, M. E., Rice, P. L., Clarke, J. R., & Ogunyemi, O. I. (2009) "A comparison of methods for assessing penetrating trauma on petrospective multi-center data." *Journal of Biomedical Informatics*, 42(2): 308–16.
Barrows, H. & G. Pickell. (1991) *Developing Clinical Problem-Solving Skills*. New York: W.W. Norton.
Berner, E.S. & M. L. Graber. (2008) "Overconfidence as a cause of diagnostic error in medicine." *The American Journal of Medicine*, 121: S2–S23.
Bossuyt, P.M. & K. McCaffery. (2009) "Additional patient outcomes and pathways in evaluations of testing." *Medical Decision Making*, 29(5), E30–E38.
Botti, S. & S.S. Iyengar. (2004) "The psychological pleasure and pain of choosing: When people prefer choosing at the cost of subsequent satisfaction." *Journal of Personal Social Psychology*, 87(3): 312–26.
Cournoyea, M. & A. Kennedy. (2014) "Causal explanatory pluralism and medically unexplained physical symptoms." *Journal of Evaluation in Clinical Practice*, 20(6): 928–33.
Emanuel, E., D. Wendler, & C. Grady. (2000) "What makes clinical research ethical?" *JAMA* May 24/31, 283(20): 2701–2711.
Fitzgerald, M., Cameron, P., Mackenzie, C., Farrow, N., Scicluna, P., Gocentas, R., Bystrzycki, A., Lee, G., Andrianopoulos, N., Dziukas, L., & Cooper, D.J. (2011) "Trauma resuscitation errors and computer-assisted decision support." *Archives of Surgery*, 146(2011): 218–25.
Greenhalgh, T. (2014) *How to Read a Paper: The Basics of Evidence-Based Medicine*. 5th ed. Chichester, West Sussex, UK: BMJ Books/Wiley-Blackwell.
Grove, W., Zald, D. H., Lebow, B. S., Snitz, B. E., & Nelson, C. (2000) "Clinical vs. Mechanical prediction: A meta-analysis." *Psychological Assessment*, 12(1): 19–30.
Howick, J. (2011) *The Philosophy of Evidence-Based Medicine*. Oxford: Wiley-Blackwell.
Katz, U., Achiron, A., Sherer, Y., & Shoenfeld, Y. (2007) "Safety of intravenous immunoglobulin therapy." *Autoimmunity Reviews*, 6(4): 257–9.
Matheny, M. E., Ogunyemi, O., Rice, P. L., & Clarke, J. R. (2005) "Evaluating the Discriminatory Power of a Computer-Based System for Assessing Penetrating Trauma on Retrospective Multi-Center Data." *AMIA Annual Symposium proceedings / AMIA Symposium. AMIA Symposium* (2005): 500–4

Mebius A., A. Kennedy & J. Howick. (forthcoming) "Research gaps in the philosophy of evidence-based medicine." *Philosophy Compass*.

Reyna, V. (2008) "A theory of medical decision making and health: Fuzzy trace theory." *Medical Decision Making*, 28(6), 850–865.

Smith, R. and D. Francesca. (2007) "Classification and diagnosis of patients with medically unexplained symptoms." *Journal of General Internal Medicine*, 22(5): 685–91.

Sox, H., M.A. Blatt, M.C. Higgins & K.I. Marton. (2007) *Medical Decision Making*. Philadelphia, PA: American College of Physicians.

Whitney, S. (2003) "A new model of medical decisions: Exploring the limits of shared decision making." *Medical Decision Making*, 23(4): 275–80.

Wood, P., Stanworth, S., Burton, J., Jones, A., Peckham, D. G., Green, T., . . . & Chapel, H. (2007) "Recognition, clinical diagnosis and management of patients with primary antibody deficiencies: A systematic review." *Clinical and Experimental Immunology*, 149(3): 410–23. Epub 2007 Jun 12.

Further Reading

"Bayes' Theorem," *Stanford Encyclopedia of Philosophy*: http://plato.stanford.edu/entries/bayes-theorem/

Efron, B. (2013). "Bayes' theorem in the 21st century." *Science*, 340(6137): 1177–8. doi:10.1126/science.1236536

Kattan, M.W. & M.E. Cowen. (2009) *Encyclopedia of Medical Decision Making*. Thousand Oaks, CA: Sage.

Shindo, T., Takahashi, T., Okamoto, T., & Kuraishi, T. (2012). "Evaluation of diagnostic results by Bayes' theorem." *IEEJ Transactions on Electrical and Electronic Engineering*, 7(5): 450–3. doi:10.1002/tee.21756

Sox, H. C., M/ C. Higgins, and D. K. Owens. (2013) *Medical Decision Making*, 2nd ed. Hoboken, NJ; Chichester, West Sussex, UK: Wiley-Blackwell.

Part V
VARIABILITY AND DIVERSITY

36
PERSONALIZED AND PRECISION MEDICINE

Alex Gamma

In Search of a Definition

There are two prevalent characterizations of "personalized" or "precision" medicine (PM; the abbreviation "PM" will refer to both of these terms, as they describe the same or very similar ideas, and whatever difference might exist between them will not matter for the purposes of this chapter). Their common core is the idea that medical treatment should be tailored to the individual characteristics of each patient in order to be maximally effective, precise, and free of side effects. Where they differ is in their view regarding which kind of patient characteristics should determine the choice of treatment. There is a broad definition that considers any kind of feature of an individual—biological, psychological, environmental—as potentially valuable in guiding treatment decisions (Anonymous, 2012; Buford and Pahor, 2012; Wolpe, 2009). The more typical view, however, is that it is essentially the molecular biology and, within that, the genetic profile of a person, which determines the optimal treatment (Ginsburg, 2001; Meyer, 2012; Personalized Medicine Coalition, 2015; Sweet and Michaelis, 2011). In this view, medical interventions can be tailored to individuals by tailoring them to their genetic constitution.

The difference between these two characterizations matters because the "genetic" version makes a strong assumption that the broad version doesn't: that a person's genome in principle contains all the disease information necessary to derive an optimal treatment. Because this assumption may not hold, the definition of PM is sometimes extended into the broad version, which includes all kinds of non-genetic disease predictors. However, when asking which version of PM should be the subject of analysis, we must consider the fact that, whereas broad PM may sometimes be advertised, it is genetic PM that is practiced. In other words, the broad version may be upheld as the image of PM, but what is actually done in labs around the world is primarily molecular biology and bioinformatics, the software-based management and analysis of genomic data (Ginsburg, 2001; Lesko, 2007; Schleidgen et al., 2012). PM should be evaluated according to the principle "actions speak louder than words." If PM is what PM does, we should go with the genetic version.

This raises the question of the ontological status of (genetic) PM. Some have described it as a new *model* (Hudis, 2007; Wikipedia, 2015), *philosophy* (Anonymous, 2012; Buford and Pahor, 2012; Wolpe, 2009), or even *paradigm* (Ginsburg, 2001; Lesko, 2007) of medicine. These terms seem a bit overblown. The best that can be said about PM is that it is a *vision* for an improved medicine, whose plausibility is to be evaluated. It must therefore be treated as something other than a medical reality, as is for example the treatment of bacterial infections by antibiotics or the prevention of viral diseases by vaccination. In its present state, it consists of a few successes and many more promises. Thus, the worst that can be said about it is that it is a *hype*.

The next two sections will compare the vision, or hype, to reality. What does PM promise and what can it deliver?

The Vision

Today's PM has arisen from a discipline called "pharmacogenetics," which is based on the fact that interindividual differences in drug response depend on which variants of drug-metabolic enzymes are present in the body. There are slow and rapid metabolizing variants that are coded for by genetic polymorphisms. "Slow metabolizers" are at risk of being overdosed by a particular drug because they keep more of the drug in the body for a longer time, while "rapid metabolizers" are at risk of being underdosed. Both conditions can have potentially dangerous consequences. Knowing a patient's specific gene variant can therefore help determine their optimal dose regimen. Today, much is known about which drugs are metabolized by which enzymes and whether slow or rapid metabolic variants exist and how frequently they occur. This information is usually included in drug labels to potentially allow doctors to make better decisions about what drugs to give in what doses.

Although pharmacogenetics has raised hopes for an increased personalization of medicine, one success story in particular inspired the vision of PM. In the late 1990s, the biotech firm Genentech developed the breast cancer drug trastuzumab (trade name "Herceptin"), for which it later gained approval from the U.S. Food and Drug Administration (FDA). Trastuzumab was different from the typical pharmacogenetic drug in two regards: its use was not determined by a patient's drug metabolic genotype, but by a different molecular feature, and this feature allowed for a stratification of patients into subgroups of responders and non-responders. Specifically, it turned out that trastuzumab only worked in a subgroup of women, whose breast cancer cells expressed high levels of a receptor molecule called HER2, which is involved in regulating cell growth. Thus, overexpression of HER2 could be used as what has come to be called a "biomarker" that indicated the likely effectiveness of treatment with trastuzumab. Indeed, clinical trials went on to show that women with HER2-positive breast cancer treated with this drug had a lower overall mortality and higher chances of disease-free survival than comparable women without the treatment (Hudis, 2007). To many, this seemed like a proof of principle for a vision of personalized medicine, in which the reality of individualized drug dosing based on drug-metabolic enzyme genotypes was extended into the possibility of individualizing *any kind* of medical treatment based on *any kind* of genetic or molecular information.

When, for example, Danielle seeks treatment for her type-2 diabetes mellitus, the genetic version of PM implies that a genome scan should, in principle, suffice to "read out" the specifics of her illness and find a tailored treatment, perhaps in the form of a drug that targets her body's insulin receptors. When Danielle goes to the doctor today, he or she won't be able to help her in this way, but for proponents of genetic PM, this is only a temporary limitation. A future Danielle, in a world where PM is perfected, would just give a drop of blood (or saliva) on a microchip, wait a few hours, and then pick up medicine targeted precisely to her specific variant of the disease.

But the emerging vision for PM is even more ambitious: it includes a vastly increased ability not just to tailor treatments to patients but also to understand the causes of disease, to predict and prevent disease in (as yet) unaffected individuals, and to make treatments more precise, effective, and less harmful. Medicine would shift to a new paradigm: from reactive to preventive, and from crude treatments found by trial-and-error to precise interventions on the molecular level (Chan and Ginsburg, 2011; Henney, 2012).

Excitement about the potential of a genomic medicine already started with the Human Genome Project. On the day of the announcement of the completed first draft of the human genome, then U.S. president Bill Clinton asserted that

> with this profound new knowledge, humankind is on the verge of gaining immense, new power to heal. . . . Genome science . . . will revolutionize the diagnosis, prevention, and treatment of most, if not all, human diseases.
>
> (cited in Collins, 2010)

Ever since, and especially with the unfolding success story of trastuzumab, the rhetoric of promised revolutions and imminent breakthroughs has never ceased to surround PM (Desiere and Spica, 2012; Ginsburg, 2001; Personalized Medicine Coalition, 2015). Not only the sensation-hungry mass media, but also researchers and university public relations offices have fueled the hype of the coming transformation of health care. Given this excitement, it can be a bit of a shock to learn that none of these fruits of PM have materialized. Instead, there is a modest number of personalized treatments, none of which have revolutionized anything.

The Reality

Most of what PM is today is still pharmacogenetics. The doses of a number of medications can be adapted to a patient's specific enzyme variants, but the extent to which actual drug responses are predictable from the genotype is highly limited because they are also affected by many non-genetic factors (Shah and Shah, 2012). Even the most celebrated success of pharmacogenetics, a drug with the romantic name "warfarin," is not an unambiguous example of tailoring a drug to a patient's genetics. Warfarin is widely used as a blood-thinning drug, but the dose needed to achieve the required clinical effect varies greatly, and apparently unpredictably, from patient to patient. This unpredictable variability is very problematic, as too much or too little warfarin can lead to dangerous over- or undercoagulation. Consequently, much research has looked into the genetic basis of its pharmacology. As it turns out, however, genetic factors do not explain more than 50% of the variation in drug response, and genotype-based dosing offers little added benefit to clinically guided dosing (Kimmel et al., 2013; Shah and Shah, 2012).

Subsequently, the same has been found for many other drugs, showing that the seemingly simple causal pathway "gene → drug enzyme → drug blood level → drug effect" is neither simple nor deterministic (Shah and Shah, 2012). A patient's drug response is determined not only by his or her genotype but also by a complicated network of metabolic pathways and by interactions with non-genetic factors such as age, sex, body weight, lifestyle, time of intake, amount of food in the stomach, renal/hepatic clearance, treatment duration, and ethnicity (Gamma, 2013). A comprehensive individualization of drug dosing using genetic information alone is therefore not possible.

The most advanced applications of PM are found in cancer research. A procedure called immunotherapy currently offers the greatest hope for effective treatment. It requires the administration of drugs such as trastuzumab, which are antibodies that block certain molecular receptors involved in the proliferation of cancer cells. These drugs are typically given in combination with conventional radio- or chemotherapy. They can be very effective in certain settings, but they are not the magic bullets they are often portrayed to be.

Trastuzumab, for example, reduces death and recurrence rates in early-stage HER2-positive breast cancer by 40% over a period of three years (Moja et al., 2012). In metastatic cancer, corresponding rates are 20% and 40%, which translates into a prolongation of life by an average of half a year (Balduzzi et al., 2014). However, the drug helps only about 25% of women with breast cancer, and of those, fewer than 50% respond to the treatment (Bartsch et al., 2007). In fact, even in responders, resistance to the treatment develops rapidly, usually within a year (Nahta and Esteva, 2006).

Although these facts qualify the efficacy of the drug, others challenge another of PM's claims: that new personalized treatments will have fewer adverse effects. In the case of trastuzumab, the opposite is true. The drug has been found to cause rare but serious cardiac dysfunction, sometimes leading to death by congestive heart failure (Dahabreh et al., 2008). This happens in about 2% of treated patients. In general, existing personalized therapeutics can have side effects as severe as those of any other drugs. Although the idea that more specific treatments will tend to have fewer side effects is plausible, the biological basis of undesired effects is not understood well enough to support general claims about a systematic reduction in adverse drug effects.

PM is a field of many activities and ambitions, but its overall clinical impact is minimal. Available treatments are mostly for rare diseases, and their benefits are rather small. PM has little to offer for the major health issues facing the world today, including common, complex diseases such as diabetes, heart disease, obesity, dementia, stroke, and affective disorders. Although countless studies searching the genome for gene-disorder associations (so-called GWAS = genome-wide association studies) have been performed, findings have been disappointing: many results have not been reproducible, and most genes that turn up contribute only a tiny risk to the overall phenotype (Hirschhorn and Gajdos, 2011). Disease prevention and prediction have generally not benefited from PM. Most illnesses are determined by a complex interplay of genetic and non-genetic factors, so that genetic tests are usually not of clinical value. Only for (rare) monogenic disorders, in which single genetic defects predict disease outcomes relatively reliably, have genetic tests been useful (e.g., screening for PKU or for BRCA variants in breast cancer). Add to that the important fact that diagnostic tests are of questionable use without concomitant therapeutic options, which are in many cases lacking.

When Danielle researches her disease on the internet, she finds that genetic information currently improves the prediction of diabetes risk achieved by non-genetic information by less than 1% (Vassy and Meigs, 2012). Like many diabetes patients, she was diagnosed late, when substantial damage to the cells in her pancreas had already occurred. But since the disease progressed imperceptibly slowly and there are no biomarkers for reliable early detection, she did not know until she was 50. Since she had normal cholesterol and blood pressure, the doctor first prescribed her a diet low in sugar and saturated fat plus moderate daily exercise to reduce her body weight. These changes helped a lot while she stuck to them, but maintaining them proved difficult. Not only was her self-discipline not the best, but there were few places in her neighborhood that made exercising really fun, and she had to drive quite far to get some of the foods she was supposed to eat. When her blood sugar levels increased, her doctor prescribed her an oral drug called Metformin, which worked well in the beginning, but finally the disease progressed to a point where her feet started hurting and feeling numb. Unfortunately, there are no known pharmacogenetic or other biomarkers to help personalize her drug treatment and reduce its side effects.

Comparing the hype of genetic PM with its clinical reality reveals a huge mismatch. How did it arise? The remainder of this chapter focuses on the epistemic challenges that generated the current situation. It will be argued that due to the particular history of genetics and the nature of modern knowledge economies in the life sciences, PM makes promises without having a principled strategy to keep them.

The Information Metaphor

The privileged position of genes in ontogeny (an organism's development from conception to death) is a basic tenet of biomedicine that hardly ever requires justification. When the question comes up, the only explanation usually given is a vague reference to the information-carrying function of DNA, what philosophers of science have called the "information metaphor." The information metaphor is the belief that genes carry information in the form of instructions that control the development of an organism (Griffiths, 2001). They are the main determinants of a creature's traits and, particularly in humans, also of its disposition for disease.

This belief in the primacy of genes in development has dominated the course of biomedicine into the 21st century. But where does it come from, and is it true? Let's tackle these questions one at a time.

A one-sentence summary of the history of biology as it relates to genes might run as follows. Scientists had always believed that seeds and germs carry the potential to form a new organism, and by the middle of the 20th century they had convinced themselves that this potential was located in the DNA of biological cells. The concept of "information" that was afloat at that time in communications theory was quickly appropriated by molecular biologists as a convenient description of what they believed genes do: passing instructions to the cell for making proteins and regulating the expression of other genes. Soon enough, the term "information" had assumed a life of its own, divorced from its original meaning of the amount of signal transmissible by a communications device. The information in genes came to be understood in semantic terms, as instructions, programs, and blueprints, and later even as a history book, user manual, and operating system of the human organism (Keller, 1995).

The deep belief that genes represent and determine traits, whereas non-genetic factors mostly provide unspecific variation, has ultimately been the driver behind the big biology projects of our time, starting with the Human Genome Project (HGP) and continuing to today's Personalized Medicine. Among other things, it has given us the idea that most diseases are ultimately genetic (see Chapter 14) and that, therefore, the cleanest, most specific, and effective treatment would be to correct the faulty genes. Another corollary is that genes hold the information about future risk for disease that can be used to predict and possibly prevent its occurrence.

Unfortunately, there is no reason to think that this belief is true. Both empirical evidence as well as conceptual considerations militate against it (Griffiths, 2001; Nijhout, 1990; Sarkar, 1996). On the conceptual side, the information metaphor posits a rich semantics of genes and their molecular entourage that has been adopted from the realm of human minds without warrant. The concept of information as meaning (e.g., conversations, messages, news), the ability to both *give* instructions *and* to understand and execute them, and the capacity to represent things (objects, states, goals) are all inherent to creatures with minds and their interactions. That is where these phenomena have their natural homes and the only place we encounter them. To ascribe them to inanimate objects like molecules is simply unjustified, unless there is compelling evidence to do so, but there is not.

A first clue is the simple but powerful observation that the DNA of genes is just an ordinary molecule, whose effects, like those of any protein, sugar, or fatty acid, depend on the specificity of its spatial, mechanical, and electromagnetic interactions with other molecules. It is hard to see where any special informational powers of genes should come into the picture.

Next, studies of the relationship between genes and phenotypic outcomes show, without exception, that genes alone do not determine outcomes. We have seen this already in the case of pharmacogenetics, where genes for drug-metabolic enzymes contribute to, but do not determine, drug response. Another example are so-called monogenic disorders, where a single gene defect is the origin of a disease. Even in such cases, one and the same mutation does not systematically "bring forth" the same disease outcome in different individuals. This is true, for example, for diseases such as phenylketonuria (PKU; Bercovich et al., 2008; Kayaalp et al., 1997; Pérez et al., 1993) and cystic fibrosis (Cutting, 2006; Schaedel et al., 2002). There are countless examples like these, and interested readers can find a rich literature on the subject (e.g., Neumann-Held and Rehmann-Sutter, 2006).

In the context of ontogeny, genes are not privileged. There is a fundamental parity of the causal roles and effects of genes and other developmental resources, such as an organism's non-genetic biochemical constituents and its physical, biological, and social environments (Griffiths, 2001). Thus, while genes may turn out *empirically* to contribute more to a given trait or disease (e.g., by causing more of its observable variability), they do not contribute more *in principle*. There is therefore no *a priori* reason to assume that diseases are ultimately genetic or that an intervention at the genetic level would always be the ideal treatment. Biological research has time and again confirmed the view that development is a matter of complex interactions between many players, genetic and non-genetic, on a level playing field. Modulation at any causal nexus can alter the outcome, and it is not usually possible to identify an unambiguous beginning of a developmental process. Genes are neither privileged origins of diseases nor privileged intervention points. It is a purely empirical question what kind of causal factor gives us the most useful information for diagnosis, prediction, prevention, or treatment, in any given situation.

For type-2 diabetes, likely causes are age, body weight, birth weight, family history, genes, diet, physical activity, and possibly serum biomarkers such as adiponectin, HDL-Cholesterol, CRP-, and HBA1c-levels. A positive family history alone more than doubles the odds of developing type-2 diabetes (Chan and Ginsburg, 2011), which is more than what single known susceptibility genes contribute (Vassy and Meigs, 2012). The same is true for other non-genetic factors such as diet and exercise. The general point is that PM is never going to be able to extract from the genome all the information it needs for its diagnostic and therapeutic vision, because this information is simply not there.

Finally, there is one important point to make about what Germans call *Etikettenschwindel* ("false labeling"). Therapeutic successes, such as new drugs against cancer and cystic fibrosis (Wainwright et al., 2015), may be claimed for personalized medicine, but that doesn't mean they are the result of its presumed success strategy, that of transforming genetic information into treatments. In fact, the opposite will usually be the case: the mechanism of action of the cancer drug Gleevec, for example, is to inhibit an enzyme in one of the cell's signaling pathways; it could not have been discovered without substantial knowledge of non-genetic metabolic processes. The same is true for other "personalized" drugs, which usually work by binding to certain receptor molecules in the cell. There is a danger that PM will allow therapeutic successes to be attributed to its gene-centered vision when in fact they are indicators of its shortcomings. This would unfortunately make it hard to learn from the present situation in biomedicine.

The Challenge of Scientific Inference

PM also faces formidable challenges of scientific inference, and it has yet to come up with a systematic and realistic way of addressing them.

Epistemically, knowledge relevant to PM sits uneasily between the opposing goals of generalization and individualization of inference. On the one hand, PM wants evidence that is valid beyond the contingent samples from which it is obtained; on the other hand, it wants evidence that, in the limit, is valid only for the individual patient from whom it is obtained. These two goals are conflicting in ways that are not yet fully understood, and each of them also has its own challenges.

The problems of the generalizability of evidence are relatively well known but nevertheless seemingly intractable. The first problem concerns representativeness: theoretically, a perfectly representative sample, obtained by random sampling from a population, would allow an accurate estimation of population effects. The effect of a drug, or the association between a gene and a disorder, found in the sample could be generalized to the population that it represents. However, in the real world, random sampling almost never holds, and samples are almost never representative. This means that drug effects and gene-disease associations will often be unreliable and replicate poorly.

Genome-wide association studies (GWAS) have been struck particularly hard by this problem, because, in their case, it is compounded by two further issues: first, by their nature, these studies test huge numbers of genes as possible predictors of disease, thereby incurring a severe multiple-testing problem (i.e., the fact that many gene-phenotype associations will reach statistical significance by mere chance). Second, because the number of predictors is so much larger than the number of subjects, the "curse of dimensionality" strikes. This term refers to the fact that in such "high-dimensional" settings, it will almost always be possible to find a combination of predictors (genes) that can discriminate well between healthy and diseased individuals, but the particular combination found in one study will not be the same as that found in the next (James et al., 2013). This is probably the main reason why many GWAS, particularly in the early years, failed to be replicated (Hirschhorn et al., 2002).

Proponents of PM often enthusiastically embrace bioinformatics and big data as the solution to the problem of extracting diagnostically or therapeutically relevant knowledge from large amounts of patient data (Hood and Flores, 2012). The sheer mass of data in combination with advanced software algorithms is seen to virtually guarantee a greater understanding of disease. However, data science, while one of the fastest growing academic and industrial markets, is still in its infancy and has yet to live up to the hopes placed on it. Sober assessments indicate that the dream of gaining knowledge by brute data force will not generally work. Crunching numbers without well-motivated hypotheses is destined to produce a flood of false leads and spurious connections. Big data will drown in noise (Jordan, 2014). As in the case of GWAS, the perpetrators are the curse of dimensionality, multiple testing, and overfitting (i.e., the situation in which a statistical model "mistakes" the random noise in a data set as true effects, which severely limits the generalizability of its results). The push toward more data and computing power will be fruitless unless analyses are based on a solid foundation of empirical and theoretical work.

As one takes a step toward drawing scientific inferences about subgroups of—or even single—patients, different sets of problems arise. PM's main epistemic approach to personalizing treatments has been to find drugs that are effective only in subgroups of patients exhibiting certain biomarkers. Although looking for subgroup-specific drugs may sound counterintuitive ("why not find drugs that work for *all* people?"), from the perspective of PM the excitement arose when trastuzumab, which initially was shown ineffective in the study sample as a whole, was later discovered to be effective, but only in a subgroup of patients. This success was seen as a proof of principle that the same strategy could generalize to many drugs and many diseases. Although this may or may not be true, subgroup analyses as currently practiced create as many problems as they solve because they are usually underpowered, lacking the necessary sample sizes to reliably detect subgroup effects (Pocock et al., 2002). The consequence is effects that are either spurious (false positives or type-I errors) or missed (false negatives, type-II errors). Widespread bad statistical practices aggravate the situation. Researchers do not always use the correct statistical methods, nor do they pre-plan their subgroup analyses, which invites post-hoc fishing for and selective reporting of statistically significant effects (Pocock et al., 2002; Smith, 2005). As about 50% of randomized clinical trials (RCTs) use subgroup analyses, the problem is substantial.

Note that despite constituting a step toward individualization, subgroup analysis still looks for results that are generalizable, namely to the entire population of patients exhibiting a certain biomarker. Proponents of PM have begun to advocate an even more individualistic methodological approach (Schork, 2015). In so-called n-of-1 trials, single patients go through several cycles of different active treatments and appropriate control conditions. This design offers some statistical advantages, such as the ability to isolate patient-by-treatment interactions (DEcIDE Methods Center N-of-1 Guidance Panel, 2014; Senn, 2001). Participants have more palpable benefits than in other kinds of trial by being able to compare different therapeutic options directly relevant to them on outcome measures directly relevant to them. The design's participatory nature is a welcome step toward a more patient-focused medicine, where treatment decisions are shared and centered on a patient's preferences.

On the down side, n-of-1 trials are not feasible for treatments whose outcomes manifest in the long term. An individual cannot both receive a treatment and wait for decades for the outcome *and* also receive a placebo and again wait for decades to see the outcome. These trials will also be difficult to conduct in terminal illnesses like end-stage cancers, where time is limited. This makes n-of-1 trials unsuitable for some of the leading causes of morbidity and mortality.

A final epistemic challenge is risk prediction. PM promises the routine use of prognostic biomarkers to the extent of turning health care from a reactive into a preventive system. At the same time, it offers little clue about how to achieve this transformation, beyond just assuming that the few existing successes will generalize. The basic problem is that, due to the complexity of human biological, psychological, and social life, single, highly predictive markers for diseases will generally not be available (and even less so if restricted to genetic markers). The norm will rather be many weakly determining causes acting in non-additive ways. In such a situation, inferential statistics hits its limits. As discussed above, results from RCTs will not generalize, most variables will only be accessible in observational studies, which suffer from problems of confounding, and sample sizes affording enough power to avoid false positive and negatives become prohibitively large.

With the insight that there are no special instructive powers of genes, prediction becomes even harder. Genes represent a relatively stable aspect of organisms that at least potentially permits long-term predictions, as demonstrated by genetic tests for monogenic disorders. When the role of genes in ontogeny is appropriately qualified, the role of non-genetic, and potentially more variable, causal factors comes to the fore. Aspects of lifestyle and environment, for example, are crucial determinants of common, complex disorders such as diabetes, but are hard-to-impossible to predict. Accordingly, the predictive power achievable from single markers will be negligible, and even combining several markers may leave substantial portions of the variance unexplained. In type-2 diabetes, for example, reliable prediction would be highly valuable in order to detect the many non-symptomatic early cases. However, even a combination of known susceptibility genes explains only about 10% of the heritable risk for the disease (Imamura and Maeda, 2011).

A Biased Knowledge Economy

The specific nature of PM cannot be understood without reference to the cultural context in which it has grown. In the past three decades, the face of biomedicine has significantly changed. The HGP has brought the commercialization of biological research into the mainstream. A new breed of scientist-entrepreneur with stakes in the biotech and pharma industry has come to dominate the field, to the point where commentators have remarked that it is hard to find leading researchers without any economic investment in their field. Many stand to gain from consulting and collaborating with pharma companies and from patenting and marketing diagnostic and therapeutic products.

The invasion of the profit motive into biomedicine has radically altered the field. Researchers and universities alike have come under the rule of the market, with pressures to compete for financial and intellectual resources intensifying. High-ranking biomedical journals have become conduits for the pharmaceutical industry by uncritically publishing their trial data, which, not surprisingly, are overwhelmingly in favor of the industry's products (Smith, 2005).

Essential to the mechanics of this new knowledge economy are media communications. Their logic of attention-grabbing headlines and dramatized reporting fits the needs of the industry to advertise its products. The result is a constant hype over ever-new claims of medical breakthroughs, with little pause to reflect critically on their basis and outcomes. The usual and almost inevitable failure of such claims therefore remains unacknowledged by most, lessons cannot be learned, and the cycle is bound to repeat.

In this context, PM can be seen as the latest in a sequence of gene-related hypes that started with the promises of the HGP and continued with the buzz surrounding gene therapy. Both of these eventually engendered disappointment (the HGP did not produce the predicted leap in our understanding and treatment of diseases, as evidenced by the failure of early gene therapy trials), but business interests were quickly restored with the vision of a genetics-driven personalized medicine.

In an environment driven by big business and the media, biomedical knowledge has become extremely biased. The strongest distortion stems from the heavy tilt toward genetics supported by the information metaphor. Then, most research on personalized treatments is industry-sponsored and biased toward exaggerating the efficacy of its products by selectively publishing positive findings or by subtly choosing outcome variables so as to make a positive result more likely (Smith, 2005). Due to their business entanglements, researchers and universities have become part of the problem as incentives shift toward competitiveness and profitability and away from finding the truth and serving patients' needs.

Danielle has heard a lot about so-called personalized medicine and hopes its promises will one day come true. It would be nice to have a pill that just cures it all. But she also knows that there could be vast improvements in diabetes care if only people could be made to maintain a healthy lifestyle. She thinks that maybe, instead of pursuing expensive high-tech research, biomedicine should focus more on how to provide better opportunities and incentives for diabetics to lead healthy lives in ways that are sustainable.

Conclusion

Genetic PM is not a realistic vision, because it has inherited the faulty belief that genes contain all of the information about an organism and direct its translation into a phenotype. It has no answers to a number of problems that severely bias scientific inference and knowledge acquisition. PM as it is currently practiced therefore lacks any principled and credible strategy to achieve its goals. If it is to have any role in the future of medicine, it will have to greatly expand its purview in a way that takes it well beyond medicine's current focus on the causal and predictive powers of genes.

References

Anonymous, 2012. What happened to personalized medicine? *Nat Biotechnol* 30, 1–1.

Balduzzi, S., Mantarro, S., Guarneri, V., Tagliabue, L., Pistotti, V., Moja, L., D'Amico, R., 2014. Trastuzumab-containing regimens for metastatic breast cancer. *Cochrane Database Syst Rev.* 6, Art. No.: CD006242.

Bartsch, R., Wenzel, C., Steger, G.G., 2007. Trastuzumab in the management of early and advanced stage breast cancer. *Biologics: Targets & Therapy* 1, 19–31.

Bercovich, D., Elimelech, A., Zlotogora, J., Korem, S., Yardeni, T., Gal, N., Goldstein, N., Vilensky, B., Segev, R., Avraham, S., Loewenthal, R., Schwartz, G., Anikster, Y., 2008. Genotype–phenotype correlations analysis of mutations in the phenylalanine hydroxylase (PAH) gene. *J Hum Genet* 53, 407–418.

Buford, T.W., Pahor, M., 2012. Making preventive medicine more personalized: Implications for exercise-related research. *Preventive Medicine* 55(1), 34–36.

Chan, I.S., Ginsburg, G.S., 2011. Personalized medicine: Progress and promise. *Annu Rev Genomics Hum Genet* 12, 217–244.

Collins, F., 2010. Has the revolution arrived? *Nature* 464, 674–675.

Cutting, G.R., 2006. Causes of variation in the cystic fibrosis phenotype. *Annales Nestlé* (English ed.) 64, 111–117.

Dahabreh, I.J., Linardou, H., Siannis, F., Fountzilas, G., Murray, S., 2008. Trastuzumab in the adjuvant treatment of early-stage breast cancer: A systematic review and meta-analysis of randomized controlled trials. *Oncologist* 13, 620–630.

DEcIDE Methods Center N-of-1 Guidance Panel, 2014. *Design and Implementation of N-of-1 Trials: A User's Guide.* AHRQ publications, Rockville, MD.

Desiere, F., Spica, V.R., 2012. Molecular diagnostics and personalised medicine. *New Biotechnology* 29(6), 611–612.

Gamma, A., 2013. The role of genetic information in personalized medicine. *Perspectives in Biology and Medicine* 56, 485–512.

Ginsburg, G., 2001. Personalized medicine: Revolutionizing drug discovery and patient care. *Trends in Biotechnology* 19, 491–496.

Griffiths, P.E., 2001. Genetic information: A metaphor in search of a theory. *Philosophy of Science* 68, 394–412.

Henney, A.M., 2012. The promise and challenge of personalized medicine: Aging populations, complex diseases, and unmet medical need. *Croat Med J* 53, 207–210.

Hirschhorn, J.N., Gajdos, Z.K.Z., 2011. Genome-wide association studies: Results from the first few years and potential implications for clinical medicine. *Annu. Rev. Med.* 62, 11–24.

Hirschhorn, J.N., Lohmueller, K., Byrne, E., Hirschhorn, K., 2002. A comprehensive review of genetic association studies. *Genetics in Medicine* 4, 45–61.

Hood, L., Flores, M., 2012. A personal view on systems medicine and the emergence of proactive P4 medicine: Predictive, preventive, personalized and participatory. *N Biotechnol* 29, 613–624.

Hudis, C.A., 2007. Trastuzumab—mechanism of action and use in clinical practice. *N. Engl. J. Med.* 357, 39–51.

Imamura, M., Maeda, S., 2011. Genetics of type 2 diabetes: The GWAS era and future perspectives. *Endocr J* 58, 723–739.

James, G., Witten, D., Hastie, T., Tibshirani, R., 2013. *An introduction to statistical learning,* Vol. 6. Springer Science & Business Media, New York.

Jordan, M., 2014. Machine-Learning Maestro Michael Jordan on the Delusions of Big Data and Other Huge Engineering Efforts. URL: https://web.archive.org/web/20150603161808/http://spectrum.ieee.org/robotics/artificial-intelligence/machinelearning-maestro-michael-jordan-on-the-delusions-of-big-data-and-other-huge-engineering-efforts (accessed 3.6.2015).

Kayaalp, E., Treacy, E., Waters, P.J., Byck, S., Nowacki, P., Scriver, C.R., 1997. Human phenylalanine hydroxylase mutations and hyperphenylalaninemia phenotypes: A meta-analysis of genotype-phenotype correlations. *American Journal of Human Genetics* 61, 1309.

Keller, E.F., 1995. *Refiguring life.* Columbia University Press, New York.

Kimmel, S.E., French, B., Kasner, S.E., Johnson, J.A., Anderson, J.L., Gage, B.F., Rosenberg, Y.D., Eby, C.S., Madigan, R.A., McBane, R.B., Abdel-Rahman, S.Z., Stevens, S.M., Yale, S., Mohler, E.R., III, Fang, M.C., Shah, V., Horenstein, R.B., Limdi, N.A., Muldowney, J.A.S., III, Gujral, J., Delafontaine, P., Desnick, R.J., Ortel, T.L., Billett, H.H., Pendleton, R.C., Geller, N.L., Halperin, J.L., Goldhaber, S.Z., Caldwell, M.D., Califf, R.M., Ellenberg, J.H., 2013. A pharmacogenetic versus a clinical algorithm for warfarin dosing. *N. Engl. J. Med.* 369, 2283–2293.

Lesko, L.J., 2007. Personalized medicine: Elusive dream or imminent reality? *Clin Pharmacol Ther* 81, 807–816.

Meyer, U.A., 2012. Personalized medicine: A personal view. *Clin Pharmacol Ther* 91, 373–375.

Moja, L., Tagliabue, L., Balduzzi, S., Parmelli, E., Pistotti, V., Guarneri, V., D'Amico, R., 2012. Trastuzumab containing regimens for early breast cancer. *Cochrane Database Syst Rev* 4, Art. No. CD006243.

Nahta, R., Esteva, F.J., 2006. Molecular mechanisms of trastuzumab resistance. *Breast Cancer Res* 8, 215.

Neumann-Held, E.M., Rehmann-Sutter, C. (Eds.), 2006. *Genes in Development: Re-reading the Molecular Paradigm*. Duke University Press, Durham, NC.

Nijhout, H.F., 1990. Metaphors and the role of genes in development. *Bioessays* 12, 441–446.

Pérez, B., Desviat, L.R., García, M.J., Ugarte, M., 1993. Different phenotypic manifestations associated with identical phenylketonuria genotypes in two Spanish families. *Journal of Inherited Metabolic Disease* 17, 377–378.

Personalized Medicine Coalition, 2015. The Age of Personalized Medicine. URL: http://www.personalizedmedicinecoalition.org (accessed 5.6.2015).

Pocock, S.J., Assmann, S.E., Enos, L.E., Kasten, L.E., 2002. Subgroup analysis, covariate adjustment and baseline comparisons in clinical trial reporting: Current practice and problems. *Stat Med* 21, 2917–2930.

Sarkar, S., 1996. Decoding "Coding": Information and DNA. *Bioscience* 4, 857–864.

Schaedel, C., de Monestrol, I., Hjelte, L., Johannesson, M., Kornfelt, R., Lindblad, A., Strandvik, B., Wahlgren, L., Holmberg, L., 2002. Predictors of deterioration of lung function in cystic fibrosis. *Pediatr. Pulmonol.* 33, 483–491.

Schleidgen, S., Klingler, C., Bertram, T., Rogowski, W.H., Marckmann, G., 2012. What is personalized medicine: Sharpening a vague term based on a systematic literature review. *BMC Med Ethics* 14, 55–55.

Schork, N.J., 2015. Time for one-person trials. *Nature* 520, 609–611.

Senn, S., 2001. Individual therapy: New dawn or false dawn? *Drug Information Journal* 35, 1479–1494.

Shah, R.R., Shah, D.R., 2012. Personalized medicine: Is it a pharmacogenetic mirage? *Br J Clin Pharmacol* 698–721.

Smith, R., 2005. Medical journals are an extension of the marketing arm of pharmaceutical companies. *Plos Med* 2, e138.

Sweet, K.M., Michaelis, R.C., 2011. *The Busy Physician's Guide To Genetics, Genomics and Personalized Medicine*. Springer Verlag, New York.

Vassy, J.L., Meigs, J.B., 2012. Is genetic testing useful to predict type 2 diabetes? *Best Pract Res Clin Endocrinol Metab* 26, 189–201.

Wainwright, C.E., Elborn, J.S., Ramsey, B.W., Marigowda, G., Huang, X., Cipolli, M., Colombo, C., Davies, J.C., De Boeck, K., Flume, P.A., Konstan, M.W., 2015. Lumacaftor–ivacaftor in patients with cystic fibrosis homozygous for Phe508del CFTR. *N Engl J Med* 373, 220–231.

Wikipedia, 2015. Personalized Medicine. URL: http://en.wikipedia.org/wiki/Personalized_medicine (accessed 6.5.15).

Wolpe, P.R., 2009. Personalized medicine and its ethical challenges. *World Medical & Health Policy* 1(1), 47–55.

Further Reading

Evans, I., Thornton, H., Chalmers, I., Glasziou, P., 2011. *Testing Treatments*. Pinter & Martin, London.

Moore, D.S., 2003. *The Dependent Gene*. Macmillan, New York.

Oyama, S., 1985. *The Ontogeny of Information*. Cambridge University Press, Cambridge.

HealthNewsReview: http://www.healthnewsreview.org

The BMJ blogs: http://blogs.bmj.com/bmj/feed/ (RSS feed)

37
GENDER IN MEDICINE

*Inmaculada de Melo-Martín
and Kristen Intemann*

Introduction

Increasingly, a significant amount of research calls attention to the influence of sex and gender in health and disease. For example, various studies suggest that disparities between men and women exist in the diagnoses and treatment of a variety of health conditions (IOM 2010, Kent, Patel, and Varela 2012). To some extent, health disparities may be the result of biological differences between groups, such as anatomical, physiological, hormonal, and genetic differences. Disparities may also be partly explained by differences in behaviors and lifestyle choices. However, gender also significantly affects differences in the power and control that men and women have over socioeconomic resources, their social position, their treatment in communities, their access to education and health care, and their susceptibility and exposure to particular health risks. Thus, gender health disparities can be the result of complex social factors, including discriminatory norms and practices related to health and disease within households, differential vulnerabilities to disease, and biases in health systems and health research (Sen and Ostlin 2008). Failure to be attentive to differences related to sex and gender can itself also create or exacerbate health disparities.

The purpose of this chapter is to show some of the ways in which attending to potential biological and social differences along sex and gender lines can be important for improving health outcomes in women and addressing gendered health inequalities. Using cardiovascular diseases (CVDs) as an example, we discuss how utilizing gender as a category of analysis in medicine can help uncover a variety of relevant phenomena crucial in researching, diagnosing, and treating a disease, and how failing to do so may obscure such phenomena. Differences between men and women in relation to CVD have become well known (Collins et al. 2011, Stock and Redberg 2012, Wenger 2012, Wyant and Collett 2015), thus this case provides an opportunity to examine the relevance of using sex and gender as categories of analyses in order to improve health outcomes.

Although in ordinary language use, "sex" and "gender" are often used interchangeably, feminist scholars have historically distinguished sex (being female or male) from gender (being a woman or a man) and have associated sex with biological characteristics and gender with social ones (Warnke 2011). Nonetheless, the categories of sex and gender, like that of race, are contested ones (Spelman 1988, Fausto-Sterling 2000, Butler 2004, Alcoff 2006, Warnke 2011, Witt 2011, Haslanger 2012) and should not be assumed to track simple binary categories. In this chapter we do not take a position on the sex/gender distinction (i.e., whether there is one and what, if anything, it amounts to). We do, however, take the category gender/sex to be a relevant category of analysis in medicine, one that can be helpful in calling attention to unjust heath disparities and needed in order to decrease existing health inequalities. In what follows, we use "gender" to refer to differences that can result from biological, psychological, and social aspects and from the interplay of these aspects. Thus, it captures, but is not limited to, traditional understandings of sex categories. Although here we focus on how gender affects health outcomes for women, notice that gender is not a category that applies only to women, and thus gender differences are relevant also to the health of all human beings. For example, masculinity norms can lead men to risk-taking behaviors that are likely to negatively affect their health (Fleming and Agnew-Brune 2015). Of course, one of the challenges of using gender as a category of analysis is that it does not track a monolithic group. There are significant differences, for instance, among women of different races, ages, sexual orientation, and socioeconomic classes. Despite these differences, our claim is that gender, along with other axes of difference, is often important to understanding and improving health problems.

In the first section we briefly introduce CVD and present evidence for the various ways in which inattention to gender and gender bias can influence medical research and clinical practice in this area and ultimately contribute to unjust gender health disparities: from animal studies, to the design of clinical trials, to clinical care. In the last section we point out different ways in which gender can be used as a relevant category of analysis throughout research and diagnosis so as to minimize gender biases and achieve better health outcomes for all.

Gender Biases in Medicine

An increasing number of studies are focusing on the effects of gender in health and disease and many have shown the existence of important gender health disparities. Gender differences in CVD in particular have received a significant amount of attention. CVDs are a class of disorders that involve the heart and blood vessels and include coronary artery disease, heart failure, stroke, and hypertension (WHO 2015). Despite significant declines in CVD mortality rates during the last decade, these diseases are currently the leading cause of death in women and men globally (WHO 2015). However, more women than men who have CVDs die from these diseases, and there has been an increase in mortality among young women with ischemic heart disease, despite a higher prevalence of this disease in men (Wyant and Collett 2015). Studies also show that a greater proportion of women with heart attacks die before hospital arrival, and a greater proportion also die within 1 year of having the disease compared with men (Stock and Redberg 2012). Similarly, women experience greater morbidity and mortality than do men after having coronary artery bypass. Additionally, although risk factors for CVD are the same in women and men, there are gender differences in the prevalence of risk factors such as high blood pressure and diabetes mellitus (Mosca, Barrett-Connor, and Wenger 2011).

Some CVD conditions that occur in both men and women have difference prevalence rates. For example, there is a predominance of pulmonary hypertension and atrial fibrillation in females and a predominance of systemic hypertension in males (Miller and Best 2011). Gender differences also exist in disease presentation (Collins et al. 2011, Stock and Redberg 2012). For instance, chest pressure spreading to the left arm is usually thought to be the distinctive symptom of coronary heart disease, but women tend to present more often with neck, jaw, or back pain, shortness of breath, nausea, indigestion, and fatigue (Collins et al. 2011). Evidence suggests that gender-specific genetic, anatomic, and physiological differences also contribute to CVD (Mercuro et al. 2010).

A variety of factors likely contribute to the observed gender differences in cardiovascular and other diseases and disorders, from physiological, anatomic, and genetic factors to delays in recognizing symptoms, to lack of knowledge of underlying pathologic processes, to underutilization of diagnostic tests and treatments (Collins et al. 2011, Mosca, Barrett-Connor, and Wenger 2011, Stock and Redberg 2012). As we show below, some of these health disparities are the result of gender biases that fail to recognize salient differences between men and women. Thus, taking gender into account as a relevant category of analysis in medical research and practice can contribute to reducing gender health disparities. That is, attention to gender can influence how research is done when trying to understand CVD and develop potential interventions to treat those diseases. It can also impact how CVD is diagnosed and treated in ways that may contribute to improved health outcomes in women. In what follows, we discuss how this is the case at different research and clinical stages.

Animal studies are usually the first step in the acquisition of knowledge about disease mechanisms and drug or medical intervention effects. Often, however, what are clearly encouraging results encountered in preclinical investigations fail to translate into efficacy in clinical studies with humans. Indeed, evidence shows that this is the case for around 90% of preclinical studies showing safety and efficacy (van der Worp and Macleod 2011). Clearly, the biology of human beings shares commonalities with rodents and nonhuman primates, but inferences from animal models to humans are fraught with difficulties. Reasons for this disconnect between preclinical and clinical outcomes include problems with the animal models used, poor methodological quality of animal studies, and publication bias (de Jong and Maina 2010, Jucker 2010, Berton, Hahn, and Thase 2012, van der Worp and Macleod 2011, Mergenthaler and Meisel 2012). Of importance for gender disparities is the fact that researchers often use exclusively male animals in preclinical research (Beery and Zucker 2011, Rice 2012, Kokras and Dalla 2014, Yoon et al. 2014, Franconi, Rosano, and Campesi 2015). Rodent studies used to evaluate the effects of drugs on behavior use males nearly exclusively, despite the evidence showing sex differences in drug metabolism (Hughes 2007). The underrepresentation of females in animal models exists even in studies for diseases that predominantly affect women. For instance, although diagnoses for depressive and anxiety disorders, stroke, and thyroid diseases are significantly more common in women than in men, the majority of research with animals to study these disorders actually used male animals (Zucker and Beery 2010). Similarly, the majority of basic and preclinical studies conducted on experimental animals studying physiological, pharmacological, and endocrinological aspects of cardiovascular disease mostly use male animals (Beery and Zucker 2011).

The predominant use of males in animal models has occurred in part because it has been assumed that hormonal variability during the estrous cycle of females presented confounding variables that made it difficult to establish clear causal relationships necessary for understanding the mechanisms of disease and for testing the efficacy of interventions. This assumption has nonetheless come under critical scrutiny (Prendergast, Onishi, and Zucker 2014). Moreover, it seems that to the extent that the hormonal cycles of females present confounding variables, it is not clear that this should count as a reason to prefer male-only animal models for diseases that affect both sexes and much less for those that predominantly affect women. After all, if differences in hormones play a role in the etiology of a disease, excluding females is likely to result in incomplete or inadequate knowledge of the mechanisms of disease. Similarly, if hormone cycles affect drug metabolism, excluding females can produce interventions that might be unsafe or ineffective when used by women. Indeed, to the extent that females are relevantly different, studying these differences will be crucial to understanding disease mechanisms, recognizing symptom presentation, and ensuring the efficacy and safety of medical interventions.

But underrepresentation of females in animal models is not the only problem in basic research that can affect gender disparities. Some studies have also shown that cells from male and female mice have different properties. For example, muscle stem cells taken from female mice regenerate new muscle faster than those taken from males (Deasy et al. 2007). Cells from male and female have likewise been found to respond differently to stress and exhibit different concentrations of many metabolites (Penaloza et al. 2009). These differences in cell characteristics have potential implications for the development of effective treatments for both men and women for conditions such as muscular dystrophy. Nonetheless, even when female animals are included, sex is not often used as a category of analysis, and thus researchers fail to track it. In fact, publications of preclinical studies rarely include sex-based analyses (Beery and Zucker 2011). This lack of attention to gender in basic research and animal studies led the National Institutes of Health (NIH) in October 2014 to implement new policies. Such policies now require NIH grant applicants to report plans for the balance of male and female cells and animals in preclinical studies or to justify that sex-specific inclusion is unwarranted (Clayton and Collins 2014).

Clinical trials, biomedical or behavioral research studies involving human participants that evaluate the efficacy and safety of medical interventions, are now ahead of animal studies with respect to female inclusion. A reason for this progress was the NIH Revitalization Act passed by Congress in 1993 in order to increase the representation of women and minorities in clinical trials. Subsequently, the NIH issued guidelines that required the inclusion and evaluation of gender/sex differences in clinical trials so as to ensure that results about safety and efficacy of drugs and other interventions will be generalizable. As a consequence of this requirement, the number of women participating in clinical research has increased significantly. In fact, evidence indicates that since 1993 more women than men have been enrolled in NIH-sponsored phase 3 trials (Kim, Tingen, and Woodruff 2010). Nonetheless, this increase in women's participation seems to be primarily the result of a few large single-sex studies of reproductive cancer trials (Kim, Tingen, and Woodruff 2010). Thus, despite the improvement in women's participation, clinical trial research often also fails to be sufficiently attentive to gender (IOM 2012). Today, women are still generally underrepresented in clinical trials, even in those studying conditions that have higher prevalence in women (Blauwet et al. 2007, Hoel et al. 2009, Jagsi et al. 2009, Weinberger, McKee, and Mazure 2010). Recent evidence shows that a significant number of clinical trials for depression, for instance, fail to report the gender composition of their sample, do not analyze outcomes by gender, or do not analyze for gender differences (Weinberger, McKee, and Mazure 2010).

Inadequate attention to gender in clinical trials can have devastating effects on women's health. This can be observed when assessing interventions for CVD approved by the FDA despite the lack of sex-specific analysis (Dhruva, Bero, and Redberg 2011). One of such interventions, the implantable cardioverter-defibrillator, was approved despite very low enrollment of women in the major trials, but a meta-analysis has shown a lack of device efficacy in primary prevention trials in women with heart failure (Ghanbari et al. 2009). Similarly, between 1997 and 2001, the FDA withdrew ten prescription drugs, eight of which were more dangerous to women than to men, and in 2000 the FDA decided to remove a component of many over-the-counter medications, phenylpropanolamine, because of a reported increased risk of bleeding into the brain in women but not in men (Pollitzer 2013).

Inattention to gender can also affect diagnosis and treatment of diseases. A variety of studies have shown gender differences in the use of diagnostic procedures to evaluate CVD and in the treatment of these diseases. These studies indicate that women are less likely to receive both invasive and noninvasive testing when they present to an emergency department with chest pain (Miracle 2010). Some research has shown that men presenting with chest pain are usually treated more aggressively than women with similar complaints and are more likely to be admitted to the intensive care unit, receive cardiac catheterization, and initial electrocardiogram. Women, however, are more likely to receive drugs, including anxiolytics (Kent, Patel, and Varela 2012). Some studies have also shown that women suffering from certain coronary problems are less likely to receive recommended prescription of beta-blockers when discharged from the hospital (Birkemeyer et al. 2014). Furthermore, some diagnostic tests have been found to be less able to correctly identify CVD in women, such as exercise electrocardiography and radionuclide myocardial perfusion imaging used to diagnose coronary artery disease (Kim, Tingen, and Woodruff 2010). All of these factors may contribute toward women having higher morbidity and mortality rates from CVD.

Gender differences in treatment side effects also exist (Mosca, Barrett-Connor, and Wenger 2011). For instance, women with acute coronary syndromes receiving antiplatelet pharmacotherapy are more likely than men to have bleeding problems, although they have similar benefits from these medications. Low-dose aspirin to prevent CVD seems to also have less beneficial effects on women than men. Although statin therapies reduce global cardiovascular risk in women as in men, evidence indicates that the absolute benefit in women who do not have established coronary disease is small (Miller and Best 2011). Thus, ignoring gender as a category of analysis can obscure the fact that the risks and benefits of various interventions can be different for men and women.

Some evidence indicates that these gender disparities in diagnosis and treatment of CVD might be the result of physicians' misinterpretations or neglect of women's symptoms because such symptoms do not fit with those that men have and because of the belief that CVDs are primarily male diseases. Other studies indicate that disparities in diagnosis and treatment intervention might result from other gender biases. In particular, because women with CVDs present with stress and anxiety, these symptoms mislead physicians toward mental health alternative diagnoses (Chiaramonte and Friend 2006). In any case, gender stereotypes can affect how physicians interpret patients' symptoms, and thus ignoring gender can conceal unintentional biases.

Making Gender Visible in Medicine

We have aimed to show that gender matters in a variety of ways to our understanding of health and disease. When researchers or physicians are not attentive to potential relevant gender differences, it can lead to unconscious biases in experimental designs, the kinds of interventions developed, the interpretation of clinical trial results, and diagnostic and treatment decisions. As we have seen, these biases can lead to incorrect diagnoses, unequal access to experimental interventions, an inattention to differences in the effectiveness of resulting treatments, and in the frequency or severity of side effects. Ultimately, a failure to take gender into account can contribute to poor health outcomes for women.

Addressing these problems, however, is not easy, particularly as it is not always possible to know a priori whether gender differences exist in a particular case or to what extent they matter for the condition studied. In addition, gender biases are generally unconscious, such that, even when they might be well-intentioned and conscientious, it is difficult for individual researchers or physicians to prevent such biases from happening. However, a variety of practices may help improve medical research and practice in ways that can contribute to reducing health disparities and improving health outcomes for both women and men. In particular, addressing the problems discussed here requires using gender as a category of analysis in both biomedical research and patient care so as to make gender and its effects more visible.

What does it mean to "use gender as a category of analysis"? In general, this means that medical researchers and practitioners ought to consider how gender differences may or may not be relevant to the decisions they make. This can be done in a variety of ways throughout the research process and in medical practice.

First, gender should be used as a category of analysis in the framing of research questions, selection of methodology, or development of interventions. In investigating a particular health problem, researchers should seek to examine similarities and differences between males and females in a variety of aspects, such as disease causes, risk factors, likely age of onset, and symptoms. In developing interventions, investigators should determine whether, and if so which, biological and social differences exist that might make specific treatments more or less effective for one sex or another. For example, as we have seen, it is important for biomedical researchers and physicians to consider questions such as whether CVD operates differently in men and women, whether risk factors are the same, and whether there are differences in the presentation of symptoms. Similarly, researchers should consider whether there are social, political, economic, or biological differences among men and women that might be relevant to understanding the epidemiology of a disease, or whether there are differences in social conditions that present challenges or require trade-offs in addressing the disease burden. For example, nearly six in ten poor adults are women, and nearly six in ten poor children live in families headed by women (DeNavas-Walt and Proctor 2015). Poverty rates are especially high for single mothers, women of color, and elderly women living alone. Socioeconomic status can affect disease risks, health outcomes, and morbidity and mortality. Being attentive to these differences makes it more likely that the framing of research problems, choice of methodologies, selection of data, and development of interventions will result in more effective treatments for both men and women.

Second, studying similarities and differences along gender lines obviously calls for increasing the participation of females in preclinical and clinical research. As noted earlier, the NIH has now developed new requirements to ensure increased inclusion of female animals. Grant applicants are obligated to present plans to appropriately balance the use of male and female cells and of animals in preclinical studies or justify that such balance is unwarranted (Clayton and Collins 2014). NIH's Office of Research on Women's Health now has supplemental funding for existing grants to incorporate animals, tissue, cells, or subjects of an underrepresented sex in order to allow for sex-based comparisons.

In human subject testing, medical researchers and practitioners should be particularly attentive to possible obstacles that might hinder the participation of women in clinical trials. For example, not uncommonly, pregnancy is an exclusion criterion in clinical trials. Such a criterion will obviously result in the exclusion of women but not men. Also, older patients are often excluded from clinical trials because of comorbid conditions, but this might result in a problematic underrepresentation of women, who develop some diseases, such as heart ones, later in life when they are more likely to have comorbid conditions (Bucholz and Krumholz 2015). Moreover, even when researchers may seek to include more women in clinical trials, social barriers may make achieving this goal difficult. For instance, because women are more likely to take care of their children, schedules chosen for clinical interventions in trials can unintentionally limit the participation of many women. Similarly, the costs often incurred in clinical trial participation (e.g., transportation, child care, etc.) might present significant burdens for many women.

Furthermore, physicians' biases might affect their beliefs about patients' preferences, anticipated logistical problems, patients' ability to understand or comply with study requirements, or the benefits of participation, and such beliefs might lead physicians to offer study participation to some patients but not to others (Howerton et al. 2007, Denson and Mahipal 2014). Thus, understanding the different barriers that can discourage the participation of women in clinical trials will also be important in achieving the goal of increasing women's participation in clinical trials.

Researchers, physicians, and health care organizations should also adopt strategies that actively encourage such participation. The creation of health registry programs that inquire about health and lifestyle and that then can serve as bases for clinical trial recommendations might be helpful in this respect. But, as we have indicated, these measures are unlikely to end the underrepresentation of women in clinical trials. More attention needs to be given to both investigating and trying to address institutional obstacles that make the participation of women difficult. Not only should inclusion criteria be scrutinized for its differential effects on women, but care should also be given to social factors that limit women's participation.

Yet, while it is clearly important to increase the participation of females in preclinical and clinical studies, this is not sufficient to reduce gender health disparities. As noted earlier, even in animal and human studies that include females, gender analyses are often not conducted. Some studies rely on cells coming from both males and females, but they do not track origin and thus gender comparisons cannot be made. Therefore, it is not just ensuring an appropriate participation of female animals and women that is important but also making sex and gender visible in experimental design and selection of methodologies. Doing so obviously calls for the tracking of sex of cells, animals, and humans used in research so as to make comparative investigations possible. Failing to do so might also contribute to the long-standing, well-known, and prevalent difficulty to reproduce research findings (Begley and Ioannidis 2015). Replicability failures and lack of predictability from animal to human studies might in some cases be exacerbated by inattention to gender differences that lead researchers to fail to track sex or to include appropriate numbers of females in both preclinical and clinical studies. Using models that allow for gendered analyses and comparisons would thus be helpful.

Fourth, gender should be used as a category of analysis in diagnosis and treatment decisions. Health care professionals involved in diagnosing and treating CVD, for example, should be attentive to their patients' gender and the ways in which gender and gender bias play a role in disease diagnosis and treatment recommendations. Physicians must be aware of the ways in which CVD presents differently for women and men and must adopt similarly aggressive testing and intervention strategies when appropriate. When relevant, treatments should be justified not only in relation to being the right course of action for a particular disease, but also for a particular disease in a patient of a particular gender. Likewise, there must be greater awareness of any differences along gender lines in terms of what side effects might result from a particular treatment. Patients should be counseled not only about what treatments are effective for a particular disease in general but also the risks and likelihoods of benefits given their sex, age, and other relevant categories. This would require medical education and training that includes a more attentive gender-specific curriculum. Although the desirability of such curriculum has been recognized for quite some time, few U.S. medical schools have fully achieved this goal (Kim, Tingen, and Woodruff 2010). The U.S. Food and Drug Administration could also help by mandating that sex-specific drug reactions be made clear to both physicians and patients.

Of course, ensuring that gender is properly used as a category of analysis throughout the research process is not only the responsibility of individual investigators and practitioners but of the health care community as a whole, including journal editors, reviewers, granting agencies, and medical educators. It should be given consideration and weight in the peer review process. Reviewers and editors should evaluate the extent to which gender analysis may be relevant and whether the presence or absence of such analysis has been appropriately accounted for in methodological choices, experimental design, and interpretation of results.

As noted earlier, one of the challenges of being attentive to gender and preventing bias is that such biases are often unintentional. As many have argued, it is very difficult for individuals to catch their own biases when they have failed to take into account potentially relevant gender differences (Longino 1990). One strategy for addressing this problem is increasing diversity among health care practitioners and within research communities. Our background assumptions and interests are significantly shaped by our experiences, and such experiences can be influenced in greater or lesser degree by a variety of factors such as our gender, race, class, or nationality. Having a diverse research community can increase the likelihood that implicit assumptions and biases of individual researchers will be identified and critically evaluated (Solomon 2001, Longino 2002, Wylie and Nelson 2007, Harding 2008, Intemann 2009, de Melo-Martín and Intemann 2011).

Given the still common underrepresentation of women in biomedical research, a call for diversity is particularly relevant. Recent statistics indicate that although in the U.S. during the last decade nearly half of all MD degrees have been awarded to women, they still constitute only about one-third of all actively practicing physicians (Lautenberger et al. 2014). Likewise, women make up just over one-third of full-time academic medicine faculty, and they constitute 34% of associate and 21% of full professors (Lautenberger et al. 2014). Moreover, the percentage of permanent women department chairs (15%) and deans (16%) remains alarmingly low (Lautenberger et al. 2014). Similarly, although more than half of the doctorates in the life sciences are currently awarded to women, they hold only 18% of full professorships (Sheltzer and Smith 2014). Encouraging strategies that would bring more women to biomedical research and medical practice, particularly in some specialties, is important because it would increase diversity in research labs, physicians' offices, faculty at medical schools, and leadership positions.

This is not to say that female researchers or female physicians are necessarily more attentive to the issues and interests of women or less susceptible to gender biases. However, insofar as background assumptions often reflect shared experiences or convention, having a research community with diverse backgrounds, experiences, values, and interests makes it more likely that erroneous assumptions will be corrected. It might also increase the possibility that new research directions, explanations, models, methods, or interventions will be explored (Longino 2002, Solomon 2006, Intemann and de Melo-Martín 2014).

Conclusion

In this chapter we have aimed to examine some of the ways in which inattention to gender can play a role in researching, diagnosing, and treating certain diseases, and ultimately in causing or exacerbating health disparities along gender lines. It is a truism that no two patients are the same, and clearly a variety of biological and social factors affect disease presentation, diagnosis, and treatment. When researchers and health care professionals are not attentive to these differences, it can lead to poorer health outcomes for certain groups, even though such disparities are not intended. We have seen how this has occurred in the use of predominantly male-only animal studies, the underrepresentation of females in clinical trials, and inattention to physician biases in the diagnosis and treatment of particular diseases. A significant effort has been devoted to trying to increase the participation of females in biomedical research, and such efforts are certainly a necessary and important step toward addressing gender health disparities. Nonetheless, this is unlikely to be sufficient for addressing these health inequities. Researchers must use gender as a category of analysis in more robust ways that attend to research questions and methodologies, as well as the effects of gender biases in diagnosing and treating diseases.

References

Alcoff, Linda. 2006. *Visible identities: Race, gender, and the self, studies in feminist philosophy*. New York: Oxford University Press.

Beery, A. K., and I. Zucker. 2011. "Sex bias in neuroscience and biomedical research." *Neuroscience and Biobehavioral Reviews* 35 (3):565–72.

Begley, C. G., and J. P. Ioannidis. 2015. "Reproducibility in science: Improving the standard for basic and preclinical research." *Circ Res* 116 (1):116–26.

Berton, O., C. G. Hahn, and M. E. Thase. 2012. "Are we getting closer to valid translational models for major depression?" *Science* 338 (6103):75–9.

Birkemeyer, R., H. Schneider, A. Rillig, J. Ebeling, I. Akin, S. Kische, L. Paranskaya, W. Jung, H. Ince, and C. A. Nienaber. 2014. "Do gender differences in primary PCI mortality represent a different adherence to guideline recommended therapy? A multicenter observation." *BMC Cardiovasc Disord* 14:71.

Blauwet, L. A., S. N. Hayes, D. McManus, R. F. Redberg, and M. N. Walsh. 2007. "Low rate of sex-specific result reporting in cardiovascular trials." *Mayo Clin Proc* 82 (2):166–70.

Bucholz, E. M., and H. M. Krumholz. 2015. "Women in clinical research: What we need for progress." *Circ Cardiovasc Qual Outcomes* 8 (2 Suppl 1):S1–3.

Butler, Judith. 2004. *Undoing gender*. New York: London: Routledge.

Chiaramonte, G. R., and R. Friend. 2006. "Medical students' and residents' gender bias in the diagnosis, treatment, and interpretation of coronary heart disease symptoms." *Health Psychol* 25 (3):255–66.

Clayton, J. A., and F. S. Collins. 2014. "NIH to balance sex in cell and animal studies." *Nature* 509 (7500):282–3.

Collins, P., C. Vitale, I. Spoletini, and G. Barbaro. 2011. "Gender differences in the clinical presentation of heart disease." *Curr Pharm Des* 17 (11):1056–8.

Deasy, B. M., A. Lu, J. C. Tebbets, J. M. Feduska, R. C. Schugar, J. B. Pollett, B. Sun, K. L. Urish, B. M. Gharaibeh, B. Cao, R. T. Rubin, and J. Huard. 2007. "A role for cell sex in stem cell-mediated skeletal muscle regeneration: Female cells have higher muscle regeneration efficiency." *J Cell Biol* 177 (1):73–86.

de Jong, M., and T. Maina. 2010. "Of Mice and humans: Are they the same? Implications in cancer translational research." *Journal of Nuclear Medicine* 51 (4):501–4.

de Melo-Martin, Inmaculada, and Kristen Intemann. 2011. "Feminist resources for biomedical research: Lessons from the HPV vaccines." *Hypatia* 26 (1):79–101.

DeNavas-Walt, Carmen, and Bernadette D Proctor. 2015. *Income and poverty in the U.S.: 2014*. Washington, DC: U.S. Government Printing Office.

Denson, A. C., and A. Mahipal. 2014. "Participation of the elderly population in clinical trials: Barriers and solutions." *Cancer Control* 21 (3):209–14.

Dhruva, S. S., L. A. Bero, and R. F. Redberg. 2011. "Gender bias in studies for Food and Drug Administration premarket approval of cardiovascular devices." *Circ Cardiovasc Qual Outcomes* 4 (2):165–71.

Fausto-Sterling, Anne. 2000. *Sexing the body: Gender politics and the construction of sexuality*. 1st ed. New York, NY: Basic Books.

Fleming, P. J., and C. Agnew-Brune. 2015. "Current trends in the study of gender norms and health behaviors." *Curr Opin Psychol* 5:72–77.

Franconi, F., G. Rosano, and I. Campesi. 2015. "Need for gender-specific pre-analytical testing: The dark side of the moon in laboratory testing." *International Journal of Cardiology* 179:514–35.

Ghanbari, H., G. Dalloul, R. Hasan, M. Daccarett, S. Saba, S. David, and C. Machado. 2009. "Effectiveness of implantable cardioverter-defibrillators for the primary prevention of sudden cardiac death in women with advanced heart failure: A meta-analysis of randomized controlled trials." *Arch Intern Med* 169 (16):1500–6.

Harding, Sandra G. 2008. *Sciences from below: Feminisms, postcolonialities, and modernities*. Durham, NC: Duke University Press.

Haslanger, Sally Anne. 2012. *Resisting reality: Social construction and social critique*. New York: Oxford University Press.

Hoel, A. W., A. Kayssi, S. Brahmanandam, M. Belkin, M. S. Conte, and L. L. Nguyen. 2009. "Underrepresentation of women and ethnic minorities in vascular surgery randomized controlled trials." *J Vasc Surg* 50 (2):349–54.

Howerton, M. W., M. C. Gibbons, C. R. Baffi, T. L. Gary, G. Y. Lai, S. Bolen, J. Tilburt, T. P. Tanpitukpongse, R. F. Wilson, N. R. Powe, E. B. Bass, and J. G. Ford. 2007. "Provider roles in the recruitment of underrepresented populations to cancer clinical trials." *Cancer* 109 (3):465–76.

Hughes, R. N. 2007. "Sex does matter: Comments on the prevalence of male-only investigations of drug effects on rodent behaviour." *Behavioural Pharmacology* 18 (7):583–9.

Institute of Medicine (IOM). Committee on Women's Health Research. 2010. *Women's health research progress, pitfalls, and promise*. Washington, DC: National Academies Press. http://www.nap.edu/openbook.php?record_id=12908.

Intemann, K. 2009. "Why diversity matters: Understanding and applying the diversity component of the National Science Foundation's broader impacts criterion." *Social Epistemology* 23 (3–4):249–66.

Intemann, K, and I. de Melo-Martin. 2014. "Addressing problems in profit-driven research: How can feminist conceptions of objectivity help?" *European Journal for Philosophy of Science* 4 (2):135–51.

Jagsi, R., A. R. Motomura, S. Amarnath, A. Jankovic, N. Sheets, and P. A. Ubel. 2009. "Underrepresentation of women in high-impact published clinical cancer research." *Cancer* 115 (14):3293–301.

Jucker, M. 2010. "The benefits and limitations of animal models for translational research in neurodegenerative diseases." *Nature Medicine* 16 (11):1210–14. doi: 10.1038/nm.2224.

Kent, J. A., V. Patel, and N. A. Varela. 2012. "Gender disparities in health care." *Mt Sinai J Med* 79 (5):555–9.

Kim, A. M., C. M. Tingen, and T. K. Woodruff. 2010. "Sex bias in trials and treatment must end." *Nature* 465 (7299):688–9.

Kokras, N., and C. Dalla. 2014. "Sex differences in animal models of psychiatric disorders." *British Journal of Pharmacology* 171 (20):4595–619.

Lautenberger, D. M., V. M. Dandar, C. L. Raezer, R. A. Sloane, and Association of American Medical Colleges. 2014. *The state of women in academic medicine: The pipeline and pathways to leadership 2013–2014*. Washington, DC: Association of American Medical Colleges.

Longino, Helen E. 1990. *Science as social knowledge: Values and objectivity in scientific inquiry*. Princeton, NJ: Princeton University Press.

———. 2002. *The fate of knowledge*. Princeton, NJ: Princeton University Press.

Mercuro, G., M. Deidda, A. Piras, C. C. Dessalvi, S. Maffei, and G. M. Rosano. 2010. "Gender determinants of cardiovascular risk factors and diseases." *J Cardiovasc Med (Hagerstown)* 11 (3):207–20.

Mergenthaler, P., and A. Meisel. 2012. "Do stroke models model stroke?" *Dis Model Mech* 5 (6):718–25.

Miller, V. M., and P. J. Best. 2011. "Implications for reproductive medicine: Sex differences in cardiovascular disease." *Sex Reprod Menopause* 9 (3):21–8.

Miracle, V. A. 2010. "Coronary artery disease in women: The myth still exists, unfortunately." *Dimens Crit Care Nurs* 29 (5):215–21.

Mosca, L., E. Barrett-Connor, and N. K. Wenger. 2011. "Sex/gender differences in cardiovascular disease prevention: What a difference a decade makes." *Circulation* 124 (19):2145–54.

Penaloza, C., B. Estevez, S. Orlanski, M. Sikorska, R. Walker, C. Smith, B. Smith, R. A. Lockshin, and Z. Zakeri. 2009. "Sex of the cell dictates its response: Differential gene expression and sensitivity to cell death inducing stress in male and female cells." *FASEB J* 23 (6):1869–79.

Pollitzer, E. 2013. "Biology: Cell sex matters." *Nature* 500 (7460):23–4.

Practice, Institute of Medicine (US) Board on Population Health and Public Health. 2012. *Sex-specific reporting of scientific research*. Washington, DC: National Academy Press.

Prendergast, B. J., K. G. Onishi, and I. Zucker. 2014. "Female mice liberated for inclusion in neuroscience and biomedical research." *Neuroscience and Biobehavioral Reviews* 40:1–5.

Rice, J. 2012. "Animal models: Not close enough." *Nature* 484 (7393):S9.

Sen, G., and P. Ostlin. 2008. "Gender inequity in health: Why it exists and how we can change it." *Glob Public Health* 3 (Suppl 1):1–12.

Sheltzer, J. M., and J. C. Smith. 2014. "Elite male faculty in the life sciences employ fewer women." *Proc Natl Acad Sci U S A* 111 (28):10107–12.

Solomon, Miriam. 2001. *Social empiricism*. Cambridge, MA: MIT Press.

———. 2006. "Norms of epistemic diversity." *Episteme: A Journal of Social Epistemology* 3 (1):23–36.

Spelman, Elizabeth V. 1988. *Inessential woman: Problems of exclusion in feminist thought*. Boston: Beacon Press.

Stock, E. O., and R. Redberg. 2012. "Cardiovascular disease in women." *Curr Probl Cardiol* 37 (11):450–526.

van der Worp, H. B., and M. R. Macleod. 2011. "Preclinical studies of human disease: Time to take methodological quality seriously." *J Mol Cell Cardiol* 51(4), 449–450.

Warnke, Georgia. 2011. *Debating sex and gender, fundamentals of philosophy series*. New York: Oxford University Press.

Weinberger, A. H., S. A. McKee, and C. M. Mazure. 2010. "Inclusion of women and gender-specific analyses in randomized clinical trials of treatments for depression." *J Womens Health (Larchmt)* 19 (9):1727–32. doi: 10.1089/jwh.2009.1784.

Wenger, N. K. 2012. "Women and coronary heart disease: A century after Herrick: Understudied, underdiagnosed, and undertreated." *Circulation* 126 (5):604–11.

Witt, Charlotte. 2011. *The metaphysics of gender, Studies in feminist philosophy*. New York: Oxford University Press.

———. 2015. Cardiovascular diseases (CVDs). Fact sheet 317. http://www.who.int/mediacentre/factsheets/fs317/en/.

Wyant, A. R., and D. Collett. 2015. "Identifying and managing chest pain in women." *JAAPA* 28 (1): 48–52.

Wylie, Alison, and Lynn Hankinson Nelson. 2007. "Coming to terms with the values of science: Insights from feminist science studies scholarship." In *Value-free science? Ideals and illusions*, edited by Harold Kincaid, John Dupre and Alison Wylie, 58–86. Oxford: Oxford University Press.

Yoon, D. Y., N. A. Mansukhani, V. C. Stubbs, I. B. Helenowski, T. K. Woodruff, and M. R. Kibbe. 2014. "Sex bias exists in basic science and translational surgical research." *Surgery* 156 (3):508–16.

Zucker, I., and A. K. Beery. 2010. "Males still dominate animal studies." *Nature* 465 (7299):690.

Further Reading

Fox Keller, E., and H. Longino (eds.). 1996. *Feminism and science*. New York: Oxford University Press. Collection of essays that summarizes the state of the art in the feminist perspective on the philosophy of science.

Harding, S. 2015. *Objectivity and diversity*. Chicago: Chicago University Press. A defense of standpoint feminism and strong objectivity.

Hochleitner, M., U. Nachtschatt, and H. Siller. 2013. "How do we get gender medicine into medical education?" *Health Care Women Int.* 34 (1):3–13. An example of incorporation of gender aspects into a medical curriculum.

Oertelt-Prigione, S., and V. Regitz-Zagrosek (eds.). 2011. *Sex and gender aspects in clinical medicine*. London: Springer-Verlag. An exploration of gender differences in a variety of diseases.

Richardson, S. 2013. *Sex itself: The search for male and female in the human genome*. Chicago: Chicago University Press. An examination of the interaction between cultural gender norms and genetic theories of sex from the beginning of the 20th century to the present, postgenomic age.

38
RACE IN MEDICINE
Sean A. Valles

Introduction

Race has an unsettled status in contemporary medicine. Genetic interpretations of race are undergoing something of a renaissance as a result of the proliferation of large-scale genomic studies of human populations. Meanwhile, examinations of health and disease patterns among populations of self-identified Whites, American Indians, etc. reveal abundant evidence of disparities. Still, it remains unclear whether these are good enough reasons to use race concepts in medicine, given race's roots in antiquated biology and its long history of abusive uses (e.g., justifying slavery). This has inspired a debate. At two poles of the race-in-medicine debate are eliminativism and conservationism. Eliminativists see race's problems as being sufficient grounds to eject race from biomedicine. Conservationists see race's usefulness as being sufficient to justify its continued use in biomedicine. There is a broad spectrum of nuanced positions between these two poles. This chapter reviews the current status of the race-in-medicine debate and the related philosophy of medicine literature.

Race has many facets and, accordingly, so does philosophy of race, including identity (Shelby 2005), law (Darby 2009), epistemology (Sullivan and Tuana 2007), moral philosophy (Blum 2002), and more. Each of these areas has some amount of overlap with philosophy of medicine, but philosophy of medicine's concerns with the topic still only represent a portion of the race discussion. It is beyond the scope of this chapter to even fully review the various positions taken on race in medicine (and philosophy of medicine), or their detailed historical development. Instead, the chapter will review the key arguments and issues at stake in the debate, as a means of guiding readers through the landscape of the topic. Because race is a complex topic that is complicated by its history, scholars of the race-in-medicine debate have faced extraordinary pressures to be painstakingly precise in how they phrase their arguments; it is a debate driven by fine distinctions. Accordingly, this chapter will often directly quote key passages from relevant authors, in order to preserve important nuances in their positions.

In order to keep a single thread running through the many different subtopics covered in the chapter, it will pay particular attention to one case study that has inspired substantial race-in-medicine research and critique. The disparity in cardiovascular disease burden between White populations and Black populations is the U.S. is widely known, even outside of the medical community. Included in that cluster of cardiovascular diseases are disparities in rates of congestive heart failure (CHF): "among adults aged 45 to 84, blacks have approximately twice the incidence of CHF as whites" (Will et al. 2012). A public controversy broke out in the 2000s when a new drug for CHF in Black patients, BiDil (hydralazine hydrochloride and isosorbide dinitrate), became the first drug the U.S. Food and Drug Administration approved exclusively for use by a single race of patients (Ellison et al. 2008; Rusert and Royal 2011).

Bidil's race-based prescribing rule has raised a number of concerns, largely related to its apparent effect of reinforcing the concept of biological race essentialism—the notion that all members of a racial grouping have some set of distinguishing traits, deeply dividing their biological makeup from that of other races. Life scientists and social scientists in the 18th and 19th centuries sought, and claimed to have found, distinctive physiological, intellectual, and behavioral characteristics in each race (Morning 2011). These purported characteristics, such as inferior intelligence in Blacks, were used to justify White supremacy and Black slavery. Today, there are still attempts to uncover modern scientific justifications for race essentialism, searching for behavioral genetics differences in features such as propensity to violence (Duster 2006; Perbal 2013) or molecular biological differences in DNA sequences (Morning 2011). Thus, a central concern is that using race concepts in medicine reinforces race essentialism and gives the false appearance of scientific objectivity to traditional racist prejudices. How should we responsibly reconcile these worries with the evidence of high rates of CHF in the Black population and the data suggesting BiDil helps to reduce that burden? These sorts of questions are rooted in philosophical judgments of ethics and evidence, as well as clinical practice.

What Is Race?

Race's significance for medicine is ambiguous partly because it is unclear what race is. Is the White race a biological reality that has been discovered or a social phenomenon that has been invented? Although many social scientists hold race to be entirely "socially constructed"—i.e., a concept imagined into existence by human minds—the biological and anthropological communities each remain internally divided over whether race has a biological basis independent of human imagination (Morning 2011).

Whatever race may be, it is a common misconception that there is some stable and universally recognized set of races. In fact, how many races there are and who qualifies for each race category are both historically and locally contingent. In the United States, the number of races named on the census has grown and shrunk, with the current standards set by the Office of Management and Budget: a minimum of five races (e.g., "Native Hawaiian or Other Pacific Islander"), which are allowed to be further subdivided (e.g., "Samoan" and "Guamanian or Chamorro") (Humes et al. 2011). The system has changed drastically over the years, with races on the census being revised ("Indian" became "American Indian" in 1960), added ("Chinese" first appeared in 1880), and subtracted ("Mexican" first appeared in 1930 then disappeared in 1940) until reaching 12 named races in 2010 (Humes et al. 2011; Nobles 2000). Meanwhile, the census also includes a question about respondents' Hispanic ethnicity, entirely independent of race. By contrast, the UK and New Zealand censuses only recognize ethnicities (Valles et al. 2015), having the effect of emphasizing culture and community over just ancestry. Taking a third approach, Brazil recognizes "colors" instead of geographical races, explicitly referring to one's appearance instead of descent (Nobles 2000). The net effect is that even a single term can have vastly different official meanings in different national contexts. In the U.S., "Black" is a race, in England it is an ethnicity, and in Brazil it is a racialized color (Nobles 2000; Valles et al. 2015).

Official national categorizations of race have so much power that they end up shaping everyday uses of race concepts. Spencer contends in his philosophical analysis of race's meaning, "no racial discourse in the United States is more widely used than census racial discourse," since Americans are so regularly exposed to them in government documents, loan applications, etc., and gradually learn "how to pigeonhole themselves into census races" (Spencer 2014a: 1027).

Variations in unofficial local race concepts can cause great confusion and distress for migrants traveling from one national context to another, such as Brazilian immigrants to the U.S. who are forced to adapt to American official and unofficial standards, including the widespread notion that having "one drop" of African ancestry means that a person cannot qualify as simply "White," regardless of skin color (Fritz 2015).

Those hoping to get an overview of some of the various scholarly positions taken on the meaning of race can get an introduction in a chapter by Charles Mills, which offers a detailed typology of race concepts (Mills 1998). Mills acknowledges, though, that his use of certain terminology is sometimes at odds with other authors' uses. For example, there are multiple senses in which race can be thought of as a "construction" (Mills 1998). To help clarify matters, Michael Hardimon distinguishes among four different race concepts that are deployed in medicine: "the ordinary concept of race," "the populationist concept of race," "the racialist concept of race," and "SOCIALRACE" (Hardimon 2013).

Hardimon's ordinary concept of race is a minimalist version of how "race" is used in day-to-day conversation, committed only to the view that groups called human races share sets of physical features, ancestries, and geographic origins. The populationist concept of race is a "scientized" version of the ordinary concept, reformulated to use technical biomedical terms and concepts to specify "phenotypic traits," genetic ancestry, and "isolated founding populations" (Hardimon 2013). Both of those race concepts are different from the insidious racialist concept of race, which has the additional features of being essentialist (believing in strict criteria that make members of one race distinct from all other races) and "evaluatively hierarchical" (holding some races to be superior to others based on their essential traits). Finally, Hardimon offers SOCIALRACE as a new term for referring to races as social groups only, without committing to any biological assumptions. It stresses (in its name) that social features of race are quite real and can have real medical repercussions (Hardimon 2013). For example, a study of emergency room analgesia prescription patterns in the U.S. found an 8% gap between the percentage of White patients vs. racial/ethnic minority patients receiving opioid painkillers (Pletcher et al. 2008), a troubling disparity in how members of different races are treated by society.

What Is the Current Status of Race in Biomedicine?

Although physical anthropology and population genetics once shared authority over "race" in the scientific community, population genetics has recently gained unofficial pride of place as a result of genomic science's rapid expansion in the wake of the Human Genome Project. In an account of teaching race in a medical school, Warwick Anderson lucidly narrates that genetics has acquired disciplinary primacy in medical students' understandings of race, with anthropology and other social sciences pushed into the margins (Anderson 2008). Phenotype (skin color, physiological measures) has been deemphasized, and clustering of genotypes has become the center of disputes over the science of race. In population genetics, there were disputes during the mid-20th century about the precise technical meaning of race (e.g., to what extent are species and race distinctions arbitrary conveniences used by researchers, as opposed to real structural boundaries in nature?), and echoes of those debates remain now (Gannett 2013; Spencer 2013).

Philosophers are now vigorously debating the status of race's biological meaning in light of recent genomics research. That debate is captured in the ongoing dialogue among Quayshawn Spencer, Jonathan Kaplan, Rasmus Winther, and Adam Hochman (among others) (Hochman 2014; Kaplan and Winther 2014; Spencer 2014a). They dispute how to interpret genomic evidence that purports to show human populations can be clustered into genetic groups that—at least in some studies—approximately line up with the five races in the U.S. federal guidelines (Spencer 2014b). Similar to Kaplan and Winther (2014), Hochman finds such data to be weak supporting evidence for the view that races are real biological entities (Kaplan and Winther 2014: 1046–1047). By contrast, Spencer judges it to be stronger evidence (Spencer 2014b). All of these authors face the daunting challenge of articulating what precisely it would mean for race to be objective, real, or biological in the first place. Interpreting the meaning of the genomic evidence is made more difficult by common confusions about the biological concepts involved.

Lisa Gannett is critical of the view that the biological and the social are polar opposites. She surmises that there are indeed correlations between racial categories and certain traits:

> . . . but we can expect such correlations to be statistical not universal, local not global, contingent not necessary, and accidental not lawful, and expect their corresponding cuts in nature to be interest-relative not mind-independent, dynamic not static, indeterminate not determinate, many not few, overlapping not nonoverlapping, and superficial not deep. The dichotomous choice of race as either "underlying" biological reality or "mere" social construction thereby becomes meaningless.
>
> (Gannett 2010: 382–383)

In sum, the current status of the genetic basis of race is muddled. This lack of clarity translates into confusion for biomedical scholars and practitioners.

In the biomedical community, by 2005, "ethnicity" had overtaken "race" as the most frequently used term in the MEDLINE database, while a new pair of hybrid terms, "race-ethnicity" and "race/ethnicity" began growing in popularity. Unfortunately, interview data indicate that both "race" and "ethnicity" have ambiguous meanings for the medical researchers who so clearly need to understand them (Baer et al. 2013; Hunt and Megyesi 2008).

Why Use "Race" in Medicine?

Given the many problems associated with using race in medicine, from its historical and philosophical foundations to its application in biomedicine, any argument in favor of using race in medicine despite those problems must somehow be more compelling. The most compelling motivator for keeping race in biomedicine is the enormous body of evidence showing "health disparities" among racial groups:

> Black infants in the United States are more than twice as likely as white infants to die before their first birthday.
>
> (Woolf and Aron 2013: 40)

> Injury victims are more likely to die at hospitals with a large percentage of minority patients, and this risk is compounded if they are uninsured.
>
> (Woolf and Aron 2013: 120)

> U.S. assault victims brought to high-level trauma centers were more likely to die if they were black, even after adjusting for other variables.
>
> (Woolf and Aron 2013: 120)

Given such disparities, Raj Bhopal, a public health scholar, contends that judicious use of race concepts in medicine is a necessary means of achieving the central goals of biomedicine:

> The goal of improving the health and well-being of minority groups and therefore the population as a whole is central to the responsible use of the racial and ethnic categories. When that goal is embraced politically and socially, it is irresponsible not to acquire and use race and ethnicity data to tackle the nation's need to improve health in both absolute and relative terms.
>
> (Bhopal 2006: 505)

Under this view, rather than putting advocates of race in medicine on the spot for needing to defend their use of dubious scientific concepts, it instead presents the use of racial concepts as an ethical necessity that overrides worries about the concepts themselves. Priority is given to people's health needs, and concerns about race concepts are made secondary. Bhopal demonstrates openness to viewing race-in-medicine debate through the lens of social and ethical considerations, but other advocates of such work wish to separate the research itself from its social and ethical interpretations.

A team of scholars of race/ethnicity in genetic epidemiology, Neil Risch, Esteban Burchard, and Hua Tang, insist that differences among races are biological facts worthy of study, and they only become socially problematic when those facts are used in pernicious value-laden ways for political ends (Risch et al. 2002).

> We believe that identifying genetics differences between races and ethnic groups, be they for random genetic markers, genes that lead to susceptibility or variation in drug response, is scientifically appropriate. What is not scientific is a value system attached to any such findings. Great abuse has occurred in the past with notions of "genetic superiority" of one group over another. The notion of superiority is not scientific, only political, and can only be used for political purposes.
>
> (Risch et al. 2002: 11)

The authors' attempt at a clean separation between facts and values clashes with contemporary philosophy of science, particularly as it seemingly manifests the so-called value-free ideal (of science that can be and should be unspoiled by interference from human values), which is contested by some philosophers specializing in science and values (Hicks 2014).

Jay Cohn, the cardiologist who owned the original methodology patent for using BiDil to treat heart failure and led the clinical trials testing it (Roberts 2011: 169), calls upon the pragmatic nature of medicine to justify the use of race despite its scientific weaknesses:

> Observed racial differences in disease frequency may be genetically, rather than racially or geographically determined, but in the absence of more refined technology, the racial designation, crude as it is, serves as a useful and available surrogate.
>
> (Cohn 2006, 552)

As noted by Katikireddi and Valles, the choice of proxy variable (e.g., using race as a variable that stands in for genetic and environmental variables) to be used in research practice raises epistemic (evidentiary) questions and ethical questions that are tightly intertwined (Katikireddi and Valles 2015). The BiDil case illustrates this amply.

To Cohn, the existence of clinical trial data suggesting that the identified "African American" population (one often neglected by biomedicine) would benefit from the drug is sufficient reason to use that race data in the clinical practice.

> The benefit of BiDil ultimately shown in the A-HeFT Trial was so profound that it would be irresponsible to deny the favorable effect and deprive a population historically underserved by our medical system of the resultant improvement in medical management.
>
> (Cohn 2006: 553)

Despite this evidence of benefits, criticism arose over the implicit aspects of how the benefits were calculated and communicated. As Kahn notes in his book chronicling the BiDil story, the drug's owner at the time (NitroMed) had specifically sought narrow FDA approval for only African American patients after the strongest data supporting BiDil's merits came from A-HeFT, the study cited above by Cohn, which included only African American subjects (Kahn 2013: 48–49). This study was pursued after earlier BiDil clinical trials, which included Black and White patients, failed to convince the FDA of the drug's effectiveness, so Cohn and other researchers reanalyzed those trials' data to determine if the effects appeared stronger within individual racial groups. It indeed appeared more effective in African Americans and led to the A-HeFT follow-up study of only African American patients (Rusert and Royal 2011). Given that there is a long history of selling pharmaceuticals to African American patients after testing them on primarily or exclusively White subjects, the FDA's 2005 decision to grant NitroMed's African American–specific labeling troublingly implied that somehow African American health benefits cannot be conversely generalized to White populations (Kahn 2013: 49–50). Thus, even the process of investigating and responding to the health needs of racial minority populations with underserved needs—seemingly the most compelling cases of appropriate use of race in medicine—can still raise serious concerns about race's appropriateness in medicine.

Why Not Use "Race" in Medicine?

Judged on the basis of its effects, race has been a disastrous concept in biomedicine. It has overstated, over-extrapolated, and essentialized. It served as a rationalization for hundreds of years of colonialism, institutional deprivations, marriage restrictions, and a host of other *de jure* and *de facto* atrocities (Washington 2006). However, as such a powerful feature of social life (setting aside the contested genetic aspects), it is also woven into individual and group identities. For example, the ugly history of the "one-drop rule" for delineating who qualifies as "Black" (previously used to distinguish those who could be owned as slaves) continues to influence the way many people form their "Black" identities, and those identities must be acknowledged and respected regardless of that troubling history (Khanna 2010). Race-in-medicine eliminativists take into account these historical and contemporary facts, ultimately judging that race must be dropped from biomedical research and practice.

As understood by contemporary social scientists of race, race is social and fluid, rather than innate and stable (Aspinall and Song 2013). This fluidity is a source of skepticism about its utility in medicine (Rotimi 2004). In one illustration of race's fluidity, Aliya Saperstein and Andrew Penner found in the U.S. that not only does one's race impact one's social status, but conversely, one's social status also impacts one's race. That is, "nonwhites who achieve high status are more likely to subsequently be seen as and identify as white" (Saperstein and Penner 2012: 701), while:

having been unemployed for a long spell, in poverty, incarcerated, and on welfare all have statistically significant effects, such that people who have had each of these experiences are more likely to be seen by others and identify themselves as black and less likely to be seen by others and identify themselves as white.

(Saperstein and Penner 2012: 698)

Extending beyond the US, across Latin American countries, socioeconomic status affects racial self-identification (Telles and Paschel 2014). People of mixed racial or ethnic backgrounds often shift identities over their lifetimes, including in response to the day-to-day demands of social situations (Aspinall and Song 2013).

Evolutionary biologists Joseph Graves and Michael Rose have decried race as an impediment to medicine obtaining precise understandings of environmental factors and genetic factors affecting health: "medicine should take both social environment and population genetics into account, not spurious 'human races' that inappropriately conflate the two" (Graves and Rose 2006: 481). Somewhat similarly, Mildred Cho argues that the combination of the fluidity and the varying definitions of race mean that race is hopeless as a biomedical tool, except for the purpose of studying perceptions of race (Cho 2006). Keita et al. (2004) take a similar position.

BiDil has served as a lightning rod for critiques of the use of racial categories in differentiating medical treatments. The aforementioned fluidity of race categories has drawn BiDil criticism like that articulated by Rotimi, a genetic epidemiologist:

The label used to designate the African American population in studies like the clinical trial for BiDil is too imprecise to be relevant for individual therapy. Some members of this population "supergroup" with heart failure will benefit from this drug, and others will not.

(Rotimi 2004: S45)

In a detailed analysis of the process of testing and approving BiDil for only African Americans, Ellison et al. make clear that the FDA failed to live up to its own goals and interests (e.g., when it encouraged the development of the single-race A-HeFT clinical trial for BiDil) (Ellison et al. 2008). It is no surprise that the FDA makes at least sporadic errors during the long process of approving drugs. This fallibility opens the door for financial interests to drive the process, with race serving as a means to earn profits while patients' interests get deprioritized. In this case, BiDil's first patent as a heart failure treatment was filed in 1989, with no mention of race (Roberts 2011: 171). Filing new patent paperwork 10 years later for BiDil as a treatment specifically for heart failure in African Americans—technically a new use—allowed BiDil to get 13 additional years of intellectual property protection (Roberts 2011: 171). In Dorothy Roberts' assessment, "the reason why BiDil was marketed according to race has more to do with its commercial appeal than its medical benefits" (Roberts 2011: 168).

Why Have Restricted Uses of "Race" in Medicine?

Virtually all scholars who advocate for the continued use of race concepts in medicine (conservationists, taken broadly) seem to favor cautious deployment of such concepts, for ethical reasons, evidentiary reasons, or both. Biomedicine's powerful influence over race discourse makes caution necessary.

Almost every aspect of racism that can be cognitively expressed originated in earlier ideas of race that were accepted on the scientific authority of their day. The biological sciences have constantly revised themselves in this regard, although common sense has lagged.

(Zack 2010: 883)

The enduring influence of biomedical science, including the echoes of outdated biomedical science, is one of the motivators for seeking particular rules or guidelines for how that biomedical science does its work.

Some critics of the use of race in medicine have targeted overly simplistic uses of race concepts, rather than seeking to eliminate racial terms altogether. Valles (2012) illustrates how racial descriptions of high-risk populations can sometimes inappropriately replace more nuanced descriptions of which specific people are at risk: "the privileging of data at the level of broad racial categories is an important obstacle to pursuing more nuanced representations of the heterogeneity of risk within racial categories" (Valles 2012: 406). This is demonstrated in two case studies where public health guidelines are made for the dietary salt intake of "African Americans" and for cystic fibrosis gene carrier screening of "Caucasians." In both cases, the guidelines ignore long-standing evidence of known subpopulations that have entirely different health risks, and hence that more fine-grained (but still simple) population descriptions would be more accurate and ethically responsible: *U.S.-born* African Americans in the former case and *non-Finnish* Whites in the latter (Valles 2012). It is not the race concepts that are inherently unacceptable; it is their incautious use despite contradictory evidence and ethical considerations.

The above position's encouragement of pursuing further specificity, when feasible, is similar to the recommendations of a 2009 report from the Institute of Medicine, which advised the collection of not only broad race/ethnicity data but also locally relevant "granular ethnicity" (Ulmer et al. 2009). For example, a hospital might collect data showing it has almost exclusively White patients, but such data would not allow differentiation between the potentially vastly different health experiences and needs of the hospital's large subpopulation of ethnically Italian patients and its large subpopulation of ethnically Albanian patients; the race data by itself would be inadequate.

Race-in-medicine debates struggle with a series of empirical claims: (1) race is a social reality; (2) race is a genetic reality; and (3) race is a biological reality. Echoing the rationale for Hardimon's aforementioned SOCIALRACE concept, Kaplan cautions against conflating (2) and (3), reconsiders the meaning of (3), and argues that the causal relationship between (1) and (3) is the reverse of what racialists contend (Kaplan 2010).

> ... taking self-identified race into account in medical decision making might make sense locally, given that self-identified race *is* a good predictor of one's experience with racial discrimination, prejudice, and racism more generally. The objection that to do so reifies race is, in essence, correct—self-identified race is *real* (it is socially contingent, but no less causally powerful for that) and, if the above is correct, it is also biological. But to be biological is not to be genetic, nor does biology make race. Race is biological because racism (and more generally a society organized by race) has profound biological effects.
>
> (Kaplan 2010)

Race can be biological without being genetic. A social environment constructed to include racist patterns of health benefits detriments (e.g., access to affordable mortgages and safe housing) creates biological differences among racial groups.

Krieger, a public health scholar, similarly focuses on racism during conversations about the utility of race in medicine, arguing that the responsible use of "race" concepts in biomedicine includes a requirement to specifically study racism in biomedicine.

> Neglecting study of the health impact of racism means that explanations for and interventions to alter population distributions of health, disease, and well-being will be incomplete and potentially misleading, if not outright harmful. Of course, work in this field will, inevitably, be fraught with controversy, because the exposure raises important themes of accountability, agency, and human rights.
>
> (Krieger 2003: 197)

Adequately explaining or responding to health disparities between races is simply not possible without examining the underlying problem that Kaplan identifies as a root cause of those disparities: racism. Even if examining racism is controversial, it remains necessary nonetheless.

As this chapter nears it conclusion, it seems appropriate to include a detailed set of rules for the responsible use of race in medicine, an abridged version of the rules laid out by Koffi Maglo.

"The Cluster Stability Rule": Researchers can investigate particular populations in medical research, but they must not assume that findings in one subgroup of a continental grouping (e.g., the Hmong ethnic group in southeast Asia) applies to all subgroups in that continental grouping (e.g., all Asians) (Maglo 2010: 366–367). In the same vein as Kahn's above point about BiDil and Valles's above point about heterogeneous subpopulations, it is essential to be cautious when making race-level (or continental population–level) generalizations based on data sampled from only a subset of patients (Kahn 2013: 49–50, Valles 2012).

"The Patient Standpoint Rule": Individual patients' needs and interests deserve the highest priority, unless serving those interests would hurt other individual patients' interests (Maglo 2010: 367). Since medicine's goal is to serve the needs of patients, considerations of patient care supersede any qualms researchers or practitioners might have with using race concepts.

"The Excluded Beneficiary Rule": When one must choose between alternative models for health patterns or different potential designs of clinical trials, preference should be given to options that benefit subpopulations that would otherwise be neglected (Maglo 2010: 367). As noted by Bhopal above (2006), medicine should be concerned with not only the heath of the whole population but also the health of oft-neglected minority populations that might have atypical health needs. When we do not know how to proceed, we ought to err on the side of helping groups that otherwise tend to be ignored. See, for example, Valles et al.'s recommendations for altering medical research design and analysis practices that (unintentionally) tend to under-report and under-analyze data from mixed-race and mixed-ethnicity patients (Valles et al. 2015).

"The Permissibility Principle": It is acceptable to prevent a subpopulation of patients from getting a treatment that would improve their health, in the sense of "species-typical normal functioning," if getting that treatment would harm the "species-typical normal functioning" of other patients (Maglo 2010: 367). Medical resources are limited, and the provision of resources to one population will often lead to fewer resources for other populations. Given the above criticisms that BiDil was approved for treating African Americans on the basis of very limited clinical research, and given that African Americans have such a disproportionate burden of heart failure, some redistribution of resources may be ethically necessary. Some of the clinical resources being directed to less burdened populations could be ethically shifted over to treating the more burdened populations, and in some cases this would indeed mean less/worse care for the former population. Because this chapter is in a philosophy of medicine handbook, it is worth noting that this last principle incorporates a philosophical commitment to a naturalist concept of health, a disputed position featured in Part I of this volume (see also: Carel and Cooper 2014). In this principle, Maglo takes health to be an objective fact of the world and human biology, not something that is open to social negotiation and value judgments. The race-in-medicine debate does not exist in a vacuum; it intersects with other philosophy of medicine debates, just as the BiDil controversy intersects with questions about the appropriateness of treating medicine as a commodity (see Chapter 47).

Before Bidil's approval was secured, sociologist of race Troy Duster had recommended taking a long view on the situation. In his view, it was not just a matter of whether the FDA approved BiDil, but also a matter of what future steps would be taken to move past the reliance on race as a proxy for unknown biological mechanisms.

> If the FDA approves BiDil, it should do so only under the condition that further research be conducted to find the markers that have the actual functional association with drug responsiveness—thus assuring that the drug be approved for everyone with those markers, regardless of their ancestry, or even of their ancestral informative markers.
>
> (Duster 2005: 1051)

In this way, Duster's position is surprisingly not all that different from that of Cohn, who affirmed that more research on heart failure is desperately needed to help guide treatment decisions (Cohn 2006). In the race-in-medicine debate, one finds interesting convergences of positions as often as one finds interesting divergences.

Conclusion

This chapter has provided a brief overview of a topic that is expansive and always changing. Given that race is embedded in social life and in biomedicine, it is perhaps not so surprising that the debate is multifaceted and evolving. The biomedical race research conducted in any time and place is influenced powerfully, even if subtly, by its social context, as is phrased elegantly in Duster's (2006) prediction of the future:

> ... the next decade will witness an outburst of behavioral genetics research, buttressed by the molecular reinscription of race tying crime to biological processes, and then correlating those biological processes to race. It is not beyond conjecture that it will be an African-American who will lead the charge, fully supported by the Pioneer Fund or some equivalent well-funded, conservative think tank or funding source. The banner will be the academic and intellectual freedom to fearlessly pursue a topic wherever it may lead. Most people will fail to recognize that such work will be driven by the prevailing winds, the Zeitgeist. Those winds will be perceived as natural and normal. "The spirit of the times" will be taken for granted.
>
> (Duster 2006: 495)

And so it has come to pass that in recent years, behavioral geneticists have been drawn to research on the gene coding for monoamine oxidase A (MAOA; Tabery 2014), variants of which have been associated with violent or antisocial behavior (sometimes given the media-friendly label "the warrior gene"), which were then used to explain minority crime rates (Perbal 2013). Journalist Nicholas Wade features MAOA research prominently in his controversial new book on racial differences (Wade 2014). Individual researchers might wish to believe that their decisions about if/when/how/why to research race, based on the above considerations, are ultimately their choice. As Duster illustrates in his prediction, the prevailing winds of the *zeitgeist*—the spirit of the times—will push the ship no matter what one does below deck. Interestingly, the zeitgeist may be shifting once again since Wade's book was forcefully rejected by the very biologists whose work he drew from to make his case (Coop et al. 2014). Meanwhile, even though Black and White U.S. primary care physicians (unsurprisingly) differ in their comfort with discussing racism and racial health disparities, they share an "excitement and anticipation regarding genomic medicine" (Bonham et al. 2009: 284).

The current status of the race-in-medicine debate can be interpreted either optimistically or pessimistically. Optimistically, a plurality of different views are now on the table, being debated by a large multidisciplinary group of scholars who care deeply about having an exacting dialogue and about the implications of the dialogue for patients. Pessimistically, the debate is problematic for both scholars and laypeople: scholars remain divided into camps and subcamps, and laypeople face a daunting challenge if they wish to formulate an educated judgment about their own stances. Sociologist Ann Morning's (2011) book offers a detailed review of current race views through a combination of theoretical analysis, interviews with biology and anthropology faculty and students, as well as analysis of high school textbooks. One key finding is, "social and biological scientists hold a wide range of beliefs about the nature of racial difference contrary to some scholars' expectations, they are far from any consensus, either within or between disciplines" (Morning 2011: 221). Such is the current state of the debate over race's status in biomedicine: unsettled.

References

Anderson, W. (2008) Teaching 'Race' at medical school: Social scientists on the margin. *Social Studies of Science* 38, 785–800.

Aspinall, P. J. and Song, M. (2013) *Mixed Race Identities*, Basingstoke: Palgrave Macmillan.

Baer, R. D., Arteaga, E., Dyer, K., Eden, A., Gross, R., Helmy, H., Karnyski, M., Papadopoulos, A. and Reeser, D. (2013) Concepts of race and ethnicity among health researchers: Patterns and implications. *Ethnicity and Health* 18, 211–225.

Bhopal, R. S. (2006) Race and ethnicity: Responsible use from epidemiological and public health perspectives. *Journal of Law, Medicine and Ethics* 34, 500–507.

Blum, L. A. (2002) *"I'm Not a Racist, But . . .": The Moral Quandary of Race*, Ithaca: Cornell University Press.

Bonham, V. L., Sellers, S. L., Gallagher, T. H., Frank, D., Odunlami, A. O., Price, E. G. and Cooper, L. A. (2009) Physicians' attitudes toward race, genetics, and clinical medicine. *Genetics in Medicine* 11, 279–286.

Carel, H. and Cooper, R. (2014) *Health, Illness and Disease: Philosophical Essays*, New York: Routledge.

Cho, M. K. (2006) Racial and ethnic categories in biomedical research: There is no baby in the bathwater. *Journal of Law, Medicine and Ethics* 34, 497–499.

Cohn, J. N. (2006) The use of race and ethnicity in medicine: Lessons from the African-American heart failure trial. *Journal of Law, Medicine and Ethics* 34, 552–554.

Coop, G., Eisen, M. B., Nielsen, R., Przeworski, M. and Rosenberg, N. (2014, September 8) Letters: A troublesome inheritance. *The New York Times*, BR6.

Darby, D. (2009) *Rights, Race, and Recognition*, Cambridge: Cambridge University Press.

Duster, T. (2005) Race and reification in science. *Science* 307, 1050–1051.

Duster, T. (2006) Lessons from history: Why race and ethnicity have played a major role in biomedical research. *Journal of Law, Medicine and Ethics* 34, 487–496.

Ellison, G. T. H., Kaufman, J. S., Head, R. F., Martin, P. A. and Kahn, J. D. (2008) Flaws in the US Food and Drug Administration's rationale for supporting the development and approval of BiDil as a treatment for heart failure only in black patients. *The Journal of Law, Medicine & Ethics* **36**, 449–457.

Fritz, C. (2015) Redefining racial categories: the dynamics of identity among Brazilian-Americans." *Immigrants and Minorities: Historical Studies in Ethnicity, Migration and Diaspora* 33(1), 45–65.

Gannett, L. (2010) Questions asked and unasked: How by worrying less about the 'really real' philosophers of science might better contribute to debates about genetics and race. *Synthese* **177**, 363–385.

Gannett, L. (2013) Theodosius Dobzhansky and the genetic race concept. *Studies in History and Philosophy of Biological and Biomedical Sciences* **44**, 250–261.

Graves, J. L. and Rose, M. R. (2006) Against racial medicine. *Patterns of Prejudice* **40**, 481–493.

Hardimon, M. O. (2013) Race concepts in medicine. *Journal of Medicine and Philosophy* **38**, 6–31.

Hicks, D. J. (2014) A new direction for science and values. *Synthese* **191**, 3271–3295.

Hochman, A. (2014) Unnaturalised racial naturalism. *Studies in History and Philosophy of Biological and Biomedical Sciences* **46**, 79–87.

Humes, K. R., Jones, N. A. and Ramirez, R. R. (2011) *Overview of Race and Hispanic Origin: 2010*, Washington, DC: United States Census Bureau.

Hunt, L. M. and Megyesi, M. S. (2008) The ambiguous meanings of the racial/ethnic categories routinely used in human genetics research. *Social Science and Medicine* **66**, 349–361.

Kahn, J. (2013) *Race in a Bottle: The Story of BiDil and Racialized Medicine in a Post-Genomic Age*, New York: Columbia University Press.

Kaplan, J. M. (2010) When socially determined categories make biological realities: Understanding black/white health disparities in the US. *The Monist* **93**, 281–297.

Kaplan, J. M. and Winther, R. G. (2014) Realism, antirealism, and conventionalism about race. *Philosophy of Science* **81**, 1039–1052.

Katikireddi, S. V. and Valles, S. A. (2015) Coupled ethical-epistemic analysis of public health research and practice: Categorizing variables to improve population health and equity. *American Journal of Public Health* **105**, e36–e42.

Keita, S. O. Y., Kittles, R. A., Royal, C. D. M., Bonney, G. E., Furbert-Harris, P., Dunston, G. M. and Rotimi, C. N. (2004) Conceptualizing human variation. *Nature Genetics* **36**, S17–S20.

Khanna, N. (2010) If you're half black you're just black: Reflected appraisals and the persistence of the one-drop rule. *The Sociological Quarterly* **51**, 96–121.

Krieger, N. (2003) Does racism harm health? Did child abuse exist before 1962? On explicit questions, critical science, and current controversies: An ecosocial perspective. *American Journal of Public Health* **93**, 194–199.

Maglo, K. N. (2010) Genomics and the conundrum of race: Some epistemic and ethical considerations. *Perspectives in Biology and Medicine* **53**, 357–372.

Mills, C. W. (1998) But what are you really? The metaphysics of race. In: Mills, C. W. (ed.) *Blackness Visible: Essays on Philosophy and Race*, Ithaca, NY: Cornell University Press, pp. 41–66.

Morning, A. (2011) *The Nature of Race: How Scientists Think and Teach about Human Difference*, Berkeley: University of California Press.

Nobles, M. (2000) History counts: A comparative analysis of racial/color categorization in US and Brazilian censuses. *American Journal of Public Health* **90**, 1738–1745.

Perbal, L. (2013) The 'warrior gene' and the Māori people: The responsibility of the geneticists. *Bioethics* **27**, 382–387.

Pletcher, M. J., Kertesz, S. G., Kohn, M. A. and Gonzales, R. (2008) Trends in opioid prescribing by race/ethnicity for patients seeking care in US emergency departments. *Journal of the American Medical Association* **299**, 70–78.

Risch, N., Burchard, E., Ziv, E. and Tang, H. (2002) Categorization of humans in biomedical research: Genes, race and disease. *Genome Biol* **3**, 1–12.

Roberts, D. (2011) *Fatal Invention: How Science, Politics, and Big Business Re-Create Race in the Twenty-First Century*, New York: The New Press.

Rotimi, C. N. (2004) Are medical and nonmedical uses of large-scale genomic markers conflating genetics and 'race'? *Nature Genetics* **36**, S43–S47.

Rusert, B. M. and Royal, C. D. M. (2011) Grassroots marketing in a global era: More lessons from BiDil. *The Journal of Law, Medicine & Ethics* **39**, 79–90.

Saperstein, A. and Penner, A. M. (2012) Racial fluidity and inequality in the United States. *American Journal of Sociology* **118**, 676–727.

Shelby, T. (2005) *We Who Are Dark: The Philosophical Foundations of Black Solidarity*, Cambridge: Harvard University Press.

Spencer, Q. (2013) Introduction to is there space for race in evolutionary biology? *Studies in History and Philosophy of Biological and Biomedical Sciences* **44**, 247–249.

Spencer, Q. (2014a) A radical solution to the race problem. *Philosophy of Science* **81**, 1025–1038.

Spencer, Q. (2014b) The unnatural racial naturalism. *Studies in History and Philosophy of Biological and Biomedical Sciences* **46**, 38–43.

Sullivan, S. and Tuana, N. (2007) *Race and Epistemologies of Ignorance*, Albany: SUNY Press.

Tabery, J. (2014) *Beyond Versus: The Struggle to Understand the Interaction of Nature and Nurture*, Cambridge, MA: MIT Press.

Telles, E. and Paschel, T. (2014) Who is black, white, or mixed race? How skin color, status, and nation shape racial classification in Latin America. *American Journal of Sociology* **120**, 864–907.

Ulmer, C., McFadden, B. and Nerenz, D. R. (2009) *Race, Ethnicity, and Language Data: Standardization for Health Care Quality Improvement*, Washington, DC: The National Academies Press.

Valles, S. A. (2012) Heterogeneity of risk within racial groups, a challenge for public health programs. *Preventive Medicine* **55**, 405–408.

Valles, S. A., Bhopal, R. S. and Aspinall, P. J. (2015) Census categories for mixed race and mixed ethnicity: Impacts on data collection and analysis in the US, UK and NZ. *Public Health* **129**, 266–270.

Wade, N. (2014) *A Troublesome Inheritance: Genes, Race and Human History*, New York: Penguin Books.

Washington, H. A. (2006) *Medical Apartheid: The dark History of Medical Experimentation on Black Americans from Colonial Times to the Present*, New York: Doubleday.

Will, J. C., Valderrama, A. L. and Yoon, P. W. (2012) Preventable hospitalizations for congestive heart failure: Establishing a baseline to monitor trends and disparities. *Preventing Chronic Disease* **9**, 110260.

Woolf, S. H. and Aron, L. (2013) *US Health in International Perspective: Shorter Lives, Poorer Health*, Washington, DC: National Academies Press.

Zack, N. (2010) The fluid symbol of mixed race. *Hypatia* **25**, 875–890.

Further Reading

Andreasen, R. O. (2008) The concept of race in medicine. In: Ruse, M. (ed) *The Oxford Handbook of Philosophy of Biology*, New York: Oxford University Press, pp. 478–503. (A chapter on race in medicine, in the *Oxford Handbook of Philosophy of Biology*)

James, M. (2012) Race. In: Zalta, E. N. (ed) *The Stanford Encyclopedia of Philosophy* (Winter 2012 Edition). (The *Stanford Encyclopedia of Philosophy*'s entry on race)

Roberts, D. (2011) *Fatal Invention: How Science, Politics, and Big Business Re-Create Race in the Twenty-First Century*, New York: The New Press. (An introductory book on race in medicine, including discussion of BiDil)

Washington, H. A. (2006) *Medical Apartheid: The Dark History of Medical Experimentation on Black Americans from Colonial Times to the Present*, New York: Doubleday. (An accessible history of racism in American biomedicine)

http://onlinelibrary.wiley.com/doi/10.1111/jlme.2006.34.issue-3/issuetoc (A special issue of *The Journal of Law, Medicine & Ethics* on the race-in-medicine debate)

39
ATYPICAL BODIES IN MEDICAL CARE

Ellen K. Feder

The "Problem" (or Problems) of Atypical Bodies

Consideration of atypical bodies in medicine requires that we investigate commonsense views about bodily differences. We may understand assessments of normality in the context of medicine to be an objective measure of what is "typical," but such measures may, as disability theorists have argued, carry unexamined judgments of value that reflect, or perhaps promote, social views of bodily difference that can, in turn, shape medical judgments concerning the meaning of these categories.

The first part of this chapter historicizes the division between normality and abnormality in medicine. The late 20th-century medical treatment of abnormal height in children illustrates gaps in the correspondence between medical views of normality and abnormality on the one hand, and health and disease on the other hand. Turning from the example of unusual height to unusual sex anatomies, the second part of the chapter looks more closely at the division between the medical and the social, and raises questions concerning the demands of conformity enforced by stigma and shaming. The complex work of what is now called "medicalization" of atypical bodies that these examples illuminate may, I conclude, spur further reflection on the ways that values and desires can make the lines between normal and abnormal, health and disease, less clear than we might have otherwise appreciated.

Normal and Abnormal

At the center of contemporary medical knowledge lies the distinction between "normal" and "abnormal." Recognizing normal function of the complex processes of the human organism forms the bedrock for medicine's role in maintaining health; indeed, it could be regarded as a standard that has been constant throughout the history of medicine. What has changed is the way that standard is figured.

Historians tell us that, for physicians in ancient Greece, health was figured as "harmony" (e.g., Plato 1989: 187a-e) or "correct proportion" of the elements (Lloyd 1983: 262). In medical terms, what we think of as "normal" (i.e., "healthy") was assessed relative to the individual: what is healthy or normal was the harmonious functioning of a particular organism.

Ancient conceptions of health certainly differ in important respects from those that we hold today; if we experience pain, discomfort, or do not "feel like ourselves," our medical providers might ask questions and order tests to assess whether individual substances fall within what is now regarded as a "normal range." This range is defined not by what is optimal for an individual but by what has been determined to be the norm or average for the species. Furthermore, examination of individual substances apart from others differs from the ancient medical view of achieving a "correct proportion" among substances.

We may nevertheless see areas of continuity between past and present views. For example, when a physician investigates whether pain or discomfort signals an infection, and tries to determine the number of white blood cells circulating in our blood, that physician's assessment is based on a prevailing view of the "species-typical norm"; at the same time, that assessment could be described in the ancient medical terms of "excess" and "deficiency." Perhaps some organ or system is not functioning as expected, and tests to measure levels of different hormones, for example, can indicate a problem that requires intervention to right that balance, to ensure that there is neither an excess nor a deficiency that signals an individual malady. These are cases in which the distinction between normal and abnormal aligns neatly with the distinction between health and illness, and they appear to be shared by ancient and contemporary physicians alike.

But observations about normality and abnormality do not always involve judgments about wellness and pathology. That a condition or capacity falls outside of the norm or average (the definition of "abnormal") may have little bearing on the health of an individual (see Chapter 1, The concept of disease). Take double-jointedness or multicolored irises, what is sometimes expressed genetically as having two different-colored eyes (heterochromia): anomalies such as these are unremarkable, not only medically but also socially. Some abnormalities that have no bearing on an individual's health can nevertheless carry significant social weight. Extraordinary height, for example, is a boon to athletes for whom it can confer an advantage, and indeed, being taller than average seems generally desirable for men, if not for women. Judgments such as these hardly seem to be matters of medicine, and yet, it was not so long ago that distressed parents of tall girls sought medical advice in order to protect them from the consequences of the stigma attached to a trait's statistical abnormality, even though this abnormality has no effect on function.

Beginning in the 1950s, concern about the negative social judgment associated with tallness in women prompted physicians to prescribe the synthetic estrogen diethylstilbestrol (DES) to stunt the growth of tall girls. Cohen and Cosgrove report in their 2009 book, *Normal at Any Cost*, that DES had been in use for at least 20 years before it was prescribed to inhibit growth in girls. Though there was no evidence of its effectiveness as an antiabortifacient, millions of pregnant women took DES to prevent miscarriage starting in the late 1930s. In 1971, the U.S. government issued a safety warning following the revelation of a correlation between DES exposure in utero and vaginal and cervical clear-cell adenocarcinoma, an extremely rare form of cancer that had previously only been diagnosed in postmenopausal women. The government warning effectively ended the prescription of DES to prevent miscarriage. It curtailed, but did not end, the prescription of estrogen to stunt growth in taller-than-average girls (Cohen and Cosgrove 2009: 46–47).

Recommendations of growth-restricting treatment for girls appear to have continued for several more years, until the publication of successive articles in *The New York Times* in 1976. The reports provoked public awareness of the medical treatment of tall girls, and also led the Lawson Wilkins Pediatric Endocrine Society and the National Institute of Child Health and Human Development to hold a conference to consider the social and medical questions raised by the practice of treating preadolescent and adolescent girls with hormones to restrict their growth. In its recommendations, the conference committee acknowledged that there was as yet no evidence that the treatment of tall girls yielded any psychosocial advantage to the girls who had been prescribed DES, and urged that studies be undertaken to determine whether confidence in the benefits imagined by physicians and the parents they counseled could be demonstrated (Cohen and Cosgrove 2009: 51). It furthermore acknowledged that the unstated risks of harm were in fact known: girls born with Turner syndrome (a condition in which the second X chromosome in females is completely or partially missing) had long been prescribed estrogen to compensate for the fact that their ovaries do not produce sufficient quantities of hormone to support typical pubertal development and to protect bone density through the lifespan. Although estrogen treatment had been effective in combating significant medical problems faced by individuals with Turner syndrome, doctors had seen an alarming incidence of uterine cancer in their patients, something that had also been noted in postmenopausal women taking estrogen replacement (Cohen and Cosgrove 2009: 51). The committee recommended that follow-up studies be undertaken to learn the outcomes of girls who had been prescribed estrogen for height. Although such studies were never undertaken, it is likely, given changes in society's views toward tall women, that there is much less treatment occurring today. According to Cohen and Cosgrove, however, this treatment nevertheless continues to be prescribed, "especially when parents insist" upon it (Cohen and Cosgrove 2009: 454).

In 2003, the U.S. Food and Drug Administration approved the use of Human Growth Hormone (hGH) for "growth hormone deficiency," an umbrella diagnosis that included different forms of dwarfism, but that also, controversially, included those with "idiopathic" short stature (i.e., shortness absent known pathology). A decade earlier, studies demonstrated the success of hGH in increasing height in otherwise healthy children (Cohen and Cosgrove 2009: 126). Leading up to and following approval, there was considerable discussion concerning the difficulty of distinguishing between interventions that were unquestionably warranted and those that were not, and whether, in the latter case, physicians should recommend them. One pediatric endocrinologist specializing in growth disorders put it this way: "The pathology you're gonna do something about. The physiology you *may* do something about" (quoted in Cohen and Cosgrove 2009: 302 original emphasis). A disease or disorder impeding human function ("pathology") unquestionably calls for medical intervention; while medical expertise may be effective in normalizing a variation in the body's appearance that does not impede function ("physiology"), such intervention would, the endocrinologist suggests, be something one might characterize as "elective."

With hindsight, the hazards involved in physicians' efforts to prevent girls from being taller than average are too apparent; the panic that led physicians to prescribe powerful hormones to arrest growth in girls may now seem farcical. How could doctors have not anticipated the dangers and ignored relevant evidence of risk? Why was there so much anxiety about tallness in girls, and why is it the business of medicine to fix that anxiety? What story will be told about the current medical treatment of short stature in boys? The administration of DES and other hormones offers an exemplary case of making a social problem (like tall stature in women) into a "medical problem," a phenomenon that sociologist Peter Conrad calls "medicalization" (Conrad 2007).

If tall stature in girls and short stature in boys are good examples of what most would agree to characterize as social, and not medical, disorders, then there are any number of conditions in which unusual stature is a feature of pathology in the ordinary sense. Serious risks of heart failure come with Marfan syndrome, characterized by unusual tallness, for example; unexpected tallness may also signal the presence of a pituitary tumor. Those with achondroplasia, the most common form of dwarfism, have a heightened risk of being born with hydrocephalus and spinal stenosis. In the case of height, history has demonstrated the extent to which questions concerning "function" may be conflated with negative social views (what in hindsight we see as different kinds of prejudice). Things get more complicated still when we consider that medical interventions intended to address features of a disorder can be prescribed as (medicalized) enhancements, as we see today in the prescription of hGH for shorter boys.

Medical and Social

Though the distinction between "pathology" and "physiology," as the pediatric endocrinologist put it, may seem obvious, we have already seen how the distinction between the two can become blurred in medical treatment of "abnormal conditions." Why, we should ask, is pathological difference yoked to certain kinds of (nonpathological) physical variation? And why is (nonpathological) variation a "problem" for medicine to solve? In his investigation of the history of European medicine, Michel Foucault noted an important shift in the conception of "health" and "normality" between the 18th and 19th centuries that suggests that the identification of pathology and variation have not always been understood in the terms they are today. Understanding more about this history may help us better appreciate the challenges we encounter in thinking about atypical bodies in medical care.

According to Foucault, standards of health in the 18th century focused on what could be understood as specific to a particular individual—"vigour, suppleness, and fluidity, which were lost to illness and which it was the task of medicine to restore" (Foucault 1994 [1963]: 35). Early modern medical standards in this sense shared with ancient Greek medicine a view of health as specific to the individual. Although physicians identified general standards, it was still possible for individuals to be their "own physicians" who judged, and managed, their health. A subtle but important change comes in the 19th century when, Foucault observes, medicine "formed its concepts and prescribed its interventions in relation to a standard of functioning and organic structure" (35) that was imposed from without. The medicine that previously took as its object "the structure of the organized being" was transformed into "*the medical bipolarity of the normal and the pathological*" (35, original emphasis). It was no longer the judgment of the individual that mattered most, but that of "experts" authorized to evaluate and treat the individual as prevailing standards dictated. Moreover, medicine comes to see its role as identifying a "single" norm that would be applied to the population as a whole.

The recent history of the medical treatment of height highlights the powerful operation of medicalization with respect to the understanding of atypical bodies. However, there is more at stake in Foucault's analysis. In the shift in focus from attending to health and disease occurring in the individual body, to assessing the individual against a standard cast in terms of the normal and abnormal, a transformation of the object of medicine occurs, as well. The health of the individual body remains a consistent object of concern, but, as Foucault's analysis suggests, individual health is understood as a component also of social health. Where previously disease had been figured principally as a threat to an individual's health and well-being, the changes in medicine marked a shift in understanding of disease, disorder, and "abnormality" as a threat to the social body, as well. Another example will better illustrate what we may understand as a deeper change in the conception of medicine that Foucault describes.

The treatment of atypical sex anatomies in the 19th century exemplifies the legacy of the change Foucault observes. As Alice Dreger recounts in her *Hermaphrodites and the Medical Invention of Sex*, physicians developed toward the end of the 19th century a taxonomic system to classify hermaphroditic "types," a nomenclature that persisted for more than 100 years. It defined maleness and femaleness in terms of standard female and male anatomy; deviations from standard anatomies were described as "male and female pseudohermaphrodites" and "true" hermaphrodites, each of which presented different mixtures of male and female anatomy (Dreger 1998: 35–40). These categorizations were not simply medico-scientific descriptions of human anatomical differences and their variations; these categories also contained prescriptions for social life and organization.

Interest in hermaphroditic bodies was entirely in line with the modernization of medicine that took place in Western Europe in the late 19th and early 20th centuries, interested as it was in carefully measuring, classifying, and mastering knowledge of the human body and its functions, normal and aberrant. However, the interest in hermaphroditic bodies was also distinct from general medical curiosity about the human body: hermaphroditism provided an inordinately rich object of scientific and medical interest that combined medical and scientific curiosity and concern with managing what was perceived as a social menace. The cases connected to the developments Dreger traces at the end of the 19th and beginning of the 20th centuries were not cases of mere physical variation, but of "mistaken sex" (i.e., cases where a person's anatomy could not be easily categorized as male or female), or more importantly at this time, where the secondary sex characteristics (e.g., body hair, musculature) and corresponding behavior (especially including the object of sexual desire) of someone who had been raised as one sex assumed the characteristics of the other sex. One example may be found in the 19th-century memoir of Herculine Barbin, made famous when 20th-century French philosopher Michel Foucault discovered the manuscript in the 1970s (Foucault 1980). Cases such as Barbin's, and others that Dreger describes, posed challenges to a social order that depended, as Dreger writes, "on there being (only) two sexes" (Dreger 1998: 8). Cases that did not present such challenges, one expects—because they were not written up and disseminated, or because people did not consult doctors or surgeons to the extent that they eventually would come to—were not extraordinary, and would not cause the sort of consternation that the birth of children with intersex anatomies did beginning in the mid-20th century.

Today, the prevailing standard of care for the medical management of intersex cases is based on work carried out by psychologist John Money and his colleagues starting in the early 1970s at Johns Hopkins University (see, e.g., Downing, Morland, & Sullivan 2015). Together with Joan and John Hampson, Money worked out what was called the "optimal gender theory," which took "gender roles" as produced not simply by nature but in interaction with culture. This interaction, Money and his colleagues proposed, suggested a certain sort of malleability that could allow children born with atypical sex anatomies to be assigned as "boys or girls" without concern (Money et al. 1955).

Money's work came to be famously discredited in the late 1990s, when it was revealed that the renowned experiment in child sex reassignment (the case of John/Joan) turned out to be a fraud (see Colapinto 1997). Despite this revelation, normalizing treatment of children with atypical sex anatomies has continued, to the distress and bewilderment of many who see the treatment of children with atypical sex anatomies as a straightforward matter of what Erik Parens, seeking to refine Conrad's term, calls "bad medicalization." Whereas medicalization takes a nonmedical problem—"life or human" problems—and casts these in medical terms, bad medicalization, Parens proposes, should be understood as the commission of a category error (i.e., taking something to be a medical issue or problem that is genuinely a social problem for which medicine intervention is inappropriate or ineffective). Parens suggests that rather than see medicalization as "bad" *per se*, it is important to recognize that there are forms of medicalization—such as that resulting in birth control—that critics of medicalization would likely celebrate (Parens 2011: 6).

The history of the management of atypical sex is complicated, in the terms Parens introduces, by the combination of forms of medicalization it has entailed. Focused on removing sexual ambiguity, medical management of atypical sex anatomies has, since the 1950s, emphasized medical (and especially surgical) fixes for what might otherwise be understood in contemporary terms as social, political, or psychological matters of sexual identity. As critics have repeatedly demonstrated since the 1990s, these interventions have entailed significant harms (see, e.g., Chase and Coventry 1997); today, a number of national and international statements have acknowledged gross violations of humans associated with normalizing atypical sex anatomies (German Ethics Council 2012; Swiss National Advisory Commission on

Bioethics 2012; United Nations Office of the High Commissioner for Human Rights 2015). One might reasonably ask why it wouldn't be simpler to take intersex conditions out of medicine altogether, to "demedicalize" conditions that might instead count as ordinary human variations. The case for understanding differences in genital appearance as matters of variation is undeniably convincing, and yet there is an equally compelling case that some of the conditions with which genital variation are associated bring genuine health challenges that require not less, but substantially more, medical attention than has been afforded them. Perhaps the best example is the case of congenital adrenal hyperplasia (CAH), a condition that in some forms poses grave dangers to an individual's health.

CAH is a genetic disorder that involves malfunction of the adrenal glands. It affects males and females in equal numbers. What was early termed the "androgen push" associated with CAH, both in utero and after birth, prompts premature development of bone growth, which ultimately results in adult short stature—a particular social concern, we know, for affected boys—and virilization of sex anatomy, which can make a male look like an "infant Hercules" (Eder 2012: 71), and can make a genetic female appear more like a male. Both males and females with CAH may suffer from serious metabolic problems soon after birth (when it is usually detected) and throughout one's life. In the "classic," salt-losing form, it results in vomiting and dehydration and, if left untreated, can lead to death. Increased vulnerabilities—caused by ordinary illness and injury that can exacerbate various sorts of imbalances associated with CAH—are among the problems that require careful attention in early childhood and beyond.

Lawson Wilkins, "the father of pediatric endocrinology," developed a detailed clinical understanding of the disease and its effects. This work led, in 1949, to the first attempt to determine whether CAH would be responsive to cortisone therapy. Wilkins refined the administration of cortisone over several years. It did not provide a hoped-for cure of the condition, but when administered in doses appropriate to the patient, cortisone could control the effects of CAH, which encompassed a range of what most would regard without controversy as ailments. Indeed, Sandra Eder's history of Wilkins's clinic (Eder 2012) suggests that we may find there an exemplary case of "good medicalization," the identification of malady or disorder that had not been previously identified or had been misunderstood. In the wake of Wilkins's work, physicians no longer theorized, as they had only some time earlier, about the "psychic influence on the part of the mother," whose thoughts were believed capable of malforming the developing fetus in her womb (Dreger 1998: 70); the physicians in Wilkins's clinic saw conditions that could alter the expected course of sex(ual) development as natural defects.

Though today it is clear that physicians and families see a continuity between the genuine medical problems caused by CAH and the social problem atypical sex is taken to present, this was not always the case. There is little evidence that "hermaphroditism" during the period that Dreger studies was associated with any condition we would regard as a disease or malady, as most forms of CAH are now recognized to be. (The important exception was the work of the French surgeon Jean Samuel Pozzi, who noted that undescended testes in those with the condition now termed "androgen insensitivity syndrome" could become cancerous [Dreger 1998: 64–65].) In other words, it was not the fact that atypical sex posed a medical danger that brought these bodies under special scrutiny (and display) in the Victorian period; instead, it is their deviation from "the norm" that is taken to be a threat to the social body (166). Rather than a concern with health, Dreger notes that the fascination with hermaphroditism was likely connected to challenges to sexual boundaries coming from new directions, particularly from first-wave feminists and homosexuals (26). Even as it appears that the continuing imperative to normalize intersex bodies is informed by the sorts of "social panic" that may be traced to its contemporary origins in the 1950s (Fausto-Sterling 2000: 72; Sandberg et al. 2004), the recognition of the genuine threat to health and life posed by some forms of CAH has led to universal newborn testing in most developed countries.

It is crucial to see how, today, these histories of atypical sex anatomies as threats to the social order, on the one hand, and to individual health, on the other hand, are today intertwined, and why there remains so much controversy about the identification and treatment of atypical sex. Most cases of what are called Disorders of Sex Development (DSD; Hughes et al. 2006) do not pose challenges to individuals' health. Understanding something of the history of the treatment of intersex goes some distance to helping us to appreciate why normalizing interventions for atypical sex anatomies in children remain the standard of care, intended as they are to "correct" anomalies that become equated with "pathology." But instead of a risk to healthy function, normalizing interventions are taken as a means to "protect" children who are regarded as—perhaps excessively or abnormally—"vulnerable" by virtue of their anatomical difference. There are today a number of robust ethical discussions of the problems entailed by the complex conceptions of health and illness, normality and abnormality, involved in the medical treatment of children and adults with atypical sex anatomies. But prior to questions of what is right or wrong, clinicians must ask themselves about the medicalization in which their practices are involved, and find ways to test the certainty with which medical decisions to normalize children's bodies are made.

Knowledge and Doubt, Certainty and Denial

Research conducted by pediatrician and ethicist Jürg Streuli and his colleagues in 2011–2012 at the University of Zurich highlights problems concerning medical authority and patient (and proxy) choice in the context of care for children with atypical sex anatomies. This work should serve to promote the asking of questions concerning the medicalization of intersex bodies that the history of treatment of atypical sex anatomies has foreclosed. It may suggest, too, how reflection on the problem, not of atypical sex anatomies themselves, but of our understanding of the need to see the problem of atypical anatomies as medical problems, may be helpful in reflecting on the phenomenon of medicalization, good and bad.

In their study, medical students were shown a short video "asking them to imagine that they had just become parents of a child with ambiguous genitalia, whose (future) gender midwives and doctors were unable to identify by looking. They were told they would watch a second video, which would be "their first contact with a specialist able to counsel and inform them about their child's condition."

There were two versions of this second video, both about six minutes long. One featured an endocrinologist's presentation of this first contact and the other a psychologist's. The scripts for both were reviewed by specialists from each discipline, ensuring that the information and presentation accurately replicated their respective approaches. The difference between the two reflected the distinctive perspectives on care that are by now well recognized. The authors reported:

> The medical professionals tended to medicalize, defining the child and its condition or behavior as a medical problem or illness that mandated or licensed the medical professional to offer a specific treatment. The parent, patient, activist, or psychologist, on the other hand, tended to demedicalize the issue by stressing the importance of the child's social world, seeing the main task as offering the child professional support in its environment.
>
> (Streuli et al. 2013: 1955)

Of the 89 students who participated in the study, 43% indicated in the questionnaire that they would choose normalizing surgery. There were considerable differences between the groups who saw the two counseling videos, with 66% of the group viewing the endocrinologist's video favoring normalizing surgery, as compared with 23% of the group viewing the psychologist's video. This point is perhaps not especially surprising; most of us would probably recognize that the position a medical specialist occupies would likely affect that individual's recommendations and our view of these recommendations.

Perhaps more interesting than this basic difference in the proportion of respondents favoring surgery was what Streuli and his colleagues report regarding the difference in the "conviction in their decision": "[T]hose informed by the endocrinologist were more secure and convinced in deciding for than against surgery . . . whereas those informed by the psychologist were more secure in deciding against than for surgery." (Gender and prior knowledge of intersex conditions did not have a significant impact on participants' decisions, though the authors note that the majority of those who rated their prior knowledge as "detailed" tended—though not overwhelmingly—to favor a nonsurgical approach [Streuli et al. 2013: 1956].)

The most startling finding was that, "*[a]ll participants believed their decision was based mainly on their own values, opinion, and attitude*." Participants saw their "personal attitude" as the main influence of their decision, and believed the content of the video had "little or no influence" (Streuli et al. 2013: 1956, emphasis added). No matter the group in which individuals participated, and regardless of the decisions they indicated they would make following their viewing of the videos, participants took their decisions to be independent of the information they had received. "Contrary to the majority participant perception," however, the authors found in no uncertain terms that "it was not just their attitude that made them decide for or against surgery but primarily a 6-minute slot of information" (Strueli et al. 2013: 1956–1957).

What is novel about the study by Strueli and his colleagues is not so much the evidence demonstrating the influence of counseling that participants received, but the evidence of *the lack of awareness* of this influence. This study helps to clarify why, as authors of a study concluded a few years earlier, "parents might see surgery as obvious and necessary, without experiencing it as something that involved a decision-making process" (quoted in Streuli et al. 2013: 1954).

Streuli and his colleagues do not explicitly address the role of culture in decision making; however, their study provides evidence of the ways that our decisions cannot be understood outside of a cultural context ordered by rules. Only some of the rules by which we think and understand, that organize our knowledge and guide our actions, are of the sort that we can name (in the realm of ethics or law, for example). So many others of these rules are captured by what social theorist Pierre Bourdieu terms the "habitus" (Bourdieu 1990), a kind of implicit normative order that tells us "how things are," how, at the same time, they are *supposed* to be. It is very difficult to describe habitus; for Bourdieu, the difficulty of making it an object of analysis is essential to its functioning as a background for knowledge. Threats to our understanding of sexual difference, and perhaps especially challenges to "the fact" of sexual difference taken as a natural division between male and female, may threaten the habitus. If, following Foucault, we see the importance of medicine and medical authority to our cultural understandings of the normal and the abnormal, we may better appreciate how the medical students, acting as parents in the Swiss study, commit themselves to act in ways that confirm and reinforce what seems "common sense," and furthermore see decisions they project as their own.

The tensions evident in the very different responses of the two groups do not contradict Bourdieu's analysis of habitus, but demonstrate how changes are adapted, and themselves made possible within its terms. Of special importance for thinking about the medical treatment of atypical bodies, we should note, with the authors, the possibility that parents charged with making decisions see their choices as "obvious and necessary," and may not therefore understand themselves to be engaged in the process of reflection and questioning that we may otherwise associate with the work of making decisions (Streuli et al. 2013: 1954). At the same time, the stark difference in the responses of the two groups suggests something of the tension the operation of habitus must involve. Respondents favoring surgical normalization, the authors report, judged autonomy less important than normality (Streuli et al. 2013: 1956). Presumably, both groups would see value in each, but the emphasis of one value over another indicates just the sort of flexibility that permits stability in systems of belief and value that habitus works to guarantee, and provides the means of working with changes it must tolerate or perhaps even promote. Taking seriously the results of the study conducted by Streuli and his colleagues, we may conclude that the sorts of certainty associated with normalizing interventions may not simply merit scrutiny but skepticism.

Conclusion

Both in medical practice and social life, atypical bodies demonstrate the sorts of changes—whether expressed as accommodation, assimilation, suppression, concealment, or celebration—required by challenges these bodies may be understood to pose. These examples illustrate significant changes in the ways that medicine and health are figured, together with the ongoing transformations of the categories of health and disease, normality and health. Examination of the medical treatment of abnormal bodies requires that we consider how exactly we understand "the problem"—or perhaps the problems—they pose. In some cases, such as simple tallness in girls, social conventions particular to a time and place may function to make of an atypical anatomy a risk to an individual's flourishing. Medicalization of tallness could be understood with respect to the individual body of the tall girl; physicians who prescribed DES to tall girls saw the problem as the body of the girl. And yet, following Conrad, we could see the problem of tallness in girls as a social problem concerning gender norms rather than a medical problem as our common sense—itself shaped by social forces—would see it. Here the problem is not her body, but rather a political and ethical problem of intolerance and the heightened vulnerability that intolerance promotes. Investigating the medical treatment of these bodies could also raise questions about the relationship between medical authority and social norms, expertise and common sense. Where we locate "the problem" of tallness is no easy matter. Investigation of the medical treatment of atypical bodies is arguably more complicated in cases where medical risks posed by differences may coincide with social risks of unusual appearance. These are more confounding when the focus of medical attention appears to be concerned more with normalizing interventions to address social consequences rather than threats to health these conditions may pose.

References

Bourdieu, P. (1990). *Logic of Practice*. Translated by Richard Nice, Stanford, CA: Stanford University Press.

Chase, C. and Coventry, M. eds. (1997). "Intersex Awakening," Special issue, *Chrysalis: The Journal of Transgressive Identities*, 2(5), 1–56.

Cohen, S. and Cosgrove, C. (2009). *Normal at Any Cost: Tall Girls, Short Boys, and the Medical Industry's Quest to Manipulate Height*, New York: Penguin.

Colapinto, John. (1997). "The True Story of John/Joan." *Rolling Stone* December 11: 54–97.

Conrad, P. (2007). *The Medicalizations of Society: On the Transformation of Human Conditions into Treatable Disorders*, Baltimore: The Johns Hopkins University Press.

Downing, L., Morland, I., and Sullivan, N. (2015). *Fuckology: Critical Essays on John Money's Diagnostic Concepts*, Chicago, IL: University of Chicago Press, pp. 19–40.

Dreger, A. D. (1998). *Hermaphrodites and the Medical Invention of Sex*, Cambridge, MA: Harvard University Press.

Eder, S. (2012). "From 'Following the Push of Nature' to 'Restoring One's Proper Sex'—Cortisone and Sex at Johns Hopkins's Pediatric Endocrinology Clinic," *Endeavor*, 36(2): 69–76.

Fausto-Sterling, A. (2000). *Sexing the Body: Gender Politics and the Construction of Sexuality*, New York: Basic Books.

Foucault, M. (1980 [1978]). *Herculine Barbin (Being the Recently Discovered Memoirs of a Nineteenth Century French Hermaphrodite)*. Translated by Richard McDougall, New York: Vintage.

———. (1994 [1963]). *The Birth of the Clinic: An Archaeology of Medical Perception*. Translated by A.M. Sheridan Smith, New York: Vintage.

German Ethics Council. (2012). "Intersexualität: Stellungnahme." *Deutscher Ethikrat* February 23. http://www.ethikrat.org/dateien/pdf/stellungnahme-intersexualitaet.pdf, accessed March 28, 2013.

Hughes, I. A., Houk, C., Ahmed, S. F., Lee, P. A., and Society, L. W. P. E. (2006). "Consensus Statement of the Management of Intersex Disorders," *Archives of Disease in Childhood*, 91: 554–563.

Lloyd, G. E. R., ed. (1983). *Hippocratic Writings*. Translated by J. Chadwick. New York: Penguin.

Money, J., Hampson, J. G. and Hampson, J. L. (1955). "Hermaphroditism: Recommendations Concerning Assignment of Sex, Change of Sex, and Psychologic Management," *Bulletin of the Johns Hopkins Hospital*, 97(4): 284–300.

Parens, E. (2011). "On Good and Bad Forms of Medicalization," *Bioethics*, 27(1): 28–35.

Plato. (1989). *Symposium*. Translated by Alexander Nehamas and Paul Woodruff, Indianapolis: Hackett.

Sandberg, D. E., Berenbaum, S. A., Cuttler, L., Kogan, B. and Lee, P. A. (2004). "Intersexuality: A Survey of Clinical Practice," *Pediatric Research*, 55(4): abstract 869.

Streuli, J. C., Vayena, E., Cavicchia-Balmer, Y. and Huber, J. (2013). "Shaping Parents: Impact of Contrasting Professional Counseling on Parents' Decision Making for Children with Disorders of Sex Development," *Journal of Sexual Medicine*, 10: 1953–1960.

Swiss National Advisory Commission on Biomedical Ethics. (2012). *On the Management of Differences of Sex Development: Ethical Issues Relating to "Intersexuality."* Opinion No. 20/2012. Berne, Switzerland: Swiss National Advisory Commission on Biomedical Ethics.

United Nations Office of the High Commissioner for Human Rights. (2015). *Discrimination and Violence Against Individuals Based on their Sexual Orientation and Gender Identity*, Geneva, Switzerland: United Nations Human Rights Council.

Further Reading

Canguilhem, G. (1991 [1966]). *The Normal and the Pathological*. Translated by C. R. Fawcett with R. S. Cohen, New York: Zone Books. (Among the most influential texts in shaping philosophers' critical perspectives on the division between normal and abnormal.)

Hacking, I. (1986). "Making Up People," *Reconstructing Individualism: Autonomy, Individuality, and the Self in Western Thought*, eds. T. C. Heller, M. Sosna, and D. E. Wellbery with A. I. Davidson, A. Swidler, and I. Watt, Stanford, CA: Stanford University Press, 222–236.

Parens, E. ed. (2006). *Surgically Shaping Children: Technology, Ethics, and the Pursuit of Normality*, Baltimore: Johns Hopkins University Press. (This collection, which looks at the cases of limb-lengthening for achondroplasia, surgical repair for craniofacial anomalies, and normalizing surgeries for atypical sex anatomies includes essays by philosophers, physicians, and individuals affected by these conditions.)

Part VI
PERSPECTIVES

40
THE BIOMEDICAL MODEL AND THE BIOPSYCHOSOCIAL MODEL IN MEDICINE

Fred Gifford

Introduction

The "biomedical model" (BMM) and the "biopsychosocial model" (BPSM) are contrasting conceptual models or frameworks that organize our thoughts, knowledge, and experience in medicine and clinical practice. The BMM has been the dominant model, understanding disease and clinical knowledge and practice in terms of biomedical science, and the BPSM was put forward, especially in the writings of George Engel (Engel 1977, 1981), as a criticism and corrective, urging more serious inclusion of psychological and social factors into our thinking. This chapter discusses both models, with emphasis on the latter.

The BPSM has had substantial resonance and impact on discussion concerning medicine and health care. Yet proponents often bemoan the fact that this shift in thinking has not been more robust, leaving the BMM too dominant (Smith et al. 2013). Coming to clear consensus about the reasonableness of the proposed shift to the BPSM and what is at stake is complicated both by these models having several components, and by ambiguity or uncertainty about just what these models are and what is involved in their adoption. Thus care must be taken in characterizing these models.

The BMM and BPSM are often said to be "models of *disease*," structuring how we think about disease, primarily in the sense of what the *causes* of a disease are, but also concerning whether the disease itself is seen or defined in biomedical or biopsychosocial terms. The BMM sees disease in terms of underlying biochemical or physiological alteration. In contrast, the BPSM sees disease as resulting from the interaction of these biomedical factors with psychological and social factors. The psychological factors brought into play include occurrent psychological features such as emotions, anxiety levels, and behaviors, as well as more permanent features of personality and lifestyle. The social factors can include personal and family relationships and support systems, cultural views, community, environment, and socio-economic status. (This dovetails with recent important work concerning the "social determinants of health" [Marmot and Wilkinson 2005].)

One point of ambiguity concerns what these models are models *of*: the models are referred to sometimes as models of *disease* and other times as models of *illness*. Just how one distinguishes disease and illness will complicate the analysis here, but typically the latter is seen as including the patient's subjective experience and is also more likely to involve its evaluative assessment. (For example, a person may have a disease process going on in one, but one that is not (yet) noticed; this would not constitute the person being *ill*.) Whether we mean a model of disease or of illness, and how we define these two, will affect the meaning and significance of the claim that there are psychosocial factors as causes.

Other questions that arise concerning the meaning and implications of these models are the following. The BMM's focus on the biomedical causes of disease and the question of whether higher-level causes are taken seriously raise certain broad philosophical questions. For example, the BMM is likely to be viewed as reductionistic, and so raises questions about reductionism, holism, and emergent properties. Is our medical science and our knowledge of medical practice ultimately reducible to or best understood in terms of underlying physical and chemical sciences? Or does an understanding exclusively in terms of such lower-level sciences provide a distorted picture? Should we endorse a "holism," positing laws, explanations, entities, or properties that are not reducible to or explainable by reference to those of the lower-level sciences? In addition, the BMM, focusing on physiology and biochemistry, might be seen as the more "scientific" view; this raises conceptual and evaluative questions about what counts as science, and as good science.

Central to the importance of this topic is, of course, the fact that which model one embraces has an impact on a number of matters of practice: diagnosis, explanation, prognosis, treatment, prevention, patient management more generally, and consequently on research about all of these things. The BPSM proponents argue that "concentration on the biomedical and exclusion of the psychosocial distorts perspectives and even interferes with patient care" (Engel 1977: 131) and that the BPSM encourages a more nuanced and perhaps more humanistic approach to the doctor-patient encounter, information gathering, and communication. Specifically, the two models suggest different implications for the data one needs to collect in order to make decisions about diagnosis and patient management. From the BMM perspective, lab tests etc. are seen as the most objective, as probing deeper into the medical reality, and thus as providing the most reliable and useful information. The BPSM perspective emphasizes more the reports from patients, including reports of subjective experiences.

Note that clinicians need to explain not just the presence or absence of disease but also such things as susceptibility to the disease, presentation of a disease in an individual patient, and severity and course of the illness. The BPSM-trained physician will also seek psychological and social explanations of certain matters that are themselves psychosocial, such as the patient's "adoption of the sick role" or visit to the physician in the first place (and the timing of these things). These will help in interpreting patients' statements, in judging how much of the patient's story has been told, and in encouraging patients to provide information that will be useful for diagnosis and management.

For the BPSM, "(t)he most essential skills of the physician involve the ability to elicit accurately and then analyze correctly the patient's verbal account of his illness experience" (Engel 1977: 132), whereas the BMM "encourages bypassing the patient's verbal account by placing greater reliance on technical procedures and laboratory measurements" (Engel 1977: 132). The BPSM will also emphasize that we should become more conscious of and indeed study scientifically the very doctor-patient encounter process itself, with the end of making this process of diagnosis and patient care more effective.

It is worth noting that the two different models also have different implications for areas of research to be pursued more generally. Areas that the BPSM would promote include research on psychosomatic illness, psychoneuroimmunology, the placebo effect, health psychology, and the interconnection with research on the social determinants of health.

Origins of the Biomedical Model

The BMM has its origins in the success of reductionistic medicine, especially in the 19th century, including the germ theory. The resulting approach to medical science, focusing on underlying biological processes, led to dramatic success in theoretical understanding and in the ability both to detect and treat disease and illness and to improve health and life expectancy. This provided confidence in the view that proper scientific understanding and effective cures will require going to lower and lower (more mechanical) levels of explanation. It was encouraged further by reforms in medical education prompted by the Flexner Report in 1910, the emphasis of which was the importance of strengthening the scientific basis of medical training.

The biomedical model generalizes from this and "assumes disease to be fully accounted for by deviations from the norm of measureable biological (somatic) variables" (Engel 1977: 130). The hope is that biomedicine, as the appropriate "scientific" approach, will eventually get to the root of and resolve all of our health problems.

George Engel and the Biopsychosocial Model

Over the years there have often been critiques of contemporary medicine as being too focused on biological or reductive science, with too little focus on the doctor-patient relationship. These ideas were given an especially important articulation in a series of articles in the 1970s and 1980s by George Engel (1913–1999), professor of psychiatry and medicine at the University of Rochester (Engel 1977, 1981). Typically, people have this set of influential writings and discussions in mind when they discuss the BPSM in the medical literature, and this set of ideas is the focus of this chapter.

But similar ideas are discussed in the broader literature, often without reference to this specific work, not necessarily conceiving of things just as Engel does. After all, critiques have been made of a number of trends in modern medicine and medical practice (and by critics with several different perspectives): the failure to understand and take seriously the responses and concerns of patients, insufficient attention to the humanistic side of clinical care, inflexibility concerning alternative medicine practices, the unreflective overuse of technology and lab tests, the use of high-tech and invasive technologies, failure to give sufficient thought to quality of life, paternalistic practice and attitudes, and even the profit motive. (Engel does note, for example, the complaint that "physicians are lacking in interest and understanding, are preoccupied with procedures, and are insensitive to the personal problems of patients and their families," as well as that they are seen as "cold and impersonal" [Engel 1977: 134].) Thus the critiques of modern, scientific medicine have a variety of aspects, much but perhaps not all captured by Engel's discussion.

Again, the value of the psychosocial aspects has been recognized over the years, and there have been changes in both education and practice. There is nevertheless frustration over the limited nature of these changes.

Initial discussion of the BPSM arose in part from a crisis in psychiatry in the 1960s about whether it is in fact on firm scientific footing, and concerning whether psychiatry should attempt to ensure its scientific status by dealing exclusively with diseases rooted in defects of the brain, or whether it should instead admit that it is something distinct from medicine *per se* (Engel 1977: 129–130). But Engel was quite explicit that medicine as a whole was in crisis. For him, the tensions in psychiatry were not to be resolved by deciding between molding it to the medical model or allowing it to go its own way. Rather, the BMM was "no longer adequate for the scientific tasks and social responsibilities of either medicine or psychiatry" (Engel 1977: 129). All of medicine should be re-conceptualized and take more seriously the psychosocial factors at play. (Although much discussion continues to occur about the BPSM within psychiatry, the present chapter focuses on the case of medicine generally.)

Engel challenges the sharp line between psychiatry and the rest of medicine. In a discussion comparing diabetes and schizophrenia, he emphasizes that we should not think of the one as a somatic disease and the other as a mental disease; both sets of factors permeate both. Concerning diabetes, he notes that

> while the diagnosis of diabetes is first suggested by certain core clinical manifestations, for example, polyuria, polydipsia, polyphagia and weight loss, and is then confirmed by laboratory documentation of relative insulin insufficiency, how these are experienced and how they are reported by any one individual, and how they affect him, all require consideration of psychological, social and cultural factors, not to mention other concurrent or complicating biological factors.
>
> (Engel 1977: 131–132)

A question arises of whether one might understand the above claim (with its emphasis on how things are *experienced*) as really being about *illnesses* rather than *diseases*. Perhaps disease could remain within the BMM and illness be the domain of the BPSM. One response to this Engel would give is that proper medical practice simply does require attention not just to disease but also to illness. No one should practice medicine by focusing only on disease and not illness.

Perhaps the BMM is appropriate for medical *science* rather than medical *practice*, and the BPSM could reign in the latter realm. (It can be noted that medicine is not itself "a science" (which does not mean that it isn't done scientifically), but instead a "practice," one that appropriately *makes use of* various sciences, like biochemistry, virology, and neurophysiology [Munson 1981].) But how well can science and practice be separated? Medical research really has to be clinical (i.e., tied to practice). If attention to psychosocial causal factors is necessary for clinical practice, then it is crucial for the research that aims to aid that practice.

In any case, psychosocial factors play crucial roles not just in the *experience* of illness—as if this could be set aside as not part of medicine *per se*. They act as causes of disease, its course, and its severity.

What Sort of Things Are These Models?

Another source of ambiguity concerns the kind of thing the BMM and (especially) the BPSM are. They are usually called "models," but sometimes "paradigms," sometimes just "approaches" or even understandings of clinical practice. The terms "model" and "paradigm" have particular technical meanings that some might say are not met by the BMM and BPSM; this can be misleading and be a source of criticism. It does seem correct to say that the BPSM is *not* to be conceived as an attempt to be a scientific model to be tested and confirmed, though one could try to gain evidence about whether its adoption leads to better outcomes. It is also worth noting that the BMM and BPSM are ideal types, and that what approach a physician takes can be a matter of degree.

Engel describes the BPSM as providing "a blueprint for research, a framework for teaching, and a design for action in the real world of health care" (Engel 1977: 135). A different and broader sense of biopsychosocial could emphasize the interdisciplinarity of a given *field of study*. For instance: "Gerontology is an interdisciplinary field of study of the elderly, using a biopsychosocial approach to study aging from the perspectives of biology, medicine, nursing, social work, the social sciences and humanities" (McCullough 2004).

Epstein and Borrell-Carrio (2005) argue that while others have sometimes supposed the BPSM to be "an empirically verifiable theory, a coherent philosophy, or a clinical method," in fact it should instead be viewed simply as "a vision and an approach to practice." Then, when getting down to their recommendations, what these authors emphasize is a number of habits of mind that need to be inculcated. These include awareness of context, attentiveness, peripheral vision, curiosity, and informed flexibility. Epstein and Borrell-Carrio take themselves to be proponents of the BPSM tradition, but they are endeavoring to place things more concretely in a clinical context, and to avoid the implication that there is "an expectation that clinicians explore each level of the biopsychosocial hierarchy in each moment of each encounter" (430).

Scientific Basis and General Systems Theory

As suggested above, the BMM may appear to be the more "scientific" model, based in "hard sciences" such as biochemistry and physiology, and with more focus on empirically objective data. Yet the BPSM, particularly as articulated and developed by Engel, emphasizes its own scientific credentials. For one thing, by including these additional variables, the BPSM provides a more complete and accurate picture and thus improved understanding. Further, as mentioned earlier, the BPSM aims to *expand* the targets of scientific investigation to include the doctor-patient relationship.

It might be objected that this broadening of the scope of investigation cannot plausibly be said to involve a broader *scientific* project, on the grounds that the psychosocial factors at issue are themselves not adequately susceptible to such a rigorous or scientific approach. Perhaps the clinical encounter is more artful than scientific, requiring a humanistic element and the reliance on "clinical judgment," something we cannot expect to be analyzed in terms of algorithms or science. Some BPSM proponents might endorse this as part of what this new model brings. The term "humanistic" is indeed often connected with BPSM, and the idea of an unanalyzable clinical judgment does come to mind in a context of integrating complex information from human interactions. But Engel's central response to this dismissal of the BPSM's scientific project would be that this comes from too narrow a view of what it is to study things scientifically. It is true that inclusion of the psychosocial factors makes things more complex, and the regularities concerning them will be less universal and more "patchy." But this does not mean that they cannot be usefully or scientifically addressed, or that we are left simply with a humanistic intuition. After all, similar things can be said of biomedical science relative to physics and chemistry, and yet this gives us no reason to conclude that biomedical science is not scientific (Schaffner 1993: chapter 3). Of course, this shows that there is no "in principle" argument against adopting the BPSM along these lines; it remains possible to argue that, for pragmatic reasons, this extra complication is not warranted.

But there is another way in which Engel stresses that the BPSM is scientific, namely by placing it within the framework of "general systems theory" (von Bertalanffy 1968), thus applying to medicine a theoretical development from biological science more generally. General systems theory highlights the fact that nature is organized into a hierarchy of levels, and emphasizes the importance of seeing both the way in which these levels are relatively autonomous and how they are interconnected. This perspective lets us represent and study both the component parts and the more complex organized wholes, as well as their interactions.

The following provides the hierarchy or system of organizational levels, organized from highest (or most complex and encompassing) to lowest (Engel 1981: 105):

BIOSPHERE
SOCIETY-NATION

CULTURE-SUBCULTURE (age, sex, class, religion, education, economic, etc.)
COMMUNITY (health care, work, neighborhood, social, recreational, etc.)
FAMILY (nuclear, extended)
TWO-PERSON (doctor, family member, coworker, friend, etc.)
PERSON (experience and behavior)
NERVOUS SYSTEM
ORGANS/ORGAN SYSTEMS
ORGANELLE
CELL
TISSUE
MOLECULE

Levels higher up are composed of those lower down. The BMM uses only the lower half of the continuum, from the person (or individual organism) down through its component parts. This reductive stance of the BMM emphasizes that these lower levels adequately explain that which occurs at the higher levels, that this is what proper explanations consist in, and that this is where we should focus our attention. The BPSM—in line with the general systems theory approach—takes seriously that the whole range of levels in the hierarchy is at least potentially relevant to explanations of something at any particular level, and it emphasizes that there is interaction; features at one level help explain the higher levels, but also higher levels sometimes explain things at lower levels (as with biological function and contribution to goals). As a result, understanding and effective practice require that one pay attention to, understand, and explore each of the levels.

Note that this view does not require endorsing holism—nor does Engel endorse this, worrying that it too can lead to dogma. In particular, it doesn't require rejecting an ontological or metaphysical thesis about reductionism—that the structure of the world really is such that the higher levels are fully constituted by and can be given an explanation in terms of the lower levels. For instance, one could hold that in principle there could exist an account, in terms of underlying biology, of psychiatric disorders as well as of psychological precursors to (somatic) diseases. But even if so, this would not change what is important for clinical practice or even genuine scientific understanding. Explanation in terms of lower levels of organization can sometimes be enlightening, but it is not always better: sometimes it will not be possible, sometimes it will be too complicated to be intelligible, sometimes it will not lead to useful questions to explore, and sometimes it will distract us.

Myocardial Infarction Case

Engel illustrates a number of themes with an extended discussion of a relatively ordinary medical case involving a patient with a myocardial infarction (MI), illustrating psychosocial factors that physicians need to track and understand, as well as the ways in which knowledge about such things as the patient's personality traits is crucial in interpreting what he says about his cardiac pain.

Here is a very brief description of the case (Engel 1981: 107 ff.): a 55-year-old man is brought to the ER with symptoms similar to those he had six months prior when he had an MI. We learn about the present event that the man is in denial about having a heart attack as he first experiences the symptoms, that an employer has persuaded him to allow her to take him to the hospital, and that his behavior and decisions are affected by a number of worries, including concerning his family, related to his felt need to be in control.

As Engel notes, "(t)he information that the patient resisted acknowledging illness and had to be persuaded to seek medical attention, especially in the face of a documented heart attack six months earlier, reveals something of this man's psychological style and conflicts" (Engel 1981: 108). Subsequent features of the case illustrate how this is important: after an apparent recovery, the patient goes into ventricular fibrillation. Engel makes the case that this is not just part of the natural course of what was going to happen (even if he had not gone to the hospital—with the implication that he was very lucky to have gotten to the hospital), but instead resulted from his losing confidence in the staff and his control over the situation when, after staff members have difficulty carrying out an arterial puncture, they run out to look for help without communicating with him.

Engel describes the course of the case in detail, using the biopsychosocial hierarchy to analyze the causal processes and the reasoning behind diagnosis and patient management: He presents a series of diagrams placing, at various levels in the hierarchy, the events and processes that were causal or that provide information from which to make (clinical) judgments about diagnosis, prognosis, and patient management. (The central nervous system level plays a key role, mediating several of the processes.)

The information about the patient's concerns and psychological dispositions potentially relevant to the stability of his cardiovascular system—and in particular the *meaning* for the patient of the illness or symptoms—is shown to serve both proper understanding and patient prognosis and management. "Alert to the patient's reluctance to submit to medical care, the physician would carefully monitor the patient's reactions to the coronary care procedures" (Engel 1981: 120). Again, this has implications for what sorts of things doctors need to have knowledge of, notice, and explore—and this has obvious implications for training.

Biomedical Model Responses to the Biopsychosocial Model

One can hardly deny that there are psychosocial as well as biological aspects to the causation of disease and illness. Nonetheless, it is worth considering reasons for reluctance to give up the BMM. First, one might feel that, even if its description of the causal processes is incomplete, the BMM works well enough, because the further factors addressed by the BPSM have a relatively small impact, and that therefore, given the BMM's advantages in terms of simplicity and scientific rigor, it is legitimate to leave these further details aside. Possibly this is arguable in some subfields or contexts more than others: for instance, the BMM may work well enough in diagnosis in oncology, but not in the case of diagnosis and patient management in the chronic illness that is increasingly part of health care.

Another line of thought raises doubts about how significant the new insights from the BPSM are, seeing it as mostly just offering advice about strategies in relation to the doctor-patient relationship. On this view, what's valuable about the BPSM is not really about medical reasoning *per se*, and thus does not justify a whole new model or a major shift in thinking. Now, it is true that important parts of the motivation and argument for the BPSM are related to the doctor-patient relationship. For instance, it is recommended for its role in ensuring better communication with patients, both because the patient can then be a good source of diagnostic information and be more willing to cooperate, and also perhaps because doing so involves respectful and humane treatment of the patient (recalling the "humanistic" element). This, combined with the "habits of mind" suggestion above from Epstein and Borrell-Carrio, might suggest that the shift to the BPSM is centrally about stressing such things as empathy and compassion. These are valuable things, to be sure, but not, so this objection goes, so centrally about medicine or medical reasoning as to warrant a reconceptualization of our understanding of disease.

But it would not be a fair interpretation of the BPSM to see it as at root about things like compassion, empathy, and humaneness. For one thing, one could be compassionate and humane and yet fail to take up more specific tasks of examining carefully (indeed, scientifically) the psychosocial causes and evidence in the medical realm. These latter are basic to the BPSM. One is not supposed to fill in the gaps by "being nice," but rather to take seriously that there is much to be gained (in accurate diagnosis, in patient management and outcome) if clinicians (and clinical researchers) pay genuine attention to the psychosocial aspects of the situation.

A related critique might emphasize as the real value of the BPSM its ability, through improved communication and trust, to contribute to improved patient compliance, and this again might be seen as peripheral advice about the doctor-patient encounter, but not really something about medical reasoning. But here what should be pointed out is that, while such "compliance" (or better, "adherence") is in fact noted as a benefit of the BPSM, this is not something peripheral but instead a crucial part of good medical practice. We want to carry out medical practice in a way that is effective toward medicine's goals (such as improved health outcomes). The specific regimens and actions chosen, and just how they are used, are crucial. Medical research may have developed a drug with the right chemical content to be known to be efficacious toward a certain health outcome, but matters such as the dosage and pharmacokinetics can make a difference to its effectiveness (and safety). Further, a drug regimen that requires patients to take pills fewer times a day may well result in patients being better able to adhere to the regimen, and this may result in better health outcomes. Similarly, strategies of patient communication that enhance diagnostic accuracy and adherence to treatment regimens can be seen to be part of scientific medical practice.

The physician's decision to choose one or the other of these strategies (perhaps based on the knowledge of her particular patient), and the effort put into research and development to generate a knowledge base about these alternatives, take into account psychological features of the process in the way recommended by the BPSM, and are rational in relation to the goals of attaining better patient outcomes. They are part of good medicine, and they can and should be taken into account by practitioners and examined scientifically by researchers.

Some further components to the debate can be seen by examining a recent defense of the BPSM. Smith et al. (2013) review concerns they see as explaining the BPSM's limited uptake. One critique they discuss is that the BPSM is not a scientific model on grounds of its being too vague and not operationalizable. This critique alleges that the BPSM does not provide us with helpful advice, but, quite to the contrary, opens up the whole psychosocial realm as potentially relevant, leaving us completely unfocused, with the impossible task of looking at all information.

One general response could be that this critique incorrectly assumes that we must look at all psychosocial data indiscriminately, as if we have no theory or background knowledge to guide us in selection (rather as in the case of "narrow inductivism," a naïve attempt to be objective by eschewing all theory and assumptions criticized by Hempel in Hempel 1966). But such a methodology does not follow from the BPSM. Especially as we do the research about the impact of various components of the doctor-patient interaction that the BPSM recommends, we will have a great deal of background knowledge that will aid us in this endeavor.

But while this response avoids the in-principle problem here, there remains an important challenge to specify just what advice the BPSM can provide. To this, Smith et al. (2013) provide a more concrete response, taking up the challenge to "operationalize" the model. They describe work done in creating and evaluating what has been called the "patient-centered interview." Methods are specified for aiding practitioners both in gathering information and in handling emotion. These patient-centered interview techniques are grounded in assumptions about how psychosocial factors have substantial influence on the disease and illness of the patients. Smith et al. cite studies demonstrating the effectiveness of these interview methods.

If these studies prove correct, then this can show one way to explore whether and in what ways the BPSM might be further or more explicitly integrated into the science, practice, and teaching of medicine. Leaving aside whether their specific proposals are adequate, this line of thought shows promise for making progress on this question of the operationalization, testing, and evaluation of aspects of the BPSM, and for following the BPSM in a concrete and scientifically grounded way. It is a worthy goal to try to gain evidence about the impact of various doctor-patient interaction techniques. The BMM and BPSM, and their comparison, can serve as guides for how to carry out this research, and this research can provide evidence of the utility of these models in different contexts.

Conclusion

The "biomedical model" and the "biopsychosocial model" offer different ways of organizing our understanding and experience in medicine and clinical practice. They differ in their conceptions of the causation of disease and illness, in their views about reduction and the relation of different sciences (such as how psychological and social causal processes connect with "lower-level" processes), and in how they understand the scope of science. And they provide different recommendations concerning such practical actions as information gathering, diagnosis, and patient management in clinical practice.

The BPSM was put forth as a response and corrective to the BMM, on the grounds both that it would more accurately capture medical reality and that it would improve patient care in terms of both effectiveness and ethical treatment of patients. On its face, the BPSM's insistence on the relevance of the psychological and social causal factors is simply correct, and the ways the BPSM directs our thinking in clinical practice count as further reasons to adopt it. As noted, it has indeed been very influential. Yet proponents complain that the BMM remains entrenched in clinical practice, training, and research, that clinicians still give too little thought to psychosocial factors and view the *real* work to be at the biomedical level.

Such incomplete adoption might be explained by reference to unreflective inertia in medical training, involving a prejudice in favor of hard sciences as more prestigious or more scientific, along with a technical conception of medicine. Still, the reluctance and incomplete adoption might also be seen as understandable, in light of some of the considerations discussed above—the claim that the BMM could sometimes be a useful heuristic, and the fact that there is a certain amount of uncertainty about just what is entailed by adopting the new model, as well as precisely what advice it provides about specific practices. Hence, there is value to further thought about these things, as well as empirical research concerning the specific impacts of taking such particular concrete steps. Included here would be the proposals about the "patient-centered interview" referred to above by Smith et al. (2013). This is in line with Engel's general suggestion that we study the processes involved in the doctor-patient relationship and their impacts.

References

Engel, G. (1977) "The need for a new medical model: A challenge for biomedicine," *Science* 196: 129–136.
Engel, G. (1981) "The clinical application of the biopsychosocial model," *Journal of Medicine and Philosophy* 6: 101–123.
Epstein, R. and Borrell-Carrio, F. (2005) "The biopsychosocial model: Exploring six impossible things," *Families, Systems, & Health* 23(4): 426–431.

Hempel, C. (1966) *The Philosophy of Natural Science*, Princeton: Princeton University Press.
Marmot, M. and Wilkinson, R., eds. (2005) *Social Determinants of Health*, 2nd ed., New York: Oxford University Press.
McCullough, L. (2004) "Geroethics", in G. Khushf (ed.), *Handbook of Bioethics* 507–523. The Dordrecht, Netherlands: Kluwer Academic Publishers.
Munson, R. (1981) "Why medicine cannot be a science," *Journal of Medicine and Philosophy* 6(2): 183–208.
Schaffner, K. F. (1993) *Discovery and Explanation in Biology and Medicine*, Chicago: University of Chicago Press.
Smith, R. C., Fortin, A. H., Dwamena, F., & Frankel, R. M. (2013) "An evidence-based patient-centered method makes the biopsychosocial model scientific," *Patient Education and Counseling* 91: 265–270.
von Bertalanffy, L. (1968) *General System Theory: Foundations, Development, Application*, Revised ed., New York: George Braziller.

Further Reading

Engel, George L. (1978) "The biopsychosocial model and the education of health professionals," *Annals of the New York Academy of Science* 310: 169–187. (An early article discussing the question of education of health care professionals)
Ghaemi, S. Nassir (2012) *The Rise and Fall of the Biopsychosocial Model: Reconciling Art and Science in Psychiatry*, Baltimore: Johns Hopkins University Press. (A recent book-length analysis, focusing on the BPSM as a model within psychiatry, and providing a critique)
White, Peter, ed. (2005) *Biopsychosocial Medicine: An Integrated Approach to Understanding Illness*, Oxford: Oxford University Press. (An edited volume drawn from a conference in 2002, covering a variety of topics but focusing not on psychiatry but on chronic medical disease)

41
MODELS OF MENTAL ILLNESS

Jacqueline Sullivan

1. Introduction

Each and every one of us, at some point in our lives, will be touched by mental illness—our own, or that of a parent, grandparent, aunt, uncle, sibling, spouse, child, grandchild, friend, or coworker. Close to 450 million people worldwide suffer from mental or neurological disorders, and the numbers continue to rise (World Health Organization 2011). In order to address the current global mental health crisis, a conceptual-explanatory framework or model adequate for investigating the causes of, diagnosing, explaining, and treating mental illness is required. Yet, what kind of model will do? Should it include psychological factors like emotions, thoughts, and memories, or social factors like income or living conditions? Would a model that understood mental illness as exclusively brain-based be sufficient? Will a single model prove adequate for understanding mental illness or are different models necessary? Providing some preliminary answers to these questions is the aim of this chapter.

2. Models of Mental Illness in the Clinic

Many different conceptual-explanatory frameworks of mental illness have been put forward in the scientific and philosophical literature (e.g., Freudian psychoanalytic theory, psychodynamic theory, cognitive-behavioral theory)—too many in fact to consider in a single chapter. However, there are three even more general frameworks that we may evaluate and in so doing learn lessons relevant for assessing the pros and cons of other models. These include (1) the folk psychological model, (2) the biopsychosocial model, and (3) the medical model. How these three models differ from each other and their merits and failings in clinical contexts may best be illustrated by means of an example.

Rebecca, a 19-year-old woman who has just entered her sophomore year of college, begins midway through the semester to miss classes, mealtimes in the cafeteria, and social events with her close group of friends. One of her friends, Tom, begins to worry and invites Rebecca out for a coffee to see how she is doing. A few minutes into their conversation, he asks her if everything is okay. In response, Rebecca becomes teary-eyed but does not say anything. Concerned, Tom gently asks her some additional questions: What are you sad about? Did something happen over the summer? Is everything all right at home? Rebecca gradually reveals to Tom that she has felt bad since the beginning of the semester, only wanting to sleep, and not really wanting to eat, but that she cannot point to anything specific that prompted this change in mood. Tom finds this answer strange, particularly because he has known Rebecca to be a basically happy person who has only been sad in the past for legitimate reasons like failing an exam or ending a friendship or romantic relationship. While Tom is concerned, he realizes that Rebecca may not feel comfortable telling him what is wrong. So, he gently suggests that she should talk to a school counselor, who he thinks might be able to help.

Tom's response to Rebecca's sadness is not uncommon. When we see a friend whose behavior is unusual, and this behavior persists and is debilitating, we tend to assume that the sources of these behavioral changes are *psychological* or "in the head" and that the best approach to helping them is to try to talk to them or encourage them to see a professional. You may not have realized it, but you use a conceptual-explanatory framework on a daily basis. Specifically, you have learned from a very young age to believe that human beings, as well as some non-human animals have some special quality—a mind, consciousness, awareness—that other kinds of things—such as rocks stars and trees—lack. You regularly describe yourself as having beliefs, desires, feelings, and intentions. You comfortably ascribe similar internal states to other human beings and some non-human animals. You appeal to these states to explain your own behavior and to make sense of the behavior of others. When you are asked why you are sad, you often put forward reasons—beliefs, feelings, unrealized desires that you have—to explain why.

Because the vast majority of us are not professional psychologists, but rather, ordinary folk, psychologists and philosophers have come to refer to this conceptual-explanatory framework as "folk" or "commonsense" psychology (e.g., Churchland 1981). When we use this framework, we are assuming what philosopher Daniel Dennett (1987) has dubbed "the intentional stance." In other words, we posit abstract mental states rather than concrete physical states (e.g., changes in the nervous system) to explain human behavior. Sometimes the ontology that we appeal to is "mixed" insofar as we talk about both mental states and mental processes (e.g., attention, memory). Some of us may be inclined to import more advanced scientific concepts into our folk-psychological theorizing (e.g., Freud's id, ego, and superego) or to be "folk neuroscientists" insofar as we may explain behavior by appeal to a rudimentary understanding of the brain. In trying to understand Rebecca's sadness, Tom was being a folk psychologist and adopting the intentional stance. However, as Tom himself recognizes, this conceptual-explanatory framework does not offer any treatment options over and above talking to a person about their feelings or suggesting that they talk to someone who is a professional at understanding the mind and mental disorders.

Suppose Rebecca takes Tom's advice and makes an appointment with the university's Center for Counseling and Personal Growth. She begins to see a licensed mental health counselor (LMHC) with a PhD in psychology (PsyD) on a weekly basis. During these sessions, the counselor asks Rebecca questions about her feelings, her recent experiences, her life growing up, her friends and family, her schoolwork, her romantic relationships, her diet and exercise habits. Over the course of these sessions, the counselor attempts to formulate hypotheses about the causes of Rebecca's sadness and to make suggestions about strategies Rebecca might implement to improve her mood. The counselor suggests yoga, mindfulness and exercise classes, social activities such as going out with friends and joining clubs, and changes in diet to include more vitamin-rich foods. He recommends that she see a medical doctor to make certain her thyroid is functioning properly and her blood-sugar levels are normal.

Implicit in such suggestions is the counselor's appeal to an explanatory framework to explain Rebecca's behavior that differs from that of folk psychology but is inclusive of it. This model has been referred to as the "biopsychosocial model" (Engel 1977) (see Gifford, Chapter 40), because those health care professionals who use it consider a variety of different kinds of causal factors that may result in changes in human health and behavior, including biological/physical factors (e.g., diet and exercise, neurobiological and physiological changes), psychological factors (feelings and thoughts or thought patterns), and social factors (social relationships and activities). American psychiatrist George Engel (1977) put forward this framework in the late

1970s because he believed that the successful diagnosis and treatment of patients with disease or illness required doctors to appeal to a multidimensional causal model. He intended it as a superior alternative to the medical or biomedical model (described below and in Chapter 40), which he regarded as interested exclusively in biological causes.

As is illustrated in Rebecca's case, the counselor believes that psychological, biological, and social factors may all be contributing to her depressed mood. Furthermore, the kinds of causal interventions that he is proposing correspond to all three of these different types of causes. For example, the counselor regards talking to him to be insufficient for improving Rebecca's mood, and he acknowledges that the causes of her sadness may be complex and that the different causal factors he has identified may be acting independently or in concert with one another. He also appeals to this framework to identify a variety of different kinds of intervention strategies that he thinks may, either independently or in combination, make Rebecca feel better. It is clear to see how this conceptual-explanatory framework differs from the folk-psychological one in terms of both the causes it posits to explain Rebecca's sadness and the kinds of treatment strategies to which it points.

Suppose Rebecca continues to see the counselor for two months. Because she does not appear to be improving, the counselor thinks that something else might be wrong—that the causes of her sadness may include neurobiological and biochemical causes in addition to psychological and social causes. Without a license to prescribe drugs or other forms of treatment to Rebecca, he is unable to address directly these other kinds of causes. However, the university's counseling center has a contract with a psychiatrist who can. The counselor suggests that in addition to seeing him on a regular basis, Rebecca should see this psychiatrist. He obtains Rebecca's permission to forward his notes on her case to the psychiatrist, and Rebecca schedules an appointment with her.

The psychiatrist begins by asking Rebecca how she is feeling. They talk for 15 minutes, and the psychiatrist indicates that based on the counselor's notes and some of the answers Rebecca has provided to her questions (e.g., how is your mood these days? your memory? your energy level?), her hypothesis is that Rebecca has Major Depressive Disorder (APA 2013, 160–168). She removes the fifth edition of the *Diagnostic and Statistical Manual of Mental Disorders* (DSM-5) from her bookshelf, explains to Rebecca what the symptoms of the disorder are, and identifies the reasons why she thinks Rebecca satisfies the criteria for this diagnosis. She notes that Rebecca exhibits the requisite set of symptoms insofar as she has been depressed for over 2 weeks, has lost interest in daily activities, and is experiencing hypersomnia, fatigue, and an inability to think and concentrate. The psychiatrist claims that research suggests that the etiology of the disorder is complex, involving genetic, developmental, and environmental factors. In other words, a complex combination of causal factors interacting throughout Rebecca's lifetime (and even before, if we consider genetic mechanisms) likely contributed to Rebecca exhibiting the symptoms of Major Depressive Disorder. The psychiatrist explains that although it is impossible to intervene effectively in this complex array of historical causes, it is possible to intervene in those causes potentially and currently contributing to Rebecca's depressed mood. One such set of causes, she claims, is neurobiological, and she identifies several drug therapies that have proved successful in intervening at the neurobiological level to treat depression. The psychiatrist tells Rebecca that counseling is an additional effective intervention for addressing past traumas and coping with day-to-day life events.

Rebecca heeds the psychiatrist's advice. She begins taking an antidepressant medication. She continues to see both the counselor and the psychiatrist during the next month. After taking the drug regularly for several weeks, Rebecca's mood begins to improve, and she finds she has more energy and more interest in her studies and spending time with her friends. Her life gradually begins to return to normal.

In contrast to Tom and the counselor, the psychiatrist adopted what we will here refer to as a *sophisticated* "medical or biomedical model" of mental illness (see Chapter 16, Chapter 40). This is to be contrasted with a "strong interpretation" of the medical model (Murphy 2009) that might understand mental illness as caused primarily by brain abnormalities (e.g., imbalance in neurotransmitters, presence of lesion, abnormal neural connectivity) and treatable exclusively by intervening in the brain or nervous system (via, e.g., surgery, electroconvulsive therapy, pharmacology). Engel introduced the biopsychosocial model in the late 1970s because he was concerned that this strong interpretation of the medical model of illness had become dominant in medicine and was not only problematic for the reasons mentioned above (see Chapter 40) but also because it led to a failure on the part of medical practitioners to treat their patients as psychological and social beings in addition to biological beings. Engel thought this dehumanized patients in ways that were antithetical to the goal of promoting their health. The biopsychosocial model was intended to encourage doctors to engage with their patients as psychological and social beings and to discover additional avenues for treatment that were not suggested by the strong interpretation of the medical model. The psychiatrist's use of a "sophisticated medical model" in Rebecca's case may be considered as a modern answer to Engel's (1977) plea for a better conceptual-explanatory framework for clinical medicine.

For the sake of highlighting the advantages of the sophisticated medical model of mental illness over that of the strong interpretation of the medical model, let's imagine an alternative scenario in which Rebecca goes *only* to see a psychiatrist who upholds the latter model. This psychiatrist assumes that the approach used to diagnose somatic illnesses is equally as effective for diagnosing and treating mental illness (Black 2005). He thus takes Rebecca's medical history but does not ask her to go into any specific details about her family history (apart from medical history), recent life experiences, or interpersonal relationships. This is because the psychiatrist believes that non-biological causes (e.g., social or psychological causes) are not relevant for diagnosing and treating mental illness. Rather, he is committed to the idea that mental disorders like depression are brain-based and that "an increased understanding of the physiology of the brain will eventually improve the care of patients with mental illness" (Black 2005, 5). Apart from the drug prescription he gives to Rebecca, a brief itemization of its side effects, and the request for a follow-up visit, the psychiatrist suggests no other modes of treatment (e.g., counseling, other lifestyle changes).

The limitations of the strong interpretation of the medical model in the clinical context are to some extent revealed by differences in the kind of care Rebecca receives from the two psychiatrists. In the first case, the psychiatrist presented Rebecca with a complex explanatory framework in which to understand her depression and how to restore her health, whereas the second psychiatrist provided a single treatment option. Given that a variety of different hypotheses exist to explain the causes of Major Depressive Disorder, and given that there are different available treatments, these differences in the two approaches make sense. Although it is an empirical question whether any clinical psychiatrists today endorse a strong version of the medical model of mental illness, there are good grounds for thinking it has few proponents. For example, psychiatrist Kevin Black (2005, 8) claims that the medical model "regards psychotherapy" as "an important part of psychiatric practice" and that few psychiatrists today "condone uncritical acceptance of pharmacological treatments, or uncritical rejection of psychotherapy."

Before we move on to consider conceptual-explanatory frameworks of mental illness operative in research contexts and their respective usefulness for guiding mental health research, let's consider some additional implications of the first three conceptual-explanatory frameworks that we have considered.

Consider first the folk-psychological model of mental illness. Although it is common to think of mental disorders as exclusively disorders of the mind, folk psychology leaves it unclear what the mind is and how it fits into the physical world. In fact, it is consistent with a philosophical view introduced by French philosopher René Descartes known as *mind-body dualism*. Descartes believed that the mind and body are separate substances; whereas minds are thinking things that are immaterial, bodies are material things that are extended in space and time. Mind-body dualism essentially jettisons the mind and, thus, disorders of the mind from the physical world. Yet insofar as science can only investigate phenomena that are observable, mind-body dualism and, thus, folk psychology seem to place minds and mental disorders beyond the realm of scientific understanding.

Yet, if the mind cannot be understood by science, a lot of negative consequences result. First, persons with mental illness will be considered beyond the hope of science insofar as their illnesses cannot be investigated or cured in the same way that somatic illnesses are. Second, if we think science is irrelevant for understanding and treating mental illness, we may be inclined to believe that persons with mental illness can simply get better on their own and overcome their illnesses with or without the help of their families and friends. This unfairly places the responsibility of having mental illnesses and the burden of overcoming them on those who have them. Yet, in many cases, talking to another person about one's mental states or experiences or trying to overcome mental illness on one's own are not viable strategies for getting better. We see this clearly in Rebecca's case, and there are many other cases. For example, persons diagnosed with schizophrenia, who suffer from hallucinations and delusions, cannot get rid of these experiences simply by talking about them or willing them away. Although talking about their experiences and having others understand what they are going through provides some relief, pharmacological interventions (e.g., antipsychotics or neuroleptics) are often required to alleviate some of their symptoms.

Notice that the medical model (both the sophisticated and strong interpretations) has a positive feature insofar as it eliminates some of the stigma associated with mental illness. People with mental illness often feel that they are beyond hope because the causes of mental illness are intangible and may remain unknown. Knowing that something tangible in their brains beyond their control may be causally contributing to their illness provides them with some psychological or emotional relief. The medical model allows for the possibility that advances in science will improve our understanding of mental illness and point the way toward viable strategies for intervention. By assuming mental illnesses have tangible, physical brain-based causes, they also afford the hope of a cure.

So where does this leave the framework of folk psychology for understanding mental illness? Neurophilosopher Paul Churchland (1981) has argued that even ordinary folk/non-scientists should abandon this conceptual-explanatory framework for a mature neuroscientific theory of mental phenomena (including mental illness) in a move that he dubs "eliminative materialism." Philosopher of psychiatry Dominic Murphy advocates for a similar position in claiming that "folk thought may be a poor guide to" individuating different kinds of mental illness (Murphy 2014, 105) and that the mind-brain sciences may serve as a better guide. One problem with abandoning folk psychology in the clinic is that even if we acknowledge that mental disorders are brain-based, research science has not provided us with complete or successful psychological or biological explanations of them. Consider neurobiological explanations of Major Depressive Disorder. Depression has historically been explained by appeal to imbalances in neurotransmitters, including catecholamines like norephinephrine ("the catecholamine hypothesis"), serotonin, and dopamine. Most recently, imbalances in the neurotransmitter, glutamate, are also thought to be involved in depression. Such shifts in hypotheses have been common for other DSM categories as well, including Schizophrenia and Substance Related and Addictive Disorders. Scientists know that in a certain percentage of the population that suffers from depression, antidepressant medications (e.g., selective serotonin reuptake inhibitors, SSRIs) seem to have positive effects, and also that neuroleptics and antipsychotics may be used to control hallucinations and delusions that accompany schizophrenia, but we currently do not have what might be considered adequate explanations of depression, schizophrenia, and the vast majority of other phenomena identified in the DSM as mental disorders.

It is also important to recognize that folk psychology will remain (at least for the foreseeable future) the dominant framework that patients and their families and friends use to explain the ways in which they are suffering to each other and to medical practitioners. It also will remain at least part of the conceptual-explanatory framework that patients use to understand themselves and their illness. If health care professionals devalue the conceptual-theoretical framework that laypeople use to understand their health and well-being, in the way that Churchland suggests we ought to, they run the risk of alienating people to the extent that they may not seek treatment. That some talk therapies are effective in treating mental illness is a compelling reason that the folk-psychological model in some form ought to continue to play a role in how medical practitioners understand, explain, and talk about mental illness to patients and their loved ones.

Insofar as the biopsychosocial model of mental illness regards psychological factors as causally relevant for explaining and treating mental illness, it is inclusive of the folk-psychological model. More specifically, because it views the mind and body as causally interacting, it is superior to the folk-psychological model. Given that application of the strong interpretation of the medical model to mental illness excludes folk psychology and the mind from both the diagnosis and treatment of mental disorders, it may be regarded as *eliminative*—because it essentially eliminates the mind from the discussion—or *reductive*—insofar as it reduces the mind to the brain—or takes the mind to be nothing over and above the brain. A sophisticated medical model, in contrast, would be closer to the biopsychosocial model.

Now that we have considered some of the basic features of different models of mental illness and some of the implications of using them for the diagnosis and treatment of persons with mental illness in clinical contexts, let's evaluate the use of these conceptual-explanatory frameworks in research contexts.

3. Models of Mental Illness in Research Contexts

Research contexts differ from clinical contexts in many ways—too many to itemize in a single chapter. However, one important difference is that when scientists conduct research into the causes of mental illness, they cannot consider the complex causal nexus in which mental disorders are situated all at once. Rather, different areas of science decide which causes they want to investigate—sociological, environmental, psychological, neurobiological, and genetic (to name only a handful)—and they use different kinds of experimental methods that carve the world up in different ways. Sociologists do field work to assess the impact of social factors on mental illness, whereas neurobiological experiments are invasive and involve the use of animal models. Within different areas of science studying the causes of mental illness, like sociology, psychology, neurobiology, and genetics, different scientists investigate different kinds of causes. For example, geneticists may look at different kinds of genes implicated in a mental illness like schizophrenia (e.g., DISC1, COMT, DTNBP1, PPP1R1B). Similarly, neurobiologists may investigate the role of different neurotransmitters (e.g., dopamine, glutamate, serotonin) involved in mental illnesses. What this means is that, at best, each area of science will yield only piecemeal explanations of a given mental illness, and within each area of science (e.g., neurobiology) there will be different models (e.g., the dopamine hypothesis of schizophrenia or the glutamate hypothesis of schizophrenia) on offer to explain it.

Although we currently have a lot of piecemeal explanations for different kinds of mental illness and such explanations have shed light on avenues for therapeutic interventions—some of which have been successful—we still have no cures. Such piecemeal explanations are not considered to be ultimately satisfactory (see Kincaid 2008; Sullivan 2013; Wimsatt 2007). Yet, can the results from different areas of science that study mental illness be fit together in ways that allow us to approximate toward better causal understandings of mental illness and the development of successful treatments? As we learned in Section 2 above, there are good reasons for medical practitioners to cast their nets widely when diagnosing and treating persons with mental illness. Is it similarly necessary for researchers investigating the causes of mental illness and who work in different areas of science to situate their results within the broader causal nexus when providing explanations for mental illness?

The short answer is "yes." In fact, there is widespread consensus that "integrative" models of mental illness are required if we want to find effective treatments for them (see, e.g., Albus et al. 2007; Cuthbert and Insel 2013; Insel et al. 2010; Sanislow et al. 2010). However, the same researchers who advocate for integrative explanations of mental illness have argued that current systems of psychiatric classification like the DSM and ICD have to date impeded the development of such explanations, and that before we make any real progress, we need to develop better alternative frameworks. Critics identify several reasons for thinking the DSM is an obstacle to progress in understanding the causes of mental illness. First, the group of scientists responsible for putting forward and fixing DSM diagnostic categories are not identical to the group of scientists conducting the research, and the two groups have different taxonomic aims. The DSM-5 is supposed to offer a *reliable* system for psychiatric diagnosis. Its authors think that just so long as most practicing clinical psychiatrists diagnose persons exhibiting the same sets of observable symptoms similarly ("interrater reliability"), this is sufficient for operationally defining a given mental disorder category. This is in contrast to the criterion of *validity* that is often operative in research contexts (e.g., Cronbach and Meehl 1955), which is intended to guarantee that scientists revise their classification systems in light of empirical discoveries. As applied to categories of mental illness, validity is supposed to ensure that diagnostic categories do not "lump together" phenomena (e.g., hallucinations, delusions, feelings of worthlessness) that do not belong in the same category or "split apart" phenomena that do (see Craver 2009).

Critics of the DSM claim that committing lumping errors is precisely what current DSM categories likely do, insofar as they "may erroneously place individuals who share superficial similarities but whose pathology springs from different sources into the same diagnostic category" (Lilienfeld 2014, 129). For example, the kinds of persons who satisfy the relevant criteria for being diagnosed as having a Major Depressive Episode comprise a diverse group of individuals who, while sharing a set of symptoms in common, likely do not share the causes of those symptoms in common (see Chapter 16). The causes of Rebecca's depression, for example, may be similar or different from other individuals diagnosed with depression. This is in part why, in the clinical context, it is important to use integrative explanatory models like the biopsychosocial model, so that we might approximate toward successful therapeutic interventions. However, research scientists advocating for the development of integrative explanations of mental illness think that they can only be attained if we "release the research community from the shackles of the DSM/ICD categorical system" and start categorizing phenomena in ways that reflect what scientists already know about "fundamental circuit-based behavior dimensions" of mental illness (First 2014, 53). In other words, investigators want to use evidence about commonalities and differences in the brain circuits disrupted in mental illness to develop what they regard as valid taxonomies of mental illness. Ideally, they want people like Rebecca to be grouped together with like individuals who share the same underlying pathology and whose illness may be treatable using the same methods.

Research scientists are not alone in criticizing the DSM. Debates about psychiatric classification in philosophy of psychiatry have centered on the question of whether or not mental disorders are natural kinds. To put it another way, philosophers have been concerned with the question of whether current systems of psychiatric classification pick out true divisions in kinds of phenomena in nature (see, e.g., Kincaid and Sullivan 2013). Some critics have argued that DSM categories fail to detect natural kinds in part because they have been historically shaped by different theoretical considerations that have impeded scientific progress. Murphy (2013), for example, claims that the concept of "delusion," which remains part of the DSM-5 definition of schizophrenia, is a folk-psychological concept. The problem with this, he claims, is that "we want to explain, taxonomize, and conceptualize mental illnesses without being inhibited by folk categories if they impede the search for power, generality, and progress" (Murphy 2006, 62) in understanding mental disorders.

To overcome the limitations of current diagnostic taxonomies, a subset of investigators at the U.S. National Institute for Mental Health (NIMH), who have advocated for the development of integrative explanations of mental illness, have put forward a new framework for thinking about mental illness known as the Research Domain Criteria (RDoC) Project. In development since 2010, RDoC constitutes a research reorientation at NIMH to direct funding away from DSM and ICD "consensus-based clusters of clinical symptoms" and toward research that conceives of current mental disorder categories as "complex combination[s] of disturbances in more fundamental processes, or dimensions of function, that do not necessarily align with currently identified categories of disorder" (Carter, Kerns, and Cohen 2009, 181). The RDoC Matrix (see, e.g., Lilienfeld 2014, 131) is essentially a table for organizing and inputting findings from current and future psychopathological research. The basic assumption upon which the matrix is based is that mental illness is caused by disruptions in discrete domains of psychological functioning. There are five such domains identified in the rows of the matrix: (1) positive and (2) negative valence systems and (3) cognitive, (4) social processing, and (5) arousal/modulatory systems. The columns of the matrix are intended to reflect the fact that research on domains of psychological functioning spans multiple "levels of organization"—from genes to cells to networks to behavior to self-reports—and that different areas of science investigate different "units" that each may causally contribute to psychological function and dysfunction. The different areas of science represented in the matrix, however, may be understood to share a taxonomy of functions/functional domains in common. According to Charles Sanislow and colleagues, RDoC "encourages integration of clinical and experimental findings from multiple approaches, including, for example, behavioral, neurophysiological, and genetic discoveries" (Sanislow et al. 2010, 3). Its advocates thus regard integrative explanatory models as fundamental for explaining mental illness.

So, in answer to the question of what kind of conceptual-explanatory framework is optimal for advancing our understanding of mental illness in research contexts, the resounding answer seems to be a framework that integrates information emanating from a wide variety of different areas of science that have historically studied the causes of mental illness. Whether RDoC is the correct approach for reconceptualizing mental illness in ways that facilitate causal discovery and the development of successful therapeutic interventions is something that perhaps only time will tell. It may be that the RDoC is too restrictive in terms of the types of causes of mental illness its proponents are willing to consider. It seems that the categories of psychological functions put forward by RDoC proponents are equally as consensus-based as the DSM categories. It is also important to recognize that, for all of its purported faults, the DSM diagnostic categories have been "serviceable for clinicians of varying [theoretical] perspectives" (Horowitz and Wakefield 2007, 97) and have allowed for researchers coming from a wide variety of different scientific backgrounds to share targets of empirical inquiry in common. It is not clear that RDoC categories will be similarly serviceable and whether the implications of adopting it will be positive or negative. Again, perhaps only time will tell. For our purposes it is simply important to recognize that discovering the causes of mental illness will continue to require collaborative efforts.

References

Albus, J.S., Bekey, G.A., Holland, J.H., Kanwisher, N.G., Krichmar, J.L., Mishkin, M., Dharmendra, S.M., Raichle, M.E., Shepard, G.M. and Tononi, G. (2007). "A Proposal for a Decade of the Mind Initiative", *Science* 317: 1321.

American Psychiatric Association. (2013). *Diagnostic and Statistical Manual of Mental Disorders* (5th ed., DSM-5). Arlington, VA: American Psychiatric Association

Cronbach, Lee and Paul Meehl. (1955). "Construct Validity in Psychological Tests", *Psychological Bulletin* 52: 281–302.

Cuthbert, B.N. and Insel, T. (2013). "Toward the Future of Psychiatric Diagnosis: The Seven Pillars of RDoC", *BMC Medicine* 11:126.

Black, K. (2005). "Psychiatry and the Medical Model", in *Adult Psychiatry*, E. Rubin and C. Zorumski (eds.), 2nd edition. Malden, MA: Blackwell, 3–15.

Carter, C., Kerns, J. and Cohen, J. (2009). "Cognitive Neuroscience: Bridging Thinking and Feeling to the Brain, and Its Implications for Psychiatry", in *Neurobiology of Mental Illness*, D. Charney and E. Nestler (eds.), 3rd edition. New York: Oxford University Press, 168–178.

Churchland, P. (1981). "Eliminative Materialism and the Propositional Attitudes", *Journal of Philosophy* 78(2): 67–90.

Craver, C. F. (2009). "Mechanisms and Natural Kinds", *Philosophical Psychology* 22(5), 575–594.

Dennett, D. (1987). *The Intentional Stance*. Cambridge, MA: MIT Press.

Engel, G. L. (1977). "The Clinical Application of the Biopsychosocial Model", *The American Journal of Psychiatry* 137: 535–44.

Horowitz, A. and Wakefield, J. (2007). *The Loss of Sadness: How Psychiatry Transformed Normal Sorrow into Depressive Disorder*. Oxford: Oxford University Press.

Insel, T.R., Cuthbert, B., Garvey, M., Heinssen, R., Kozalk, M., Pine, D.S., Quinn, K., Sanislow, C. and Wang, P. (2010). "Towards a New Classification Framework for Research on Mental Disorders", *American Journal of Psychiatry* 167: 748–751.

Kincaid, H. (2008). "Do We Need a Theory to Study Disease? Lessons from Cancer Research and Their Implications for Mental Illness", *Perspectives in Biology and Medicine* 51(3): 367–378.

Lilienfeld, S. (2014). "The Research Domain Criteria (RDoC): An Analysis of Methodological and Conceptual Challenges", *Behaviour Research and Therapy* 62: 129–139.

Murphy, D. (2006). *Psychiatry in the Scientific Image*. Cambridge, MA: MIT Press.

Murphy, D. (2009). "Psychiatry and the Concept of Disease as Pathology," in M. Broome and L. Bortolotti (eds.), *Psychiatry as Cognitive Neuroscience: Philosophical Perspectives*. New York: Oxford University Press, 103–117.

Murphy, D. (2013). "The Medical Model and the Philosophy of Science", in K. Fulford, M. Davies, R. Gipps, G. Graham, J. Sadler, G. Stanghellini, & T. Thornton (Eds.), *The Oxford Handbook of Philosophy and Psychiatry*. Oxford: Oxford University Press, 103–117.

Murphy, D. (2014). "Natural Kinds in Folk Psychology and in Psychiatry", in Harold Kincaid and Jacqueline A. Sullivan (eds.), *Classifying Psychopathology: Mental Kinds and Natural Kinds*. Cambridge: MIT Press, 105–122.

Sanislow, C., Pine, D., Quinn, K., Kozack, M., Garvey, M., Heinssen, R. Wang, P. and Cuthbert, B. (2010). "Developing Constructs for Psychopathology Research: Research Domain Criteria", *Journal of Abnormal Psychology*, doi: 10.1037/a0020909.

Sullivan, J. (2013). "Stabilizing Mental Disorders: Prospects and Problems", in Harold Kincaid and Jacqueline A. Sullivan (eds.), *Classifying Psychopathology: Mental Kinds and Natural Kinds*. Cambridge: MIT Press, 257–281.

Wimsatt, W. (2007). *Reengineering Philosophy for Limited beings: Piecewise Approximations to Reality*. Cambridge: Harvard University Press.

World Health Organization. (2011). *Mental Health Atlas*. Geneva, Switzerland: WHO. Available at: http://apps.who.int/iris/bitstream/10665/44697/1/9799241564359_eng.pdf

Further Reading

Kincaid, H. and J. Sullivan. (2014). *Classifying Psychopathology: Mental Kinds and Natural Kinds*. Cambridge: MIT Press.

Murphy, D. (2015). "Philosophy of Psychiatry." *Stanford Encyclopedia of Philosophy*. Available at: http://plato.stanford.edu/entries/psychiatry/

42
PHENOMENOLOGY AND HERMENEUTICS IN MEDICINE

Havi Carel

Introduction

Phenomenology enables us to focus on the experience of illness itself, while bracketing the causal and ontological assumptions that accompany the biomedical model. This chapter gives an overview of phenomenology and hermeneutics, focusing on a core phenomenological distinction between the objective "biological" body and the subjective "body as lived." It also links this distinction to the hermeneutical emphasis on the different horizons of patient and health professional. The distinction between the objective body and the body as lived is mapped onto the disease/illness distinction. This enables us to (1) differentiate between the different dimensions of the body and thus provide a detailed account of embodied illness experiences; (2) conceptualize the complexity of the intersubjective encounter in the clinic; and (3) understand particular embodied experiences that characterize illness, such as alienation, uncanniness, and bodily doubt (Carel 2014).

The chapter then turns to hermeneutics, focusing on the work of Fredrik Svenaeus (2000a, 2000b, 2001) to explain the usefulness of this approach. Using the notion of illness as "unhomelike being in the world," I discuss the role of medicine in this context, which is to find a way to return patients to a more homelike being.

Emma's Story

Emma, who lives in Manchester, England, is 29. Until recently, things were looking well for Emma. She was a fit and active young woman, about to start a family with her husband, George, and working in a job she loved that involved lots of travel and excitement. Emma loves horse riding, hiking, and travel. She loves seeing the world and being part of an exciting start-up business.

But this is all over now. Emma has recently been diagnosed with a chronic, progressive lung disease called lymphangioleiomyomatosis (LAM, see http://www.thelamfoundation.org). The diagnosis shattered her life. She was forced to give up her job, because air travel caused her repeated lung collapses, requiring lengthy hospitalizations, and travel insurance costs have become prohibitive. She was advised not to get pregnant, as a pregnancy will accelerate the rate of disease progression and cause other, potentially life-threatening problems, like renal bleeding. There is a treatment she could try, which at best would halt the deterioration of her lungs but will not restore the lung function she has already lost. Approximately 30% of LAM patients do not respond to the treatment. Emma's prognosis is bleak. If she does not respond to the treatment and without a lung transplant, Emma would—say the doctors—be at risk of respiratory failure within two years.

Emma sinks into despair. She questions the benevolence of the world in which she lives. Why did this happen to her? Apparently, it was just bad luck: LAM is caused by a somatic mutation, causing cell proliferation that creates cysts in the lungs; it could happen to anyone. What can she do about it? Well, she could carry on exercising gently, she could use ambulatory oxygen to increase her ability to do things, she could go on the lung transplant waiting list, and she could hope that the treatment works for her. But there are no guarantees. Emma is paralyzed with fear and unable to think clearly; she can barely bring herself to talk about her illness or what she should do next.

Emma's despair is exacerbated by the way she feels she is treated by the medical and health professionals in her local hospital. Although they are courteous, they are also cold, she finds. She thinks they are mortified to be dealing with such a difficult case and are sad that they are unable to offer her a cure. She has no one to talk to about her diagnosis who she feels can really understand her. Her parents are devastated. She feels alienated from her friends, who seem disgustingly lucky to her. She is worried that her partner may not want to take on the long-term prospect of living with severe illness and disability. Emma feels envy, bitterness, and confusion. More than anything, she is scared of what the future holds in store for her: how will she choose between the Scylla of lung transplants, with their modest survival rates, and the Charybdis of respiratory failure and premature death?

She notes the spiraling decline of her lung function; it scares her beyond words. Each week her world shrinks and things have to be given up. One day, she gives up cycling to work. A few months later, she gives up work altogether, as she is made so breathless by talking or walking down the corridor and also suffers from fatigue. Eventually, she is confined to a wheelchair and can't even tie her own shoelaces. Bending down makes her too breathless. She can no longer climb stairs; she can no longer talk without needing to pause for a breath. She loses weight as eating makes her so breathless that she subsists mainly on protein shakes. A narrow, unchosen life descends on her. She feels trapped in a body and a life she never wanted.

Although she has lots of medical appointments, Emma doesn't know how to explain her anguish and despair, as well as the ways in which her world has shrunk both geographically and temporally, to the health professionals involved in her care. What will they understand and how will they incorporate that knowledge into their interactions with her? How can Emma convey the life-changing impact of her illness on her?

There is an impasse between the medical staff's focus on her disease, symptoms, and lung function, and Emma's first-hand experience of these. The impasse is part of a broader problem: Emma needs concepts and language with which to describe what has happened to her. Her confusion and suffering might subside, but only if she is given tools with which to order, discern, and describe her illness experience. And she then needs to find a way to share it with her family, carers, and doctors. A phenomenological approach to illness offers a framework with which she can do that.

What Is Phenomenology?

Phenomenology is a descriptive philosophical method, developed in the first half of the 20th century in Germany and France. It aims to be a practice rather than a system (Moran 2000: 4). The goal of phenomenology is to accurately describe the relationship between a perceiving consciousness and the world. It has been used to describe aesthetic experiences; for example, the experience of looking at Van Gogh's 1886 painting, *Peasant Shoes*

(Heidegger 1993: 158–161). It can be used to describe how something appears from a particular point of view, in a certain environment, as Merleau-Ponty's analysis of Cezanne's paintings does (1964: 9–25). Or it can be used to analyze the experience of listening to a melody (Husserl 1990).

Because of its sensitivity to acts of perception and its rich account of the relationship between perceiver and perceived, phenomenology is uniquely suited to describing the experience of illness as it is lived by the ill person. A particular approach within phenomenology, developed by the French philosopher Maurice Merleau-Ponty (1908–1961) in *Phenomenology of Perception* (1962 [1945]) has special utility here. In Merleau-Ponty's view, the body is the locus of subjectivity, the "null centre" from which consciousness radiates toward the world; he coined the term "body-subject" to capture the mind-body unity that is core to his view.

Merleau-Ponty rejects the mind/body separation traditionally espoused by philosophers and theologians. For Merleau-Ponty the body is "the origin of the rest, expressive movement itself, that which causes things to begin to exist as things, under our hands and eyes" (1962: 146, translation modified). This is not just an empirical claim about perceptual activity, but a transcendental view that posits the body as the condition of possibility of perception and action. As Gallagher and Zahavi write, ". . . the body is considered a constitutive or transcendental principle, precisely because it is involved in the very possibility of experience" (2008: 135). In other words, without the body, we would not be able to experience ourselves, the world, and others. Having the kind of body we have makes us the kind of beings we are.

In Merleau-Ponty's view, perceptual experience is the foundation of subjectivity. The kind of creature we are is circumscribed by the types of experiences we have and the kinds of actions we perform, which are shaped by our bodies and brains. Any attempt to understand human nature would have to begin with the body and perception as the foundations of personhood (Merleau-Ponty 1962: 146). To think of a human being is to think of a perceiving, feeling, and thinking animal, rooted in a meaningful context and interacting with things and people in its environment. To be is to be a body that perceives the world. This body is situated and intends toward objects around it. Human existence takes place within the horizons opened up by perception.

The body, for Merleau-Ponty, is the locus of human existence. Thus, when we become ill, this is not simply a biological dysfunction but a pervasive disturbance of our being in the world. On a phenomenological view, illness is not a localized dysfunction, although it may be located in a specific organ or system, but a pervasive concern. Thus, we see the habits that anchor our everyday routines disrupted in illness; Emma is no longer able to run for the bus. She can no longer cook for herself or do her own grocery shopping.

This disruption of habits is not a superficial disturbance. The habitual body, as Merleau-Ponty calls it, is very much at the core of lived experience. The ease and expertise with which we perform everyday actions leads us to view them as trivial tasks. But in illness the tasks that form a seamless part of our everyday life become demanding and require planning and attention. It is this kind of disruption to our plans and our ability to act in the world that changes both the ill person and her experience of the world.

Illness as Life-Transforming

The first insight phenomenology provides is that our agency, our ability to operate in the world, is restricted when our bodies are damaged. Emma's agency and freedom are radically curtailed by her failing lungs. She can no longer do any of the things she enjoyed or took for granted previously, and she is no longer independent. Her agency, ability to pursue her goals, and independence are restricted by illness.

An extreme example of such restriction would be Jean-Dominique Bauby's account of a stroke, which left him in complete paralysis, or "locked-in syndrome." In the space of a few minutes, Bauby turns from being a successful man in the prime of life to lying helplessly in bed, unable to communicate or eat. Although his mind is alert, the total paralysis of his body imposes a complete halt on all the activities he previously enjoyed. His account of his illness was painstakingly dictated using the batting of his one functioning eyelid. (An assistant would read out the alphabet and Bauby would blink when she got to the letter he wanted—a process that took many months.) The laboriously produced account was published as a novel, *The Diving Bell and the Butterfly* (Bauby 1998), and later made into a film by Julian Schnabel.

It would be impossible for us to describe the changes to Emma's or Bauby's existence merely in terms of physiological changes. These changes have to be understood not just biologically, but also existentially (i.e., as impacting on one's entire existence, including one's psychological, social, and temporal being). We can begin to appreciate these changes by thinking about simple things, like going for a meal with friends, or having a shower, which can become impossible in illness. Illness, as opposed to disease, is not a physiological dysfunction but a shutting down of horizons and closure of possibilities.

Illness changes our relationship to the world, or more specifically, our relationship to the environment, to other people and to possibilities. The geography of our world changes with illness, when old invitations (a stairway leading somewhere) become new limitations (Carel 2013: 25). Toombs describes how a bookcase in her house was initially a place to store books, then became an object to hang on to as her walking became less steady, and eventually turned into an obstacle she had to wheel her wheelchair around (1995: 16). For Emma, the gym was once a place to exercise and enjoy physical activity and fitness. It is now off-bounds for her; a meaningless location.

Similarly, the social world is transformed by illness. As described above, Emma's relationship to her parents, friends, and husband have to be renegotiated. Some relationships become less natural, or weighed by guilt, awkwardness, and other responses to illness. Illness, especially if visible, may mark the ill person out, or put her in a "sick role," in which she is expected to behave in particular ways (e.g., be grateful for her medical care, and want to get better) (Parsons 1991: 436–8). Heidegger's notion of being-with (*Mitsein*) captures the magnitude of the change brought about by illness. By being-with, Heidegger expresses the inherent sociality that lies at the core of a human being (1962: 149–50). Anything that modifies our ways of being-with will have far-reaching consequences, stretching beyond the physiological process of disease. In particular, Emma's relationship to her husband has changed. He will quickly become her carer, and the new restrictions on her life will also be his. He will witness from a second-person perspective how illness changes Emma's way of being.

Finally, Emma's relationship to herself, in terms of her possibilities, goals, and her experience of time and of her future, is also modified by illness. Emma is faced with a poor prognosis, substantial limitations on work and leisure, and a pressing need to change habits and to rethink plans for the future. Illness impacts on every dimension of her life. Heidegger views the human being as a temporal synthesis of past, present, and future; as a temporal creature whose actions are informed by her past and directed toward her future (Heidegger 1962: 376, 418). Emma feels that her future is tainted by fear, uncertainty, and grave prognosis. Thus, her experience of time also changes.

This temporal dimension also includes finitude, as plans for the future are always constrained by our finite existence, as a stretch from birth to death (Carel 2006: 70). When faced with a poor prognosis like Emma's, we need to rethink our life plans and to adjust our expectations to what remains possible. Again, this process of adjustment is reflexive and time-consuming.

Emma needs to invest considerable time and effort into adapting to her illness. But adaptation is possible, although it never fully compensates for the freedom that is lost (Carel 2007: 104).

Varying Perspectives on Illness: The Objective Body and the Body as Lived

In medicine we encounter two perspectives. First, we have the experiences of the patient and of her family and carers, which are first- and second-person perspectives, respectively. They are the lived experience of illness. Second, we have the experiences of health professionals for whom the perspective of the patient may be quite alien. Health professionals see Emma for short appointments, but they do not know what impact her illness has on different areas of her life, like her social world and her self-identity, unless they inquire or Emma shares this with them.

Health professionals focus on lung function measurements, rate of decline, and the secondary physical symptoms, such as a lung collapse. They have an interest in quality of life, but their understanding of the disease is theoretical, professional, and characteristically takes on the third-person perspective. The patients live the illness, and therefore primarily view it as a lived experience. As Toombs (1987) notes, the two partners in the conversation in the clinical meeting are talking about two different dimensions of the body. A second-person perspective is also possible in medicine, but health professionals are educated to primarily seek and occupy the third-person perspective.

The phenomenological literature distinguishes between the objective body and the body as lived. The objective body is the physical body, the object of medicine: it is what becomes diseased. Sartre calls this body the "body of Others": it is the body as viewed by others, not as experienced by me (Sartre 2003). The body as lived is the first-person experience of this objective body, the body as experienced by the person whose body it is. And it is on this level that illness, as opposed to disease, appears. This distinction is fundamental to any attempt to understand the phenomenon at hand: the ill person is only and ever the one who experiences the illness from within (although others may have an experience of someone else's illness, as second-perspective witnesses).

Only they can say if they feel pain or fatigue, or what a medical procedure or a particular symptom *feels like*. This is a source of significant and medically relevant knowledge, but also contains an element of unshareability. Thus, the experience of illness contains a measure of communication difficulties that should be acknowledged (Carel 2013). Or as Sartre put it more strongly: "the existed body is ineffable" (Sartre 2003: 377). Disease, on the other hand, is a process in the objective body that may be observed by any other person and may yield information that is not available through first-person reports. For example, one may have elevated cholesterol while having no experience of this. Often such knowledge comes from medical tests that yield objective facts with no experiential correlate.

The relationship between illness and disease is not simple: the two aspects do not just mirror one another. Illness may precede one's knowledge of disease: disease is commonly, but not always, diagnosed following the appearance of symptoms experienced by the patient. These symptoms are part of her illness experience and are lived by the patient. Disease may appear without illness, as in asymptomatic disease such as high blood pressure. Or often we have both illness and disease, but the two do not perfectly cohere. For example, severe disease or disability (e.g., quadriplegia, COPD) may give rise to an illness experience that is tolerable, due to adaptation (Carel 2009). So although the disease may be clinically "severe," the illness experience is not as correspondingly negative as might be expected. In fact, it has been well-documented that there is a surprising lack of correlation between disease severity and level of subjective well-being (happiness) that patients report (Angner et al. 2009; Carel 2007, 2009; Riis et al. 2005).

Another difference that emerges in illness is the difference between the habitual body and the body as it is in this moment. Routine actions can be performed expertly and efficiently because they have become habit, and they form what Merleau-Ponty calls the "habitual body" (2012). The ease with which we perform habitual tasks often disappears in illness, where the body as it is in this moment is incapable of performing routines familiar to the habitual body. Illness thus reveals the difference between the habitual body and the body as it is at this moment (i.e., as no longer able to continue its habits).

Returning to Emma, we can see that a large part of her communicative difficulties stem from health professionals' focus on the objective body and disease process, while for Emma it is the body as lived that is both experienced and central. The distinction between the objective and the lived body makes clear the fundamental difference between the two perspectives. The physician's perspective limits the physician, who can only ever perceive the disease through objective observation. The illness experience in its first-person form is not accessible to the physician, by definition, other than via Emma's account. This means that Emma is the expert on her own experienced illness, and this expertise should be taken into account in medical epistemic practices.

Taking the objective perspective may lead the physician to seek to treat the disease (sometimes with inadequate understanding of the illness) or to have little understanding of the impact of the disease on the patient's life as a whole. The patient, on the other hand, can observe the objective indicators of disease (e.g., look at blood test results or an x-ray) but also has unique access to the lived experience of the disease—namely, illness.

In this sense the patient may have, at least in principle, an epistemic advantage of having access to her own illness experience *and* to the objective knowledge about the disease. This double epistemic advantage, of both having direct access to the illness experience and of having both subjective and objective knowledge, often goes unacknowledged, and the patient experience may be subsumed under the medical view or discounted (Carel and Kidd 2014). The unique ability to oscillate between the two perspectives gives the patient a deeper understanding of the illness experience, and potentially to the dual nature of the body, but this may also cause confusion and miscommunication. As Toombs (1987) notes, the physician's focus on disease may clash with the patient's primary interest in her illness, so although they may seem to speak of the same entity, they in fact refer to two different entities (disease vs. illness), and therefore have a communicative and interpretative gap that must be addressed before effective communication becomes possible (Toombs 1987).

Hermeneutics of Medicine

A closely related approach to illness is hermeneutics. It shares with phenomenology the fundamental premise that human existence is inherently meaning-making and meaning-seeking. Therefore, the role of medicine is not merely that of repairing physiological dysfunction, but of creating sense, what Fredrik Svenaeus (2000a, 2000b) calls "homelike being in the world," in an existence that has become unhomelike, alienated, due to illness.

Svenaeus developed a hermeneutical account of medicine. He writes: "medical practice is not only essentially a meeting, it is also interpretation: clinical hermeneutics" (Svenaeus 2001: 148). This interpretation takes place through language, and more specifically, through the "fusion of horizons" in the clinical encounter, a fusion that serves as a meeting point for patient and physician interpretations. This particular kind of interpretative work has a specific productive goal—namely, the restoration of homelike being in the world, helping patients feel at home in their bodies and environment, despite changes to embodiment and despite the alienation and objectification of their bodies that typically occurs in illness. Svenaeus also draws on Drew Leder's hermeneutic analysis (Leder 1990, and discussed in Svenaeus 2001), according to which the patient constitutes a primary text and is interpreted in a process of hermeneutic circling, moving between the primary text and secondary texts (experiential, narrative, physical, and instrumental texts), which are derived from the primary text or stand in some relationship to it.

Medical hermeneutics pays close attention to the language of the body and to the process of inscription, documented also by Getz et al. (2011). On this view, what a patient has lived through and experienced is etched into her body and gives rise to her health state, in ways that demand careful interpretation, and a holistic-therapeutic approach. As Getz et al. document, particular forms of childhood trauma, sexual abuse, and disturbed domestic environment can give rise to permanent changes to one's body (e.g., elevated cortisol levels) that then give rise to further illness and suffering that are only treatable in the context of understanding the trauma as the causal source of the illness.

In *The Enigma of Health* (1996), Hans Georg Gadamer suggests that medicine is a dialogue aimed at reaching a mutual understanding of why a patient is ill. The health professional must practice empathy in her effort to understand the patient's situation. The health professional is here the reader or interpreter, while the patient is the text, albeit a text that can question and engage in dialogue. Here hermeneutics is not understood as a method for analyzing patient data but an attitude of curiosity and openness, aimed at securing an improved health state for the patient. This work in hermeneutics of medicine remains theoretically very close to the phenomenological approach described above. I now turn to survey recent work on the phenomenology and hermeneutics of illness.

Work on Phenomenology of Illness

A growing number of philosophers have in recent years turned their attention to illness, using a phenomenological or hermeneutical lens. This has generated a small but rapidly growing literature, starting with S. Kay Toombs' seminal article, "Illness and the Paradigm of Lived Body" (1988). Toombs' article applies Merleau-Ponty's distinction between the body as lived and the biological body to the case of illness, demonstrating the problems and limitations arising from understanding illness as merely a disruption of biological function. Rather, Toombs argues, illness disrupts the lived experience of one's body, leading to an overarching disruption of the ill person's way of being in the world and their lifeworld.

Toombs' work also explores temporal changes in illness using Sartre and Husserl, as well as examining the patient-clinician encounter through a phenomenological lens (1990, 1987). She also uses phenomenology to characterize the general features of chronic illness and disability, weaving together examples from her life with multiple sclerosis and phenomenological analysis (1995, 1993).

Toombs' trailblazing work was followed by Fredrik Svenaeus, Matthew Ratcliffe, Luna Dolezal, Jenny Slatman, Darian Meacham, and Havi Carel, among others. As discussed above, Svenaeus published a series of influential articles developing a Heideggerian and Gadamerian analysis of illness as an unhomelike experience. He describes medicine's role as showing the patient the way home, back from an uncanny experience (Svenaeus 2000a, 2000b). His work was further developed in *The Hermeneutics of Medicine and the Phenomenology of Health*. In this book Svenaeus provides a novel account of medicine's aim, using a hermeneutic phenomenological approach to describe medicine as an interpretive practice (Svenaeus 2001). This emphasis on hermeneutic aspects of the patient-clinician encounter, as well as on the interpretative work involved in diagnosis and in other epistemic aspects of medical work, draws on Gadamer's account to provide a view of illness as based in social and interpretative practices of generating meaning.

Within philosophy of psychiatry, Matthew Ratcliffe (2008) developed a novel account of what he calls "existential feelings" and how these underpin our sense of belonging in the world, a sense that is disturbed in a variety of ways in mental disorder. In addition, Ratcliffe has in recent years written a number of influential papers using phenomenology to describe mental disorders such as depression. Giovanni Stanghellini (2004) draws primarily on Merleau-Ponty in his work *Disembodied Spirits and Deanimated Bodies*, also examining a phenomenological understanding of mental disorder.

Recently, Luna Dolezal (2014) has written on the phenomenology of shame in the medical encounter. Jenny Slatman's (2014) work has systematically explored the phenomenological alternative to historical dualistic notions of embodiment. Darian Meacham has edited a volume on phenomenology, politics, and health (2015). Finally, Havi Carel's *Illness* uses Merleau-Ponty and Heidegger to provide a comprehensive description of the first-person experience of illness (2013). The book confronts the tendency of philosophy to work from a third-person perspective and criticizes the central debate in the philosophy of medicine, between those advancing a naturalistic value-free description of disease and those claiming that disease is fundamentally a social and normatively laden concept. Carel argues that this debate excludes the experience of illness, which is highly relevant, and intersects with, the concept of disease. She suggests augmenting the debate by providing a phenomenological account of the first-person experience of illness, examining the personal, social, physical, and temporal dimensions of illness.

Phenomenology as Research Method

Phenomenology is primarily a philosophical method, but it has also been applied as a framework for qualitative research. It is used as an interpretative technique in the social sciences, which helps researchers distill salient themes emerging from interviews. Van Manen's work has been influential in this regard, and many researchers have used his description of the six activities involved in phenomenological research in the social sciences, including in health care research. On van Manen's account, phenomenology can be used to select a phenomenon of interest to the researcher (namely, lived experience); investigate this experience as we live it, not as it is conceptualized; reflect on essential themes characterizing the experience; describe the experience through writing and rewriting; maintain an oriented relationship to the experience; and finally, balance the research context by considering how the parts relate to the whole (1990: 31–32).

Another research method used in qualitative interviews in nursing and health care research is Interpretative Phenomenological Analysis (IPA). IPA is a qualitative research method with an idiographic emphasis. Its aim is to offer insight on how a particular person or small group (usually between 5 and 15 interviewees), in a particular context, experience a certain phenomenon (e.g., their illness or another aspect of their lives). IPA normally uses interviews, focus groups, or diaries to gather data. IPA is a unique method in that it does not set out to test a hypothesis, but rather is a more self-reflexive and open-ended method, in which the researchers acknowledge their own biases and preconceptions and attempt to bracket these in order to produce an account of the experiential world of their subjects. IPA is influenced by hermeneutic phenomenology and is ultimately aimed at understanding meaning-making processes, using both a phenomenological description of an experience and a multi-layered interpretation of that experience (Larkin et al. 2016).

There is scope for developing further phenomenological research tools. Existing tools include "walking with" exercises and a host of research techniques that go beyond the traditional qualitative interview. For example, some have argued that researchers need to notice the body language and facial gestures of interviewees, rather than merely their words, in order to understand the meaning they are trying to convey. Miczo (2003) recommends the use of video recordings, rather than merely using transcripts of interviews, as is commonly done. The phenomenological methods used to understand and report the experience of illness can be developed and taken beyond the existing paradigm of questionnaires and interviews, to include a host of nonverbal embodied methods.

Phenomenology in Teaching and Training

Phenomenology is also starting to be used as a teaching and training tool. Basic phenomenological concepts, like the distinction between the biological and lived body, motor intentionality and habitual body, are used to instruct health care professionals and trainees about the changes to their patients' lives. Phenomenology as a pedagogical tool has proved useful in pilot workshops and in medical school teaching (Carel, unpublished reports). Providing health care professionals with an understanding of basic phenomenological concepts of embodiment would enable them to understand the holistic and embodied nature of illness. By understanding their patients as body-subjects, clinicians would be able to appreciate the impact illness has on patients' lives, not just as a secondary effect of the biological disease, but as a primary phenomenon.

Another future application is the development of a "phenomenological toolkit," which would enable patients to systematically and comprehensively describe their experience (Carel 2012). Such a toolkit would enable patients to take a fractured set of experiences and to make sense of them through describing and ordering. This information can then be presented to the clinician, as well as aiding the patient's self-understanding. Bringing to light the different perspectives on illness can help construct a shared meaning of illness. This would improve communication and understanding in patient-clinician dialogue, which could in turn improve patient trust in physicians and compliance.

References

Angner, E., Ray, M. N., Saag, K. G., & Allison, J. J. (2009) Health and happiness among older adults a community-based study. *Journal of Health Psychology* 14(4), 503–512.
Bauby, J.D. (1998) *The Diving Bell and the Butterfly*. New York: Fourth Estate.
Carel, H. (2006) *Life and Death in Freud and Heidegger*. New York & Amsterdam: Rodopi.
_____. (2007) "Can I be Ill and Happy?" *Philosophia* 35(2), 95–110.
_____. (2009) "'I Am Well, Apart from the Fact that I Have Cancer': Explaining Wellbeing within Illness." In L. Bortolloti (ed.), *The Philosophy of Happiness*. Basingstoke: Palgrave, pp. 82–99.
_____. (2012) "Phenomenology as a Resource for Patients." *Journal of Medicine and Philosophy* 37(2), 96–113.
_____. (2013) *Illness*. Durham: Acumen.
_____. (2014) "Bodily Doubt." *Journal of Consciousness Studies* 20(7–8), 178–197.
Carel, H., & Kidd, I. J. (2014). Epistemic injustice in healthcare: A philosophial analysis. *Medicine, Health Care and Philosophy* 17(4), 529–540.
Dolezal, L. (2014) *The Body and Shame: Phenomenology, Feminism and the Socially Shaped Body*. New York: Lexington Books.
Gadamer, H.-G. (1996) *The Enigma of Health: The Art of Healing in a Scientific Age*. Stanford, CA: Stanford University Press.
Gallagher, S. & D. Zahavi. (2008) *The Phenomenological Mind*. New York and London: Routledge.

Getz, L., Kirkengen, A. A., & Ulvestad, E. (2001) "The human biology: saturated with experience." *Tiddskr Nor Legeforen* 7, 683–687.

Heidegger, M. (1962/1927) *Being and Time*. Oxford: Blackwell.

———. (1993) "The Origin of the Work of Art." In *Basic Writings*. New York and London: Routledge, pp. 143–212.

Husserl, E. (1990 [1928]) On the Phenomenology of the Consciousness of Internal Time. Dordrecht: Kluwer.

Leder, D. (1990) *The Absent Body*. Chicago, IL: University of Chicago Press.

Meacham, D. E. (ed.). (2015) *Medicine and Society, New Perspectives in Continental Philosophy*. Dordrecht: Springer.

Merleau-Ponty, M. (1962) *Phenomenology of Perception*. London: Routledge and Kegan Paul.

Merleau-Ponty, M. (1964) "Cezanne's Doubt." In *Sense and Nonsense*. Evanston, IL: Northwestern University Press, pp. 9–25.

———. (2012 [1945]) *Phenomenology of Perception*. New York & London: Routledge.

Miczo, N. (2003) "Beyond the 'Fetishism of Words': Considerations on the Use of the Interview to Gather Chronic Illness Narratives." *Qualitative Health Research* 13(4), 469–490

Moran, D. (2000) *Introduction to Phenomenology*. Oxford, UK: Routledge Press.

Parsons, T. (1991 [1951]) *The Social System*. London: Routledge.

Ratcliffe, M. (2008) *Feelings of Being: Phenomenology, Psychiatry and the Sense of Reality*. Oxford: Oxford University Press.

Riis, J., Loewenstein, G., Baron, J., Jepson, C., Fagerlin, A., & Ubel, P. A. (2005) Ignorance of Merleau-Ponty, M. _Phenomenology of Perception_ London and Henley: Routledge and Kegan Paul hedonic adaptation to hemodialysis: A study using ecological momentary assessment. *Journal of Experimental Psychology: General* 134(1), 3.

Sartre, J.-P. (2003) *Being and Nothingness: An Essay on Phenomenological Ontology*. Oxford, UK: Routledge.

Slatman, J. (2014) *Our Strange Body: Philosophical Reflections on Identity and Medical Interventions*. Amsterdam: Amsterdam University Press.

Stanghellini, G. (2004) *Disembodied Spirits and Deanimated Bodies*. Oxford: Oxford University Press.

Svenaeus, F. (2000a) "Das Unheimliche—Towards a Phenomenology of Illness." *Medicine, Health Care and Philosophy* 3, 3–16.

———. (2000b) "The Body Uncanny—Further Steps Towards a Phenomenology of Illness." *Medicine, Health Care and Philosophy* 3, 125–137.

Toombs, S. K. (1987) "The Meaning of Illness: A Phenomenological Approach to the Patient-Physician Relationship." *The Journal of Medicine and Philosophy* 12, 219–240.

———. (1988) "Illness and the Paradigm of Lived Body." *Theoretical Medicine* 9, 201–226.

———. (1990) "The Temporality of Illness: Four Levels of Experience." *Theoretical Medicine* 11, 227–241.

———. (1993) "The Metamorphosis: The Nature of Chronic Illness and Its Challenge to Medicine." *Journal of Medical Humanities* 14(4), 223–230.

———. (1995) "The Lived Experience of Disability." *Human Studies* 18, 9–23.

Van Manen, M. (1990) *Researching Lived Experience: Human Science for an Active Sensitive Pedagogy*. Albany, NY: State University of New York Press.

Further Reading

Carel, H. (2013) *Illness*. London: Routledge.

Carel, H. (2016) *Phenomenology of Illness*. Oxford: Oxford University Press.

Dolezal, L. (2014) *The Body and Shame: Phenomenology, Feminism and the Socially Shaped Body*. New York: Lexington Books.

Slatman, J. (2014) *Our Strange Body: Philosophical Reflections on Identity and Medical Interventions*. Amsterdam: Amsterdam University Press.

Toombs, S. K. (1999) *The Meaning of Illness: A Phenomenological Account of the Different Perspectives of Physician and Patient*. Amsterdam: Kluwer.

43
EVOLUTIONARY MEDICINE
Michael Cournoyea

We are so frequently concerned with *how* we get sick that we often neglect to ask *why* we get sick. Not simply why we get sick compared to our colleagues or younger selves, but why we get sick at all. Although this might be posed as a deeply philosophical question about the human condition, it can also be considered empirically. An anthropologist might point to the norms and politics of socio-cultural institutions; a geneticist may look to family trees and hereditary mutations that persist between generations; a physician might cite the pathophysiology of disease and curious cases from clinical experience. This chapter is about the evolutionary biologist, who tells a different story: we get sick because we are evolved beings, riddled with evolutionary trade-offs and maladapted to modern lifestyles. In the light of evolution, so the story goes, sickness and health need to be reinterpreted.

Evolutionary medicine attempts to answer why we get sick by exploring the ultimate, evolutionary causes of health and disease. This exploration aims to "transform the way patients and doctors see disease" (Nesse and Williams 1994: 245) and "bring the full power of evolutionary biology to bear on problems of human health" (Nesse and Stearns 2008: 43). With the help of evolutionary thinking, bodily vulnerabilities are envisioned as evolutionary trade-offs or the result of a mismatch between ancestral lifestyles and the modern world. A recent review surveys the diversity of medically relevant issues that would be reconceptualized by evolutionary thinking: "bottle-feeding, caesarian sections, infection, cleanliness, fever, exercise, diet, mate choice, contraception, semen sampling, and body odor suppression" (Gallup et al. 2014: 69). Breadth is heralded as a strength of the field.

Evolution also helps organize human physiology and anatomy into a coherent but historically contingent model of the body. Just as anatomy and chemistry are core subjects in the medical classroom, evolutionary biology would serve as a foundational subject that unified the seemingly disconnected volumes of medical knowledge. Learning tools focused on evolutionary medicine have been developed for high school, undergraduate, and medical students in the last 10 years (Antolin et al. 2012, Hidaka et al. 2015).

Before evolutionary medicine began to ask why we get sick, evolutionary theory helped us to understand the evolution of pathogens and the genetic dynamics of contemporary human populations. These applications have saved countless lives and remain essential to public health and clinical practice. Only in the last 25 years have evolutionary biologists begun to theorize about human ancestral biology and what it might mean for health today. This chapter explores such theories, their assumptions, and their potential role in clinical practice.

A Brief History of Evolution and Medicine

Throughout the 20th century, the relationship between evolution and medicine had a troubled history. Most striking was the rise of eugenics, before and during World War II, which led to a post-war distancing of evolutionary theory from medicine. Even today, proponents of evolutionary medicine must clarify that their recommendations do not advocate genetic purification or racial typologies, the outcomes of oversimplified views on genetic determinism (e.g., Gluckman et al. 2009, Nesse et al. 2010). Evolutionary biologist George C. Williams and psychiatrist Randolph Nesse's foundational book, *Why We Get Sick: The New Science of Darwinian Medicine* (1994), clearly stated that evolutionary medicine is neither eugenic nor aligned with turn-of-the-century ideologies about Social Darwinism that "helped to justify withholding medical care from the poor" (11). Modern evolutionary medicine, like clinical medicine, aims to help those that suffer rather than the species as a whole.

Contemporary evolutionary medicine became possible because of 20th-century developments in evolutionary theory, genetics, and their intersection. Quantitative advances in our knowledge of human population genetics, host-pathogen interactions, and the phylogenetics of bacteria and viruses drew attention to the importance of natural selection in medical contexts (Zampieri 2009). When the Flexner Report of 1910 highlighted the importance of biomedicine in medical training and revamped medical professionalism, evolutionary thinking was in a lull and so neglected (Nesse and Stearns 2008, Duffy 2011). Fisher had yet to quantify population genetics, modern evolutionary theory had yet to be built, and E. O. Wilson's controversial *Sociobiology* (1975) had yet to incite discussion about the application of evolution and genetic selectionism to human populations.

In the last 25 years, there has been growing interest in evolutionary explanations for health and disease. In 1991, Williams and Nesse urged physicians to embrace evolution as a foundation of medical knowledge in "The Dawn of Darwinian Medicine." Embracing evolution meant modifying medical school curricula, re-evaluating clinical practice, and creating interdisciplinary institutes to advance evolutionarily informed medical research. If only physicians and researchers "were as attuned to Darwin as they have been to Pasteur" (1991: 2), they claimed, would medical knowledge progress at an unprecedented pace. Despite their optimism, the dawn started slowly.

The vast majority of evolutionary medicine's theoretical and educational contributions have occurred in the last decade. There are a growing number of university courses, two dedicated journals (*Evolution, Medicine, and Public Health*; and the *Journal of Evolutionary Medicine*), collaborative research centers (e.g., the Triangle Center for Evolutionary Medicine [TriCEM] and the Arizona State University Center for Evolution and Medicine), and a curriculum supplement for U.S. high school science classes produced by the National Institutes of Health (Pennisi 2009, 2011). A recent survey of U.S. medical school deans indicated that evolution is more integrated into the curricula of some schools than it was a decade ago, and medical faculty are more likely to be trained in evolutionary biology; however, nearly half of those that responded to the survey worried that controversy might erupt should evolution be taught at their schools (Hidaka et al. 2015). This is unsurprising given the controversial place of evolution in the U.S. classroom (Glaze and Goldston 2015). Physicians have also been cautious to adopt evolutionary explanations because their role in clinical diagnostics, prognostics, and treatments remains unclear.

Evolutionary Explanations in Medicine

Before exploring evolutionary medicine critically, I need to outline the kinds of explanations it claims to offer. I mentioned above that evolutionary medicine explores the *ultimate* causes of health and disease. Ultimate, here, does not mean final or decisive, but refers to evolutionary causes as opposed to proximal causes. The Harvard evolutionary biologist Ernst Mayr clearly distinguished between ultimate and proximate causes in his "Cause and Effect in Biology" (1961), arguing that both kinds of causes (and their associated explanations) were necessary to fully understand biological phenomena. He contrasted the functional biologist with the evolutionary biologist: the functional biologist asks how the mechanisms within organisms operate, whereas the evolutionary biologist asks why organismal traits were adaptive. The former asks proximate questions, the latter ultimate. These two question types establish typical disciplinary boundaries and an "explanatory asymmetry" that distinguishes "not between past and present causes, but between different *types* of causes or processes" (Scholl and Pigliucci 2015: 655). The ultimate cause of fever, for instance, might be that it helped our ancestors to ward off pathogens (e.g., by boosting our immune responses) and so fosters a selective advantage; this evolutionary explanation need not appeal to the proximate physiological causes of fever, such as its cellular mechanisms or hormonal regulations. The merits of the proximate-ultimate distinction continue to be debated by philosophers of biology (e.g., Ariew 2003, Amundson 2005, Laland et al. 2013, Scholl and Pigliucci 2015); nevertheless, it remains essential to evolutionary thinking in medicine.

Not long after Mayr's article, Tinbergen (1963) went on to distinguish four questions that, when answered, offer a complete explanation for an organismal trait. Tinbergen did not map these onto Mayr's proximate-ultimate distinction, but subsequent work mapped the parallel lines of inquiry, for better or worse (Laland et al. 2013). Integrating the frameworks might look like the following:

Proximate questions (the "how" questions)

1. How does the trait's causal mechanism work?
2. How does the trait develop in the organism?

Evolutionary questions (the "why" questions)

3. Why is the trait adaptive (in its selective environment)?
4. Why did the trait evolve as it did?

Evolutionary medicine aims to bring the latter two "why" questions into medical reasoning. Why do we sneeze? Why do infants reflexively hold their breath underwater? Why do certain populations have Tay-Sachs disease? The hope is that posing these questions (and searching for their answers) will lead to lines of inquiry that more firmly ground medicine in biological evolution. With a firmer grasp of ultimate and proximate explanations, we can begin to tease apart the themes often found in evolutionary medicine and how these questions address them. The "why" questions in evolutionary medicine tackle four broad themes, as I have noted elsewhere (Cournoyea 2013):

the macro-domain (long-term human evolution)

(1) the origin and adaptive function of physiological processes (like breastfeeding or yawning)
(2) the adaptive use of supposed dysfunctions mismatched to modern lifestyles (like fever or autism)

the micro-domain (short-term microorganism or human evolution)

(3) the evolutionary mechanisms that shape modern host-pathogen interactions (like HIV/AIDS)
(4) the genetic dynamics of human populations (like heterozygote advantage with sickle-cell anemia).

Although both domains use evolutionary reasoning, each domain is unique in its subject matter, methodology, epistemic standards, and implications. Tinbergen's "why" questions can be applied to both domains, but answers in the macro-domain remain speculative—theorizing about the adaptive significance of ancient bodily traits. The micro-domain is not as epistemologically or methodologically problematic, since experimental evolution and genotyping, for example, can be conducted and tested on contemporary organisms. When the evolutionary medicine literature lumps these domains together (without noting their unique styles of reasoning, evidence, and implications), it becomes possible to conflate success in the micro-domain for success in both domains. Unfortunately, this has been common practice.

A similar distinction has been made by Pierre-Olivier Méthot (2011) between forward-looking evolutionary medicine and backward-looking Darwinian medicine—these seem to align with the micro- and macro-domains, respectively. While the terms "evolutionary" and "Darwinian" are often used interchangeably in the literature, Méthot draws a methodological distinction between these research traditions and the role they play in research and clinical medicine. Forward-looking evolutionary medicine "tries to predict the effects of ongoing evolutionary processes on human health and disease in contemporary environments (e.g., hospitals)" while backward-looking Darwinian medicine "typically applies evolutionary principles from the vantage point of the evolutionary past of humans (here, the Pleistocene epoch) in order to assess present states of health and disease among populations" (76). Only Darwinian medicine is epistemologically and methodologically cohesive; evolutionary medicine is a disciplinary umbrella under which diverse research agendas share their use of evolutionary thinking. This distinction is apt but may lead to some confusion. Both evolutionary and Darwinian research traditions employ backward-looking ultimate explanations and forward-looking predictive models, at times—the central difference is one of time-scale. For example, the evolutionary processes and outcomes of antibiotic resistance, a topic typically explored in evolutionary medicine, may require backward-looking evolutionary thinking to trace the genetic ancestors of bacterial populations; a researcher may need to look "back" through ancestral lineages to understand the mutational process that has led to resistant strains. Méthot is aware of this potential confusion and qualifies that these explanatory styles are not used absolutely.

These nuances about naming may seem trivial, but they remain important to critical debates about what evolutionary medicine is and how it should be used. Although the micro-domain may be epistemologically and methodologically challenging, these challenges are acknowledged and debated by microbiologists, geneticists, and even physicians. The macro-domain remains problematic, and its relevance to medical research and practice remains unclear. For the remainder of the chapter, I use the term *evolutionary medicine* to refer exclusively to the macro-domain, which has been the focus of recent critiques. Having looked at these preliminaries, let me begin to sketch my two central criticisms of evolutionary medicine.

Adaptationism and Ancestral Ideals of Health

Adaptationism is an approach to evolutionary thinking that privileges natural selection over non-selective evolutionary forces, such as developmental constraints (which constrain variability) or drift (change in populations due to random sampling). To the adaptationist, fever evolved because it gave organisms a selective advantage: the ability to survive pathogens and have more offspring with the capacity for fever themselves. This style of reasoning offers a powerful approach for understanding why organisms are so well adapted to their environments. In general, evolutionary medicine is committed to the following adaptationist stances: (1) adaptations give us insight into why our bodies are so vulnerable to disease (the vulnerability stance);

(2) adaptations elucidate the mismatches between our ancestral biology and modern lifestyles (the diseases of civilization stance); and (3) adaptationism is important because it unifies disparate pieces of medical knowledge about the evolved body (the unificationist stance).

The adaptationism in these stances is not singular. Godfrey-Smith (2001) has outlined three kinds of adaptationism: explanatory, methodological, and empirical. Explanatory adaptationism claims that the uncanny fit between organism and environment should be the central puzzle in biology, and this puzzle is most effectively solved by appeal to natural selection. Methodological adaptationism is a pragmatic approach to conceptualizing organismal design that recommends adaptationist thinking as an organizing concept. Williams and Nesse (1991, 1994) are committed to methodological adaptation, although they do not always hold explanatory adaptationism (Méthot 2015). When hypothesizing the function of fever, we test evolutionary hypotheses (methodologically) by carefully considering all adaptive and non-adaptive possibilities. This does not imply that *all* traits are best explained by adaptationism; the strongest hypothesis for some traits may be that they have resulted from non-selective forces (Nesse 2011).

Empirical adaptationism asserts that natural selection is the most influential and important force in biological change. Evolutionary medicine's strongest advocates appear to be committed to empirical adaptationism, specifically a pan-selectionist view of biological change where non-adaptive forces have little influence (Valles 2012). This is not to say that they advocate a view of the body in which all traits are adaptive, but that these traits were likely adaptive (or the trade-offs of other adaptations) in the environment in which we evolved. Many theorize that the human body's adaptations took place during the Pleistocene epoch (approximately 2 million years ago until about 12,000 years ago), during what John Bowlby (1969) has called the environment of evolutionary adaptedness. Since nearly every aspect of our lives has been altered since that time—from diet to housing to social structures—evolutionary medicine argues that our bodies are not fit for modern environments. Yet the environment of evolutionary adaptedness has been challenged both for its oversimplification of our ancestral environment and because we can do little more than speculate about its conditions or impact on ancestral populations (Buller 2005). There is some debate and discussion of these critiques, but the vast majority of evolutionary medicine's proponents appear committed to empirical adaptationism. Empirical adaptationism is taken for granted in two textbooks (Trevathan et al. 2008, Gluckman et al. 2009), though Gluckman et al. at least note that adaptive arguments are hypothetical. Although empirical adaptationism is assumed in Nesse and William's early work, Nesse (2011) does warn students about accepting adaptive hypotheses without careful and thorough consideration. But even with careful consideration, empirical adaptationism remains speculative. These concerns were the thrust of Gould and Lewontin's (1979) profound critique of the adaptationist program, and most biologists have heeded their call to consider alternative explanations.

There may also be a more nuanced critique of empirical adaptationism, articulated both by Canguilhem's contextualism (Sholl 2014) and Walsh's situated adaptationism (Walsh 2012). These views emphasize the dynamic interplay between organisms and their environments, putting the organism back into naturalist accounts of health, such that "health and disease are not to be found in the separate parts or matter comprising organisms, but in their total organization" (Sholl 2014: 142). This view of health undermines the importance of the environment of evolutionary adaptedness because "organisms are not only shaped by their past, but construct their present, healthy behavior is that which allows organisms to offset potentially problematic 'mismatches' or evolved constraints" (161). These "ecological" accounts are also echoed by approaches to health in disability studies and health promotion, which attempt to situate health in the dynamic organism-environment relationship.

Without a contextual account of health, we run the risk of idealizing the lives of our Stone Age ancestors. Our modern environment is likely responsible for many "diseases of civilization," but these diseases are not simply the result of an environmental mismatch. We cannot so simply naturalize health. Since, as Nesse (2011) asserts, evolutionary medicine is only one approach to medicine among many, it must adhere to the ethical and personal idiosyncrasies of a patient's socio-cultural context and wider normative beliefs about health and illness (Lewis 2008). If suffering is the root of illness, then a patient might be healthy despite adaptive dysfunctionality or unhealthy while the body is functionally aligned with its evolutionary history. It is too simple to say that "medicine is based on biology and biology is based on evolution" (Nesse 2008: 416). Evolutionary medicine might even commit the naturalistic fallacy of suggesting that the natural should be normal, especially when naturalizing health/disease using speculative hypotheses about our evolutionary past. We should be wary of claims that certain biological "facts" are natural; naturalness is never a good enough reason to suggest that something should be normal. These concerns about adaptationism underscore my central concern: are ultimate explanations ever relevant to clinical practice? I am not optimistic, and I turn to this issue now.

The Clinical Irrelevance of Evolutionary Medicine

Clinical medicine aims to alleviate suffering through pragmatic, patient-centered, interventionist treatments. This aim is as old as medicine, enshrined in the Hippocratic Oath. Medical research that has no applicability to clinical practice may be interesting but irrelevant, details better suited to a basic science. This is not to say that research must always have direct applications to the clinic—developments in the basic sciences may eventually lead to advances in practice, even if we do not know when or in what capacity. The clinical applicability of evolutionary explanations has been a perennial concern, with Nesse and Williams (1994) admitting that "medicine is a practical enterprise, and it hasn't been immediately obvious how evolutionary explanations might help us prevent or treat disease" (241). In the last five years, critiques from the philosophy of medicine have challenged the epistemic and pragmatic usefulness of such ultimate explanations to clinical medicine, with varied degrees of optimism (Ruse 2012, Valles 2012, Cournoyea 2013, Méthot 2015).

At the heart of these concerns is whether ultimate explanations can offer clinical guidelines independent of proximate explanations. Even though a complete explanation of a biological trait would include both proximate and ultimate explanations, these two explanations are epistemically independent (Griffiths 2009). Ultimate explanations are essential to the biological sciences, but they are merely heuristic in medicine. Proximate explanations should take precedence because they offer the possibility of intervention. This point is reinforced by the confidence we can have in proximate explanations that we cannot have in ultimate explanations. The need for practical certainty in medicine may even ethically preclude the speculative claims of adaptationist, ultimate explanations. Ruse (2012) briefly notes this point, citing Schaffner's (1993) thorough discussion of how functional teleological language is merely heuristic and eventually superseded by mechanistic explanations.

Ultimate explanations do not always point us in the right direction. Knowing the evolutionary history and adaptive value of fever, for example, does not necessarily lead us to understand its proximate causes. Indeed, Nesse and Stearns (2008) acknowledge that despite clinically relevant recommendations,

> [. . .] does understanding evolution change dramatically what a physician does in her day-to-day work? In general it does not, and should not. Clinical decisions based

on theory alone are notoriously suspect. Treatment should, whenever possible, be based on controlled studies of treatment outcomes. However, lack of evolutionary understanding among physicians fosters misunderstanding about issues as important as aging, diet, and when it is wise to use medications to block defensive responses. While there is a trend for doctors to just carry out protocols, we want doctors to have a deep knowledge base so their decisions are informed by understanding the body and disease. Better decisions come from doctors who understand the ecology of immune responses, the evolutionary reasons for polygenic diseases, the phylogeny of cancer cells, and the origins of antibiotics.

(41–42)

Yet these "controlled studies" explore the proximate causes of treatment options, and the "better decisions" come from the micro-domain: immune ecologies, intergenerational genetic diseases, the adaptability of cancer cells, and the development of antibiotic resistance. So what is the place of macro-domain evolutionary medicine in clinical practice? It remains unclear. Whether macro-domain evolutionary theory can influence interventionist medicine remains an open (and controversial) question—one that might need to be answered one speculative hypothesis at a time (Nesse 2011).

Consider Nunn et al.'s (2015) review of the Inaugural Meeting of the International Society for Evolution, Medicine and Public Health (ISEMPH) held in March 2015. They review some of the notable speakers who discussed putative evolutionary mismatches: shoes weaken our feet and predispose us to injury; unnecessary caesarian deliveries may lead to obesity and immune disorders; in-vitro fertilization may harm newborns "by bypassing the evolutionary norm of postcopulatory cryptic female choice such as choosing among particular sperm" (127). These are speculative hypotheses about the impacts of modern environments that seem to offer the potential for direct clinical applications, but before any of these hypotheses lead to clinical recommendations, we need proximate explanations to validate such ultimate theories.

It may be that an evolutionary perspective offers us a more complete "feeling for the organism" (Nesse 2008, Nesse et al. 2010), unifying physiology's "hodgepodge of unconnected facts" (Nesse 2008: 427). This unificationist stance is largely educational, with the potential to encourage certain lines of inquiry and foster a "deep knowledge" of the human body (as Nesse and Stearns note above). Yet evolutionary medicine is not the only field vying for curricular time in medical schools, and it remains unclear that ultimate explanations would have a positive impact on clinical decision-making. A deep knowledge of evolution may help us in answering "What does it mean to be a human organism?" (Gluckman and Bergstrom 2011), but clinical medicine has little use for such speculations, as we see with the example of fever.

An Example: The Adaptive Value of Fever

Body temperature is dynamically maintained in a complex homeostasis. This is critical for physiological reactions, which are dependent on temperature, and varies from bodily core to periphery and during the time of day; these variations are also crucial in regulating tissue functions and maintaining circadian rhythms. Temperature is dynamically held at a set-point, and when actual core temperature differs from this point, a negative feedback loop (via the anterior hypothalamus) induces heat dissipation mechanisms such as sweating (Cannon 2013).

Fever occurs when body temperature rises above an individual's healthy (or normal) range (usually 35.6°C–38.2°C when taken orally), a process triggered by a wide variety of infectious and non-infectious agents. These agents are either endogenous (e.g., cytokines or prostaglandins) or exogenous (e.g., bacterial proteins) and stimulate the hypothalamus to increase the homeostatic set-point. Fevers differ from other, non-pathological elevations in temperature (i.e., hyperthermia, caused by exercise, heatstroke, drug reactions, etc.), because of this change in set-point (Cannon 2013).

The metabolic and behavioral changes that induce fever are costly—febrile animals have significant increases in both metabolism and oxygen consumption (Manthous et al. 1995). Fevers may even lead to collateral tissue damage or morbidities caused by metabolic demands that exceed the capabilities of its host (Hasday et al. 2000). Fevers can be fatal, especially in young children, and both environmental adjustments and medications are used to treat them. Paracetamol (acetaminophen) is the most commonly administered analgesic and antipyretic (i.e., fever preventative) to relieve the discomfort of fever and reduce temperatures; it acts by antagonizing prostaglandins, but it may ease discomfort because it helps with pain, rather than reducing one's fever (Best and Schwartz 2014). Intensive care units (ICUs) commonly prescribe paracetamol for its analgesic and antipyretic effects (Jefferies et al. 2012, Suzuki et al. 2015).

The adaptive value of fever has been studied for more than 40 years. Fever in response to infection occurs in a wide variety of endothermic and ectothermic animals, including mammals, birds, ectothermic vertebrates, arthropods, and even annelids (Hasday et al. 2000). When infected with bacterial pathogens, both desert iguanas and goldfish move to warmer locations; mortality increases when these species are kept at afebrile temperatures (Covert and Reynolds 1977, Kluger 1978). In a fascinating example of convergent evolution, honeybees also generate hive-wide fever in response to a fungal, heat-sensitive pathogen (Starks et al. 2000). The widespread incidence of fever strongly suggests that it is an ancient, adaptive response to infection that may have evolved over 600 million years ago; this persistence is remarkable given fever's metabolic cost and potential harms (Hasday et al. 2000).

The importance of fever for the human defense system is more obscure, with evidence tending to indicate that antipyretic treatments are sometimes harmful. Clinical recommendations remain highly specific to the circumstances of a patient's condition. A multicenter prospective observational study conducted by Lee et al. (2012) found that the administration of non-steroidal anti-inflammatory drugs (NSAIDs) or acetaminophen to septic patients increased mortality. They hypothesize four potential explanations for the increase: (1) lowered body temperatures might prevent anti-viral and anti-bacterial effects of fever; (2) NSAIDs and acetaminophen may sometimes be toxic; (3) patients that failed to develop a fever had worse outcomes, or (4) some combination of these factors. This supports the conclusion that fever may be "naturally protective" (9), but only with cases of infection. Schulman et al. (2005) similarly found that aggressive antipyretic treatment was correlated with higher mortality in a randomized, prospective study of 82 patients. Animal models infected with influenza are also at increased risk of mortality with antipyretic treatment (Eyers et al. 2010).

In contrast to these results, several recent studies have concluded that antipyretic treatment is clinically neutral or beneficial. A meta-analysis of three randomized control trials showed that antipyretic treatment was not associated with better or worse outcomes in critically ill patients (Neto et al. 2014). The recently completed HEAT trial (a long-anticipated multicenter, randomized, placebo-control trial in Australia and New Zealand) concluded that ICU patients, who were suspected to have infections, did not spend less time in the ICU when treated with acetaminophen and exhibited no difference in "28-day mortality, 90-day mortality, or survival time to day 90" (Young et al. 2015: 2223). In the largest multicenter retrospective, observational study of paracetamol use in the ICU, results from four Australian ICUs (and 15,000 patients) found that antipyretic paracetamol therapy actually lowered mortalities, contrary to the hypothesis; however, survival varied with the severity of illness, suggesting the need for further research (Suzuki et al. 2015).

Despite the evidence that some fevers have adaptive significance, even the most thorough reviews conclude that clinical guidelines are tough to formulate broadly, since "the complex interactions among immunological and homeostatic mechanisms in critically ill patients precludes an accurate prediction of the ultimate effect of fever in such patients" (Hasday et al. 2000: 1900). Despite the potentially detrimental effects of antipyretics and "a greater understanding of evolved defence mechanisms" like fever, researchers who advocate adaptive explanations conclude that "[c]urrent [evolutionary] research is insufficient to warrant changing clinical practice but indicates the urgent need for further studies" (Best and Schwartz 2014: 92). Although evolutionary medicine has emphasized the naturalness and adaptive efficacy of fever, it offers little more than encouragement for further research. We are still gathering evidence about the best uses of antipyretics, and the burden should not fall on evolutionary medicine to provide a rationale for clinical guidelines.

Fever does seem to be an excellent candidate for adaptive thinking: unifying our knowledge of infection and bodily defenses, and explaining why antipyretics may sometimes be harmful. Williams and Nesse (1991) mention this in their pioneering article, noting that when fever is pharmaceutically blocked, it "may" be easier to resist infection—a conclusion that, for them, "clearly illustrates use of the adaptationist program to make medically important and heuristically useful predictions" (6–7). Explanatory and methodological adaptationism also seem well suited to an exploration of fever, especially considering its prevalence throughout the animal kingdom.

There is one final concern. Evidence used to evaluate whether fever is functionally adaptive is the same as that used to evaluate whether fever should be treated. Inferences from homology and a comparative analysis of other bodily vulnerabilities give us some insight into the adaptiveness of febrile responses, but the details of these inferences are discovered in clinical trials and physiological research like those outlined above. We have good reason to believe that fever is adaptive, but whether fever is adaptive only during certain bacterial infections, or below a certain temperature threshold, or in certain patient populations is to be established by clinical research. And this very research allows physicians to construct the standards of clinical practice. Ultimate explanations, then, may be simply redundant.

Proximate explanations gleaned from controlled studies remain essential to guiding physicians in treating fevers. Antipyretic treatment must be evaluated in light of the infection's progression, the fever's intensity, and the patient's capacity to withstand increased metabolic demands. The adaptive value of fever does not suggest that it should always be left to take its course or that patient outcomes (or comfort) would not be better aided with medications. Allowing most fevers to run their course may very well be the ideal clinical recommendation, but this needs to be based on clinical trials rather than evidence from evolution.

The Evolutionary Biologist's Tale

Where does this leave the evolutionary biologist's story in medicine? On optimistic but dubious terrain, I'm afraid. As Anne Gammelgaard (2000) argues, the natural goals of evolutionary theory and medicine are distinct and potentially incompatible. The goals of clinical medicine are specific to the suffering and life history of each patient as they aim to live a healthy life; however, macro-domain evolutionary explanations offer little more than speculative hypotheses about our ancestral past. Such hypotheses may lead us to problematically naturalize evolved functions and "exemplify the naturalistic fallacy since the healthy functioning of the individual cannot be derived from the evolutionary history of that individual" (115). These challenges must be balanced against the need for medical students and physicians to appreciate the complex history of our biologies. Ultimately, evolutionary medicine must itself adapt, taking such trade-offs and philosophical vulnerabilities seriously.

References

Amundson, R. (2005) *The changing role of the embryo in evolutionary thought: Roots of Evo-Devo*, Cambridge: Cambridge University Press.

Antolin, M. F., Jenkins, K. P., Bergstrom, C. T., Crespi, B. J., De, S., Hancock, A., Hanley, K. A., Meagher, T. R., Moreno-Estrada, A, Nesse, R. M., Omenn, G. S., & Stearns, S. C. (2012) "Evolution and Medicine in Undergraduate Education: A Prescription for All Biology Students," *Evolution* 66(6): 1991–2006.

Ariew, A. (2003) "Ernst Mayr's 'Ultimate/Proximate' Distinction Reconsidered and Reconstructed," *Biology and Philosophy* 18: 553–65.

Best, E. V., and M. D. Schwartz (2014) "Fever," *Evolution, Medicine, and Public Health* 1(92).

Bowlby, J. (1969) *Attachment: Vol. 1 of attachment and loss*, London: Hogarth Press.

Buller, D. J. (2005) *Adapting minds: Evolutionary psychology and the persistent quest for human nature*, Boston: MIT Press.

Cannon, J. G. (2013) "Perspective on Fever: The Basic Science and Conventional Medicine," *Complementary Therapies in Medicine* 21(SUPPL.1): S54–60.

Cournoyea, M. (2013) "Ancestral Assumptions and the Clinical Uncertainty of Evolutionary Medicine," *Perspectives in Biology and Philosophy* 56(1): 36–52.

Covert, J. B., and W. W. Reynolds (1977) "Survival Value of Fever in Fish," *Nature* 267: 43–5.

Duffy, T. P. (2011) "The Flexner Report—100 Years Later," *Yale Journal of Biology and Medicine* 84(3): 269–76.

Eyers, S., Weatherall, M., Shirtcliffe, P., Perrin, K., & Beasley, R. (2010) "The Effect on Mortality of Antipyretics in the Treatment of Influenza Infection: Systematic Review and Meta-Analysis," *Journal of the Royal Society of Medicine* 103(10): 403–11.

Gallup Jr., G. G., Reynolds, C. J., Bak, P. A., & Aboul-Seoud, F. (2014) "Evolutionary Medicine: The Impact of Evolutionary Theory on Research, Prevention, and Practice," *The Journal of the Evolutionary Studies Consortium* 6(1): 69–79.

Gammelgaard, A. (2000) "Evolutionary Biology and the Concept of Disease," *Medicine, Health Care, and Philosophy* 3(2): 109–16.

Glaze, A. L., and M. J. Goldston (2015) "U.S. Science Teaching and Learning of Evolution: A Critical Review of the Literature 2000–2014," *Science Education* 99(3): 500–18.

Gluckman, P. D., A. Beedle, and M. Hanson (2009) *Principles of evolutionary medicine*, Oxford: Oxford University Press.

Gluckman, P. D., and C. T. Bergstrom (2011) "Evolutionary Biology within Medicine: A Perspective of Growing Value," *British Medical Journal* 343: d7671.

Godfrey-Smith, P. (2001) "Three kinds of adaptationism," In S. H. Orzack and E. Sober (eds.) *Adaptationism and optimality*, Cambridge: Cambridge University Press. 335–57.

Gould, S. J., and R. Lewontin (1979) "The Spandrels of San Marcos and the Panglossian Paradigm: A Critique of the Adaptationist Program," *Proceedings of the Royal Society of London* (Series B, Biological Sciences) 205(1161): 581–98.

Griffiths, P. E. (2009) "In What Sense Does 'Nothing Make Sense except in the Light of Evolution'?," *Acta Biotheoretica* 57(1–2): 11–32.

Hasday, J. D., Fairchild, K. D., & Shanholtz, C. (2000) "The Role of Fever in the Infected Host," *Microbes and Infection* 2(15): 1891–1904.

Hidaka, B. H., Asghar, A., Aktipis, C. A., Nesse, R. M., Wolpaw, T. M., Skursky, N. K., Bennett, K. J., Beyrouty, M. W., & Schwartz, M. D. (2015) "The Status of Evolutionary Medicine Education in North American Medical Schools," *BMC Education* 15: 38.

Jefferies, S., M. Saxena, and P. Young (2012) "Paracetamol in Critical Illness: A Review," *Critical Care and Resuscitation* 14(1): 74–80.

Kluger, M. J. (1978) "The Evolution and Adaptive Value of Fever," *American Scientist* 66(1): 38–43.

Laland, K. N., Odling-Smee, J., Hoppitt, W., & Uller, T. (2013) "More on How and Why: Cause and Effect in Biology Revisited," *Biology and Philosophy* 28(5): 719–45.

Lee, B. H., Inui, D., Suh, G. Y., Kim, J. Y., Kwon, J. Y., Park, J., ... & Koh, Y. (2012) "Association of Body Temperature and Antipyretic Treatments with Mortality of Critically Ill Patients with and without Sepsis: Multi-Centered Prospective Observational Study," *Critical Care* 16(1): 1–13.

Lewis, S. (2008) "Evolution at the intersection of biology and medicine," In W. R. Trevathan, E. O. Smith, and J. J. McKenna (eds.) *Evolutionary medicine and health: New perspectives*, New York: Oxford University Press. 399–416.

Manthous, C. A., Hall, J. B., Olson, D., Singh, M., Chatila, W., Pohlman, A., ... & Wood, L. D. (1995) "Effect of Cooling on Oxygen Consumption in Febrile Critically Ill Patients," *American Journal of Respiratory and Critical Care Medicine* 151(1): 10–4.

Mayr, E. (1961) "Cause and Effect in Biology," *Science* 134(3489): 1501–6.

Méthot, P.-O. (2011) "Research Traditions and Evolutionary Explanations in Medicine," *Theoretical Medicine and Bioethics* 32(1): 75–90.

Méthot, P.-O. (2015) "Darwin, evolution, and medicine: Historical and contemporary perspectives," In T. Heams, P. Huneman, G. Lecointre, & M. Silberstein (eds.) *Handbook of evolutionary thinking in the sciences*, Netherlands: Springer. 587–617.

Nesse, R. M. (2008) "The importance of evolution for medicine," In W. R. Trevathan, E. O. Smith, and J. J. McKenna (eds.) *Evolutionary medicine and health: New perspectives*, New York: Oxford University Press. 416–33.

Nesse, R. M. (2011) "Ten Questions for Evolutionary Studies of Disease Vulnerability," *Evolutionary Applications* 4: 264–77.

Nesse, R. M., and S. C. Stearns (2008) "The Great Opportunity: Evolutionary Applications to Medicine and Public Health," *Evolutionary Applications* 1: 28–48.

Nesse, R. M., and G. C. Williams (1994) *Why we get sick: The new science of Darwinian medicine*, New York: Vintage.

Nesse, R. M., Bergstrom, C. T., Ellison, P. T., Flier, J. S., Gluckman, P., Govindaraju, D. R., Niethammer, D., Omenn, G. S., Perlman, R. L., Schwartz, M. D., Thomas, M. G., Stearns, S. C., & Valle, D. (2010) "Making Evolutionary Biology a Basic Science for Medicine," *Proceedings of the National Academy of Sciences USA* 107(suppl. 1): 1800–07.

Neto, A. S., Pereira, V. G. M., Colombo, G., Scarin, F. C. D., Pessoa, C. M. S., & Rocha, L. L. (2014) "Should We Treat Fever in Critically Ill Patients? A Summary of the Current Evidence from Three Randomize," *Einstein (São Paulo)* 12(4): 518–23.

Nunn, C. L., Wallace, I., and C. M. Beall (2015) "Connecting Evolution, Medicine, and Public Health," *Evolutionary Anthropology* 24: 127–9.

Pennisi, E. (2009) "Darwin Applies to Medical School," *Science* 324(5924): 162–3.

Pennisi, E. (2011) "Darwinian Medicine's Drawn-Out Dawn," *Science* 334(6062): 1486–7.

Ruse, M. (2012) *The philosophy of human evolution*, Cambridge: Cambridge University Press.

Schaffner, K. F. (1993) *Discovery and explanation in biology and medicine*, Chicago: University of Chicago Press.

Sholl, J. (2014) "Evolution and Normativity," Doctoral Dissertation. KU Leuven, Belgium.

Scholl, R., and M. Pigliucci (2015) "The Proximate-Ultimate Distinction and Evolutionary Developmental Biology: Causal Irrelevance versus Explanatory Abstraction," *Biology and Philosophy* 30: 653–70.

Schulman, C. I., Namias, N., Doherty, J., Manning, R. J., Li, P., Elhaddad, A., Lasko, D., Amortegui, J., Dy, C. J., Dlugasch, L., Baracco, G., & Cohn, S. M. (2005) "The Effect of Antipyretic Therapy Upon Outcomes in Critically Ill Patients: A Randomized, Prospective Study," *Surgical Infections* 6(4): 369–75.

Starks, P. T., C. A. Blackie, and T. D. Seeley (2000) "Fever in Honeybee Colonies," *Naturwissenschaften* 87(5): 229–31.

Suzuki, S., Eastwood, G. M., Bailey, M., Gattas, D., Kruger, P., Saxena, M., ... & Bellomo, R (2015) "Paracetamol Therapy and Outcome of Critically Ill Patients: A Multicenter Retrospective Observational Study," *Critical Care* 19(162): 1–10.

Tinbergen, N. (1963) "On Aims and Methods of Ethology," *Z. Tierpsychologie* 20: 410–33.

Trevathan, W. R., E. O. Smith, and J. J. McKenna (2008) *Evolutionary medicine and health: New perspectives*, New York: Oxford University Press.

Valles, S. A. (2012) "Evolutionary Medicine at Twenty: Rethinking Adaptationism and Disease," *Biology and Philosophy* 27(2): 241–61.

Walsh, D. (2012) "Situated adaptationism," In W. P. Kabasenche, M. Rourke, and M. H. Slater (eds.) *The environment: Philosophy, science, and ethics*, Boston: MIT Press. 89–116.

Williams, G. C., and R. M Nesse (1991) "The Dawn of Darwinian Medicine," *Quarterly Review of Biology* 66(1): 1–22.

Young, P., Saxena, M., Bellomo, R., Freebairn, R., Hammond, N., Van Haren, F., Holliday, M., Henderson, S., Mackle, D., McArthur, C., McGuinness, S., Myburgh, J., Weatherall, M., Webb, S., & Beasley, R. (2015) "Acetaminophen for Fever in Critically Ill Patients with Suspected Infection," *The New England Journal of Medicine* 373(23): 2215–2224.

Zampieri, F. (2009) "Medicine, Evolution, and Natural Selection: An Historical Overview," *Quarterly Review of Biology* 84(4): 333–355.

Further Reading

Gluckman, P. D., A. Beedle, and M. Hanson (2009) *Principles of evolutionary medicine*, Oxford: Oxford University Press. (the most recent textbook on evolutionary medicine)

Méthot, P.-O. (2015) "Darwin, evolution, and medicine: Historical and contemporary perspectives," In T. Heams, P. Huneman, G. Lecointre, & M. Silberstein (eds.), *Handbook of evolutionary thinking in the sciences*, Netherlands: Springer. 587–617. (a comprehensive historical and theoretical overview of the field)

Nesse, R. M., and G. C. Williams (1994) *Why we get sick: The new science of Darwinian medicine*, New York: Times Books. (the book that began evolutionary medicine as a distinct field of inquiry)

Valles, S. A. (2011) "Evolutionary medicine at twenty: Rethinking adaptationism and Disease," *Biology and Philosophy* 27(2): 241–61. (a critique of evolutionary medicine's pan-selectionist adaptationism)

Zampieri, F. (2009) "Medicine, evolution, and natural selection: An historical overview," *Quarterly Review of Biology* 84(4): 333–355. (a detailed history of Darwinian and evolutionary medicine)

44
PHILOSOPHY OF NURSING: CARING, HOLISM, AND THE NURSING ROLE(S)

Mark Risjord

1. Introduction

One of the central philosophical questions about nursing is: Who is a nurse? In other words, is the nursing role defined by distinctive responsibilities, abilities, or domains of knowledge? And if so, what are they? This is a question of identity for practicing nurses. New nurses will often wonder whether a problem falls within their domain of expertise or responsibility. Should I respond directly, they ask themselves, or do I need to refer this problem to the physician (or social worker, or nutritionist, or housekeeping, or . . .)? Questions of identity are not only about how one regards oneself. Identity is also a matter of how a person is regarded by others. What a nurse is asked to do by a patient or by other members of a health care team depends in part on what they think "nurses" ought to do. The question of identity is also important to the nursing profession. Nursing is regulated and licensed. Decisions must be made about the content and assessment of nursing curricula, as well as the legal boundaries of professional responsibility. Such decisions depend, at least in part, on judgments about the proper responsibilities, abilities, and expertise of nurses: they depend on who nurses *are*.

The question of identity is made difficult by the enormous variety of health care activities in which people with nursing credentials engage. Nurses are stereotypically employed in hospitals, but nurses also conduct research, teach, serve as administrators, or work in community health settings. Even if we restrict our concern to nurses who have direct contact with patients, there is wide variation in responsibilities and expertise. Consider the following three examples.

Example 1: James is a Licensed Practical Nurse (LPN) in a small hospital. He received his training in one year of community college education, leading to the National Council Licensure Examination (NCLEX-PN). He works with hospital patients recovering from surgery. This means that he makes rounds to check bandages and vital signs, change dressings when needed, and deliver prescribed medication. When patients need assistance, he will help them eat or use the bathroom. Because he spends a lot of time with patients, he often hears their personal stories. He is also often the first to recognize status changes, such as increased body temperature, pain, or bleeding. His work is supervised by a staff nurse, who was educated with a four-year Bachelor of Nursing Science (BSN) degree and who has a Registered Nurse (RN) license. While the patient treatments he implements are ordered by a physician, he has little direct interaction with the physician. Important information about the patients is communicated to James by the staff nurse.

Example 2: Lakshmi is an RN who works in the behavioral health division of a large teaching hospital. She obtained her BSN degree through four years of work at a university-affiliated nursing school. While an undergraduate, she took a number of courses in psychology and biology. After passing her licensure exam and practicing for several years, she went back to school and received a Master of Science in Nursing, specializing in psychiatric nursing. Many of the patients she sees have experienced an acute psychological crisis. She conducts evaluations and consults with a psychiatrist (MD) to develop care plans that respond to the individual patient's needs, complementing pharmacological interventions or helping to manage their side effects. While the laws of the state where Lakshmi works do not permit advanced practice nurses to prescribe medication, she has colleagues in other states who may do so. As a part of a teaching hospital, her unit receives medical students and residents. Her relationship with young physicians is ambiguous and often difficult. She has much more experience than they do, but the physicians have authority to sign off on prescriptions and care plans. Her duties include supervision of several LPNs and non-nursing hospital staff who help feed, bathe, toilet, and, when necessary, restrain patients. While not an official part of her job description, she works more directly with the patient's families than do the psychiatrists in her unit. She educates families about the patient's condition and treatment options, listens to their concerns, and uses this information to shape the patients' care plans.

Example 3: Keisha is a nurse practitioner (NP) in a rural clinic. She passed her licensure exam for the RN after completing her BSN in a four-year nursing school. She went on to get a Master in Public Health (MPH) and worked for several years in the county health department. As a public health nurse, she conducted home visits for the elderly, developed and implemented educational programs, and conducted research for the health department. Seeing a need for more clinics in her area, she worked toward a Doctor of Nursing Practice degree and took the exam to get her NP license. She is now one of three NPs who staff the clinic. She sees patients in the clinic, treating the full spectrum of health needs from the treatment of minor ailments to the management of chronic disease. She diagnoses patient conditions and recommends treatments. State law requires that clinics be supervised by an MD. The physician who supervises her clinic rarely interacts directly with patients. Her primary responsibility is to sign off on the prescriptions, care plans, or referrals recommended by the NPs from two clinics in the area. Patients in the area also have access to clinics staffed by physicians, but many choose a nurse practitioner as their caregiver, saying that they feel like they get a more holistic approach to their health care.

These three examples are a partial representation of the wide variety of roles that are counted as "nursing" in contemporary health care. (The variety is even broader than portrayed here, since all three of these examples are drawn from the American context; nursing varies internationally as well.) While there is variety, there are patterns too. All three examples highlight the way that nurses in clinical practice tend to be on the first line of patient care. In hospital settings, nurses often spend more time interacting with patients and families than do physicians or other health professionals. This is partly because duties like changing bandages, taking vital signs, or educating patients require lengthy personal interactions. The character of their typical interactions with patients has several consequences for nurses.

First, nurses are often the nexus of a patient's care. This is particularly true of nurses who are trained to the RN level and who work in hospital settings. These nurses often find themselves acting as the link between the physician, the patient, and the patient's family. The nurse must communicate changes in the patient's health status or patient concerns to the physician. S/he must help the patient and his or her family understand the diagnosis and treatment. In a hospital's complex environment, the nurse often ends up with the responsibility of making sure that all of the pieces of patient care fall into place. Nurses, then, must understand the full range of factors that influence a patient's health and the elements of the health care system that address them. A further consequence of nurses' front-line status is that nurses are educated in a distinctive way. While the duration of their nursing education may vary from two to eight or more years, nurses are educated in specially accredited schools and must pass specialized exams. A Bachelor of Nursing Science degree invariably requires substantial clinical practice. Coursework will include basic science—anatomy, physiology, pharmacology, genetics—and it will also emphasize the cognitive, emotional, and social aspects of disease and health.

2. Philosophical Questions and the Nursing Role(s)

Against this background, we can ask again: What, if anything, is distinctive of the nursing role? Nursing occupies a place in the division of labor, but is there any good reason why health care roles are divided as they are? If we look across the range of societies, we see health care roles divided in a variety of ways. In traditional societies, childbirth was often attended by specialists, typically female, while the treatment of non-pregnancy-related conditions was delegated to others. By contrast, in 18th-century America, midwives were the only source of health care for those not rich enough to afford the physician or too ill to wait for him to arrive on his occasional rounds. The contemporary European and North American articulation of roles in health care dates largely to the development of modern hospitals and clinics in the 19th century. Florence Nightingale proposed a model of nursing that remains central to contemporary thinking (Nightingale 1969 [1860]). Nurses were to be responsible for the environment of the patient, including the layout of the patient's room, nutritional needs, bandages, and medication. Nightingale's conceptualization of nursing was strongly gendered and hierarchical. Nursing was to be a profession for unmarried, middle-class women, and nurses generally acted at the behest of physicians. As hospitals became more complex, the nursing role became more specialized, more varied, and often more autonomous. Nurses began to specialize in things like surgical care or community health. Some of their earlier responsibilities, such as cooking, doing laundry, or cleaning floors, were allocated to others.

Clearly, many aspects of the nursing role depend on the immediate context. Where physicians are rare and expensive, nursing action must expand to cover a variety of patient needs. In a hospital, where a large number of patients must be served, it makes sense to divide roles more finely. One might conclude that our main question—what is the nursing role?—does not have an interesting general answer. To understand the nursing role, we need only to look at a particular time and place, and to describe the responsibilities actually taken up by those called "nurses." We might call this answer to our question the social-relativist answer: the nursing role is determined by the social environment. The nursing role is *relative* to the social environment in the sense that its character depends on the environment. In the three examples, a social-relativist would say that the duties of each nurse are fixed by the hospital or clinic, and that's all there is to it.

The social-relativist position is probably correct as far as it goes, but is it an answer to our question? When individual nurses ask "What is the nursing role?" they are often concerned with the propriety of demands made upon them: is this the sort of duty that I *ought* to be responsible for? There are "oughts" embedded in the questions of nursing policy, management, education, and licensure as well. The question of the nursing role is "normative" in the sense that it involves judgments about what is good, valuable, or right for nurses to do. This is what makes it a philosophical question. The environment in which a nurse acts is relevant to answering the normative questions about the nursing role, but it is not the full answer. Because of differences in the environment, the duties of a staff nurse in a teaching hospital *should* be different from a nurse practitioner in a rural clinic, and both *ought* to be distinct from the responsibilities of the physician. But more needs to be said to explain *why* the nurses ought to be responsible for one dimension of patient care rather than another.

When nurses reflect on what their role ought to be, their accounts invoke special characteristics of either the nurse or the nurse-patient interaction. In the sections below, we will explore two commonly mentioned ideas. First, caring is central to nursing. Caring is an attitude that might be distinctive of a nurse's comportment, and it may also refer to the way in which a nurse cares *for* the patient. What these ideas of caring entail, and whether caring can serve as a justification for the nursing role, will be explored in Section 3. Second, nurses possess specialized knowledge or expertise. Section 4 will turn to the character of nursing knowledge and give particular attention to the idea of "holism," which has been used to characterize both nursing knowledge and nurse-patient interactions.

3. Care and Self-Care

In trying to characterize what the nursing role ought to be, a number of nurse scholars have used the idea that nurses help a patient care for him or herself. Virginia Henderson used this idea to create a well-known and influential definition of nursing:

> The unique function of the nurse is to assist the individual, sick or well, in the performance of those activities contributing to health or its recovery (or to peaceful death) that he would perform unaided if he had the necessary strength, will, or knowledge. And to do this in such a way as to help him gain independence as rapidly as possible.
> (Henderson 1966: 15)

This conception of nursing puts the patient and his or her abilities in the center of the picture. If we think of a nurse like James in Example 1, it is easy to see the attraction of this definition. If I cut my finger with a knife, I clean it, bandage it, and keep an eye out for infection. Patients who have undergone surgery have temporarily lost this ability to care for themselves. They may need to be fed, bathed, and have their clothing changed. In addition, they have incurred a host of new needs for which they do not have the training: monitoring vital signs, inserting IVs, administering medication. The role of the nurse is to fill the gap between what patients can currently do for themselves and what they would do if they had the "strength, will, or knowledge." Dorthea Orem called this gap the "self-care deficit" (Orem 1971), and similar ideas have played a role in other theories of nursing (Orlando 1961; Wiedenbach 1964).

An argument in favor of the self-care conception of nursing is that it both distinguishes the domain of the nurse from that of the physician and explains the relationship between them. Physicians are, arguably, charged with responsibility to diagnose and treat disease. This is only possible if the patient continues to receive adequate nutrition, bathes, has a clean environment, receives medications at the right times, and so on. In ordinary conditions, we do these things for ourselves. Disease or injury can compromise a person's ability to care for him or herself, and a self-care deficit arises. A necessary health care role, then, is to support the medical treatment prescribed by the physician by ensuring that all of the patient's self-care activities are carried out. This is the nurse's role.

While the idea that nurses address the self-care needs of the patient has been popular, it has also been subject to objections. A first concern is that such a definition puts the domain of nursing responsibility squarely within a hierarchical relationship to the physicians. While there may be contexts where a central responsibility for treatment and recovery is appropriate—surgery or emergency care might be examples—there are many domains of health care where more autonomous action by nurses is necessary. Nurses who conduct visits to home-bound patients encounter a wide range of conditions, some of which are not under the direct care of a physician. In the case of Keisha, the Nurse Practitioner (Example 3), she has substantial professional autonomy and needs it to carry out her responsibilities. While the physician has a supervisory role, it does not make sense to think of the nurse practitioner's role as simply keeping the patient properly fed and cleaned so that the physician's treatment can take effect.

The case of Lakshmi (Example 2) adds additional layers of complexity. Patients with a severe mental illness may never be independent. Moreover, some of her patients have never been competent to make their own decisions. What does it mean for the nurse to perform for such a patient an activity that he or she would have done if s/he "had the necessary strength, will, or knowledge"? In response to this sort of problem, some nursing scholars have argued that to care for patients is to respond to their needs, whether or not these are part of an individual's ordinary self-care activities (Edwards 2001).

Thinking of caring-for as a matter of responding to needs raises another concern about the self-care conception of the nursing role. How are a patient's needs determined? Health is not simply a biological matter. Life aspirations, family considerations, and other values also determine a person's health care needs. Health care providers are obligated to respect the health- care choices of competent patients; to respect their "autonomy." The idea of doing for a competent patient what s/he cannot do for him or herself presupposes, one might argue, that the nurse, not the patient, is determining the patient's needs. When the self-care conception of nursing is used to guide nursing for competent adults, it may fail to respect the patient's autonomy; it is "paternalistic" (Mitchell and Cody 1992). Thinking through this objection may require us to reevaluate our understanding of autonomy and paternalism in relation to the nursing role (Risjord 2014).

The conception of nursing as addressing self-care deficits emphasizes the relational aspect of the concept of care: to care is to care *for* someone or something. Caring is also an attitude or stance. Henderson's definition of nursing, quoted above, is notable for the absence of any caring attitude on the part of the nurse. The "unique function of nursing" could apparently be carried out by machines. To be more charitable, we might read Henderson as presupposing that the nurses have a caring attitude and that such an attitude is necessary to provide good care *for* the patient. The next question, then, is about this implicit notion of caring. What sort of attitude or stance do "caring" nurses exhibit, and why is it necessary for good patient care?

Kristen Swanson has argued that a nurse's therapeutic actions (as described by Henderson, for instance) are supported by three other dimensions of caring, what she calls "maintaining belief," "knowing," and "being with" (Swanson 1993). (We will discuss the characteristics of nursing knowledge in more detail below, and thus focus on the other two dimensions here.) "Maintaining belief" denotes an attitude of support for the patient as he or she confronts or copes with a health problem. It is a faith that the patient can get through the crisis, or if the condition is chronic or terminal, find ways to come to grips with it. Maintaining belief is an attitude shared by teachers and coaches: if the coach didn't think I could get better at my tennis backhand, there would not be much point in her instructions. "Being with" is a matter of being "emotionally present" to the patient, including "not just the side-by-side physical presence but also the clearly conveyed message of availability and ability to endure with" the patient (Swanson 1993: 355). "Being with" captures the affective dimensions of nursing care, but it is more than empathy. The attitude must be communicated to the patient (and perhaps the family). And there must be limits to emotional availability. Human life is full of tragedy, and asking nurses to share unconditionally in their patients' suffering would be an impossible demand.

Adding Swanson's analysis of the nurse's attitude or stance to Henderson's characterization of nursing's therapeutic domain clearly provides a more complete picture of nursing care. It also might be defended in a way that does not depend on a hierarchy between nurse and physician. Steven Edwards argues that caring is grounded in the fact that humans are vulnerable; we have the capacity to feel pain and to suffer (Edwards 2001). Human vulnerabilities give rise to the needs, and these are only partly met by medical diagnostic and treatment regimes. Faith in the patient, specialized knowledge, and emotional support may not be biologically necessary (in the way that nutrition or infection control are), but they too contribute to the patient's health and ability to heal. This sort of support requires a health care professional with direct and extended contact with the patient, and this would be the nursing role.

One might object that using this concept of care to define the nursing role either fails to be unique to nursing or excludes some appropriate activities. It is certainly true that each of the proposed aspects of care is or should be shared by other health care professionals. A physician who was emotionally disengaged or who exhibited no faith that the patient could cope would be a poor physician. Hospital cooks and nutritionists also meet the patients' needs. On the other hand, this conception of caring might seem too demanding. Arguably, at least some nursing activities can be conducted without the emotional engagement described above (Seedhouse 2000). Nurses who are responding to acute crises in an emergency room or on a battlefield might be examples. Moreover, many nurses do not have direct responsibility for patient care: they run research programs, administer staff, teach, or analyze policy. It is not clear how the dimensions of caring discussed above would figure in their roles. Are we to conclude that such heath care practitioners are not nurses, despite their titles, credentials, and training?

The foregoing objections arise insofar as caring is taken to be a *description* of nurses' attitudes and activities. In the discussion of the social-relativist conception of the nursing role, we noted that a mere description of nurses' duties does not answer our question. Similarly, a mere description of nurses' attitudes will not answer our question either. The question is a normative one: how should we understand the nursing role? In Edwards' argument that the nursing role is grounded in human vulnerability to pain and suffering, he notes:

> Nursing can be understood as a moral response to this [vulnerability]. Thus nursing is exposed as an essentially moral enterprise.
>
> (Edwards 2001: 112)

Caring, then, is not a description of nurses, it is a moral dimension of nursing practice. It is an ideal or value embedded in the call to nursing. Treating caring as a value goes some way toward mitigating the foregoing objections. There may be particular nursing activities that do not require the attitude of caring or that do not respond directly to patient needs. But insofar as they are within the appropriate scope of a nurse's duties, they should be consistent with promoting or achieving a caring response to human vulnerability. And as a value, caring should be shared by physicians and other health care personnel. The difference is not in the possession of the value, but the range and type of needs to which nurses respond.

4. Holism and the Nursing Perspective

One way in which nurses make a distinctive contribution to health care is through their particular knowledge or expertise. Florence Nightingale was instrumental in the creation of a nursing profession because she was able to synthesize and express the knowledge that nurses bring to patient care. Nurses, on her view, understood the larger context in which patient healing could take place. This included the light and air in the patient's room, the patient's emotional, cognitive, and spiritual needs, as well as wound care, nutrition, and so on. Later, nurses recognized the social environment of the patient—his or her family, community, and the larger cultural context—as important too. In the early 20th century, nurse scholars argued that nursing was a profession, because it encompassed a unique domain of knowledge.

What is the knowledge distinctive of nursing? Clearly, nurses share much knowledge with other health care providers—knowledge of anatomy, disease process, pharmacology, and psychology, for example—and this is entirely appropriate, even necessary. Many have argued that this shared knowledge must fall within a distinctively nursing perspective on health, healing, and health care. The challenge, then, is to articulate what this distinctive perspective amounts to.

Some contrast the nursing perspective with the medical perspective. Medical research has made great strides through better understanding of the biological underpinnings of normal physiological function and its disruption. This understanding informs medical diagnosis and treatment. If nurses adopt a "medical model" it would, arguably, shape nursing interactions:

> The medical model forces nursing to view health-illness manifestations as organic phenomenon where emphasis is upon disorders in the structure and function of the body. With this disease-oriented approach to clients, the nurse is concerned with underlying defects or structural aberrations . . . The utilization of the medical model compels a person to view disease as the failure of the body as a physiochemical machine, and patients are helped by interventions in bodily processes. . . .
>
> (Phillips 1977: 5)

While the medical model is helpful in organizing knowledge about patients, according to Phillips, it limits nursing. Nursing needs to integrate knowledge of the patient's larger context, such as the patient's emotional or psychological state, relationship to his or her family, or social position. Phillips argues that this larger context should not be understood as mere extension of a mechanistic perspective to emotions or social relationships. If nursing is to be a distinctive profession, it needs a unique framework within which to develop nursing expertise. Phillips suggests that the unique framework for nursing is a model that "views man in his totality in his interaction with the environment" (Phillips 1977: 7). While Phillips does not use the term, many who characterized the nursing perspective in this way call it "holism." The nursing perspective sees the patient and the patient's health care as something more than a mechanistic process. The parts combine irreducibly into a whole; in this sense, it is "holistic."

One answer, then, to the question of what makes nursing knowledge distinctive locates nursing knowledge within a particular paradigm or conceptual framework.

> The new paradigm in nursing has been characterized in multiple ways, with the major emphasis on the irreducible unity of human beings and their worlds, the dynamic unpredictable unfolding of multidimensional life patterns, and the human freedom to choose direction in life based on personal values and meanings. The human is seen as a living unity and a self-interpreting free agent. Health is seen, not as a condition to be characterized as good, bad, more, or less, but as a life pattern experienced qualitatively by humans. . . .
>
> (Cody 2000: 96)

Understood in this way, holism is a body of substantive commitments. The propositions Cody articulates—e.g., that humans are free to choose their direction in life, or that health is a life pattern not to be characterized as good or bad—are not statements that might be confirmed or refuted by careful observation. They are assumptions or presuppositions; statements that should be accepted by nurses to guide their practice and their empirical inquiry. The language of a "paradigm" expresses the idea that scientific knowledge is not based on observation alone. It is guided by fundamental ideas. Big differences in scientific perspectives, sometimes exhibited in dramatic historical shifts in scientific theory, are partly differences about fundamental presuppositions. On this kind of view, then, the distinctive feature of nursing knowledge is found in a unique set of assumptions.

The idea that a holistic paradigm should characterize nursing knowledge has been criticized by nurse scholars on several grounds. First, it is not clear that assumptions of the sort identified by Cody preclude a mechanistic understanding of disease processes. It is possible, for instance, for a nurse or physician to agree that a human being is a "living unity and a self-interpreting free agent" and still deploy their knowledge of pharmacology when exploring treatment options. Arguably, the best care requires both an understanding of biological micro-processes and a broad appreciation of the particular patient and his or her situation. Moreover, if a commitment to the holistic paradigm really did preclude a nurse's use of biological knowledge, then nurses are blocked from finding or using any effective health interventions. Any action that is effective in addressing a patient's health concern must be *causally* effective. To develop such interventions and understand their limitations, we need to know about causes, and this means understanding mechanisms involving organs, tissues, cells, genes, hormones, and so on.

A somewhat different conception of holism was expressed by Rosemary Ellis:

> Holism, if used as the appropriate view for aiding a patient, requires that one be concerned with any factor, be it physiological, social or any other, which affects the patient's health. It requires that the factors be treated in combination, not in isolation. It also means that the combination is not the same as the sum over each factor. Nursing requires the recognition of the inseparability and interdependence of many factors.
>
> (Ellis 1968: 218)

The first point to notice is that this sort of holism concerns practice. Ellis is not treating holism as an assumption, presupposition, or anything else a nurse might believe or disbelieve. Holism, according to Ellis, is something that might be "the appropriate view for aiding a patient," and it requires a specific kind of concern. It is a practical commitment to treat the factors influencing a patient's health in combination. This means that holism is a characteristic of a nurse's *practical* knowledge; her theoretical knowledge need not be holistic.

In her essay, Ellis is implicitly opposing the "reductionism" found in medicine. Physicians were portrayed in this period (and often since) as narrowly concerned with the patient's disease or dysfunction. The nurse, by contrast, is charged with a broad responsibility for the patient and his or her environment. A reductionist approach in this domain would consider a patient's disease or dysfunction in abstraction from the other factors. Because of her role in health care, a nurse cannot be so limited. The nursing role demands that a nurse attend to any and all factors that affect a patient's health. To limit nursing concern to one organ system, or to one dimension of psychological or social dysfunction, would be to ignore the broader context of patient health that has been central to nursing since Nightingale's time.

In response to Ellis's argument for practical holism, one might point out that it relies on a false contrast between medicine and nursing. Just as physicians should not be portrayed as uncaring while nurses are caring, physicians should not be portrayed as reductionist while nurses are holistic in their practice. Practical holism should be part of good medical care too. That said, it remains true that the typical range of a nurse's holistic concern in patient care is broader than a physician's. As the examples in Section 1 illustrate, nurses often have a much richer engagement with the patient's family and personal history. Arguably, then, the scope of holistic care in nursing is broader than holistic care in medicine.

A further objection might be that while a holistic practical attitude describes many nursing situations, it does not describe all of them. While stanching bleeding, one would probably do best to attend just to the wound; the patient's story can wait. And like the use of caring to define the nurse's role, practical holism privileges nurses involved with direct patient care. The parallel with the difficulty we saw in using caring to define nursing suggests that the root of the problem might be the same. In the approaches we have discussed in this section, holism is taken to describe something about nurses or nursing. Perhaps we should understand practical holism as "an ideal, an abstraction mean to orient our awareness, not to depict a tangible outcome" (Thorne 2001: 261). Practical holism is an ideal or value orientation. Thinking about holism this way blunts the objection that holism does not include all nurses or nursing activities. There may well be situations where a holistic approach is irrelevant or may even be counterproductive. Nonetheless, it can remain one of the values that informs the full range of a nurse's activities.

5. Conclusion: Thinking the Nursing Yet to Come

We have seen in the previous two sections that "caring" and "holism" are plausible ways of identifying the nursing role, subject to two provisos. First, neither caring nor holism should be thought of as a description of nursing attitudes, actions, or knowledge. Rather, they are values or ideals embedded in the practice of nursing. Second, neither should be used as an all-or-nothing contrast (e.g., caring versus curing or holism versus reductionism) with other health care providers. Both are values that ought to be exhibited to some degree by any health care provider. The difference lies in the content or scope of the value.

The astute reader will have noticed that this second proviso gives rise to a potential circularity. We began with the question of how to understand the proper scope of nursing. The partial answer developed so far is that nursing practice involves the values of caring and holism. We have insisted that the difference between nurses and other health care providers does not lie in holding caring and holism as values, but in the scope or content of the value. The kinds of situations with which they are confronted mandates the kind of caring and holism valued by nurses. But the situations confronting nurses depend on the appropriate domain of nursing action. In other words, the proper nursing role is defined by . . . the proper nursing role.

Were our project simply one of definition, circularity would be devastating. It sheds no light on the meaning of "inspiring" to define it as "having the effect of inspiring someone," as the online Oxford English Dictionary does (Oxford Dictionaries 2015). But as the first part of this essay indicated, the question about how to understand the nursing role is not a request for a definition. It is a question that arises for practicing nurses when considering the appropriate scope of their duties. The current roles occupied by nurses have a history, and they have been formed by forces both within and outside of nurses' control. At any point in that history, nurses can (and did) ask, is our role what it *ought* to be? Should we resist or advocate for changes in our role? The apparent circularity identified above is a tension between the "is" and the "ought" of the nursing role, between what the nursing role currently is and what the ideals of nursing demand of it. To understand who a nurse *is*, we must adjust the current nursing role(s) in the light of what they *ought* to be.

The deeper question to which we have come, then, is: what should the ideals or values of nursing practice be? We have seen arguments for including caring and holism among the values of nursing, but this is only the beginning of a much larger inquiry. The important point, and the one with which we must close, is that a philosophical question lies at the heart of nursing practice. Nursing practice is implicitly committed to a variety of values. Reflecting on these values is a part of nursing, but it is not something that a typical nurse will do on a typical day in the clinic. Nursing philosophy serves, in John Drummond's phrase, as an "avant-garde" for nursing practice (Drummond 2004). Like the avant-garde in the arts, it explores the space of possibilities. Good practitioners need to tack back and forth between the avant-garde and the mainstream, using explorations of novelty to expand horizons of the familiar. An essential part of the call to nursing, then, is philosophical: to thoughtfully engage "the nursing yet to come" (Drummond 2004).

References

Cody, W. K. (2000) "Paradigm Shift or Paradigm Drift? A Meditation on Commitment and Transcendence," *Nursing Science Quarterly* 13 (2): 93–102.
Drummond, J. S. (2004) "Nursing and the Avant-Garde," *International Journal of Nursing Studies* 41 (5): 525–533.
Edwards, S. D. (2001) *Philosophy of Nursing: An Introduction*, New York: Plagrave Macmillan.
Ellis, R. (1968) "Characteristics of Significant Theories," *Nursing Research* 17 (3):217–222.
Henderson, V. (1966) *The Nature of Nursing*, New York: The Macmillan Company.
Mitchell, G. J., and W. K. Cody (1992) "Nursing Knowledge and Human Science: Ontological and Epistemological Considerations," *Nursing Science Quarterly* 5 (2): 54–61.
Nightingale, F. (1969) *Notes on Nursing: What It Is and What It Is Not*, New York: Dover Publications. Original edition, 1860.
Orem, D. E. (1971) *Nursing: Concpets of Practice*, New York: McGraw-Hill.
Orlando, I. J. (1961) *The Dynamic Nurse-Patient Relationship: Function, Process, and Principles*, New York: G. P. Putnam and Sons.
Oxford Dictionaries (2015) *Inspiring*, Oxford University Press. Accessed September 6, 2015. Available from http://www.oxforddictionaries.com/definition/english/inspiring.
Phillips, J. R. (1977) "Nursing Systems and Nursing Models," *Image* 9 (1): 4–7.
Risjord, M. (2014) "Nursing and Human Freedom," *Nursing Philosophy* 15 (1): 35–35.
Seedhouse, D. (2000) *Practical Nursing Philosophy: The Universal Ethical Code*, London: John Wiley & Sons.
Swanson, K. M. (1993) "Nursing as Informed Caring for the Well-Being of Others," *IMAGE: Journal of Nursing Scholarship* 25 (4): 352–357.
Thorne, S. (2001) "People and Their Parts: Deconstructing the Debates in Theorizing Nuring's Clients," *Nursing Philosophy* 2 (3): 259–262.
Wiedenbach, E. (1964) *Clinical Nursing: A Helping Art*, New York: Springer-Verlag.

Further Reading

Comprehensive introductions to philosophical issues in nursing are found in S. D. Edwards, *Philosophy of Nursing: An Introduction* (New York: Palgrave Macmillan, 2001), and D. Seedhouse, *Practical Nursing Philosophy: The Universal Ethical Code* (London: John Wiley & Sons, 2000).
For a more thorough, but still introductory, discussion of caring, see D. Sellman, "Nursing as Caring," in *Handbook of the Philosophy of Medicine*, edited by T. Schramme and S. Edwards (Dordrecht: Springer, 2017).
Clear discussions of the holism issue are found in P. S. Hawley et al., "Reductionism in the Pursuit of Nursing Science: (In)congruent with Nursing's Core Values?" *Canadian Journal of Nursing Research* 32 (2): 75–88 (2000) and R. Kolcaba "The Primary Holisms in Nursing" *Journal of Advanced Nursing* 25 (2): 290–296 (1997).

45
CONTEMPORARY CHINESE MEDICINE AND ITS THEORETICAL FOUNDATIONS

Judith Farquhar

In 1982, philosopher Renzong Qiu published a review article in the first number of the new journal *Metamedicine*. His topic was "Philosophy of Medicine in China, 1930–1980" (Qiu 1982). As a professional philosopher working for the Chinese government's major social research institute, the Chinese Academy of Social Sciences (CASS), he sought to bring to the attention of an English-reading academic public the theoretical and policy debates that had swirled around the field of traditional Chinese medicine (TCM) during the 20th century. Of necessity, and recognizing the recent international interest in "medicine and philosophy," he distinguished between the natural philosophy that had informed the specialized knowledge of medicine in China for at least 2,000 years and the recently formalized discipline of philosophy of medicine (Qiu 1982: 36). (I will return to "natural philosophy" below.)

Qiu's article shows that the philosophy of medicine in China, which included both academic philosophy and medical theory-building, was an especially lively field in 1982. There was a complex history of intellectual and institutional struggle, dating from the 1920s, between those who sought to replace indigenous medical practices with modern biomedicine and the experts of "traditional" medicine who fought to maintain the value of their field (Lei 2014). Debates had turned at mid-century from merely factual or natural scientific matters to include "philosophical foundations." The journal *Yixue yu zhexue* (*Medicine and Philosophy*) began publishing in Chinese in 1980, just two years after the journal *Medicine and Philosophy* was inaugurated in the United States. Many of the Chinese journal's articles explored the logical and ontological character of TCM through readings of specific bodies of literature, drawn from a vast archive spanning several thousand years.

Colleges of medicine were teaching the history and philosophy of science and medicine to medical students, often working from departments of "Natural Dialectics," a phrase adopted from an obscure philosophy of science work by Friedrich Engels. Introductory textbooks of TCM were revised and reissued with larger and more classically oriented sections on "theoretical foundations." Leading doctors turned their attention to the production of rich histories of ideas in medicine, and critical clinicians advanced new terms to talk about the deep conceptual differences they found between their practice and that of biomedical physicians.

"Epistemology" (*renshilun*), "methodology" (*fangfalun*), and "style of thought" (*siwei fangshi*) were keywords for senior Chinese doctors and graduate students alike, and scholarly books were published on the aesthetics, phenomenology, and metaphysics of Chinese medicine. This philosophical labor undertaken in the 20th century by the modernizing users of an effective and historically deep medical practice was integral to the modern constitution of the clinical and research field of TCM.

As Renzong Qiu pointed out, Chinese medicine, ancient and modern, is nothing if not philosophical. It is, however, quite a recent phenomenon for specialists to abstractly and systematically analyze the "philosophical foundations" of Chinese medicine. And though some of the best philosophers concerned with TCM have been privately dismissed by full-time clinicians as "not very good doctors," and thus unreliable as philosophers or theorists as well, many thinkers and writers have successfully shown deep currents of connection between ancient metaphysics and modern "traditional" medical practice. TCM, notwithstanding all its clinical and scientific modernizations, is one of the few active fields of human endeavor in the contemporary world that keeps a radically different world of things and forces in play, for thought, for healing, and for use in everyday life.

As I will argue in the remainder of this discussion, TCM works with natural processes and entities that are not available for thought in the terms provided by the modernist natural sciences. The fact that an "other world" of things and forces can ground an effective medicine in the 21st century should at least, to adopt Isabelle Stengers' advice, "slow down thought" (Stengers 2005). Modern Chinese philosophers have argued, critically and correctly, that all medicine everywhere adopts epistemological and metaphysical habits from wider fields of thought. Perhaps through engaging with Chinese medicine as philosophy, as some writers in Chinese have been doing for the last half-century, it will be possible to see medicine in general with fresh eyes. At the very least, after coming to terms with the radically different discursive possibilities of the classical Chinese scientific past, it should be easier to perceive some of the "tenacious [metaphysical] assumptions" informing the perceptions and strategies of doctors of all kinds (see Gordon 1988).

What Is TCM?

First it will be necessary to describe the modern field known as traditional Chinese medicine (TCM), especially as it exists in China today. The modern institutional complex that is TCM came into being during the Chinese nation's struggle toward modern nation-statehood. To call the field "traditional" is thus something of a misnomer (and in fact, it is not usually called "traditional" in spoken Chinese). During much of the 20th century, as political modernization advanced, many great and little traditions of medicine in China were lumped together for the sake of argument and referred to as "national medicine" (*guoyi*) or "old medicine" (*jiuyi*). Because, as Sean Lei has shown, there was vocal opposition in government and scientific circles to the continuance of East Asian traditions of medical practice, starting as early as the turn of the 20th century, a number of reform movements were initiated within Chinese medicine. Imagining an integrated national medicine, for example, some traditionally trained scholar-doctors sought to demonstrate their understanding of anatomy, bacteriology, surgery, and other biomedical forms of knowledge, but they also asserted a form of Chinese empiricism, insisting that national medicine was "experiential" and thus required a philosophical foundation that was more phenomenological and pragmatic than positivist (Zhao 1989, Farquhar 1994, Lei 2014).

Even before the 1949 founding of the People's Republic, reformed and expanded medical education, scientific research on natural pharmaceuticals, textbook editing projects, and speculative philosophy drawing on the theoretical heritage of Chinese medicine were becoming important; they were accompanied by efforts in the Ministry of Health to impose a narrowed vision of what national medicine could be in a modern and later revolutionary China (Croizier 1968, Lampton 1977, Zhao 1989, Andrews 2014). Most historians date the birth of a modern system of TCM from the mid-1950s, when Mao Zedong declared in several speeches that "our motherland's medicine" had intrinsic value and should be studied even by those trained in biomedicine (Taylor 2005). From 1956 onward, the institutions of Chinese medicine proliferated with Communist Party support, with a slowing of growth only in the Cultural Revolution years between 1966 and 1976. Schools, hospitals, clinics, professional associations, and many publications offered a national terrain in which a particular version of Chinese medicine could flourish (Scheid 2007). Edited and systematized, technically complex and clinically effective, TCM became a national treasure.

In the early decades of the development of TCM as a distinct scientific endeavor, scholars often spoke of the need to "salvage, sort, systematize, and elevate" the knowledge of the field. Modernization of a 2,000-year tradition, and its public health institutionalization, were thus recognized by many as requiring significant epistemological and metaphysical (including ontological) work. Arguing that Chinese medicine was very often an effective clinical practice, as well as being less invasive than biomedical interventions—a fact acknowledged even by its critics—was not enough to secure continuing official recognition and support, however. As one medical historian said to me, echoing the debates discussed by Renzong Qiu, "Unless you can show that there is theory, there's no hope of the field being acknowledged as medicine." He was referring in part to a persistent tendency in health policy circles to separate the tools of TCM—needles and herbs, manipulation techniques, and pharmaceutical formulary—from its systematic knowledge (the epistemology [*renshilun*] and methodology [*fangfalun*] that provide theory [*lilun*]). Biomedically oriented policy makers, by contrast, have often suggested that needling techniques and individual medicinal herbs could, eventually, be assessed in clinical trials and added one-by-one to the arsenal of "world" biomedicine, making no theoretical alteration to the modernist scientific foundations of biomedicine. There is still no shortage of critics who believe that the conceptual world of TCM, thus "scientized," would then naturally fall by the wayside, superseded by modern science and technology and remembered as a different *system* only by historians.

But clinicians and teachers of medicine know that no medicine is just its tools and techniques; rather, within the world of TCM, many have insisted that a "style of thought" (*siwei fangshi*) particular to TCM can be articulated. Any effort to understand what TCM in modern China is, then, must attend to the philosophical work undertaken by experts in the field in the 20th century, as they drew on the speculative philosophies of a great many ancient writings. These skilled and thoughtful designers of a modern *cum* traditional medicine sought to "preserve the essence and discard the dross" they found in the vast and heterogeneous medical-philosophical archive, which spanned 2,000 years and included tens of thousands of texts. How they decided what was essence and what was dross, how they sorted their heritage for modern use in public health, is a story of philosophy in action. Indeed, TCM may be the only world arena of "applied science" in which a non-modern metaphysical system is still hard at work, grappling with the gritty reality of *things* that go unacknowledged by modern science and other objectivities (Heidegger 1971).

Philosophy: A Metaphysics of Processes

Chinese medicine heals in a world of unceasing transformation. This condition of constant change, this fluidity of material forms, stands in sharp contrast to a (modern Western) commonsense world of discrete entities characterized by fixed essences, which seem to be exhaustively describable in structural terms. Mathematical and physical theories of relativity and indeterminacy notwithstanding, in our everyday life we still assume a Newtonian world of inertial masses, a world in which motion and change result from causes external to entities. Events must be accounted for in a logic of cause and effect, an ultimately mechanical relationship that requires the radical reduction of the plenum of phenomena to objects and forces. Think billiard balls: for one mass to go into motion, it must be directly contacted by another mass already in motion, which has been put into motion by another thing that is also being impelled from outside itself. For a commonsense modernist metaphysics, changes in the nature of an object, not to mention life and death, remain ultimate mysteries that escape mechanical and structural explanations. What drives a viral mutation or the development of an el Niño season? What is the objective difference between being alive (perhaps on a ventilator, or in a coma) and being dead? Such questions have presented challenges to natural scientists, but for the traditional Chinese sciences, questions like this are beside the point. Constant change in form and essence is taken for granted. The challenge is to perceive the patterns of transformation rather than to identify original causes.

The early Chinese sciences were indebted to what we might call a vitalist metaphysics. Generation and transformation are intrinsic to existence; nature in all its diversity is thought of as that which comes about spontaneously, of itself (*ziran*). It is stability and fixity, rather, that call for explanation: indeed, in TCM, stasis is usually diagnosed as a pathology. Things are held in place only through concerted action; motion and change, on the other hand, are a given and seldom need to be explained with reference to their causes. Working with TCM clinicians in the early 1980s, I found that they had little interest in what they called "disease causes" (*bingyin*). Patterns of pathological process (*binglixue*), on the other hand, interested them greatly.

One consequence of this dynamic bias in Chinese medicine is that the body and its organs (i.e., anatomical structures) appear as effects or by-products of the more fundamental physiological processes that are always taking place throughout the body and beyond. Air breathed converts to energy, food and drink become flesh and blood, through processes that can be understood by reading the body's expressive signs. Contenting oneself, as diagnostician, with locating and visualizing an internal structure, such as a tumor or a lesion, would be like closing the barn door after the horse has bolted. For the scholar-doctor, as for the classic philosopher, paying attention to the ongoing patterns of phenomena, or "the myriad things" (*wanwu*), as they become manifest in the world is the only effective way of discerning and intervening in a disordered natural process, that is to say, disease. The seasoned practitioner notes the qualities and forms of illness signs and the changing time and space relationships among them, eventually combining the vital powers of known therapies (such as the efficacies of herbal drugs) to influence developments in a more wholesome direction.

Philosophy: Cosmogony and Transformation

Theorists of Chinese medicine tend to consider the nature of transformation and existence, illness and health, in a cosmogonic framework. Medical practice is informed by a logic of becoming: let's call this a transformative cosmogony. The process through which the world emerged at the origin of everything, and through which it continues to happen and generate myriad transforming things, helps readers and doctors understand how bodies work in the present.

Imagine yourself, then, as a TCM physician in a modern clinic, with blood pressure cuff, prescription pad, and acupuncture needles at the ready. Most of the patients coming to see you, especially in recent years, suffer from conditions that both you and they consider to be "puzzling, recalcitrant, multiple disorders" (*yinan zabing*). This sort of problem, known to English-speaking biomedical practitioners as multiple comorbidity, is often encountered in clinics around the world, and TCM is widely acknowledged in China to be good at managing such problems.

Suppose that a patient comes to you with a chief complaint of radical digestive system breakdown. Almost everything she eats produces stomach pain, diarrhea, and sometimes vomiting. A biomedical physician has diagnosed the condition as celiac disease, and rheumatologists have tried to control her food allergies (which go well beyond gluten intolerance). Nothing has helped, and her condition has only worsened over several years.

As this patient's TCM physician, you "seek the root" (*qiuben*) of her overall condition. This root is not only the original "cause" of the disorder (stress? early malnutrition?) but also the ongoing, deep-seated deviation from healthy physiological processes. Direct action to pacify and regulate discrete parts of the digestive tract will not work; stomach and intestinal malfunction are only symptoms of disorder at the root. More important are the entities known to TCM: yin and yang qi, and dynamic relations between physiological systems (which can be analyzed as the five phases, see below) affecting the activity of the whole body.

In meeting with your patient, you need to collect more observations than she thinks (at first) to share with you. As you inquire into her history, you find that as a world traveler working far from her old home, she has often had to change her diet. When you ask, she tells you that she is able to eat dates, honey, some grains, and fruit, but mainly when she visits her family in the arid far west of China. With more questioning you determine that she sleeps little, works long hours, sometimes has disturbing dreams, and sometimes inexplicably weeps. She has had several broken bones in recent years. You also note in talking with her that she is of a very animated disposition, with sparkling eyes and a ready laugh.

Is there one process that can be identified as responsible for all of these symptomatic expressions? In theory, for modern TCM, the answer would almost always be yes. Even though the body known to Chinese medicine is differentiated, spatially organized in systems, networks, and regions, most clinicians believe it is profoundly interactive, or, in their constant refrain, "a dynamic whole." As you search for the root in this case, you work your way "back" through the signs of disease to put them together as expressions of a particular "holistic" process. Only if you understand or have a theory of this underlying unitary process can you design an effective intervention.

Suppose you decide that the problems all stem from a rather severe instance of "rising yang qi," resulting from an overactive and inharmonious liver system, which is disrupting the smooth function of the heart system and depriving the spleen system (responsible for the downward movement of digestion) of the energy it would need to function properly. This discrimination of a deep pattern (*bianzheng*) would have important therapeutic implications. Avoiding, perhaps, oral medicines that would not easily be incorporated by the patient's malfunctioning systems, you might decide to intervene with acupuncture in a system that may not yet have been disturbed, yet is closer to the root and origin of all physiology. You might, in other words, choose to bolster the generative and regulatory activity of the kidney system, home of "original qi" (*yuanqi*) and source of much of the energy that usually makes the liver system thrive. If the liver system, in other words, can be returned to its normal function of distributing yang qi smoothly throughout the body (rather than just "flaming upward"), then the heart will be less disturbed and the spleen will have what it needs to return digestion to normal.

This will take time. Rowing upstream toward the cosmogonic source of illness manifestations is not easy work, especially when a pathological process is long established as a "multiple comorbidity." Moreover, it is not necessary for you, our clinician, to be a theorist or to explain the diagnosis to your patient in a metaphysical language. By the time you have been working for a decade or two, you know, in a bodily way, the TCM things that require tweaking and the TCM habits of the world.

This clinical vignette is not easy to read as scientific or even rational; the systems I have invoked here, for example, are clearly not the organs of human anatomy. But looked at as generative processes, as cosmogonic, they make sense in TCM. Cosmogony matters because doctors must trace the roots of disorder. Here is a very ancient explanation of why, paraphrased from a text called *Plain Questions*:

> Gui Yuqu said: Your servant has studied the *Notebooks on the Ultimate Beginnings of the Heavenly Origins* at great length. In this work it says: the vast and limitless heavenly void is the foundation of the root and origin of the generation and transformation of matter, and it is the beginning of the production of the myriad things. The five movements pass through the Dao of Heaven, ending only to begin anew, distributing the steady original qi of heaven and earth, epitomizing and comprehending the root and origin of generation and transformation upon the great earth, among the nine stars twinkling in the sky, and among the seven planets that revolve according to the degrees of heaven. Consequently, ceaseless change results from the myriad things having yin and yang aspects, having different characters of hard and soft, while opaque darkness and clear brightness emerge according to a definite positional order, and cold and heat come and go according to certain seasons. These mechanisms of ceaseless generation, this Way (*Dao*) of inexhaustible transformation, and the differing forms and manifestations of the world's myriad things all in this way come out and are manifest.

What would we have to know to find a passage like this legible in any way? Perhaps the most intuitive idea invoked here is that of the seasons, or natural temporality: in the course of a year, cold gives way to warmth; in the course of a day, darkness gives way to brightness. Time, in other words, has patterns. It is an oscillation of qualities (light and dark, warmth and cold) across a polarized continuum. Nature, or the spontaneous Way (Dao), is a patterned process of material transformation that has a certain predictability. And perhaps the good doctor comes to "know" this patterning in his bones.

The fundamental polarizing movements in ontogenesis are called yin and yang. So it is also important to grasp the philosophical or metaphysical character of yinyang—that is, their relationship—if we are to understand how "the world's myriad things all in this way come out and are manifest." Further, what are "the five movements"? What in the world is "qi"?

Students of Chinese medicine approach the metaphysics of their craft when they are learning the theoretical foundations of TCM: they must learn to think with qi, yinyang, and the five phases. "Qi" (pronounced "chee") is an indispensible term for Chinese thought, so fundamental that writers on medicine have hesitated to define qi as if it were an ordinary noun referring to a discrete thing. Instead, think of qi as the "stuff that makes things happen" in processes (Sivin 1987: 47). Though qi is quite substantial, it is at the same time a form of action. It is physiological and pathological processes of flow—driven by and formed from qi—that generate the concrete body and its (dis)abilities. *Qi is the activity and substance of the natural processes that produce the real world.*

Popular writers in English sometimes call qi "vital energy," but the first major study of TCM epistemology in English, by Manfred Porkert (1974), translates it as "configurative force." Porkert's term seeks to capture the inseparability of qi ("the substance and vitality of which the universe is made") from "pattern" (*li*, a central concern of neo-Confucian philosophy of the 11th and 12th centuries; see Scheid 2007). Qi is a force that manifests only in its configured results: the things of this world (e.g., hair and toenails, spleens and circulation channels, rivers and stars) are qi-patterned *gatherings* of qi-substance. Rather than translating or defining qi, then, most scholars now writing in Western languages use the term untranslated, as qi or *ch'i*, and writers in Chinese presume that readers have a sense of the term without needing a definition.

Armed with qi, then, the similarly ambiguous meanings of yin and yang might be more accessible. Modern discussions, like the TCM teaching materials I am quoting here, feel impelled to define these ancient metaphysical terms, but often do so more by putting them to use in situations than by anatomizing them. Here is a 1982 theoretical foundations textbook, for example:

> Yin and yang are the Dao [the Way] of Heaven and Earth. They are the network of the myriad things, the father and mother of alteration and transformation, the root and beginning of life-giving and death-bringing, the abode of vitality and intelligence. The treatment of illness must trace this root. Thus it is that gathered yang is heaven and gathered yin is earth; yang ends life, yin begins life in latency. Yang transforms qi; yin brings forms to maturity. When cold reaches an extreme, it gives rise to heat; when heat reaches an extreme, it gives rise to cold. Cold qi generates what is turbid [yin]; hot qi generates what is clear [yang]. When clear qi is in the lower [parts of the body, which are relatively yin], it gives rise to "rice-gruel diarrhea"; when turbid qi is in the upper [part of the body, which are relatively yang], it produces swelling and distention. This is the opposed action of yin and yang, the counter-movement and following movement of illness.
> (Cheng Shide et al. 1982: 68–69)

In this fragment, which is both poetic and medical, and which evokes the "rising yang" syndrome described above, the seasonal or fundamentally temporal character of yin and yang is quite clear. Natural, spontaneous oscillations between polar qualities can be named and even analyzed with a yinyang logic. An important feature of this passage, however, is the relationship it suggests between *classification* of phenomena as relatively yin or yang and the dynamic *interaction* of the phenomena so classified. Cool and turbid are yin, and they interact with yang-classified things and qualities to produce effects (illness, for example). The things of the world, or products of the great cosmogonic flow of the Dao, such as heaven and earth, life and death, the soft and the hard, are named as yin or yang and in the same moment placed in a polar and interactive relationship to each other.

The polarity of yin and yang thus allows the doctor to perceive numerous positions on the continuum of possibilities between its extreme points; just as hot shades into cold, and clear fluids can become turbid by degrees, the difference between life and death is discernible as infinite particularities of the yinyang relationship. Because all manifest phenomena can be placed on a continuum of effects, the yinyang of the bodily person and of the medical techniques through which her or his symptoms can be read is not much different from that of the cosmos as a whole. When we say, for instance, that "the female (or the cool, dark, junior) is yin and the male (or the warm, bright, senior) is yang," producing perhaps a two-column list of paired qualities—hot/cold, male/female—we are classifying phenomena as *relatively* yin or yang. There is no pure yang-ness or yin-ness, but there is a polarized continuum along which there is a process of ceaseless change: "This is the opposed action of yin and yang, the counter-movement and following movement of illness" (Cheng Shide et al. 1982).

The Powers of Correlation

Let's explore the classificatory dynamics to be found in yinyang analysis a little farther by taking up the "five phases" system. Philosophers of TCM have sometimes argued that TCM is a correlative science (Porkert 1974, Farquhar 1994). The ability of a yinyang metaphysics to both classify and interrelate phenomena is an instance of correlative logic, a method that groups things together according to their qualities or characteristics. This method is perhaps even more clear in the system of the five phases than in the very generalizing relativities of yin and yang. The "five phases" are conventionally listed in TCM as "wood, fire, earth, metal, and water." Consider this exuberant and puzzling first-century CE text on the correlations of the "wood" phase, for example:

> The eastern quarter generates a wind, [this] wind generates wood, wood generates sour, sour generates the liver, the liver generates muscle, muscle generates the heart, the liver [system] rules the eyes. This [process of generation (*sheng*)] in Heaven is dark generative potential (*xuan*), in man is the Dao, on earth is transformation (*hua*). Transformation generates the five flavors, the Dao generates wisdom, dark potential generates vitality. Vitality in heaven is wind, on earth is wood, in the human frame is muscle, among the viscera is the liver, among the colors is blue-green, among the musical notes is *jue*, among the inflected tones is *hu*, in movement is grasping, among the orifices is the eyes, among the flavors is sour, and among the intentions is anger. Anger injures liver and sorrow overcomes anger, wind injures muscle and dry overcomes wind, sour injures muscle and pungent overcomes sour.
>
> (Cheng Shide et al. 1982: 82–83)

This passage just begs to be made into a chart! Supplemented by four more lists of five phases correlations, we could make a five-column table, listing under each rubric all manner of things: here, "wood" is the heading under which the east wind, sour flavors, the liver (as organ and as system of functions), muscle, the eyes, blue-green, the *jue* note, the *hu* inflection, grasping, and anger are grouped together. Once the full five-part table was assembled, things like the heart, sorrow, wind, dryness, and pungent flavors, also mentioned here, would have their place under other rubrics.

Contrary to some early translations of five phases texts, the array of things and qualities gathered under the five terms wood, fire, earth, metal, and water is not a materialism of "elements" or "humors," which could only be seen as a kind of error about the ingredients of the human body. It is not as if ancient Chinese scientists believed that sorrow and pungent flavors and the liver were all made from the basic element of wood. Rather, in keeping with the processual emphasis discussed above, wood is a convenient column heading for a grouping of things that all manifest the same style of transformation.

Moreover, these things and qualities can "generate, injure, and overcome" each other. In therapeutics, herbal medicines classified as yang and warming can be used to treat a cold stasis; medicines classified as salty, and affiliated with the water phase, can be used to drain a local swelling. Perhaps the fives phases are not much more than mnemonics, reminding the practitioner what is classically (and arbitrarily) aligned with what. But the passage I quote here (from the *Plain Questions*, 1st century CE) argues that there are vast cosmological processes involved in the way things fall into groups and underlying the order of their systematic interaction: "This [process of generation (*sheng*)] in Heaven is dark generative potential (*xuan*), in man is the Dao, on earth is transformation (*hua*). Transformation generates the five flavors, the Dao generates wisdom, dark potential generates vitality." The tendency of spontaneous vital genesis to fall into patterns; the obvious fact that patterns (of climate, of dying and being born, of waxing and waning) have a regularity; and the active capacity of humans to make use of the "counter-movement and following movement" of phenomena, in our efforts to bring about better health—these are metaphysical assumptions that are prior to all challenge and all proof.

Conclusion: Sources and Manifestations

In the second classical quote above, which argues that yin and yang are the Way of heaven and earth, the author links cosmogony to medicine: presuming that things arise from a certain "root and origin of generation and transformation," which flows from "the vast and limitless heavenly void," as the Dao, or the Way of nature, "the treatment of illness must trace this root." The ancient metaphysics that still informs the field of TCM speculated about sources, and in order to understand manifestations it sought to trace the hidden, primordial, but still active roots of the phenomena experienced by patients and doctors alike. Arguably, it is only by intervening close to the source of a disorder that a lasting cure can be achieved. How to "trace this root," as they say in China, is thus a strategic question even for a modern medical practitioner: it is a matter of knowing the right moment in an etiological chain of mostly hidden events, expertly judging where and when to interrupt the spontaneous processes that continuously produce pathological symptoms.

Clearly, cosmogony, this world-generating process, is not only about the past origins of present entities. As was pointed out, it works as well for imagining the continuing processes of generation and transformation that operate in the present. Consequently, as medicine in treating an illness traces its roots (*zhibing qiuben*), it considers continuing transformations of the manifest—the pathological transformation, for example, of easy breathing into gasping and coughing, of unfelt digestion into heartburn, of comfortable walking into painful knees and feet. In the hope of altering a pathological development, or arresting it as it develops, doctors must link the "counter-movements" that they can control—drugs, acupuncture, massage, and more—to the "following movements" of the ensemble of processes that is a living person.

I began this article with the help of a 1982 literature review by philosopher Renzong Qiu. His report was written in the midst of controversy about the institutional forms of TCM, and he reported considerable thought about the epistemological and ontological features of this "traditional" form of applied philosophy. He proposed "medicine and philosophy" as a new field, and he contrasted the academic writings he reviewed with the deep and usually inarticulate "natural philosophy" characteristic of China's medical heritage.

More than 20 years later, a similarly well-placed philosopher, Qicheng Zhang of the Beijing University of Chinese Medical Sciences, wrote a comprehensive textbook for a series now used throughout China, titled *Philosophical Foundations of Chinese Medicine*. This survey text reflects a certain progress toward consensus that was made during the decades since the end of the Maoist period, when Qiu was writing. But even Zhang's essays on the work of the most ancient philosophers carry a trace of recent polemic. Take the three pages on Wang Chong, for example: Wang Chong (27–104 CE) is known to moderns as the first ancient "skeptic." He is thought of as opposing myth and mysticism, insisting on a natural uncreated universe, encouraging scientific investigation, and placing humans at the center of history. Zhang discusses these features of Wang Chong's philosophical writings in order to make a link to the contentious present: he is shown to be a secular modernist *avant la lettre*, interested in science and practical about causes and techniques.

This Wang Chong would thus be acceptable as a proto-scientist to the Natural Dialectics theorists presented by Qiu in 1982, as well as to the more recent "scientizers" of TCM, those who would like to incorporate the effective tools of Chinese medicine into a biomedical arsenal while scuttling "theory" and metaphysics. But Zhang and his TCM colleagues understand that the natural philosophy underpinning medicine (everywhere) is not so easily separated from its tools and techniques. In this textbook, then, which aims to teach the philosophical foundations of medicine, Wang Chong is characterized as propagating a "natural philosophy of the Way of Heaven" (Zhang 2004: 76). His "view of nature" (*ziranguan*) is stated as follows:

> Wang Chong took original qi (*yuanqi*) to be the root and source of the cosmos, considering that the myriad things of heaven and earth naturally come into being through the gathering together and condensing of original qi. . . . He posited that "heaven and earth unite qi, and the myriad things are spontaneously born of this."

The qi of heaven and earth, of course, are yang and yin qi, and even in this very simple (and often-quoted) formula, it is clear that the things of the world result from a gathering process. The gathering and condensing of the empirical world into a myriad of forms happens prior to all observation, and yet it is always happening. If the things that interest medicine—symptoms, lesions, organs, pathogens, drug properties, pulse qualities—are among the myriad things (and how could they be otherwise?), then they are always coming into existence. Their powers and dangers can be understood and managed through a highly technical analysis of normal and pathological patterns. Qi, yinyang, and the five phases are only examples of the many logical systems available to help Chinese medical doctors analyze and intervene in illness. There is no doubt that effective medical thought and action are deeply indebted to a frequently challenged and much tinkered-with, but always understandable, metaphysical "style of thought." Perhaps a fuller engagement with the "theory" of TCM can help us see that any medical practice in any era bolsters its efficacy with philosophy.

References

Andrews, Bridie, 2014, *The Making of Modern Chinese Medicine, 1850–1960*. Vancouver: UBC Press.
Cheng Shide et al., eds, 1982 (1st C. CE), *Suwen Zhushi Huicui (The Compiled and Annotated Basic Questions)* juan 2 section 5, p. 69. Beijing, China: People's Health Press.
Croizier, Ralph, 1968, *Traditional Medicine in Modern China: Science, Nationalism, and the Tensions of Cultural Change*. Berkeley: University of California Press.
Farquhar, Judith, 1994, *Knowing Practice: The Clinical Encounter of Chinese Medicine*. Boulder, CO: Westview Press.
Gordon, Deborah R., 1988, Tenacious Assumptions in Western Medicine. In Margaret Lock and Deborah R. Gordon, eds., *Biomedicine Examined*. Dordrecht, Netherlands: Kluwer Academic, pp. 19–56.
Heidegger, Martin, 1971, The Thing. Tr. Albert Hofstadter, *Poetry, Language, Thought*. New York: Harper and Row, pp. 163–180.
Lampton, David M., 1977, *The Politics of Medicine in China: The Policy Process 1949–1977*. Boulder, CO: Westview Press.
Lei, Sean Hsiang-lin, 2014, *Neither Donkey Nor Horse: Medicine in the Struggle Over China's Modernity*. Chicago: University of Chicago Press.
Porkert, Manfred, 1974, *The Theoretical Foundations of Chinese Medicine: Systems of Correspondence*. Cambridge, MA: MIT Press.
Qiu, Renzong, 1982, Philosophy of Medicine in China (1930–1980). *Metamedicine* 3: 35–7
Scheid, Volker, 2007, *Currents of Tradition in Chinese Medicine, 1626–2006*. Seattle, WA: Eastland Press.
Shandong College of Chinese Medicine and Hebei Medical College, eds, 1982, *Huangdi Neijing Suwen Jiaoshi (The Yellow Emperor's Inner Canon, Plain Questions, with Commentary)*, 2 vols. Beijing: Renmin Weisheng Chubanshe.
Sivin, Nathan, 1987, *Traditional Medicine in Contemporary China*. Ann Arbor, MI: Center for Chinese Studies.
Stengers, Isabelle, 2005, The Cosmopolitical Proposal. In Bruno Latour and Peter Weibel, eds., *Making Things Public: Atmospheres of Democracy*. Cambridge, MA: MIT Press, pp. 994–1003.
Taylor, Kim, 2005, *Chinese Medicine in Early Communist China, 1945–1963: A medicine of revolution*. London/New York: Routledge Curzon.

Zhang, Qicheng, 2004, *Zhongyi Zhexue Jichu* (*The Philosophical Foundations of Chinese Medicine*). Beijing, China: Chinese Medicine and Pharmacy Press.

Zhao, Hongjun, 1989, *Jindai Zhongxiyi Lunzhengshi* (*History of Modern Controversies Between Chinese and Western Medicine*). Fuyang, Anhui: Anhui Science and Technology Press.

Further Reading

J. Farquhar, *Knowing Practice: The clinical encounter of Chinese medicine* (Boulder: Westview Press,1994) is a book-length and case-based discussion that expands upon ideas in this article.

D. L. Hall and R. T. Ames, *Thinking Through Confucius* (Albany: State University of New York Press, 1987) lucidly introduces much of early Chinese metaphysics and political philosophy.

M. Porkert, *The Theoretical Foundations of Chinese Medicine: Systems of correspondence* (Cambridge, MA: MIT Press, 1974) is a work that remains one of the most thorough philosophical introductions to classical Chinese medicine.

V. Scheid, *Chinese Medicine in Contemporary China: Plurality and Synthesis* (Durham, NC: Duke University Press, 2002) provides an anthropological and practitioner's views of TCM's recent history.

Q. Zhang, *Zhongyi Zhexue Jichu* (*The Philosophical Foundations of Chinese Medicine*) (Beijing: Chinese Medicine and Pharmacy Press, 2004) is a well-written introduction to this article's themes for readers of modern Chinese.

46
DOUBLE TRUTHS AND THE POSTCOLONIAL PREDICAMENT OF CHINESE MEDICINE

Eric I. Karchmer

Traditional Medicine and the Postcolonial Condition

It was a gray and chilly November morning in 2002, two years after my graduation from the Beijing University of Chinese Medicine. I had returned to Dongzhimen Hospital, the main teaching hospital of my alma mater, to spend two months studying with several senior physicians, trying to master the clinical skills I needed to become a doctor of Chinese medicine. I was walking briskly to get to the hospital before the outpatient clinic opened at 8 a.m. Because the outpatient clinic is first-come first-served, throngs of patients were already filling the waiting room and hallways of the clinic. As I approached the hospital, I began to prepare myself for the intense focus I would need for the next four hours. I was going to be shadowing a senior clinician during his morning shift. It would take my full concentration to follow, understand, and take good notes on his clinical work, as he efficiently worked through two dozen or more patients. I was determined to make the most of this opportunity. I had completed a five-year medical school degree in Chinese medicine, but as a foreign expatriate, it probably would have been legally impossible for me to work in a Chinese medicine hospital and follow the standard career trajectory of a doctor through residency, attending physician, senior physician, and so on. My opportunities to learn the clinical craft of Chinese medicine from experienced physicians were now limited to occasional short visits to China like this one. I envied my Chinese classmates, who would be able to gradually hone their skills through such mentoring relationships in the hospitals and clinics where they were now working.

While I was concerned about my development as a physician, I knew that many of my Chinese classmates were even more apprehensive about their own futures. A significant number, in fact, wished to *gaihang* "change professions." I had recently caught up with Chen Yao, a classmate who was working for the multinational medical nutrition company, Nutricia, in their Beijing office. She told me that in the two and a half years since graduation, almost one-third of our 60 Chinese classmates were now working for pharmaceutical companies as drug representatives. Chen Yao did not dislike Chinese medicine. Indeed, like most of our classmates, she had a certain affinity for the profession after having devoted an entire college career to studying it. But she clearly preferred the financial benefits of working for a global pharmaceutical firm over the difficult and poorly compensated work of a doctor of Chinese medicine. The other classmates that I caught up with during my visit shared Chen Yao's ambivalence about Chinese medicine. They seemed to be either reluctant doctors or were in search of other career opportunities. Some of the graduate students that I met at the Beijing University of Chinese Medicine took another path out of the profession, transitioning from an undergraduate degree in Chinese medicine to a graduate program in Western medicine. Wang Bo, the teaching assistant for my biochemistry class, was one such example. She and her husband had both studied Chinese medicine as undergraduates. She was able to test into a Master's program in Biochemistry at the Beijing University of Chinese Medicine. Not long after completing her M.S., she followed her husband to New York City in 2003, when he got accepted into a Ph.D. program for oncology research at New York University.

This ambivalence about pursuing a career in Chinese medicine is widespread, particularly among students and young doctors. I consider it one of the defining traits of what I call the "postcolonial condition" of Chinese medicine. In postcolonial studies, scholars such as Dipesh Chakrabarty have defined the "postcolonial condition" as a state in which the West continues to exercise a cultural dominance over the formerly colonized, making the West the necessary point of reference for any historical, sociological, or scientific claim made about the East (Chakrabarty 2000: 8). In other words, the power inequalities of European colonization have often persisted in the contemporary period, even though the vast majority of formerly colonized societies achieved political independence by the early 1960s. Instead of overt political domination, these societies struggle with colonial-like power inequalities that take subtle, cultural forms. Taking contemporary China as an example, the unquestioned prestige of Western medicine is reminiscent of Frantz Fanon's analysis of how settler values were privileged over native ones in colonial Africa (Fanon 1963). The ambivalence of young students and doctors toward Chinese medicine also resonates with Fanon's insights into the psychology of colonized.

> All colonized people . . . position themselves in relation to the civilizing language, i.e. the metropolitan culture. The more the colonized has assimilated the cultural values of the metropolis, the more he will have escaped the bush.
> Fanon 2008 (1952)

As we will see in the ethnographic vignette below, physicians of Chinese medicine must also position themselves in relation to the dominant practice of biomedicine in order to engage in clinical work.

Historically, we can also trace how Chinese intellectuals positioned themselves in relation to the "civilizing language" of European and Japanese imperialism (1840–1949) and the effects of this relationship on the Chinese medicine profession in general. In the early 20th century, a vocal opposition to Chinese medicine emerged among China's most educated elite, the first generation of Chinese intellectuals to study the Western sciences and humanities, often at European, American, or Japanese universities. For these elites, Chinese medicine was not only false and incorrect, but a superstitious practice that made the populace at large resistant to modernization. Although some of these intellectuals became important figures in the Nationalist government (1927–1948) and sought to use their political position to promote Western medicine and restrict Chinese medicine, they had little effect on the general prestige of Chinese medicine. Skilled physicians of Chinese medicine remained highly sought after in both rural and urban communities (Karchmer and Scheid 2015). A critical mass of trained doctors of Western medicine did emerge during this period, creating the foundation for the later development of a national health care system under the Communists, but they were vastly outnumbered by doctors of Chinese medicine (Karchmer 2015: 199–201). On the eve of the Communist Revolution in 1949, there were an estimated 500,000 doctors of Chinese medicine compared to 38,000 doctors of Western medicine, who practiced almost exclusively in urban centers (Cui 1993: 67).

In the early years of the People's Republic, the fate of the Chinese medicine profession was in doubt. The Chinese state was more centralized and powerful than it had been in decades. Colonial possessions had been returned; the political opposition had been destroyed. The Chinese Communist Party (CCP) could turn to the urgent task of building a modern Chinese nation, including a national health care system. Within the CCP, not unlike the Nationalist Party (KMT) that preceded it, there was strong opposition to Chinese medicine, particularly among the technocrats in the Ministry of Health, who were almost all trained in biomedicine. Following an internal party struggle that led to the purge of the He Cheng, the Vice Minister of Health, in 1955, the promotion of both Chinese medicine and Western medicine became official CCP policy (Croizier 1968; Lampton 1977). Although the place of Chinese medicine in China's modern health care system was ensured from this point going forward, the state has nonetheless given clear priority to development of Western medicine throughout the Communist era. When I finished my medical training in Chinese medicine in 2000, the Ministry of Health was reporting that there were 1.33 million doctors of Western medicine, compared to 337,200 doctors of Chinese medicine. The number of doctors of Western medicine expanded more than thirty-five-fold in the Communist era, while their Chinese medicine counterparts had actually declined (Editorial Committee of the China Medical Yearbook 2001).

The ambivalence of contemporary students and young doctors of Chinese medicine begins with the structure of the national health care system, which contains two parallel but unequal systems of medicine. There are state-run schools, hospitals, and research institutions for both Western medicine and Chinese medicine, but the former are more numerous, more respected, more authoritative, and considered more scientific than the latter. The hegemonic place of Western medicine in contemporary Chinese society leads to other important asymmetries, despite some government policies to address this imbalance. For example, students of Western medicine are generally required to take a semester-long introduction to Chinese medicine, but this training, even in the rare case that a doctor goes on to use some Chinese medicine in his or her clinical practice, is irrelevant to the doctor's professional advancement. By contrast, students of Chinese medicine have a curriculum that devotes nearly 50% of allotted course time to topics in Western medicine. Moreover, as we will see below, a good command of Western medicine is an essential requirement for practicing Chinese medicine today.

Students of Chinese medicine must take Western medicine seriously from the start of their education. The significance of this dual training becomes apparent in the third year of medical school, when clinical clerkships begin and students observe doctors using both medical practices in their everyday clinical work. Under these unusual training conditions, perhaps it should not be surprising that many students gravitate toward Western medicine and struggle to embrace Chinese medicine. As my classmates often reminded me, the language, concepts, and principles of Chinese medicine were almost as foreign to them as they were to me. Their high school education was focused on biology, chemistry, and physics—the foundations of Western medicine—not on Chinese philosophy and other disciplines more useful to the study of Chinese medicine. One of my classmates, who graduated third in our class, thereby earning an automatic admission to the Master's program of her choosing, told me quite unabashedly after our graduation that she preferred the apparently straightforward logic of Western medicine because she couldn't "grasp the Chinese medicine way of thinking."

What have these power inequalities meant for the actual practice of Chinese medicine? In the vignette below, I will show that postcolonial power inequalities are an unavoidable feature of every clinical encounter and have profound consequences for the theory and practice of Chinese medicine. Doctors of Chinese medicine cannot practice a "pure" Chinese medicine, willfully ignoring the world of Western medicine around them. Rather, they must adjust their clinical work to the standards of Western medicine, an intellectual burden that does not apply to their Western medicine counterparts. These power inequalities are further institutionalized in the structures of China's health care system. Thus, the Chinese state has created a separate institutional sphere for the practice of Chinese medicine, but it has also constrained the profession by imposing regulations on Chinese medicine practice that have no parallel in Western medicine institutions. For example, hospitals of Chinese medicine require that medical records for all admitted patients contain a "double diagnosis," one for each medical system. There is no such requirement for biomedical institutions, where it would be scandalous if the state were to try to impose one. The result of the "double diagnosis" is that doctors of Chinese medicine also frequently use both medical practices to treat their patients, whereas doctors of Western medicine rarely treat with two types of medicine.

As doctors move back and forth between these two medical systems, they must also negotiate an epistemological quandary: what is the truth status of Chinese medicine claims about the body, illness, and healing? The answer is fraught. Chinese medicine is defined by its difference with Western medicine; difference marks its "Chinese-ness." At the same time, this difference is inherently problematic. Any deviation from the standards of Western medicine can be interpreted as error and evidence that Chinese medicine is not scientific. Doctors of Chinese medicine resolve this postcolonial dilemma by embracing a position of "double truths": Western medicine is scientific, they concede, but "Chinese medicine is scientific too." By insisting on this adverb "too," the all-important supplement that makes this claim work, doctors are implicitly recognizing the power inequalities between Western medicine and Chinese medicine. Chinese medicine can only be true if it is a lesser truth.

The Contemporary Clinical Encounter

I hustled through the waiting room of the outpatient clinic just a few minutes before 8 a.m., pulling off my winter jacket and slipping into my white doctor's coat, standard attire for all doctors and hospital technicians, just before striding into the consultation room. On this day, I had arranged to work with Dr. Sun, who had been the main lecturer for our important fourth-year class, Chinese Internal Medicine. In a pedagogic environment where most professors stayed close to the textbook materials, Dr. Sun had stood out, with his carefully researched lectures, dynamic speaking style, and memorable anecdotes from his own clinical cases. I was hoping that his clinical skills would match his rhetorical talents.

Established in 1958, Dongzhimen Hospital is one of the older hospitals of Chinese medicine in China. The well-worn state of its buildings belies the clinical excitement that sometimes transpires inside. Dr. Sun's consultation hours were being held in a long, narrow room on the second floor that had almost surely been converted from some other use to a consultation space. Dr. Sun sat at a yellow desk, the same basic wooden desk that could be found in all of the consultation rooms, positioned halfway between the hallway door and a tiny window on the far wall. The room was so narrow that when Dr. Sun sat at his desk, I would have to awkwardly squeeze between the wall and his chair to get past him. The desk was where all the action took place. Dr. Sun would spend the entire morning seated in front of it, conducting consultations and writing prescriptions, too busy on most days to even stand up for a break. The patient would enter from his left and take a seat at a small, three-legged stool at the side of the desk nearest the entrance. Students, such as myself, would sit to his right, huddling around the far side of the desk as we took notes.

On this morning, I was sharing the far end of the desk with another medical student, who turned out to be a distant cousin of Dr. Sun. We were participating in the time-honored tradition of "copying prescriptions *chao fangzi*," in which a student follows a senior doctor, making notes about the consultation and recording the doctor's prescription for later study. As this expression suggests, the prescription is central to this training method. Far more than a record of the doctor's treatment for an individual patient, the prescription is a condensation of the doctor's therapeutic strategy, both with respect to a specific disorder and his overall clinical style. Unlike Western medicine prescriptions, Chinese medicine prescriptions often contain a dozen or more herbs that the patient usually cooks together in water to make a decoction. Doctors of Chinese medicine assert that the clinical efficacy of a prescription depends not so much on the properties of any single item but on the collective action of the herbs together. Moreover, prescriptions are not standardized for medical conditions. Indeed, physicians generally try to individualize the prescription to the patient's unique presentation to the greatest degree possible. Writing a prescription is therefore an art, based on the physician's interpretation of the patient's underlying condition, drawing on a mastery of hundreds of Chinese medicinal herbs and centuries of formulary scholarship on how to best combine them. By copying these prescriptions, the student hopes to inscribe and ultimately embody the teacher's art.

On most days, the rush of patients is so overwhelming that doctor and student may have little opportunity to discuss prescriptions and treatment strategies. But on this day, a light drizzle had begun, thinning out the usual morning crowd, giving us occasional opportunities to talk. Around 9:30 a.m., an 84-year-old woman shuffled into the room, her daughter supporting her as she took a seat. The daughter opened her purse and pulled out her mother's outpatient record book, a worn and folded yellow notebook, the size of an elongated index card. Dr. Sun took the notebook and placed it on the desk, pushing aside the blood pressure cuff he had used for the last consultation. He scanned the notes from previous consultations and then looked at the patient, asking, "What's bothering you today?" "My whole body aches," she said in the Beijing patois, as she put her hand to her chest.

While she spoke, Dr. Sun flipped through the many laboratory tests and other exam results that had been folded and stapled into the record book. They included an electrocardiogram from a month ago with a depressed ST section, indicating mild cardiac ischemia. Blood work from a visit two weeks ago showed that her white blood cell count had been high (14.3×10^9 cells/liter) and her neutrophil distribution elevated (84%), both signs of infection. A biochemical panel did not indicate conclusively any one problem, but Dr. Sun declared it "chaotic" with eight abnormal results. The daughter handed Dr. Sun a recent chest X-ray. Holding it up to the light and angling it toward us, the students, Dr. Sun pointed out cobweb-like interstitial markings caused by a pulmonary infection and drew our attention to the increased spacing of the ribs indicative of emphysematous changes. Putting down the X-ray, he then showed us the notes from her last hospital visit, in which a different doctor had diagnosed her with coronary heart disease, chronic nephritis, and interstitial pneumonia.

Turning to a fresh page in the record book, Dr. Sun began writing today's entry, asking the patient questions as he wrote. He noted complaints about heart palpitations and back pain and then asked her to stick out her tongue. The tongue exam is one of the distinctive features of the Chinese medicine exam. Doctors consider it one of the most important and reliable ways to assess the patient's overall condition. Dr. Sun carefully noted the shape and color of the tongue, as well as the texture and color of the tongue coating. Next he gestured toward the patient's wrist to begin the pulse exam, another distinctive feature of a Chinese medicine consultation. She extended her arm. Dr. Sun put three fingers on her radial artery, letting his fingertips gently roll over the artery, sensing its resilience as he varied the pressure, recording the texture in terms of the 28 basic pulse presentations recognized in Chinese medicine. He repeated this process with the other wrist. In Chinese medicine pulse taking, the three sites on both wrists, six positions all together, are significant because minute differences in pulse presentation at each site might indicate the pathological changes of a particular region or organ of the body (Kuriyama 1999; Farquhar 2014). Like the tongue exam, the pulse is considered an excellent indicator of the patient's overall condition and an essential part of any consultation. The pulse exam is so iconic to Chinese medicine clinical work that some patients will silently extend their wrist at the beginning of a consultation, expecting, or sometimes challenging, the doctor to make a diagnosis based on the pulse exam alone.

Having completed his exam, Dr. Sun looked up from his notes and addressed the two women. He recommended that the patient be admitted to the hospital. Her condition was too complicated and unstable to be treated on an outpatient basis. Since Dr. Sun has recently joined the Nephrology Department, he suggested that the patient be admitted to this ward. It would be permissible with her chronic nephritis, and he could personally care for her in that department. They quickly agreed to this plan, and the daughter gathered up the record book, the X-rays, and other belongings and escorted her mother out of the room to begin the admissions process.

The next patient did not enter right away, so Dr. Sun turned toward his two students to discuss this case with the excitement that only a devoted teacher might have. "What formula would you use for that patient?" he quizzed. Dr. Sun's cousin and I looked back at him blankly. I felt overwhelmed by the complexity of the case. Could a single formula address the patient's heart, lung, and kidney problems? Each one alone would be difficult to treat. "First of all," Dr. Sun broke the silence, "the patient should be diagnosed as having Chest Blockage *xiongbi*, due to Cold and Phlegm. In the sixth edition of the Chinese Internal Medicine textbook, Chest Blockage was misleadingly renamed Chest Blockage and Heart Pain *xiongbi xintong*. But interstitial pneumonia corresponds perfectly to Chest Blockage, which can also account for the patient's mild cardiac ischemia as well. The proper formula should be Trichosanthes Fruit, Chinese Garlic, and Pinellia Decoction *gualou xiebai banxia tang* to 'invigorate Chest yang *zhenfen xiongyang*.'" Dr. Sun was asserting that the patient's interstitial pneumonia was her most urgent issue and needed to be addressed first.

I was instantly intrigued by this explanation that went to the crux of the patient's condition and was also critical of standardizing conventions in Chinese medicine, particularly as represented by the sixth edition of the national textbook. Dr. Sun continued his explanation, demonstrating his mastery over both Chinese medicine and Western medicine, moving nimbly between these two different medical systems and showing us how to navigate the potential pitfalls that awaited the inexperienced physician. "This formula is an excellent choice for this patient. Antibiotics are generally not very effective in treating interstitial pneumonia. In Western medicine, one might also consider steroids. But this approach compromises the immune system and could actually exacerbate the infection. In a similar fashion, we must not use the related formula Unripe Bitter Orange, Chinese Garlic, and Cinnamon Twig Decoction *zhishi xiebai guizhi tang*, because Cinnamon Twig is too warming and might also worsen the infection. We could consider replacing it with Ephedra *mahuang*, which is also warming but won't intensify the infection because of its strong Lung dispersing properties."

Although Dr. Sun had proposed a unique solution to the patient's complex medical condition, his approach was typical of the contemporary Chinese medicine clinical encounter in many ways. Doctors of Chinese medicine continually tack back and forth between "Chinese medicine" and "Western medicine" as they determine a diagnosis and design a treatment. Western medicine has become so intertwined with the contemporary practice of Chinese medicine that one professor at the Beijing University of Chinese medicine told me: "Chinese medicine today cannot exist without Western medicine." Hybrid medicine is utterly mundane.

As the vignette above suggests, these hybrid practices are complex. On the one hand, doctors use both medical systems in tandem, as if they have little in common with each other. Dr. Sun's examination of the patient could be considered a dual exam, producing two diagnoses appropriate to each medical system. And once the patient had been admitted to the hospital, it is highly likely that she would have also been prescribed some form of Western medicine treatment. Even though Dr. Sun seemed to suggest that Chinese medicine treatment would be superior, Western medicine therapies are widely used in the inpatient wards. During my training at Dongzhimen Hospital, I did not observe, or hear about, any admitted patients that were treated with Chinese medicine therapies exclusively. (Outpatients are much more likely to be treated with only Chinese medicine therapies.) At the same time, doctors are not just practicing two types of medicine at once. They are also continually strategizing about how to integrate the two medical systems. For example, Dr. Sun equated the Chinese medicine disorder of "Chest Blockage" with the Western medicine diagnosis of interstitial pneumonia. He argued that the Chinese medicine treatment principle of the "invigorating Chest yang" could cure the Western medicine pathology of a lung infection. He cautioned against using certain herbs that, according to Chinese medicine classifications, would be too "warming" and might exacerbate the infection.

Purification and Hybridization

The clinical encounter described above challenges us with a very basic question: "What is Chinese medicine?" Is it possible to define Chinese medicine as the specific moments in the above clinical encounter—the tongue and pulse exam, the unfamiliar terms of diagnosis (Chest Blockage, the pattern of Cold and Phlegm obstruction), the formula and herbs—that are distinct from the other, more familiar moments—EKGs, X-rays, blood work, interstitial pneumonia readily recognized as Western medicine? Or should we consider the entire mélange of practices to be Chinese medicine or perhaps some third form of medicine?

Most contemporary scholars have not written explicitly about the hybridity of contemporary practice (Sivin 1987; Farquhar 1994; Hsu 1999). But most doctors do not write about this phenomenon either, even though they are deeply enmeshed in it through their clinical work. In my many years of studying Chinese medicine, I have never encountered a theory of medical integration. Doctors tend to talk about a "pure" Chinese medicine, while tacitly blending it with Western medicine in practice. These two tendencies—to speak about a purified medicine while producing a hybrid one—are reminiscent of Bruno Latour's analysis of modernity. Latour has argued that we are continually trapped in the predicament of modernity because we give great credence to work of purification—in which "Nature" is opposed to "Culture"—and pay little attention to the work of hybridization—in which "Nature" and "Culture" are always being intermingled (Latour 1993). In considering the contemporary condition of Chinese medicine, I argue that a similar dynamic defines its "postcoloniality": we can't understand the hybridity of contemporary Chinese medicine—the "integration of Chinese medicine and Western medicine"—without also understanding its opposing purifications—the constructions of "Chinese medicine" and "Western medicine" that doctors attempt to combine.

Returning to the diagnosis of Dr. Sun, we can now better understand how these two processes of purification and hybridization operate in clinical practice and what it means for the truth claims of Chinese medicine. While Dr. Sun could have made his diagnosis of "Chest Blockage" without any reference to Western medicine, he clearly felt the need to relate it to the patient's known biomedical diagnosis. This move, the "double diagnosis" mentioned above, would be required for any admitted patient and is rarely omitted in outpatient care either. The first step in this process is to equate a biomedical disease category with a similar nosological category in Chinese medicine, known as *bing*. In fact, this term also means "disease" in the context of Western medicine, but I leave it untranslated here to distinguish its different connotations in Chinese medicine. Each Chinese medicine *bing* category is defined by a loose cluster of symptoms, which may overlap with some disease categories of Western medicine but frequently do not correspond exactly. Taking "Chest Blockage" as an example, it is defined in the popular fifth edition of the Chinese Internal Medicine textbook as: "tightness and pain in the chest, with severe cases producing pain that pierces to the back, shortness of breath, wheezing, and an inability to lie down" (Zhang, Dong, and Zhou 1985: 108). Because of the postcolonial imperative to relate Chinese medicine concepts to Western medicine ones, many doctors take this definition, drawn from classic sources and using primarily the original language of those texts, as a close approximation of coronary heart disease. But Dr. Sun criticized this kind of reductionist thinking, which was given a prominent place in the sixth edition of the national textbook, the one he was required to teach to my class. In this edition of the national textbook, "Chest Blockage" was renamed "Chest Blockage and Heart Pain," and its definition was rewritten in the biomedical language of coronary heart disease. Although Heart Pain is a traditional term, Dr. Sun was suggesting that the editors appended it to Chest Blockage to strengthen the implied correlation to coronary heart disease.

When a Chinese medicine *bing* category is equated with a single biomedical disease, the epistemological value of the Chinese medicine term is diminished, even erased. Dr. Sun's critique of the sixth edition textbook was addressing precisely this problem. In fact, the editors of the national textbooks, first published in 1960 under the auspices of the Ministry of Health and then periodically revised every 5 to 10 years, have grappled with this very issue over the past 50 years. The most popular editions, the second and the fifth, scrupulously avoided any direct reference to Western medicine concepts, even though they were not impervious to more indirect influences. Perhaps inspired by the growing role of Western medicine in clinical practice in the mid-1990s, the editors of the sixth edition decided to incorporate limited elements of Western medicine, such as the new definition of Chest Blockage. Dr. Sun's critique was not just an academic one either. By suggesting that "Chest Blockage" could also be considered an equivalent for interstitial pneumonia, Dr. Sun was also arguing that a series of formulas from the second-century canon, *Essentials from the Golden Chamber*, where the term "Chest Blockage" originates, can be used to effectively treat interstitial pneumonia. That this formula treats certain presentations of coronary heart disease made it even more appropriate for this case. Whether Dr. Sun was correct or not, his approach was certainly unique. In an historical moment where working across the two medical systems is an imperative, most contemporary doctors would probably treat interstitial pneumonia with formulas thought to address lung function rather than framing the problem in terms of Chest Blockage.

Dr. Sun's creative use of "Chest Blockage" was only the first step in his clinical assessment. He also cautioned us to be aware of the problems of Cold and Phlegm. These terms refer to what contemporary practitioners call the patient's "pattern" *zheng*, a general description of the patient's underlying pathological condition, as understood within the principles of Chinese medicine. Despite Dr. Sun's emphasis on *bing* in this particular case, the pattern is usually the physician's main focus. Typically, it would be stated more robustly, with a four- to eight-character phrase that describes the quality of the condition (in terms of hot or cold, excess or deficiency), the nature of the pathogen (which might refer to external factors, like wind, heat, cold, dampness, summer heat, dryness, or internal ones, such as qi constraint, blood stasis, phlegm), and the location of the problem (such as, in the organs, meridians, qi, or blood) (Deng and Guo 1987: 94–141). Taking Dr. Sun's recommended formula, Trichosanthes Fruit, Chinese Garlic, and Pinellia Decoction as an example, this formula would be indicated for patients presenting with a pattern of "flagging chest yang, qi constraint and phlegm obstruction *xiongyang buzhen, qizhi tanzu*" (Duan, Li, and Shang 1995: 180). The treatment principle for this formula is to reverse these pathological tendencies, "opening yang and dispersing clumps, moving qi and eliminating phlegm." This clinical approach is called "pattern recognition and treatment determination" *bianzheng lunzhi*, and it is universally celebrated as the central methodology of Chinese medicine.

In order to explore the problem of "double truths," we need to dig a little deeper into the relationship between disease, *bing*, and pattern, which can only be fully understood through the hospital medical record and the practice of "double diagnosis." When making a "double diagnosis," doctors actually make two levels of comparison. First, the *bing* is matched to disease, as discussed above. At the second level, a pattern is determined and added to the Chinese medicine side of the diagnosis as a subcategory of *bing*. If the disease presents in several common forms or has well-known complications, then the pattern will be matched to the disease variant or sequelae, although equivalency at this level is not considered essential for writing a proper medical record. For example, "Wasting and Thirst" *xiaoke* is widely recognized as the Chinese medicine *bing* that most closely matches the biomedical disease of diabetes mellitus. But depending on whether the patient has type I or type II diabetes or presents with complications, such as neuropathy, retinopathy, arteriosclerosis, nephropathy, and so on, the admitting physician will try to match the pattern to these disease variants or complications. Elements of this relationship are also visible in the structure of the Chinese Internal Medicine textbook. This textbook presents about 50 of the most commonly seen *bing*. Treatment for each *bing* is broken down into roughly five or six commonly seen patterns; each pattern will indicate a treatment with a well-known formula.

Although the nuances of the textbooks and medical record writing are beyond the scope of this chapter, the key point is that the relationship among disease, *bing*, and pattern is a product of the postcolonial moment. Prior to the 1950s, most doctors of Chinese medicine had little to no knowledge of Western medicine, and among the small elite that knew both medical systems, there were no established conventions on how to integrate them. Because the vast majority of doctors practiced in private clinics, there were no legal or institutional requirements to do a double diagnosis in the extremely small number of hospitals of Chinese medicine that did exist. Moreover and most importantly, due to these vastly different circumstances, the Chinese terms of *bing* and *zheng* had very different connotations. Before the 1950s, it would be incorrect to translate *zheng* as pattern, because it was neither a term of diagnosis nor a subcategory of *bing*. Rather, I believe *zheng* is best translated as "presentation," which explains why it was sometimes used interchangeably with the new biomedical term for symptom during this time period (Karchmer and Daidoji forthcoming). Further evidence for this translation can be found in the first dictionary of Chinese medicine, published in 1921 by the respected physician and scholar Xie Guan, where *zheng* is defined as the "external expression of an internal *bing*" (Xie 1994 [1921]).

With the rapid growth of the biomedical profession in the Communist era, Western medicine soon, perhaps by the late 1950s, became the dominant form of medical practice in China. With this shift in power relations, doctors of Chinese medicine became increasingly oriented to the epistemological standards of Western medicine. The meaning and use of concepts, such as *bing* and *zheng*, began to shift in an emerging era of hybrid medical practices. Because *bing* could be equated to disease, it facilitated the process of hybridization, but it did so at significant cost. The proximity of the two concepts weakened *bing*, making it expendable, except when a skilled physician like Dr. Sun can demonstrate creative new applications. Because *zheng* took on the meaning of "pattern" (of pathological process), which had no corresponding term in Western medicine, it contributed to the process of purification. Doctors frequently assert, and the national textbooks acclaim, that the methodology of "pattern recognition and treatment determination" *bianzheng lunzhi* is one of the defining characteristics of Chinese medicine (Yin and Zhang 1984: 8). As a result, when comparing the two medical systems, most doctors will make the following claim: "Western medicine differentiates disease; Chinese medicine differentiates pattern" (Karchmer 2010).

Although these epistemological shifts are quite profound, contemporary doctors are generally not aware of them. Or more accurately, they have been "forgotten" through a process of normalization that Thomas Kuhn argues follows every scientific revolution (Kuhn 1970). What is evident to students and young doctors is that Chinese medicine can only be practiced in a world of "double truths." While pattern is not an expendable truth like *bing*, it remains a lesser one. I suspect that these power inequalities and the conundrum of double truths contribute significantly to the ambivalence of young doctors discussed above. They easily recognize the relative weakness of Chinese medicine vis-à-vis Western medicine, and the disappointed ones seek out other professional opportunities. However, if we take Dr. Sun at his word, this postcolonial form of medical practice is not ineffective and far from impotent. When a doctor develops the degree of clinical experience and proficiency to see beyond the conundrum of double truths, to rethink the conventions around hybrid medical practice that can delegitimize the claims of Chinese medicine, Chinese medicine may indeed (once again) become a potent therapy. In short, the practice of Chinese medicine is deeply constrained by the hegemony of Western medicine in contemporary China. But Dr. Sun's clinical case above suggests that the first step out of this postcolonial predicament may be a critical examination of these power inequalities, so that doctors may more effectively deploy the conceptual tools they already possess.

References

Chakrabarty, Dipesh (2000). *Provincializing Europe: Postcolonial Thought and Historical Difference*. Princeton, NJ: Princeton University Press.

Croizier, Ralph C. (1968). *Traditional Medicine in Modern China: Science, Nationalism, and the Tensions of Cultural Change*. Cambridge: Harvard University Press.

Cui, Yueli (1993). *Founder of the Chinese Medicine Profession in New China: The Collected Writings of Lu Bingkui's Sixty Years in Medicine*. Beijing: Huaxia Press.

Deng, Tietao, and Zhenqiu Guo (1987). *Chinese Medicine Diagnosis*. In *Educational Reference Series for Higher Education Chinese Medicine Institutes*. Beijing: The People's Medical Press.

Duan Fujin, Fei Li, and Chichang Shang (1995). *Formulary*. In *Standardized Textbooks for General Higher Education in Chinese Medicine and Drugs*. Shanghai: Shanghai Science and Technology Press.

Editorial Committee of the China Medical Yearbook (2001). *The China Medical Yearbook 2001*. Beijing: People's Medical Press.

Fanon, Frantz (1963). *The Wretched of the Earth*. New York: Grove Press.

Fanon, Frantz (2008 [1952]). *Black Skin, White Masks*. New York: Grove Press.

Farquhar, Judith (1994). In *Knowing Practice: The Clinical Encounter of Chinese Medicine*. edited by John Comaroff, Pierre Bourdieu and Maurice Bloch. Boulder: Westview Press.

Farquhar, Judith (2014). "Reading Hands: Pulse Qualities and the Specificity of the Clinical." *East Asian Science, Technology, and Society: An International Journal* no. 8 (1):9–24.

Hsu, Elizabeth (1999). *The Transmission of Chinese Medicine*. Cambridge: Cambridge University Press.

Karchmer, Eric I. (2010). "Chinese Medicine in Action: On the Postcoloniality of Medicine in China." *Medical Anthropology* no. 29 (3):1–27.

Karchmer, Eric I. (2015). "Slow Medicine: How Chinese Medicine Became Efficacious Only for Chronic Conditions." In *Historical Epistemology and the Making of Modern Chinese Medicine*, edited by Howard Chiang (pp. 188–216). Manchester: Manchester University Press.

Karchmer, Eric I., and Keiko Daidoji (forthcoming). "The Case of the Suzhou Hospital of National Medicine (1939–1941): War, Medicine, and Eastern Civilization." *East Asian Science, Technology, and Society* no. 11 (2).

Karchmer, Eric I., and Volker Scheid (2015). "The History of Chinese Medicine, 1890–2010." In *Modern Chinese Religion II: 1850–2015*, edited by Vincent Goossaert, Jan Kiely and John Lagerway (pp. 141–194). Leiden: Brill.

Kuhn, Thomas S. (1970). *The Structure of Scientific Revolutions*. 2nd ed. Chicago: University of Chicago Press.

Kuriyama, Shigehisa (1999). *The Expressiveness of the Body and the Divergence of Greek and Chinese Medicine*. New York: Zone Books.

Lampton, David M. (1977). *The Politics of Medicine in China: The Policy Process, 1949–1977*. Boulder: Westview Press.

Latour, Bruno (1993). *We Have Never Been Modern*. Cambridge: Harvard University Press.

Sivin, Nathan (1987). *Traditional Medicine in Contemporary China: A Partial Translation of Revised Outline of Chinese Medicine (1972) with an Introductory Study on Change in Present Day and Early Medicine*. Ann Arbor: University of Michigan Center for Chinese Studies.

Xie, Guan (1994/1921). *Comprehensive Dictionary of Chinese Medicine*. Beijing: Chinese Medicine Publishing House. Original edition, 1921.

Yin, Huihe and Bo'ne Zhang (1984). *Basic Theory of Chinese Medicine*. In *Textbooks for Higher Education Medical Institutions*. Shanghai: Shanghai Science and Technology Publishing House.

Zhang, Boyu, Jianhua Dong, and Zhongying Zhou (1985). *Chinese Internal Medicine*. In *Textbooks for Higher Education Medical Institutions*. Shanghai: Shanghai Science and Technology Press.

Further Reading

Farquhar, Judith (1994). *Knowing Practice: The Clinical Encounter of Chinese Medicine*. Boulder: Westview Press.

Karchmer, Eric I. (2015). "Slow Medicine: How Chinese Medicine Became Efficacious Only for Chronic Conditions." In *Historical Epistemology and the Making of Modern Chinese Medicine*, edited by Howard Chiang (pp. 188–216). Manchester: Manchester University Press.

Lei, Sean Hsiang-lin (2014). *Neither Donkey Nor Horse: Medicine in the Struggle over China's Modernity*. Chicago: University of Chicago Press.

Scheid, Volker (2002). *Chinese Medicine in Contemporary China: Plurality and Synthesis*. Durham, NC: Duke University Press.

Zhan, Mei (2009). *Other Wordly: Making Chinese Medicine through Transnational Frames*. Durham: Duke University Press.

47
MEDICINE AS A COMMODITY
Carl Elliott

In 1971, *The New England Journal of Medicine* published a classic defense of medical care as a market commodity. "Medical care is neither a right nor a privilege," wrote Robert Sade, a cardiovascular surgeon, "it is a service that is provided by doctors and others to people who wish to purchase it." In Sade's view, sick people do not have a right to demand medical care; doctors do not have an obligation to provide it; and it is improper for the state to interfere with services that doctors produce and own. Sade wrote: "In a free society, man exercises his right to sustain his own life by producing economic values in the form of goods and services that he is, or should be, free to exchange with other men who are similarly free to trade with him or not" (Sade 1971).

Sade's article appeared only six years after the introduction of Medicare and Medicaid in the Social Security Amendments of 1965, which greatly expanded access to medical care but which many American doctors at the time viewed as a threat to their autonomy as independent practitioners. What made Sade's article controversial was how explicitly he dismissed the traditional view of medicine as a profession. In the traditional view, far from being a commodity, medicine is a profession that carries duties and obligations beyond those of pure market exchanges. The most important of these duties is for doctors to ensure that their decisions and actions serve the welfare of their patients, even if there is a cost to doctors themselves. The fact that medicine carries ethical duties to patients beyond those of ordinary market transactions is what has traditionally exempted it from formal regulation as a business. The public generally grants professions a certain degree of trust to govern themselves.

That trust extends to medical research. Medical publishing operates on a kind of honor system; journal editors and other members of the scientific community simply trust medical researchers to conduct their research ethically and present data honestly. Of course, the scientific community has checks to ensure the quality of medical research—peer review, replication of results, financial disclosure statements, and so on—but there is little external oversight, and rarely are there any legal penalties for violations. The foundations of medical publishing rest on the integrity of medical researchers.

Yet the line between medicine and business has never been entirely clear, at least not in the United States. Most developed countries guarantee basic health care for all of their citizens, regardless of the ability of a citizen to pay. But the United States has embraced market-based medicine with open arms. If anything, American medicine looks far more like a commodity today than it did in 1971. Pharmaceuticals are advertised directly to consumers on television. Access to most of the medical literature is controlled by large, for-profit publishing companies. Hospitals are owned and managed by large corporations. Clinical trials are conducted by multinational Contract Research Organizations. The line between patient and consumer has been blurred by cosmetic surgery practices, fertility clinics, and weight-loss centers. And, of course, American doctors are very well-paid.

For and Against Commodification

What does it mean to say that medicine is a commodity? To commodify something, in Michael Sandel's phrase, is to transform it into an "instrument of profit and use" (Sandel 2013). It is to assign market values to a good, service, or idea that was previously thought to be outside the sphere of the market. An extreme example of commodification would be the slave trade, in which human beings were owned, bought, and sold. Of course, slavery is almost universally condemned, but the commodification of some other goods and services is genuinely controversial. Reasonable people disagree, for instance, about whether it is a good idea to buy and sell sex, embryos, or gestational surrogacy.

The most common arguments for commodification in controversial cases are libertarian. Why should the state prohibit the market exchange of a good or service, as long as the exchange does not harm anyone? Just as the state should not prohibit anyone from selling their furniture, their vegetables, or their services as a computer programmer, neither should the state prohibit them from selling their kidney, their sperm, or their services as a surrogate mother.

But there is also a broader, less individualistic argument for commodification, which holds that market exchanges improve our collective well-being (Sandel 2013). This argument follows from the basic argument for capitalism as an economic system, dating to Adam Smith. If there is a demand for a product or service, people will compete to provide it, and the rest of us are free to buy it from wherever we get the best bargain. In this way, capitalism increases the production of goods and services and allocates them to whoever values those goods and services most highly. Both buyers and sellers are made better off.

In medicine, of course, there is always a demand for goods and services, because people always get sick or disabled. So, in theory, we could simply submit medicine to market forces, and then let people buy medical goods and services wherever they like. Market forces should generate the supply. A market-based system would allow all of us to sell goods and services that are useful to others in the medical marketplace, such as our services as an organ donor or a research subject.

Should there be any limits on what goods and services are placed in the market sphere? Objections to commodification generally fall into one of a few broad categories. The first concern is about fairness and inequality. Once goods and services are moved into the realm of the market, they are generally distributed according to a buyer's willingness—or ability—to pay. This system of distribution is the basic mechanism behind the law of supply and demand. Yet while this may seem like an acceptable way to distribute luxury items, or even basic consumer goods and services, it seems morally problematic as a way of addressing genuine human needs, such as nutrition, police and fire protection, or medical care. It feels unfair to withhold crucial, life-sustaining goods or services from people who are unable to pay.

This concern often arises when there is controversy over the distribution of (arguably) public goods, such as water, public lands, or scientific knowledge. For instance, the vast majority of the medical literature today is published in journals owned and managed by for-profit publishers. From 2000 to 2005, Elsevier, the largest publisher of scientific journals, earned profits of close to $10 billion, largely by charging university libraries extraordinarily high electronic licensing fees (Smith 2006). Access to scientific publications is thus largely limited to people who are affiliated with universities or specialized corporations, or who are wealthy enough to pay out of pocket for access. A growing number of critics argue that this arrangement is unjust and that science should be seen as a public good. Richard Smith, the former editor of the *British*

Medical Journal, has written, "The whole business of medical journals is corrupt because owners are making money from restricting access to important research, most of it funded by public money" (Smith 2006).

A second, somewhat related concern is about constraints on voluntariness (Sandel 2013; Walzer 1983). Many people can be persuaded to do things that are risky, uncomfortable, or degrading, as long as they are paid enough. An economist might argue that if people do such things, it is because they value the payment enough to justify taking the risks. This may well be true, but it is also true that the voluntariness of these choices depends not just on the value of what is being offered, but on the background conditions of the choice. If a person provides a service that is dangerous or degrading for money, it might well be a result of the fact that the person is desperately poor. To many people, it seems wrong to take advantage of that desperation.

Both of these concerns emerge in debates over the legalization of markets for transplant organs. The current system of harvesting organs for transplantation relies on volunteers—either people who agree to donate their organs after they die or living organ donors who volunteer to donate while they are still alive. Once harvested, these organs are generally distributed according to a complex algorithm based on factors such as the patient's medical need, the likelihood of medical benefit, and the length of time the patient has been on a waiting list.

The problem is that the need for organs far outstrips the supply. Critics of the volunteer system argue that the shortage could be remedied with a market-based solution, in which patients in need of transplantation would be permitted to pay for organs and potential donors would be paid to donate. In this way, at least in theory, the supply of organs would increase to meet the demand (Cherry 2005).

Defenders of a volunteer system usually object to a market-based solution on two related grounds. First, it seems unfair to distribute organs based on wealth, rather than need. Why should wealthy patients be able to jump the queue ahead of those who may be in greater need or who may have been waiting longer? Second, it is argued that financial incentives would lure potential organ donors into risking their own health, especially potential living donors who are very poor and do not have access to the high-quality, follow-up medical care that is necessary after donation (Scheper-Hughes 2002). Whether these problems could be prevented or mitigated by regulatory mechanisms—such as, say, a futures market for cadaveric organs—is a matter of some debate (Cohen 1993).

A third general concern about commodification is more elusive. According to many critics, some goods become corrupted or degraded when they are moved into the market sphere (Sandel 2013). This type of concern is very different from those based on fairness or on problems with voluntariness. It relies on the notion that some institutions, practices, and attitudes must be insulated from market forces in order to preserve their integrity. For example, some people object to the legalization of sex work not simply because they believe it exploits the poor, whose economic circumstances force them into work they would otherwise avoid, but because they believe it degrades women and encourages harmful social attitudes toward sex. Similar arguments lie behind the widespread prohibition against buying or selling babies for adoption. It is argued that children ought to be valued for themselves as human beings, rather than as instruments of profit.

Sometimes commodification may erode the value of a good by changing it into something else entirely (Andre 1992; Walzer 1983). For example, we do not allow the buying and selling of awards and prizes, such as the Nobel Prize or the Congressional Medal of Honor. Partly this is based on grounds of fairness: we think the distribution of these awards should be based on merit, not wealth. But another reason is because a market in these awards would corrupt their value. The very meaning of the Nobel Prize would vanish if receiving one were no longer an honor.

A related concern is sometimes raised about the corrosion of medicine as a profession, especially in the context of so-called enhancement technologies, such as cosmetic surgery, sexual enhancements, or performance-enhancing drugs for athletes (Parens 1998). Traditionalists argue that the purpose of medicine is to treat illnesses and disabilities and to relieve suffering, not to improve a person's appearance or performance. Of course, many people are more than willing to pay for medical enhancements such as cosmetic surgery, and many physicians are equally willing to provide them. But traditionalists argue that the willingness of physicians to take part in a market for medical enhancements has degraded their professional status as healers, transforming them into mere vendors of medical goods and services.

Markets may also generate potentially perverse financial incentives. Many people believe this has occurred with the privatization of clinical research, as will be illustrated by the case below.

Example: The Commodification of Drug Research

Until the early 1990s, most clinical research on pharmaceuticals took place in academic settings. Pharmaceutical companies provided the funding, but the companies would generally partner with academic physicians in medical schools and teaching hospitals to carry out the research. These arrangements began to change in the early 1990s, when pharmaceutical companies started outsourcing clinical research to the private sector. This change was driven largely by economic factors. Clinical trials were becoming larger and more complex, and the business model of the pharmaceutical industry was shifting toward the production of so-called blockbuster drugs, such as Prozac, Prilosec, and Lipitor. (Blockbuster drugs generate at least $1 billion in annual revenue and are often aimed at common medical problems such as depression, diabetes, asthma, or high cholesterol.) In 1994, about 70% of clinical researchers were affiliated with academic health centers, but by 2006, that figure had decreased to 36% (Getz 2007).

This shift to the private sector has created a niche for businesses specializing in various aspects of the research enterprise. The largest and most important are Contract Research Organizations (CROs), such as Covance, Quintiles, and Parexel, which manage and organize clinical trials (Fisher 2008; Petryna 2009). But other, smaller companies have emerged to take advantage of more specialized tasks. Today, for example, clinical trials are likely to be conducted in private clinical trial sites, overseen by contract researchers. The research subjects might be recruited by specialized patient recruitment companies. The publication of the results of these trials may be managed by a team of medical writers and "publication planners" working for a specialized medical communications company. And the ethical oversight of the trials will likely be conducted by a for-profit Institutional Review Board (IRB), which is paid by the sponsor of the study (Emmanuel et al. 2006). (The Food and Drug Administration requires that clinical trials submitted in support of marketing approval of a new drug be approved by an IRB. Traditionally, IRBs were staffed by volunteers and located in universities and hospitals, but today many IRBs are free-standing, for-profit entities that charge fees to research sponsors for reviewing their protocols.)

In theory, transforming medical research into a market commodity should have produced a faster, more efficient system of drug development, resulting in better products. And it is true that privatization has helped the pharmaceutical industry generate extraordinary profits. By the mid-2000s, the pharmaceutical industry was easily the most profitable business in the world (Angell 2005). But the changes of the 1990s and 2000s also introduced an unprecedented series of scandals, many of them involving the manipulation of research results to promote drugs of dubious benefit. By 2010, according to Public Citizen, the pharmaceutical industry had surpassed the defense industry as the leading defrauder of the federal government. Over a 20-year period, pharmaceutical companies were forced to pay $19.8 billion in federal penalties (Public Citizen 2010).

One of the most alarming consequences of this period was the corruption of the medical literature. As Richard Horton, the editor of *The Lancet*, has pointed out, a market-based research system forces medical researchers to compete with one another in a market economy and treat scientific data as a trade secret. And because success is measured by profits, a market-based system gives industry-backed researchers financial incentives to produce results that are favorable to their employers. In Horton's view, market forces have corrupted the very institution of medical publication. He writes, "Journals have devolved into information laundering operations for the pharmaceutical industry" (Horton 2004).

Over the past two decades, study after study has found that scientific articles published by researchers with pharmaceutical industry funding are more likely than independently funded articles to contain results favoring the industry sponsor (Flacco et al. 2015; Lexchin et al. 2003; Lundh et al. 2012). In 1994, for instance, a meta-analysis of studies of nonsteroidal anti-inflammatory drugs for arthritis found that not a single industry-funded article presented results that were unfavorable to the sponsor (Rochon et al. 1994). Similarly, a 2006 meta-analysis in the *American Journal of Psychiatry* examined head-to-head comparisons of three different atypical antipsychotic drugs for the treatment of schizophrenia. It found that 90% of those studies favored the antipsychotic produced by the company that funded and designed the trial (Heres et al. 2006).

Findings such as this can be partly explained by the suppression of unfavorable results. A 2008 analysis in *The New England Journal of Medicine* found that the manufacturers of antidepressants buried the results of the vast majority of studies that reflected poorly on their products. The drug makers published only 14% of studies with poor or ambiguous results, as compared to 94% of the studies with positive results. Yet the suppression of unfavorable studies only explains part of the problem. According to the same 2008 analysis, many of the antidepressant studies that did wind up in print misrepresented their results, reporting findings that were positive for the sponsor's drug even when such a conclusion was not warranted by the data (Turner et al. 2008).

As Richard Smith has explained, the easiest way for pharmaceutical companies to get the results they want is by rigging the design of the studies in advance. Some of these tricks are simple. A study sponsor might choose a low dosage of a competitor drug so that it appears less effective. Or it might choose an excessively high dosage so that the competitor drug appears toxic and causes unpleasant or dangerous side effects. It might also underpower the trial by choosing a sample size unlikely to demonstrate a statistically significant difference between the two drugs and then claim equivalence. If a company designs its study in the right way, using these or other more statistically sophisticated tricks, it will not need to worry about burying unfavorable results. A positive result is all but guaranteed (Smith 2005).

Many pharmaceutical companies outsource the publication of their studies to specialized medical communications agencies, which integrate marketing goals into the "publication planning" for new drugs (Matheson 2008). Publication planners will explore the ways in which a new drug differs from other drugs on the market, which journals and specialists should be targeted, and the ways in which publication can be integrated into academic conferences, public relations efforts, and Continuing Medical Education events. A medical writer—or "ghostwriter"—will write up the publications. Finally, the agency will often recruit an academic physician to serve as "author" for the study, sometimes in exchange for a fee. The academic affiliation of the "author" of the study gives it a veneer of objectivity and helps disguise the fact that the article was produced by a pharmaceutical company working in concert with a communications agency.

Although medical ghostwriting has played a part in many of the fraud scandals of the past 15 years (and has been widely condemned as a result), it is hard to know just how often it occurs. In 2003, David Healy and Dinah Cattell tracked all of the scientific publications on the antidepressant Zoloft (sertraline) produced by a medical communications agency called Current Medical Directions, on behalf of Pfizer, the manufacturer of Zoloft, over a three-year period. They found that over half of all Zoloft publications during that period had been produced by the agency. The agency-produced articles were published in more prestigious journals and were cited at a rate five times more often than the "traditionally authored" articles (Healy and Cattell 2003).

The distortion of the medical literature is clearly a problem for practicing physicians, who rely on honest studies in order to make prescribing decisions for their patients. But it is also a serious problem for research subjects, many of whom take risks to their health when they sign up for research studies. Would these subjects enroll in research studies if they knew that the results were likely to be buried or manipulated? If so, under what conditions would they be willing to enroll?

The Emergence of Paid Research Subjects

For many years it was seen as ethically troubling for researchers to pay potential subjects to sign up for studies. Critics worried that payment might tempt people to risk their health in studies that were painful, unpredictable, or dangerous. On the occasions where subjects were offered money, it was usually only a small amount, often to reimburse them for expenses such as meals, transportation, or parking.

As clinical research moved into the private sector, however, the amount of money available for subjects began to increase. By the mid-2000s, it was not unusual to see clinical trial sites advertising studies with payments of $6,000 and up. The highest-paying studies today are Phase 1 clinical trials, which are generally conducted at the beginning of drug development. Their main aim is generally to determine if an experimental compound is safe. In contrast to later-stage trials, which are conducted on ill patients in search of a new treatment, Phase 1 trials are usually conducted on healthy subjects. These trials usually require subjects to check into an inpatient unit for at least three weeks, so that they can be carefully monitored. Some trials also include invasive procedures, such as biopsies and endoscopies, in which case the payment is generally higher. In some areas, this approach has generated an underground economy of semi-professional research subjects, who enroll in one study after another in exchange for a fee (Abadie 2010; Elliott 2014).

The range of studies offering payment has broadened as well. Until recently, patients who were ill and looking for new treatments were generally not paid, largely because it was felt that sick people were too vulnerable to potential manipulation. Today, however, many trial sites offer payment to patients with illnesses such as asthma, diabetes, anxiety disorders, and even schizophrenia. In some areas, for instance, clinical trial sites recruit mentally ill subjects from recovery houses and homeless shelters (Elliott 2014).

The arguments in favor of paying subjects are straightforward and pragmatic. For many types of trials, it is hard to imagine many people enrolling without a considerable financial incentive. Unlike later-stage clinical trials, for instance, which can offer sick patients the possibility of effective therapy, Phase 1 trials offer subjects nothing but risk and discomfort. Their very purpose is to determine how toxic an experimental drug is. And since most Phase 1 trials are sponsored by pharmaceutical companies, which are generally ranked among the world's least trusted industries, it is unlikely that many subjects would volunteer purely for altruistic reasons.

The primary objection to payment is related to constraints on voluntariness. If a desperately poor person tests the safety of an experimental drug for money, to what extent is the decision truly voluntary? The standard conceptual framework for addressing such concerns is that of "undue influence" (Jones et al. 2010). The motivating idea behind the concept of undue influence is that payments should not be so large as to persuade people to take risks that they would not otherwise take. But measuring how much money constitutes an undue influence is conceptually fraught. When a subject is desperately poor, even a small amount of money might persuade him or her to enroll in a dangerous study, whereas for a rich person, no amount of money would be enough.

If payment is conceptualized as compensation for work, then keeping payment low seems unfair (Elliott and Abadie 2008). Why shouldn't subjects be well-compensated for what is, after all, an unpleasant and possibly risky endeavor that results in a social good? Concern about undue influence merely magnifies the problem. Why should payment rates be designed so that the poorer the subject, the less he or she is paid?

Research subjects may be paid, but they get few of the protections and benefits that come with other kinds of employment, such as worker's compensation, health insurance, and the right to unionize. In fact, if a subject is injured in a research study in the United States, chances are that he or she will be responsible for his or her own medical expenses (Elliott 2012). (The United States is the only developed country that does not require sponsors to pay for the medical care of injured subjects.)

Although no central agency tracks the number of deaths and injuries in clinical trials, there have been a number of high-profile cases in which paid subjects in clinical trials have died or been severely injured. The most notorious recent case was the Northwick Park disaster near London in 2006, where six healthy subjects nearly died in a "first-in-man" study of a monoclonal antibody named TGN1412. In 2004, a paid volunteer named Tracy Johnson committed suicide in an Eli Lilly–sponsored, Phase 1 study of its antidepressant Cymbalta. And in 2007, Walter Jorden, a mentally ill man living in a Philadelphia recovery house, died after being paid to test an experimental antipsychotic drug produced by Astellas Pharma (Elliott 2014).

It is also worth pointing out that while oversight bodies are instructed to ensure that payments do not unduly influence subjects, there are no such restrictions on the amount of money paid to contract researchers for conducting the trials. In fact, payment to trial sites is often on a per-head basis: the more subjects a trial site recruits, the more money it receives. For contract researchers seeing patients who might otherwise receive standard treatment, there is another perverse incentive, according to sociologist Jill Fisher: any given medical service will generate two to five times the amount of money when a pharmaceutical company is paying the bill, rather than a health insurance company or a government agency (Fisher 2008). So if a private physician is conducting a trial, it is in his or her financial interests to enroll patients in a trial rather than simply to treat them.

Federal guidelines permit research sponsors to pay subjects, although they do require selection of subjects of clinical trials to be "equitable" and stipulate that special protection must be provided to subjects who are "economically disadvantaged." Yet the very structure of many inpatient trials, such as Phase 1 and bioequivalence studies, generally excludes most people with regular jobs. Not many people are able to check into an inpatient facility for three or four weeks at time. The only populations easily able to do that are generally students, unemployed people, contract workers, and people prohibited from taking other jobs, such as undocumented immigrants (Abadie 2010; Elliott and Abadie 2008).

A better framework for conceptualizing this issue might be exploitation, rather than undue influence. To exploit someone is to take unfair advantage of them, usually in a relationship of unequal power. A common example of exploitation is sweatshop labor, in which large corporations take advantage of the poverty of people in the developing world by offering them poorly paid jobs under dangerous conditions. A similar case could be made for paid subjects in clinical trials. The ethical issue would then be whether a research sponsor is taking unfair advantage of a subject's poverty and desperation (Elliott 2014).

Conclusion

An old philosophical debate about the costs of commodification has become gradually more important for American medicine. On the one hand, a capitalist economy makes the production of goods and services far more efficient by introducing the principles of mass production. In theory, this should be as true of medical practice as it was for Adam Smith's pin factory. Yet the problem, as Marx famously pointed out, is that these principles also alienate workers from their labor and its products. If a job requires little skill or creativity, and has no larger moral purpose, its meaning is drained away.

In many ways, such a transformation has been well under way in American medicine for decades. While the figure of the solo family practitioner may be idealized in the popular imagination, in reality that figure has been replaced by teams of specialists working in group settings owned by large corporations, where they often practice according to rigid treatment guidelines and reimbursement plans developed by distant managers and experts. Whether this transformation has produced more efficient delivery of medical care is a matter of some debate, but it has clearly demoralized doctors, who have grown increasingly unhappy and more alienated in their work. A 2008 survey of more than 12,000 physicians found that a mere 6% thought that the morale of the profession was positive. Approximately 60% said they would not recommend medicine as a career. Only 22% said the practice of medicine was highly rewarding (Physicians' Foundation 2008).

There are many reasons for this demoralization—declining reimbursement rates, pressure to limit the amount of time spent with patients, the crushing burden of paperwork—but it is hard to separate these things from the transformation of medicine into a business. Many doctors went into medicine precisely because they believed that it was not simply a business selling goods and services, but rather, a profession with a larger moral purpose. Yet that purpose seems contrary to the spirit motivating much of American medicine today. In a recent survey, nearly one-third of American medical school faculty members said that their institution discourages altruism. More than half felt that medical school administrators are "only interested in me for the revenue I generate" (Pololi et al. 2012). The collective sense of alienation is summed up by a physician quoted by Dr. Sandeep Janduhar, author of the memoir *Doctored*: "I feel like a pawn in a moneymaking game for hospital administrators. There are so many other ways I could have made my living and been more fulfilled. The sad part is we chose medicine because we thought it was worthwhile and noble, but from what I have seen in my short career, it is a charade" (Jauhar 2014).

References

Abadie, R. (2010) *The Professional Guinea Pig: Big Pharma and the Risky World of Human Subjects*. 1st ed. Durham, NC: Duke University Press.

Andre, J. (1992) Blocked Exchanges: A Taxonomy. *Ethics* 103:1, 29–47.

Angell, M. (2005) *The Truth About the Drug Companies: How They Deceive Us and What to Do About It.* 1st ed. New York: Random House.

Cherry, M. (2005) *Kidney for Sale by Owner: Human Organs, Transplantation, and the Market.* 1st ed. Washington, DC: Georgetown University Press.

Cohen, L. (1993) A Futures Market in Cadaveric Organs: Would It Work. *Transplantation Proceedings* 25:1, 60–61.

Elliott, C., Abadie, R. (2008) Exploiting a Research Underclass in Phase 1 Clinical Trials. *New England Journal of Medicine* 358, 2316–2317.

Elliott, C. (2012) Justice for Injured Research Subjects. *New England Journal of Medicine* 367, 6–8.

Elliott, C. (2014) The Best-Selling, Billion-Dollar Pills Tested on Homeless People, Matter, July 27. Available at: https://medium.com/matter/did-big-pharma-test-your-meds-on-homeless-people-a6d8d3fc7dfe [Accessed 22 July 2015].

Emmanuel, E., Lemmens, T., Elliott, C. (2006) Should Society Allow Research Ethics Boards to be Run As For-Profit Enterprises? *PLoS Medicine* [Online] 3:9, e391. Available at: http://journals.plos.org/plosmedicine/article?id=10.1371/journal.pmed.0030309 [Accessed 22 July 2015].

Fisher, J. (2008) *Medical Research for Hire: The Political Economy of Pharmaceutical Clinical Trials*, 1st ed. New Brunswick, NJ: Rutgers University Press.

Flacco, M., Manzoli, L., Boccia, S., Capasso, L., Aleksovska, K., Rosso, A., Scaioli, G., De Vito, C., Siliquini, R., Villari, P., Ioannidis, J. (2015) Head-to-Head Randomized Trials are Mostly Industry Sponsored and Almost Always Favor the Industry Sponsor. *Journal of Clinical Epidemiology* 68:7, 811–820.

Getz, K. (2007) Industry Trials Poised to Win Back Academia. *Applied Clinical Trials* 16(4), 35–38.

Healy, D., Cattell, D. (2003) Interface between Authorship, Industry and Science in the Domain of Therapeutics. *British Journal of Psychiatry* 183, 22–27.

Heres, S., Davis, J., Maino, K., Jetzinger, E., Kissling, W., Leucht, S. (2006) Why Olanzapine Beats Risperidone, Risperidone beats Quetiapine, and Quetiapine Beats Olanzapine: An Exploratory Analysis of Head-to-Head Comparison Studies of Second-Generation Antipsychotics. *American Journal of Psychiatry* 163:2, 185–194.

Horton, R. (2004) The Dawn of McScience. *New York Review of Books* 51, 7–9.

Jauhar, S. (2014) *Doctored: The Disillusionment of an American Physician.* 1st ed. New York: Farrar, Straus and Giroux.

Jones, E., Liddell, K., Saunders, J. (2010) Should Healthy Volunteers in Clinical Trials be Paid According to Risk? *British Medical Journal* 340:7738, 130–131.

Lexchin, J., Bero, L. A., Djulbegovic, B., Clark, O. (2003) Pharmaceutical Industry Sponsorship and Research Outcome and Quality. *British Medical Journal* 326, 1167–1170.

Lundh, A., Sismondo, S., Lexchin, J., Busuioc, O. A., Bero, L. (2012) Industry Sponsorship and Research Outcome, Cochrane Database Systematic Reviews 12. Available at: http://onlinelibrary.wiley.com/doi/10.1002/14651858.MR000033.pub2/full [Accessed 22 July 2015].

Matheson, A. (2008) Corporate Science and the Husbandry of Scientific and Medical Knowledge by the Pharmaceutical Industry. *BioSocieties* 3, 355–382.

Parens, E. (ed.) (1998) *Enhancing Human Traits.* 1st ed. Washington, DC: Georgetown University Press.

Petryna, A. (2009) *When Experiments Travel: Clinical Trials and the Global Search for Human Subjects.* 1st ed. Princeton, NJ: Princeton University Press.

Pololi, L., Krupat, E., Civian, J., Ash, A., Brennan, R. (2012) Why Are a Quarter of Faculty Considering Leaving Academic Medicine? A Study of Their Perceptions of Institutional Culture and Intentions to Leave at 26 Representative U.S. Medical Schools. *Academic Medicine* 87:7, 859–869.

Public Citizen. (2010) Pharmaceutical Industry Is Biggest Defrauder of the Federal Government under the False Claims Act, New Public Citizen Study Finds. Available at: https://www.citizen.org/Page.aspx?pid=4734 [Accessed 22 July 15].

Rochon, P., Gurwitz, J., Simms, R., Fortin, P., Felson, D., Minaker, K. L., Chalmers, T. C. (1994) A Study of Manufacturer-Supported Trials of Nonsteroidal Anti-Inflammatory Drugs in the Treatment of Arthritis. *Archives of Internal Medicine* 154, 157–163.

Sade, R. (1971) Medical Care as a Right: A Refutation. *New England Journal of Medicine* 285, 1288–1292.

Sandel, M. (2013) *What Money Can't Buy*. 1st ed. New York: Farrar, Straus and Giroux.
Scheper-Hughes, N. (2002) The Ends of the Body: Commodity Fetishism and the Global Traffic in Organs. *SAIS Review* 22:1, 61–80.
Smith, R. (2005) Medical Journals Are an Extension of the Marketing Arm of Pharmaceutical Companies. *PLoS Med* 2(5): e138. Available at: http://journals.plos.org/plosmedicine/article?id=10.1371/journal.pmed.0020138 [Accessed 22 July 2015].
Smith, R. (2006) *The Trouble with Medical Journals*. London: Royal Society of Medicine Press.
Turner, E., Matthews, A., Linardatos, E., Tell, R., Rosenthal, R. (2008) Selective Publication of Antidepressant Trials and its Influence on Apparent Efficacy. *New England Journal of Medicine* 358: 3, 52–60.
Walzer, M. (1983). *Spheres of Justice*. 1st ed. New York: Basic Books.

Further Reading

Sandel, Michael (2013) *What Money Can't Buy* (New York: Farrar, Straus and Giroux), is a lively, sophisticated overview of the costs of what Sandel calls "a market society."

My book [Elliott, Carl] (2010) *White Coat, Black Hat* (Boston: Beacon Press), examines how the market has generated a range of new quasi-medical jobs, from professional research subjects to medical ghostwriters.

For a journalistic introduction to the types of body parts that can be bought and sold, see the excellent Carney, Scott (2011) *The Red Market: On the Trail of the World's Organ Brokers, Bone Thieves, Blood Farmers, and Child Traffickers* (New York: William Morrow), and Fisher, Jill. (2008) *Medical Research for Hire* (New Brunswick, NJ: Rutgers University Press), an authoritative, readable overview of the contract research industry.

Angell, Marcia (2005) *The Truth about the Drug Companies* (New York: Random House), remains the most persuasive critique available of pharmaceutical industry corruption.

INDEX

Page numbers for figures are in *italics*.

abdominal obstruction 158
abdominal pain 231–2
ab initio predictions 82
abnormalities 11, 23, 31, 144, 432–41; behavioral 164, 170–2, 175; biological 177–8; genetic diseases and 153
abortion 40–1, 105
academic health centers 522
accidents 129
acetaminophen 254–5, 482
achondroplasia 434
ACP Journal Club 209
action 17–18, 104, 126, 391
actuarial judgment 364, 370
actuarial tables 162
acupuncture 258–60, 264–5, 501
ADA (Americans with Disabilities Act) 10, 37–8, 43–5
adaptationism 478–80
ADD (attention deficit disorder) 10, 156, 165
addiction 130–1, 460
Addyi (flibanserin) 161–2
adenine 145
ADHD (attention deficit hyperactivity disorder) 20, 343
adoption, of children 107, 521
Adorno, T. 379
adrenal glands 437
advertising 166, 405, 519
Advisory Committee to the Surgeon General of the Public Health Service 251
aeipatheia 98
aestheticization 111
affect 126–30
affiliation 376–7
Africa 509
African Americans 424–8
African ancestry 421
after-baby body 111
ageism 339
Agency for Healthcare Research and Quality (AHRQ) 330, 358
agent-based models 277–8
aging 16–18
Aguilera, C. 111
A-heFT clinical trials 425
aided reasoning 364, 370

AIDS 149, 172, 298, 311–15, 374
airflow obstruction 184–5
albinism 11
alcohol 19, 131
alendronic acid (Fosamax) 348
Alexander, E.: "Haircut" 376
algorithms 275, 293, 366–7, 388, 403
alienation 6, 373
alleles 86, 146–9, 298
allergic reactions 243
allergies 385, 387
allocation 156, 199–203, 212–13, 219, 228, 339
alpha-synuclein 86
alpha values 220
alternative medicine 257–67
Alzheimer's disease 11, 19, 72, 118
American Cancer Society 359
American College of Cardiology/American Heart Association 224–5
American Heritage Medical Dictionary, The 39
American Indians 419
American Journal of Psychiatry 523
American Psychiatric Association: *Diagnostic and Statistical Manual of Mental Disorders* (DSM) 12, 97, 159–62, 170–1, 174–6, 457–63
American Recovery and Reinvestment Act 330
Americans with Disabilities Act (ADA) 10, 37–8, 43–5
amino acids 82, 145, 150
amoebas 116–17
amputation 253
Amundson, R. 41–2
analgesia 140–1
analogy 53, 59, 251
Analytical Language of John Wilkins, The (Borges) 182–3
analytic training 326
anatomical criteria 184–5
ancestral ideals 478–80
ancient Greece 432
Andersen, H. 213–14
Anderson, E. 326
Anderson, W. 421
androcentrism 109
anecdotal reports 357–8
anemia 94

angina 28
angina pectoris 135
angiosarcoma 58–66
Anglophone tradition 2
angst 108
animalism 118–22
animals: classifying 182; dairy 148–9; disease and 8; models and studies 54, 59, 62–3, 252, 271–3, 281–2, 345, 410–11, 460
animations 277
anistreplase 229–33
Ankeny, R. 1, 273
ankle fractures 364
ankylosing spondylitis 159
anomalies 291–2
anonymity 108
anterior cingulate 127–30
Anthropology from a Pragmatic Point of View (Kant) 6
antiabortifacients 433
antibiotic-resistant organisms 390
antibiotics 138, 243, 300–3, 385, 390, 397–8
anticipation of pain 128–9
antidepressants 12, 171, 221, 343, 349–50, 457, 460
anti-discrimination law 44
anti-fat industry 164
anti-narrative 374
antiplatelet pharmacotherapy 412
antipsychotics 58, 460
antipyretic treatment 482–3
antiquity 5–6
anti-realism 90–9, 176
antisocial behavior 165, 428
anxiety 6, 17, 20, 85, 129–30, 156, 257, 410; classifications of 159; meditation and 258; placebo treatments for 141
anxiolytics 411
aphasia 38
Apkarian, V. 126–7
apodicticity 105
apoptosis 277–8
appearance, physical 164
appendicitis 231–2
appetite 129
appraisals 33, 210
Aquinas 6–8
AR (Attributable Risk) 152
Arendt, H. 104
Aristotelian notions 182
Aristotle 5
Arnold, T. 157
arousal 130
arthritis 159
asbestos 67, 243
Ashkenazi Jews 147, 298
aspirin 412
ASSENT-2 trial 347
assessments: of disabilities 36; of medical treatments 52; subjective outcomes and 221; of treatment efficacy 219
association models 271–5, 278–82
Astellas Pharma 525
asthma 254–7, 387
asymmetric relationships 107
asymptomatic individuals 8, 11–13, 20–2, 64, 152
asymptotic reassurances 243
atherosclerosis 245
atoms 84
atrial fibrillation 409
at-risk states 11
attention 17, 376–7
attention deficit disorder (ADD) 10, 156, 165
attention deficit hyperactivity disorder (ADHD) 20, 343
Attributable Risk (AR) 152
attributes 77
attrition bias 61
atypical bodies 432–41
auditory sensations 109
auditory system 32
AURORA (statin trial) 225
Australia 483
authority, argument from 353
authority figures 210
autism 12
autoimmune diseases 185, 314, 387, 391
automation 293–4
autonomy 31, 381, 491
autopsies 296–7
autosomal dominant alleles 147–8
autosomal genes 146–8
Avandia (rosiglitazone) 343–50
avoidance of pain 127
awareness 17, 118–20, 165–6

Bachelor of Nursing Science (BSN) 487–9
background knowledge 50, 263–4
back pain 21–2, 36, 124–9, 409
bacteria 48–9, 249, 305, 476
bacterial infections 302, 370, 387–90, 397–8
bacterial resistance 302–3
bacterial sinusitis 385–8
Bain, D. 125–6
Barbin, H. 436
Barthes, R. 378–9
Bauby, J-D.: *The Diving Bell and the Butterfly* 468
Bayesian network models 274–8, 281
Bayesian trials 199, 219–20, 228–36
Bayes, T. 229
Bayes' Theorem 231–2, 384–5, 389
Beauchamp, T. 381
Beck, A. 2
becoming 104
Beecher, H. 134–6, 139–41, 373; "The Powerful Placebo" 135

INDEX

behaviors 10–12, 20, 23, 408, 436; disease categories and 162–4; medicalization of 175–6; risky 68; studies of 411
Beijing University of Chinese Medical Sciences 505
Beijing University of Chinese Medicine 508–9, 513
being 103–5
Benedetti, F. 140
benzene 281–2
bereavement 12
Bernard, C. 6, 273
beta-blockers 412
Bextra (valdecoxib) 348
Bh$_4$ (tetrahydrobiopterin) 151
Bhabha, H. 374
Bhaktin, M. 375
Bhopal, R. 423, 427
bianzheng 501
bianzheng lunzhi 517
biases 50, 55, 218–22, 225, 228, 233–5, 261–2, 339, 386; allocation 212–13; consensus and 357; disabilities and 40–1, 44; gender 408–18; knowledge and 405; observational studies and 238–41; of physicians 413; publication 248, 345, 349–50, 410; randomized controlled trials (RCTs) and 195–201; selection 61; treatment of trial participants and 204; unaided reasoning and 364; unintentional 415; values and 321–4
BiDil (hydralazine hydrochloride and isosorbide dinitrate) 419–20, 423–8
big data 305, 403
Bîgu, D. 292
bile 5–6
bilious type 305
bimatoprost 338
bing 515–17
binglixue 500
bingyin 500
bioassays 59
biochemical causes 457
biochemical defects 23
bioequivalence studies 525
bioethics 41, 373–4, 381
bioinformatics 397, 403
biological gradient 59, 251
biological inheritance 40
biological maturation 109
biological/physical factors 456
biological processes 447
biological race essentialism 420, 424
biological research, commercialization of 404–5
biological statistics 39
biomarkers 11, 19, 160, 282, 398–400, 403
biomedical corporations 10
biomedical information management 186–7
biomedical journals 405

biomedical mode 458–60
biomedical model (BMM) 10, 445–54
biomedical research 59, 68–9, 411, 415
biomedicine 6, 65, 153, 401, 405; race in 421–9; reductionism in 83–7
biopsies 49–51
biopsychosocial model (BPSM) 11, 23, 373, 445–62
biostatistical theory 7–8, 65
biotech industry 404
bipolar disease 6, 10–12, 176
Bird, A. 2, 297
birth 22, 103–14, 129, 215, 331–9
birth centers 103
birth control pills 66
Birth Day (television show) 111–12
Birth Night Live (television show) 111–12
Birthplace in England Research Programme 331–6
birth preparation classes 106
black bile 5–6, 183
black-box approaches 304
Black, K. 458
Black populations: biological race essentialism and 420, 424; cardiovascular diseases (CVDs) and 419; CHF (congestive heart failure) and 419–20
Blackwell, E. 381
bladder cancer 346
bleeding 412
blinding 239
blindness 10, 38–40, 92, 338
blockbuster drugs 522
blood 5–6, 183, 391
blood-brain barrier 151
blood circulation 276, 300–1
blood clots 390
blood glucose 92, 96, 162
blood pressure 162, 240, 400
blood sugar 400
blood supply 118
blood tests 314
blood-thinning drugs 399
blood transfusions 312
blood vessels 92, 409
Bluhm, R. 214
BMI (body mass index) 162–4, 253
BMM (biomedical model) 10, 445–54
Bo, W. 509
bodies: atypical 432–41; categorization of 162–4; disease and 6; habituated 108, 467, 470; reductionism and 85
bodily differences 432–41
bodily functioning 28
bodily integrity 125–6, 130
bodily memory 108–10
bodily processes view 92–3
body as lived 465, 469–70

531

body mass index (BMI) 162–4, 253
body-subject 467
body temperature 481–3
body weight 400
Bognar, G. 28–30
bonding 106–7
bone density 434
bone growth 437
bone marrow transplant 149
bone problems 150
bones, broken 257
Boorse, C. 7–8, 10, 173
Borgerson, K. 213, 264–5
Borges, J.: *The Analytical Language of John Wilkins* 182–3
Borrell-Carrio, F. 449–51
Boston 288–9
Bourdieu, P. 439–40
Bowlby, J. 479
Boyd, R. 93
BPSM (biopsychosocial model) 11, 23, 373, 445–62
brachial plexus injury 332
BRAF gene 185
BRAF inhibitors 185
brain: consciousness and 118; damage to 118, 121–2; death 120–2; disease and 10–12; firing patterns of 128; imaging 174; neural development of 104; pain and 126–9; Parkinson's disease and 86; persistent vegetative state (PVS) 118–19; reward mechanism of 140; sciences 48, 176; states 17
brain-mind, function of 12
Brazil 420
Brazilian immigrants 421
BRCA-deficient tumors 55
breast cancer 398–400; asymptomatic 22; hormone replacement therapy and 62; mechanisms and 49–56; screening 358–9
Breast Cancer Screening (NIH Consensus Conference) 355
Britain 111
British Medical Journal 61, 520–1
Broadbent, A. 2, 68, 304–5, 334
Brocklehurt, P. 332
Brody, H. 313
Brooks, C. 378
Broome, J. 29–30
Brown, R. 286
Bruner, J. 380
BSN (Bachelor Of Nursing Science) 487–9
Buck v. Bell (1927) 40
Burchard, E. 423
burden 39–40, 111
bureaucratization 372
Buse, J. 350
Butler, J.: *Giving an Account of Oneself* 379
bypass, coronary artery 409

cadaveric material 297
CAH (congenital adrenal hyperplasia) 437–8
calcium pyrophosphate crystals 93–4
calculus 81
California 40, 312
CAM (complementary/alternative medicine) 257–67
Canada 210, 355
Canadian Preventative Services Task Force 355
cancer 173, 185, 249–50, 282, 314, 402; advanced-stage 239; asymptomatic 172; causation 53, 56, 64–5, 72; cells 6, 49, 53, 399; end-stage 404; health care costs for 124; immunization genes 65; patients 129; rates 60; research 399–400; susceptibility 61, 387
candida 19–20
candidiasis 314
Canguilhem, G. 8–9, 41, 479
Cannon, W. 6
capitalism 520, 526
Caplan, A. 2
carcinogenic mechanisms 245
carcinogens 52, 58–68
cardiac dysfunction 400
cardiac events 224–5
cardiac toxicity 345
cardiovascular diseases (CVDs) 64, 195–9, 224, 243–4, 344, 408–14, 419
cardiovascular harms 350
cardiovascular risk 225, 237, 238, 245, 412
cardioverter-defibrillator 411
care 17, 107, 156, 490–2
caregiving 135–6, 374
Carel, H. 2, 380, 471; *Illness* 472
Carey, L. 55–6
Carnap, R. 299–300
Carnegie Foundation 373
carnivores 242
Carpenter, D. 223
Carter, K. 63
Cartesian mind-body dualism 373, 381
Cartwright, N. 65, 205–6, 255
Cartwright, S. 185
case-controlled studies 59, 252, 253
case reports 59, 211
case studies 211, 310–18
caspases coordinate demolition 277
CASS (Chinese Academy of Social Sciences) 497
Cassell, E. 130, 373
casual trials 260–1
cataloguing 187, 251
categories 6, 10, 52; diagnostic 156–69; making of 160–4; race and 420; use of 164–5
Cattell, D. 524
Caucasians 347, 426
causal Bayesian networks 275–6, 281
causal effects 238–41

causal factors 73, 79–86, 173–4, 255, 300–1
causal inference 58–70, 214; epidemiology and 249–54; observational studies and 239, 241–6; randomized controlled trials (RCTs) 202–3, 219
causality 58–80, 214, 240–1, 244–5, 279, 297, 300, 316
Causality Probability Connection Principle (CPCP) 75–9
causal models 71–80, 271, 275–82
causal pathways 152–3
causal realism 94
causal relationships 271–2, 275, 278–9, 304, 307
causation 214, 242, 249, 254, 304–5
cause and effect 49, 500
Cause and Effect in Biology (Mayr) 477
cause-of-death statistics 184, 187
Cavarero, A. 379
CCP (Chinese Communist Party) 499, 509–10
CCR5 Δ32 allele 149
CD4+ helper cells 314
C-E correlation 55, 68, 303
celecoxib (Celebrex) 348
cell death 53, 277–8
cell experiments 315
cell generation 115
cell receptors 140
cells: DNA repair mechanisms and 53; reductionism and 84–7; typing of 49–51
cellular differences 51
cellular mechanisms 48–9
cellular/molecular level 92–3
cellular pathology 183
cellular processes 145
C-E mechanism 55
census 420
cerebral palsy 159
cerebrum 121
ceremonies of birth 103
cervical cancers 58
cervical clear-cell adenocarcinoma 433
cervix, dilation of 106
cesarean birth 103, 109–10
ceteris paribus 239, 337–8
Ceusters, W. 181, 189
Cezanne, P. 467
CFS (chronic fatigue syndrome) 16, 20, 96, 160–1, 172
CFTR allele 147
Chakrabarty, D. 509
Chambers, T. 2, 381
change 499–500
chaperone systems 151
Charon, R. 314, 372, 375–7; *Narrative Medicine: Honoring the Stories of Illness* 374
Chauhan, C. 338
checklists 159–60
chemicals 83

chemosynthesis 115
chemotherapy 53, 238–9, 314, 399
Chervenak, F. 334–5
chest pain 411
chest pressure 409
CHF (congestive heart failure) 400, 419–20
chi 264–5
ch'i 503
Chicago 138
chickenpox 71, 257
childbed fever 286–7
childbirth *see* birth
children: abnormal height in 432; adoption of 107, 521; atypical sex anatomies and 438–40; bipolar disease in 176; birth of 104, 109–12; cancer in 185; development of 107; disabilities and 40; fever and 369–70; hGH (Human Growth Hormone) and 434; poverty and 413; sex reassignment and 436; stunted growth and 348; surgery on 263
Childress, J. 381
China 20, 497–9, 505
Chinese Academy of Social Sciences (CASS) 497
Chinese Communist Party (CCP) 499, 509–10
Chinese Internal Medicine textbook 513–16
Cho, M. 425
choice 30–1, 335
cholera 64, 250
cholesterol 224–5, 241, 245, 331, 400
Christidi, G. 348–9
Christie, W. 162–3
chromosomes 42, 144–6, 282
chronic fatigue syndrome (CFS) 16, 20, 96, 160–1, 172
chronic, non-communicable diseases (CNCDs) 249–50
chronic obstructive pulmonary disease (COPD) 19
chronotope 375
Churchland, P. 459–60
CI (confidence interval) 220, 230–4
cigarettes 254, 354
Cimino, J. 186–9
circulatory system 117, 120
citalopram 216
citizenship 164
civil rights movement 373
claims, causal 58–61, 64–5, 68, 279–80, 303–4
claims, testing of 75
Clarke, B. 52, 279
class domination 97
classes 158, 183, 186
classical (frequentist) trials 199–201, 219–20, 228–36
classicists 5–6
classification manuals 170–1
classification-of-classifications debate 7

classifications 5–6; of disabilities 38–9; of diseases 180–91, 238–9; of models 272; principles of 157–8; psychopathology and 170–9; systems of 158–60, 461–3; of triple-negative breast cancer (TNBC) 55–6
classifier models 275
clavicle, fractured 332
clear-cell adenocarcinoma 433
Cleveland, C. 37–8, 43
Cleveland v. Policy Management Systems Corporation (1999) 37–8, 44–5
climate change 354
clinical effect 263
clinical improvement 135–6
clinical judgment 363–71, 449
Clinical Judgment (Feinstein) 366
clinical medicine 18–20, 71, 134, 198–9, 304, 480–1
clinical psychology 170
clinical reasoning 367
clinical research 195, 310, 413; harm detection and 342–52; privatization of 522–3
clinical safety 343
clinical trials 53, 56, 59–62, 185, 209–10, 214, 234; for breast cancer screening 359; gender and 411–14; paid research subjects and 524–6; placebos and 134–41; race and 411, 424; random allocation in 203; registry 326, 350; reliability of 218–27; tracking of 324
clinician-reported questionnaires 331
clinicians 6–7, 173, 364–5
clinico-pathological classifications 188–9
clinimetrics 367
Clinimetrics (Feinstein) 366
Clinton, B. 399
cliques 274
close reading 372, 375–9
Clouser, K. 8–9
clustering 158
"Cluster Stability Rule, The" (Maglo) 427
CMV (cytomegalovirus) 312–14
CNCDs (chronic, non-communicable diseases) 249–50
coagulation 246
cochlear implants 31
Cochrane Collaboration 209–17
Cochrane Database of Systematic Reviews 222
coding 145, 159
codons 145
Cody, W. 493–4
cognition 118–20, 128–30, 150, 306–7
cognitive development 151
cognitive diagnoses 38
Cohen, S.: *Normal at Any Cost* 433–4
coherence 59, 251, 303–7
Cohn, J. 423–4
cohort studies 59, 197, 252–3, 334, 338
Colditz, G. 196–7

collectives 77–8
colonialism 424, 509
color blindness 148
Columbia University 374–5; College of Physicians and Surgeons 381
comfort 17
commercial interests 218, 225
commodity, medicine as 519–28
common cold 19, 22–3, 135, 299–300
common good 40
commonsense psychology 456
Common Variable Immune Deficiency (CVID) 387–91
Communist era 516–17
Communist Revolution 510
community health centers 373
community interests 40
comorbidities 216, 413
comparative studies 237–8
competitiveness 405
complementary/alternative medicine (CAM) 257–67
complex diseases 144–6, 153
complexity 54–6, 81, 86–8
complex-systems mechanisms 276–8, 279
compliance 60
compulsive behaviors 130–1, 137
computer-aided prognosis 391
computerized tomography (CT) 51, 385–7
computer simulations 278
computer systems 180, 186
concatenated models 87
conception 104
conceptual developments 251
conceptual-explanatory framework 455–60
concrete realism 92–5
Condillac, É. 6
conditional probability 232
confidence interval (CI) 220, 230–4
confirmation theory 315
confirmatory trials 220
conformity 42, 432
confounders 55, 61–4, 196–203, 212–14, 220, 238–44
congenital adrenal hyperplasia (CAH) 437–8
congestive heart failure (CHF) 400, 419–20
connectedness 119
connections, causal 12, 23, 220–1, 275–6, 279
Conrad, P. 434–6, 440
conscientiousness 369
consciousness 105, 118–22, 466–7
consensus 37, 160, 203, 353–60
conservationism 419
consistency 59, 251
constellations 96
constitutional hypothesis 61, 64–5, 253
constitutive mechanisms 49
constructivism 90–100, 175–7

INDEX

construct theories 337
contemporary Chinese medicine 497–507
contested narratives 377
context 182, 189, 369
contextualism 479
contextual unanimity 65, 68–9
Continuing Medical Education events 523
continuity 104, 109, 118–20
continuous parameters 343
contraceptives 76–9, 246
contractile structures 300
contractions 106
contract researchers 519, 522, 525
Contract Research Organizations (CROs) 519, 522
contrastivism 241
control groups 50, 60, 135, 212–13, 228, 237–8, 241–2, 248, 260, 263, 337–8
conventional medicine 258, 261, 264
convergence 307
Conway, K. 380
COPD (chronic obstructive pulmonary disease) 19
copyright 161
CORONA (statin trial) 224–5
coronary artery 274
coronary artery disease 409–12
coronary heart disease 249–50, 409
corporeal narratives 381
corpses 117
correlations 55, 68, 214, 279–80, 297, 503–4
cortex 126
corticosteroids 215
Cosgrove, C.: *Normal at Any Cost* 433–4
cosmetic surgery 520–2
Cosmides, L. 302
cosmogony 500–5
costs 330, 339, 354
coughs 135
counter-examples 75–9
counterfactual theory 63–7, 241–6, 253
Covance 522
Coxibs 243–4
CPCP (Causality Probability Connection Principle) 75–9
Craver, C. 48; "Thinking about Mechanisms" 181
creation, divine acts of 103
creative capacities 108–9
creative writing 372
creativity 377
creeping reductions 83–7
Crick, F. 48
crime rates 428
criminal acts 10
criteria 59–62, 96–8, 120–2
critical appraisal 210
critical race theory 381
critical theorists 9–10

critical thinking 326, 379
Critique of Pure Reason, The (Kant) 71–2
Croft, D. 294
CROs (Contract Research Organizations) 519, 522
crosstalk, nervous system and 126
CRPD (United Nations' Convention on the Rights of People with Disabilities) 44
crystals 93–5
CT (computerized tomography) 51, 385–7
Cultural Revolution 499
culture 105, 157, 439–40
Culver, C. 8–9
Cummins, R. 301
cure 17, 48–51, 130
Current Medical Directions 524
Currie, A. 273
CVDs (cardiovascular diseases) 64, 195–9, 224, 243–4, 344, 408–14, 419
CVID (Common Variable Immune Deficiency) 387–91
Cymbalta 525
cystic fibrosis 147, 152, 358, 402, 426
cytomegalovirus (CMV) 312–14
cytosine 145

dairy consumption 148–9
Daly, J. 209
Daniels, N. 42–3
Danish Board of Technology 355
Danish Medical Research Council 355
Dao 503–5
Darby, G. 279
Darden, L. 48; "Thinking about Mechanisms" 181
Darrason, M. 305
Darwin, C. 476
Darwinian medicine 302–3, 478
Darzi, A.: *High Quality Care for All* 330
DasGupta, S. 374
data: acquisition 391; analysis 62; biomedical 180; causal claims and 58–63, 66; management of 180, 186; mining 281; out-of-protocol cleaning of 221–2; science 403; sets 273
data-based methods 198
databases 210, 310
Davies, M.: *The Oxford Handbook of Philosophy and Psychiatry* 1
"Dawn of Darwinian Medicine, The" (Nesse and Williams) 476
DCIS (ductal carcinoma in situ) 19–21
deafness 10, 27–33, 38–40
death 12, 103–5, 115–23, 211, 239, 343–7, 400; autosomal recessive mutations and 147; causes of 19, 158, 325, 409; early 124; fear of 24; statins and 224
death certificates 325
deathlessness 116–17
debiasing 218, 221–2

INDEX

decapitation 120
decision making: causality and 68; clinical 209, 364–7; close reading and 376; culture and 439–40; medical 383–93; treatment priorities and 389
decoctions 512
deconstructionism 374
de Crespigny, L. 332–6
deduction 297, 299
Deductive-Nomological (D-N) model 296–302, 307
deductive-statistical (D-S) model 298–300, 307
deep vein thrombosis (DVT) 66
defects 23, 40, 144
defense mechanisms 301–2
deficiency 433
dehumanization 373, 458
delinquency 20
Deliver Me! (television show) 111–12
delivery, birth and 106
demedicalization 437
de Melo-Martin, I. 335
dementia 115, 118–22, 370
demographics 212
demoralization 526
dendritic cells 180, 184
Denmark 134
Dennett, D. 124–5, 456
deoxyribose sugar 145
depression 6, 10–12, 156, 171–8, 216, 370, 410, 457–8; classifications of 159, 176; clinical trials for 411; overdiagnosis of 174; placebo treatments for 141
deprivation 37–8, 130
derangement 6
Derridean Deconstructionism 378
DES (diethylstilbestrol) 433–5
Descartes, R. 6, 459
descriptions 18, 238, 322, 366–7
Deshauer, D. 216
desiderata 186
design, of experiments 210–12, 218–22, 234–5, 253–4
detection technologies 11
deterioration 83, 330
determinism 36–7, 64, 476
deterministic causality 71–3
development: cellular processes and 145–6; constraints 478; continuing 104; resources 148
diabetes 12, 44, 68, 73, 78, 90–6, 144, 239–42, 350, 400, 405, 448; causal claims and 58–9; descriptive studies and 238; health care costs for 124
diabetes mellitus 186–7, 409
diabetes type-2 58, 344
diabetic ketoacidosis 92
diabetic patients 186–7

diagnosis 6, 36–8, 50–1, 92, 96–8; categories of 156–69; causal 386; codes 159; differential 316, 383, 386–7; double 511; error/mismeasurement of 61; expansion of 10; gender and 411–12, 414; medical decision making (MDM) and 383–93; physical 38, 160; prognosis and 165; Tibetan medicine and 158; traditional Chinese medicine and 158; uncertainty of 368, 388
Diagnosis Related Groups 159
Diagnostic and Statistical Manual of Mental Disorders (DSM; American Psychiatric Association) 12, 97, 159–62, 170–1, 174–6, 457–63
diagnostic apps 156
diagnostic groups 91
diagnosticians 61
diagnostic products 404
diagnostic tests 388, 412
diarrhea 250, 312
didactic roles 157, 165
diet 58–60, 162–4, 253, 400
diethylstilbestrol (DES) 433–5
diet industry 163–4
difference: birth and 108; bodily 432–41; classification and 157–8; means and 331
differential equations 278
differentiation 81
digestive functioning 119
digital automation 293–4
dilution 257
dimensionality, curse of 403
dioxin 325
directed acyclic graphs 274–6
direct relations 64–5
disabilities 31–2; accommodations for 43–4; biomedical model of 10; defined 39, 43; experiences of 374; history of 38–9; medicalization of 39–41, 44; moral model of 39–41; narratives of 377, 380; normality and 36–47; social model of 10, 42–4
disability laws 12, 37–41
disability pensions 40
disability rights 373
disability studies 372, 381
disability theorists 432
disadvantage 36–43
discontinuation study design 216
discourse, medical 161–2, 165–6
discovery 285–95
discrimination 17, 24, 37–8, 43–4, 164, 408
disease: acceptable 96; categorizing and classifying of 156–69, 180, 238–9, 249–51; causation 50, 53; chronic 64; concepts of 5–15, 144; environmental exposures and 281–2; explanation schema 306; exposure 271; genetic 83, 144–55, 401; illness, sickness, and 16–26, 172–5, 380, 448, 465, 469–70; mechanisms and 48–56; mongering 10; as multifactorial 249–51;

natural history of 135, 139–41; nature of 90, 96, 184; nomenclature 184; persistence of 94; physical 180–91; predicting and controlling 271–9, 303, 397; progression of 50–1; purpose of 302; race and 419; reductionism and 81–7; risk 24, 146; tokens and types 91–8; untreatable 384; variants of 82–3; vulnerabilities to 408
disease-related states 238–9
Disembodied Spirits and Deanimated Bodies (Stanghellini) 472
disharmony 158
disjointedness 183
disorders 11–13, 128; categorizing and classifying of 156–69, 173–4; impulse control 130; open-ended definitions of 170–1
Disorders of Sex Development (DSD) 438
dissatisfaction 20–2
dissent 354, 357–9
dissidence 20–2
distractibility 156
distress 130, 156, 159
disunity of medicine 307
Disunity of Science, The (Dupré) 182
diversity 415
divine order 159
Diving Bell and the Butterfly, The (Bauby) 468
dizziness 78
Djulbegovic, B. 368
DNA 49; abnormalities or defects in 144; double helix of 145; double-strand breaks in 53; information-carrying function of 401; mechanisms of 85; mitochondrial 145, 148; race and 420; repair mechanisms 53; structure of 48–9; X-ray crystallography photography of 48
D-N (Deductive-Nomological) model 296–302, 307
Doctored (Janduhar) 526
doctor-patient relationship 313, 373, 446–7, 451–3, 471
documentation 105, 180, 186–7
Dolezal, L. 471–2
Doll, R. 72, 252
Donald, A. 380
Dongzhimen Hospital 508, 511, 514
dormancy 116
dose-response curve 59
dosing 50, 321, 398–9, 446
double-jointedness 433
double masking 261, 319
double truths 511, 517
Douglas, H. 324–7
Douglas, M. 166
downstream consequences 152
drapetomania 9, 20–1, 185
Dreger, A. 2; *Hermaphrodites and the Medical Invention of Sex* 435–7
drug companies 52–3, 161–2; *see also* pharmaceutical industry

drug labels 398
drug metabolism 398, 401–2, 410
drugs: blockbuster 522; consumption of 131; dependence on 131; designing of 49; development of 203; dosages of 50; evaluation of 134; experimental 342–52; market withdrawal of 223, 411; responses to 398; safety, efficacy, and cost of 52, 203, 219–23, 343; targets of 53
Drugs and Insomnia (NIH Consensus Conference) 355
drug therapies 457
Drummond, J. 496
DSD (Disorders of Sex Development) 438
D-S (deductive-statistical) model 298–300, 307
DSM (*Diagnostic and Statistical Manual of Mental Disorders*; American Psychiatric Association) 12, 97, 159–62, 170–1, 174–6, 457–63
duality, of pain 125–7
ductal carcinoma in situ (DCIS) 19–21
Dumit, J. 160
duodenum 305
Dupré, J. 65, 185, 188; *The Disunity of Science* 182
Duster, T. 428–9
DVT (deep vein thrombosis) 66
dwarfism 10, 434
dying 115–17, 373; *see also* death
dysfunctions 6–11, 32–3, 39–41, 173–5; activity-limiting biological 44; diagnostic categories and 156–69; mind-brain 174; reductionism and 81–2, 86
dysphoria 128–9

ears 20
East Asia 498
eastern Europe 344
Ebers Papyrus 158
EBM (evidence-based medicine) 209–17, 228, 235, 296, 303, 314, 349, 356–7, 364–9
ECG (electrocardiogram) 274, 291
economic interests: birth and 110–11; disease categories and 159–60; personalized and precision medicine (PM) and 404–5; value-laden decisions and 325
economic rights 17
economic support 17–19, 22
Eder, S. 437
education: cases studies and 313–16; disabilities and 40; grants for 161; medical 373
Edwards, S. 492
effectiveness 54, 61, 203, 228, 232, 260–3, 354–5
Effectivex 319–20, 323–4
effect size 54, 232
efficacy 52, 203, 218–27, 343, 359; animal models and 273; of anistreplase 229–33; complementary/alternative medicine (CAM) and 258–60, 263–6; of conventional medicine 258

537

efficiency 330
egg cells 148
egosyntonic disorders 6
elderly people 347, 413
elective interventions 335
electrocardiogram (ECG) 274, 291
electromagnetic hypersensitivity 16
Electronic Health Record 186–8
electrons 91, 95, 177
electroshock treatment 171
elementary particles 95
Eli Lilly 161, 350, 525
eliminative abduction 297
eliminative materialism 459
eliminativism 419, 424, 460
Eliot, T. 378
Ellis, B. 93
Ellis, R. 494–5
Ellison, G. 425
Elsevier 520
embryos 104–5, 111, 115–16
emergency settings 385, 411
emergent properties 446
emotion 124–30
empathy 135–6, 139
empirical adaptationism 479–80
empirical observation 6–8, 343
empirical trials 388
empiricism 96, 159, 498
employer discrimination 37–8, 43–4
encouragement, of caregivers 135–6
endocrine disorders 391
endocrine system 85
endorphins 140
Engel, G. 11, 373, 445–58
Engelhardt, H. 97–8, 365–7
Engelhardt, T. 122
Engels, F. 497
engineering 97–8, 303–4
England 72, 258, 331–6
enhancement technologies 522
Enigma of Health, The (Gadamer) 471–2
Enkin, M. 321
enrichment strategies 346
environmental diseases 82, 144, 149
environmental exposures 281–2
environmental insults 11
Environmental Protection Agency (EPA) 58
environmental reasoning 367
environment, health states and 29, 32
enzymatic function 151
enzymes 287
enzyme variants 399
EORTC QLQ (European Organization for Research and Treatment of Cancer Quality of Life Questionnaire) 337
Eosinophilic granuloma 188–9
EPA (Environmental Protection Agency) 58

epidemics 184
epidemiology 58, 64, 72, 198–9, 245, 281–2, 315; analysis 187; clinical 195–6, 209–17, 248; counterfactual reasoning in 241–3; methods 209–17; molecular 282; philosophy of 248–56; studies 62–3, 152, 213–14; traditional view of 304; transition 250
epigenetic traits 145
epilepsy 9–11, 94
episiotomy 335–7
epistemic causality 67, 214
epistemic challenges 20–2
epistemic values 322–3
epistemological thesis 279
epistemology 1, 5, 105, 419
Epistulae morales ad Lucilium (Seneca) 24
epithelial lung cells 245
epochal studies 245
Epstein, R. 449–51
EQ-5D (EuroQol) 338–9
equilibrium 125, 147
erectile dysfunction 10, 161
Ereshefsky, M. 7
errors 61, 86, 195–9, 203, 223–4, 326
essence 7
essential properties 182
Essentials from the Golden Chamber 515
estrangement 17
estrogen 198, 434
estrous 410
ethicists 6–7
ethics 1, 33, 104, 134, 374
ethnicity 163–6, 420–2, 426
Etikettenschwindel (false labeling) 402
etiology 21–3, 49, 63–4, 86, 184–5, 239–41, 251, 289, 315, 410
eugenics 40, 111, 476
eukaryotes 145
Europe 120, 162, 349, 355, 489
European medicine 435
European Organization for Research and Treatment of Cancer Quality of Life Questionnaire (EORTC QLQ) 337
European Union (EU) 44
EuroQol (EQ-5D) 338–9
euthanasia 40
evaluations 8, 348
evaluative views 28–9
evidence: case studies and 314–15; complementary/alternative medicine (CAM) and 257–67; conventional medicine and 258; diagnostic categories and 156; of efficacy 264–6; empirical 232, 322, 401; epidemiology and 248; harms and 351; hierarchies of 62, 209–17, 314, 357; integrating 245–6; mechanisms and 48–56, 62–3; misleading 262; panels 52–3; secrecy surrounding 342–3, 349–51; social processes and 176; standards of 259, 263–4;

triangulation of 246, 252–4; uncertainty and 368; of values 320–1
evidence-based label 59
evidence-based medicine (EBM) 209–17, 228, 235, 296, 303, 314, 349, 356–7, 364–9
Evidence-Based Medicine: How to Practice and Teach EBM (Sackett) 210
Evidence-Based Medicine Working Group 210
evil spirits, diseases as 92
evolution 41–2, 175, 262, 300–1, 301–3, 477
evolutionary ancestry 273
evolutionary function account of disease 173–5
evolutionary medicine 475–86
evolutionary psychiatry 302
evolutionary theory 273, 475
"Excluded Beneficiary Rule, The" (Maglo) 427
exclusion criteria 324, 346–7
executive function 128
exercise 58, 128–9, 162–4, 253, 400
exercise electrocardiography 412
exertion 287
existence 105–8, 116–22, 130
expectations 139, 243, 290–2
experiment: causation and 59, 62–3, 251
experimental groups 228
experimentalism 59–62
experimental studies 50, 237, 244–5, 253
experimental treatments 134, 204
experimenter bias 61
experts 160, 211, 223, 353–60, 388
explanans and *explanandum* 297–302, 307
explanations 249, 476–8; locus of 82; in medicine 296–309; models for 278
explanatory adaptationism 479–80
explanatory frameworks 456
explanatory pattern 306
explanatory trials 260–1
explanatory values 307
exploitation 526
exposure 60, 238, 248, 271–2
external agents 238
external causes of disease 6, 10
external interventions 242
external substance nosologies 6
external validity 51, 195–208, 220, 223, 255
extra-patient considerations 390
extrapolations 273
extrauterine 106
extremal diseases 94
eye color 338, 433
eyesight issues 11
eyes, opening of 119

failures, scientific 94–5
fairness 29, 520–1
fair tests 220, 225
fallibility 50–4, 59, 246, 357
false labeling (*Etikettenschwindel*) 402

Falun Gong 20
families 105–6, 110–11
family planning 111
fangfalun 498
Fanon, F. 509
fashion industry 163
fast food 72–5
father-centered approach 109
fathers 106, 246
fatigue 20, 128–9, 287, 385, 409
fat people 162–4
FDA (Food and Drug Administration) 51–2, 161–2, 220, 223–5, 338, 343, 348–50, 398, 411, 414, 419, 424–5, 428, 434
fear 17, 24, 103, 108–9, 129, 174
feasibility conditions 242
fecal soiling 137
feeblemindedness 38
feelings 17, 128–9
Feinstein, A. 209–11, 366–70; *Clinical Judgment* 366; *Clinimetrics* 366
Feldman, F. 116–17
Felski, R. 378–80
female animals, for research 410–11
femaleness 435
female sexual interest/arousal disorder 161–2
feminism 103, 371, 408–9, 437
fenfluramine (Pondimin) 348
fertility clinics 520
fetal celebrity 111–12
fetuses 40, 104–11
fever 12, 95–6, 231–2, 255, 312, 369–70, 481–3
fever-reducing medication 255
fibromyalgia 17, 20, 23
film 377
financial disclosure 519
financial incentives 248, 262, 323, 425, 521–6
first-in-human studies 345–6, 349, 525
first-wave feminists 437
Fish, S. 378
Fisher, J. 525
Fisher, R. 61, 64–5, 199–200, 219, 476
five phases 502–6
flat distributions 233
Fleck, L.: *Genesis and Development of a Scientific Fact* 183–4
Fleischman, S. 156
Fletcher, R. 195, 209
Fletcher, S. 209
Flexner, A. 373
Flexner Report 447, 476
flibanserin (Addyi) 161–2
flu 305
fMRI (functional magnetic resonance imaging) 11
folk psychological model 455–60

Food and Drug Administration (FDA) 51–2, 161–2, 220, 223–5, 338, 343, 348–50, 398, 411, 414, 419, 424–5, 428, 434
Fordyce, W. 125, 128–9
forebrain 131
foreign fetus model 106–7
forgetting 105–9
formal consensus 353–4
formal definitions 186
formalized norms 24
Fosamax (alendronic acid) 348
Foucault, M. 9, 379, 435–6
founder effect 147
four-dimensional ultrasound 111
fractures 332, 357, 364
frameshift mutations 150
France 162, 257–8, 466
Frank, A. 377, 380
Franklin, R. 48–9
fraud 218
freedom 31
free will 175
Freidson, E. 165
frequency 71–80
frequentist trials 199–201, 219–20, 228–36
Freud, S. 108–9
frogs 272
Frontiers of Justice (Nussbaum) 41
fruit flies 273
Fryback, D. 339
function 8, 173–5, 300–2
functional biology 477
functional categories 161
functional determinism 41–3
functional/dysfunctional difference 301–2
functional efficiency, health as 31
functional explanations 300–2
functional magnetic resonance imaging (fMRI) 11
funding 161, 185, 330
funding agencies 324
fusion 87, 117
Future of Disability in America, The (Institute of Medicine [IOM]) 44

Gadamer, H.-G. 380; *The Enigma of Health* 471–2
gag clauses 349
Galen 5, 381
Galenic tradition 158
Gallagher, S. 467
Gammelgaard, A. 483
Gannett, L. 422
Gao, Z. 147
Garden, R. 375
gasoline 281
gasses 83–4
gatifloxacin (Gatiflo) 348
Gaucher disease 298

Gawande, A. 358, 381
gay-related immunodeficiency disease (GRID) 313
Geertz, C. 380
gender 245, 408–18; binaries 381, 408–9; categories of 408–9; discrimination 79; dysphoria 171–3, 178; identity 171; labels 170
gendered groups 165–6
gender/sex distinction 408–11
gender-specific curricula 414
genealogy 372–5
Genentech 398
generalizability 203, 320, 323, 402–3, 411
generalizations 99, 156–8, 311, 315
general probability 231–2
general systems theory 449–50
generic causality 71, 74
genes 65, 245, 298, 404–5; Alzheimer's disease and 72; coding 145; combination 82; as difference makers 146; dominant 148; expression 145, 345; gene concept and 145; genetic diseases and 144–55; interactions 82; loci 185; mutant 146; Parkinson's disease and 87–8; phenotypic outcomes and 401–2; regulation of 145; sickle cell anemia and 85–6; traits and 401; variation in 42, 146–7, 253
Genesis and Development of a Scientific Fact (Fleck) 183–4
genes-RNA-proteins 306
gene therapy 405
genetic abnormalities 23
genetic classifications 182
genetic code 145
genetic diseases 83, 144–55, 401
genetic drift 147
genetic epidemiology 423
geneticists 460
genetic mapping 185
genetic mutations 20, 86
genetic polymorphisms 240, 398
genetic prenatal diagnostics 110
genetic purification 476
genetics 12, 292, 476
genetic testing 22, 107, 110–11, 152, 400, 404
genetic variability 305
Genette, G. 375
genomes 42, 145–6, 149, 397–8
genome-wide association studies (GWAS) 304–5, 403
genomics 397–9, 419–22
genotypes 145, 152, 399, 421
genotypic variance 305
Germany 172, 373, 466
germ theory 6, 250, 447
Germ Theory Explanation Schema 306
Gert, B. 8–9
Gesundheitsurlaub (health vacation) 172
Getz, L. 471

INDEX

ghostwriters 326, 523–4
Gifford, F. 2
Gillies, D. 52, 279; "Varieties of Propensity" 77
girls 433–4
GISSI-HF (statin trial) 225
Giving an Account of Oneself (Butler) 379
glaucoma 338
GlaxoSmithKline 344, 347, 350
Gleevec 402
Global Burden of Disease Study 124
Gluckman, P. 479
glucose 92–3, 96, 239–41, 287–9
glyceraldehyde phosphate dehydrogenase (GPD) 288–93
glycogen 287–9
Glymour, C. 242
Glymour, M. 242
goals 9, 32–3
Godfrey-Smith, P. 479
gods 103
Goesaert v. Cleary (1948) 38
Goldacre, B. 321
Goldenberg, M. 214
good life 9–10, 27, 31–3
Gottlieb, M. 312–15
Gøtzsche, P. 134–41, 221–3
Gould, S. 479
gout 93–5
GPD (glyceraldehyde phosphate dehydrogenase) 288–93
GRADE Working Group 211, 214, 357
graphs, causal 275–6
Gräsbeck, R. 97
gratification 130
Graunt, J.: *Natural and Political Observations* 158; "Reflections on the Weekly Bills of Mortality" 158
Graves, J. 425
GREAT Group 229
Greenhalgh, T. 2, 380; *Narrative Based Medicine: Dialogue and Discourse in Clinical Practice* 374
Greenland, S. 368
Grene, M. 98–9
GRID (gay-related immunodeficiency disease) 313
grief 12, 16, 130–1, 373
grief-related depression 178
Gronenborn, A. 87
Grossman, J. 213
group comparisons 248
groupings 91
growth 145
growth hormone deficiency 434
growth-restricting treatment 433–4
guanine 145
guideline syntheses 359
guoyi 498
Guyatt, G.: *The Users' Guides to the Medical Literature* 210–14

Guy's Hospital 287
GWAS (genome-wide association studies) 304–5, 403
gym industry 163

habituated body 108, 467, 470
habitus 439–40
Hacking, I. 161, 178; *Rewriting the Soul* 176
Hackshaw, A. 219
"Haircut" (Alexander) 376
Haitian migrants 312
Hamilton Rating Scale for Depression (HAMD) 344
Hampson, J. 436
H&P (history and physical examination) 377
hand washing 297
Hanson, N. 286–7, 290–4
haphazard allocation 200
Haraway, D. 374
Hardie, J. 205–6
Hardimon, M. 421, 426
hard trial outcomes 223
Hardwig, J. 380
Harlem 376
harmful dysfunction model 10
harms, measuring of 342–52
Harvard Medical School 288–9
Harvard University 160–1
Harvey, W. 6
Hawkins, A. 380
Hawthorne effect 138–40
HbS allele 147
Head, H. 126
headache 19, 135, 159, 343
healing 125, 130, 265
health: aesthetic of 164; concepts of 6–9, 435–6; defined 5, 27–8; disparities 408–10, 419–31; economists 28; insurance 20, 97, 159–62, 198, 326, 330; justice 390; policy 6, 29–33, 40–2, 245, 256; psychology 446; registry programs 414; vacations 172; well-being and 27–35
health care: costs 124, 330–1, 339; education 372; professionals 16–17, 20–3, 136–9
health-measurement literature 28
health outcomes *see* outcomes
health-related quality of life 28
Healy, D. 524
heart 276, 299–301, 409
heart attack 92, 204, 211, 274, 346–7, 409
heart disease 72–5, 79, 144, 241; health care costs for 124; multi-causal fork for 74
heart failure 409–11, 425
HEAT trial 482
hedonism 27
Hegel, G. 379
hegemonic narratives 374
Heidegger, M. 103, 380, 468, 472
Heideggerian analysis 471

541

height, abnormal 432–4
Heiserman, A. 377
Helgesson, C. 221–2
Helicobacter Pylori 305
helixes 145
hemodialysis 225
hemoglobin 147
Hempel, C. 286–9, 293, 297–300
Henderson, V. 490–2
Henle, J. 250–1
HER2-positive breast cancer 398–400
herbal medicine 258
Herceptin (trastuzumab) 398–400, 403
heritability 305
Hermaphrodites and the Medical Invention of Sex (Dreger) 435–7
hermaphroditism 435–8
hermeneutics 104, 465–74
Hermeneutics of Medicine and the Phenomenology of Health, The (Svenaeus) 471–2
herpes zoster 158
Herrick, S. 162
Hesslow, G. 75–9
Hesslow example 76, 79
heterogeneity: of beliefs 306–7; complementary/alternative medicine (CAM) and 258; of conditions 134–5; genetic diseases and 149–50; health states and 28; placebos 134–9; of reasoning tasks 365; of subpopulations 427
heterozygotes 146–7
heuristic roles 157, 165
hGH (Human Growth Hormone) 434
HGP (Human Genome Project) 399–401, 404–5, 421
HIE (hypoxic-ischemic encephalopathy) 332
high blood pressure 10–11, 19, 73, 94, 409
higher brain criterion 121–2
High Quality Care for All (Darzi) 330
high-risk populations 426
Hildegard of Bingen 381
Hill, A. B. 58–63, 251–2
Hill criteria for causation 58–63
Hippocrates 6, 165
Hippocratic Corpus 5, 313
Hippocratic descriptions 94
Hippocratic medicine 300
Hippocratic Oath 480
hip replacements 335
Hirose, I. 28–30
hirsuteness 11
Hispanic ethnicity 420
histiocytosis 184
Histiocytosis X 180; *see also* Langerhans Cell Histiocytosis (LCH)
historiography 374
history and physical examination (H&P) 377
HIV 315, 387
HIV/AIDS 8, 149, 387

Hobart, J. 337
Hobbes, T. 103
Hochman, A. 421–2
Hofmann, B. 7
holism 8, 11, 260, 446, 450, 490–5
holistic-therapeutic approach 471
Holland, N. 378
Holleb, A. 356
Hollowell, J. 332, 335
Holmes, O. 40
home birth 103, 106, 331–2, 335–9
Home Birth Data collection form 333
home health care practitioners 390
homeopathy 52, 257–64
homeostasis 6
homogenous populations 64–5
homosexuality 9, 17–22, 170, 173
homosexuals 311–14, 437
homozygotes 146, 147
horizontal mechanisms 49
hormonal cycles 410
hormonal variability 410
hormone replacement therapy 62, 195–201
hormones 433–4
Horton, R. 523
Horwitz, A.: *The Loss of Sadness* 174
hospice movement 373
hospital births 103, 106, 110
hospitalization 22, 344
hospital management 519
host data 367
host-pathogen interactions 476
House, MD (television show) 311
how-explanations 302
Howick, J. 137–8, 213–14
HPV (human papilloma virus) 19, 58
H. Pylori 305–7
Hróbjartsson, A. 134–41
HSDD (hypoactive sexual desire disorder) 162
Human Genome Project (HGP) 399–401, 404–5, 421
Human Growth Hormone (hGH) 434
humanistic approaches 372–3, 449
humanistic data 367
human papilloma virus (HPV) 19, 58
human subjects 60, 272, 373, 413
Hume, D. 71
humerus, fractured 332
humors 5–6, 158, 183–4
hunger 125
Hunt, S. 337
Hunter, K. 313, 379–80
hunter-gatherers 175, 302
Huntington's disease 147
Hurst, M. 375
Hursthouse, R. 368
Hurwitz, B. 2, 380; *Narrative Based Medicine: Dialogue and Discourse in Clinical Practice* 374

Husserl, E. 471
Husserlian phenomenology 105
hybridization 7, 10–11, 513–17
hydralazine hydrochloride and isosorbide dinitrate (BiDil) 419–20, 423–8
hydrocephalus 434
hydrogen bomb 103
hypercholesterolemia 11, 22, 245
hyperglycemia 20–2
hypertension 20–2, 94, 137, 245, 409
hypervigilance 129
hyperviscosity 390
hypoactive sexual desire disorder (HSDD) 162
hypothesis testing 220, 323, 324–5
hypothetico-deductive view 297
hypothyroidism 237–8, 391
hypoxic-ischemic encephalopathy (HIE) 332

IARC (International Agency for Research on Cancer) 52, 58, 61, 64; *IARC Monographs on the Evaluation of Carcinogenic Risk* 63
IBE (inference to the best explanation) 94, 252, 297
IBS (irritable bowel syndrome) 19–20
ICD (*International Classification of Diseases*; World Health Organization [WHO]) 159, 170, 180, 187–9, 461–2
ICE (individual causal effect) 65–6
ICU patients 482
IDDM (insulin-dependent diabetes mellitus) 92–3, 96–7
ideal genetic disease 152
ideas, history of 109
identity: diagnosis and 156–7; gender 171; nursing and 487; race 419, 425
idiopathic diseases 82
IgG 388–92
Illari, P. 48
Illich, I. 97
illness: concepts of 7–8, 11; disease, sickness, and 16–26, 172–5, 380, 448, 469–70; experiences of 374; narratives of 377, 380; pain and 129; suffering and 130; as transforming 467–9
Illness (Carel) 472
"Illness and the Paradigm of Lived Body" (Toombs) 471
imaging technology 22, 162
imbalances 5–6, 183–4
immune deficiency 387
immune-reactive diseases 184
immune-suppressing drugs 314
immune systems 172, 313; genetic diseases and 149; pregnancy and 106–7
immunity 65, 390
immunocompromised groups 312–14
immunodeficiency diseases 387
immunoglobulin 391
immunological categories 161

implantation, of embryos 104
impulses 109
impulsivity 130–1
inactivity 44
inborn autoimmune disorder 314
incidentalomas 19
inclusion criteria 189, 324, 346–7
incompetence 20–2
indeterministic causality 71–6, 79
indexing 186
index measures 336–7
index patients 313
Indiana 40
indigenous medical practices 497
indigestion 409
indispensability 98–9
individual causal effect (ICE) 65–6
individual characteristics 397
individualistic methodological approach 404
individuality 104
individualization 158
individualized drug dosing 398–9
individualized morbidity outcomes 332
individual rights 29
Indonesia 165
inductive inference 219
inductive logic 287, 290, 299
Inductive-Statistical (I-S) model 298–300, 307
inductivism 286
industry-sponsored research 221–3, 325
ineffective treatment 262
inequality, socioeconomic 249, 520
infant mortality 215
infants 109–10, 118
infections 12, 31, 52, 72, 296–7, 347, 385–7, 390
infectious agents 6–8, 92, 170
infectious diseases 40, 146, 149, 184, 250, 304–5
inferences 233–4, 249, 402–4; incorrect 187; randomized trials and 202; reliable and reproducible 366
inference to the best explanation (IBE) 94, 252, 297
inferentialism 59–63, 67
inferential statistics 404
infertility 8, 22, 28–9
infertility clinics 115
inflammation 6, 93–5, 387
influenza 165, 305
informal norms 24
informatics 180, 186–7
information bias 196–8
information, meaning of 390
information metaphor 400–2
infusion centers 390
infusions, placebo 140
inheritance 148, 298
inherited immunodeficiency diseases 149–52
iniparib 51–5

injections, placebo 137, 140
injury 31–2, 109, 129–30
insanity 165
insomnia 16
instantiation 189
Institute of Medicine (IOM) 160–1, 355, 426; *The Future of Disability in America* 44
institutional incentives 218
institutionalization 40
Institutional Review Board (IRB) 522
instrumentalism 96
insula 126–30
insulin 12, 92–3, 96
insulin-dependent diabetes mellitus (IDDM) 92–3, 96–7
insulin receptors 398
insulin treatment 241
insurance industry 20, 159–62, 326, 330
insurance reimbursement 97, 159–60, 198
integrated cognitive-social explanation schema 306
integration 81–2
integrative models 461
integrity 521
intellectual disabilities 40
intellectual virtues 369
Intemann, K. 335
intensity, of pain 126
intentionality 105–6
interest groups 170, 173
interface consensus 355
Intergovernmental Panel on Climate Change (IPCC) 354
interindividual differences 398
internal validity 195–208, 220–3
International Agency for Research on Cancer (IARC) 52, 58, 61, 64; *IARC Monographs on the Evaluation of Carcinogenic Risk* 63
international classification 159
International Classification of Diseases (ICD; World Health Organization [WHO]) 159, 170, 180, 187–9, 461–2
International Consensus Development Conference on Female Sexual Dysfunction 161
International Philosophy of Medicine Roundtable 1
International Society for Evolution, Medicine and Public Health (ISEMPH) 481
International Society for Pharmacoeconomics and Outcomes Research (ISPOR) 339
International Society of Quality of Life Research 337
interpretability 337–8
Interpretative Phenomenological Analysis (IPA) 472
intersex cases 436–7
inter-subjectivity 17
interventionism 63–7, 241

interventions 36, 195–6, 211, 244–5; benefits of 343, 384; causal 457; disabilities and 40; elective 335; gender in 410, 413; locus of control and 82–3; locus of explanation and 82, 85; models and 273–6; observational research and 237, 244–5; personalized and precision medicine (PM) and 397; pregnancy, birth, and 110–11, 215; randomized controlled trials (RCTs) 215; reductionism and 82, 86–8; strategies 62; variables 59–60; very large treatment effects (VLEs) of 222–3, 234
interviews 385
intestinal flora 390
Intraocular Lens Implantation (NIH Consensus Conference) 355
intrapartum care guidelines 334
intrapartum events 332
intrauterine 106, 109
intravenous IgG 390
introspection 17
intuition 373
INUS conditions 63–4
invasive testing 110–11, 411
in-vitro fertilization (IVF) 110, 116
in vivo evidence 54
Ioannidis, J. 222–4, 348–9
iodoacetate poisoning 288
IOM (Institute of Medicine) 160–1, 355, 426; *The Future of Disability in America* 44
IPA (Interpretative Phenomenological Analysis) 472
IPCC (Intergovernmental Panel on Climate Change) 354
I-QOL (Urinary Incontinence Quality of Life Scale) 335–7
IRB (Institutional Review Board) 522
irreversibility 120
irritable bowel syndrome (IBS) 19–20
Irvine, C. 374, 379
ischemic heart disease 19, 409
ISEMPH (International Society for Evolution, Medicine and Public Health) 481
I-S (Inductive-Statistical) model 298–300, 307
isolation 43–4
ISPOR (International Society for Pharmacoeconomics and Outcomes Research) 339
itch 125
IVF (in-vitro fertilization) 110, 116

Jadad, A. 321
JAMA: The Journal of the American Medical Association 210
Janduhar, S.: *Doctored* 526
Jansen, L. 182–3, 186–9
Japanese imperialism 509
jaw pain 409
Jews 147, 298

jiuyi 498
Johansson, I. 185
Johns Hopkins University 436
Johnson, T. 525
joint probability distribution 274–5, 278
joints 93–5
Jones, A. 381
Jones Criteria 98
Jorden, W. 525
journal editors 519
JUPITER (statin trial) 224–5
Jurecic, A. 380
justice, theories of 41

Kadane, J. 199
Kahane, G. 36–7
Kahn, J. 424, 427
Kant, I.: *Anthropology from a Pragmatic Point of View* 6; *The Critique of Pure Reason* 71–2
Kaplan, J. 421–2, 426–7
Kaposi sarcoma 311–12
Katikireddi, S. 423
Kearns–Sayre syndrome 148
Keeper of the Word, A (Stringfellow) 372
keywords 187
kidneys 28, 347
Kiene, H.: "The powerful placebo effect: fact or fiction?" 135–6
Kienle, G.: "The powerful placebo effect: fact or fiction?" 135–6
Kincaid, H. 1, 176; *Value-Free Science? Ideals and Illusions* 322–3
kinds 181
King, L. 97
King, M. 149–50
King's Fund Forum 358
Kirk, S. 176
Kirsch, I. 137
Klein, C. 125–6
Kleinman, A. 380
Klum, H. 111
KMT (Nationalist Party) 510
knee replacements 335
knowledge, causal 205–6, 254–5
knowledge economy 404–5
Koch, R. 63–4, 72, 250–1
Koch's postulates 63–4
Krieger, N. 426
Kripke, S. 93
Kübler-Ross, E. 373
Kuhn, T. 287–93, 517
Kukla, R. 2
Kutchens, H. 176

labeling 165, 173–6, 223
labor 103, 106
laboratory evidence 53
laboratory experiments 59

laboratory tests 162, 223
labor interventions 110
labor pains 106
lactase persistence 148–9
lactate 288, 291
lactose intolerance 21, 148–9
lactose pills 136
LaFollette, H. 273
Laing, R. 10
Lamping, D. 337
Lancet, The 311, 523
Lancet Neurology 337
Langerhans Cell Histiocytosis (LCH) 180–1, 184–5, 188–9
language 17; of the body 471; controlled medical vocabularies 186–7; of diagnosis 156
large pragmatic trials 204–5
late-onset diseases 147
lateral pathway, pain and 126
Latin 134
Latin America 425
Latour, B. 514
Laudan, L. 94–5
Lawlor, D. 198
laws: disabilities and 12, 37–41, 44–5; of nature 95, 298; race and 419
Lawson Wilkins Pediatric Endocrine Society 433
LCH (Langerhans Cell Histiocytosis) 180–1, 184–5, 188–9
LDL (low-density lipoprotein) cholesterol level 225
leadership 357
Leboyer, F. 109–10
Leder, D. 471
Lee, B. 482
legitimacy 160
Lei, S. 498
Lemoine, M. 307
Leonelli, S. 273
Leray, J. 162
lesions 6–7, 10–12, 51, 170
leukemia 149, 243, 281–2, 300
Leven, G. 162
Levinas, E. 379
Levinasian paradox 379
levothyroxine 391
Levy, A. 273
Lewis, D. 65
Lewontin, R. 479
Lewy bodies 86–7
LGBT studies 381
libertarianism 520
Licensed Practical Nurses (LPNs) 487–8
licensing fees 520
life: beginning of 104–5; existence and 116–21; meanings of 104, 115–16
life expectancy 339
life sciences 48, 180–1

INDEX

lifestyles 239, 259, 265, 408
life support 121
lighting, productivity and 138
like with like comparisons 220
limbic system 126–8
limbs, loss of 92
Lindemann, H. 2, 381
linguistics 156, 186–7
Linnaean hierarchy 181
Lipitor 522
Lipton, P. 252, 297
listening 379
literary theory and criticism 372–8
Literature and Medicine curricula 373–5
liveliness 109
liver angiosarcomas 58, 61
liver cancer 68, 325
liver damage 346
liver tissue 325
location, of pain 126
Locke, J. 6
Lockean persons 118–21
locked-in syndrome 468
logical inference 366
logical positivism 285–7
London 287, 525
Longino, H. 324, 354
looping 176
Los Angeles 288
loss 12, 24, 128–30
Loss of Sadness, The (Horwitz and Wakefield) 174
loveliness 252, 297
low-density lipoprotein (LDL) cholesterol level 225
lower back pain 21–2, 124
lower respiratory infections 19
LPNs (Licensed Practical Nurses) 487–8
lung cancer 61, 64, 67, 72–7, 243–5, 250–4
lungs 184–5, 276
Lyme disease 16, 19–20, 160
lymphadenopathy 385
lymphangioleiomyomatosis (LAM) 465–6
lymph nodes 184–5, 312
lymphocytes 314
Lynøe, N. 185

Ma, J. 163
Machamer, P. 48; "Thinking about Mechanisms" 181
machine learning 275
Mackenzie, F. 213
macromolecules 84–5
Madhava-kara 381
madness 38
magic bullets 251
Maglo, K.: "The Cluster Stability Rule" 427; "The Excluded Beneficiary Rule" 427; "The Patient Standpoint Rule" 427; "The Permissibility Principle" 427–8
magnetic resonance 51
magnitude, of pain 126–7
Maimonides 6
major depressive disorders 12, 171, 174, 177–8, 457–62; classifications of 159; overdiagnosis of 174
Major Diagnostic Groups 159
malady 8–9, 16–19, 24
malaria 21, 147
male animals, for research 410–11
male gods 103
maleness 435
male physicians 103
malfunctions 170–7
malignant cancers 391
mammals 272
mammography 358–9
managed care 330
mania 6
manipulability approaches 316
many-many relationships 151–2
MAOA (monoamine oxidase A) 428–9
Maoist period 505
Mao Zedong 499
mapping, genetic 185
Marcum, J. 369
Marcus, E. 374
Marfan syndrome 434
marginalization 336, 381
Marie Claire 111
marital discord 137
market-based medicine 519–28
market exchanges 519
market withdrawals 223
Markov network models 274
married women 38
Marshall, B. 305
Marx, K. 526
masculinity norms 409
masking 54–6, 199–200, 203, 221, 228, 261
mass-based classification systems 162
Master in Public Health (MPH) 488
Master of Science in Nursing 488
masturbation 9, 20–1
materialist analysis 43
maternal-fetal relation 106–7
maternal inheritance 148
mathematical statistics 219
mathematics 98, 286
maturation 109
Mayr, E.: *Cause and Effect in Biology* 477
McArdle, B. 287–9
McArdle disease 285–93
McClellan, J. 149–50
McMaster University 210
MDM (medical decision making) 383–93

INDEX

ME (myalgic encephalomyelitis) 16, 20–2
Meacham, D. 471–2
meaning, information as 401
meaning-making 379, 470
meaning-seeking 470
means, difference and 331
measurement outcomes 334–5
measuring: of diseases 238–9; errors in 198; of harms 342–52; of health and well-being 28–33; instruments and indications 333–5; of outcomes 62, 156, 330–41; of weight 162–3
mechanical dysfunction 6
mechanisms 48–57, 62–7, 82–5, 181, 184, 213–14, 245–6, 271–3, 276, 300–5, 346
meconium aspiration syndrome 332
MECP2 genes 148
media 22, 311, 399, 405
medial pathway 126
medial prefrontal cortex (mPFC) 127–8
mediating proteins 52
Medicaid 519
medical access 110–11
medical anthropology 16
medical communications companies 522–3
medical consensus conference movement 354–8
medical decision making (MDM) 383–93
medical disease model 171–2
medicalization 9, 12, 21, 175–6; atypical bodies and 432–40; bad and good 436–7; of birth 110–11; disabilities and 39–41, 44
medical journals 222, 310–11
medical knowledge resources 358
medical literature 210, 331, 523
medically unexplained physical symptoms (MUPS) 21
Medical Microbiology 285
medical models 170–2, 455, 458–60, 493
medical praxis 110
medical professionals 97
medical publishing 519–21, 524
medical research: reliability of 218–27; secrecy and 342–3, 349–50; values in 319–29
Medical Subject Headings (MeSH) 187
medical terminology 22
Medicare 519
Medicare Coverage Advisory Committee 355–7
Medicare Evidence Development Coverage Advisory Committee 357
medication 103, 110, 248
medicine: authority of 156; case study in 310–18; complementary/alternative medicine (CAM) 257–67; discovery in 285–95; disunity of 307; explanation in 296–309; models in 271–84; nosology of 180–91; pessimistic meta-induction and 95; placebos and 137; race in 419–31; reductionism in 81–7; as science of diseases 90
Medicine and Philosophy (*Yixue yu zhexue*) 497
medieval theology 5–6

meditation 258
Mediterranean descent 312
MEDLINE database 422
Meehl, P. 364
meiosis 298
melancholia 6
melanoma 11
melody 467
Melzack, R. 127
memory 105–10, 128, 174
men: cardiovascular diseases (CVDs) and 196, 408–11; clinical trials and 410–11; gender in medicine and 408–18; masculinity norms and 409; medical discourse and 165; sex-linked genetic diseases and 147–8; sexual dysfunction in 161
meningitis 370
menopause 62
mental disease 448
mental disorders 6, 10–13, 174
mental functioning 28
mental health 162
mental illness 9–11, 146, 170–4, 455–64
mentalism 118–22
meridians 158
merit 521
Merleau-Ponty, M. 108, 470–2; *Phenomenology of Perception* 467
MeSH (*Medical Subject Headings*) 187
mesotheliomas 243
meta-analysis 135–9, 209–17, 221, 257, 344, 347–50
metabolic activity levels 150
metabolic genotypes 398
metabolic phenotypes 151–2
metabolism 92–3
metabolites 150, 411
Metamedicine 497
metaphysics 1, 5, 189, 365, 498–500
metastatic cancer 399–400
Metformin 400
methodological adaptationism 479–80
methodologies: choices and 49, 413; for classification 182–3; developments and 251; for disease classification 180; placebos and 134–41; reduction as 85–7; of research 210, 342
Méthot, P.-O. 302–3, 478
methylphenidate (Ritalin) 343, 348
Metropolitan Life 162
MI (myocardial infarction) 17, 224–5, 229, 242, 450–1
mice 272–3, 281, 410–11
Michigan 39
microbes 306
microbiology 183
microorganisms 63–4
Miczo, N. 473
mid-brain 86
middle-aged adults 44

middle range theories 298
MID (minimal important difference) 337–8
midwives 103, 106, 110, 296–7, 333–4, 489
migraines 12, 129
migrants 421
migration 147
milk 149
Mill's harm principle 265
Mills, C. 421
mind-body dualism 459
mind/body separation 467
mind-brain dysfunction 174
mindism 118–22
minds 118
mind sciences 48
minimal important difference (MID) 337–8
minimally conscious state 119
Ministry of Health (China) 499
minority crime rates 428
miscarriage 433
misclassification 198, 238–9
misfolded proteins 82, 86
misleading evidence 262
missense mutations 150–1
mistaken sex 436
Mitchell, S. 87
mitochondrial diseases 148
mitochondrial DNA (mtDNA) 145, 148
Mitteleuropa 286
mixed racial or ethnic backgrounds 425
models 59, 62–3; gender and 414; in medicine 271–84; of mental illness 455–64; navigating 87–8
modernity 514
modernization 435–6, 499
Modified Evidence Requirement 263–4
molecular biological differences 420
molecular biology 84, 397
molecular disease 85–7
molecular epidemiology 282
molecular level 92–3
molecularly targeted oncological drug 234
molecular medicine 42
molecules 83–6, 345
Money, J. 436
monoamine oxidase A (MAOA) 428–9
mono-causality 73
monogenic disorders 400, 404
monogenic Parkinson's disease 82, 86–8
monohierarchy 188–9
mononucleosis 312
mononucleosis-like syndrome 312
Montgomery, K. 2
moral considerations 29
moral failing 6, 164
morale, of doctors 526
morality 8

moral model of disability 39–41
moral norms 21
moral philosophy 419
moral virtues 369
morbidity 186–7, 331–5, 339, 409
Morgan, L. 111
Morgangni, G. 6
Morning, A. 429
morphine 135, 140
morphological classifications 182
morphology 23
mortality 103–4, 186, 223, 240–2, 253–4, 330–5, 339; hypothyroidism and 237; rates 72, 224, 229–30, 237, 296–7; risks 238–9, 245; statistics 187
mosaicism 11
mother–child relations 109
mothers 105–9, 246, 413
motivation 125–31
motor control 86–7
mourning 130
movements 86, 125
mPFC (medial prefrontal cortex) 127–8
MPH (Master in Public Health) 488
mtDNA (mitochondrial DNA) 145, 148
multi-causality 72–6, 79
multicolored irises 433
multifactorialism 249–51
multifocal classes 188
multigenic Parkinson's disease 86–8
multiple hierarchies 189
multiple inheritance 186, 189
multiple personality disorder 176
multiple sclerosis 67
multiple-testing problem 403
multisystem classes 188
mumps 94
MUPS (medically unexplained physical symptoms) 21
muraglitazar 346
Murphy, D. 173–8, 459, 462
muscle contraction 272
muscle cramps 287–8, 291
muscle metabolism 288
muscle relaxants 125
muscles 125
muscle stem cells 410–11
muscle weakness 347
muscular dystrophy 411
muscularity 163
mutant genes 146
mutations 8, 20, 86, 306, 401–2; autosomal recessive 147; cystic fibrosis and 152; genetic diseases and 149–50; mitochondrial DNA (mtDNA) and 148; phenylketonuria (PKU) and 150–2; sickle cell anemia and 85–6; tobacco smoke and 245; X chromosomes and 148

myalgia 347
myalgic encephalomyelitis (ME) 16, 20–2
myocardial infarction (MI) 17, 224–5, 229, 242, 450–1
myopathy 347
myophosphorylase 287–93
myopia 21
Myth of Mental Illness, The (Szasz) 9–10, 175–6

NAc (nucleus accumbens) 126–31
nail biting 137
naloxone 140
naming 111
Narrative Based Medicine: Dialogue and Discourse in Clinical Practice (Greenhalgh and Hurwitz) 374
narrative medicine 372–82
Narrative Medicine: Honoring the Stories of Illness (Charon) 374
narratives 310, 314
narratology 374–6
narrowest reference classes 300
narrow inductivist view 293
nasal swelling 385
natality 104–6
National Association of the Deaf 31–2
National Cancer Institute 185
National Cancer Institute Thesaurus (NCIT) 187
National Council Licensure Examination (NCLEX-PN) 487
national groups 165–6
National Guidelines Clearinghouse 358, 359
National Health Interview Survey 258
National Health Service (NHS) 52, 330
National Heart, Lung and Blood Institute 163–4
National Institute for Care and Health Excellence 52
National Institute for Clinical Excellence (NICE) 334, 339
National Institute of Child Health and Human Development 433
National Institutes of Health (NIH) 355–6, 411–13, 476
National Institute of Mental Health (NIMH) 12, 462
Nationalist government 509
Nationalist Party (KMT) 510
National Party (New Zealand) 165
nation-specific modifications 159
Natural and Political Observations (Graunt) 158
natural birth 103; videos of 112
natural causes 63
Natural Dialectics 505
Natural Dialectics departments 497–8
natural disaster 147
natural history, of diseases 135, 139–41
naturalism 7–11, 18, 28–9, 176, 428

natural kinds 90–9, 181, 462
natural philosophy 497
natural selection 302, 476, 478
nature: categories and 157, 161; laws of 95; as orderly 159
naturopathic approach 265
nausea 27, 409
Nazi researchers 373
NCIT (*National Cancer Institute Thesaurus*)187
NCLEX-PN (National Council Licensure Examination) 487
near-death experience 108
NEC (Not Elsewhere Classified) classes 186–8
neck pain 409
negative reinforcement mechanisms 130–1
Nelson, H. 381
nematode worms 273
neonatal encephalopathy 332–4
neonatal mortality and morbidity 332, 338–9
neoplasms 68
neoplastic diseases 180, 184–5
nerve cells 86
nervous system 85, 86, 126
Nesse, R. 302, 479–83; "The Dawn of Darwinian Medicine" 476; *Why We Get Sick: The New Science of Darwinian Medicine* 476
networks 214, 275–6, 315
neural development 104
neurobiology 126–7, 131, 174, 298, 457, 460
neurochemistry 12
neuroendocrine dysregulation 128
neuroleptics 460
neurological circuitry 12
neurological functioning, disturbed 332
neurological illnesses 11
neurological systems 12
neurons 126–7
neuropathy 241
neuroscience 126, 176
neurotic symptoms 108–9
neurotransmitters 11–12
neutrinos 90–1
newborns 105–10; moral states of 104; phenylketonuria (PKU) screening for 150; thalidomide and 310
New Criticism 378
New England Journal of Medicine, The 350, 519, 523
New Historicism 378
Newtonianism 500
New York 312, 374, 381
New York Times, The 433
New York University 509
New Zealand 159–60, 165, 420
NICE (National Institute for Clinical Excellence) 334, 339
Nietzsche, F. 379
Nightingale, F. 489, 492–3

INDEX

NIH (National Institutes of Health) 355–6, 411–13, 476
NIH Consensus Development Conference Program 355–8
NIH Revitalization Act 411
NIMH (National Institute of Mental Health) 12, 462
19th Amendment 38
NIPD (noninvasive genetic prenatal testing) 111
Nissen, S. 350
nitrite inhalant drugs (poppers) 315
NitroMed 424
Nobel Prize 285
nociception 125–7
n-of-1 trials 404
nomenclature 184, 435
nominal essence 7
nominalists 98
no-miracles argument 93–5
nomological view 298
non-choleric fever 250
non-epistemic values 322–5
non-exercisers 162
non-experimental treatments 204–5
non-genetic metabolic processes 402
nonhierarchical lists 184
nonhuman primate studies 410
non-inheritable injury 40
noninvasive genetic prenatal testing (NIPD) 111
noninvasive tests 110–11, 411
non-masked assessments 221
non-obstetric units 331–2, 335–6
non-random allocation 199
nonrandomized research 211–12, 215, 348–9; *see also* observational research
non-reciprocity 107
non-replicability 218–22
nonsense mutations 150
non-sex chromosomes 146
non-smokers 75, 250
non-steroidal anti-inflammatory drugs (NSAIDs) 482
non-study populations 51
non-utility measures 336–7
non-vegetarians 239–42
nonvital processes 116
Nordenfelt, L. 8, 98–9
Normal at Any Cost (Cohen and Cosgrove) 433–4
normal birth 106, 109
normality 432–4; concepts of 435–6; disabilities and 36–47; genetic diseases and 144, 153
normalization 41–2
normativism 7–11, 18–22, 36–7, 248, 343, 489
normotension 94
North America 170, 355, 369, 489
Northwick Park disaster 525
Norton, J. 225
Norway 17

Norwegian Institute for Hospital Research 358
nosology 6–13, 23, 180–91, 515
Not Elsewhere Classified (NEC) classes 186–8
nouveau roman 374
novelistic fiction 374
novel pharmaceuticals 349
noxious stimuli 125
NP (nurse practitioner) 488, 491
n-pronged multi-causal fork 73, 79
NSAIDs (non-steroidal anti-inflammatory drugs) 482
nuclear magnetic resonance spectroscopy 82
nucleotides 145–6, 150–3
nucleus accumbens (NAc) 126–31
null centre 467
null hypothesis 200, 219, 230–4
nulliparous women 334
numbness 400
Nunn, C. 481
Nuremburg Trials 373
nurse practitioner (NP) 488
nursing homes 44
nursing, philosophy of 487–96
nursing roles 489–90
Nussbaum, M.: *Frontiers of Justice* 41
Nutricia 508
nutrients 306

obesity 16–17, 21–2, 44, 68, 73, 162–4, 245, 253–4
"Obesity as Disease" (Obesity Society) 164
obesity-reduction program 254
Obesity Society: "Obesity as Disease" 164
objective Bayesian network approach 281
objective body 465, 469–70
objective data 366–7
objective interpretations 178, 229; probability and 76–9
objective judgment 364
objective values 9–11, 17
objectivity 77–8, 170, 173, 177, 199
object of interest 331; *see also* outcomes
observational research 50–1, 62, 196–9, 203–5, 211, 237–47, 248, 253, 263–4, 314
observational studies 348, 391
observation, clinical 366
observer effect 138–40
obstetricians 106
obstetric units 331–9
Ocana, A. 231
Occam's Razor 157
O'Connor, M.-F. 130
octogenarians 299
odds ratio (OR) 229–30, 232
Odegaard, C. 373
Office of Management and Budget 420
official documents 105
olanzapine (Zyprexa) 348–50

older adults 44, 370, 413
oncological trials 231
online checklists 156
ontic evils 8–9
ontogeny 402, 404
ontological status: of personalized and precision medicine (PM) 397–8
ontology 1, 111; classification and 180–3, 186–9; prenatal existence and 106–7
open-access journals 326
open science movement 326
operationalizing 452–3
opiates 140
Oppenheim, P. 297; "The Unity of Science as a Working Hypothesis" 83–7
OR (odds ratio) 229–30, 232
oral contraceptives 246
oral history 372
oral thrush 311
ordinal-level information 337
ordinary concept of race 421
Orem, D. 490
Oreskes, N. 354
organ donors 521
organelles 148
organic chemicals 281
organism-environment interactions 304
organisms 63–4, 83–4; life and 115–16
organization: mechanisms and 49
organizational levels 449–50
organs: disease and 6; reductionism and 84–5
organ-specific complications 391
organ transplant recipients 312–13
organ transplants 521
Orion 96
oseltamivir (Tamiflu) 350
Osimani, B. 349
Osler, W. 368
osseous lesions 6–7
osteoarthritis pain 140
osteopathy 258
osteoporotic spinal fractures 357
Other, the 379
others, as classifications 182–3
Ottawa Ankle Rules 364
outcomes 211, 344; animal studies and 410; categories and 157; causal claims and 58–61, 64–5, 68; classification and 158; defined 331; developmental 146; empathy and 135–6; epidemiology and 248–9; fatal 223; gender and 408; label changes and 223; measuring of 62, 156, 330–41; participation and 197–8; patient-important 211, 228; randomized controlled trials (RCTs) and 215, 220; replicability crisis and 218; reporting bias and 136; selection bias and 221–2; surrogate 211; tests and 384
out-of-protocol data cleaning 221–2
ovaries 434

overcoagulation 399
overdiagnosis 21, 174
overdosing 398
overprescription: of statins 218, 224–5
overseas populations: studies and 51
oversight 519
oversight bodies 525
over-the-counter medications 411
overweight 162–4
Oxford English Dictionary 495
Oxford Handbook of Philosophy and Psychiatry, The (Davies) 1
oxygen deprivation 334
Ozo, I. 368

pacemakers 276
Pacific Islanders 164
Paget, J. 285
PAH enzyme (phenylalanine hydroxylase enzyme) 150–3
PAH gene (phenylalanine hydroxylase gene) 150–3
paid research subjects 524–6
pain 17, 27, 68, 262; accounts of 379; acupuncture and 258; acute and chronic 124–33; mental states and attitudes 135; placebo effects and 134–6, 140–1; processing 125–7; reflex 125
painkillers 135, 140
pain on exertion 287
pain-relief medication 103
pancreas 92–3, 400
pancreatic islet transplantation 241
panic 108
Papanikolaou, P. 348–9
Paracelsus 6
paraclinical domains 366
paradigms 494
Parallel Chart 377
paraplegia 10
parasites, fetuses as 107
parasitic infections 390
Parens, E. 436
parental depressive episodes 178
parenthood 111
Parexel 522
Parfit, D. 118–19
Parkinson's disease 82–3, 86–8
paroxetine 216
PARP (poly(adenosine diphosphate–ribose) polymerase) 53–5
Parsons, T. 165
participation 17, 197–8
particle physics 83
partus/parturition 106–7
passive surveillance 343, 348
Pasteur, L. 72, 93, 476
paternalism 109, 491

INDEX

pathogenesis 184–5
pathographies 380
pathological difference 435
pathologies 9, 28, 172, 184, 434
pathology textbooks 189
pathophysiology 156–7, 160, 357, 386
patient: advocacy 373; autonomy 22, 390; characteristics 397; history 367, 387; interests 391–2; monitoring 204; perspectives 314; populations 51, 54–5, 136; response 51; satisfaction 390; welfare 369
patient-by-treatment interactions 404
patient-centered care 373
patient-centered criteria 97
patient-centered interview techniques 452–3
Patient-Centered Outcomes Research Institute (PCORI) 330
patient-important outcomes 211, 228
Patient Protection and Affordable Care Act 330
patient recruitment companies 522
Patient-Reported Outcome Measures (PROMs) 330, 335–9
patient's rights movement 373
"Patient Standpoint Rule, The" (Maglo) 427
pattern recognition 311–13
patterns 515–17
Pauling, L. 85–7
PCORI (Patient-Centered Outcomes Research Institute) 330
PDQ database 185
Peabody, F. 373
Pearl, J. 77
Peasant Shoes (van Gogh) 466–7
pedagogy, progressive 372, 376
pediatric hospital admissions 144
peer pressure 354, 357, 519
peer review 326, 350, 414
PEF (Population Etiologic Fraction) 152–3
Pellegrino, E. 6–7, 365–6
pelvic girdle pain 17
penicillin 138, 299–300
penicillin-resistant streptococcus 299
Penner, A. 424–5
Pentech 161
People's Republic of China 498–9, 510
peptic ulcer 305–7
perception 108, 467
Pereira, T. 223
perfect health 98
performance-enhancing drugs 522
peri-acquiductal gray regions 127, 130
perimenopausal symptoms 198
"Perinatal and maternal outcomes for planned place of birth for healthy women with low risk pregnancies: the Birthplace in England national prospective cohort study" 332
perinatal separation 109–10
peripheral neuropathy 68

"Permissibility Principle, The" (Maglo) 427–8
peroxisome proliferator activated receptor (PPAR) 343–6
persimals 117–22
persistent vegetative state (PVS) 118–19
personal change 130
personal interests 40
personality structures 11
personalized and precision medicine (PM) 397–407
personal loss 11
personism 118–22
pertussis 187
pessimistic meta-induction 94–5
pesticides 82
Peto, R. 72
Pfizer 161, 524
p-hacking 321
pharmaceutical industry 10, 165–6, 173; African American population and 424–5; categorization and 161–2; clinical research and 522–3; management agencies 159–60; overweight and 163–4; paid research subjects and 524–6; personalized and precision medicine (PM) and 404–5; profits of 522; randomized controlled trials (RCTs) and 220–2; scandals involving 522–3
pharmaceuticals 83, 342–52
pharmacogenetics 398–402
pharmacokinetics 452
pharmacotherapeutic approach 159
phase transitions 83–4
phenomenology 104–5, 381, 465–74
Phenomenology of Perception (Merleau-Ponty) 467
phenotypes 42, 145, 150–2, 305, 401–2, 421
phenylalanine (Phe) 150
phenylalanine hydroxylase enzyme (PAH enzyme) 150–3
phenylalanine hydroxylase gene (PAH gene) 150–3
phenylketonuria (PKU) 145, 150–3, 401–2
phenylpropanolamine 411
Philadelphia 525
Phillips, J. 493
Philosophical Foundations of Chinese Medicine (Zhang) 505
philosophy 372–3; birth and 103; of discovery 285–7; of epidemiology 248–56; history of 181; of mathematics 98; of medicine 16, 104, 110–11, 285, 293, 419; of psychiatry 1; of race 419; of science 285–7, 293
"Philosophy of Medicine in China, 1930–1980" (Qiu) 497–8
Philosophy of Psychiatry: A Companion, The (Radden) 1
phlegm 5–6, 183
phosphate 145
phosphorylase 289

photopigments 148
photosynthesis 115
phylogenetics 273, 476
physical anthropology 421
physical environments 85
physical exams 385
physical functioning 129
physical health 162
physical organizations 84
physical performance 128
physical-state view 92–3
physical variation 435
physicians 7, 97, 103, 413
physics 64, 83–7, 95–8
physiological: aberrations 23; adaptation 41; biomarkers 11; entities and events 17; explanations 301; maturation 109; mechanisms 213–14; research 209–11
physio-pathological mechanisms 225
pie charts 64
Pinel, P. 6
pink ribbons 165
pink Viagra 161–2
pioglitazone 346
pituitary tumors 434
PKU (phenylketonuria) 145, 150–3, 401–2
placebo 134–8, 219–21, 237, 260, 316, 404
placebo analgesia 140–1
placebo-controlled RCTs 215
placebo effects 134–43, 257, 446
placenta delivery 106
Place of Birth study 331
Plain Questions 502
Plante, D. 375
plants, disease and 8
plasticity 41–2
Plato 5, 103
plausibility 232–3; causation and 59, 63, 251; complementary/alternative medicine (CAM) and 258, 264; constructivism and 99; of personalized and precision medicine (PM) 397–8
pleasure 126–7, 130–1
Pleistocene epoch 302, 479
pluralism 182, 185, 305–7
pluralistic realism 11
PM (personalized and precision medicine) 397–407
pneumonia 138, 238, 311, 314, 370
POA (potential outcomes approach) 253–5
Pocock, S. 232–3
poetry 375–8
point conception of discovery 289–3
poisonous causes 6
polio 165
political dissidence 21
political factors 42–4, 104, 159–62
political values 9–10

pollution 72, 276
poly(adenosine diphosphate–ribose) polymerase (PARP) 53–5
polydactylism 21
polyhierarchy 186–8
polymorphisms 398
polyvinyl chloride (PVC) 58
Pondimin (fenfluramine) 348
Popper, K. 77, 286
poppers (nitrite inhalant drugs) 315
population drift 478
population effects 403
Population Etiologic Fraction (PEF) 152–3
population genetics 421, 476
population health 248–9
populationist concept of race 421
population-level associations 62
population-level research 214
populations, differences in 305
population size 147
population-specific statistics 164
Porkert, M. 502
Porphyrian Tree 181
positive reinforcement mechanisms 130–1
positivism 285–93, 498
postclassical narrative theory 374–5
postcolonialism 374, 508–11
posterior distribution 232–3
posterior insula 127
post hoc ergo procter hoc fallacy 135–7
post-market studies 348
postmenopausal women 195–9, 433
postnatal existence 107–12
postoperative pain 135
post-test probabilities 387–9
posture 125
potato blight 8
potentiality 104
potential outcomes approach (POA) 253–5
poverty 413
"Powerful Placebo, The" (Beecher) 135
"The powerful placebo effect: fact or fiction?" (Kienle and Kiene) 135–6
power relations 165
Pozzi, J. 437
PPAR (peroxisome proliferator activated receptor) 343–6
practical holism 494–5
practice-based observations 310
pragmatic trials 204–5, 260–4
pragmatism 97–8, 307
praxis 104, 110
precision medicine (PM) 397–407
preclinical animal trials 272–3, 410
preclinical evidence 54
preclinical stages 12
preclinical trials 50, 53, 411–14
pre-diabetes 19

prediction 177, 255; causality and 68; epidemiology and 249, 254–5; models for 271–9, 303
predictive testing 19, 22
predictors 403
preemption 242
preferences 30
prefrontal cortex 131
pregnancy 18–22, 76–9, 105–7, 110–12, 245–6, 331–5; diethylstilbestrol (DES) and 433–5; immunological paradox of 106–7; prevention of 66; social and medical perspectives of 110–11; termination of 40–1; thalidomide and 310
pre-ictal seizures 11
prejudices 24, 104
premature birth 215
premenopausal women 162
premenstrual dysphoric disorder 10
prenatal education 111
prenatal existence 105–12
prenatal testing 40, 110–11
prenatal X-ray exposure 243
pre-reflective experiences 108
prescriptions 512
prestige 17, 22–4, 354
presuppositions 104
pre-test probabilities 387–9
pre-trial run-in periods 321
prevalence 231–2, 409
prevention 50, 224–5
preventive medicine 398
price, of health care 330
Prilosec 522
primary elemental relation 107
primary immunodeficiency 387
primary prevention 224–5
primate studies 410
prior hypothesis 233–5
private clinical trial sites 522
privatization 522–3
probabilistic independence 274–5
probabilistic methods 383
probabilistic theory 63–7
probability 200, 229–32, 273–4; causality and 71–80; distribution 219, 274–6
Procter and Gamble 161
prodromal states 11–12
production, of models 280–1
productivity 138
profile measures 336–7
profit motive 405, 425, 523
prognosis 49, 53–5, 62, 243; association models for 274; biomarkers for 404; considerations for 391; defined 384; descriptive studies and 238; diagnosis and 165; medical decision making (MDM) and 383–93; observational studies and 239; uncertainty and 368
programmed cell death 53

prominent ears 20
promiscuous realism 182, 185
PROMs (Patient-Reported Outcome Measures) 330, 335–9
propensity, causality and 71–80
propensity to violence 420
properties 77, 183
property rights 38–9
property universals 181
prophylactic antibiotic therapy 390
prose 378
prospective cohort studies 197
prostate cancer screening 165
protections 12, 125–8
protein-bound medications 391
proteins 49, 52, 145–6, 150; Alzheimer's disease and 118; folding of 82, 85–7; sickle cell anemia and 85–6
protein synthesis 48–9
protocols 50, 157, 220–2
protons 91
proximate questions 477
Prozac 522
pseudogout 93–4
pseudohermaphrodites 435
Psillos, S. 95
psychiatric classifications 172–3
psychiatric diagnoses 38, 97
psychiatric disorders 5–6, 12–13, 40, 162, 170, 180
psychiatric interventions 347
psychiatric nosologies 11–13
psychiatrists 97, 457–8
psychiatry 1, 170, 447–8
psychoanalysis 108
psychoanalytic theory 372
psychological adaptation 41
psychological disorders 170–9
psychological maturation 109
psychological-neurobiological terms 174
psychological placebos 136
psychological trauma 129
psychology 127, 176
psychoneuroimmunology 446
psychopathology 6, 170–9
psycho-physiological connections 109
psychosis 6, 11, 58
psychosocial factors 447–52
psychosomatic disorders 160, 446
PTSD 165
puberty 103, 434
publication 62, 222, 310, 405
publication bias 248, 345, 349–50, 410
publication planning 522–3
public awareness 321–2
Public Citizen 523
public health 156, 255; classifications and 158; complementary/alternative medicine

(CAM) and 258–9, 265–6; emergencies 52–3; guidelines 426; interventions 224; policies 58
publicity 326
public participation 355
public policy 38, 40–2, 44, 51–3
published research, appraisal of 210
publishing companies 519–21, 524
puerperal fever 296–7
pulmonary hypertension 409
pulmonary LCH 188–9
pulse exams 512–14
purchasing choices 262
purification 514–17
Putnam, H.: no-miracles argument of 93; "The Unity of Science as a Working Hypothesis" 83–7
putrefaction 120
p-values 140, 218–22, 230–1
PVC (polyvinyl chloride) 58
PVS (persistent vegetative state) 118–19

QALY (Quality Adjusted Life Year) 331, 336–9
qi 158, 502–3, 506
Qiu, R. 499, 505; "Philosophy of Medicine in China, 1930–1980" 497–8
qiuben 501
qualitative approaches 252–3, 372, 472
qualitative mechanistic models 277–8
Quality Adjusted Life Year (QALY) 331, 336–9
Quality and Outcomes Framework 330
quality controls 218
quality of life 28–30, 141, 331–9
Quality of Life Research 337
quantitative contrast-thinking 246
quantitative genetics 304–5
quantitative mechanistic models 277–8, 281
quantum mechanics 95
quarks 93
quasi-religious communities 265–6
quaternary protein structures 82, 87
questionnaires 331, 337
Quintiles 522

rabies 93
race 96, 347, 419–31
race-ethnicity and race/ethnicity 422–3, 426
race-in-medicine debate 419–31
racial descriptions 426
racialist concept 421
racialized color 420
racial self-identification 425–6
racial typologies 476
racism 425–9
Radden, J.: *The Philosophy of Psychiatry: A Companion* 1
radiation levels 50
radiographs 364
radionuclide myocardial perfusion imaging 412

radiotherapy 399
random allocation 199–203, 219, 228
random errors 195, 203
randomization 60–1, 199–203, 209–16, 221–2, 243
randomized controlled trials (RCTs) 50–1, 54–6, 59–65, 140, 209–12, 237, 243–6, 304, 313–14, 319, 321, 357; Bayesian vs. frequentist 228–36; complementary/alternative medicine (CAM) and 258–66; epidemiology and 248; experimental drugs and 342, 346–9; homeopathy and 257; internal and external validity 195–208; medical decision making (MDM) and 383; misclassification in 239; reliability of 218–27
random sampling 403
Rank, O. 108–9
Ransom, J. 378
rapid metabolizers 398
rare conditions 144, 311
Ratcliffe, M. 471–2
rational persons 8
rats 252–3, 272, 325
Raz, J. 32
RCTs *see* randomized controlled trials (RCTs)
RDoC (Research Domain Criteria) project 462–3
reactive diseases 100
reactive medicine 398
Read Codes 159
reader-response criticism 379
real essence 7
realism: constructivism and 90–100, 176; metaphysical 189; pluralistic 11; promiscuous 182, 185
reboxetine 349–50
receptor molecules 398, 402
recessive genes 148
reclassification 63
RECORD trial 344, 347
recovery 55, 125, 134–6
recursive Bayesian networks 278, 281
red blood cells 117
reductionism 81–9, 96, 446–7, 450, 460, 494
redundant causation 67
reference classes 78–9, 299–301
"Reflections on the Weekly Bills of Mortality" (Graunt) 158
reflective memory 107–8
reflexology 258
Registered Nurses (RNs) 487–8
registration systems 324
regularity theory 63–7
regulators 222–3, 234–5, 348
regulatory mechanisms 146, 220–3, 319, 342, 519–21
Reichenbach, H. 286
Reiss, J. 2
relationship-centered care 373

relevance 299
reliabilist virtues 369
Reliance Weighing Machine 162
relief 127–31
religion 20, 104
religious communities 265–6
religious values 9
remission 135
Renaissance 6
Renggly, F. 108
Rennie, D.: *The Users' Guides to the Medical Literature* 210–14
renshilun 498
repeatables 77–8, 181
replacement IgG 388
replicability 218–23, 414, 519
reporting 62, 136
representation 376
representational memory 108
representativeness 402–3
reproductive cancer trials 411
reproductive medicine 110–11
reproductive process 40, 145
reproductive technologies 107
reprogenetics 111
Research Domain Criteria (RDoC) project 462–3
researchers 519
research questions 413
research trials 319–20
resemblances 157–8
residual confounding 198
resilience 42
resistance 302–3, 400
resources, allocation of 156, 339
respiration 115, 119–20
respiratory distress 109
respiratory infections 388
respiratory system 184–5
responsibilist virtues 369
responsibility attribution 58
rest 125
retroduction 290–2
Rett syndrome 12, 148
reversibility 120–2
revisionary approaches 37
rewards 126–31
Rewriting the Soul (Hacking) 176
Reznek, L. 9, 93–4
rhabdomyolysis 347
rheumatic fever (RF) 98
ribosomes 49
Richards, I. 378
Ricoeur, P. 374, 380
rights-driven social progress 44
Risch, N. 423
risk 11, 24, 144, 239–45, 251, 255, 384; absolute 334–5; causal claims and 58–68; childbirth and 103, 110–11; descriptive studies and 238; extraneous factors 200–2; first-in-human studies and 345–6; gender and 408; genes and 401; prediction 254, 404; random allocation and 200–2; relative 331, 334–5
risk-reward calculations 127
risk-taking behaviors 409
risperidone (Risperdal) 348
Ritalin (methylphenidate) 343, 348
rituals 103, 138
RNA 49, 85
RNs (Registered Nurses) 487–8
Roberts, D. 425
rodents 281, 345, 410
rofecoxib (Vioxx) 321, 348
Rorty, R. 374
Rose, M. 425
Rosenberg, J. 116–17
Rosenblatt, L. 378
rosiglitazone (Avandia) 343–50
rosuvastatin 225
Rothman, K. 196, 205
Rotimi, C. 425
routine care 204–5
Routledge Press 2
rubella 257
ruminations 129
run-in periods 321
Ruse, M. 480
Russell, B. 286
Russo, F. 52, 214, 279
Russo-Williamson thesis 214, 303

Sackett, D. 211, 364; *Evidence-Based Medicine: How to Practice and Teach EBM* 210
Sacks, O. 311
sacred texts 378
sacroiliitis 159–60
Sade, R. 519
sadness 18–20, 156
safety 52, 203, 218–27, 343
Salmon, W. 299–300
salt intake 426
sameness 41–2
sample size 61, 222, 225, 231
Sandel, M. 520
Sanislow, C. 462
Saperstein, A. 424–5
Sartre, J.-P. 469–71
saturated fat 400
Saunders, C. 373
Savulescu, J. 36–7, 332–6
Scadding, J. 7
scales 162–3
Scandinavian consensus conferences 355
Scanlon, T. 32
Scarry, E. 379
Schaffner, K. 84–7, 297–300, 480
Schering-Lough 161

Schiffer, C. 111
schizophrenia 9–12, 448, 459–60
Schulman, C. 482
SCIDS (Severe Combined Immunodeficiency Syndrome) 149
science, classification in 180–2
sciences, vocabulary of 83
scientific inference 402–4
scientific models 271
scientific pluralism 182
scientific realism 176
scientific reasoning 286
scientist-entrepreneurs 404
screening 19, 22
screening mammography 358–9
Scriver, C. 151
seasickness 22
secondary immunodeficiency 387
secondary prevention 224–5
secondary sex characteristics 436
second opinions 354
secrecy 342–3, 349–50
seeds 115–16
Seidenfeld, T. 199
seizures 11, 40, 210, 334, 347
selection bias 61, 196–8, 221–2, 349
selectionism 476
selective abortion 40–1
selective serotonin reuptake inhibitors (SSRIs) 216, 349–50
self-awareness 115, 118–22
self-care 490–2
self-consciousness 108
self-diagnosis 156, 163–4
self-empowerment 359
self-help industry 163–4
self-infusion 391
self, intactness of 130
self-mutilation 344
self-reported questionnaires 331–5
Semmelweis, I. 286, 296–7
Seneca: *Epistulae morales ad Lucilium* 24
Senn, S. 200
sensory cortices 128
sensory diagnoses 38
sensory encoding 126
separation relationships 274–5
sertraline 216
serum cholesterol 245
serum creatinine 346
serum immunoglobulins 387
Severe Combined Immunodeficiency Syndrome (SCIDS) 149
Severinsen, M. 97
sex anatomies 432, 435–40
sex-based analyses 411
sex categories 408–9
sex chromosomes 147–8

sex/gender distinction 408–9
sex-linked genetic diseases 147–8
sex reassignment 436
sexual behaviors 9
sexual dysfunction 161–2
sexual enhancements 522
sexually transmitted diseases (STDs) 184
sex work 521
Shakespeare, T. 10
sham acupuncture 261
shaming 432
Shanks, N. 273
shared decision making 389–90
Sharpless, N. 55–6
Sherlock Holmes mysteries 379–80
Sherrington, C. 126
Shields, B. 111
shift-work sleep disorder 10
shortness of breath 409
short stature 434, 437
shyness 18, 156
sick leave 17, 20, 24
sickle cell anemia 85–6, 147
sickle cell trait 21
sickness 16–26, 156, 165, 172–5, 465
side effects 78–9, 224, 243, 321, 412–14
significance test 199–200, 233–4
significance threshold 230
signs-and-symptoms view 92–3
sildenafil 161
silence 108
Silvers, A. 43
similarity 157–8
Simon, H. 293–4
simplification 81
Simpson's paradox 79
simulated behavior 278
single-case causality 71
single-gene disorders 144–6, 149–50, 292
single mothers 413
single-race A-heFT clinical trials 425
single RCTs 211
singularity 41–2
sinusitis 385, 390
situated adaptationism 479
situational depressed mood 175
siwei fangshi 498–9
size scale 82, 85–7
skeletal muscles 287–9
Skinner v. State of Oklahoma, ex. rel. Williamson (1942) 40
skin problems 150
Skyrms, B. 65
Slatman, J. 471–2
sleep 10, 129, 159
sleep/wake cycle 118–19
slenderness 162–4
slow metabolizers 398

small intestine 250
small treatment effects 222
Smith, A. 520, 526
Smith, B. 181, 186–9
Smith, K. 152–3
Smith, R. 452–3, 520–3
smoking 58, 61, 64, 67, 72–9, 242–54
smoking cessation 141
SNOMED (*Systematized Nomenclature of Medicine*) 159, 187
SOAP (subjective, objective, assessment, and plan) notes 377
social anxiety disorder 156
social birth 111
social change 130, 381
social constructivism 172–8
social conventions 38
social Darwinism 476
social determinants 446
social disapproval 170–3
social disarray 165
social expectations 156, 164
social factors 159–62, 456
social harm 11
social interactions 17, 343
social isolation 43–4
socialized medicine 159
social justice 372
social model of disability 10, 42–4
social norms 20, 106, 170–3
social perspectives 16, 110–11
social phobia 156
social policy 29–31, 42
SOCIALRACE 421, 426
social-relational worries 130
social-relativism 489, 492
social responsibilities 165
social science studies 170–1
Social Security Amendments of 1965 519
Social Security Disability Insurance (SSDI) 37–8, 43
social status 18–23
social values 29, 178, 343
sociobiology 175
Sociobiology (Wilson) 476
socioeconomic inequality 249
socioeconomic status 197–201, 413, 425
sociological factors 170
sociologists 460
sociology 16, 372
sociopolitical power 97
sociopolitical values 9–10, 185, 423
soft endpoints 223
software algorithms 403
solipsism 353
Solway Pharmaceuticals 161
somatic illnesses 448, 458
somatic sensations 109

somatosensory system 126
sophisticated mode of mental illness 458–60
sore throat 385
sorrow 21–2
Southern Cross 96
Sparling, F. 350
spatio-temporal gaps 62
spatio-temporally contiguous processes 276–8
specialization 372
species atypicality 11
species norms 7, 39, 41–3, 95, 175, 433
species persistence 41–2
specificity 59, 251
Spencer, Q. 420–2
Spiegel, M. 374, 377
Spiegelhalter, D. 229, 232–3
spinal stenosis 434
spine 125–6, 184–5
spinothalamic tracts 126
Spitzak, C. 164
sponsored trials 221–3
spontaneous improvement 135
spores 115–16
spurious associations 58
S-R (Statistical Relevance) model 299–300, 307
SSDI (Social Security Disability Insurance) 37–8, 43
SSRIs (serotonin reuptake inhibitors) 216, 349–50
stability, of causal relations 68
Stampfer, M. 196–7
standardizations 110, 373
standardized assessment tools 370
Standard Nomenclature of Diseases and Pathological Conditions, Injuries, and Poisonings for the United States (U.S. Department of Commerce) 184
Stanghellini, G.: *Disembodied Spirits and Deanimated Bodies* 472
stars 96
stasis 500
state policy 30–1, 265–6
states of being 242
statins 218, 224–5, 412
statistical analysis 156, 187, 199, 203, 221–2
statistical discrepancy 231
statistical evidence 218–27
statistical explanations 305
statistical frameworks 228–36
statistical frequencies 78
statistical independence 61
statistical judgment 364, 367–8
statistical literacy 222
statistical models 298–300
statistical power 222, 332
Statistical Relevance (S-R) model 299–300, 307
statistical research 39
statistical significance 218, 221, 230
statistics 275

status 24
STDs (sexually transmitted diseases) 184
Stearns, S. 480–1
Stegenga, J. 214–16, 223, 334
Stengers, I. 498
stereotyping 37
sterility 147
sterilization 40
stiffness 287
stigma 11, 17, 24, 40, 164, 173, 380, 432
stillbirth 103, 332
stomach 305
Stone Age 480
stop codons 145
storytelling 372–4
strength 59–61, 251
strep infection 299–300
stress 85, 129–30, 305
stressors 11
Streuli, J. 438–40
Stringfellow, W.: *A Keeper of the Word* 372
stroke 19, 37–8, 92, 204, 211, 224–5, 242, 346, 409–10, 468
structural abnormalities 387
structuralism 374
structure 183
studies: fallibility of 50; timing of 62
stunted growth 348
subcellular levels 81–6
subcellular organelles 148
subclinical hydrocephaly 42
subclinical hypothyroidism 237
subconsciousness 108–9
subcutaneous IgG 390–1
subgroup-specific drugs 403–4
subjective body 465–7
subjective interpretations 67, 220, 231–3, 364; classifications as 177–8; disease categories and 162–4; probability and 76–7
subjective, objective, assessment, and plan (SOAP) notes 377
subjective outcomes 221
subjective states 30
subjective values 9, 17
subjectivity 467
subjects, of research 524–6
submechanisms 81–2
sub-optimal dosing 321
sub-organism level 85
subpopulations 51, 55–6, 65, 205, 298, 426–7
substance related and addictive disorders 460
substantia nigra 86–7
subtypes 183
suffering 8, 12, 17, 22, 124–33, 160, 172–3
sugar 400
sugar pills 134–7, 141
sugary drinks 68
suicidality 344

suicide 171
superclasses 186
Suppes, P. 199
supply and demand 520
suppression 109, 523
Supreme Court 37–40, 45
surgery 110, 263
surrogate endpoints 321
surrogate motherhood 107
surrogate outcomes 211
surveillance 342–3
susceptibility 61, 64, 73
suspended animation 115–16
Svenaeus, F. 465, 470; *The Hermeneutics of Medicine and the Phenomenology of Health* 471–2
Swanson, K. 491–2
sweat 125
sweat chloride levels 152
Sweden 221
sweeping reductions 83–5
Swiss mice 60
Sydenham, T. 6, 159
symbiotic relations 107
symbolic logic 286
symptomatology 367
symptom-based diagnostic standards 160–1
symptoms 8, 12, 21, 71, 96; of cardiovascular diseases (CVDs) 409; categorization of 160, 184–5; causality and 61; classifications of 170; diagnosis and 156; homeopathy and 257; of insulin-dependent diabetes mellitus (IDDM) 92; mild 152; pain and 128; reductionism and 81–6; of suffering 129–30
syndromes 172, 175
synecdoche 156
syphilis 183–4
Syria 165
systematic allocation 200
systematic errors 195–6, 203
systematic reviews 209–17, 262
systematization 272
Systematized Nomenclature of Medicine (SNOMED) 159, 187
systemic hypertension 409
Szasz, T. 178; *The Myth of Mental Illness* 9–10, 175–6

tactile sensations 109
Tal, E. 331
talk therapies 460
Tamiflu (oseltamivir) 350
Tang, H. 423
Tannock, I. 231
Tap Pharmaceuticals 161
tar 252–4
target systems 81, 271–3, 277–8
taxonomy 23, 435, 462–3
Tay-Sachs disease 147

TCM (traditional Chinese medicine) 158, 264–5, 380, 497–518
teaching 473
tea-drinker experiment 199–200
technical consensus 355
technologies 23, 40–1, 44, 48, 156–7
Teira, D. 2
teleological functioning 6
television 111–12
Teller, E. 103
telling 379
Temkin, O. 93, 98
temperature 125, 481–3
temporality 59, 62, 251, 468–71
temporal sequences 126–7
tension 130
terminologies 83, 183–4, 343
tesaglitazar 346
test populations 60
tests 218, 221–2
tetanus 52
tetrahydrobiopterin (Bh_4) 151
TGN1412 525
Thagard, P. 294, 305–7, 315
thalidomide 310
theoretical foundations textbook 503
theoretical medicine 71, 74, 87
theoretical models 271–5, 278–82
theory neutral 175
therapeutic capacity 23
therapeutic interventions 211
therapeutic products 404
therapeutic reasoning 367
therapy research 108, 211
thermal painful stimulus 126–7
thinking 128–9
"Thinking about Mechanisms" (Machamer, Darden, and Craver) 181
thirst 92, 125
Thomas, J. 162
Thompson, R. 296, 304
thoracic surgery 140
threats, severity of 127
three-armed trials 137
three poisons and three humors 158
thrombolytic therapy 229–30
thrombosis 76–9
thrush 311–12
thymine 145
thyroid diseases 118, 410
thyroid hormone substitute 237
Tibetan medicine 158
ticks 20
time pressure 357
time scales 84–5
time trade-off technique 336–8
time-trend data 252–4
Tinbergen, N. 477–8
tissue 6, 84–5, 124–7

tissue macrophages 184
T-lymphocytes 314
TNBC (triple-negative breast cancer) 49–56
tobacco smoke 245
token causality 71
tokens 90–6
tongue exams 512–14
Tooby, J. 302
Toombs, S. 380, 468–72; "Illness and the Paradigm of Lived Body" 471
toothache 129
tooth decay 21
topography, critical 378–81
total evidence 50
toxicity 49, 54
toxicology testing 272–3
toxins 82
tractability 272
trade secrets 523
traditional Chinese medicine (TCM) 158, 264–5, 380, 497–518
training 473
traits 248, 401
transcription errors 86
transformation 105–6, 500–3
transgressive conduct 40
trans* identity 381
transition 109–12
transition-rating index (TRI) 337
translational medicine 2
transmission 315
transparency 111, 326
trastuzumab (Herceptin) 398–400, 403
trauma 108–9, 124, 129
trauma care 391
treatment 12, 17–22, 50–2, 55–6, 241–2; causality and 65–8; decisions 58, 243, 391–2; diagnosis and 156; effects 137, 233–5, 243; efficacy of 219; exaggeration of benefits of 135; experimental 134; gender and 411–14; groups 50, 60, 212–13, 248; humors and 158; ineffective 262; medical decision making (MDM) and 383–93; placebo effects and 134–7; progression 204; randomized trials and 203; responses to 49, 216; status 60–1; withdrawal 204; withholding of 60–1
tremors 86
TRI (transition-rating index) 337
trial industry 220
trials 50–1, 56, 134–41, 324–7; duration of 348; participants in 203–4
triangulation 246, 252–4
Trichinella spiralis 285
triggering conditions 86
triglycerides 92–3
triple-negative breast cancer (TNBC) 49–56
triplets, mapping of 145
troglitazone 346
troublesome discoveries 290–2

trust 353–4, 519
tubercle bacilli 72
tuberculosis 72, 90, 96, 184–5
tumors 31, 49–51, 55, 325, 345
Turner, B. 160
Turner syndrome 434
Tuskegee syphilis study 373
Twaddle, A. 16–17
two-pronged multi-causal fork 74, 79
type causality 71, 74
types 91–8
type-2 diabetes 58, 344, 402
type-2 diabetes mellitus 398
tyrosine 150

ultra-rare diseases 144
ultrasound photos 111–12
ultrasound screening 107
ultrastuctural abnormalities 23
umbilical cord 109
unaided reasoning 364, 370
unassisted birth 112
uncanniness, feelings of 17
uncertainty 325, 367–8
unconsciousness 19
undercoagulation 399
underdosing 398
underestimation, of harms 342–4
unification 83, 305
unifocal classes 188
Uniform Determination of Death Act 120
uniformity 183–4
unifying models 81–3
unintended effects 243–4
unintentional biases 415
unisystem classes 188
United Kingdom 44, 51–2, 72, 120, 162, 210, 287, 330–5, 339, 358, 420
United Nations' Convention on the Rights of People with Disabilities (CRPD) 44
United States 1, 36, 40, 51–2, 58, 110–12, 120, 124, 159–62, 176, 220, 224, 258, 287, 321, 330–1, 355–8, 373, 398, 414, 419–21, 433, 462, 489, 497, 519, 525–6
United States Congress 330, 411
"Unity of Science as a Working Hypothesis, The" (Oppenheim and Putnam) 83–7
universal causes 63
universal coverage 21
universal determinism 64
universals 181, 189
University of Alabama at Birmingham 1
University of California, Berkeley 79
University of California, Los Angeles 288
University of Rochester 447
University of Zurich 438
unknown cofounders 212
unmarried women 38
unmasked trials 221

unpleasantness 17, 20, 126
unpredictably 399
unreason 9
unresponsive wakefulness syndrome 119
unscrupulous research 318–19, 322
unsystematic clinical observations 211
untreatable disease 384
untreated groups 137–9
unviable processes 116
upper respiratory infections 91
Upshur, R. 369
urate crystals 93–5
Urbach, P. 202
urinary incontinence 335
Urinary Incontinence Quality of Life Scale (I-QOL) 335–7
urination 92
urine 96
U.S. Census Bureau 36, 44–5
U.S. Congress 45
U.S. Department of Commerce: *Standard Nomenclature of Diseases and Pathological Conditions, Injuries, and Poisonings for the United States* 184
use-mention mistakes 187
Users' Guides to the Medical Literature, The (Guyatt and Rennie) 210–14
U.S. Preventative Services Task Force 355
uterine cancer 434
uterus 106, 111–12
utilitarian constructivism 97
utility measures 336–9

vaccination 398
vaginal birth 110
vaginal clear-cell adenocarcinoma 433
vagueness 287
valdecoxib (Bextra) 348
validity 51, 195–208, 220–3, 238–41, 255
Valles, S. 423, 426–7
value choices 259, 265
value-free classifications 173
value-free ideal 7–8
value-free research 319, 322–7
Value-Free Science? Ideals and Illusions (Kincaid) 322–3
value, of health care 330
values: corruption of 521; defined 322; disabilities and 37; domain-specific 225; epidemiology and 248; health states and 29–33; in medical research 319–29; mental illnesses and 10; objective and universal 8–11; socio-political 9–10, 185, 423; subjective 9, 17
Vandenbroucke, J. 349
van Gogh, V.: *Peasant Shoes* 466–7
van Manen, M. 472
variability 203–4, 211, 214, 344–6
variable immune deficiency 387
variables 50, 56, 59–61, 75, 273

INDEX

variation 42, 330
varicella zoster virus (VZV) 71
"Varieties of Propensity" (Gillies) 77
vascular events 224
VC (vinyl chloride) 58–68
Veatch, R. 122
vegetarian diet 239–42
vegetative state (VS) 118–19
ventricular fibrillation 451
vertebra 159–60
vertebroplasty 357
vertical mechanisms 49
vertigo 27
very large treatment effects (VLEs) 222–3, 234
Vetlesen, A. 107
viability account of life 115–17, 122
Viagra 161–2
vibrio cholerae 250
Vichow, R. 6
Victorian period 437
video 112
Vienna Circle 286–7
Vienna General Hospital 296–7
vinyl chloride (VC) 58–68
violence 109–10, 420, 428
Vioxx (rofecoxib) 321, 348
viral diseases 398
viral infections 311, 385, 390
viral upper respiratory tract infection 385
Virginia 40
virtue-based frameworks 367–70
viruses 48–9, 72, 172, 249, 305, 476
visualization of birth 112
vital force 264–5
vitalist metaphysics 500
vital processes 115–17
vitamin C 134, 299–300
vitamin-D deficiency 67
vitamin deficiencies 118
VLEs (very large treatment effects) 222–3, 234
vocabularies 83, 186–7
Vogt, E. 2
Vogue 111
voices, hearing of 170
voluntariness 521, 525
voluntary behavior 118–19
vomiting 78, 301–2
von Mises' principle 77–80
von Mises, R. 77–80
voting rights 38
VS (vegetative state) 118–19
vulnerabilities to disease 408
VZV (varicella zoster virus) 71

Wade, N. 428–9
Wakefield, J. 10, 173–8; *The Loss of Sadness* 174

Walach, H. 260–2
Wales 331
walking with exercises 473
Walsh, D. 479
Wang Chong 505–6
wanwu 500
warfarin 399
Warren, R. 305, 378
Washington 40
waste removal 115
water 52
water bears 115
Waters, P. 151
Watson, J. 48
weak associations 61–2
Weber, M. 273
weight 58, 400; gain 350; loss 92, 164, 312
weigh tables 162–3
weight-loss centers 520
welfare 40
well-being 9, 262, 330–1; disabilities and 36; health and 27–35
Western Electric Company 138
Western Europe 435–6
Western medicine 158, 183–4, 380, 508–17
"What Evidence in Evidence-Based Medicine" (Worrall) 212–13
which-explanations 302
whiplash 19–22
Whitbeck, C. 316
White, H. 374
Whitehead, A. 286
White populations: biological race essentialism and 420, 424; cardiovascular diseases (CVDs) and 419; CHF (congestive heart failure) and 419–20; as research subjects 424
WHO *see* World Health Organization (WHO)
whooping cough 187
why-questions 302, 477–8
Why We Get Sick: The New Science of Darwinian Medicine (Williams and Nesse) 476
Wilkins, L. 437
Williams, G. 302, 479–80, 483; "The Dawn of Darwinian Medicine" 476; *Why We Get Sick: The New Science of Darwinian Medicine* 476
Williamson, J. 48, 52, 214, 279
Willowbrook hepatitis experiments 373
"Will the QALY Survive?" panel 339
Wilson, E.: *Sociobiology* 476
Winnicott, D. 109–10
Winther, R. 421–2
withdrawal 105
womb 106, 109–11, 121
women: abnormal height and 433–4; AIDS and 312; birth and 103–14, 215, 331–9; birth without 103; cardiovascular diseases (CVDs)

and 195–9, 408–11; clinical trials and 410–11; of color 413; contraceptive pills and 76–9; disparities between men and 408–10; gender in medicine and 408–18; health outcomes and 408–9; married and unmarried 38; medical discourse and 165; perimenopausal 198; postmenopausal 195–9, 433; premenopausal 162; puerperal fever and 296–7; randomized controlled trials (RCTs) and 347; right to choose by 40–1; sex-linked genetic diseases and 147–8; sexual dysfunction in 161–2; smoking and 245–6; thalidomide and 310; triple-negative breast cancer (TNBC) and 49
women's health care movement 359, 373
Women's Health Initiative Study 197–202
Woodward, J. 59–60, 66
work absence 17
World Health Organization (WHO) 9, 19, 27, 258, 387; *International Classification of Diseases* (ICD) 159, 170, 180, 187–9, 461–2
World War II 135, 476
worms 273
Worrall, J. 2, 202–3, 347; "What Evidence in Evidence-Based Medicine" 212–13
worst-case scenarios 41
wounds 52, 135
writing prompts 376
wrongful birth 111

X chromosome 147–8
xiaoke 516

X-linked recessive disorders 147–8
X-ray crystallography 48, 82
X-rays 238, 243

yang 506
yang qi 501
Yao, C. 508–9
Y chromosomes 148
yearning 130
yellow bile 5–6, 183
yin 501
yin and yang 158, 503–5
yinan zabing 501
yin qi 506
yinyang 502–6
Yixue yu zhexue (Medicine and Philosophy) 497
Y-linked diseases 148
younger adults 44
YouTube 111
yuanqi 501

Zahavi, D. 467
Zaner, R. 381
Zarkovich, E. 369
Zerubavel, E. 157
Zhang, Q.: *Philosophical Foundations of Chinese Medicine* 505
zheng 516–17
ziran 500
Zoloft 524
Zonagen 161
Zyprexa (olanzapine) 348–50